MyMathLab®:
Support You Need, When You Need It

MyMathLab is the world's leading online program in mathematics, integrating homework with support tools and tutorials in an easy-to-use format. MyMathLab helps you get up to speed on course material, visualize the content, and understand how math will play a role in your future career.

Review Prerequisite Skills

Integrated Review content identifies gaps in prerequisite skills and offers help for just those skills you need. With this targeted practice, you will be ready to learn new material.

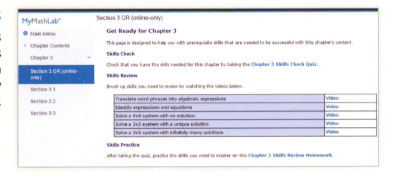

Tutorial Videos

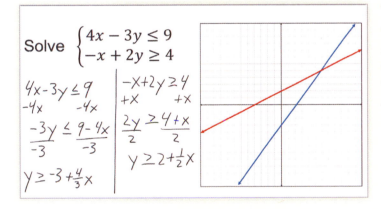

Tutorial videos are available for every section of the textbook and cover key examples from the text. These videos are especially handy if you miss a lecture or just need another explanation.

Interactive Figures

Interactive Figures illustrate key concepts and help you visualize the math. MyMathLab includes assignable exercises that require use of Interactive Figures and instructional videos that explain the concept behind each figure.

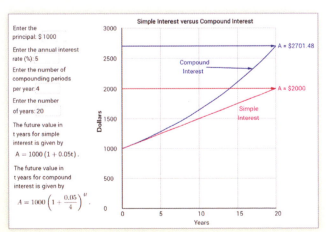

twelfth edition

Finite Mathematics

& ITS APPLICATIONS

Larry J. Goldstein
Goldstein Educational Technologies

David I. Schneider
University of Maryland

Martha J. Siegel
Towson University

Steven M. Hair
The Pennsylvania State University

330 Hudson Street, NY, NY 10013

Director, Portfolio Management: *Deirdre Lynch*
Executive Editor: *Jeff Weidenaar*
Editorial Assistant: *Jennifer Snyder*
Content Producer: *Patty Bergin*
Managing Producer: *Karen Wernholm*
Media Producer: *Stephanie Green*
MathXL Content Manager: *Kristina Evans*
Product Marketing Manager: *Claire Kozar*
Marketing Assistant: *Jennifer Myers*

Senior Author Support/Technology Specialist: *Joe Vetere*
Rights and Permissions Project Manager: *Gina Cheselka*
Manufacturing Buyer: *Carol Melville, LSC Communications*
Associate Director of Design: *Blair Brown*
Composition: *iEnergizer Aptara®, Inc.*
Text Design, Production Coordination, Composition,
  and Illustrations: *iEnergizer Aptara®, Inc.*
Cover Design: *Cenveo*
Cover Image: *Doug Chinnery/Getty Images*

Library of Congress Cataloging-in-Publication Data
Names: Goldstein, Larry Joel. | Schneider, David I. | Siegel, Martha J. |
  Hair, Steven M.
Title: Finite mathematics & its applications.
Other titles: Finite mathematics and its applications
Description: Twelfth edition / Larry J. Goldstein, Goldstein Educational
  Technologies, David I. Schneider, University of Maryland, Martha J.
  Siegel, Towson State University, Steven M. Hair, Pennsylvania State
  University. | Boston: Pearson Education, [2018] | Includes indexes.
Identifiers: LCCN 2016030690 | ISBN 9780134437767 (hardcover) | ISBN
  0134437764 (hardcover)
Subjects: LCSH: Mathematics—Textbooks.
Classification: LCC QA39.3 .G65 2018 | DDC 511/.1—dc23
LC record available at https://lccn.loc.gov/2016030690

Student Edition ISBN-13: 978-0-134-43776-7
Student Edition ISBN-10:   0-134-43776-4

Contents

The book divides naturally into four parts. The first part consists of linear mathematics: linear equations, matrices, and linear programming (Chapters 1–4); the second part is devoted to probability and statistics (Chapters 5–7); the third part covers topics utilizing the ideas of the other parts (Chapters 8 and 9); and the fourth part explores key topics from discrete mathematics that are sometimes covered in the modern finite mathematics curriculum (Chapters 10–12).

KV 12.31.2018 1622

Preface

This work is the twelfth edition of our text for the finite mathematics course taught to first- and second-year college students, especially those majoring in business and the social and biological sciences. Finite mathematics courses exhibit tremendous diversity with respect to both content and approach. Therefore, in developing this book, we incorporated a wide range of topics from which an instructor may design a curriculum, as well as a high degree of flexibility in the order in which the topics may be presented. For the mathematics of finance, we even allow for flexibility in the approach of the presentation.

The Series

This text is part of a highly successful series consisting of three texts: *Finite Mathematics & Its Applications*, *Calculus & Its Applications*, and *Calculus & Its Applications, Brief Version*. All three titles are available for purchase in a variety of formats, including as an eBook within the MyMathLab online course.

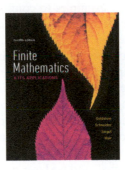 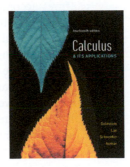 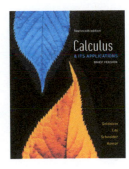

Topics Included

This edition has more material than can be covered in most one-semester courses. Therefore, the instructor can structure the course to the students' needs and interests. The book divides naturally into four parts:

- Part One (Chapters 1–4) consists of linear mathematics: linear equations, matrices, and linear programming.
- Part Two (Chapters 5–7) is devoted to counting, probability, and statistics.
- Part Three (Chapters 8 and 9) covers topics utilizing the ideas of the other parts.
- Part Four (Chapters 10–12) explores key topics from discrete mathematics that are sometimes included in the modern finite mathematics curriculum.

Minimal Prerequisites

Because of great variation in student preparation, we keep formal prerequisites to a minimum. We assume only a first year of high school algebra, and we review, as needed, those topics that are typically weak spots for students.

New to This Edition

We welcome to this edition a new co-author, Steven Hair from Penn State University. Steve has brought a fresh eye to the content and to the MyMathLab course that accompanies the text.

We are grateful for the many helpful suggestions made by reviewers and users of the text. We incorporated many of these into this new edition. We also analyzed aggregated student usage and performance data from MyMathLab for the previous edition of this text. The results of this analysis helped improve the quality and quantity of exercises that matter the most to instructors and students. Additionally, we made the following improvements in this edition:

- **Help-Text Added.** We added blue "help text" next to steps within worked-out examples to point out key algebraic and numerical transitions.
- **Updated Technology.** We changed the graphing calculator screen captures to the more current TI-84 Plus CE format. The discussions of Excel now refer to Excel 2013 and Excel 2016.
- **Additional Exercises and Updated Data.** We have added or updated 440 exercises and have updated the real-world data appearing in the examples and exercises. The book now contains 3580 exercises and 370 worked-out examples.
- **Technology Solutions.** We added technology-based solutions to more examples to provide flexibility for instructors who incorporate technology. For instance, the section on the method of least-squares (1.4) now relies more on technology and less on complicated calculations. In Section 7.6, several examples now demonstrate how to compute the area under a normal curve using a graphing calculator, in addition to the table-based method. In the finance chapter, many TI-84 Plus TVM Solver screen captures accompany examples to confirm answers. Instructors have the option of using TVM Solver for financial calculations instead of complicated formulas.
- **Linear Inequalities Section Relocated.** We moved this section from 1.2 (in the 11e) to the beginning of the linear programming chapter (Ch. 3) in this edition. The move places the topic in the chapter where it is used. Also, the move allows us to use conventional names (such as *slope-intercept form*) in the section.
- **Improved Coverage of Counting Material.** In Chapter 5, we added several definitions and discussions to aid student comprehension of counting problems. We moved the definition of *factorials* to 5.4 and rewrote the permutation and combination formulas in 5.5 in terms of factorials. In 5.6, the *complement rule* for counting is now formally defined, and we have added a discussion of when addition, subtraction, and multiplication is appropriate for solving counting problems.
- **Section Added to the End of the Finance Chapter.** Titled "A Unifying Equation," this new section shows that the basic financial concepts can be described by a difference equation of the form $y_n = a \cdot y_{n-1} + b$, y_0 given, and that many of the calculations from the chapter can be obtained by solving this difference equation. Examples and exercises show that this difference equation also can be used to solve problems in the physical, biological, and social sciences. This section can be taught as a standalone section without covering the preceding sections of the finance chapter.
- **Revision of Logic Material.** We substantially revised Chapter 11 on logic to better meet student needs. We moved the definition of *logical equivalence* and De Morgan's laws from 11.4 to 11.2. By stating key ideas related to truth tables and implications in terms of logical equivalence, students will be better equipped to understand these concepts. To remove confusion between the inclusive and exclusive "or" statements, we removed the word "either" from inclusive "or" statements in English. In 11.4, we added the definition of the *inverse* of an implication. This is a key concept in the topic of implications and logical arguments. To help students understand when a logical argument is invalid, we expanded 11.5 to include more discussion of invalid arguments. Additionally, we added the fallacies of the inverse and converse, and two new examples where arguments are proven to be invalid.

- **Difference Equation Chapter Moved Online.** We moved former Chapter 11 online (relabeling it Chapter 12 in the process). The chapter is available directly to students at www.pearsonhighered.com/mathstatsresources and within MyMathLab. All support materials for the chapter appear online within MyMathLab. *Note:* The new section at the end of the finance chapter contains the fundamental concepts from the difference equation chapter.

New to MyMathLab

Many improvements have been made to the overall functionality of MyMathLab (MML) since the previous edition. However, beyond that, we have also invested in increasing and improving the content specific to this text.

- Instructors now have more exercises than ever to choose from in assigning homework. There are approximately 2540 assignable exercises in MML.
- We heard from users that the Annotated Instructor Edition for the previous edition required too much flipping of pages to find answers, so MML now contains a downloadable Instructor Answers document—*with all answers in one place*. (This augments the downloadable Instructor Solutions Manual, which contains all *solutions*.)
- Interactive Figures are now in HTML format (no plug-in required) and are supported by assignable exercises and tutorial videos.
- An Integrated Review version of the MML course contains pre-made quizzes to assess the prerequisite skills needed for each chapter, plus personalized remediation for any gaps in skills that are identified.
- New Setup & Solve exercises require students to show how they set up a problem as well as the solution, better mirroring what is required of students on tests.
- StatCrunch, a fully functional statistics package, is provided to support the statistics content in the course.
- MathTalk and StatTalk videos highlight applications of the content of the course to business. The videos are supported by assignable exercises.
- Study skills modules help students with the life skills that can make the difference between passing and failing.
- 110 new tutorial videos by Brian Rickard (University of Arkansas) were added to support student learning.
- Tutorial videos involving graphing calculators are now included within MML exercises to augment videos showing "by hand" methods. If you require graphing calculator usage for the course, your students will find these videos very helpful. (If you do not use calculators, you can hide these videos from students.)
- Graphing Calculator and Excel Spreadsheet Manuals, specific to this course, are now downloadable from MML.

Trusted Features

Though this edition has been improved in a variety of ways to reflect changing student needs, we have maintained the popular overall approach that has helped students be successful over the years.

Relevant and Varied Applications

We provide realistic applications that illustrate the uses of finite mathematics in other disciplines and everyday life. The variety of applications is evident in the Index of Applications at the end of the text. Wherever possible, we attempt to use applications to motivate the mathematics. For example, the concept of linear programming is introduced in Chapter 3 via a discussion of production options for a factory with labor limitations.

Plentiful Examples

The twelfth edition includes 370 worked examples. Furthermore, we include computational details to enhance comprehension by students whose basic skills are weak.

Knowing that students often refer back to examples for help, we built in fidelity between exercises and examples. In addition, students are given Now Try exercise references immediately following most examples to encourage them to check their understanding of the given example.

Exercises to Meet All Student Needs

The 3580 exercises comprise about one-quarter of the book—the most important part of the text, in our opinion. The exercises at the ends of the sections are typically arranged in the order in which the text proceeds, so that homework assignments may be made easily after only part of a section is discussed. Interesting applications and more challenging problems tend to be located near the ends of the exercise sets. Exercises have odd-even pairing, when appropriate. Chapter Review Exercises are designed to prepare students for end-of-chapter tests. Answers to the odd-numbered exercises, and all Chapter Review Exercises, are included at the back of the book.

Check Your Understanding Problems

The Check Your Understanding problems are a popular and useful feature of the book. They are carefully selected exercises located at the end of each section, just before the exercise set. Complete solutions follow the exercise set. These problems prepare students for the exercise sets beyond just covering simple examples. They give students a chance to think about the skills they are about to apply and reflect on what they've learned.

Use of Technology

We incorporated technology usage into the text in ways that provide you with flexibility, knowing that the course can vary quite a bit based on how technology is incorporated. Our basic approach in the text is to assume minimal use of technology and clearly label the opportunities to make it a greater part of the course. Many of the sections contain Incorporating Technology features that show how to use Texas Instruments graphing calculators, Excel spreadsheets, and Wolfram|Alpha. In addition, the text contains appendixes on the use of these technologies. Each type of technology is clearly labeled with an icon:

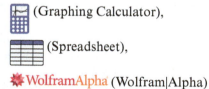

(Graphing Calculator),

(Spreadsheet),

WolframAlpha (Wolfram|Alpha)

In our discussions of graphing calculators, we specifically refer to the TI-84 Plus models, since these are the most popular graphing calculators. New to this edition, screen shots display the new color versions of the TI-84. Spreadsheets refer to Microsoft Excel 2016. The web application discussed is Wolfram|Alpha, which is an exceptionally fine and versatile product that is available online or on mobile devices for free or at low cost. We feel that Wolfram|Alpha is a powerful tool for learning and exploring mathematics, which is why we chose to include activities that use it. We hope that by modeling appropriate use of this technology, students will come to appreciate the application for its true worth.

End-of-Chapter Study Aids

Near the end of each chapter is a set of problems entitled Fundamental Concept Check Exercises that help students recall key ideas of the chapter and focus on the relevance of these concepts as well as prepare for exams. Each chapter also contains a two-column grid giving a section-by-section summary of key terms and concepts with examples. Finally, each chapter has Chapter Review Exercises that provide more practice and preparation for chapter-level exams.

Chapter Projects

Each chapter ends with an extended project that can be used as an in-class or out-of-class group project or special assignment. These projects develop interesting applications or enhance key concepts of the chapters.

Technology and Supplements

MyMathLab® Online Course (access code required)

Built around Pearson's best-selling content, MyMathLab is an online homework, tutorial, and assessment program designed to work with this text to engage students and improve results. MyMathLab can be successfully implemented in any classroom environment—lab-based, hybrid, fully online, or traditional. **By addressing instructor and student needs, MyMathLab improves student learning.**

Used by more than 37 million students worldwide, MyMathLab delivers consistent, measurable gains in student learning outcomes, retention, and subsequent course success. Visit www.mymathlab.com/results to learn more.

Preparedness

One of the biggest challenges in Finite Mathematics courses is making sure students are adequately prepared with the prerequisite skills needed to successfully complete their course work. Pearson offers a variety of content and course options to support students with just-in-time remediation and key-concept review.

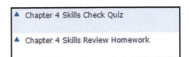

- **Integrated Review Courses** can be used for just-in-time prerequisite review. These courses provide additional content on review topics, along with pre-made, assignable skill-check quizzes, personalized homework assignments, and videos integrated throughout the course.

Motivation

Students are motivated to succeed when they're engaged in the learning experience and understand the relevance and power of mathematics. MyMathLab's online homework offers students immediate feedback and tutorial assistance that motivates them to do more, which means they retain more knowledge and improve their test scores.

- **Exercises with immediate feedback**—over 2540 assignable exercises—are based on the textbook exercises, and regenerate algorithmically to give students unlimited opportunity for practice and mastery. MyMathLab provides helpful feedback when students enter incorrect answers and includes optional learning aids including Help Me Solve This, View an Example, videos, and an eText.

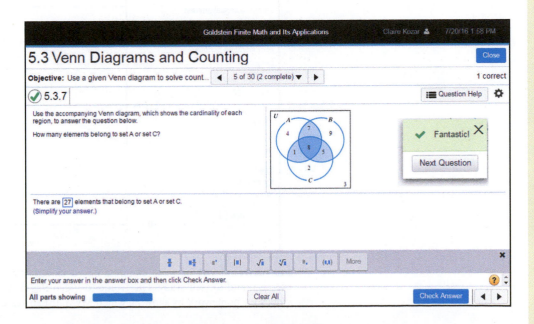

- **Setup and Solve Exercises** ask students to first describe how they will set up and approach the problem. This reinforces students' conceptual understanding of the process they are applying and promotes long-term retention of the skill.
- **MathTalk and StatTalk videos** connect the math to the real world (particularly business). The videos include assignable exercises to gauge students' understanding of video content.
- **Learning Catalytics™** is a student response tool that uses students' smartphones, tablets, or laptops to engage them in more interactive tasks and thinking. Learning Catalytics fosters student engagement and peer-to-peer learning with real-time analytics.

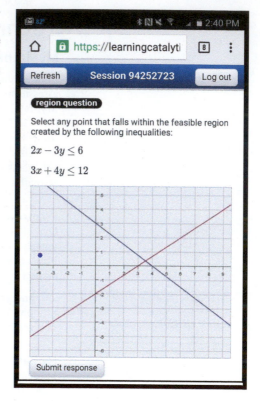

Learning and Teaching Tools

- **Interactive Figures** illustrate key concepts and allow manipulation for use as teaching and learning tools. MyMathLab includes assignable exercises that require use of figures and instructional videos that explain the concept behind each figure.

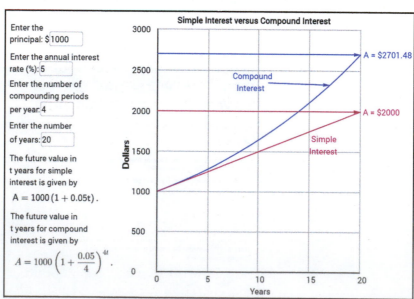

- **Instructional videos**—238 example-based videos—are available as learning aids within exercises and for self-study. The Guide to Video-Based Assignments makes it easy to assign videos for homework by showing which MyMathLab exercises correspond to each video.

- **Graphing Calculator videos** are available to augment "by hand" methods, allowing you to match the help that students receive to how graphing calculators are used in the course. Videos are available within select exercises and in the Multimedia Library.
- **Complete eText** is available to students through their MyMathLab courses for the lifetime of the edition, giving students unlimited access to the eText within any course using that edition of the textbook.
- **StatCrunch**, a fully functional statistics package, is provided to support the statistics content in the course.
- **Skills for Success Modules** help students with the life skills that can make the difference between passing and failing. Topics include "Time Management" and "Stress Management."
- **Excel Spreadsheet Manual,** specifically written for this course.
- **Graphing Calculator Manual,** specifically written for this course.
- **PowerPoint Presentations** are available for download for each section of the book.
- **Accessibility** and achievement go hand in hand. MyMathLab is compatible with the JAWS screen reader, and enables multiple-choice and free-response problem types to be read and interacted with via keyboard controls and math notation input. MyMathLab also works with screen enlargers, including ZoomText, MAGic, and SuperNova. And, all MyMathLab videos have closed-captioning. More information is available at http://mymathlab.com/accessibility.
- **A comprehensive gradebook** with enhanced reporting functionality allows you to efficiently manage your course.
  - **The Reporting Dashboard** provides insight to view, analyze, and report learning outcomes. Student performance data is presented at the class, section, and program levels in an accessible, visual manner so you'll have the information you need to keep your students on track.

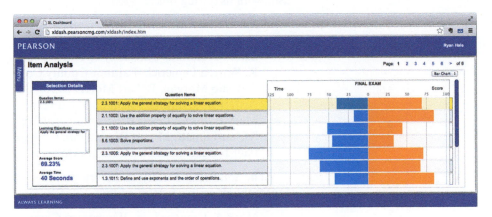

- **Item Analysis** tracks class-wide understanding of particular exercises so you can refine your class lectures or adjust the course/department syllabus. Just-in-time teaching has never been easier!

MyMathLab comes from an experienced partner with educational expertise and an eye on the future. Whether you are just getting started with MyMathLab, or have a question along the way, we're here to help you learn about our technologies and how to incorporate them into your course. To learn more about how MyMathLab helps students succeed, visit www.mymathlab.com or contact your Pearson rep.

MathXL® is the homework and assessment engine that runs MyMathLab. (MyMathLab is MathXL plus a learning management system.) MathXL access codes are also an option.

Student Solutions Manual

ISBN-10: 0-134-46344-7 | ISBN-13: 978-0-134-46344-5
Contains fully worked-out solutions to odd-numbered exercises. Available in print and downloadable from within MyMathLab.

Instructor Answers / Instructor Solutions Manual (downloadable)

ISBN-10: 0-134-46343-9 | ISBN-13: 978-0-134-46343-8
The Instructor Answers document contains a list of answers to all student edition exercises. The Instructor Solutions Manual contains solutions to all student edition exercises. Downloadable from the Pearson Instructor Resource Center www.pearsonhighered.com/irc, or from within MyMathLab.

TestGen (downloadable)

ISBN-10: 0-134-46346-3 | ISBN-13: 978-0-134-46346-9

TestGen enables instructors to build, edit, print, and administer tests using a bank of questions developed to cover all objectives in the text. TestGen is algorithmically based, allowing you to create multiple but equivalent versions of the same question or test. Instructors can also modify testbank questions or add new questions. The software and testbank are available to qualified instructors for download and installation from Pearson's online catalog www.pearsonhighered.com and from within MyMathLab.

PowerPoints

ISBN-10: 0-134-46407-9 | ISBN-13: 978-0-134-46407-7

Contains classroom presentation slides for this textbook featuring lecture content, worked-out examples, and key graphics from the text. Available to qualified instructors within MyMathLab or through the Pearson Instructor Resource Center www.pearsonhighered.com/irc.

Acknowledgments

While writing this book, we have received assistance from many people, and our heartfelt thanks go out to them all. Especially, we should like to thank the following reviewers, who took the time and energy to share their ideas, preferences, and often their enthusiasm, with us during this revision:

Jeff Dodd, Jacksonville State University
Timothy M. Doyle, University of Illinois at Chicago
Sami M. Hamid, University of North Florida
R. Warren Lemerich, Laramie County Community College
Antonio Morgan, Robert Morris University
Arthur J. Rosenthal, Salem State University
Mary E. Rudis, Great Bay Community College
Richard Smatt, Mount Washington College
Paul J. Welsh, Pima Community College

The following faculty members provided direction on the development of the MyMathLab course for this edition:

Mark A. Crawford, Jr., Waubonsee Community College
Cymra Haskell, University of Southern California
Ryan Andrew Hass, Oregon State University
Melissa Hedlund, Christopher Newport University
R. Warren Lemerich, Laramie County Community College
Sara Talley Lenhart, Christopher Newport University
Enyinda Onunwor, Stark State College
Lynda Zenati, Robert Morris University

We wish to thank the many people at Pearson who have contributed to the success of this book. We appreciate the efforts of the production, design, manufacturing, marketing, and sales departments. We are grateful to Lisa Collette for her thorough proofreading and John Morin and Rhea Meyerholtz for their careful and thorough checking for accuracy. Our sincere thanks goes to Erica O'Leary for her assistance throughout the revision of the book. Content Producer Patty Bergin did a fantastic job keeping the book on schedule. The authors wish to extend special thanks to editor Jeff Weidenaar.

If you have any comments or suggestions, we would like to hear from you. We hope you enjoy using this book as much as we have enjoyed writing it.

Larry J. Goldstein
larrygoldstein@predictiveanalyticsshop.com

Martha J. Siegel
msiegel@towson.edu

David I. Schneider
dis@math.umd.edu

Steven M. Hair
smh384@psu.edu

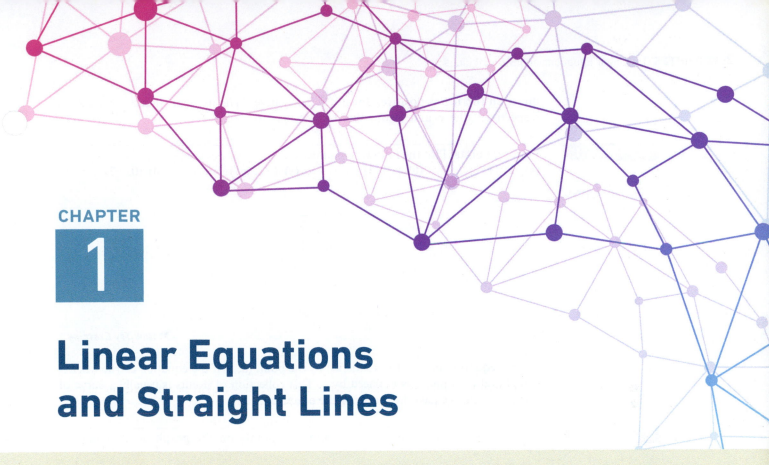

Linear Equations and Straight Lines

Many applications considered later in this text involve linear equations and their geometric counterparts—straight lines. So let us begin by studying the basic facts about these two important notions.

1.1 Coordinate Systems and Graphs

Often, we can display numerical data by using a **Cartesian coordinate system** on either a line or a plane. We construct a Cartesian coordinate system on a line by choosing an arbitrary point O (the **origin**) on the line and a unit of distance along the line. We then assign to each point on the line a number that reflects its directed distance from the origin. Positive numbers refer to points on the right of the origin, negative numbers to points on the left. In Fig. 1, we have drawn a Cartesian coordinate system on the line and have labeled a number of points with their corresponding numbers. Each point on the line corresponds to a number (positive, negative, or zero).

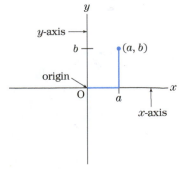

Figure 2

Figure 1

In a similar fashion, we can construct a Cartesian coordinate system to numerically locate points on a plane. Each point of the plane is identified by a pair of numbers (a, b). See Fig. 2. To reach the point (a, b), begin at the origin, move a units in the x direction (to the right if a is positive, to the left if a is negative), and then move b units in the y

direction (up if b is positive, down if b is negative). The numbers a and b are called, respectively, the **x-** and **y-coordinates** of the point.

EXAMPLE 1   **Plotting Points** Plot the following points:

(a) $(2, 1)$ (b) $(-1, 3)$ (c) $(-2, -1)$ (d) $(0, -3)$

SOLUTION

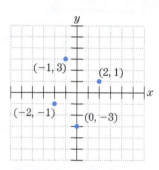

>> *Now Try Exercise 1*

An equation in x and y is satisfied by the point (a, b) if the equation is true when x is replaced by a and y is replaced by b. This collection of points is usually a curve of some sort and is called the **graph of the equation**.

EXAMPLE 2   **Solution of an Equation** Are the following points on the graph of the equation $8x - 4y = 4$?

(a) $(3, 5)$ (b) $(5, 17)$

SOLUTION   (a) $8x - 4y = 4$ Given equation

$8 \cdot 3 - 4 \cdot 5 \overset{?}{=} 4$ $x = 3, y = 5$

$24 - 20 \overset{?}{=} 4$ Multiply.

$4 = 4$ Subtract.

Since the equation is satisfied, the point $(3, 5)$ is on the graph of the equation.

(b) $8x - 4y = 4$ Given equation

$8 \cdot 5 - 4 \cdot 17 \overset{?}{=} 4$ $x = 5, y = 17$

$40 - 68 \overset{?}{=} 4$ Multiply.

$-28 \overset{?}{=} 4$ Subtract.

The equation is not satisfied, so the point $(5, 17)$ is *not* on the graph of the equation.   >> *Now Try Exercises 11 and 13*

Linear Equations

A linear equation is an equation whose graph is a straight line. Figure 3 shows four examples of linear equations, along with their graphs and some points on their graphs.

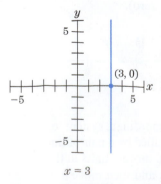

$x = 3$

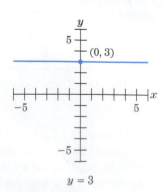

$y = 3$

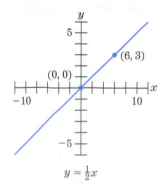

$y = \frac{1}{2}x$

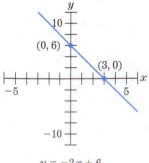

$y = -2x + 6$

Figure 3 Four linear equations and their graphs

Intercepts

The intercepts of a line are the points where the line crosses the x- and y-axes. These points have 0 for at least one of their coordinates. For the graph of $y = -2x + 6$ in Fig. 3, the x-intercept is the point $(3, 0)$ and the y-intercept is the point $(0, 6)$.* The y-intercept of a line having an equation of the form $y = mx + b$ is the point $(0, b)$, since setting x equal to 0 gives y the value b. The x-intercept is the point having the solution of the equation $0 = mx + b$ as the first coordinate and 0 as the second coordinate.

Table 1 shows how to draw the graphs of the four types of linear equations shown in Fig. 3. The equations $y = b$ and $y = mx$ are actually special cases of $y = mx + b$.

Table 1 **Graphs of Linear Equations**

Equation	Description of Graph	How to Draw Graph
$x = a$	Vertical line through the point $(a, 0)$	Plot $(a, 0)$ and draw the vertical line through the point.
$y = b$	Horizontal line through the point $(0, b)$	Plot $(0, b)$ and draw the horizontal line through the point.
$y = mx$	Line through the origin	Draw the line through the origin and any other point on the graph.
$y = mx + b$; $m \neq 0, b \neq 0$	Line having two different intercepts	Draw the line through any two points (often the two intercepts) of the line.

> **General Form of a Linear Equation** Any equation whose graph is a straight line can be written in the **general form**
>
> $$cx + dy = e$$
>
> where c, d, and e are constants and c and d are not both zero.

An equation in general form having $d \neq 0$ (that is, an equation in which y appears) can be solved for y. The resulting equation will have the form of one of the last three equations in Table 1. An equation in which y does not appear can be solved for x and the resulting equation will have the form of the first equation in Table 1.

EXAMPLE 3

Graph of an Equation Write the equation $x - 2y = 4$ in one of the forms shown in Table 1 and draw its graph.

SOLUTION Since y appears in the equation, solve for y.

$$x - 2y = 4 \qquad \text{Given equation}$$
$$-2y = -x + 4 \qquad \text{Subtract } x \text{ from both sides.}$$
$$y = \tfrac{1}{2}x - 2 \qquad \text{Divide both sides by } -2.$$

Since the equation $y = \tfrac{1}{2}x - 2$ has the form of the last equation in Table 1, it can be graphed by finding its two intercepts and drawing the straight line through them.

---

*Intercepts are sometimes defined as numbers, such as x-intercept 3 and y-intercept 6. In this text, we define them as pairs of numbers, such as $(3, 0)$ and $(0, 6)$.

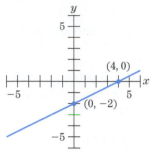

Figure 4 Graph of $x - 2y = 4$

The y-intercept is the point $(0, -2)$ since setting x equal to 0 gives y the value -2. The x-intercept is found by setting y equal to 0 and solving for x.

$$y = \tfrac{1}{2}x - 2 \qquad \text{Given equation}$$
$$0 = \tfrac{1}{2}x - 2 \qquad \text{Set } y \text{ equal to 0.}$$
$$2 = \tfrac{1}{2}x \qquad \text{Add 2 to both sides.}$$
$$x = 4 \qquad \text{Multiply both sides by 2. Rewrite.}$$

Therefore, the x-intercept is the point $(4, 0)$.

The graph in Fig. 4 was obtained by plotting the intercepts $(4, 0)$ and $(0, -2)$ and drawing the straight line through them. **»» Now Try Exercise 27**

EXAMPLE 4 **Graph of an Equation** Write the equation $-2x + 3y = 0$ in one of the forms shown in Table 1 and draw its graph.

SOLUTION Since y appears in the equation, solve for y.

$$-2x + 3y = 0 \qquad \text{Given equation}$$
$$3y = 2x \qquad \text{Add } 2x \text{ to both sides.}$$
$$y = \tfrac{2}{3}x \qquad \text{Divide both sides by 3.}$$

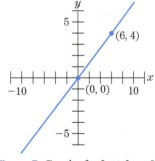

Figure 5 Graph of $-2x + 3y = 0$

Because the graph of the equation $y = \tfrac{2}{3}x$ passes through the origin, the point $(0, 0)$ is both the x-intercept and the y-intercept of the graph. In order to draw the graph, we must locate another point on the graph. Let's choose $x = 6$. Then $y = \tfrac{2}{3} \cdot 6 = 4$. Therefore, the point $(6, 4)$ is on the graph. The graph in Fig. 5 was obtained by plotting the points $(0, 0)$ and $(6, 4)$ and drawing the straight line through them.

»» Now Try Exercise 19

The next example gives an application of linear equations.

EXAMPLE 5 **Linear Depreciation** For tax purposes, businesses must keep track of the current values of each of their assets. A common mathematical model is to assume that the current value y is related to the age x of the asset by a linear equation. A moving company buys a 40-foot van with a useful lifetime of 5 years. After x months of use, the value y, in dollars, of the van is estimated by the linear equation

$$y = 25{,}000 - 400x.$$

(a) Draw the graph of this linear equation.
(b) What is the value of the van after 5 years?
(c) When will the value of the van be $15,000?
(d) What economic interpretation can be given to the y-intercept of the graph?

SOLUTION (a) The y-intercept is $(0, 25{,}000)$. To find the x-intercept, set $y = 0$ and solve for x.

$$0 = 25{,}000 - 400x \qquad \text{Set } y = 0.$$
$$400x = 25{,}000 \qquad \text{Add } 400x \text{ to both sides.}$$
$$x = 62.5 \qquad \text{Divide both sides by 400.}$$

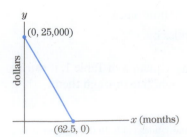

Figure 6

The x-intercept is $(62.5, 0)$. The graph of the linear equation is sketched in Fig. 6. Note how the value decreases as the age of the van increases. The value of the van reaches 0 after 62.5 months. Note also that we have sketched only the portion of the graph that has physical meaning—namely, the portion for x between 0 and 62.5.

(b) After 5 years (or 60 months), the value of the van is

$$y = 25{,}000 - 400(60) = 25{,}000 - 24{,}000 = 1000.$$

Since the useful life of the van is 5 years, this value represents the *salvage value* of the van.

(c) Set the value of y to 15,000, and solve for x.

$15{,}000 = 25{,}000 - 400x$	Set $y = 15{,}000$.
$400x + 15{,}000 = 25{,}000$	Add $400x$ to both sides.
$400x = 10{,}000$	Subtract 15,000 from both sides.
$x = 25$	Divide both sides by 400.

The value of the van will be $15,000 after 25 months.

(d) The y-intercept corresponds to the value of the van at $x = 0$ months—that is, the initial value of the van, $25,000.

» *Now Try Exercise 41*

INCORPORATING

TECHNOLOGY

Appendix B contains instructions for TI-84 Plus calculators. (For the specifics of other calculators, consult the guidebook for the calculator.) The appendix shows how to obtain the graph of a linear equation of the form $y = mx + b$, find coordinates of points on the line, and determine intercepts. Vertical lines can be drawn with the Vertical command from the DRAW menu. To draw the vertical line $x = k$, go to the home screen, press [2nd] [DRAW] **4** to display the word Vertical, type in the value of k, and press [ENTER].

❋WolframAlpha Appendix D contains an introduction to Wolfram|Alpha.

Straight lines can be drawn with instructions of the following forms:

plot ax + by = c;   plot y = ax + b;   plot x = a

If a phrase of the form **for x from x₁ to x₂** is appended to the instruction, only the portion of the line having x-values from x_1 to x_2 will be drawn.

An equation of the form $ax + by = c$, with $b \neq 0$, can be converted to the form $y = mx + b$ with the instruction **solve ax + by = c for y.**

The intercepts of an equation can be found with an instruction of the form **intercepts [equation]**. An expression in x can be evaluated at $x = a$ with an instruction of the form **evaluate [expression] at x = a.** For instance, the instruction

evaluate 2500 − 400x at x = 5

gives the result 500.

Check Your Understanding 1.1

Solutions can be found following the section exercises.

1. Plot the point (500, 200).

2. Is the point $(4, -7)$ on the graph of the linear equation $2x - 3y = 1$? Is the point $(5, 3)$?

EXERCISES 1.1

In Exercises 1–8, plot the given point.

1. $(2, 3)$

2. $(-1, 4)$

3. $(0, -2)$

4. $(2, 0)$

5. $(-2, 1)$

6. $\left(-1, -\frac{5}{2}\right)$

7. $(-20, 40)$

8. $(25, 30)$

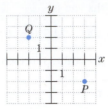

Figure 7

9. What are the coordinates of the point Q in Fig. 7?

10. What are the coordinates of the point P in Fig. 7?

In Exercises 11–14, determine whether the point is on the graph of the equation $-2x + \frac{1}{3}y = -1$.

11. $(1, 3)$ 12. $(2, 6)$ 13. $\left(\frac{1}{2}, 3\right)$ 14. $\left(\frac{1}{3}, -1\right)$

In Exercises 15–18, each linear equation is in the form $y = mx + b$. Identify m and b.

15. $y = 5x + 8$ 16. $y = -2x - 6$

17. $y = 3$ 18. $y = \frac{2}{3}x$

In Exercises 19–22, write each linear equation in the form $y = mx + b$ or $x = a$.

19. $14x + 7y = 21$ 20. $x - y = 3$

21. $3x = 5$ 22. $-\frac{1}{2}x + \frac{2}{3}y = 10$

In Exercises 23–26, find the x-intercept and the y-intercept of each line.

23. $y = -4x + 8$ 24. $y = 5$

25. $x = 7$ 26. $y = -8x$

In Exercises 27–34, graph the given linear equation.

27. $y = \frac{1}{3}x - 1$ 28. $y = 2x$ 29. $y = \frac{5}{2}$

30. $x = 0$ 31. $3x + 4y = 24$ 32. $x + y = 3$

33. $x = -\frac{5}{2}$ 34. $\frac{1}{2}x - \frac{1}{3}y = -1$

35. Which of the following equations describe the same line as the equation $2x + 3y = 6$?
(a) $4x + 6y = 12$ (b) $y = -\frac{2}{3}x + 2$ (c) $x = 3 - \frac{3}{2}y$
(d) $6 - 2x - y = 0$ (e) $y = 2 - \frac{2}{3}x$ (f) $x + y = 1$

36. Which of the following equations describe the same line as the equation $\frac{1}{2}x - 5y = 1$?
(a) $2x - \frac{1}{5}y = 1$ (b) $x = 5y + 2$
(c) $2 - 5x + 10y = 0$ (d) $y = .1(x - 2)$
(e) $10y - x = -2$ (f) $1 + .5x = 2 + 5y$

37. Each of the lines L_1, L_2, and L_3 in Fig. 8 is the graph of one of the equations (a), (b), and (c). Match each of the equations with its corresponding line.
(a) $x + y = 3$ (b) $2x - y = -2$ (c) $x = 3y + 3$

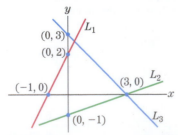

Figure 8

38. Which of the following equations is graphed in Fig. 9?
(a) $x + y = 3$ (b) $y = x - 1$ (c) $2y = x + 3$

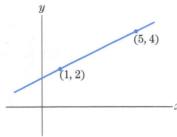

Figure 9

39. **Heating Water** The temperature of water in a heating tea kettle rises according to the equation $y = 30x + 72$, where y is the temperature (in degrees Fahrenheit) x minutes after the kettle was put on the burner.
(a) What physical interpretation can be given to the y-intercept of the graph?
(b) What will the temperature of the water be after 3 minutes?
(c) After how many minutes will the water be at its boiling point of 212°?

40. **Life Expectancy** The average life expectancy y of a person born x years after 1960 can be approximated by the linear equation $y = \frac{1}{6}x + 70$.
(a) What interpretation can be given to the y-intercept of the graph?
(b) In what year did people born that year have an average life expectancy of 75 years?
(c) What is the average life expectancy of people born in 1999?

41. **Cigarette Consumption** The worldwide consumption of cigarettes has been increasing steadily in recent years. The number of trillions of cigarettes, y, purchased x years after 1960, is estimated by the linear equation $y = .075x + 2.5$.
(a) Draw the graph of this linear equation.
(b) What interpretation can be given to the y-intercept of the graph?
(c) When were there 4 trillion cigarettes sold?
(d) If this trend continues, how many cigarettes will be sold in the year 2024?

42. **Ecotourism Income** In a certain developing country, ecotourism income has been increasing in recent years. The income y (in thousands of dollars) x years after 2000 can be modeled by $y = 1.15x + 14$.
(a) Draw the graph of this linear equation.
(b) What interpretation can be given to the y-intercept of this graph?
(c) When was there $20,000 in ecotourism income?
(d) If this trend continues, how much ecotourism income will there be in 2022?

43. **Insurance Rates** Yearly car insurance rates have been increasing steadily in the last few years. The rate y (in dollars) for a small car x years after 1999 can be modeled by $y = 23x + 756$.
(a) Draw the graph of this linear equation.
(b) What interpretation can be given to the y-intercept of this graph?
(c) What was the yearly rate in 2007?
(d) If this trend continues, when will the yearly rate be $1308?

44. Simple Interest If $1000 is deposited at 3% simple interest, the balance y after x years will be given by the equation $y = 30x + 1000$.
   (a) Draw the graph of this linear equation.
   (b) Find the balance after two years.
   (c) When will the balance reach $1180?

45. College Freshmen The percentage, y, of college freshmen who entered college intending to major in general biology increased steadily from the year 2000 to the year 2014 and can be approximated by the linear equation $y = .2x + 4.1$ where x represents the number of years since 2000. Thus, $x = 0$ represents 2000, $x = 1$ represents 2001, and so on. (*Source: The American Freshman: National Norms.*)
   (a) What interpretation can be given to the y-intercept of the graph of the equation?
   (b) In 2014, approximately what percent of college freshmen intended to major in general biology?
   (c) In what year did approximately 5.5% of college freshmen intend to major in general biology?

46. College Freshmen The percentage, y, of college freshmen who smoke cigarettes decreased steadily from the year 2004 to the year 2014 and can be approximated by the linear equation $y = -.46x + 6.32$ where x represents the number of years since 2004. Thus, $x = 0$ represents 2004, $x = 1$ represents 2005, and so on. (*Source: The American Freshman: National Norms.*)
   (a) What interpretation can be given to the y-intercept of the graph of the equation?
   (b) In 2014, approximately what percent of college freshmen smoked?
   (c) In what year did approximately 2.6% of college freshmen smoke?

47. College Tuition Average tuition (including room and board) for all institutions of higher learning in year x can be approximated by $y = 461x + 16,800$ dollars, where $x = 0$ corresponds to 2004, $x = 1$ corresponds to 2005, and so on. (*Source: U.S. National Center of Education Statistics.*)
   (a) Approximately what was the average tuition in 2011?
   (b) Assuming that the formula continues to hold, when will the average tuition exceed $25,000?

48. Bachelor's Degrees The number of bachelor's degrees conferred in mathematics and statistics in year x can be approximated by $y = 667x + 12,403$, where $x = 0$ corresponds to 2003, $x = 1$ corresponds to 2004, and so on. (*Source: U.S. National Center of Education Statistics.*)
   (a) Approximately how many bachelor's degrees in mathematics and statistics were awarded in 2007?
   (b) Assuming that the model continues to hold, approximately when will the number of bachelor's degrees in mathematics and statistics awarded exceed 25,000?

49. Find an equation of the line having x-intercept $(16, 0)$ and y-intercept $(0, 8)$.

50. Find an equation of the line having x-intercept $(.6, 0)$ and y-intercept $(0, .9)$.

51. Find an equation of the line having y-intercept $(0, 5)$ and x-intercept $(4, 0)$.

52. Find an equation of the line having x-intercept $(5, 0)$ and parallel to the y-axis.

53. What is the equation of the x-axis?

54. Can a line other than the x-axis have more than one x-intercept?

55. What is the general form of the equation of a line that is parallel to the y-axis?

56. What is the general form of the equation of a line that is parallel to the x-axis?

In Exercises 57–60, find a general form of the given equation.

57. $y = 2x + 3$

58. $y = 3x - 4$

59. $y = -\frac{2}{3}x - 5$

60. $y = 4x - \frac{5}{6}$

61. Show that the straight line with x-intercept $(a, 0)$ and y-intercept $(0, b)$, where a and b are not zero, has $bx + ay = ab$ as a general form of its equation.

62. Use the result of Exercise 61 to find a general form of the equation of the line having x-intercept $(5, 0)$ and y-intercept $(0, 6)$.

In Exercises 63–70, give the equation of a line having the stated property. *Note:* There are many answers to each exercise.

63. x-intercept $(9, 0)$

64. y-intercept $(0, 10)$

65. passes through the point $(-2, 5)$

66. passes through the point $(3, -3)$

67. crosses the positive part of the y-axis

68. passes through the origin

69. crosses the negative part of the x-axis

70. crosses the positive part of the x-axis

71. The lines with equations $y = \frac{2}{3}x - 2$ and $y = -4x + c$ have the same x-intercept. What is the value of c?

72. The lines with equations $6x - 3y = 9$ and $y = 4x + b$ have the same y-intercept. What is the value of b?

TECHNOLOGY EXERCISES

In Exercises 73–76, (a) graph the line, (b) use the utility to determine the two intercepts, (c) use the utility to find the y-coordinate of the point on the line with x-coordinate 2.

73. $y = -3x + 6$

74. $y = .25x - 2$

75. $3y - 2x = 9$

76. $2y + 5x = 8$

In Exercises 77 and 78, determine an appropriate window, and graph the line.

77. $2y + x = 100$

78. $x - 3y = 60$

Solutions to Check Your Understanding 1.1

1. Because the numbers are large, make each hatchmark correspond to 100. Then the point $(500, 200)$ is found by starting at the origin, moving 500 units to the right and 200 units up (Fig. 10 on the next page).

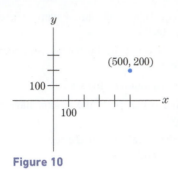

Figure 10

2.

$$2x - 3y = 1 \quad \text{Given equation}$$
$$2(4) - 3(-7) \overset{?}{=} 1 \quad x = 4, y = -7$$
$$29 \overset{?}{=} 1 \quad \text{False}$$

Since the equation is not satisfied, $(4, -7)$ is not on the graph.

$$2x - 3y = 1 \quad \text{Given equation}$$
$$2(5) - 3(3) \overset{?}{=} 1 \quad x = 5, y = 3$$
$$1 = 1 \quad \text{True}$$

Since the equation is satisfied, $(5, 3)$ is on the graph.

1.2 The Slope of a Straight Line

In this section, we consider only lines whose equations can be written in the form $y = mx + b$. Geometrically, this means that we will consider only nonvertical lines. Slope is not defined for vertical lines.

> **DEFINITION** Given a nonvertical line L with equation $y = mx + b$, the number m is called the **slope** of L. That is, the slope is the coefficient of x in the equation of the line. The equation is called the **slope–intercept** form of the equation of the line.

EXAMPLE 1 **Finding the Slope of a Line from its Equation** Find the slopes of the lines having the following equations:
(a) $y = 2x + 1$ **(b)** $y = -\frac{3}{4}x + 2$ **(c)** $y = 3$ **(d)** $-8x + 2y = 4$

SOLUTION **(a)** $m = 2$.
(b) $m = -\frac{3}{4}$.
(c) When we write the equation in the form $y = 0 \cdot x + 3$, we see that $m = 0$.
(d) First, write the equation in slope–intercept form.

$$-8x + 2y = 4 \quad \text{Given equation}$$
$$2y = 8x + 4 \quad \text{Add } 8x \text{ to both sides.}$$
$$y = 4x + 2 \quad \text{Divide both sides by 2.}$$

Thus, $m = 4$.   **» Now Try Exercise 1**

The definition of the slope is given in terms of an equation of the line. There is an alternative equivalent definition of *slope*.

> **DEFINITION Alternative Definition of Slope** Let L be a line passing through the points (x_1, y_1) and (x_2, y_2), where $x_1 \neq x_2$. Then, the slope of L is given by the formula
>
> $$m = \frac{y_2 - y_1}{x_2 - x_1}. \tag{1}$$

That is, the slope is the difference in the y-coordinates divided by the difference in the x-coordinates, with both differences formed in the same order. *Note:* x_1 is pronounced "x sub 1."

Before proving this definition equivalent to the first one given, let us show how it can be used.

EXAMPLE 2	**Finding the Slope of a Line from Two Points** Find the slope of the line passing through the points $(1, 3)$ and $(4, 6)$.

SOLUTION   We have

$$m = \frac{[\text{difference in } y\text{-coordinates}]}{[\text{difference in } x\text{-coordinates}]} = \frac{6 - 3}{4 - 1} = \frac{3}{3} = 1.$$

Thus, $m = 1$. Note that if we reverse the order of the points and use formula (1) to compute the slope, then we get

$$m = \frac{3 - 6}{1 - 4} = \frac{-3}{-3} = 1,$$

which is the same answer. The order of the points is immaterial. The important concern is to make sure that the differences in the x- and y-coordinates are formed in the same order.   **»** *Now Try Exercise 7*

The slope of a line does not depend on which pair of points we choose as (x_1, y_1) and (x_2, y_2). Consider the line $y = 4x - 3$ and two points $(1, 1)$ and $(3, 9)$, which are on the line. Using these two points, we calculate the slope to be

$$m = \frac{9 - 1}{3 - 1} = \frac{8}{2} = 4.$$

Now, let us choose two other points on the line—say, $(2, 5)$ and $(-1, -7)$—and use these points to determine m. We obtain

$$m = \frac{-7 - 5}{-1 - 2} = \frac{-12}{-3} = 4.$$

The two pairs of points give the same slope.

Justification of Formula (1)   Since (x_1, y_1) and (x_2, y_2) are both on the line, both points satisfy the equation of the line, which has the form $y = mx + b$. Thus,

$$y_2 = mx_2 + b$$
$$y_1 = mx_1 + b.$$

Subtracting these two equations gives

$$y_2 - y_1 = mx_2 - mx_1 = m(x_2 - x_1).$$

Dividing by $x_2 - x_1$, we have

$$m = \frac{y_2 - y_1}{x_2 - x_1},$$

which is formula (1). So the two definitions of slope lead to the same number.   **«**

Let us now study four of the most important properties of the slope of a straight line. We begin with the **steepness property**, since it provides us with a geometric interpretation for the number m.

Steepness Property   Let the line L have slope m. If we start at any point on the line and move 1 unit to the right, then we must move m units vertically in order to return to the line (Fig. 1 on the next page). (Of course, if m is positive, then we move up; and if m is negative, we move down.)

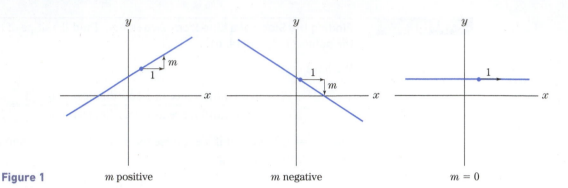

Figure 1 m positive m negative $m = 0$

EXAMPLE 3 **Steepness Property of a Line** Illustrate the steepness property for each of the lines.
(a) $y = 2x + 1$ (b) $y = -\frac{3}{4}x + 2$ (c) $y = 3$

SOLUTION (a) Here, $m = 2$. So starting from any point on the line, proceeding 1 unit to the right, we must go 2 units up to return to the line (Fig. 2).
(b) Here, $m = -\frac{3}{4}$. So starting from any point on the line, proceeding 1 unit to the right, we must go $\frac{3}{4}$ unit down to return to the line (Fig. 3).
(c) Here, $m = 0$. So going 1 unit to the right requires going 0 units vertically to return to the line (Fig. 4).

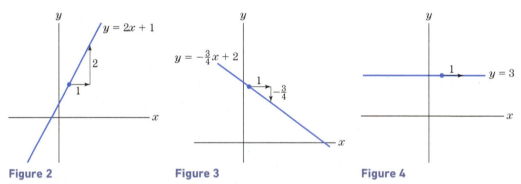

Figure 2 **Figure 3** **Figure 4**

>> *Now Try Exercise 59*

In the next example, we introduce a new method for graphing a linear equation. This method relies on the steepness property and is often more efficient than finding two points on the line (e.g., the two intercepts).

EXAMPLE 4 **Using the Steepness Property to Graph a Line** Use the steepness property to draw the graph of $y = \frac{1}{2}x + \frac{3}{2}$.

SOLUTION The y-intercept is $\left(0, \frac{3}{2}\right)$, as we read from the equation. We can find another point on the line by using the steepness property. Start at $\left(0, \frac{3}{2}\right)$. Go 1 unit to the right. Since the slope is $\frac{1}{2}$, we must move vertically $\frac{1}{2}$ unit to return to the line. But this locates a second point on the line. So we draw the line through the two points. The entire procedure is illustrated in Fig. 5.

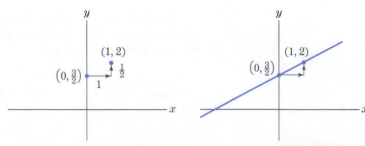

Figure 5

>> *Now Try Exercise 13*

Actually, to use the steepness property to graph an equation, all that is needed is the slope plus *any* point (not necessarily the *y*-intercept).

EXAMPLE 5   **Using the Steepness Property to Graph a Line** Graph the line of slope −1, which passes through the point (2, 2).

SOLUTION   Start at (2, 2), move 1 unit to the right and then −1 unit vertically—that is, 1 unit down. The line through (2, 2) and the resulting point is the desired line. (See Fig. 6.)   ≪

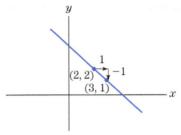

Figure 6

Slope measures the steepness of a line. That is, the slope of a line tells whether it is rising or falling, and how fast. Specifically, lines of positive slope rise as we move from left to right. Lines of negative slope fall, and lines of zero slope stay level. The larger the magnitude of the slope, the steeper the ascent or descent will be. These facts are directly implied by the steepness property. (See Fig. 7.)

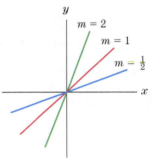 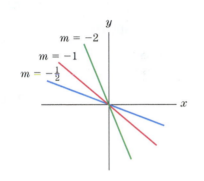

Figure 7

Justification of the Steepness Property   Consider a line with equation $y = mx + b$, and let (x_1, y_1) be any point on the line. If we start from this point and move 1 unit to the right, the first coordinate of the new point will be $x_1 + 1$, since the *x*-coordinate is increased by 1. Now, go far enough vertically to return to the line. Denote the *y*-coordinate of this new point by y_2. (See Fig. 8.) We must show that to get y_2, we add m to y_1. That is, $y_2 = y_1 + m$. By equation (1), we can compute m as

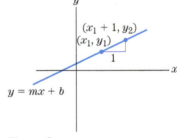

Figure 8

$$m = \frac{[\text{difference in } y\text{-coordinates}]}{[\text{difference in } x\text{-coordinates}]} = \frac{y_2 - y_1}{1} = y_2 - y_1.$$

In other words, $y_2 = y_1 + m$, which is what we desired to show.   ≪

Often, the slopes of the straight lines that occur in applications have interesting and significant interpretations. An application in the field of economics is illustrated in the next example.

EXAMPLE 6   **Slope of the Cost Line** A manufacturer finds that the cost *y* of producing *x* units of a certain commodity is given by the equation $y = 2x + 5000$. What interpretation can be given to the slope of the graph of this equation?

SOLUTION   Suppose that the firm is producing at a certain level and increases production by 1 unit. That is, *x* is increased by 1 unit. By the steepness property, the value of *y* then increases by 2, which is the slope of the line whose equation is $y = 2x + 5000$. Thus, each additional unit of production costs $2. The graph of $y = 2x + 5000$ is called a **cost curve**. It relates the size of production to total cost. The graph is a straight line, and economists call its slope the **marginal cost of production**. The *y*-coordinate of the *y*-intercept is called the **fixed cost**. In this case, the fixed cost is $5000, and it includes costs such as rent and insurance, which are incurred even if no units are produced.

≫ *Now Try Exercise 35*

In applied problems having time as a variable, the letter t is often used in place of the letter x. If so, straight lines have equations of the form $y = mt + b$ and are graphed on a ty-coordinate system.

EXAMPLE 7

Straight-Line Depreciation The federal government allows businesses an income tax deduction for the decrease in value (or **depreciation**) of capital assets (such as buildings and equipment). One method of calculating the depreciation is to take equal amounts over the expected lifetime of the asset. This method is called **straight-line depreciation**. Suppose that, for tax purposes, the value V of a piece of equipment t years after purchase is figured according to the equation $V = -100{,}000t + 700{,}000$ and the expected life of the piece of equipment is 5 years.
(a) How much did the piece of equipment originally cost?
(b) What is the annual deduction for depreciation?
(c) What is the *salvage value* of the piece of equipment? (That is, what is the value of the piece of equipment after 5 years?)

SOLUTION

(a) The original cost is the value of V at $t = 0$, namely

$$V = -100{,}000(0) + 700{,}000 = 700{,}000.$$

That is, the piece of equipment originally cost $700,000.
(b) By the steepness property, each increase of 1 in t causes a decrease in V of 100,000. That is, the value is decreasing by $100,000 per year. So the depreciation deduction is $100,000 each year.
(c) After 5 years, the value of V is given by

$$V = -100{,}000(5) + 700{,}000 = 200{,}000.$$

The salvage value is $200,000.   «

We have seen in Example 5 how to sketch a straight line when given its slope and one point on it. Let us now see how to find the equation of the line from this data.

> **Point-Slope Equation** The equation of the straight line passing through (x_1, y_1) and having slope m is given by $y - y_1 = m(x - x_1)$.

EXAMPLE 8

Finding the Equation of a Line from Its Slope and a Point on the Line Find the slope–intercept equation of the line that passes through $(2, 3)$ and has slope $\frac{1}{2}$.

SOLUTION

Here, $x_1 = 2$, $y_1 = 3$, and $m = \frac{1}{2}$. So the point–slope equation is

$$y - 3 = \tfrac{1}{2}(x - 2)$$
$$y - 3 = \tfrac{1}{2}x - 1 \qquad \text{Perform multiplication on right side.}$$
$$y = \tfrac{1}{2}x + 2 \qquad \text{Add 3 to both sides.}$$

» *Now Try Exercise 49*

EXAMPLE 9

Finding the Equation of a Line Find the slope–intercept equation of the line through the points $(3, 1)$ and $(6, 0)$.

SOLUTION

We can compute the slope from equation (1).

$$m = \frac{y_2 - y_1}{x_2 - x_1} = \frac{1 - 0}{3 - 6} = -\frac{1}{3}.$$

Now, we can determine the equation from the point–slope equation with $(x_1, y_1) = (3, 1)$ and $m = -\frac{1}{3}$.

$$y - 1 = -\tfrac{1}{3}(x - 3) \quad \text{Point-slope equation}$$
$$y - 1 = -\tfrac{1}{3}x + 1 \quad \text{Perform multiplication on right side.}$$
$$y = -\tfrac{1}{3}x + 2 \quad \text{Add 1 to both sides.}$$

[*Question:* What would the equation be if we had chosen $(x_1, y_1) = (6, 0)$?]

» *Now Try Exercise 55*

EXAMPLE 10 **Sales Generated by Advertising** For each dollar of monthly advertising expenditure, a store experiences a 6-dollar increase in sales. Even without advertising, the store has $30,000 in sales per month. Let x be the number of dollars of advertising expenditure per month, and let y be the number of dollars in sales per month.
(a) Find the equation of the line that expresses the relationship between x and y.
(b) If the store spends $10,000 in advertising, what will be the sales for the month?
(c) How much would the store have to spend on advertising to attain $150,000 in sales for the month?

SOLUTION (a) The steepness property tells us that the line has slope $m = 6$. Since $x = 0$ (no advertising expenditure) yields $y = \$30,000$, the y-intercept of the line is $(0, 30,000)$. Therefore, the slope–intercept equation of the line is

$$y = 6x + 30,000.$$

(b) If $x = 10,000$, then $y = 6(10,000) + 30,000 = 90,000$. Therefore, the sales for the month will be $90,000.
(c) We are given that $y = 150,000$, and we must find the value of x for which

$$150,000 = 6x + 30,000.$$

Solving for x, we obtain $6x = 120,000$, and hence, $x = \$20,000$. To attain $150,000 in sales, the store should invest $20,000 in advertising.   » *Now Try Exercise 45*

Verification of the Point–Slope Equation Let (x, y) be any point on the line passing through the point (x_1, y_1) and having slope m. Then, by equation (1), we have

$$m = \frac{y - y_1}{x - x_1}.$$

Multiplying through by $x - x_1$ gives

$$y - y_1 = m(x - x_1). \tag{2}$$

Thus, every point (x, y) on the line satisfies equation (2). So (2) gives the equation of the line passing through (x_1, y_1) and having slope m. «

Perpendicular and Parallel Lines

The next property of slope relates the slopes of two perpendicular lines.

Perpendicular Property When two nonvertical lines are perpendicular, their slopes are negative reciprocals of one another. That is, if two lines with nonzero slopes m and n are perpendicular to one another, then

$$m = -\frac{1}{n}.$$

Conversely, if two lines have slopes that are negative reciprocals of one another, they are perpendicular.

A proof of the perpendicular property is outlined in Exercise 88. Let us show how it can be used to help find equations of lines.

EXAMPLE 11   **Perpendicular Lines**   Find an equation of the line perpendicular to the graph of $y = 2x - 5$ and passing through $(1, 2)$.

SOLUTION   The slope of the graph of $y = 2x - 5$ is 2. By the perpendicular property, the slope of a line perpendicular to it is $-\frac{1}{2}$. If a line has slope $-\frac{1}{2}$ and passes through $(1, 2)$, it has the point-slope equation

$$y - 2 = -\tfrac{1}{2}(x - 1) \quad \text{or} \quad y = -\tfrac{1}{2}x + \tfrac{5}{2}.$$

>> Now Try Exercise 21

The final property of slope gives the relationship between slopes of parallel lines. A proof is outlined in Exercise 87.

> **Parallel Property**   Parallel lines have the same slope. Conversely, if two different lines have the same slope, they are parallel.

EXAMPLE 12   **Parallel Lines**   Find an equation of the line through $(2, 0)$ and parallel to the line whose equation is $y = \frac{1}{3}x - 11$.

SOLUTION   The slope of the line having equation $y = \frac{1}{3}x - 11$ is $\frac{1}{3}$. Therefore, any line parallel to it also has slope $\frac{1}{3}$. Thus, the desired line passes through $(2, 0)$ and has slope $\frac{1}{3}$, so its equation is

$$y - 0 = \tfrac{1}{3}(x - 2) \quad \text{or} \quad y = \tfrac{1}{3}x - \tfrac{2}{3}.$$

>> Now Try Exercise 23

INCORPORATING

TECHNOLOGY

A graphing calculator can find the equation of the line through two points. Refer to the graphing calculator discussion in the Incorporating Technology feature of Section 1.4 and find the equation of the least-squares fit to the two points.

Excel can find the equation of the line through two points. Refer to the Excel discussion in the Incorporating Technology feature of Section 1.4 and find the equation of the least-squares fit to the two points.

✹WolframAlpha The following instructions produce the equation of the line described.

line through (a, b) and (c, d)

line through (a, b) with slope m

line through (a, b) perpendicular to y = mx + b

line through (a, b) parallel to y = mx + b

Check Your Understanding 1.2   Solutions can be found following the section exercises.

Suppose that the revenue y from selling x units of a certain commodity is given by the formula $y = 4x$. (Revenue is the amount of money received from the sale of the commodity.)

1. What interpretation can be given to the slope of the graph of this equation?

2. The cost curve discussed in Example 6 intersects the revenue curve at the point $(2500, 10{,}000)$. What economic interpretation can be given to the value of the x-coordinate of the intersection point?

EXERCISES 1.2

In Exercises 1–6, find the slope of the line having the given equation.

1. $y = \frac{2}{3}x + 7$

2. $y = -4$

3. $y - 3 = 5(x + 4)$

4. $7x + 5y = 10$

5. $\frac{x}{5} + \frac{y}{4} = 6$

6. $\frac{x}{7} - \frac{y}{8} = 1$

In Exercises 7–10, plot each pair of points, draw the straight line through them, and find its slope.

7. $(3, 4), (7, 9)$

8. $(-2, 1), (3, -3)$

9. $(0, 0), (5, 4)$

10. $(4, 17), (-2, 17)$

11. What is the slope of any line parallel to the y-axis?

12. Why doesn't it make sense to talk about the slope of the line between the two points $(2, 3)$ and $(2, -1)$?

In Exercises 13–16, graph the given linear equation by beginning at the y-intercept, and moving 1 unit to the right and m units in the y-direction.

13. $y = -2x + 1$

14. $y = 4x - 2$

15. $y = 3x$

16. $y = -2$

In Exercises 17–24, find the equation of line L.

17.

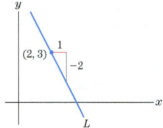

18.

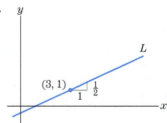

19.

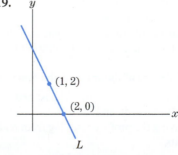

20.

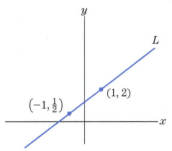

21.

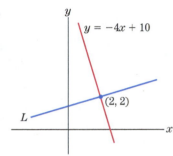

L perpendicular to $y = -4x + 10$

22.

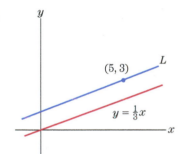

L parallel to $y = \frac{1}{3}x$

23.

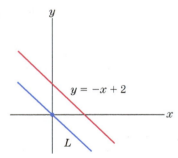

L parallel to $y = -x + 2$

24.

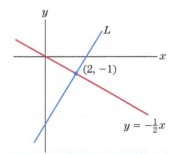

L perpendicular to $y = -\frac{1}{2}x$

In Exercises 25–28, give the slope–intercept form of the equation of the line.

25.

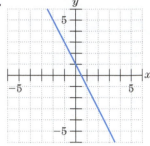

26.

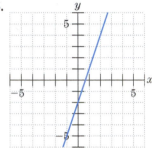

27.

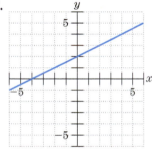

28.

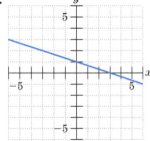

29. Find the equation of the line passing through the point $(2, 3)$ and parallel to the x-axis.

30. Find the equation of the line passing through the point $(2, 3)$ and parallel to the y-axis.

31. Find the y-intercept of the line passing through the point $(5, 6)$ and having slope $\frac{3}{5}$.

32. Find the y-intercept of the line passing through the points $(-1, 3)$ and $(4, 6)$.

33. Find the equation of the line passing through $(0, 4)$ and having undefined slope.

34. Find the equation of the line passing through the point $(1, 4)$ and having y-intercept $(0, 4)$.

35. Cost Curve A manufacturer has fixed costs (such as rent and insurance) of \$2000 per month. The cost of producing each unit of goods is \$4. Give the linear equation for the cost of producing x units per month.

36. Demand Curve The price p that must be set in order to sell q items is given by the equation $p = -3q + 1200$.
 (a) Find and interpret the p-intercept of the graph of the equation.
 (b) Find and interpret the q-intercept of the graph of the equation.
 (c) Find and interpret the slope of the graph of the equation.
 (d) What price must be set in order to sell 350 items?
 (e) What quantity will be sold if the price is \$300?
 (f) Draw the graph of the equation.

37. Boiling Point of Water At sea level, water boils at a temperature of 212°F. As the altitude increases, the boiling point of water decreases. For instance, at an altitude of 5000 feet, water boils at about 202.8°F.
 (a) Find a linear equation giving the boiling point of water in terms of altitude.
 (b) At what temperature does water boil at the top of Mt. Everest (altitude 29,029 feet)?

38. Cricket Chirps Biologists have found that the number of chirps that crickets of a certain species make per minute is related to the temperature. The relationship is very close to linear. At 68°F, those crickets chirp about 124 times a minute. At 80°F, they chirp about 172 times a minute.
 (a) Find the linear equation relating Fahrenheit temperature F and the number of chirps c.
 (b) If you count chirps for only 15 seconds, how can you quickly estimate the temperature?

39. Cost Equation Suppose that the cost of making 20 cell phones is \$6800 and the cost of making 50 cell phones is \$9500.
 (a) Find the cost equation.
 (b) What is the fixed cost?
 (c) What is the marginal cost of production?
 (d) Draw the graph of the equation.

Exercises 40–42 are related.

40. Cost Equation Suppose that the total cost y of making x coats is given by the formula $y = 40x + 2400$.
 (a) What is the cost of making 100 coats?
 (b) How many coats can be made for \$3600?
 (c) Find and interpret the y-intercept of the graph of the equation.
 (d) Find and interpret the slope of the graph of the equation.

41. Revenue Equation Suppose that the total revenue y from the sale of x coats is given by the formula $y = 100x$.
 (a) What is the revenue if 300 coats are sold?
 (b) How many coats must be sold to have a revenue of \$6000?
 (c) Find and interpret the y-intercept of the graph of the equation.
 (d) Find and interpret the slope of the graph of the equation.

42. Profit Equation Consider a coat factory with the cost and revenue equations given in Exercises 40 and 41.
 (a) Find the equation giving the profit y resulting from making and selling x coats.

(b) Find and interpret the *y*-intercept of the graph of the profit equation.

(c) Find and interpret the *x*-intercept of the graph of the profit equation.

(d) Find and interpret the slope of the graph of the profit equation.

(e) How much profit will be made if 80 coats are sold?

(f) How many coats must be sold to have a profit of $6000?

(g) Draw the graph of the equation found in part (a).

43. Heating Oil An apartment complex has a storage tank to hold its heating oil. The tank was filled on January 1, but no more deliveries of oil will be made until sometime in March. Let *t* denote the number of days after January 1, and let *y* denote the number of gallons of fuel oil in the tank. Current records show that *y* and *t* will be related by the equation $y = 30{,}000 - 400t$.

(a) Graph the equation $y = 30{,}000 - 400t$.

(b) How much oil will be in the tank on February 1?

(c) How much oil will be in the tank on February 15?

(d) Determine the *y*-intercept of the graph. Explain its significance.

(e) Determine the *t*-intercept of the graph. Explain its significance.

44. Cash Reserves A corporation receives payment for a large contract on July 1, bringing its cash reserves to $2.3 million. Let *y* denote its cash reserves (in millions) *t* days after July 1. The corporation's accountants estimate that *y* and *t* will be related by the equation $y = 2.3 - .15t$.

(a) Graph the equation $y = 2.3 - .15t$.

(b) How much cash does the corporation have on the morning of July 16?

(c) Determine the *y*-intercept of the graph. Explain its significance.

(d) Determine the *t*-intercept of the graph. Explain its significance.

(e) Determine the cash reserves on July 4.

(f) When will the cash reserves be $.8 million?

45. Weekly Pay A furniture salesperson earns $220 a week plus 10% commission on her sales. Let *x* denote her sales and *y* her income for a week.

(a) Express *y* in terms of *x*.

(b) Determine her week's income if she sells $2000 in merchandise that week.

(c) How much must she sell in a week in order to earn $540?

46. Weekly Pay A salesperson's weekly pay depends on the volume of sales. If she sells *x* units of goods, then her pay is $y = 5x + 60$ dollars. Give an interpretation to the slope and the *y*-intercept of this straight line.

In Exercises 47–58, find an equation for each of the following lines.

47. Slope is $-\frac{1}{2}$; *y*-intercept is $(0, 0)$.

48. Slope is 3; *y*-intercept is $(0, -1)$.

49. Slope is $-\frac{1}{3}$; $(6, -2)$ on line.

50. Slope is 1; $(1, 2)$ on line.

51. Slope is $\frac{1}{2}$; $(2, -3)$ on line.

52. Slope is -7; $(5, 0)$ on line.

53. Slope is $-\frac{2}{5}$; $(0, 5)$ on line.

54. Slope is 0; $(7, 4)$ on line.

55. $(5, -3)$ and $(-1, 3)$ on line.

56. $(2, 1)$ and $(4, 2)$ on line.

57. $(2, -1)$ and $(3, -1)$ on line.

58. $(0, 0)$ and $(1, -2)$ on line.

In each of Exercises 59–62, we specify a line by giving the slope and one point on the line. We give the first coordinate of some points on the line. Without deriving an equation of the line, find the second coordinate of each of the points.

59. Slope is 2, $(1, 3)$ on line; $(2, \ )$; $(0, \ )$; $(-1, \ )$.

60. Slope is -3, $(2, 2)$ on line; $(3, \ )$; $(4, \ )$; $(1, \ )$.

61. Slope is $-\frac{1}{4}$, $(-1, -1)$ on line; $(0, \ )$; $(1, \ )$; $(-2, \ )$.

62. Slope is $\frac{1}{3}$, $(-5, 2)$ on line; $(-4, \ )$; $(-3, \ )$; $(-2, \ )$.

63. Each of the lines (A), (B), (C), and (D) in Fig. 9 is the graph of one of the linear equations (a), (b), (c), and (d). Match each line with its equation.

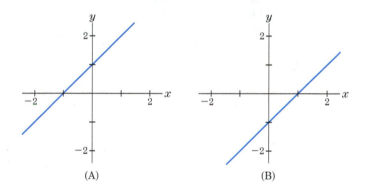

(A)   (B)

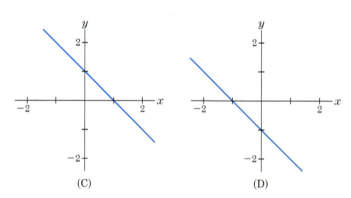

(C)   (D)

Figure 9

(a) $x + y = 1$ **(b)** $x - y = 1$

(c) $x + y = -1$ **(d)** $x - y = -1$

64. The table that follows gives several points on the line $Y_1 = mx + b$. Find *m* and *b*.

X	Y_1
4.8	3.6
4.9	4.8
5.0	6.0
5.1	7.2
5.2	8.4
5.3	9.6
5.4	**10.8**

$Y_1 = 10.8$

In Exercises 65–70, give an equation of a line with the stated property. *Note:* There are many answers to each exercise.

65. rises as you move from left to right

66. falls as you move from left to right

67. has slope 0

68. slope not defined

69. parallel to the line $2x + 3y = 4$

70. perpendicular to the line $5x + 6y = 7$

71. **Temperature Conversion** Celsius and Fahrenheit temperatures are related by a linear equation. Use the fact that $0°C = 32°F$ and $100°C = 212°F$ to find an equation.

72. **Dating of Artifacts** An archaeologist dates a bone fragment discovered at a depth of 4 feet as approximately 1500 B.C. and dates a pottery shard at a depth of 8 feet as approximately 2100 B.C. Assuming that there is a linear relationship between depths and dates at this archeological site, find the equation that relates depth to date. How deep should the archaeologist dig to look for relics from 3000 B.C.?

73. **College Tuition** The average college tuition and fees at four-year public colleges increased from \$3735 in 2001 to \$8312 in 2013. (See Fig. 10.) Assuming that average tuition and fees increased linearly with respect to time, find the equation that relates the average tuition and fees, y, to the number of years after 2001, x. What were the average tuition and fees in 2009? (*Source:* National Center for Education Statistics, *Digest of Education Statistics.*)

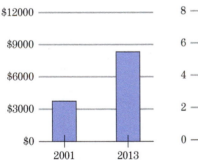

Figure 10 College Tuition

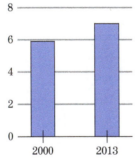

Figure 11 College Enrollments (in millions)

74. **College Enrollments** Two-year college enrollments increased from 5.9 million in 2000 to 7.0 million in 2013. (See Fig. 11.) Assuming that enrollments increased linearly with respect to time, find the equation that relates the enrollment, y, to the number of years after 2000, x. When was the enrollment 6.5 million? (*Source:* National Center for Education Statistics, *Digest of Education Statistics.*)

75. **Gas Mileage** A certain car gets 25 miles per gallon when the tires are properly inflated. For every pound of pressure that the tires are underinflated, the gas mileage decreases by $\frac{1}{2}$ mile per gallon. Find the equation that relates miles per gallon, y, to the amount that the tires are underinflated, x. Use the equation to calculate the gas mileage when the tires are underinflated by 8 pounds of pressure.

76. **Home Health Aid Jobs** According to the U.S. Department of Labor, home health aide jobs are expected to increase from 913,500 in 2014 to 1,261,900 in 2024. Assuming that the number of home health aide jobs increases linearly during that time, find the equation that relates the number of jobs, y, to the number of years after 2014, x. Use the equation to estimate the number of home health aide jobs in 2018. (*Source:* Bureau of Labor Statistics, *Occupational Projections Data.*)

77. **Bachelor's Degrees in Business** According to the U.S. National Center of Education Statistics, 263,515 bachelor's degrees in business were awarded in 2001 and 360,823 were awarded in 2013. If the number of bachelor's degrees in business continues to grow linearly, how many bachelor's degrees in business will be awarded in 2020? (*Source:* National Center for Education Statistics, *Digest of Education Statistics.*)

78. **Pizza Stores** According to *Pizza Marketing Quarterly*, the number of U.S. Domino's Pizza stores grew from 4818 in 2001 to 4986 in 2013. If the number of stores continues to grow linearly, when will there be 5100 stores?

79. **Super Bowl Commercials** The average cost of a 30-second advertising slot during the Super Bowl increased linearly from \$3.5 million in 2012 to \$4.5 million in 2015. Find the equation that relates the cost (in millions of dollars) of a 30-second slot, y, to the number of years after 2012, x. What was the average cost in 2014?

80. **Straight-Line Depreciation** A multi-function laser printer purchased for \$3000 depreciates to a salvage value of \$500 after 4 years. Find a linear equation that gives the depreciated value of the multi-function laser printer after x years.

81. **Supply Curve** Suppose that 5 million tons of apples will be supplied at a price of \$3000 per ton and 6 million tons of apples will be supplied at a price of \$3400 per ton. Find the equation for the supply curve and draw its graph. Let the units for q be millions of tons and the units for p be thousands of dollars.

82. **Demand Curve** Suppose that 5 million tons of apples will be demanded at a price of \$3000 per ton and 4.5 million tons of apples will be demanded at a price of \$3100 per ton. Find the equation for the demand curve and draw its graph. Let the units for q be millions of tons and the units for p be thousands of dollars.

83. Show that the points $(1, 3)$, $(2, 4)$, and $(3, -1)$ are *not* on the same line.

84. For what value of k will the three points $(1, 5)$, $(2, 7)$, and $(3, k)$ be on the same line?

85. Find the value of a for which the line through the points $(a, 1)$ and $(2, -3.1)$ is parallel to the line through the points $(-1, 0)$ and $(3.8, 2.4)$

86. Rework Exercise 85, where the word *parallel* is replaced by the word *perpendicular*.

87. Prove the parallel property. [*Hint:* If $y = mx + b$ and $y = m'x + b'$ are the equations of two lines, then the two lines have a point in common if and only if the equation $mx + b = m'x + b'$ has a solution for x.]

88. Prove the perpendicular property. [*Hint:* Without loss of generality, assume that both lines pass through the origin. Use the point–slope formula, the Pythagorean theorem, and Fig. 12.]

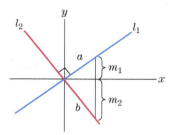

Figure 12

89. **Temperature Conversion** Figure 13 gives the conversion of temperatures from Celsius to Fahrenheit. What is the Fahrenheit equivalent of 30°C?

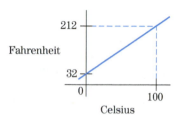

Figure 13

90. **Shipping Costs** Figure 14 gives the cost of shipping a package from coast to coast. What is the cost of shipping a 20-pound package?

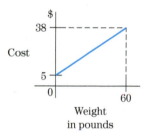

Figure 14

91. **Costs and Revenue** A T-shirt company has fixed costs of $25,000 per year. Each T-shirt costs $8.00 to produce and sells for $12.50. How many T-shirts must the company produce and sell each year in order to make a profit of $65,000?

92. **Costs and Revenue** A company produces a single product for which variable costs are $100 per unit and annual fixed costs

are $1,000,000. If the product sells for $130 per unit, how many units must the company produce and sell in order to attain an annual profit of $2,000,000?

93. **Demand and Revenue** Suppose that the quantity q of a certain brand of mountain bike sold each week depends on price according to the equation $q = 800 - 4p$. What is the total weekly revenue if a bike sells for $150?

94. **Demand and Revenue** Suppose that the number n of single-use cameras sold each month varies with the price, according to the equation $n = 2200 - 25p$. What is the monthly revenue if the price of each camera is $8?

95. **Setting a Price** During 2015, a manufacturer produced 50,000 items that sold for $100 each. The manufacturer had fixed costs of $600,000 and made a profit before income taxes of $400,000. In 2016, rent and insurance combined increased by $200,000. Assuming that the quantity produced and all other costs were unchanged, what should the 2016 price be if the manufacturer is to make the same $400,000 profit before income taxes?

96. **Setting a Price** Rework Exercise 95 with a 2015 fixed cost of $800,000 and a profit before income taxes of $300,000.

TECHNOLOGY EXERCISES

97. Graph the three lines $y = 2x - 3$, $y = 2x$, and $y = 2x + 3$ together, and then identify each line without using TRACE.

98. Graph the two lines $y = .5x + 1$ and $y = -2x + 9$ in the standard window $[-10, 10]$ by $[-10, 10]$. Do they appear perpendicular? If not, use **ZSquare** to obtain true aspect, and look at the graphs.

99. Graph the line $y = -.5x + 2$ with the window **ZDecimal**. Without pressing TRACE, move the cursor to a point on the line. Then move the cursor one unit to the right and down .5 unit to return to the line. If you start at a point on the line and move 2 units to the right, how many units down will you have to move the cursor to return to the line? Test your answer.

100. Graph the three lines $y = 2x + 1$, $y = x + 1$, and $y = .5x + 1$ together, and then identify each line without using TRACE.

101. Repeat Exercise 99 for the line $y = .7x - 2$, using *up* instead of *down* and .7 instead of .5.

Solutions to Check Your Understanding 1.2

1. By the steepness property, whenever x is increased by 1 unit, the value of y is increased by 4 units. Therefore, each additional unit of production brings in $4 of revenue. (The graph of $y = 4x$ is called a **revenue curve**, and its slope is called the **marginal revenue of production**.)

2. When 2500 units are produced, the revenue equals the cost. This value of x is called the **break-even point**. Since profit = (revenue) − (cost), the company will make a profit only if its level of production is greater than the break-even point (Fig. 15).

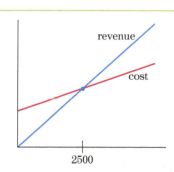

Figure 15

1.3 The Intersection Point of a Pair of Lines

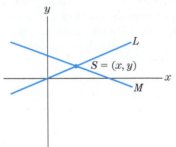

Figure 1

Suppose that we are given a pair of intersecting straight lines L and M. Let us consider the problem of determining the coordinates of the **point of intersection** $S = (x, y)$. (See Fig. 1.) We may as well assume that the equations of L and M are given in slope–intercept or vertical form. First, let us assume that both lines are in slope–intercept form—that is, that the equations are

$$L: \quad y = mx + b,$$
$$M: \quad y = nx + c.$$

Since the point S is on both lines, its coordinates satisfy both equations. In particular, we have two expressions for its y-coordinate:

$$y = mx + b = nx + c.$$

The last equality gives an equation from which x can easily be determined. Then, the value of y can be determined as $mx + b$ (or $nx + c$). Let us see how this works in a particular example.

EXAMPLE 1 **Finding the Point of Intersection** Find the point of intersection of the two lines $y = 2x - 3$ and $y = x + 1$.

SOLUTION To find the x-coordinate of the point of intersection, equate the two expressions for y and solve for x.

$$2x - 3 = x + 1 \quad \text{Equate the two expressions for } y.$$
$$x - 3 = 1 \quad \text{Subtract } x \text{ from both sides.}$$
$$x = 4 \quad \text{Add 3 to both sides.}$$

To find the value of y, set $x = 4$ in either equation—say, the first. Then,

$$y = 2 \cdot 4 - 3 = 5.$$

So the point of intersection is $(4, 5)$. See Fig. 2. **»» Now Try Exercise 1**

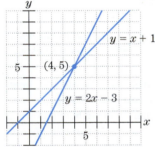

Figure 2

EXAMPLE 2 **Finding the Point of Intersection** Find the point of intersection of the two lines $x + 2y = 6$ and $5x + 2y = 18$.

SOLUTION To use the method described above, the equations must be in slope–intercept form. Solving both equations for y, we get

$$y = -\tfrac{1}{2}x + 3$$
$$y = -\tfrac{5}{2}x + 9.$$

Equating the expressions for y gives

$$-\tfrac{1}{2}x + 3 = -\tfrac{5}{2}x + 9$$
$$\tfrac{5}{2}x - \tfrac{1}{2}x + 3 = 9 \quad \text{Add } \tfrac{5}{2}x \text{ to both sides.}$$
$$2x + 3 = 9 \quad \text{Combine } x \text{ terms.}$$
$$2x = 6 \quad \text{Subtract 3 from both sides.}$$
$$x = 3 \quad \text{Divide both sides by 2.}$$

Setting $x = 3$ in the first equation gives

$$y = -\tfrac{1}{2}(3) + 3 = \tfrac{3}{2}.$$

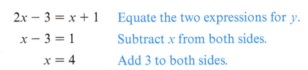

Figure 3

So the intersection point is $\left(3, \tfrac{3}{2}\right)$. See Fig. 3. **»» Now Try Exercise 3**

The preceding method works when both equations have the slope–intercept form ($y = mx + b$). In case one equation has the form $x = a$, things are much simpler. The value of x is then given directly without any work—namely, $x = a$. The value of y can be found by substituting a for x in the other equation.

EXAMPLE 3   **Finding the Point of Intersection** Find the point of intersection of the lines $y = 2x - 1$ and $x = 2$.

SOLUTION   The x-coordinate of the intersection point is 2, and the y-coordinate is $y = 2 \cdot 2 - 1 = 3$. Therefore, the intersection point is $(2, 3)$.   **≫ Now Try Exercise 5**

The method just introduced may be used to solve systems of two equations in two variables.

EXAMPLE 4   **Solving a System of Equations** Solve the following system of linear equations:

$$\begin{cases} 2x + 3y = 7 \\ 4x - 2y = 9. \end{cases}$$

SOLUTION   First, convert the equations to slope–intercept form:

$$2x + 3y = 7 \qquad \text{Given equation}$$
$$3y = -2x + 7 \qquad \text{Subtract } 2x \text{ from both sides.}$$
$$y = -\tfrac{2}{3}x + \tfrac{7}{3} \qquad \text{Divide both sides by 3.}$$

$$4x - 2y = 9 \qquad \text{Given equation}$$
$$-2y = -4x + 9 \qquad \text{Subtract } 4x \text{ from both sides.}$$
$$y = 2x - \tfrac{9}{2} \qquad \text{Divide both sides by } -2.$$

Now, equate the two expressions for y and then solve for x and y.

$$2x - \tfrac{9}{2} = -\tfrac{2}{3}x + \tfrac{7}{3}$$
$$\tfrac{8}{3}x - \tfrac{9}{2} = \tfrac{7}{3} \qquad \text{Add } \tfrac{2}{3}x \text{ to both sides.}$$
$$\tfrac{8}{3}x = \tfrac{7}{3} + \tfrac{9}{2} \qquad \text{Add } \tfrac{9}{2} \text{ to both sides.}$$
$$\tfrac{8}{3}x = \tfrac{14}{6} + \tfrac{27}{6} = \tfrac{41}{6} \qquad \text{Add fractions on right.}$$
$$x = \tfrac{3}{8} \cdot \tfrac{41}{6} = \tfrac{41}{16} \qquad \text{Multiply both sides by } \tfrac{3}{8}.$$
$$y = 2x - \tfrac{9}{2} = 2\left(\tfrac{41}{16}\right) - \tfrac{9}{2} \qquad \text{Substitute value for } x \text{ into second equation.}$$
$$y = \tfrac{41}{8} - \tfrac{36}{8} = \tfrac{5}{8} \qquad \text{Perform arithmetic.}$$

So the solution of the given system is $x = \tfrac{41}{16}$, $y = \tfrac{5}{8}$.   **≫ Now Try Exercise 9**

Supply and Demand Curves

The price p that a commodity sells for is related to the quantity q available. Economists study two kinds of graphs that express relationships between q and p. To describe these graphs, let us plot *quantity* along the horizontal axis and *price* along the vertical axis. The first graph relating q and p is called a **supply curve** (Fig. 4) and expresses the relationship between q and p from a manufacturer's point of view. For every quantity q, the supply curve specifies the price p for which the manufacturer is willing to produce the quantity q. The greater the quantity to be supplied, the higher the price must be. So supply curves rise when viewed from left to right.

The second curve relating q and p is called a **demand curve** (Fig. 5) and expresses the relationship between q and p from the consumer's viewpoint. For each quantity q,

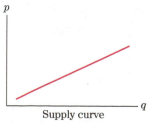

Supply curve

Figure 4

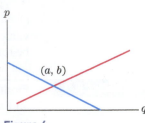

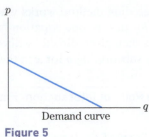

Figure 5   **Figure 6**

the demand curve gives the price p that must be charged in order for q units of the commodity to be sold. The greater the quantity that must be sold, the lower the price must be that consumers are asked to pay. So demand curves fall when viewed from left to right.

Suppose that the supply and demand curves for a commodity are drawn on a single coordinate system (Fig. 6). The intersection point (a, b) of the two curves has an economic significance: The quantity produced will stabilize at a units, and the price will be b dollars per unit. This is the *equilibrium point*.

EXAMPLE 5   **Applying the Law of Supply and Demand** Suppose that the supply curve for a certain commodity is the straight line whose equation is $p = .0002q + 2$ (p in dollars). Suppose that the demand curve for the same commodity is the straight line whose equation is $p = -.0005q + 5.5$. Determine both the quantity of the commodity that will be produced and the price at which it will sell in order for supply to equal demand.

SOLUTION   We must solve the system of linear equations

$$\begin{cases} p = .0002q + 2 \\ p = -.0005q + 5.5. \end{cases}$$

$.0002q + 2 = -.0005q + 5.5$	Equate the two expressions for p.
$.0007q + 2 = 5.5$	Add $.0005q$ to both sides.
$.0007q = 3.5$	Subtract 2 from both sides.
$q = \frac{3.5}{.0007} = 5000$	Divide both sides by $.0007$.
$p = .0002(5000) + 2$	Substitute the value for q into first equation.
$p = 1 + 2 = 3$	Perform arithmetic.

Thus, 5000 units of the commodity will be produced, and it will sell for \$3 per unit.

» *Now Try Exercise 19*

INCORPORATING

TECHNOLOGY

Graphing utilities have commands that find the intersection point of a pair of lines. Figure 7 shows the result of solving Example 4 with the intersect command of the CALC menu. Since the x-coordinate of the intersection point is assigned to ANS, the x-coordinate can be converted to a fraction by pressing MATH 1 ENTER from the home screen. See Fig. 8.

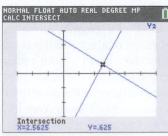

Figure 7 $[-3, 6]$ *by* $[-3, 3]$

Figure 8

❋WolframAlpha An instruction of the form

solve [first linear equation], [second linear equation]

graphs the corresponding lines and finds their intersection point. For instance, consider Example 4. The instruction

solve 2x + 3y = 7, 4x − 2y = 9

produces a graph of the two lines and displays the result "$x = \frac{41}{16} \approx 2.56250$ and $y = \frac{5}{8} \approx 0.625000$."

Check Your Understanding 1.3

Solutions can be found following the section exercises.

Figure 9 shows a type of polygon that plays a prominent role in Chapter 3; its four vertices are labeled A, B, C, and D.

1. Use the method of this section to find the coordinates of the point C.

2. Determine the coordinates of the points A and B by inspection.

3. Find the coordinates of the point D.

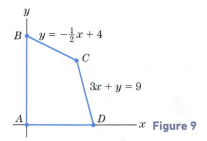

Figure 9

EXERCISES 1.3

In Exercises 1–6, find the point of intersection of the given pair of straight lines.

1. $\begin{cases} y = 4x - 5 \\ y = -2x + 7 \end{cases}$
2. $\begin{cases} y = 3x - 15 \\ y = -2x + 10 \end{cases}$
3. $\begin{cases} x - 4y = -2 \\ x + 2y = 4 \end{cases}$

4. $\begin{cases} 2x - 3y = 3 \\ y = 3 \end{cases}$
5. $\begin{cases} y = \frac{1}{3}x - 1 \\ x = 12 \end{cases}$
6. $\begin{cases} 2x - 3y = 3 \\ x = 6 \end{cases}$

7. Does $(6, 4)$ satisfy the following system of linear equations?

$$\begin{cases} x - 3y = -6 \\ 3x - 2y = 10 \end{cases}$$

8. Does $(12, 4)$ satisfy the following system of linear equations?

$$\begin{cases} y = \frac{1}{3}x - 1 \\ x = 12 \end{cases}$$

In Exercises 9–12, solve the systems of linear equations.

9. $\begin{cases} 2x + y = 7 \\ x - y = 3 \end{cases}$
10. $\begin{cases} x + 2y = 4 \\ \frac{1}{2}x + \frac{1}{2}y = 3 \end{cases}$

11. $\begin{cases} 5x - 2y = 1 \\ 2x + y = -4 \end{cases}$
12. $\begin{cases} x + 2y = 6 \\ x - \frac{1}{3}y = 4 \end{cases}$

In Exercises 13–16, find the coordinates of the labeled points.

13.

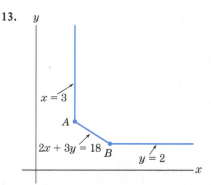

14.

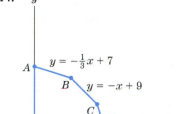
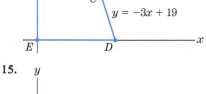
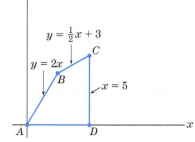

15.

16.

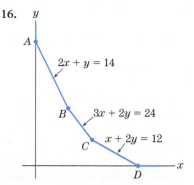

17. **Supply Curve** The supply curve for a certain commodity is $p = .0001q + .05$.
    (a) What price must be offered in order for 19,500 units of the commodity to be supplied?
    (b) What prices result in no units of the commodity being supplied?

18. **Demand Curve** The demand curve for a certain commodity is $p = -.001q + 32.5$.
    (a) At what price can 31,500 units of the commodity be sold?
    (b) What quantities are so large that all units of the commodity cannot possibly be sold no matter how low the price?

19. **Supply and Demand** Suppose that supply and demand for a certain commodity are described by the supply and demand curves of Exercises 17 and 18. Determine the equilibrium quantity of the commodity that will be produced and the selling price.

20. **Supply and Demand** A discount book seller has determined that the supply curve for a certain author's newest paperback book is $p = \frac{1}{300}q + 13$. The demand curve for this book is $p = -.03q + 19$. What quantity of sales would result in supply exactly meeting demand, and for what price should the book be sold?

21. **Supply and Demand** Suppose that the demand curve for corn has the equation $p = -.15q + 6.925$ and the supply curve for corn has the equation $p = .2q + 3.6$, where p is the price per bushel in dollars and q is the quantity (demanded or produced) in billions of bushels.
    (a) Find the quantities supplied and demanded when the price of corn is $5.80 per bushel.
    (b) Determine the equilibrium quantity of corn that will be produced and the price at which it will sell.

22. **Supply and Demand** Suppose that the demand curve for soybeans has the equation $p = -2.2q + 19.36$ and the supply curve for soybeans has the equation $p = 1.5q + 9$, where p is the price per bushel in dollars and q is the quantity (demanded or produced) in billions of bushels.
    (a) Find the quantities supplied and demanded when the price of soybeans is $16.50 per bushel.
    (b) Determine the equilibrium quantity of soybeans that will be produced and the price at which it will sell.

23. **Temperature Conversion** The formula for converting Fahrenheit degrees to Celsius degrees is $C = \frac{5}{9}(F - 32)$. For what temperature are the Celsius and Fahrenheit values the same?

24. **Temperature Conversion** The precise formula for converting Celsius degrees to Fahrenheit degrees is $F = \frac{9}{5}C + 32$. An easier-to-use formula that approximates the conversion is $F = 2C + 30$.
    (a) Compare the values given by the two formulas for a temperature of 5°C.
    (b) Compare the values given by the two formulas for a temperature of 20°C.
    (c) For what Celsius temperature do the two formulas give the same Fahrenheit temperature?

25. **Manufacturing** A clothing store can purchase a certain style of dress shirt from either of two manufacturers. The first manufacturer offers to produce shirts at a cost of $1200 plus $30 per shirt. The second manufacturer charges $500 plus $35 per shirt. Write the two equations that show the total cost y of manufacturing x shirts for each manufacturer. For what size

order will the two manufacturers charge the same amount of money? What is that amount of money?

26. **Time Apportionment** A plant supervisor must apportion her 40-hour workweek between hours working on the assembly line and hours supervising the work of others. She is paid $12 per hour for working and $15 per hour for supervising. If her earnings for a certain week are $504, how much time does she spend on each task?

27. **Calling Card Options** A calling card offers two methods of paying for a phone call. Method A charges 1 cent per minute, but has a 45-cent connection fee. Method B charges 3.5 cents per minute, but has no connection fee. Write the equations that show the total cost, y, of a call of x minutes for methods A and B, and determine their intersection point. What does the intersection point represent?

28. **Towing Fees** Sun Towing Company charges $50 plus $3 per mile to tow a car, whereas Star Towing Company charges $60 plus $2.50 per mile. Write the equations that show the total cost y of towing a car x miles for each company. For what number of miles will the two companies charge the same amount? What is that amount of money?

In Exercises 29 and 30, find the area of the shaded triangle. Each triangle has its base on one of the axes. The area of a triangle is one-half the length of its base times its height.

29.

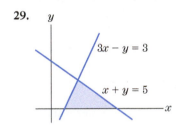

30.

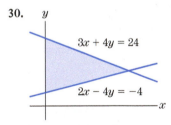

31. **Weight Determination** In a wrestling competition, the total weight of the two contestants is 700 pounds. If twice the weight of the first contestant is 275 pounds more than the weight of the second contestant, what is the weight (in pounds) of the first contestant?

32. **Sales Determination** An appliance store sells a 42″ TV for $400 and a 55″ TV of the same brand for $730. During a one-week period, the store sold 5 more 55″ TVs than 42″ TVs and collected $26,250. What was the total number of TV sets sold?

TECHNOLOGY EXERCISES

In Exercises 33–36, graph the lines and estimate the point of intersection to two decimal places.

33. $\begin{cases} y = .25x + 1.3 \\ y = -.5x + 4.1 \end{cases}$

34. $\begin{cases} y = -\frac{2}{3}x + 4.5 \\ y = 2x \end{cases}$

35. $\begin{cases} x - 4y = -5 \\ 3x - 2y = 4.2 \end{cases}$

36. $\begin{cases} 2x + 3y = 5 \\ -4x + 5y = 1 \end{cases}$

Solutions to Check Your Understanding 1.3

1. Point C is the point of intersection of the lines with equations $y = -\frac{1}{2}x + 4$ and $3x + y = 9$. To use the method of this section, the second equation must be put into its slope–intercept form $y = -3x + 9$. Now, equate the two expressions for y and solve.

$$-\tfrac{1}{2}x + 4 = -3x + 9$$
$$\tfrac{5}{2}x = 5$$
$$x = \tfrac{2}{5} \cdot 5 = 2$$
$$y = -\tfrac{1}{2}(2) + 4 = 3$$

Therefore, $C = (2, 3)$.

2. $A = (0, 0)$, because the point A is the origin. $B = (0, 4)$, because it is the y-intercept of the line with equation $y = -\frac{1}{2}x + 4$.

3. D is the x-intercept of the line $3x + y = 9$. Its first coordinate is found by setting $y = 0$ and solving for x.

$$3x + (0) = 9$$
$$x = 3$$

Therefore, $D = (3, 0)$.

1.4 The Method of Least Squares

People compile graphs of literally thousands of different quantities: the purchasing value of the dollar as a function of time, the pressure of a fixed volume of air as a function of temperature, the average income of people as a function of their years of formal education, or the incidence of strokes as a function of blood pressure. The observed points on such graphs tend to be irregularly distributed due to the complicated nature of the phenomena underlying them, as well as to errors made in observation. (For example, a given procedure for measuring average income may not count certain groups.) In spite of the imperfect nature of the data, we are often faced with the problem of making assessments and predictions based on them. Roughly speaking, this problem amounts to filtering the sources of errors in the data and isolating the basic underlying trend. Frequently, perhaps on the basis of a working hypothesis, we may suspect that the underlying trend is linear—that is, the data should lie on a straight line. But which straight line? This is the question that the **method of least squares** attempts to answer. To be more specific, let us consider the following task.

> **TASK** Given observed data points $(x_1, y_1), (x_2, y_2), \ldots, (x_N, y_N)$ in the plane, find the straight line that "best" fits these points.

In order to completely understand the statement of the task being considered, we must define what it means for a line to "best" fit a set of points. If (x_i, y_i) is one of our observed points, then we will measure how far it is from a given line $y = ax + b$ by the vertical distance, E_i, from the point to the line. (See Fig. 1.)

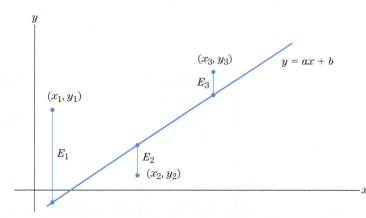

Figure 1 Fitting a line to data points

Statisticians prefer to work with the square of the vertical distance E_i. The total error in approximating the data points $(x_1, y_1), \ldots, (x_N, y_N)$ by the line $y = ax + b$ is measured by the sum E of the squares of the vertical distances from the points to the line,

$$E = E_1^2 + E_2^2 + \cdots + E_N^2.$$

E is called the **sum-of-squares error** of the observed points with respect to the line.

EXAMPLE 1 **Finding the Sum-of-Squares Error** Determine the sum-of-squares error when the line $y = 1.5x + 3$ is used to approximate the data points $(1, 6)$, $(4, 5)$, and $(6, 14)$.

SOLUTION Figure 2 shows the line, the points, and the vertical distances. The vertical distance of a point from the line is determined by finding the second coordinate of the point on the line having the same x-coordinate as the point. For instance, for the data point $(1, 6)$, the point on the line with x-coordinate 1 has y-coordinate $y = 1.5(1) + 3 = 4.5$ and, therefore, vertical distance $6 - 4.5 = 1.5$. Table 1 summarizes the vertical distances. The sum-of-squares error is $1.5^2 + 4^2 + 2^2 = 2.25 + 16 + 4 = 22.25$.

Table 1 Vertical Distances from the Line $y = 1.5x + 3$

Data Point	Point on Line	Vertical Distance
$(1, 6)$	$(1, 4.5)$	1.5
$(4, 5)$	$(4, 9)$	4
$(6, 14)$	$(6, 12)$	2

» *Now Try Exercise 1*

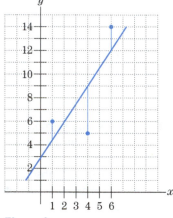

Figure 2

In general, we cannot expect to find a line $y = mx + b$ that fits the observed points so well that the sum-of-squares error E is zero. Actually, this situation will occur only if the observed points lie on a straight line. However, we can rephrase our original task as follows:

> **TASK** Given data points $(x_1, y_1), (x_2, y_2), \ldots, (x_N, y_N)$ in the plane, find the straight line $y = mx + b$ for which the sum-of-squares error E is as small as possible. This line is called the **least-squares line** or the **regression line**.

In this day and age, the task is always solved with technology. However, there is a computational method of solving the task that does not require modern technology. We will show both ways of finding the least-squares line for the data points in Example 1.

Obtaining the Least-Squares Line with a Graphing Calculator

On the TI-84 Plus graphing calculator screens in Fig. 3, the data points are entered into lists, the least-squares line is calculated with the item **LinReg(ax+b)** of the STAT/CALC menu, and the data points and line are plotted with [STAT PLOT] and $\boxed{\text{GRAPH}}$. The end of this section contains the details for obtaining least-squares lines with a graphing calculator.

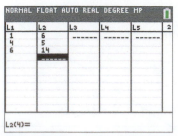  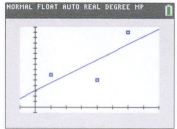

Figure 3 Obtaining a least-squares line with a TI-84 Plus

Obtaining a Least-Squares Line with an Excel Spreadsheet

Excel has special functions that calculate the slope and y-intercept of the least-squares line for a collection of data points. In Fig. 4, the least-squares line of Example 1 is calculated and graphed in Excel 2016. The steps used to obtain the graph in Fig. 4 are given at the end of this section.

	A	B
1	1	6
2	4	5
3	6	14
4	slope	1.447368421
5	y-intercept	3.026315789
6		
7		
8	Cell	Formula
9	B4	=SLOPE(B1:B3,A1:A3)
10	B5	=INTERCEPT(B1:B3,A1:A3)

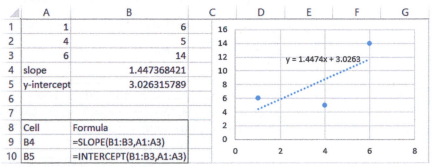

Figure 4 Obtaining a least-squares line with Excel 2016

Obtaining a Least-Squares Line with Wolfram|Alpha

Figure 5 shows how to ask for the least-squares line. Figure 6 shows part of the result after the equal sign is clicked.

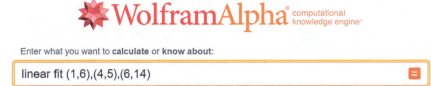

Figure 5 Ask Wolfram|Alpha to calculate a least-squares line

Least-squares best fit:

$1.44737\,x + 3.02632$

Figure 6 Output produced by Wolfram|Alpha

The Computational Method for Obtaining a Least-Squares Fit

The least-squares line for the general case given in the TASK box on page 26 can be obtained with the following formulas.

$$m = \frac{N \cdot \sum xy - \sum x \cdot \sum y}{N \cdot \sum x^2 - (\sum x)^2}$$

$$b = \frac{\sum y - m \cdot \sum x}{N},$$

where

$\sum x$ = sum of the x-coordinates of the data points

$\sum y$ = sum of the y-coordinates of the data points

$\sum xy$ = sum of the products of the coordinates of the data points

$\sum x^2$ = sum of the squares of the x-coordinates of the data points

N = number of data points.

That is,

$$\Sigma x = x_1 + x_2 + \cdots + x_N$$
$$\Sigma y = y_1 + y_2 + \cdots + y_N$$
$$\Sigma xy = x_1 \cdot y_1 + x_2 \cdot y_2 + \cdots + x_N \cdot y_N$$
$$\Sigma x^2 = x_1^2 + x_2^2 + \cdots + x_N^2.$$

EXAMPLE 2 **Finding the Least-Squares Line** Use the computational method to find the least-squares line for the data points of Example 1.

SOLUTION The sums are calculated in Table 2 and then used to determine the values of m and b.

Table 2

x	y	xy	x^2
1	6	6	1
4	5	20	16
6	14	84	36
$\Sigma x = 11$	$\Sigma y = 25$	$\Sigma xy = 110$	$\Sigma x^2 = 53$

$$m = \frac{3 \cdot 110 - 11 \cdot 25}{3 \cdot 53 - 11^2} = \frac{55}{38} \approx 1.45$$

$$b = \frac{25 - \frac{55}{38} \cdot 11}{3} = \frac{38 \cdot 25 - 55 \cdot 11}{38 \cdot 3} = \frac{345}{114} = \frac{115}{38} \approx 3.03$$

Therefore, the equation of the least-squares line is $y = \frac{55}{38}x + \frac{115}{38}$. With this line, the sum-of-squares error can be shown to be about 22.13. **» Now Try Exercise 7**

EXAMPLE 3 **Applying the Least-Squares Line** Table 3 gives the U.S. per-capita health expenditures for several years. The least-squares line for these data is $y = 311.3x + 8394$, where x is the number of years since 2010. The figures in the margin verify the computations.

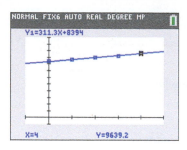

Table 3 U.S. Per-Capita Health Expenditures

Year	2010	2011	2012	2013	2014
Dollars	8428	8698	8996	9255	9706

(*Source:* Centers for Medicare and Medicaid Services, *NHE Fact Sheet.*)

(a) Assuming that the least-squares line continues to describe health expenditures, estimate the per-capita health expenditures for the year 2017.

(b) Assuming that the least-squares line continues to describe health expenditures, when will the per-capita health expenditures reach $14,000?

SOLUTION **(a)** The year 2017 corresponds to $x = 7$.

$$y = 311.3(7) + 8394$$

$$\approx 10{,}573.10$$

Therefore, an estimate of per-capita health expenditures in the year 2017 is $10,573.

(b) Set the value of y equal to 14,000, and solve for x.

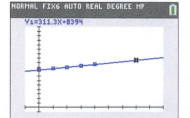

$y = 311.3x + 8394$	Least-squares line
$14{,}000 = 311.3x + 8394$	$y = 14{,}000$
$5606 = 311.3x$	Subtract 8394 from both sides.
$x = 18.0084$	Divide both sides by 311.3. Rewrite.

Expenditures are projected to reach $14,000 in approximately 18 years after 2010—that is, in the year 2028.

» Now Try Exercise 19

INCORPORATING

TECHNOLOGY

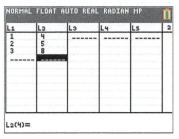

Figure 7

Figure 8

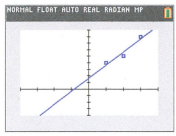

Figure 9

The following steps show how to use a calculator to find the straight line that minimizes the sum-of-squares error for the points (1,4), (2,5), and (3,8):

1. Press $\boxed{\text{STAT}}$ **1** to obtain a table to be used for entering the data.
2. If there are no data in columns labeled $\mathbf{L}_1$ and $\mathbf{L}_2$, proceed to step 4.
3. Move the cursor up to $\mathbf{L}_1$, and press $\boxed{\text{CLEAR}}$ $\boxed{\text{ENTER}}$ to delete all data in $\mathbf{L}_1$'s column. Move the cursor right and up to $\mathbf{L}_2$, and press $\boxed{\text{CLEAR}}$ $\boxed{\text{ENTER}}$ to delete all data in $\mathbf{L}_2$'s column.
4. If necessary, move the cursor left to the first blank row of the $\mathbf{L}_1$ column. Press **1** $\boxed{\text{ENTER}}$ **2** $\boxed{\text{ENTER}}$ **3** $\boxed{\text{ENTER}}$ to place the x-coordinates of the three points into the $\mathbf{L}_1$ column.
5. Move the cursor right to the $\mathbf{L}_2$ column, and press **4** $\boxed{\text{ENTER}}$ **5** $\boxed{\text{ENTER}}$ **8** $\boxed{\text{ENTER}}$ to place the y-coordinates of the three points into the $\mathbf{L}_2$ column. The screen should now appear as in Fig. 7.
6. Press $\boxed{\text{STAT}}$ $\boxed{\blacktriangleright}$ and press the number for **LinReg(ax+b)**.
7. Press $\boxed{\text{ENTER}}$ five times. The screen should now appear as in Fig. 8. The least-squares line is $y = 2x + \frac{5}{3}$. (*Note:* $\frac{5}{3} \approx 1.666666667$.)
8. If desired, the linear function can be assigned to $\mathbf{Y}_1$ with the following steps:
   (a) Press $\boxed{\text{Y=}}$ $\boxed{\text{CLEAR}}$ to erase the current expression in $\mathbf{Y}_1$
   (b) Press $\boxed{\text{VARS}}$ **5** $\boxed{\blacktriangleright}$ $\boxed{\blacktriangleright}$ and the number for **RegEQ** to assign the linear function to $\mathbf{Y}_1$.
9. The original points can be easily plotted along with the least-squares line. Assume that the linear function has been assigned to $\mathbf{Y}_1$, that all other functions have been cleared or deselected, and that the window has been set to $[-4, 4]$ *by* $[-4, 9]$. Press $\boxed{\text{2nd}}$ [STAT PLOT] $\boxed{\text{ENTER}}$ $\boxed{\text{ENTER}}$ $\boxed{\text{GRAPH}}$ to see the display in Fig. 9. *Note 1:* To turn off the point-plotting feature, press $\boxed{\text{2nd}}$ [STAT PLOT] $\boxed{\text{ENTER}}$ $\boxed{\blacktriangleright}$ $\boxed{\text{ENTER}}$. *Note 2:* The point-plotting feature can be toggled from the $\mathbf{Y}$= editor by moving the cursor to the word "**Plot1**" on the top line and pressing $\boxed{\text{ENTER}}$.

The following nine steps produce the graph in Fig. 4 on page 27 using Excel 2016.

1. Enter the six numbers in the range A1:B3, and then select the range.
2. Click on the $\boxed{\text{Insert}}$ tab, and then click on the scatter icon (⠿) on the ribbon. (A drop-down box titled *Scatter* will appear.)
3. Click on the first icon (⠿) in the drop-down box. (A coordinate system showing the three points will appear.)
4. Click on the **Chart Elements** button (+) at the upper-right corner of the coordinate system to display a **Chart Elements** list.
5. In the **Chart Elements** list, uncheck the Chart Title box and click on the Trendline box. (A dotted least-squares line will appear.)
6. Click on the arrowhead to the right of the word "Trendline," and then click on More Options in the list that appears. (A **Format Trendline** pane will appear on the right side of the screen.)
7. If the next to last line of the pane does not contain the words "Display Equation on chart," click on the histogram icon (▥) near the top of the dialog box.
8. Check the "Display Equation on chart" box.
9. Click on the Close button (×) at the upper-right corner of the dialog box.

Excel can make least-squares projections by dragging the fill handle. Fig. 10(a) contains data on college enrollments. If you select the data in column B, as shown in Fig. 10(b), and then drag the fill handle down three rows, Excel will insert values into the three empty cells in column B as shown in Fig. 10(c). Excel carries out the computation by determining the least-squares equation for the four selected data values and uses the equation to calculate the values for the three empty cells.

	A	B
		Enrollment
1	Year	(in millions)
2	2013	21.49
3	2014	21.73
4	2015	21.94
5	2016	22.19
6	2017	
7	2018	
8	2019	

(a)

	A	B
		Enrollment
1	Year	(in millions)
2	2013	21.49
3	2014	21.73
4	2015	21.94
5	2016	22.19
6	2017	
7	2018	
8	2019	

(b)

	A	B
		Enrollment
1	Year	(in millions)
2	2013	21.49
3	2014	21.73
4	2015	21.94
5	2016	22.19
6	2017	22.42
7	2018	22.65
8	2019	22.88

(c)

Figure 10 Calculate a least-squares projection

WolframAlpha The instruction

$$\text{linear fit } (x_1, y_1), (x_2, y_2), \ldots, (x_N, y_N)$$

displays an equation and a graph for the least-squares line.

Check Your Understanding 1.4

Solutions can be found following the section exercises.

1. Can a vertical distance be negative?

2. Under what condition will a vertical distance be zero?

EXERCISES 1.4

1. Suppose that the line $y = 3x + 1$ is used to fit the four data points in Table 4. Complete the table, and determine the sum-of-squares error E.

Table 4

Data Point	Point on Line	Vertical Distance
(1, 3)		
(2, 6)		
(3, 11)		
(4, 12)		

2. Suppose that the line $y = -2x + 12$ is used to fit the four data points in Table 5. Complete the table, and determine the sum-of-squares error E.

Table 5

Data Point	Point on Line	Vertical Distance
(1, 11)		
(2, 7)		
(3, 5)		
(4, 5)		

3. Find the sum-of-squares error E for the least-squares line fit to the four points in Fig. 11.

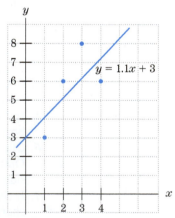

Figure 11

4. Find the sum-of-squares error E for the least-squares line fit to the five points in Fig. 12.

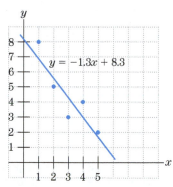

$y = -1.3x + 8.3$

Figure 12

5. Complete Table 6, and find the values of m and b for the straight line that provides the least-squares fit to the data.

Table 6

x	y	xy	x^2
1	7		
2	6		
3	4		
4	3		
$\Sigma x =$	$\Sigma y =$	$\Sigma xy =$	$\Sigma x^2 =$

6. Complete Table 7, and find the values of m and b for the straight line that provides the least-squares fit to the data.

Table 7

x	y	xy	x^2
1	2		
2	4		
3	7		
4	9		
5	12		
$\Sigma x =$	$\Sigma y =$	$\Sigma xy =$	$\Sigma x^2 =$

7. Consider the data points $(1, 2)$, $(2, 5)$, and $(3, 11)$. Find the straight line that provides the least-squares fit to these data.

8. Consider the data points $(1, 8)$, $(2, 4)$, and $(4, 3)$. Find the straight line that provides the least-squares fit to these data.

9. Consider the data points $(1, 9)$, $(2, 8)$, $(3, 6)$, and $(4, 3)$. Find the straight line that provides the least-squares fit to these data.

10. Consider the data points $(1, 5)$, $(2, 7)$, $(3, 6)$, and $(4, 10)$. Find the straight line that provides the least-squares fit to these data.

11. Consider the data points $(5, 4)$ and $(7, 3)$.
    (a) Find the straight line that provides the least-squares fit to these data.
    (b) Use the method from Section 1.2 to find the equation of the straight line passing through the two points.
    (c) Explain why we could have predicted that the straight line in (b) would be the same as the straight line in (a).

12. The data points $(2, 3)$, $(5, 9)$, and $(10, 19)$ all lie on the line $y = 2x - 1$. Explain why that line must be the least-squares fit to the three data points.

13. According to Example 2, the sum-of-squares error for the least-squares fit to the data points $(1, 6)$, $(4, 5)$, and $(6, 14)$ is $E = 22.13$.
    (a) Find the equation of the straight line through the two points $(1, 6)$ and $(6, 14)$.
    (b) What is the sum-of-squares error when the line in (a) is used to fit the three data points?

14. According to Example 2, the sum-of-squares error for the least-squares fit to the data points $(1, 6)$, $(4, 5)$, and $(6, 14)$ is $E = 22.13$.
    (a) Find the equation of the straight line through the two points $(4, 5)$ and $(6, 14)$.
    (b) What is the sum-of-squares error when the line in (a) is used to fit the three data points?

15. **Fuel Economy** The following table gives the city and highway miles per gallon for four hybrid cars: (*Source:* www.fueleconomy.gov.)

Model	City MPG	Highway MPG
2016 Toyota Prius	54	50
2016 Honda Accord	50	45
2016 Ford Fusion FWD	44	41
2016 Toyota Camry LE	43	39

(a) Obtain the least-squares line that fits these data.
(b) Use the equation from (a) to estimate the highway mpg for a hybrid car that gets 47 mpg in city driving.
(c) Use the equation from (a) to estimate the city mpg for a hybrid car that gets 47 mpg in highway driving.

16. **Pizzerias** The following table gives the number of stores and the amount of sales (in millions of dollars) for the leading U.S. pizza chains in 2013. (*Source:* PMQ.com.)
    (a) Obtain the least-squares line that fits these data. (Let x represent the number of stores in thousands.)

(b) Use the equation from (a) to estimate the 2013 sales for a pizza chain having 4000 stores.

(c) Use the equation from (a) to estimate the number of stores for a pizza chain whose 2013 sales were 2000 million dollars.

Pizzeria	Number of Stores	Sales (in Millions of Dollars)
Pizza Hut	6326	5700
Domino's	4986	3800
Little Caesars	3890	3025
Papa John's	3207	2485

17. Lung Cancer and Smoking The following table gives the crude male death rate for lung cancer in 1950 and the per capita consumption of cigarettes in 1930 in various countries. Figure 13 shows the least-squares line for the data. (*Source:* U.S. Dept. of Health, Education, and Welfare)

Country	Cigarette Consumption (Per Capita)	Lung Cancer Deaths (Per Million Males)
Norway	250	95
Sweden	300	120
Denmark	350	165
Australia	470	170

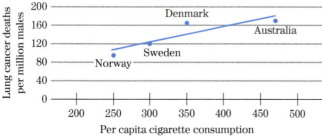

Figure 13

(a) Obtain the equation of the line in Fig. 13.

(b) In 1930, the per capita cigarette consumption in Finland was 1100. Use the equation found in part (a) to estimate the male lung cancer death rate in Finland in 1950.

18. Cigarette Use The percentage of college freshmen who smoke declined substantially from the year 2004 to the year 2014. Figure 14 shows the percentage of college freshmen who smoked during six of the years of that time period and the least-squares line for the data, where x represents the number of years after 2004. (*Source:* Higher Education Research Institute, UCLA.)

(a) Obtain the least-squares line that fits the data.

(b) Use the equation to determine the percentage of the 2009 freshmen class who smoked.

(c) According to the least-squares line, in what year did approximately 4.9% of college freshmen smoke?

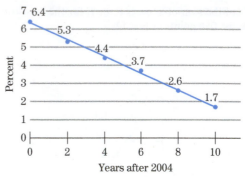

Figure 14 Percentage of College Freshmen Who Smoke

19. College Graduates Figure 15 gives the percent of persons 25 years and over who have completed four or more years of college. (*Source:* U.S. Dept. of Education.)

(a) Obtain the least-squares line that fits these data. (Let $x = 0$ correspond to 1989.)

(b) Estimate the percent for the year 2012.

(c) If the trend determined by the straight line in part (a) continues, when will the percent reach 35?

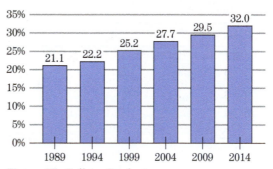

Figure 15 College Graduates

20. Average College Costs Figure 16 gives the average college costs (in thousands of dollars) for public colleges in the U.S. (*Source:* U.S. Dept. of Education.)

(a) Obtain the least-squares line that fits these data. (Let $x = 0$ correspond to 2005.)

(b) Estimate the average cost in 2010.

(c) If the trend determined by the straight line in part (a) continues, when will the average cost reach $22,000?

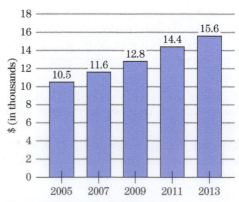

Figure 16 Average College Costs

21. Life Expectancy The following table is an abbreviated life expectancy table for U.S. males:

Current Age	Life Expectancy
0	76.4
20	77.3
40	78.7
60	81.7
80	88.3

(a) Find the straight line that provides the least-squares fit to these data.

(b) Use the straight line of part (a) to estimate the life expectancy of a 30-year-old U.S. male. *Note:* The actual life expectancy is 78 years.

(c) Use the straight line of part (a) to estimate the life expectancy of a 50-year-old U.S. male. *Note:* The actual life expectancy is 79.7 years.

(d) Use the straight line of part (a) to estimate the life expectancy of a 90-year-old U.S. male. *Note:* The actual life expectancy is 94.1 years.

22. Banking Two Harvard economists studied countries' relationships between the independence of banks and inflation rates from 1955 to 1990. The independence of banks was rated on a scale of -1.5 to 2.5, with -1.5, 0, and 2.5 corresponding to least, average, and most independence, respectively. The following table gives the values for various countries. (*Source:* Harvard University; The World Bank.)

Country	Independence Rating	Inflation Rate (%)
New Zealand	-1.4	7.6
Italy	$-.75$	7.2
Belgium	.3	4.0
France	.4	6.0
Canada	.9	4.5
United States	1.6	4.0
Switzerland	2.2	3.1

(a) Obtain the least-squares line that fits these data.

(b) What relationship between independence of banks and inflation is indicated by the least-squares line?

(c) Japan has a .6 independence rating. Use the least-squares line to estimate Japan's inflation rate.

(d) The inflation rate for Britain is 6.8. Use the least-squares line to estimate Britain's independence rating.

23. Consumer Price Index The following table gives the average price of a pound of spaghetti and macaroni in January of the given years. (*Source:* U.S. Bureau of Labor Statistics, Consumer Price Index.)

Year	Price
2000	$0.88
2004	$0.92
2008	$1.02
2011	$1.21
2015	$1.26

(a) Obtain the least-squares line that fits these data. (Let $x = 0$ correspond to 2000.)

(b) Estimate the average price of a pound of spaghetti and macaroni in January 2013.

(c) If this trend continues, when will the average price of a pound of spaghetti and macaroni be $1.45?

24. Greenhouse Gases Although greenhouse gases are essential to maintaining the temperature of the Earth, an excess of greenhouse gases can raise the temperature to dangerous levels. One of the most threatening greenhouse gases is carbon dioxide (CO_2). The Mauna Loa atmospheric CO_2 measurements constitute the longest continuous record of atmospheric CO_2 concentrations available in the world. The following table shows the concentration of CO_2 (in parts per million) at Mauna Loa, Hawaii, for three years. (*Source:* noaa.gov.)

Year	CO_2 (ppm)
1968	323
1999	368
2015	401

(a) Obtain the least-squares line that fits these data. (Let $x = 0$ correspond to 1968.)

(b) Estimate the concentration of CO_2 in the year 2008. How does this compare with the actual reading of 386?

(c) If the trend continues, in what year will the concentration of CO_2 reach 408 ppm?

Solutions to Check Your Understanding 1.4

1. No. The word *distance* implies a nonnegative number. It is the absolute value of the difference between the y-coordinate of the data point and the y-coordinate of the point on the line.

2. The vertical distance will be zero when the point actually lies on the least-squares line.

CHAPTER 1   Summary

KEY TERMS AND CONCEPTS	EXAMPLES

1.1 Coordinate Systems and Graphs

The **graph** of $ax + by = c$ is a straight line.

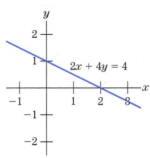

Standard forms: $y = mx + b$ (slope–intercept form)
$\qquad x = a$ (vertical form)

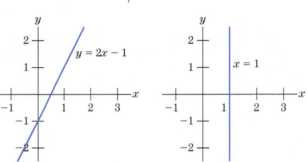

To find an *x*-**intercept:** Set $y = 0$, and solve for x.

Find the intercepts of $2x + 4y = 4$.

$$2x + 4(0) = 4 \quad \text{Set } y = 0.$$
$$2x = 4 \quad \text{Simplify.}$$
$$x = 2 \quad \text{Divide by 2.}$$

To find a *y*-**intercept:** Set $x = 0$, and solve for y.

$$2(0) + 4y = 4 \quad \text{Set } x = 0.$$
$$4y = 4 \quad \text{Simplify.}$$
$$y = 1 \quad \text{Divide by 4.}$$

1.2 The Slope of a Straight Line

Slope–intercept form
The equation of the line with slope m and y intercept $(0, b)$ is $y = mx + b$.

The line with equation $y = 2x + 5$ has slope 2 and y-intercept $(0, 5)$.

Point–slope form
The equation of the line of slope m passing through (x_1, y_1) is $y - y_1 = m(x - x_1)$.

The line of slope 2 that passes through $(3, 4)$ has equation $y - 4 = 2(x - 3)$.

The line passing through (x_1, y_1) and (x_2, y_2) has **slope** $m = \frac{y_2 - y_1}{x_2 - x_1}$.

The line through $(3, 4)$ and $(5, 8)$ has slope $\frac{8 - 4}{5 - 3} = \frac{4}{2} = 2$.

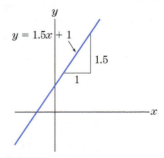

Steepness property: If we start at any point on a line of slope m and move 1 unit to the right, then we must move m units vertically to return to the line.

Parallel lines have the same slope.

The parallel lines $y = 3x + 4$ and $y = 3x - 5$ both have slope 3.

Slopes of **perpendicular lines** are negative reciprocals of each other.

A line perpendicular to $y = 5x + 2$ has slope $-\frac{1}{5}$.

KEY TERMS AND CONCEPTS	EXAMPLES

1.3 The Intersection Point of a Pair of Lines

One method for finding the **point of intersection** of a pair of lines is as follows:

1. Write each of the equations of the lines in slope–intercept or vertical form.

2. If both equations are in slope–intercept form, equate the two expressions for y.

3. Otherwise, substitute the value of x from the equation of the form $x = a$ into the other equation.

Solve $\begin{cases} 4x - 2y = 16 \\ 3x + y = 17 \end{cases}$

$\begin{cases} y = 2x - 8 \\ y = -3x + 17 \end{cases}$ Write equations in slope–intercept form.

$2x - 8 = -3x + 17$ Equate expressions for y.
$5x = 25$ Add $3x$ and 8 to both sides.
$x = 5$ Divide by 5.
$y = 2 \cdot 5 - 8 = 2$ Substitute into first equation.

Therefore, $x = 5, y = 2$.

Solve $\begin{cases} 5x - y = 8 \\ x = 3 \end{cases}$

$\begin{cases} y = 5x - 8 \\ x = 3 \end{cases}$ Write equations in slope–intercept or vertical form.

$y = 5 \cdot 3 - 8 = 7$ Substitute $x = 3$ into first equation.

Therefore, $x = 3, y = 7$.

1.4 The Method of Least Squares

Let $y = mx + b$ be the **least-squares line** for the points (x_1, y_1), $(x_2, y_2), \ldots, (x_N, y_N)$. Then,

$$m = \frac{N \cdot \Sigma xy - \Sigma x \cdot \Sigma y}{N \cdot \Sigma x^2 - (\Sigma x)^2}$$

$$b = \frac{\Sigma y - m \cdot \Sigma x}{N},$$

where

$\Sigma x = x_1 + x_2 + \cdots + x_N$

$\Sigma y = y_1 + y_2 + \cdots + y_N$

$\Sigma xy = x_1 \cdot y_1 + x_2 \cdot y_2 + \cdots + x_N \cdot y_N$

$\Sigma x^2 = x_1^2 + x_2^2 + \cdots + x_N^2.$

Find the least-squares line for the points $(2, 7)$, $(5, 9)$, and $(6, 14)$.

$$\Sigma x = 2 + 5 + 6 = 13$$
$$\Sigma y = 7 + 9 + 14 = 30$$
$$\Sigma xy = 2 \cdot 7 + 5 \cdot 9 + 6 \cdot 14 = 143$$
$$\Sigma x^2 = 4 + 25 + 36 = 65$$
$$m = \frac{3 \cdot 143 - 13 \cdot 30}{3 \cdot 65 - 13^2} = \frac{39}{26} = 1.5$$
$$b = \frac{30 - 1.5 \cdot 13}{3} = \frac{10.5}{3} = 3.5$$

Therefore, $y = 1.5x + 3.5$.

CHAPTER 1 Fundamental Concept Check Exercises

1. How do you determine the coordinates of a point in the plane?

2. What is meant by the *graph of an equation* in x and y?

3. What is the y-intercept of a line? How do you find the y-intercept from an equation of a line?

4. What is the x-intercept of a line? How do you find the x-intercept from an equation of a line?

5. Give a method for graphing the equation $y = mx + b, m \neq 0$.

6. What is the general form of a linear equation in x and y?

7. Define the slope of a line, and give a physical description.

8. What is the slope–intercept form of a non-vertical line?

9. Suppose that you know the slope and the coordinates of a point on a line. How could you draw the line without first finding its equation?

10. What is the point–slope form of the equation of a line?

11. Describe how to find the equation for a line when you know the coordinates of two points on the line.

12. What can you say about the slopes of perpendicular lines?

13. What can you say about the slopes of parallel lines?

14. Describe how to obtain the point of intersection of two lines.

15. What is the least-squares line approximation to a set of data points?

CHAPTER 1 Review Exercises

1. What is the equation of the y-axis?

2. Graph the linear equation $y = -\frac{1}{2}x$.

3. Find the point of intersection of the pair of straight lines $x - 5y = 6$ and $3x = 6$.

4. Find the slope of the line having equation $3x - 4y = 8$.

5. Find the equation of the line having y-intercept $(0, 5)$ and x-intercept $(10, 0)$.

6. Find the point of intersection of the pair of straight lines $2x - y = 1$ and $x + 2y = 13$.

7. Find the equation of the straight line passing through the point $(15, 16)$ and parallel to the line $2x - 10y = 7$.

8. Find the y-coordinate of the point having x-coordinate 1 and lying on the line $y = 3x + 7$.

9. Find the x-intercept of the straight line with equation $x = 5$.

10. Solve the system of linear equations.

$$\begin{cases} 3x - 2y = \ \ 1 \\ 2x + \ \ y = 24. \end{cases}$$

11. Find the y-intercept of the line passing through the point $(4, 9)$ and having slope $\frac{1}{2}$.

12. **Cost of Moving** The fee charged by a local moving company depends on the amount of time required for the move. If t hours are required, then the fee is $y = 35t + 20$ dollars. Give an interpretation of the slope and y-intercept of this line.

13. Are the points $(1, 2)$, $(2, 0)$, and $(3, 1)$ on the same line?

14. Write an equation of the line with x-intercept $(3, 0)$ and y-intercept $(0, -2)$.

15. If $x + 7y = 30$ and $x = -2y$, then determine the value of y.

16. Solve the system of linear equations

$$\begin{cases} 1.2x + 2.4y = \ .6 \\ 4.8y - 1.6x = 2.4. \end{cases}$$

17. Find the equation of the line through $(1, 1)$ and the intersection point of the lines $y = -x + 1$ and $y = 2x + 3$.

18. Do the three graphs of the linear equations $2x - 3y = 1$, $5x + 2y = 0$, and $x + y = 1$ contain a common point?

19. Graph the equation $x + \frac{1}{2}y = 4$, and give the slope and both intercepts.

20. Show that the lines with equations $2x - 3y = 1$ and $3x + 2y = 4$ are perpendicular.

21. Each of the lines L_1, L_2, and L_3 in Fig. 1 is the graph of one of the equations (a), (b), and (c). Match each of the equations with its corresponding line.

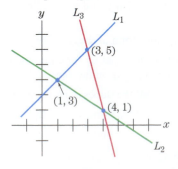

Figure 1

(a) $4x + y = 17$ (b) $y = x + 2$ (c) $2x + 3y = 11$

22. **Supply and Demand** Consider the following four equations.

$$p = .01q - 5$$
$$p = .005q + .5$$
$$p = -.01q - 10$$
$$p = -.01q + 5$$

One is the equation of a supply curve, and another is the equation of a demand curve. Identify the two equations, and then find the intersection point of those two curves.

23. **Out-of-State Students** Many of the nation's prominent public universities are shifting to admitting more out-of-state students. The number of out-of-state freshmen at the University of Alabama x years after 2004 is given approximately by the linear equation $y = 3.6x + 28$. (*Source: The Washington Post.*)
    (a) What interpretation can be given to the y-intercept of the graph of the equation?
    (b) Approximately what percent of freshmen entering the University of Alabama in 2009 were from out of state?
    (c) If this trend continues, in what year will approximately 82% of the University of Alabama freshmen be from out of state?

24. **Profit Equation** For a certain manufacturer, the production and sale of each additional unit yields an additional profit of $10. The sale of 1000 units yields a profit of $4000.
    (a) Find the profit equation.
    (b) Find the y- and x-intercepts of the graph of the profit equation.
    (c) Draw the graph of the profit equation.

25. **Car Rentals** One-day car rentals cost $50 plus 10 cents per mile from company A and $40 plus 20 cents per mile from company B.
    (a) For each company, give the linear equations for the cost, y, when x miles are driven.
    (b) Which company offers the best value when the car is driven for 80 miles?
    (c) Which company offers the best value when the car is driven for 160 miles?
    (d) For what mileage do the two companies offer the same value?

26. **Inflation** In 2003, bacon cost an average of $3.20 per pound. In 2015, bacon cost an average of $5.45 per pound. (*Source: U.S. Bureau of Labor Statistics, Consumer Price Index.*)
    (a) Assuming a linear increase in the price per pound of bacon, find the equation that relates the cost, y, to the number of years after 2003, x.
    (b) When did bacon cost an average of $4.60 per pound?

27. **Medical Assistant Jobs** According to the U.S. Department of Labor, medical assistant jobs are expected to increase from 591,300 in 2014 to 730,200 in 2024. Assuming that the number of medical assistant jobs increases linearly during that time, find the equation that relates the number of jobs, y, to the number of years after 2014, x. Use the equation to predict the number of medical assistant jobs in 2020.

28. **Sales Commission** A furniture store offers its new employees a weekly salary of $200 plus a 3% commission on sales. After one year, employees receive $100 per week plus a 5% sales

commission. For what weekly sales level will the two scales produce the same salary?

29. **Soft Drinks** According to *Beverage Digest* (March 2015), Coke Classic's percentage of the soft drink market declined from 20.4 in 2000 to 17.3 in 2014. If the percentage declined linearly during that time, estimate Coke Classic's percentage of the soft drink market in 2012.

30. **Bachelor's Degrees in Education** According to the U.S. National Center of Education Statistics, 107,238 bachelor's degrees in education were awarded in 2006 and 104,647 were awarded in 2013. If the number of bachelor's degrees in education continues to decline linearly, how many bachelor's degrees in education will be awarded in 2019?

31. **First-Choice College** The following table gives the percentage of first-year students who were accepted by their first-choice college in recent years. (*Source: The American Freshman: National Norms.*)
    (a) Use the method of least squares to obtain the straight line that best fits these data. (Let $x = 0$ represent 2012.)
    (b) If this trend continues, what percentage of first-year students in 2017 will have been accepted by their first-choice college?
    (c) If this trend continues, when will 67% of first-year students have been accepted by their first-choice college?

Year	Percent
2012	76.7
2013	75.6
2014	72.7
2015	75.5

32. **Life Expectancy** The following table gives the 2015 life expectancy at birth for several countries. (*Source: CIA World Fact Book, 2015 estimates.*)

Country	Male	Female
United States	77.3	82.0
Germany	78.3	83.0
Australia	79.7	84.7
Japan	81.4	88.3

(a) Use the method of least squares to obtain the straight line that best fits these data.
(b) In Greece, the life expectancy of men is 77.8 years. Use the least-squares line from (a) to estimate the life expectancy for women.

(c) In France, the life expectancy for women is 85.0 years. Use the least-squares line from (a) to estimate the life expectancy for men.

33. **Cancer and Diet** The following table gives the (age-adjusted) death rate per 100,000 women from breast cancer and the daily dietary fat intake (in grams per day) for various countries. (*Source: Advances in Cancer Research.*)

Country	Fat Intake	Death Rate
Japan	41	4
Poland	90	10
Finland	118	13
United States	148	21

(a) Use the method of least squares to obtain the straight line that best fits these data.
(b) In Denmark, women consume an average of 160 grams of fat per day. Use the least-squares line to estimate the breast cancer death rate.
(c) In New Zealand, the breast cancer death rate is 22 women per 100,000. Use the least-squares line to estimate the daily fat intake in New Zealand.

Conceptual Exercises

34. Consider a linear equation of the form $y = mx + b$. When the value of b remains fixed and the value of m changes, the graph rotates about the point $(0, b)$. As the value of m increases, will the graph rotate in a clockwise or counterclockwise direction?

35. Consider an equation of the form $y = mx + b$. When the value of m remains fixed and the value of b changes, the graph is translated vertically. As the value of b increases, does the graph move up or down?

36. What is the difference between a line having undefined slope and having zero slope?

37. When is the x-intercept of a line the same as the y-intercept?

38. Suppose that you have found the least-squares line for a collection of points and that you edit the data by adding a point on the line to the data. Will the expanded data have the same least-squares line? Explain the rationale for your conclusion, and then experiment to test whether your conclusion is correct.

39. Does every line have an x-intercept? A y-intercept?

40. Consider the line having the slope–intercept equation $y = 2x + 5$.
    (a) How many point–slope equations does the line have?
    (b) Find two point–slope equations for the line.

Break-Even Analysis

We discussed linear demand curves in Section 1.3. Demand curves normally apply to an entire industry or to a **monopolist**—that is, a manufacturer so large that the quantity that it supplies affects the market price of the commodity. We discussed linear cost curves in Section 1.2. In this chapter project, we combine demand curves and cost curves with least-squares lines to determine break-even points.

1. Table 1 can be used to obtain the demand curve for a monopolist who manufactures and sells a unique type of camera. The first column gives several production quantities in thousands of cameras, and the second column gives the corresponding prices per camera. For instance, in order to sell 200 thousand cameras, the manufacturer must set the price at $316 per camera. Find the least-squares line that best fits these data; that is, find a demand curve for the camera.

Table 1

q (thousands)	p (dollars)
100	360
200	316
300	288
400	236
500	200

2. Use the demand curve from part 1 to estimate the price that must be charged in order to sell 350 thousand cameras. Calculate the revenue for this price and quantity. *Note: The revenue is the amount of money received from the sale of the cameras.*

3. Use the demand curve to estimate the quantity that can be sold if the price is $300 per camera. Calculate the revenue for this price and quantity.

4. Determine the expression that gives the revenue from producing and selling q thousand cameras. *Note: The number of cameras sold will be 1000q.*

5. Assuming that the manufacturer has fixed costs of $8,000,000 and that the variable cost of producing each thousand cameras is $100,000, find the equation of the cost curve.

6. Graph the revenue curve from part 4 and the cost curve from part 5 on a graphing calculator, and determine the two points of intersection.

7. What is the break-even point–that is, the lowest value of q for which cost equals revenue?

8. For what values of q will the company make a profit?

CHAPTER 2

Matrices

We begin this chapter by developing a method for solving systems of linear equations in any number of variables. Our discussion of this method will lead naturally into the study of mathematical objects called *matrices*. The arithmetic and applications of matrices are the main topics of the chapter. We discuss in detail the application of matrix arithmetic to input–output analysis, which can be (and is) used to make production decisions for large businesses and entire economies.

2.1 Systems of Linear Equations with Unique Solutions

In Chapter 1, we presented a method for solving systems of linear equations in two variables. The method of Chapter 1 is very efficient for determining the solutions. Unfortunately, it works only for systems of linear equations having *two* variables. In many applications, we meet systems having more than two variables, as the following example illustrates.

EXAMPLE 1

Manufacturing The Upside Down Company specializes in making down jackets, ski vests, and comforters. The requirements for down and labor and the profits earned are given in the following chart:

	Down (pounds)	Time (labor-hours)	Profit ($)
Jacket	3	2	6
Vest	2	1	6
Comforter	4	1	2

Each week, the company has available 600 pounds of down and 275 labor-hours. It wants to earn a weekly profit of $1150. How many of each item should the company make each week?

SOLUTION

The requirements and earnings can be expressed by a system of equations. Let x be the number of jackets, y the number of vests, and z the number of comforters. If 600 pounds of down are used, then

$$[\text{down in jackets}] + [\text{down in vests}] + [\text{down in comforters}] = 600$$
$$3\,[\text{no. jackets}] \quad + \quad 2\,[\text{no. vests}] \quad + \quad 4\,[\text{no. comforters}] \quad = 600.$$

That is,

$$3x + 2y + 4z = 600.$$

Similarly, the equation for labor is

$$2x + y + z = 275,$$

and the equation for the profit is

$$6x + 6y + 2z = 1150.$$

The numbers x, y, and z must simultaneously satisfy a system of three linear equations in three variables.

$$\begin{cases} 3x + 2y + 4z = 600 \\ 2x + y + z = 275 \\ 6x + 6y + 2z = 1150. \end{cases} \tag{1}$$

Later, we present a method for determining the solution to this system. This method yields the solution $x = 50$, $y = 125$, and $z = 50$. It is easy to confirm that these values of x, y, and z satisfy all three equations:

$$3(50) + 2(125) + 4(50) = 600$$
$$2(50) + (125) + (50) = 275$$
$$6(50) + 6(125) + 2(50) = 1150.$$

Thus, the Upside Down Company can make a profit of $1150 by producing 50 jackets, 125 vests, and 50 comforters. ◀◀

In this section, we develop a step-by-step procedure for solving systems of linear equations such as (1). The procedure, called the **Gauss–Jordan elimination method**, consists of repeatedly simplifying the system, using so-called elementary row operations, until the solution stares us in the face!

In the system of linear equations (1), the equations have been written in such a way that the x-terms, the y-terms, and the z-terms lie in different columns. We shall always be careful to display systems of equations with separate columns for each variable. One of the key ideas of the Gauss–Jordan elimination method is to think of the solution as a system of linear equations in its own right. For example, we can write the solution of system (1) as

$$\begin{cases} 1x + 0y + 0z = 50 \\ 0x + 1y + 0z = 125 \\ 0x + 0y + 1z = 50 \end{cases} \quad \text{or} \quad \begin{cases} x \phantom{{}+{}} = 50 \\ y \phantom{{}+{}} = 125 \\ z = 50. \end{cases} \tag{2}$$

This is just a system of linear equations in which the coefficients of most terms are zero! Since the only terms with nonzero coefficients are arranged on a diagonal, such a system is said to be in **diagonal form**.

Our method for solving a system of linear equations consists of repeatedly using three operations that alter the system but do not change the solutions. The operations are used to transform the system into a system in diagonal form. Since the operations involve only elementary arithmetic and are applied to entire equations (i.e., rows of the system), they are called **elementary row operations**. Let us begin our study of the Gauss–Jordan elimination method by introducing these operations.

> **Elementary Row Operation 1** Interchange any two equations.

This operation is harmless enough. It certainly does not change the solutions of the system.

> **Elementary Row Operation 2** Multiply an equation by a nonzero number.

For example, if we are given the system of linear equations

$$\begin{cases} 2x - 3y + 4z = 11 \\ 4x - 19y + z = 31 \\ 5x + 7y - z = 12, \end{cases}$$

then we may replace it by a new system obtained by leaving the last two equations unchanged and multiplying the first equation by 3. To accomplish this, multiply each term of the first equation by 3. The transformed system is

$$\begin{cases} 6x - 9y + 12z = 33 \\ 4x - 19y + z = 31 \\ 5x + 7y - z = 12. \end{cases}$$

The operation of multiplying an equation by a nonzero number does not change the solution of the system. For if a particular set of values of the variables satisfies the original equation, it satisfies the resulting equation, and vice versa.

Elementary row operation 2 may be used to make the coefficient of a particular variable 1.

EXAMPLE 2 **Demonstrating Elementary Row Operation 2** Replace the system

$$\begin{cases} -5x + 10y + 20z = 4 \\ x \phantom{{}+ 10y} - 12z = 1 \\ x + y + z = 0 \end{cases}$$

by an equivalent system in which the coefficient of x in the first equation is 1.

SOLUTION The coefficient of x in the first equation is -5, so we use elementary row operation 2 to multiply the first equation by $-\frac{1}{5}$. Multiplying each term of the first equation by $-\frac{1}{5}$ gives

$$\begin{cases} x - 2y - 4z = -\frac{4}{5} \\ x \phantom{{}- 2y} - 12z = 1 \\ x + y + z = 0. \end{cases}$$

»› Now Try Exercise 1

Another operation that can be performed on a system without changing its solutions is to replace one equation by its sum with some other equation. For example, consider this system of equations:

$$A: \begin{cases} x + y - 2z = 3 \\ x + 2y - 5z = 4 \\ 5x + 8y - 18z = 14. \end{cases}$$

We can replace the second equation by the sum of the first and the second. Since

$$\begin{array}{r} x + y - 2z = 3 \\ + \quad x + 2y - 5z = 4 \\ \hline 2x + 3y - 7z = 7, \end{array}$$

the resulting system is

$$B: \begin{cases} x + y - 2z = 3 \\ 2x + 3y - 7z = 7 \\ 5x + 8y - 18z = 14. \end{cases}$$

If a particular choice of x, y, and z satisfies system A, it also satisfies system B. This is because system B results from adding equations. Similarly, system A can be derived from system B by subtracting equations. So any particular solution of system A is a solution of system B, and vice versa.

The operation of adding equations is usually used in conjunction with elementary row operation 2. That is, an equation is changed by adding to it a nonzero multiple of another equation. For example, consider the system

$$\begin{cases} x + y - 2z = 3 \\ x + 2y - 5z = 4 \\ 5x + 8y - 18z = 14. \end{cases}$$

Let us change the second equation by adding to it twice the first. Since

$$\begin{array}{ll} 2(\text{first}) & 2x + 2y - 4z = 6 \\ + (\text{second}) & \underline{x + 2y - 5z = 4} \\ & 3x + 4y - 9z = 10, \end{array}$$

the new second equation is

$$3x + 4y - 9z = 10$$

and the transformed system is

$$\begin{cases} x + y - 2z = 3 \\ 3x + 4y - 9z = 10 \\ 5x + 8y - 18z = 14. \end{cases}$$

Since addition of equations and elementary row operation 2 are often used together, let us define a third elementary row operation.

Elementary Row Operation 3   Change an equation by adding to it a multiple of another equation.

For reference, let us summarize the elementary row operations that we have just defined.

> **Elementary Row Operations**
>
> 1. Interchange any two equations.
> 2. Multiply an equation by a nonzero number.
> 3. Change an equation by adding to it a multiple of another equation.

The idea of the Gauss–Jordan elimination method is to transform an arbitrary system of linear equations into diagonal form by repeated applications of the three elementary row operations. To see how the method works, consider the following example:

EXAMPLE 3   **Solving a System of Equations by Gauss–Jordan Elimination** Solve the following system by the Gauss–Jordan elimination method:

$$\begin{cases} x - 3y = 7 \\ -3x + 4y = -1. \end{cases}$$

SOLUTION   Let us transform this system into diagonal form by examining one column at a time, starting from the left. Examine the first column:

$$\begin{matrix} x \\ -3x \end{matrix}$$

The coefficient of the top x is 1, which is exactly what it should be for the system to be in diagonal form. So we do nothing to this term. Now, examine the next term in the column, $-3x$. In diagonal form, this term must be absent. In order to accomplish this, we add a multiple of the first equation to the second. Since the coefficient of x in the second is -3, we add three times the first equation to the second equation in order to cancel the x-term. (Abbreviation: $R_2 + 3R_1$. The R_2 refers to the second equation—that is, the equation in the second row of the system of equations. The expression $R_2 + 3R_1$ means that we are replacing the second equation by the original equation plus 3 times the first equation.)

$$\begin{cases} x - 3y = 7 \\ -3x + 4y = -1 \end{cases} \xrightarrow{R_2 + 3R_1} \begin{cases} x - 3y = 7 \\ - 5y = 20. \end{cases}$$

The first column now has the proper form, so we proceed to the second column. In diagonal form, that column will have one nonzero term—namely, the second—and the coefficient of y in that term must be 1. To bring this about, multiply the second equation by $-\frac{1}{5}$ [abbreviation $\left(-\frac{1}{5}\right)R_2$]:

$$\begin{cases} x - 3y = 7 \\ - 5y = 20 \end{cases} \xrightarrow{\left(-\frac{1}{5}\right)R_2} \begin{cases} x - 3y = 7 \\ y = -4. \end{cases}$$

The second column still does not have the correct form. We must get rid of the $-3y$-term in the first equation. We do this by adding a multiple of the second equation to the first. Since the coefficient of the term to be canceled is -3, we add three times the second equation to the first:

$$\begin{cases} x - 3y = 7 \\ y = -4 \end{cases} \xrightarrow{R_1 + 3R_2} \begin{cases} x = -5 \\ y = -4. \end{cases}$$

The system is now in diagonal form, and the solution can be read off: $x = -5$, $y = -4$.

» Now Try Exercise 47

NOTE ▶   The abbreviation for interchanging the first and second equation is $R_1 \leftrightarrow R_2$. ◀

EXAMPLE 4   **Solving a System of Equations by Gauss–Jordan Elimination**   Use the Gauss–Jordan elimination method to solve the system

$$\begin{cases} 2x - 6y = -8 \\ -5x + 13y = 1. \end{cases}$$

SOLUTION   We can perform the calculations in a mechanical way, proceeding column by column from the left:

$$\begin{cases} 2x - 6y = -8 \\ -5x + 13y = 1 \end{cases} \xrightarrow{\ \frac{1}{2}R_1\ } \begin{cases} x - 3y = -4 \\ -5x + 13y = 1 \end{cases}$$

$$\xrightarrow{\ R_2 + 5R_1\ } \begin{cases} x - 3y = -4 \\ \quad\ -2y = -19 \end{cases}$$

$$\xrightarrow{\ (-\frac{1}{2})R_2\ } \begin{cases} x - 3y = -4 \\ \qquad\ y = \frac{19}{2} \end{cases}$$

$$\xrightarrow{\ R_1 + 3R_2\ } \begin{cases} x \quad\ = \frac{49}{2} \\ \quad\ y = \frac{19}{2}. \end{cases}$$

So the solution of the system is $x = \frac{49}{2}$, $y = \frac{19}{2}$.   **» Now Try Exercise 55**

The calculation becomes easier to follow if we omit writing down the variables at each stage and work only with the coefficients. At each stage of the computation, the system is represented by a rectangular array of numbers. For instance, the original system is written as

$$\begin{bmatrix} 2 & -6 & | & -8 \\ -5 & 13 & | & 1 \end{bmatrix}.$$

The vertical line between the second and third columns is a placemarker that separates the data obtained from the left- and right-hand sides of the equations. Each row of the array corresponds to an equation, and each column to the left of the vertical line corresponds to a variable.

The elementary row operations are performed on the rows of this rectangular array just as if the variables were there. So, for example, the first step in the preceding solution is to multiply the first equation by $\frac{1}{2}$. This corresponds to multiplying the first row of the array by $\frac{1}{2}$ to get

$$\begin{bmatrix} 1 & -3 & | & -4 \\ -5 & 13 & | & 1 \end{bmatrix}.$$

The diagonal form corresponds to the array

$$\begin{bmatrix} 1 & 0 & | & \frac{49}{2} \\ 0 & 1 & | & \frac{19}{2} \end{bmatrix}.$$

Note that this array has ones down the diagonal and zeros everywhere else on the left. The solution of the system appears on the right.

A rectangular array of numbers is called a **matrix** (plural *matrices*). Matrices (such as the one above) that are derived from systems of linear equations are called **augmented matrices**. In the next example, we use augmented matrices to carry out the Gauss–Jordan elimination method.

EXAMPLE 5   **Solving a System of Equations by Gauss—Jordan Elimination**   Use the Gauss—Jordan elimination method to solve the system

$$\begin{cases} 3x - 6y + 9z = 0 \\ 4x - 6y + 8z = -4 \\ -2x - y + z = 7. \end{cases}$$

SOLUTION The initial array corresponding to the system is

$$\begin{bmatrix} 3 & -6 & 9 & | & 0 \\ 4 & -6 & 8 & | & -4 \\ -2 & -1 & 1 & | & 7 \end{bmatrix}.$$

We must use elementary row operations to transform this array into diagonal form—that is, with ones down the diagonal, and zeros everywhere else to the left of the vertical line:

$$\begin{bmatrix} 1 & 0 & 0 & | & * \\ 0 & 1 & 0 & | & * \\ 0 & 0 & 1 & | & * \end{bmatrix}.$$

We proceed one column at a time.

$$\begin{bmatrix} 3 & -6 & 9 & | & 0 \\ 4 & -6 & 8 & | & -4 \\ -2 & -1 & 1 & | & 7 \end{bmatrix} \xrightarrow{\frac{1}{3}R_1} \begin{bmatrix} 1 & -2 & 3 & | & 0 \\ 4 & -6 & 8 & | & -4 \\ -2 & -1 & 1 & | & 7 \end{bmatrix} \xrightarrow{R_2 + (-4)R_1}$$

$$\begin{bmatrix} 1 & -2 & 3 & | & 0 \\ 0 & 2 & -4 & | & -4 \\ -2 & -1 & 1 & | & 7 \end{bmatrix} \xrightarrow{R_3 + 2R_1} \begin{bmatrix} 1 & -2 & 3 & | & 0 \\ 0 & 2 & -4 & | & -4 \\ 0 & -5 & 7 & | & 7 \end{bmatrix} \xrightarrow{\frac{1}{2}R_2}$$

$$\begin{bmatrix} 1 & -2 & 3 & | & 0 \\ 0 & 1 & -2 & | & -2 \\ 0 & -5 & 7 & | & 7 \end{bmatrix} \xrightarrow{R_1 + 2R_2} \begin{bmatrix} 1 & 0 & -1 & | & -4 \\ 0 & 1 & -2 & | & -2 \\ 0 & -5 & 7 & | & 7 \end{bmatrix} \xrightarrow{R_3 + 5R_2}$$

$$\begin{bmatrix} 1 & 0 & -1 & | & -4 \\ 0 & 1 & -2 & | & -2 \\ 0 & 0 & -3 & | & -3 \end{bmatrix} \xrightarrow{(-\frac{1}{3})R_3} \begin{bmatrix} 1 & 0 & -1 & | & -4 \\ 0 & 1 & -2 & | & -2 \\ 0 & 0 & 1 & | & 1 \end{bmatrix} \xrightarrow{R_1 + 1R_3}$$

$$\begin{bmatrix} 1 & 0 & 0 & | & -3 \\ 0 & 1 & -2 & | & -2 \\ 0 & 0 & 1 & | & 1 \end{bmatrix} \xrightarrow{R_2 + 2R_3} \begin{bmatrix} 1 & 0 & 0 & | & -3 \\ 0 & 1 & 0 & | & 0 \\ 0 & 0 & 1 & | & 1 \end{bmatrix}$$

The last array is in diagonal form, so we just put back the variables and read off the solution:

$$x = -3, \qquad y = 0, \qquad z = 1.$$

Because so much arithmetic has been performed, it is a good idea to check the solution by substituting the values for x, y, and z into each of the equations of the original system. This will uncover any arithmetic errors that may have occurred.

$$\begin{cases} 3x - 6y + 9z = 0 \\ 4x - 6y + 8z = -4 \\ -2x - y + z = 7 \end{cases} \qquad \begin{cases} 3(-3) - 6(0) + 9(1) = 0 \\ 4(-3) - 6(0) + 8(1) = -4 \\ -2(-3) - (0) + (1) = 7 \end{cases}$$

$$\begin{cases} -9 - 0 + 9 = 0 \\ -12 - 0 + 8 = -4 \\ 6 - 0 + 1 = 7 \end{cases}$$

$$\begin{cases} 0 = 0 \\ -4 = -4 \\ 7 = 7 \end{cases}$$

So we have indeed found a solution of the system. **》 Now Try Exercise 53**

NOTE Note that so far we have not had to use elementary row operation 1, which allows interchange of equations. But in some examples, it is definitely needed.

Consider this system:

$$\begin{cases} y + z = 0 \\ 3x - y + z = 6 \\ 6x - z = 3. \end{cases}$$

The first step of the Gauss–Jordan elimination method consists of making the x-coefficient 1 in the first equation. But we cannot do this, since the first equation does not involve x. To remedy this difficulty, just interchange the first two equations to guarantee that the first equation involves x. Now, proceed as before. Of course, in terms of the matrix of coefficients, interchanging equations corresponds to interchanging rows of the matrix. **«**

INCORPORATING

TECHNOLOGY

Pressing 2ND [MATRIX] presents the three menus shown in Fig. 1. You define a new matrix or alter an existing matrix with the EDIT menu. You perform operations on existing matrices with the MATH menu. You place the name of an existing matrix on the home screen with the NAMES menu. (Specifically, to display the name of a matrix on the home screen press 2ND [MATRIX], cursor down to the matrix in the NAMES menu, and press ENTER).

To define a matrix from the EDIT menu, you cursor down to one of the names **[A]**, **[B]**, **[C]**, ... for the matrix, press ENTER, specify the size of the matrix, and fill in the entries as shown in Fig. 2.

The MATRIX/MATH menu contains commands for 16 matrix operations. (Appendix B shows how to carry out the three elementary row operations from the menu.) The command **rref** carries out the complete Gauss–Jordan elimination method. See Fig. 3. *Note:* **rref** stands for "reduced row echelon form," the name given to the final form of a matrix that has been completely row reduced.

Matrix entries are normally displayed as decimals. From the home screen, you can display the entries as fractions with a command such as **[A]▶Frac** or **Ans▶Frac**. (**▶Frac** is displayed by pressing MATH **1**.) See Fig. 4.

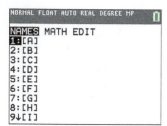

Figure 1

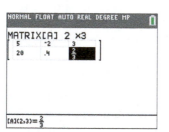

Figure 2

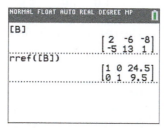

Figure 3

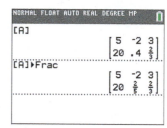

Figure 4

The systems of linear equations presented in this section can be solved with a device called **Solver**. Appendix C shows how to use Solver to find the solution to Example 5.

✳ WolframAlpha Each row of a matrix is represented as a sequence of numbers (separated by commas) inside a pair of braces. A matrix is represented as a sequence of rows (separated by commas) inside a pair of braces. Consider Example 3. The instruction

row reduce $\{\{1, -3, 7\}, \{-3, 4, -1\}\}$

performs Gauss–Jordan elimination on the matrix. The instruction

solve $x - 3y = 7, -3x + 4y = -1$

solves the system of linear equations directly.

Check Your Understanding 2.1

Solutions can be found following the section exercises.

1. Determine whether the following systems of linear equations are in diagonal form:

   (a) $\begin{cases} x & +z = 3 \\ y & = 2 \\ & z = 7 \end{cases}$ (b) $\begin{cases} x & = -1 \\ y & = 0 \\ 3z = & 4 \end{cases}$

2. Give the meaning of each of the following abbreviations for row operations on a matrix.

   (a) $R_i \leftrightarrow R_j$ (b) $cR_i, c \neq 0$ (c) $R_i + cR_j$

3. Perform the indicated elementary row operation.

   (a) $\begin{cases} x - 3y = 2 \\ 2x + 3y = 5 \end{cases} \xrightarrow{R_2 + (-2)R_1}$

   (b) $\begin{cases} x + y = 3 \\ -x + 2y = 5 \end{cases} \xrightarrow{R_2 + 1R_1}$

4. State the next elementary row operation that should be performed when applying the Gauss–Jordan elimination method.

   (a) $\begin{bmatrix} 0 & 2 & 4 & | & 1 \\ 0 & 3 & -7 & | & 0 \\ 3 & 6 & -3 & | & 3 \end{bmatrix}$ (b) $\begin{bmatrix} 1 & -3 & 4 & | & 5 \\ 0 & 2 & 3 & | & 4 \\ -6 & 5 & -7 & | & 0 \end{bmatrix}$

EXERCISES 2.1

In Exercises 1–8, perform the indicated elementary row operation and give its abbreviation.

1. Operation 2: Multiply the first equation by 2.

   $\begin{cases} \frac{1}{2}x - 3y = 2 \\ 5x + 4y = 1 \end{cases}$

2. Operation 2: Multiply the second equation by -1.

   $\begin{cases} x + 4y = 6 \\ -y = 2 \end{cases}$

3. Operation 3: Change the second equation by adding to it 5 times the first equation.

   $\begin{cases} x + 2y = 3 \\ -5x + 4y = 1 \end{cases}$

4. Operation 3: Change the second equation by adding to it $\left(-\frac{1}{2}\right)$ times the first equation.

   $\begin{cases} x - 6y = 4 \\ \frac{1}{2}x + 2y = 1 \end{cases}$

5. Operation 3: Change the third equation by adding to it (-4) times the first equation.

   $\begin{cases} x - 2y + z = 0 \\ y - 2z = 4 \\ 4x + y + 3z = 5 \end{cases}$

6. Operation 3: Change the third equation by adding to it 3 times the second equation.

   $\begin{cases} x + 6y - 4z = 1 \\ y + 3z = 1 \\ -3y + 7z = 2 \end{cases}$

7. Operation 3: Change the first row by adding to it $\frac{1}{2}$ times the second row.

   $\begin{bmatrix} 1 & -\frac{1}{2} & | & 3 \\ 0 & 1 & | & 4 \end{bmatrix}$

8. Operation 3: Change the third row by adding to it (-4) times the second row.

   $\begin{bmatrix} 1 & 0 & 7 & | & 9 \\ 0 & 1 & -2 & | & 3 \\ 0 & 4 & 8 & | & 5 \end{bmatrix}$

In Exercises 9–12, write the augmented matrix corresponding to the system of linear equations.

9. $\begin{cases} -3x + 4y = -2 \\ x - 7y = 8 \end{cases}$ 10. $\begin{cases} \frac{2}{3}x - 3y = 4 \\ y = -5 \end{cases}$

11. $\begin{cases} x + 13y - 2z = 0 \\ 2x - z = 3 \\ y = 5 \end{cases}$ 12. $\begin{cases} y - z = 22 \\ 2x = 17 \\ x - 3y = 12 \end{cases}$

In Exercises 13–16, write the system of linear equations corresponding to the matrix.

13. $\begin{bmatrix} 0 & -2 & | & 3 \\ 1 & 7 & | & -4 \end{bmatrix}$ 14. $\begin{bmatrix} -5 & \frac{2}{3} & | & 3 \\ 1 & 7 & | & -\frac{5}{8} \end{bmatrix}$

15. $\begin{bmatrix} 3 & 2 & 0 & | & -3 \\ 0 & 1 & -6 & | & 4 \\ -5 & -1 & 7 & | & 0 \end{bmatrix}$ 16. $\begin{bmatrix} \frac{6}{5} & -1 & 12 & | & -\frac{2}{3} \\ -1 & 0 & 0 & | & 5 \\ 0 & 2 & -1 & | & 6 \end{bmatrix}$

In Exercises 17–22, describe in your own words the meaning of the notation with respect to a matrix.

17. $\frac{1}{3}R_2$ 18. $R_2 + (-4)R_1$ 19. $R_1 + 3R_2$

20. $(-1)R_1$ 21. $R_2 \leftrightarrow R_3$ 22. $R_1 \leftrightarrow R_2$

In Exercises 23–28, carry out the indicated elementary row operation.

23. $\begin{bmatrix} 1 & 2 & | & 0 \\ -3 & 4 & | & 5 \end{bmatrix} \xrightarrow{R_2 + 3R_1} \begin{bmatrix} & & | & \\ & & | & \end{bmatrix}$

24. $\begin{bmatrix} -\frac{1}{2} & 2 & | & \frac{3}{4} \\ -3 & 4 & | & 9 \end{bmatrix} \xrightarrow{(-2)R_1} \begin{bmatrix} & & | & \\ & & | & \end{bmatrix}$

25. $\begin{bmatrix} \frac{1}{7} & \frac{2}{7} & | & \frac{3}{7} \\ 3 & -2 & | & 0 \end{bmatrix} \xrightarrow{\ 7R_1\ } \begin{bmatrix} & & \\ & & \end{bmatrix}$

26. $\begin{bmatrix} 1 & 3 & | & -2 \\ 4 & 4 & | & 5 \end{bmatrix} \xrightarrow{\ R_2 + (-4)R_1\ } \begin{bmatrix} & & \\ & & \end{bmatrix}$

27. $\begin{bmatrix} 0 & 1 & | & 7 \\ 1 & 3 & | & -5 \end{bmatrix} \xrightarrow{\ R_1 \leftrightarrow R_2\ } \begin{bmatrix} & & \\ & & \end{bmatrix}$

28. $\begin{bmatrix} 4 & 5 & | & 6 \\ -3 & 2 & | & 0 \end{bmatrix} \xrightarrow{\ R_1 + 1R_2\ } \begin{bmatrix} & & \\ & & \end{bmatrix}$

In Exercises 29–36, state the next elementary row operation that should be performed in order to put the matrix into diagonal form. Do not perform the operation.

29. $\begin{bmatrix} 1 & -5 & | & 1 \\ -2 & 4 & | & 6 \end{bmatrix}$

30. $\begin{bmatrix} 1 & 3 & | & 4 \\ 0 & 2 & | & 6 \end{bmatrix}$

31. $\begin{bmatrix} 1 & 2 & | & 3 \\ 0 & 1 & | & 4 \end{bmatrix}$

32. $\begin{bmatrix} 1 & -2 & 5 & | & 7 \\ 0 & -3 & 0 & | & 9 \\ 4 & 5 & -6 & | & 7 \end{bmatrix}$

33. $\begin{bmatrix} 0 & 5 & -3 & | & 6 \\ 2 & -3 & 4 & | & 5 \\ 4 & 1 & -7 & | & 8 \end{bmatrix}$

34. $\begin{bmatrix} 1 & 4 & -2 & | & 5 \\ 0 & -3 & 6 & | & 9 \\ 0 & 4 & 3 & | & 1 \end{bmatrix}$

35. $\begin{bmatrix} 1 & 0 & 3 & | & 4 \\ 0 & 1 & 2 & | & 5 \\ 0 & 0 & 1 & | & 6 \end{bmatrix}$

36. $\begin{bmatrix} 1 & 2 & 4 & | & 5 \\ 0 & 0 & 3 & | & 6 \\ 0 & 1 & 1 & | & 7 \end{bmatrix}$

In Exercises 37 and 38, two steps of the Gauss–Jordan elimination method are shown. Fill in the missing numbers.

37. $\begin{bmatrix} 1 & 1 & -1 & | & 6 \\ -3 & 7 & 5 & | & 0 \\ 2 & -4 & 3 & | & -1 \end{bmatrix} \rightarrow \begin{bmatrix} 1 & 1 & -1 & | & 6 \\ 0 & 10 & \square & | & \square \\ 0 & -6 & \square & | & \square \end{bmatrix}$

38. $\begin{bmatrix} 1 & 2 & 7 & | & -3 \\ 1 & -5 & -4 & | & 2 \\ -4 & 6 & 9 & | & 3 \end{bmatrix} \rightarrow \begin{bmatrix} 1 & 2 & 7 & | & -3 \\ 0 & -7 & \square & | & \square \\ 0 & 14 & \square & | & \square \end{bmatrix}$

The screen captures in Exercises 39–46 show a matrix, A, corresponding to a system of linear equations and the matrix **rref**(A) obtained after the Gauss–Jordan elimination method is applied to A. Write the system of linear equations corresponding to A, and use **rref**(A) to give the solution to the system of linear equations.

39.
```
NORMAL FLOAT AUTO REAL DEGREE MP
[A]
                        [1  1   7]
                        [1 -1   1]
rref([A])
                        [1 0  4]
                        [0 1  3]
```

40.
```
NORMAL FLOAT AUTO REAL DEGREE MP
[A]
                        [2  3  23]
                        [6 -4   4]
rref([A])
                        [1 0  4]
                        [0 1  5]
```

41.
```
NORMAL FLOAT AUTO REAL DEGREE MP
[A]
                        [3 -4 -27]
                        [1  2  11]
rref([A])
                        [1 0 -1]
                        [0 1  6]
```

42.
```
NORMAL FLOAT AUTO REAL DEGREE MP
[A]
                        [4 -3 18]
                        [2 -1  8]
rref([A])
                        [1 0  3]
                        [0 1 -2]
```

43.
```
NORMAL FLOAT AUTO REAL DEGREE MP
[A]
                        [2  1  3 31]
                        [1  1 -2  3]
                        [4 -2  5 17]
rref([A])
                        [1 0 0  3]
                        [0 1 0 10]
                        [0 0 1  5]
```

44.
```
NORMAL FLOAT AUTO REAL DEGREE MP
[A]
                        [2  7  4  1]
                        [3 -8  9 20]
                        [4  0  5 16]
rref([A])
                        [1 0 0  4]
                        [0 1 0 -1]
                        [0 0 1  0]
```

45.
```
NORMAL FLOAT AUTO REAL DEGREE MP
[A]
                        [ 3  7  2 5]
                        [ 7 -6 -3 4]
                        [10  9 -7 3]
rref([A])
                        [1 0 0 1]
                        [0 1 0 0]
                        [0 0 1 1]
```

46.
```
NORMAL FLOAT AUTO REAL DEGREE MP
[A]
                        [3  2  1 10]
                        [8 -1  6 16]
                        [5  3 -1  9]
rref([A])
                        [1 0 0  .5]
                        [0 1 0  3]
                        [0 0 1 2.5]
```

In Exercises 47–60, solve the linear system by the Gauss–Jordan elimination method.

47. $\begin{cases} x + 9y = 8 \\ 2x + 8y = 6 \end{cases}$

48. $\begin{cases} \frac{1}{3}x + 2y = 1 \\ -2x - 4y = 6 \end{cases}$

$$49. \begin{cases} x - 3y + 4z = 1 \\ 4x - 10y + 10z = 4 \\ -3x + 9y - 5z = -6 \end{cases}$$

$$50. \begin{cases} \frac{1}{2}x + y = 4 \\ -4x - 7y + 3z = -31 \\ 6x + 14y + 7z = 50 \end{cases}$$

$$51. \begin{cases} 2x - 2y + 4 = 0 \\ 3x + 4y - 1 = 0 \end{cases}$$

$$52. \begin{cases} 2x + 3y = 4 \\ -x + 2y = -2 \end{cases}$$

$$53. \begin{cases} 4x - 4y + 4z = -8 \\ x - 2y - 2z = -1 \\ 2x + y + 3z = 1 \end{cases}$$

$$54. \begin{cases} x + 2y + 2z - 11 = 0 \\ x - y - z + 4 = 0 \\ 2x + 5y + 9z - 39 = 0 \end{cases}$$

$$55. \begin{cases} .2x + .3y = 4 \\ .6x + 1.1y = 15 \end{cases}$$

$$56. \begin{cases} \frac{3}{2}x + 6y = 9 \\ \frac{1}{2}x - \frac{2}{3}y = 11 \end{cases}$$

$$57. \begin{cases} x + y + 4z = 3 \\ 4x + y - 2z = -6 \\ -3x + 2z = 1 \end{cases}$$

$$58. \begin{cases} -2x - 3y + 2z = -2 \\ x + y = 3 \\ -x - 3y + 5z = 8 \end{cases}$$

$$59. \begin{cases} -x + y = -1 \\ x + z = 4 \\ 6x - 3y + 2z = 10 \end{cases}$$

$$60. \begin{cases} x + 2z = 9 \\ y + z = 1 \\ 3x - 2y = 9 \end{cases}$$

61. A baked potato smothered with cheddar cheese weighs 180 grams and contains 10.5 grams of protein. If cheddar cheese contains 25% protein and a baked potato contains 2% protein, how many grams of cheddar cheese are there?

62. A high school math department purchased brand A calculators for $80 each and brand B calculators for $95 each. It purchased a total of 20 calculators at a total cost of $1780. How many brand A calculators did the department purchase?

Exercises 63 and 64 are multiple choice exercises with five possible choices. Each exercise consists of a question and two statements that may or may not provide sufficient information to answer the question. Select the response (a)–(e) that best describes the situation.

(a) Statement I alone is sufficient to answer the question, but statement II is not sufficient.

(b) Statement II alone is sufficient to answer the question, but statement I is not sufficient.

(c) Both statements together are sufficient to answer the question, but neither alone is sufficient.

(d) Each statement alone is sufficient to answer the question.

(e) Both statements together are not sufficient to answer the question.

63. A box of golf balls and a golf glove cost a total of $20. How much does the box of balls cost?

   Statement I: The golf glove costs three times as much as the box of golf balls.

   Statement II: The golf glove costs $15.

64. I have four nickels and three pennies in my pocket. What is the total weight of these coins?

   Statement I: A nickel weighs twice as much as a penny.

   Statement II: The total weight of a nickel and two pennies is 10 grams.

65. **Sales** A street vendor has a total of 350 short- and long-sleeve T-shirts. If they sell the short-sleeve shirts for $10 each and the long-sleeve shirts for $14 each, how many of each did they sell if they sold all of their stock for $4300?

66. **Sales** A grocery store carries two brands of bleach. A 30-ounce bottle of the national brand sells for $2.59, while the same-size bottle of the store brand sells for $2.09. How many

bottles of each brand were sold if a total of 82 bottles were sold for $194.88?

67. **Movie Tickets** A 275-seat movie theater charges $11.25 admission for adults and $8.50 for children. If the theater is full and $2860 is collected, how many adults and how many children are in the audience?

68. **Batting Average** A baseball player's batting average is determined by dividing the number of hits by the number of times at bat and multiplying by 1000. (Batting averages are usually, but not necessarily, rounded to the nearest whole number.) For instance, if a player gets 2 hits in 5 times at bat, his batting average is 400: $\left(\frac{2}{5} \times 1000 = 400\right)$. Partway through the season, a player thinks to himself, "If I get a hit in my next time at bat, my average will go up to 250; if I don't get a hit, it will drop to 187.5." How many times has this player batted, how many hits has he had, and what is his current batting average?

69. **Areas of Countries** The United States and Canada are the largest of the 23 countries that make up North America. The bar graph in Fig. 5 shows their areas and that of the remaining North American countries in millions of square miles. The total area of North America is 9.5 million square miles. Canada is 200,000 sq. mi. larger than the United States. The area of the United States is one-half the area of Canada plus 1.75 million sq. mi. Let x, y, and z represent the three areas shown in the figure. Use the methods of this section to determine the values of x, y, and z.

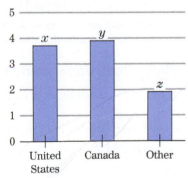

Figure 5 Areas in North America

70. **College Majors** The bar graph in Fig. 6 gives the intended majors of a group of 100 randomly selected college freshmen. (The biology category includes the biological and life sciences.) Six more students intend to major in biology than intend to major in business. The number of students intending to major in fields other than business or biology is 4 more than twice the number of students majoring in business or biology. Let x, y, and z represent the three numbers shown in the figure. Use the methods of this section to determine the values of x, y, and z.

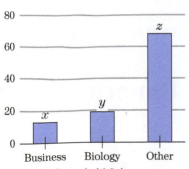

Figure 6 Intended Majors

71. Coffee Blends A one-pound blend of coffee uses Brazilian (60 ¢ per ounce), Columbian (50 ¢ per ounce), and Peruvian (70 ¢ per ounce) coffee beans and costs $9.70. The blend contains twice the weight of Brazilian beans as Columbian beans. How many ounces of each type of bean does the blend contain?

72. Nut Mixture A one-pound mixture of nuts contains cashews (70 ¢ per ounce), almonds (60 ¢ per ounce), and walnuts (80 ¢ per ounce) and costs $11.10. The mixture contains the same weight of cashews as walnuts. How many ounces of each type of nut does the mixture contain?

73. Investment Planning A bank wishes to invest a $100,000 trust fund in three sources: a bond fund paying 8%; a health sciences fund paying 7%; and a real estate fund paying 10%. The bank wishes to realize an $8000 annual income from the investment. A condition of the trust is that the total amount invested in the bond fund and the health sciences fund must be triple the amount invested in the real estate fund. How much should the bank invest in each possible category?

74. Nutrition Planning A dietitian wishes to plan a meal around three foods. Each ounce of food I contains 10% of the daily requirements for carbohydrates, 10% for protein, and 15% for vitamin C. Each ounce of food II contains 10% of the daily requirements for carbohydrates, 5% for protein, and 0% for vitamin C. Each ounce of food III contains 10% of the daily requirements for carbohydrates, 25% for protein, and 10% for vitamin C. How many ounces of each food should be served in order to supply exactly the daily requirement for each nutrient?

75. Candy Assortments A small candy store makes three types of party mixes. The first type contains 40% nonpareils and 60%

peanut clusters, while the second type contains 30% peanut clusters and 70% chocolate-covered raisins. The third type consists of 40% nonpareils, 30% peanut clusters, and 30% chocolate-covered raisins. If the store has 90 pounds of nonpareils, 100 pounds of peanut clusters, and 120 pounds of chocolate-covered raisins available, how many pounds of each type of party mix should be made?

76. Investment Planning New parents Jim and Lucy want to start saving for their son's college education. They have $5000 to invest in three different types of plans. A traditional savings account pays 1% annual interest, a certificate of deposit pays 3.6% annual interest, and a prepaid college plan pays 5.5% annual interest. If they want to invest the same amount in the prepaid college fund as in the other two plans together, how much should they invest in each plan to realize an interest income of $195 for the first year?

TECHNOLOGY EXERCISES

In Exercises 77–80, use technology to put the matrix in the specified exercise into diagonal form.

77. Exercise 31   **78.** Exercise 32

79. Exercise 33   **80.** Exercise 34

In Exercises 81–84, use technology to solve the system of linear equations in the specified exercise.

81. Exercise 55   **82.** Exercise 56

83. Exercise 57   **84.** Exercise 58

Solutions to Check Your Understanding 2.1

1. (a) Not in diagonal form, since the first equation contains both x and z.

  (b) Not in diagonal form, since the coefficient of z is not 1.

2. (a) Interchange the ith and j^{th} rows of the matrix.

  (b) Multiply the ith row of the matrix by the number c.

  (c) Add c times the jth row of the matrix to the ith row of the matrix in order to change the ith row.

3. (a) Change the system into another system in which the second equation is altered by having (-2) (first equation) added to it. The new system is

$$\begin{cases} x - 3y = 2 \\ \quad\;\; 9y = 1. \end{cases}$$

The equation $9y = 1$ was obtained as follows:

$$
\begin{array}{ll}
(-2)(\text{first equation}) & -2x + 6y = -4 \\
+ \,(\text{second equation}) & \underline{\;2x + 3y = \;\;5} \\
& \quad\quad\; 9y = \;\;1.
\end{array}
$$

(b) Change the second equation by adding to it 1 times the first equation. The result is

$$\begin{cases} x + \;y = 3 \\ \quad\;\; 3y = 8. \end{cases}$$

In general, notation of the form $R_i + kR_j$ specifies that the first row mentioned, R_i, be changed by adding to it a multiple, k, of the second row mentioned, R_j. The row R_j is not changed by the operation.

4. (a) The first row should contain a nonzero number as its first entry. This can be accomplished by interchanging the first and third rows. The notation for this operation is

$$R_1 \leftrightarrow R_3.$$

(b) The first column can be put into proper form by eliminating the -6. To accomplish this, multiply the first row by 6 and add this product to the third row. The notation for this operation is

$$R_3 + 6R_1.$$

2.2   General Systems of Linear Equations

In this section, we introduce the operation of pivoting and consider systems of linear equations that do not have exactly one solution.

Roughly speaking, the Gauss–Jordan elimination method applied to a matrix proceeds as follows: Consider the columns one at a time, from left to right. For each column, use the elementary row operations to transform the appropriate entry to a one and the remaining entries in the column to zeros. (The "appropriate" entry is the first entry in the first column, the second entry in the second column, and so forth.) This sequence of elementary row operations performed for each column is called **pivoting**. More precisely,

---

Method   To pivot a matrix about a given nonzero entry,

1. Transform the given entry into a one.
2. Transform all other entries in the same column into zeros.

Pivoting is used in solving problems other than systems of linear equations. As we shall see in Chapter 4, it is the basis for the simplex method of solving linear programming problems.

EXAMPLE 1   **Pivoting** Pivot the matrix about the circled element.

$$\begin{bmatrix} 18 & \boxed{-6} & 15 \\ 5 & -2 & 4 \end{bmatrix}$$

SOLUTION   The first step is to transform the -6 to a 1. We do this by multiplying the first row by $-\frac{1}{6}$.

$$\begin{bmatrix} 18 & -6 & 15 \\ 5 & -2 & 4 \end{bmatrix} \xrightarrow{(-\frac{1}{6})R_1} \begin{bmatrix} -3 & 1 & -\frac{5}{2} \\ 5 & -2 & 4 \end{bmatrix}.$$

Next, we transform the -2 (the only remaining entry in column 2) into a 0:

$$\begin{bmatrix} -3 & 1 & -\frac{5}{2} \\ 5 & -2 & 4 \end{bmatrix} \xrightarrow{R_2 + 2R_1} \begin{bmatrix} -3 & 1 & -\frac{5}{2} \\ -1 & 0 & -1 \end{bmatrix}.$$

The last matrix is the result of pivoting the original matrix about the circled entry.

≫ Now Try Exercise 1

In terms of pivoting, we can give the following summary of the Gauss–Jordan elimination method:

Gauss–Jordan Elimination Method to Transform a System of Linear Equations into Diagonal Form

1. Write down the matrix corresponding to the linear system.
2. Make sure that the first entry in the first column is nonzero. Do this by interchanging the first row with one of the rows below it, if necessary.
3. Pivot the matrix about the first entry in the first column.
4. Make sure that the second entry in the second column is nonzero. Do this by interchanging the second row with one of the rows below it, if necessary.
5. Pivot the matrix about the second entry in the second column.
6. Continue in this manner until the left side of the matrix is in diagonal form.
7. Write the system of linear equations corresponding to the matrix.

All of the systems considered in the preceding section had only a single solution. In this case, we say that the solution is **unique**. Let us now use the Gauss–Jordan elimination method to study the various possibilities other than a unique solution.

EXAMPLE 2   **A System of Equations That Has No Solution** Find all solutions of the system

$$\begin{cases} x - y + z = 3 \\ x + y - z = 5 \\ -2x + 4y - 4z = 1. \end{cases}$$

SOLUTION   We apply the Gauss–Jordan elimination method to the matrix of the system. (The elements pivoted about are circled.)

$$\begin{bmatrix} ① & -1 & 1 & 3 \\ 1 & 1 & -1 & 5 \\ -2 & 4 & -4 & 1 \end{bmatrix} \xrightarrow[R_3 + 2R_1]{R_2 + (-1)R_1} \begin{bmatrix} 1 & -1 & 1 & 3 \\ 0 & ② & -2 & 2 \\ 0 & 2 & -2 & 7 \end{bmatrix}$$

$$\xrightarrow[\substack{R_1 + 1R_2 \\ R_3 + (-2)R_2}]{\frac{1}{2}R_2} \begin{bmatrix} 1 & 0 & 0 & 4 \\ 0 & 1 & -1 & 1 \\ 0 & 0 & 0 & 5 \end{bmatrix}$$

We cannot pivot about the last zero in the third column, so we have carried the method as far as we can. Let us write out the equations corresponding to the last matrix:

$$\begin{cases} x = 4 \\ y - z = 1 \\ 0 = 5. \end{cases}$$

Note that the last equation is a built-in contradiction. The last equation can never be satisfied, no matter what the values of x, y, and z are. Thus, the original system has no solution. Systems with no solution can always be detected by the presence of a matrix row of the form $[0 \ \ 0 \ \ \cdots \ \ 0 \ | \ a]$, where a is a nonzero number.

≫ Now Try Exercise 25

EXAMPLE 3   **Solving a System of Equations That Has Infinitely Many Solutions** Determine all solutions of the system

$$\begin{cases} 2x + 2y + 4z = 8 \\ x - y + 2z = 2 \\ -x + 5y - 2z = 2. \end{cases}$$

SOLUTION   We set up the matrix corresponding to the system and perform the appropriate pivoting operations.

$$\begin{bmatrix} ② & 2 & 4 & 8 \\ 1 & -1 & 2 & 2 \\ -1 & 5 & -2 & 2 \end{bmatrix} \xrightarrow[\substack{R_2 + (-1)R_1 \\ R_3 + 1R_1}]{\frac{1}{2}R_1} \begin{bmatrix} 1 & 1 & 2 & 4 \\ 0 & ⟨-2⟩ & 0 & -2 \\ 0 & 6 & 0 & 6 \end{bmatrix}$$

$$\xrightarrow[\substack{R_1 + (-1)R_2 \\ R_3 + (-6)R_2}]{(-\frac{1}{2})R_2} \begin{bmatrix} 1 & 0 & 2 & 3 \\ 0 & 1 & 0 & 1 \\ 0 & 0 & 0 & 0 \end{bmatrix}$$

Note that our method must terminate here, since there is no way to transform the third entry in the third column into a 1 without disturbing the columns already in appropriate form. The equations corresponding to the last matrix read

$$\begin{cases} x + 2z = 3 \\ y = 1 \\ 0 = 0. \end{cases}$$

The last equation does not involve any of the variables and so may be omitted. This leaves the two equations

$$\begin{cases} x + 2z = 3 \\ y = 1. \end{cases}$$

Now, taking the $2z$-term in the first equation to the right side, we can write the equations

$$\begin{cases} x = 3 - 2z \\ y = 1. \end{cases}$$

The value of y is given: $y = 1$. The value of x is given in terms of z. To find a solution to this system, assign any value to z. Then, the first equation gives a value for x and thereby a specific solution to the system. For example, if we take $z = 1$, then the corresponding specific solution is

$$z = 1$$
$$x = 3 - 2(1) = 1$$
$$y = 1.$$

If we take $z = -3$, the corresponding specific solution is

$$z = -3$$
$$x = 3 - 2(-3) = 9$$
$$y = 1.$$

Thus, we see that the original system has infinitely many specific solutions, corresponding to the infinitely many possible different choices for z.

We say that the **general solution** of the system is

$$z = \text{any value}$$
$$x = 3 - 2z$$
$$y = 1. \qquad \text{» Now Try Exercise 27}$$

Gauss–Jordan Elimination Method for the Matrix of a Linear System That Cannot Be Transformed into Diagonal Form

1. Apply the Gauss–Jordan elimination method to put as many columns as possible into proper form. (A column is in proper form if one entry is 1 and the other entries are 0.) Proceed from left to right, but do not disturb columns that have already been put into proper form. As much as possible, each row should have a 1 in its leftmost nonzero entry. (Such a 1 is called a *leading 1*.) The column for each leading 1 should be to the right of the columns for the leading 1s in the rows above it.
2. If at any time one or more of the rows is of the form $[0 \ 0 \ \cdots \ 0 \,|\, a]$, where a is a nonzero number, then the linear system has no solution.
3. Otherwise, there are infinitely many solutions. Variables corresponding to columns not in proper form can assume any value. The other variables can then be expressed in terms of these variables.

EXAMPLE 4 **Solving a System of Equations That Has Infinitely Many Solutions** Find all solutions of the linear system

$$\begin{cases} x + 2y - z + 3w = 5 \\ \quad\quad y + 2z + w = 7. \end{cases}$$

SOLUTION The Gauss–Jordan elimination method proceeds as follows:

$$\begin{bmatrix} 1 & 2 & -1 & 3 & | & 5 \\ 0 & ① & 2 & 1 & | & 7 \end{bmatrix} \quad \text{(The first column is already in proper form.)}$$

$$\xrightarrow{R_1 + (-2)R_1} \begin{bmatrix} 1 & 0 & -5 & 1 & | & -9 \\ 0 & 1 & 2 & 1 & | & 7 \end{bmatrix}.$$

We cannot do anything further with the third and fourth columns (without disturbing the first two columns), so the corresponding variables, z and w, can assume any values. Writing down the equations corresponding to the last matrix yields

$$\begin{cases} x & - 5z + w = -9 \\ & y + 2z + w = 7 \end{cases}$$

or

$$z = \text{any value}$$
$$w = \text{any value}$$
$$x = -9 + 5z - w$$
$$y = 7 - 2z - w.$$

To determine a specific solution, let, for example, $z = 1$ and $w = 2$. Then, a specific solution of the original system is

$$z = 1$$
$$w = 2$$
$$x = -9 + 5(1) - (2) = -6$$
$$y = 7 - 2(1) - (2) = 3.$$ **» Now Try Exercise 33**

EXAMPLE 5   **Solving a System of Equations That Has Infinitely Many Solutions** Find all solutions of the system of equations

$$\begin{cases} x - 7y + z = 3 \\ 2x - 14y + 3z = 4. \end{cases}$$

SOLUTION   The first pivot operation is routine:

$$\begin{bmatrix} ① & -7 & 1 & | & 3 \\ 2 & -14 & 3 & | & 4 \end{bmatrix} \xrightarrow{R_2 + (-2)R_1} \begin{bmatrix} 1 & -7 & 1 & | & 3 \\ 0 & 0 & 1 & | & -2 \end{bmatrix}.$$

However, it is impossible to pivot about the zero in the second column. So skip the second column and pivot about the second entry in the third column to get

$$\xrightarrow{R_1 + (-1)R_2} \begin{bmatrix} 1 & -7 & 0 & | & 5 \\ 0 & 0 & 1 & | & -2 \end{bmatrix}.$$

This is as far as we can go. The variable corresponding to the second column—namely, y—can assume any value, and the general solution of the system is obtained from the equations

$$\begin{cases} x - 7y & = 5 \\ & z = -2. \end{cases}$$

Therefore, the general solution of the system is

$$y = \text{any value}$$
$$x = 5 + 7y$$
$$z = -2.$$ **» Now Try Exercise 37**

EXAMPLE 6   **Finding Specific Solutions** A high school music department purchased a total of 20 new clarinets, trumpets, and violins in anticipation of increased interest in these instruments. Each clarinet cost $1400, each trumpet cost $1200, and each violin cost $900. If the music department spent a total of $24,900, give two different combinations for the number of each type of instrument purchased.

SOLUTION   Let x be the number of clarinets purchased, y be the number of trumpets purchased, and z be the number of violins purchased. Since a total of 20 instruments are purchased, $x + y + z = 20$. Since the cost of each instrument is given, the total cost is given by $1400x + 1200y + 900z = 24900$.

So we need to solve the system

$$\begin{cases} x + y + z = 20 \\ 1400x + 1200y + 900z = 24900 \end{cases}$$

Applying the Gauss-Jordan elimination method to this system we get

$$\begin{bmatrix} 1 & 1 & 1 & \bigm| & 20 \\ 1400 & 1200 & 900 & \bigm| & 24900 \end{bmatrix} \xrightarrow{R_2 + (-1400)R_1} \begin{bmatrix} 1 & 1 & 1 & \bigm| & 20 \\ 0 & -200 & -500 & \bigm| & -3100 \end{bmatrix}$$

$$\xrightarrow{-\frac{1}{200}R_2} \begin{bmatrix} 1 & 1 & 1 & \bigm| & 20 \\ 0 & 1 & \frac{5}{2} & \bigm| & \frac{31}{2} \end{bmatrix} \xrightarrow{R_1 + (-1)R_2} \begin{bmatrix} 1 & 0 & -\frac{3}{2} & \bigm| & \frac{9}{2} \\ 0 & 1 & \frac{5}{2} & \bigm| & \frac{31}{2} \end{bmatrix}.$$

So the general solution to this system of equations is

$$x = \tfrac{9}{2} + \tfrac{3}{2}z, \ y = \tfrac{31}{2} - \tfrac{5}{2}z, \ z = \text{any value}.$$

However this problem is about instruments so z can't really be ANY value. Notice that if $z = 0$, then $x = \frac{9}{2} = 4.5$ which doesn't make sense since x is the number of clarinets. So we need to be careful in selecting values for z. If $z = 1$, then $x = 6$ and $y = 13$ which are all sensible values for a number of instruments. Similarly, if we let $z = 3$, then $x = 9$ and $y = 8$ are also sensible values. Therefore, the music department could have purchased 6 clarinets, 13 trumpets, and 1 violin, or 9 clarinets, 8 trumpets, and 3 violins.

» Now Try Exercise 45

At first, it might seem strange that some systems have no solution, some have one, and yet others have infinitely many. The reason for the difference can be explained geometrically. For simplicity, consider the case of systems of two equations in two variables. Each equation in this case has a graph in the xy-plane, and the graph is a straight line. As we have seen, solving the system corresponds to finding the points lying on both lines. There are three possibilities. First, the two lines may intersect. In this case, the solution is unique. Second, the two lines may be parallel. Then, the two lines do not intersect and the system has no solution. Finally, the two equations may represent the same line, as for example, the equations $2x + 3y = 1$ and $4x + 6y = 2$ do. In this case, every point on the line is a solution of the system; that is, there are infinitely many solutions (Fig. 1).

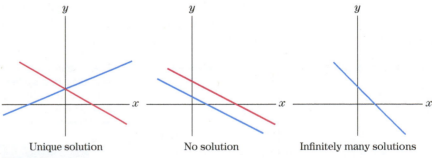

Unique solution   No solution   Infinitely many solutions

Figure 1

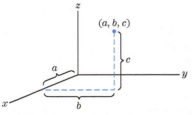

Figure 2

Systems of three equations in three variables can be examined in an analogous way by using a 3-dimensional coordinate system. The point having coordinates (a, b, c) is obtained by starting at the origin and moving a units in the x-direction, b units in the y-direction, and then c units in the z-direction (up if c is positive, down if c is negative). See Fig. 2. The collection of points that satisfy a specific linear equation in three variables forms a plane. Therefore, the solution of a system of linear equations in three variables consists of all points that are simultaneously on the planes corresponding to each equation. Figure 3 shows some possible configurations for these planes.

Unique solution

No solution

Infinitely many solutions

Figure 3

INCORPORATING

TECHNOLOGY

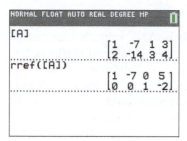

Figure 4

📈 The complete Gauss–Jordan elimination method can be carried out in one step, as shown in Fig. 4 for the matrix of Example 5. The command **rref** is found in the MATRIX/MATH menu of the TI-84 Plus calculators. *Note:* **rref** stands for "reduced row echelon form," the name given to the final form of a matrix that has been completely row reduced.

📊 When Solver is used with a system of linear equations having infinitely many solutions, only one solution is given. When Solver is used with a system of linear equations having no solution, the sentence "Solver could not find a feasible solution." is displayed.

✳**WolframAlpha** When Wolfram|Alpha is asked to solve a system of linear equations having no solution, the phrase "no solutions exist" is displayed. When systems have infinitely many solutions, the form of the answers might differ from those in the textbook.

Check Your Understanding 2.2

Solutions can be found following the section exercises.

1. Find a specific solution to a system of linear equations whose general solution is

$$w = \text{any value}$$
$$y = \text{any value}$$
$$z = 7 + 6w$$
$$x = 26 - 2y + 14w.$$

2. Find all solutions of this system of linear equations.

$$\begin{cases} 2x + 4y - 4z - 4w = 24 \\ -3x - 6y + 10z - 18w = -8 \\ -x - 2y + 4z - 10w = 2 \end{cases}$$

EXERCISES 2.2

In Exercises 1–8, pivot the matrix about the circled element.

1. $\begin{bmatrix} ② & -4 & 6 \\ 3 & 7 & 1 \end{bmatrix}$

2. $\begin{bmatrix} 1 & 2 & 3 \\ 4 & ⑧ & -12 \end{bmatrix}$

3. $\begin{bmatrix} 7 & 1 & 4 & 5 \\ -1 & 1 & ② & 6 \\ 4 & 0 & 2 & 3 \end{bmatrix}$

4. $\begin{bmatrix} 5 & 10 & -10 & 12 \\ 4 & 3 & 6 & 12 \\ 4 & ⓘ(-4) & 4 & -16 \end{bmatrix}$

5. $\begin{bmatrix} ② & 3 \\ 6 & 0 \\ 1 & 5 \end{bmatrix}$

6. $\begin{bmatrix} 2 & 1 \\ ⓘ(-1) & 0 \end{bmatrix}$

7. $\begin{bmatrix} 4 & 3 & 0 \\ \frac{2}{3} & 0 & -2 \\ 1 & 3 & ⑥ \end{bmatrix}$

8. $\begin{bmatrix} 1 & 0 & 2 \\ -1 & 1 & ⓘ(-2) \\ 1 & 2 & 6 \end{bmatrix}$

The screen captures in Exercises 9–16 show a matrix, A, corresponding to a system of linear equations and the matrix rref(A) obtained after the Gauss–Jordan elimination method is applied to A. Write the system of linear equations corresponding to A, and use rref(A) to give all solutions to the system of linear equations.

9.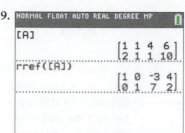

10.

```
NORMAL FLOAT AUTO REAL DEGREE MP
[A]
                    [ 2 -2  1  2]
                    [-6  6 -3  5]
rref([A])
                    [1 -1 .5 0]
                    [0  0  0 1]
```

11.

```
NORMAL FLOAT AUTO REAL DEGREE MP
[A]
                    [-5 15 -10 5]
                    [ 1 -3   2 0]
rref([A])
                    [1 -3 2 0]
                    [0  0 0 1]
```

12.

```
NORMAL FLOAT AUTO REAL DEGREE MP
[A]
                    [ 2 -6 -4 0]
                    [-3  9  6 0]
rref([A])
                    [1 -3 -2 0]
                    [0  0  0 0]
```

13.

```
NORMAL FLOAT AUTO REAL DEGREE MP
[A]
                    [ 2 -1  5 12]
                    [-1 -4  2  3]
                    [ 8  5 11 30]
rref([A])
                    [1 0  2  5]
                    [0 1 -1 -2]
                    [0 0  0  0]
```

14.

```
NORMAL FLOAT AUTO REAL DEGREE MP
[A]
                    [2 -1  2  4]
                    [3  1  1 -2]
                    [1  2 -1  5]
rref([A])
                    [1 0 .6  0]
                    [0 1 -.8 0]
                    [0 0  0  1]
```

15.

```
NORMAL FLOAT AUTO REAL DEGREE MP
[A]
                    [1 2 3 -1  4]
                    [2 3 0  1 -3]
                    [4 7 6 -1  5]
rref([A])
                    [1 0 -9  5 -18]
                    [0 1  6 -3  11]
                    [0 0  0  0   0]
```

16.

```
NORMAL FLOAT AUTO REAL DEGREE MP
[A]
                    [1 1  1 -1]
                    [1 2 -1 -6]
                    [2 1  4  3]
rref([A])
                    [1 0  3  4]
                    [0 1 -2 -5]
                    [0 0  0  0]
```

In Exercises 17–36, use the Gauss–Jordan elimination method to find all solutions of the system of linear equations.

17. $\begin{cases} 2x - 4y = 6 \\ -x + 2y = -3 \end{cases}$

18. $\begin{cases} -\frac{1}{2}x + y = \frac{3}{2} \\ -3x + 6y = 10 \end{cases}$

19. $\begin{cases} -x + 3y = 11 \\ 3x - 9y = -30 \end{cases}$

20. $\begin{cases} .25x - .75y = 1 \\ -2x + 6y = -8 \end{cases}$

21. $\begin{cases} x + 2y = 5 \\ 3x - y = 1 \\ -x + 3y = 5 \end{cases}$

22. $\begin{cases} x - 6y = 12 \\ -\frac{1}{2}x + 3y = -6 \\ \frac{1}{3}x - 2y = 4 \end{cases}$

23. $\begin{cases} 4x + 5y = 3 \\ 3x + 6y = 1 \\ 2x - 3y = 7 \end{cases}$

24. $\begin{cases} 2x + 3y = 12 \\ 2x - 3y = 0 \\ 5x - y = 13 \end{cases}$

25. $\begin{cases} x - y + 3z = 3 \\ -2x + 3y - 11z = -4 \\ x - 2y + 8z = 6 \end{cases}$

26. $\begin{cases} x - 3y + z = 5 \\ -2x + 7y - 6z = -9 \\ x - 2y - 3z = 6 \end{cases}$

27. $\begin{cases} x + y + z = -1 \\ 2x + 3y + 2z = 3 \\ 2x + y + 2z = -7 \end{cases}$

28. $\begin{cases} x - 3y + 2z = 10 \\ -x + 3y - z = -6 \\ -x + 3y + 2z = 6 \end{cases}$

29. $\begin{cases} 6x - 2y + 2z = 4 \\ 3x - y + 2z = 2 \\ -12x + 4y - 8z = 8 \end{cases}$

30. $\begin{cases} x + 2y + 3z = 4 \\ 5x + 6y + 7z = 8 \\ x + 2y + 3z = 5 \end{cases}$

31. $\begin{cases} x + 2y + 8z = 1 \\ 3x - y + 4z = 10 \\ -x + 5y + 10z = -8 \\ x + y + z = 3 \end{cases}$

32. $\begin{cases} 2x + 6y + 6z = 0 \\ -3x - 10y + z = 1 \\ -x - 4y + 3z = 1 \\ 5x + 6y + 8z = 9 \end{cases}$

33. $\begin{cases} x + y - 2z + 2w = 5 \\ 2x + y - 4z + w = 5 \\ 3x + 4y - 6z + 9w = 20 \\ 4x + 4y - 8z + 8w = 20 \end{cases}$

34. $\begin{cases} 2y + z - w = 1 \\ x - y + z + w = 14 \\ -x - 9y - z + 4w = 11 \\ x + y + z = 9 \end{cases}$

35. $\begin{cases} x - y + z + w = 1 \\ y + 3z + 2w = -7 \\ y - z - 3w = 1 \\ x + 4z + 3w = 0 \end{cases}$

36. $\begin{cases} x + y + 2w = 4 \\ x + 2y + 2z = 7 \\ 2x + y - z + 6w = 5 \\ -x + y + 2z - 6w = 2 \end{cases}$

In Exercises 37–40, find three solutions to the system of equations.

37. $\begin{cases} x + 2y + z = 5 \\ y + 3z = 9 \end{cases}$

38. $\begin{cases} x + 5y + 3z = 9 \\ 2x + 9y + 7z = 5 \end{cases}$

39. $\begin{cases} x + 7y - 3z = 8 \\ z = 5 \end{cases}$

40. $\begin{cases} x = 4 \\ y - 3z = 7 \end{cases}$

41. Nutrition Planning In a laboratory experiment, a researcher wants to provide a rabbit with exactly 1000 units of vitamin A, exactly 1600 units of vitamin C, and exactly 2400 units of vitamin E. The rabbit is fed a mixture of three foods whose nutritional content is given by the accompanying table. How many grams of each food should the rabbit be fed?

	Food 1	Food 2	Food 3
Vitamin A (units per gram)	2	4	6
Vitamin C (units per gram)	3	7	10
Vitamin E (units per gram)	5	9	14

42. **Nutrition Planning** Rework Exercise 41 with the requirement for vitamin E changed to 2000 units.

43. **Nutrition Planning** The nutritional content of three foods is given by the accompanying table. Use the methods of this section to show that no combination of these three foods can contain 72 units of B-12, 68 calories, and 60 units of iron.

	Food A	Food B	Food C
B-12 (units per gram)	3	10	15
Calories (per gram)	4	12	8
Iron (units per gram)	5	14	1

44. **Nutrition Planning** Refer to Exercise 43. Show that when the required units of iron are increased to 64 units, there are several combinations of the three foods that meet the requirement. Give two such combinations.

45. **Furniture Manufacturing** A furniture manufacturer makes sofas, chairs, and ottomans. The accompanying table gives the number of hours of labor required for the carpentry and upholstery that goes into each item. Suppose that, each day, 380 labor-hours are available for carpentry and 450 labor-hours are available for upholstery. Give three different combinations for the numbers of each type of furniture that can be manufactured each day.

	Ottoman	Sofa	Chair
Carpentry	1 hour	3 hours	6 hours
Upholstery	3 hours	6 hours	3 hours

46. **Computer Equipment** An office manager placed an order for computers, printers, and scanners. Each computer cost $1000, each printer cost $100, and each scanner cost $400. She ordered 15 items for $10,200. Give two different combinations for the numbers of each type of item that she could have purchased.

47. **Quilting** Granny's Custom Quilts receives an order for a patchwork quilt made from square patches of three types: solid green, solid blue, and floral. The quilt is to be 8 squares by 12 squares, and there must be 15 times as many solid squares as floral squares. If Granny's charges $3 per solid square and $5 per floral square, and if the customer wishes to spend exactly $300, how many of each type of square may be used in the quilt?

48. **Purchasing Options** Amanda is decorating her new home and wants to buy some house plants. She is interested in three types of plants costing $7, $10, and $13. If she has budgeted exactly $150 for the plants and wants to buy exactly 15 of them, what are her options?

49. For what values(s) of k will the following system of linear equations have no solution? Infinitely many solutions?
$$\begin{cases} 2x - 3y = 4 \\ -6x + 9y = k \end{cases}$$

50. For what value of k will the following system of linear equations have a solution?
$$\begin{cases} 2x + 6y = 4 \\ x + 7y = 10 \\ kx + 8y = 4 \end{cases}$$

51. Figure 5 shows the graphs of the equations from a system of three linear equations in two variables. How many solutions does the system have?

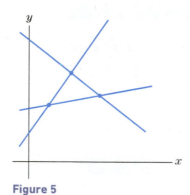

Figure 5

52. Suppose that after the Gauss–Jordan elimination method has been applied to a matrix corresponding to a system of linear equations, the matrix has a row of all zeros. Must the system have infinitely many solutions?

TECHNOLOGY EXERCISES

In Exercises 53–56, graph the three equations together and determine the number of solutions (exactly one, none, or infinitely many). If there is exactly one solution, estimate the solution.

53. $$\begin{cases} x + y = 10 \\ 2x - 3y = 5 \\ -x + 3y = 2 \end{cases}$$

54. $$\begin{cases} 2x + 3y = 5 \\ -3x + 5y = 22 \\ 2x + y = -1 \end{cases}$$

55. $$\begin{cases} 2x + y = 12 \\ 3x - y = 2 \\ x + 2y = 16 \end{cases}$$

56. $$\begin{cases} 3x - 2y = 3 \\ -2x + 4y = 14 \\ x + y = 11 \end{cases}$$

57. Apply **rref** or **row reduce** to the matrix in Example 2. How does the final matrix differ from the one appearing in the text?

58. Apply **rref** or **row reduce** to the matrix in Example 3. How does the final matrix differ from the one appearing in the text?

Solutions to Check Your Understanding 2.2

1. Since w and y can each assume any value, select any numbers—say, $w = 1$ and $y = 2$. Then, $z = 7 + 6(1) = 13$ and $x = 26 - 2(2) + 14(1) = 36$. So $x = 36$, $y = 2$, $z = 13$, $w = 1$ is a specific solution. There are infinitely many different specific solutions, since there are infinitely many different choices for w and y.

2. Apply the Gauss–Jordan elimination method to the matrix of the system.

$$\begin{bmatrix} \textcircled{2} & 4 & -4 & -4 & | & 24 \\ -3 & -6 & 10 & -18 & | & -8 \\ -1 & -2 & 4 & -10 & | & 2 \end{bmatrix}$$

$$\xrightarrow[\substack{\frac{1}{2}R_1 \\ R_2 + 3R_1 \\ R_3 + 1R_1}]{} \begin{bmatrix} 1 & 2 & -2 & -2 & | & 12 \\ 0 & 0 & \textcircled{4} & -24 & | & 28 \\ 0 & 0 & 2 & -12 & | & 14 \end{bmatrix}$$

$$\xrightarrow[\substack{\frac{1}{4}R_2 \\ R_1 + 2R_2 \\ R_3 + (-2)R_2}]{} \begin{bmatrix} 1 & 2 & 0 & -14 & | & 26 \\ 0 & 0 & 1 & -6 & | & 7 \\ 0 & 0 & 0 & 0 & | & 0 \end{bmatrix}$$

The corresponding system of equations is

$$\begin{cases} x + 2y & - 14w = 26 \\ & z - 6w = 7. \end{cases}$$

The general solution is

$$w = \text{any value}$$
$$y = \text{any value}$$
$$z = 7 + 6w$$
$$x = 26 - 2y + 14w.$$

2.3 Arithmetic Operations on Matrices

We introduced matrices in Sections 2.1 and 2.2 to display the coefficients of a system of linear equations. For example, the linear system

$$\begin{cases} 5x - 3y = \frac{1}{2} \\ 4x + 3y = -1 \end{cases}$$

is represented by the matrix

$$\begin{bmatrix} 5 & -3 & | & \frac{1}{2} \\ 4 & 2 & | & -1 \end{bmatrix}.$$

After we have become accustomed to using such matrices in solving linear systems, we may omit the vertical line that separates the left and right sides of the equations. We need only remember that the right side of the equations is recorded in the right column. So, for example, we would write the preceding matrix in the form

$$\begin{bmatrix} 5 & -3 & \frac{1}{2} \\ 4 & 2 & -1 \end{bmatrix}.$$

A matrix is *any* rectangular array of numbers and may be of any size. Here are some examples of matrices of various sizes.

$$\begin{bmatrix} 3 & 7 \\ 0 & -1 \end{bmatrix}, \quad \begin{bmatrix} 1 \\ 2 \end{bmatrix}, \quad \begin{bmatrix} 2 & 1 \end{bmatrix}, \quad \begin{bmatrix} 6 \end{bmatrix}, \quad \begin{bmatrix} 5 & 7 & -1 \\ 0 & 3 & 5 \\ 6 & 0 & 5 \end{bmatrix}.$$

Matrices can be added, subtracted, and multiplied. This section discusses these arithmetic operations and how they are used in applications. Before we can do so, however, we need some vocabulary with which to describe matrices.

A matrix is described by its **size**—that is, the number of rows and columns that it contains. For example, the matrix

$$\begin{bmatrix} 7 & 5 \\ \frac{1}{2} & -2 \\ 2 & -11 \end{bmatrix}$$

has three rows and two columns and is referred to as a *3 × 2* (read: "three-by-two") *matrix*. The matrix $\begin{bmatrix} 4 & 5 & 0 \end{bmatrix}$ has one row and three columns and is a *1 × 3 matrix*. A matrix with only one row is called a **row matrix**. A matrix, such as

$$\begin{bmatrix} 2 \\ 7 \end{bmatrix},$$

that has only one column is called a **column matrix**. If a matrix has the same number of rows and columns, it is called a **square matrix**. Here are some square matrices of various sizes:

$$[5], \quad \begin{bmatrix} 1 & 2 \\ 3 & 4 \end{bmatrix}, \quad \begin{bmatrix} 2 & -1 & 0 \\ 3 & 5 & 4 \\ 0 & 3 & -7 \end{bmatrix}.$$

The rows of a matrix are numbered from the top down, and the columns are numbered from left to right. For example, the first row of the matrix

$$A = \begin{bmatrix} 1 & -1 & 0 \\ 2 & 1 & 7 \\ -3 & 2 & 4 \end{bmatrix}$$

is $\begin{bmatrix} 1 & -1 & 0 \end{bmatrix}$, and its third column is

$$\begin{bmatrix} 0 \\ 7 \\ 4 \end{bmatrix}.$$

The numbers in a matrix, called **entries**, may be identified in terms of the row and column containing the entry in question. For example, the entry in the first row, third column, of matrix A is 0:

$$\begin{bmatrix} 1 & -1 & 0 \\ 2 & 1 & 7 \\ -3 & 2 & 4 \end{bmatrix};$$

the entry in the second row, first column, is 2:

$$\begin{bmatrix} 1 & -1 & 0 \\ 2 & 1 & 7 \\ -3 & 2 & 4 \end{bmatrix};$$

and the entry in the third row, third column, is 4:

$$\begin{bmatrix} 1 & -1 & 0 \\ 2 & 1 & 7 \\ -3 & 2 & 4 \end{bmatrix}.$$

We use double-subscripted lowercase letters to indicate the locations of the entries of a matrix. We denote the entry in the ith row, jth column, of the matrix A by a_{ij}. For instance, we have $a_{13} = 0$, $a_{21} = 2$, and $a_{33} = 4$.

We say that two matrices A and B are **equal**, denoted $A = B$, provided that they have the same size and that all of their corresponding entries are equal.

Addition and Subtraction of Matrices

We define the sum $A + B$ of two matrices A and B only if A and B are two matrices of the same size—that is, if A and B have the same number of rows and the same number of columns. In this case, $A + B$ is the matrix formed by adding the corresponding entries of A and B. For example,

$$\begin{bmatrix} 2 & 0 \\ 1 & 1 \\ 5 & 3 \end{bmatrix} + \begin{bmatrix} 5 & 4 \\ 0 & 2 \\ 2 & 6 \end{bmatrix} = \begin{bmatrix} 2+5 & 0+4 \\ 1+0 & 1+2 \\ 5+2 & 3+6 \end{bmatrix} = \begin{bmatrix} 7 & 4 \\ 1 & 3 \\ 7 & 9 \end{bmatrix}.$$

We subtract matrices of the same size by subtracting corresponding entries. Thus, we have

$$\begin{bmatrix} 7 \\ 1 \end{bmatrix} - \begin{bmatrix} 3 \\ 2 \end{bmatrix} = \begin{bmatrix} 7-3 \\ 1-2 \end{bmatrix} = \begin{bmatrix} 4 \\ -1 \end{bmatrix}.$$

Multiplication of Matrices

It might seem that, to define the product of two matrices, one would start with two matrices of like size and multiply the corresponding entries. But this definition is not useful, since the calculations that arise in applications require a somewhat more complex multiplication. In the interests of simplicity, we start by defining the product of a row matrix times a column matrix.

If A is a row matrix and B is a column matrix, then we can form the product $A \cdot B$ provided that the two matrices have the same length. The product $A \cdot B$ is the 1×1 matrix obtained by multiplying corresponding entries of A and B and then forming the sum.

We may put this definition into algebraic terms as follows. Suppose that A is the row matrix

$$A = [a_1 \quad a_2 \quad \cdots \quad a_n],$$

and B is the column matrix

$$B = \begin{bmatrix} b_1 \\ b_2 \\ \vdots \\ b_n \end{bmatrix}.$$

Note that A and B are both of the same length—namely, n. Then,

$$A \cdot B = [a_1 \quad a_2 \quad \cdots \quad a_n] \cdot \begin{bmatrix} b_1 \\ b_2 \\ \vdots \\ b_n \end{bmatrix}$$

is calculated by multiplying corresponding entries of A and B and forming the sum; that is,

$$A \cdot B = [a_1 b_1 + a_2 b_2 + \cdots + a_n b_n].$$

Notice that the product is a 1×1 matrix—namely, a single number in brackets.

Here are some examples of the product of a row matrix times a column matrix:

$$\begin{bmatrix} 3 & \frac{1}{2} \end{bmatrix} \cdot \begin{bmatrix} 1 \\ 4 \end{bmatrix} = [3 \cdot 1 + \frac{1}{2} \cdot 4] = [5];$$

$$[2 \quad 0 \quad -1] \cdot \begin{bmatrix} 6 \\ 5 \\ 3 \end{bmatrix} = [2 \cdot 6 + 0 \cdot 5 + (-1) \cdot 3] = [9].$$

In multiplying a row matrix times a column matrix, it helps to use both of your hands. Use your left index finger to point to the first element of the row matrix and your right to point to the first element of the column matrix. Multiply the elements that you are pointing to, and keep a running total of the products in your head. After each multiplication, move your fingers to the next elements of each matrix. With a little practice, you should be able to multiply a row times a column quickly and accurately.

The preceding definition of *multiplication* may seem strange. But products of this sort occur in many down-to-earth problems. Consider, for instance, the next example.

EXAMPLE 1   **Total Revenue as a Matrix Product** A dairy farm produces three items—milk, eggs, and cheese. The wholesale prices of the three items are \$1.70 per gallon, \$.80 per dozen, and \$3.30 per pound, respectively. In a certain week, the dairy farm sells 30,000 gallons of milk, 2000 dozen eggs, and 5000 pounds of cheese. Represent its total revenue as a matrix product.

SOLUTION The total revenue equals

$$(1.70)(30,000) + (.80)(2000) + (3.30)(5000).$$

This suggests that we define two matrices. The first displays the prices of the various items:

$$[1.70 \quad .80 \quad 3.30].$$

The second represents the production:

$$\begin{bmatrix} 30,000 \\ 2000 \\ 5000 \end{bmatrix}.$$

Then, the revenue for the week, when placed in a 1×1 matrix, equals

$$[1.70 \quad .80 \quad 3.30] \begin{bmatrix} 30,000 \\ 2000 \\ 5000 \end{bmatrix} = [69,100].$$

>> *Now Try Exercise 89*

The principle behind Example 1 is this: Any sum of products of the form $a_1b_1 + a_2b_2 + \cdots + a_nb_n$, when placed in a 1×1 matrix, can be written as the matrix product

$$[a_1b_1 + a_2b_2 + \cdots + a_nb_n] = [a_1 \quad a_2 \quad \cdots \quad a_n] \cdot \begin{bmatrix} b_1 \\ b_2 \\ \vdots \\ b_n \end{bmatrix}.$$

Let us illustrate the procedure for multiplying more general matrices by working out a typical product:

$$\begin{bmatrix} 2 & 1 \\ 0 & 1 \\ 1 & 0 \end{bmatrix} \cdot \begin{bmatrix} 1 & 1 \\ 4 & 2 \end{bmatrix}.$$

To obtain the entries of the product, we multiply the rows of the left matrix by the columns of the right matrix, taking care to arrange the products in a specific way to yield a matrix, as follows. Start with the first row on the left, [2 1], and the first column on the right, $\begin{bmatrix} 1 \\ 4 \end{bmatrix}$. Their product is [6], so we enter 6 as the element in the first row, first column, of the product:

$$\begin{bmatrix} 2 & 1 \\ 0 & 1 \\ 1 & 0 \end{bmatrix} \cdot \begin{bmatrix} 1 & 1 \\ 4 & 2 \end{bmatrix} = \begin{bmatrix} 6 & \\ & \end{bmatrix}.$$

The product of the first row of the left matrix and the second column of the right matrix is [4], so we put a 4 in the first row, second column, of the product:

$$\begin{bmatrix} 2 & 1 \\ 0 & 1 \\ 1 & 0 \end{bmatrix} \cdot \begin{bmatrix} 1 & 1 \\ 4 & 2 \end{bmatrix} = \begin{bmatrix} 6 & 4 \\ & \end{bmatrix}.$$

There are no more columns that can be multiplied by the first row, so let us move to the second row and shift back to the first column. Correspondingly, we move down one row in the product.

$$\begin{bmatrix} 2 & 1 \\ 0 & 1 \\ 1 & 0 \end{bmatrix} \cdot \begin{bmatrix} 1 & 1 \\ 4 & 2 \end{bmatrix} = \begin{bmatrix} 6 & 4 \\ 4 & \end{bmatrix};$$

$$\begin{bmatrix} 2 & 1 \\ 0 & 1 \\ 1 & 0 \end{bmatrix} \cdot \begin{bmatrix} 1 & 1 \\ 4 & 2 \end{bmatrix} = \begin{bmatrix} 6 & 4 \\ 4 & 2 \end{bmatrix}.$$

We have now exhausted the second row of the left matrix, so we shift to the third row and, correspondingly, move down one row in the product:

$$\begin{bmatrix} 2 & 1 \\ 0 & 1 \\ 1 & 0 \end{bmatrix} \cdot \begin{bmatrix} 1 & 1 \\ 4 & 2 \end{bmatrix} = \begin{bmatrix} 6 & 4 \\ 4 & 2 \\ 1 & \end{bmatrix};$$

$$\begin{bmatrix} 2 & 1 \\ 0 & 1 \\ 1 & 0 \end{bmatrix} \cdot \begin{bmatrix} 1 & 1 \\ 4 & 2 \end{bmatrix} = \begin{bmatrix} 6 & 4 \\ 4 & 2 \\ 1 & 1 \end{bmatrix}.$$

Note that we have now multiplied every row of the left matrix by every column of the right matrix. This completes the computation of the product:

$$\begin{bmatrix} 2 & 1 \\ 0 & 1 \\ 1 & 0 \end{bmatrix} \cdot \begin{bmatrix} 1 & 1 \\ 4 & 2 \end{bmatrix} = \begin{bmatrix} 6 & 4 \\ 4 & 2 \\ 1 & 1 \end{bmatrix}.$$

EXAMPLE 2   **Matrix Multiplication** Calculate the following product:

$$\begin{bmatrix} 1 & 5 \\ 3 & 2 \end{bmatrix} \cdot \begin{bmatrix} 1 & 2 \\ 1 & 0 \end{bmatrix}.$$

SOLUTION

$$\begin{bmatrix} 1 & 5 \\ 3 & 2 \end{bmatrix} \cdot \begin{bmatrix} 1 & 2 \\ 1 & 0 \end{bmatrix} = \begin{bmatrix} 6 & \\ & \end{bmatrix}$$

$$\begin{bmatrix} 1 & 5 \\ 3 & 2 \end{bmatrix} \cdot \begin{bmatrix} 1 & 2 \\ 1 & 0 \end{bmatrix} = \begin{bmatrix} 6 & 2 \\ & \end{bmatrix}$$

$$\begin{bmatrix} 1 & 5 \\ 3 & 2 \end{bmatrix} \cdot \begin{bmatrix} 1 & 2 \\ 1 & 0 \end{bmatrix} = \begin{bmatrix} 6 & 2 \\ 5 & \end{bmatrix}$$

$$\begin{bmatrix} 1 & 5 \\ 3 & 2 \end{bmatrix} \cdot \begin{bmatrix} 1 & 2 \\ 1 & 0 \end{bmatrix} = \begin{bmatrix} 6 & 2 \\ 5 & 6 \end{bmatrix}$$

Thus,

$$\begin{bmatrix} 1 & 5 \\ 3 & 2 \end{bmatrix} \cdot \begin{bmatrix} 1 & 2 \\ 1 & 0 \end{bmatrix} = \begin{bmatrix} 6 & 2 \\ 5 & 6 \end{bmatrix}.$$

≫ Now Try Exercise 33

Notice that we cannot use the preceding method to compute the product $A \cdot B$ of *any* matrices A and B. For the procedure to work, it is crucial that the number of entries of each row of A be the same as the number of entries of each column of B. (Or, to put it another way, the number of columns of the left matrix must equal the number of rows of the right matrix.) Therefore, in order for us to form the product $A \cdot B$, the sizes of A and B must match up in a special way. If A is $m \times n$ and B is $p \times q$, then the product $A \cdot B$ is defined only in case the "inner" dimensions n and p are equal. In that case, the size of the product is determined by the "outer" dimensions m and q. It is an $m \times q$ matrix:

$$\underset{m \times n}{A} \quad \cdot \quad \underset{p \times q}{B} \quad = \quad \underset{m \times q}{C}.$$

equal

So, for example,

$$
\begin{bmatrix} \\ \\ \\ \end{bmatrix} \begin{bmatrix} \\ \\ \\ \\ \end{bmatrix} = \begin{bmatrix} \\ \\ \\ \end{bmatrix}
$$

$$
\underset{3\times 4}{} \qquad \underset{4\times 2}{} \qquad \underset{3\times 2}{}
$$

$$
\begin{bmatrix} \\ \\ \end{bmatrix} \begin{bmatrix} \\ \\ \end{bmatrix} = \begin{bmatrix} \\ \\ \end{bmatrix}.
$$

$$
\underset{2\times 2}{} \qquad \underset{2\times 1}{} \qquad \underset{2\times 1}{}
$$

If the sizes of A and B do not match up in the way just described, the product $A \cdot B$ is not defined.

EXAMPLE 3   **Matrix Multiplication** Calculate the following products, if defined:

(a) $\begin{bmatrix} 3 & -1 \\ 2 & 0 \\ 1 & 5 \end{bmatrix} \begin{bmatrix} 1 & 0 \\ 5 & -4 \\ 2 & -1 \end{bmatrix}$ (b) $\begin{bmatrix} 3 & -1 \\ 2 & 0 \\ 1 & 5 \end{bmatrix} \begin{bmatrix} 5 & 4 \\ -2 & 3 \end{bmatrix}$

SOLUTION

(a) The matrices to be multiplied are 3×2 and 3×2. The inner dimensions do not match, so the product is undefined.

(b) We are asked to multiply a 3×2 matrix times a 2×2 matrix. The inner dimensions match, so the product is defined and has size determined by the outer dimensions—that is, 3×2.

$$
\begin{bmatrix} 3 & -1 \\ 2 & 0 \\ 1 & 5 \end{bmatrix} \begin{bmatrix} 5 & 4 \\ -2 & 3 \end{bmatrix} = \begin{bmatrix} 3\cdot 5 + (-1)\cdot(-2) & 3\cdot 4 + (-1)\cdot 3 \\ 2\cdot 5 + 0\cdot(-2) & 2\cdot 4 + 0\cdot 3 \\ 1\cdot 5 + 5\cdot(-2) & 1\cdot 4 + 5\cdot 3 \end{bmatrix}
$$

$$
= \begin{bmatrix} 17 & 9 \\ 10 & 8 \\ -5 & 19 \end{bmatrix}.
$$

≫ Now Try Exercise 35

Multiplication of matrices has many properties in common with multiplication of ordinary numbers. However, there is at least one important difference. With matrix multiplication, the order of the factors is usually important. For example, the product of a 2×3 matrix times a 3×2 matrix is defined: The product is a 2×2 matrix. If the order is reversed to a 3×2 matrix times a 2×3 matrix, the product is a 3×3 matrix. So reversing the order may change the size of the product. Even when it does not, reversing the order may still change the entries in the product, as the following two products demonstrate:

$$
\begin{bmatrix} 1 & 5 \\ 3 & 2 \end{bmatrix} \begin{bmatrix} 1 & 2 \\ 1 & 0 \end{bmatrix} = \begin{bmatrix} 6 & 2 \\ 5 & 6 \end{bmatrix}; \qquad \begin{bmatrix} 1 & 2 \\ 1 & 0 \end{bmatrix} \begin{bmatrix} 1 & 5 \\ 3 & 2 \end{bmatrix} = \begin{bmatrix} 7 & 9 \\ 1 & 5 \end{bmatrix}.
$$

EXAMPLE 4   **Investment Earnings** An investment trust has investments in three states. Its deposits in each state are divided among bond funds, real estate funds, and capital funds. On January 1, the amount (in millions of dollars) invested in each category by state is given by the matrix

	Bond funds	Real estate funds	Capital funds
State A	10	5	20
State B	30	12	10
State C	15	6	25

The current average yields are 5% for bond funds, 10% for real estate funds, and 7% for capital funds. Determine the earnings of the trust from its investments in each state.

SOLUTION   Define the matrix of investment yields by

$$
\begin{matrix}
& \text{Yield} \\
\begin{bmatrix} .05 \\ .10 \\ .07 \end{bmatrix} & \begin{matrix} \text{Bond funds} \\ \text{Real estate funds} \\ \text{Capital funds.} \end{matrix}
\end{matrix}
$$

The amount earned in state A, for instance, is

$$
\begin{bmatrix} \text{amount of} \\ \text{bond funds} \end{bmatrix} \cdot \begin{bmatrix} \text{yield of} \\ \text{bond funds} \end{bmatrix} + \begin{bmatrix} \text{amount of} \\ \text{real estate} \\ \text{funds} \end{bmatrix} \cdot \begin{bmatrix} \text{yield of} \\ \text{real estate} \\ \text{funds} \end{bmatrix}
$$

$$
+ \begin{bmatrix} \text{amount of} \\ \text{capital funds} \end{bmatrix} \cdot \begin{bmatrix} \text{yield of} \\ \text{capital funds} \end{bmatrix}
$$

$$
= (10) \cdot (.05) + (5) \cdot (.10) + (20) \cdot (.07)
$$

And this is just the first entry of the product:

$$
\begin{bmatrix} 10 & 5 & 20 \\ 30 & 12 & 10 \\ 15 & 6 & 25 \end{bmatrix} \begin{bmatrix} .05 \\ .10 \\ .07 \end{bmatrix}.
$$

Similarly, the earnings for the other states are the second and third entries of the product. Carrying out the arithmetic, we find that

$$
\begin{bmatrix} 10 & 5 & 20 \\ 30 & 12 & 10 \\ 15 & 6 & 25 \end{bmatrix} \begin{bmatrix} .05 \\ .10 \\ .07 \end{bmatrix} = \begin{bmatrix} 2.40 \\ 3.40 \\ 3.10 \end{bmatrix}.
$$

Therefore, the trust earns $2.40 million in state A, $3.40 million in state B, and $3.10 million in state C.   **>> Now Try Exercise 69(b)**

EXAMPLE 5   **Manufacturing Revenue** A clothing manufacturer has factories in Los Angeles, San Antonio, and Newark. Sales (in thousands of items) during the first quarter of last year are summarized in the production matrix

	Los Angeles	San Antonio	Newark
Coats	12	13	38
Shirts	25	5	26
Sweaters	11	8	8
Ties	5	0	12

During this period, the selling price of a coat was $100, a shirt $20, a sweater $35, and a tie $25.

(a) Use a matrix calculation to determine the total revenue produced by each of the factories.

(b) Suppose that the prices had been $120, $15, $50, and $20, respectively. How would this have affected the revenue of each factory?

SOLUTION   (a) For each factory, we wish to multiply the price of each item by the number produced to arrive at revenue. Since the production figures for the various items of clothing are arranged down the columns, we arrange the prices in a row matrix, ready for multiplication. The price matrix is

	Coat	Shirt	Sweater	Tie
Price	[100	20	35	25].

The revenues of the various factories are then the entries of the product

$$[100 \quad 20 \quad 35 \quad 25] \begin{bmatrix} 12 & 13 & 38 \\ 25 & 5 & 26 \\ 11 & 8 & 8 \\ 5 & 0 & 12 \end{bmatrix} = [\begin{matrix} \text{Los Angeles} & \text{San Antonio} & \text{Newark} \\ 2210 & 1680 & 4900 \end{matrix}].$$

Since the production figures are in thousands, the revenue figures are in thousands of dollars. That is, the Los Angeles factory has revenues of $2,210,000, the San Antonio factory $1,680,000, and the Newark factory $4,900,000.

(b) In a similar way, we determine the revenue of each factory if the price matrix had been $[120 \quad 15 \quad 50 \quad 20]$.

$$[120 \quad 15 \quad 50 \quad 20] \begin{bmatrix} 12 & 13 & 38 \\ 25 & 5 & 26 \\ 11 & 8 & 8 \\ 5 & 0 & 12 \end{bmatrix} = [\begin{matrix} \text{Los Angeles} & \text{San Antonio} & \text{Newark} \\ 2465 & 2035 & 5590 \end{matrix}].$$

The change in revenue at each factory can be read from the difference of the revenue matrices:

$$[2465 \quad 2035 \quad 5590] - [2210 \quad 1680 \quad 4900] = [255 \quad 355 \quad 690].$$

If prices had been as given in (b), revenues of the Los Angeles factory would have increased by $255,000, revenues at San Antonio would have increased by $355,000, and revenues at Newark would have increased by $690,000.

>> *Now Try Exercise 73*

Identity Matrix

There are special matrices analogous to the number 1. Such matrices are called *identity matrices*.

> **DEFINITION**   The **identity matrix** I_n is the $n \times n$ square matrix with all zeros except for ones down the upper-left-to-lower-right diagonal.

Here are the identity matrices of sizes 2, 3, and 4:

$$I_2 = \begin{bmatrix} 1 & 0 \\ 0 & 1 \end{bmatrix}; \qquad I_3 = \begin{bmatrix} 1 & 0 & 0 \\ 0 & 1 & 0 \\ 0 & 0 & 1 \end{bmatrix}; \qquad I_4 = \begin{bmatrix} 1 & 0 & 0 & 0 \\ 0 & 1 & 0 & 0 \\ 0 & 0 & 1 & 0 \\ 0 & 0 & 0 & 1 \end{bmatrix}.$$

The characteristic property of an identity matrix is that it plays the role of the number 1; that is,

$$I_n \cdot A = A \cdot I_n = A$$

for all $n \times n$ matrices A.

Scalar Multiplication

The **scalar product** of the number c and the matrix A, denoted cA, is the matrix obtained by multiplying each element of A by c. For instance, if A is the matrix $\begin{bmatrix} 3 & 0 \\ 4 & -1 \end{bmatrix}$, then $2A$ is the matrix $\begin{bmatrix} 2 \cdot 3 & 2 \cdot 0 \\ 2 \cdot 4 & 2 \cdot (-1) \end{bmatrix}$. That is,

$$2A = \begin{bmatrix} 6 & 0 \\ 8 & -2 \end{bmatrix}.$$

EXAMPLE 6   **Scalar Multiplication** Consider the price matrix $[100 \quad 20 \quad 35 \quad 25]$, and call it P, from the solution of part (a) of Example 5. Calculate the new price matrix that results from increasing each price by 5%.

SOLUTION   Since 5% is .05, each price p will be increased to $p + .05p$.

$$
\begin{aligned}
p + .05p &= 1 \cdot p + .05p \\
&= (1 + .05)p \\
&= 1.05p.
\end{aligned}
$$

Since each entry of matrix P should be multiplied by 1.05, the new matrix is

$$1.05P = [105 \quad 21 \quad 36.75 \quad 26.25].$$

» Now Try Exercise 65(c)

One of the principal uses of matrices is in dealing with systems of linear equations. Matrices provide a compact way to write and solve linear systems.

EXAMPLE 7   **Representing a System of Equations as a Matrix Equation** Write the system of linear equations

$$
\begin{cases}
-2x + 4y = 2 \\
-3x + 7y = 7
\end{cases}
$$

as a matrix equation.

SOLUTION   The system of equations can be written in the form

$$
\begin{bmatrix} -2x + 4y \\ -3x + 7y \end{bmatrix} = \begin{bmatrix} 2 \\ 7 \end{bmatrix}.
$$

So consider the matrices

$$
A = \begin{bmatrix} -2 & 4 \\ -3 & 7 \end{bmatrix}, \qquad X = \begin{bmatrix} x \\ y \end{bmatrix}, \qquad B = \begin{bmatrix} 2 \\ 7 \end{bmatrix}.
$$

Notice that

$$
AX = \begin{bmatrix} -2 & 4 \\ -3 & 7 \end{bmatrix} \begin{bmatrix} x \\ y \end{bmatrix} = \begin{bmatrix} -2x + 4y \\ -3x + 7y \end{bmatrix}.
$$

Thus, AX is a 2×1 column matrix whose entries correspond to the left side of the given system of linear equations. Since the entries of B correspond to the right side of the system of equations, we can rewrite the given system in the form

$$AX = B$$

—that is,

$$
\begin{bmatrix} -2 & 4 \\ -3 & 7 \end{bmatrix} \begin{bmatrix} x \\ y \end{bmatrix} = \begin{bmatrix} 2 \\ 7 \end{bmatrix}.
$$

» Now Try Exercise 57

The matrix A of the preceding example displays the coefficients of the variables x and y, and so it is called the **coefficient matrix** of the system.

INCORPORATING TECHNOLOGY

The arithmetic operations $+$, $-$, and $*$ can be applied to matrices in much the same way as to numbers, as Figs. 1 and 2 show. *Note:* The matrix product `[A]*[B]` also can be calculated as `[A][B]`, and the scalar product `c*[A]` also can be calculated as `c[A]`. Identity matrices are placed on the home screen with the command `identity` (in the MATRIX/MATH menu), as shown in Figure 3. The value in the *i*th row, *j*th column, of the matrix **[A]** can be displayed with **[A](i,j)**.

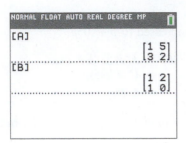

Figure 1

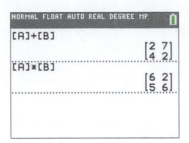

Figure 2

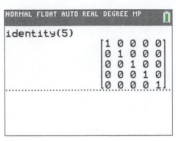

Figure 3

The entries of an $m \times n$ matrix can be typed directly into an *m* by *n* range of cells. (In order to insert a fraction, type in a formula, such as **=2/3**. A matrix is most easily used in calculations if it has been given a name, such as **A** or **B**. To name a matrix, select it, click on the Name box, type the name, and press Enter. Single-letter names can consist of any letter other than *C* or *R*.

Matrices are added, subtracted, and multiplied with the + operator, the − operator, and the MMULT function, respectively. The ∗ operator is used for scalar multiplication. The following steps carry out these operations on the matrices *A* and *B* and the number *c*:

1. Highlight a range of cells having the size of the matrix to be computed.
2. Type **=A+B**, **=A−B**, **=MMULT(A,B)** or **=c*A**.
3. Press Ctrl+Shift+Enter. *Note:* The plus signs indicate that the three keys should be held down together. The computed matrix will be displayed in the highlighted block.

☀**WolframAlpha** Two matrices are added, subtracted, and multiplied with the operators +, −, and period. An asterisk is used as the operator for scalar multiplication. For instance, the results of the instructions **{{1,5},{3,2}}.{{1,2},{1,0}}** and **2*{{1,5},{3,2}}** are

$$\begin{pmatrix} 6 & 2 \\ 5 & 6 \end{pmatrix} \text{ and } \begin{pmatrix} 2 & 10 \\ 6 & 4 \end{pmatrix}, \text{ respectively.}$$

Check Your Understanding 2.3

Solutions can be found following the section exercises.

1. Compute

$$\begin{bmatrix} 3 & 1 & 2 \\ -1 & 0 & \frac{1}{2} \\ 0 & 4 & 1 \end{bmatrix} \begin{bmatrix} 7 & -1 & 0 \\ 5 & 4 & 2 \\ -6 & 0 & 4 \end{bmatrix}.$$

2. Give the system of linear equations that is equivalent to the matrix equation

$$\begin{bmatrix} 3 & -6 \\ 2 & 1 \end{bmatrix} \begin{bmatrix} x \\ y \end{bmatrix} = \begin{bmatrix} 5 \\ 0 \end{bmatrix}.$$

3. Give a matrix equation equivalent to this system of equations:

$$\begin{cases} 8x + 3y = 7 \\ 9x - 2y = -5. \end{cases}$$

EXERCISES 2.3

In Exercises 1–6, give the size and special characteristics of the given matrix (such as square, column, row, identity).

1. $\begin{bmatrix} 3 & 2 & 4 \\ \frac{1}{2} & 0 & 6 \end{bmatrix}$

2. $\begin{bmatrix} 3 \\ -1 \end{bmatrix}$

3. $\begin{bmatrix} 2 & \frac{1}{3} & 0 \end{bmatrix}$

4. $\begin{bmatrix} 1 & 0 \\ 0 & 1 \end{bmatrix}$

5. $\begin{bmatrix} 1 & 0 \\ 0 & 0 \end{bmatrix}$

6. $[5]$

Exercises 7–10 refer to the 2×3 matrix $A = \begin{bmatrix} 2 & -4 & 6 \\ 0 & 3 & -1 \end{bmatrix}$.

7. Find a_{12} and a_{21}.

8. Find a_{23} and a_{11}.

9. For what values of *i* and *j* does $a_{ij} = 6$?

10. For what values of *i* and *j* does $a_{ij} = 3$?

In Exercises 11–26, perform the indicated matrix calculation.

11. $\begin{bmatrix} 4 & -2 \\ 3 & 0 \end{bmatrix} + \begin{bmatrix} 5 & 5 \\ 4 & -1 \end{bmatrix}$ **12.** $\begin{bmatrix} 8 \\ -3 \end{bmatrix} + \begin{bmatrix} 5 \\ 6 \end{bmatrix}$

13. $\begin{bmatrix} 1.3 & 5 & 2.3 \\ -6 & 0 & .7 \end{bmatrix} + \begin{bmatrix} .7 & -1 & .2 \\ .5 & 1 & .5 \end{bmatrix}$

14. $\begin{bmatrix} \frac{5}{6} & 10 & \frac{1}{2} \end{bmatrix} + \begin{bmatrix} \frac{2}{3} & -7 & \frac{3}{2} \end{bmatrix}$

15. $\begin{bmatrix} 2 & 8 \\ \frac{4}{3} & 4 \\ 1 & -2 \end{bmatrix} - \begin{bmatrix} 1 & 5 \\ \frac{1}{3} & 2 \\ -3 & 0 \end{bmatrix}$ **16.** $\begin{bmatrix} 1 & 0 \\ 0 & 1 \end{bmatrix} - \begin{bmatrix} .8 & .5 \\ .2 & .5 \end{bmatrix}$

17. $\begin{bmatrix} -5 \\ \frac{1}{2} \end{bmatrix} - \begin{bmatrix} 2 \\ \frac{1}{3} \end{bmatrix}$

18. $\begin{bmatrix} 1.4 & 0 & 3 \\ .5 & -1.2 & 2.5 \end{bmatrix} - \begin{bmatrix} .6 & -1 & 3 \\ .1 & .4 & 1 \end{bmatrix}$

19. $\begin{bmatrix} 5 & 3 \end{bmatrix} \begin{bmatrix} 1 \\ 2 \end{bmatrix}$ **20.** $\begin{bmatrix} 1 & 0 & 0 \end{bmatrix} \begin{bmatrix} \frac{1}{2} \\ 6 \\ 2 \end{bmatrix}$

21. $\begin{bmatrix} 6 & 1 & 5 \end{bmatrix} \begin{bmatrix} \frac{1}{2} \\ -3 \\ 2 \end{bmatrix}$ **22.** $\begin{bmatrix} 0 & 0 \end{bmatrix} \begin{bmatrix} 5 \\ -3 \end{bmatrix}$

23. $\frac{2}{3} \begin{bmatrix} 6 & 0 & -1 \\ -9 & \frac{3}{4} & \frac{1}{2} \end{bmatrix}$ **24.** $1.5 \begin{bmatrix} 4 & .5 \\ 0 & 1.2 \end{bmatrix}$

25. $2 \begin{bmatrix} \frac{1}{2} & -1 \\ 4 & 0 \end{bmatrix} + 3 \begin{bmatrix} \frac{2}{3} & 7 \\ 5 & 1 \end{bmatrix}$ **26.** $\frac{3}{5} \begin{bmatrix} 10 & 25 \end{bmatrix} - \begin{bmatrix} 6 & -3 \end{bmatrix}$

In Exercises 27–32, the sizes of two matrices are given. Tell whether or not the product AB is defined. If so, give its size.

27. $A, 3 \times 4$; $B, 4 \times 5$ **28.** $A, 3 \times 3$; $B, 3 \times 4$

29. $A, 3 \times 2$; $B, 3 \times 2$ **30.** $A, 1 \times 1$; $B, 1 \times 1$

31. $A, 3 \times 3$; $B, 3 \times 1$ **32.** $A, 4 \times 2$; $B, 3 \times 4$

In Exercises 33–52, perform the multiplication.

33. $\begin{bmatrix} 3 & 1 \\ 0 & 2 \end{bmatrix} \begin{bmatrix} 1 & 4 \\ 3 & 5 \end{bmatrix}$ **34.** $\begin{bmatrix} 4 & -1 \\ 2 & \frac{1}{2} \end{bmatrix} \begin{bmatrix} 3 \\ 2 \end{bmatrix}$

35. $\begin{bmatrix} 4 & 1 & 0 \\ -2 & 0 & 3 \\ 1 & 5 & -1 \end{bmatrix} \begin{bmatrix} 5 \\ 1 \\ 2 \end{bmatrix}$ **36.** $\begin{bmatrix} 0 & 0 \\ 0 & 0 \\ 0 & 0 \end{bmatrix} \begin{bmatrix} 1 & 2 \\ 3 & 4 \end{bmatrix}$

37. $\begin{bmatrix} 1 & 0 \\ 0 & 1 \end{bmatrix} \begin{bmatrix} 5 & 6 \\ 7 & 8 \end{bmatrix}$ **38.** $\begin{bmatrix} 1 & 2 \\ 1 & 3 \end{bmatrix} \begin{bmatrix} 3 & -2 \\ -1 & 1 \end{bmatrix}$

39. $\begin{bmatrix} .6 & .3 \\ .4 & .7 \end{bmatrix} \begin{bmatrix} .6 & .3 \\ .4 & .7 \end{bmatrix}$ **40.** $\begin{bmatrix} 0 & 1 & 2 \\ -1 & 4 & \frac{1}{2} \\ 1 & 3 & 0 \end{bmatrix} \begin{bmatrix} 3 & -1 & 5 \\ 0 & 2 & 2 \\ 4 & -6 & 0 \end{bmatrix}$

41. $\begin{bmatrix} 2 & -1 & 4 \\ 0 & 1 & 0 \\ \frac{1}{2} & 3 & -2 \end{bmatrix} \begin{bmatrix} 4 & 8 & 0 \\ 3 & -1 & 2 \\ 5 & 0 & 1 \end{bmatrix}$

42. $\begin{bmatrix} 1 & 0 & 0 \\ 0 & 1 & 0 \\ 0 & 0 & 1 \end{bmatrix} \begin{bmatrix} 1 \\ 2 \\ 3 \end{bmatrix}$ **43.** $\begin{bmatrix} \frac{1}{3} & \frac{2}{3} \\ \frac{1}{3} & \frac{2}{3} \end{bmatrix} \begin{bmatrix} \frac{1}{3} & \frac{2}{3} \\ \frac{1}{3} & \frac{2}{3} \end{bmatrix}$

44. $\begin{bmatrix} .4 & .4 & .4 \\ .4 & .4 & .4 \\ .2 & .2 & .2 \end{bmatrix} \begin{bmatrix} .4 & .4 & .4 \\ .4 & .4 & .4 \\ .2 & .2 & .2 \end{bmatrix}$

45. $\begin{bmatrix} 2 & 5 \end{bmatrix} \begin{bmatrix} 0 & 3 \\ 6 & 7 \end{bmatrix}$ **46.** $\begin{bmatrix} 4 & 0 & 1 \end{bmatrix} \begin{bmatrix} 2 & 3 \\ 4 & 5 \\ 0 & 6 \end{bmatrix}$

47. $\begin{bmatrix} 2 & 0 \\ 0 & 3 \end{bmatrix} \begin{bmatrix} 5 & 0 \\ 0 & 5 \end{bmatrix}$ **48.** $\begin{bmatrix} \frac{1}{3} & 0 \\ 0 & \frac{1}{4} \end{bmatrix} \begin{bmatrix} 6 & 0 \\ 0 & 8 \end{bmatrix}$

49. $\begin{bmatrix} 0 & 0 \\ 0 & 0 \end{bmatrix} \begin{bmatrix} 2.34 & 5.6 \\ -3.7 & .08 \end{bmatrix}$ **50.** $\begin{bmatrix} -78 & 56 \\ 312 & 23 \end{bmatrix} \begin{bmatrix} 0 & 0 \\ 0 & 0 \end{bmatrix}$

51. $\begin{bmatrix} 23 & 24 \\ 25 & 26 \end{bmatrix} \begin{bmatrix} 1 & 0 \\ 0 & 1 \end{bmatrix}$ **52.** $\begin{bmatrix} 1 & 0 \\ 0 & 1 \end{bmatrix} \begin{bmatrix} 2.4 & 5.6 \\ 7.8 & 9.9 \end{bmatrix}$

In Exercises 53–56, give the system of linear equations that is equivalent to the matrix equation. Do not solve.

53. $\begin{bmatrix} 2 & 3 \\ 4 & 5 \end{bmatrix} \begin{bmatrix} x \\ y \end{bmatrix} = \begin{bmatrix} 6 \\ 7 \end{bmatrix}$ **54.** $\begin{bmatrix} -3 & 4 \\ 0 & 1 \end{bmatrix} \begin{bmatrix} x \\ y \end{bmatrix} = \begin{bmatrix} 1 \\ 1 \end{bmatrix}$

55. $\begin{bmatrix} 1 & 2 & 3 \\ 4 & 5 & 6 \\ 7 & 8 & 9 \end{bmatrix} \begin{bmatrix} x \\ y \\ z \end{bmatrix} = \begin{bmatrix} 10 \\ 11 \\ 12 \end{bmatrix}$ **56.** $\begin{bmatrix} 1 & 0 & 0 \\ 0 & 1 & 0 \\ 0 & 0 & 1 \end{bmatrix} \begin{bmatrix} x \\ y \\ z \end{bmatrix} = \begin{bmatrix} 1 \\ 2 \\ 3 \end{bmatrix}$

In Exercises 57–60, write the given system of linear equations as a matrix equation.

57. $\begin{cases} 3x + 2y = -1 \\ 7x - y = 2 \end{cases}$ **58.** $\begin{cases} 5x - 2y = 6 \\ -2x + 4y = 0 \end{cases}$

59. $\begin{cases} x - 2y + 3z = 5 \\ y + z = 6 \\ z = 2 \end{cases}$ **60.** $\begin{cases} -2x + 4y - z = 5 \\ x + 6y + 3z = -1 \\ 7x + 4z = 8 \end{cases}$

The *distributive law* says that $(A + B)C = AC + BC$. That is, adding A and B and then multiplying on the right by C gives the same result as first multiplying each of A and B on the right by C and then adding. In Exercises 61 and 62, verify the distributive law for the given matrices.

61. $A = \begin{bmatrix} 1 & 2 \\ 0 & 3 \end{bmatrix}$, $B = \begin{bmatrix} 3 & -2 \\ 4 & 5 \end{bmatrix}$, $C = \begin{bmatrix} 1 & 6 \\ 2 & 0 \end{bmatrix}$

62. $A = \begin{bmatrix} 1 & 0 & 0 \\ 0 & 1 & 0 \\ 0 & 0 & 1 \end{bmatrix}$, $B = \begin{bmatrix} 2 & 1 & 3 \\ 0 & 5 & -1 \\ 3 & 6 & 0 \end{bmatrix}$, $C = \begin{bmatrix} 0 \\ 3 \\ -4 \end{bmatrix}$

Two $n \times n$ matrices A and B are called *inverses* (of one another) if both products AB and BA equal I_n. Check that the pairs of matrices in Exercises 63 and 64 are inverses.

63. $\begin{bmatrix} 3 & -1 \\ -1 & \frac{1}{2} \end{bmatrix}, \begin{bmatrix} 1 & 2 \\ 2 & 6 \end{bmatrix}$

64. $\begin{bmatrix} 2 & 8 & -11 \\ -1 & -5 & 7 \\ 1 & 2 & -3 \end{bmatrix}, \begin{bmatrix} 1 & 2 & 1 \\ 4 & 5 & -3 \\ 3 & 4 & -2 \end{bmatrix}$

65. Wardrobe Costs The quantities of pants, shirts, and jackets owned by Mike and Don are given by the matrix A, and the costs of these items are given by matrix B.

$$\begin{array}{c} \\ \text{Mike} \\ \text{Don} \end{array} \begin{array}{ccc} \text{Pants} & \text{Shirts} & \text{Jackets} \\ \begin{bmatrix} 6 & 8 & 2 \\ 2 & 5 & 3 \end{bmatrix} \end{array} = A$$

$$\begin{array}{c} \\ \text{Pants} \\ \text{Shirts} \\ \text{Jackets} \end{array} \begin{array}{c} \text{Cost} \\ \begin{bmatrix} 20 \\ 15 \\ 50 \end{bmatrix} \end{array} = B$$

(a) Calculate the matrix AB.
(b) Interpret the entries of the matrix AB.
(c) Calculate the matrix $1.25B$.
(d) Interpret the entries of the matrix $1.25B$.

66. Retail Sales Two stores sell the exact same brand and style of a dresser, a nightstand, and a bookcase. Matrix A gives the retail prices (in dollars) for the items. Matrix B gives the number of each item sold at each store in one month.

$$A = \begin{bmatrix} \overset{\text{Dresser}}{250} & \overset{\text{Nightstand}}{80} & \overset{\text{Bookcase}}{60} \end{bmatrix}$$

$$B = \begin{array}{c} \\ \\ \\ \end{array} \begin{bmatrix} \overset{\text{Store 1}}{40} & \overset{\text{Store 2}}{35} \\ 30 & 35 \\ 50 & 75 \end{bmatrix} \begin{array}{l} \text{Dresser} \\ \text{Nightstand} \\ \text{Bookcase} \end{array}$$

(a) Calculate AB.
(b) Interpret the entries of AB.
(c) Calculate the matrix $1.1A$.
(d) Interpret the entries of the matrix $1.1A$.

67. Retail Sales A candy shop sells various items for the price per pound (in dollars) indicated in matrix A. Matrix B gives the number of pounds of coated peanuts, raisins, and espresso beans prepared in a week. Matrix C gives the total number of pounds of white chocolate-covered, milk chocolate-covered, and dark chocolate-covered items sold each week.

$$A = \begin{array}{c} \\ \\ \\ \end{array} \begin{bmatrix} \overset{\text{White}}{3} & \overset{\text{Milk}}{3} & \overset{\text{Dark}}{5.8} \\ 2.5 & 3.5 & 6 \\ 9 & 8 & 9.5 \end{bmatrix} \begin{array}{l} \text{Peanuts} \\ \text{Raisins} \\ \text{Espresso beans} \end{array}$$

$$B = \begin{bmatrix} \overset{\text{Peanuts}}{210} & \overset{\text{Raisins}}{175} & \overset{\text{Espresso beans}}{135} \end{bmatrix}$$

$$C = \begin{bmatrix} 105 \\ 390 \\ 285 \end{bmatrix} \begin{array}{l} \text{White} \\ \text{Milk} \\ \text{Dark} \end{array}$$

Determine and interpret the following matrices.
(a) BA (b) AC (c) $.9C$

68. Wholesale and Retail Sales A company has three appliance stores that sell washers, dryers, and ranges. Matrices W and R give the wholesale and retail prices of these items, respectively. Matrices N and D give the quantities of these items sold by the three stores in November and December, respectively.

$$W = \begin{bmatrix} \overset{\text{Washers}}{300} & \overset{\text{Dryers}}{250} & \overset{\text{Ranges}}{450} \end{bmatrix}$$

$$R = \begin{bmatrix} \overset{\text{Washers}}{500} & \overset{\text{Dryers}}{450} & \overset{\text{Ranges}}{750} \end{bmatrix}$$

$$N = \begin{bmatrix} \overset{\text{Store 1}}{30} & \overset{\text{Store 2}}{40} & \overset{\text{Store 3}}{20} \\ 20 & 30 & 10 \\ 10 & 5 & 35 \end{bmatrix} \begin{array}{l} \text{Washers} \\ \text{Dryers} \\ \text{Ranges} \end{array}$$

$$D = \begin{bmatrix} \overset{\text{Store 1}}{20} & \overset{\text{Store 2}}{50} & \overset{\text{Store 3}}{30} \\ 30 & 10 & 20 \\ 10 & 20 & 30 \end{bmatrix} \begin{array}{l} \text{Washers} \\ \text{Dryers} \\ \text{Ranges} \end{array}$$

Determine and interpret the following matrices:
(a) WN (b) WD
(c) RN (d) RD
(e) $R - W$ (f) $(R - W)N$
(g) $(R - W)D$ (h) $N + D$
(i) $(R - W)(N + D)$ (j) $.95R$

69. Course Grades Three professors teaching the same course have entirely different grading policies. The percentage of students given each grade by the professors is summarized in the following matrix:

	Grade				
	A	B	C	D	F
Prof. I	25	35	30	10	0
Prof. II	10	20	40	20	10
Prof. III	5	10	20	40	25

(a) The point values of the grades are A = 4, B = 3, C = 2, D = 1, and F = 0. Use matrix multiplication to determine the average grade given by each professor.
(b) Professor I has 240 students, professor II has 120 students, and professor III has 40 students. Use matrix multiplication to determine the numbers of As, Bs, Cs, Ds, and Fs given.

70. Semester Grades A professor bases semester grades on four 100-point items: homework, quizzes, a midterm exam, and a final exam. Students may choose one of three schemes summarized in the accompanying matrix for weighting the points from the four items. Use matrix multiplication to determine the most advantageous weighting scheme for a student who earned 97 points on homework, 72 points on the quizzes, 83 points on the midterm exam, and 75 points on the final exam.

	Items			
	HW	Qu	ME	FE
Scheme I	.10	.10	.30	.50
Scheme II	.10	.20	.30	.40
Scheme III	.15	.15	.35	.35

71. Voter Analysis In a certain town, the proportions of voters voting Democratic and Republican by various age groups is summarized by this matrix:

	Dem.	Rep.	
Under 30	.65	.35	
30–50	.55	.45	$= A.$
Over 50	.45	.55	

The population of voters in the town by age group is given by the matrix

$$B = \begin{bmatrix} \underset{\substack{\text{Under} \\ 30}}{6000} & \underset{\substack{\text{30–50}}}{8000} & \underset{\substack{\text{Over} \\ 50}}{4000} \end{bmatrix}.$$

Interpret the entries of the matrix product BA.

72. Voter Analysis Refer to Exercise 71.
(a) According to the data, which party would win and what would be the percentage of the winning vote?

(b) Suppose that the population of the town shifted toward older residents, as reflected in the population matrix $B = [2000 \quad 4000 \quad 12{,}000]$. What would be the result of the election now?

73. Labor Costs Suppose that a contractor employs carpenters, bricklayers, and plumbers, working three shifts per day. The number of labor-hours employed in each of the shifts is summarized in the following matrix:

	Shift 1	2	3
Carpenters	50	20	10
Bricklayers	30	30	15
Plumbers	20	20	5

Labor in shift 1 costs $20 per hour, in shift 2 $30 per hour, and in shift 3 $40 per hour. Use matrix multiplication to compute the dollar amount spent on each type of labor.

74. Epidemiology A flu epidemic hits a large city. Each resident of the city is either sick, well, or a carrier. The proportion of people in each of the categories is expressed by the following matrix:

	Age 0–10	10–30	Over 30
Well	.70	.70	.60
Sick	.10	.20	.30
Carrier	.20	.10	.10

$= A.$

The population of the city is distributed by age and sex as follows:

		Male	Female
	0–10	60,000	65,000
Age	10–30	100,000	110,000
	Over 30	200,000	230,000

$= B.$

(a) Compute AB.
(b) How many sick males are there?
(c) How many female carriers are there?

75. Nutrition Analysis Mikey's diet consists of food X and food Y. The matrix N represents the number of units of nutrients 1, 2, and 3 per ounce for each of the foods.

$$N = \begin{bmatrix} 60 & 50 & 38 \\ 42 & 50 & 67 \end{bmatrix} \begin{matrix} X \\ Y \end{matrix}$$

with columns labeled 1, 2, 3.

The matrices B, L, and D represent the number of ounces of each food that Mikey eats each day for breakfast, lunch, and dinner, respectively.

$$B = [2 \quad 1] \quad L = [1 \quad 3] \quad D = [2 \quad 4]$$

with columns labeled $X \quad Y$.

Calculate and interpret the following:
(a) BN
(b) LN
(c) DN
(d) $B + L + D$
(e) $(B + L + D)N$

76. Bakery Sales A bakery makes three types of cookies, I, II, and III. Each type of cookie is made from the four ingredients A, B, C, and D. The number of units of each ingredient used in each type of cookie is given by the matrix M. The cost per

unit of each of the four ingredients (in cents) is given by the matrix N. The selling price for each of the cookies (in cents) is given by the matrix S. The baker receives an order for 10 type I cookies, 20 type II cookies, and 15 type III cookies, as represented by the matrix R.

$$M = \begin{bmatrix} 1 & 0 & 2 & 4 \\ 3 & 2 & 1 & 1 \\ 2 & 5 & 3 & 1 \end{bmatrix} \begin{matrix} I \\ II \\ III \end{matrix}$$

with columns labeled $A \quad B \quad C \quad D$.

$$N = \begin{bmatrix} 10 \\ 20 \\ 15 \\ 17 \end{bmatrix} \begin{matrix} A \\ B \\ C \\ D \end{matrix} \quad \text{Cost}$$

$$S = \begin{bmatrix} 175 \\ 150 \\ 225 \end{bmatrix} \begin{matrix} I \\ II \\ III \end{matrix} \quad \begin{matrix} \text{Selling} \\ \text{price} \end{matrix}$$

$$R = [10 \quad 20 \quad 15] \quad \text{Order}$$

with columns labeled $I \quad II \quad III$.

Calculate and interpret the following:
(a) RM
(b) MN
(c) RMN
(d) $S - MN$
(e) $R(S - MN)$
(f) RS

77. Revenue A community fitness center has a pool and a weight room. The admission prices (in dollars) for residents and nonresidents are given by the matrix

$$P = \begin{bmatrix} 4.50 \\ 5.00 \end{bmatrix} \begin{matrix} \text{Residents} \\ \text{Nonresidents.} \end{matrix} \quad \text{Price}$$

The average daily numbers of customers for the fitness center are given by the matrix

$$A = \begin{bmatrix} 90 & 63 \\ 78 & 59 \end{bmatrix} \begin{matrix} \text{Pool} \\ \text{Weight room.} \end{matrix}$$

with columns labeled Residents, Nonresidents.

(a) Compute AP.
(b) What is the average amount of money taken in by the pool each day?

78. Production Planning A company makes DVD players and TV sets. Each DVD player requires 3 hours of assembly and $\frac{1}{2}$ hour of packaging, while each TV set requires 5 hours of assembly and 1 hour of packaging.
(a) Write a matrix T representing the required time for assembly and packaging of DVD players and TV sets.
(b) The company receives an order from a retail outlet for 30 DVD players and 20 TV sets. Find a matrix S so that either ST or TS gives the total assembly time and the total packaging time required to fill the order. What is the total assembly time? What is the total packaging time?

79. Production Planning A bakery sells Boston cream pies and carrot cakes. Each Boston cream pie requires 30 minutes preparation time, 30 minutes baking time, and 15 minutes for finishing. Each carrot cake requires 45 minutes preparation time, 50 minutes baking time, and 10 minutes for finishing.
(a) Write a matrix T representing the required time for preparation, baking, and finishing for the Boston cream pies and the carrot cakes.
(b) The bakery receives an order for 20 Boston cream pies and 8 carrot cakes for a large party. Find a matrix S so that either ST or TS gives the total preparation, baking, and finishing times required to fill this order.
(c) What is the total baking time? What is the total finishing time?

80. Time Requirements A beauty salon offers manicures and pedicures. A manicure requires 20 minutes for preparation, 5 minutes for lacquering, and 15 minutes for drying. A pedicure requires 30 minutes for preparation, 5 minutes for lacquering, and 20 minutes for drying.
   (a) Construct a matrix T representing the time required for preparation, lacquering, and drying for manicures and pedicures.
   (b) Suppose that the salon will be giving manicures and pedicures to a large wedding party. If 15 members of the wedding party want manicures and 9 want pedicures, find a matrix S so that either ST or TS gives the total amount of time required for each of the three steps.
   (c) What is the total time required for drying?

81. Production and Revenue The J.E. Carrying Company makes two types of backpacks. The larger Huge One backpack requires 2 hours for cutting, 3 hours for sewing, and 2 hours for finishing, and sells for \$32. The smaller Regular Joe backpack requires 1.5 hours for cutting, 2 hours for sewing, and 1 hour for finishing, and sells for \$24.
   (a) Construct a matrix T representing the time required for each of the three steps in making the backpacks.
   (b) Construct a matrix S representing the sales prices for the two types of backpacks.
   (c) Suppose that the J.E. Carrying Company receives an order for 27 Huge One backpacks and 56 Regular Joe backpacks. Construct a matrix A so that either AT or TA gives the total time required to construct the backpacks in this order, and either AS or SA gives the total revenue generated by this order.
   (d) How much total time is needed for sewing to fill this order?
   (e) What is the total revenue for this order?

82. MP3 Sales A store sells three types of MP3 players. Matrix A contains information about size (in gigabytes), battery life (in hours), and weight (in ounces) of the three MP3 players. Matrix B contains the sales prices (in dollars) of the MP3 players, while matrix C contains the number of each type of player sold in one week.

$$A = \begin{matrix} \text{Type I} & \text{Type II} & \text{Type III} \\ \begin{bmatrix} 4 & 8 & 16 \\ 25 & 30 & 30 \\ 1 & 1.9 & 1.1 \end{bmatrix} & & \begin{matrix} \text{Size} \\ \text{Battery Life} \\ \text{Weight} \end{matrix} \end{matrix}$$

$$B = \begin{matrix} \text{Type I} & \text{Type II} & \text{Type III} \\ [\,\, 40 & 75 & 150 \,\,] \end{matrix}$$

$$C = \begin{bmatrix} 25 \\ 16 \\ 32 \end{bmatrix} \begin{matrix} \text{Type I} \\ \text{Type II} \\ \text{Type III} \end{matrix}$$

Calculate and interpret the following:
   (a) BC **(b)** AC
   (c) the row 2, column 1 entry of AC

83. Make up an application whose answer is that the total cost is given by

$$[20 \quad 30] \begin{bmatrix} 600 \\ 700 \end{bmatrix}.$$

84. Find the values of a and b for which $A \cdot B = I_3$, where

$$A = \begin{bmatrix} 3 & 2 & 0 \\ 1 & 1 & 0 \\ 0 & 0 & 1 \end{bmatrix} \quad \text{and} \quad B = \begin{bmatrix} a & b & 0 \\ -1 & 3 & 0 \\ 0 & 0 & 1 \end{bmatrix}.$$

In Exercises 85 and 86, determine the matrix B on the basis of the screen shown.

85.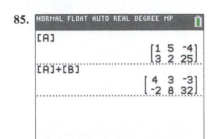

86.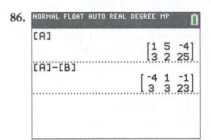

87. If A is a 3×4 matrix and $A(BB)$ is defined, what is the size of matrix B?

88. If B is a 3×5 matrix and $(AA)B$ is defined, what is the size of matrix A?

89. School Enrollments Table 1 gives the number of public school teachers (elementary and secondary) and the average number of pupils per teacher for three mid-Atlantic states in a recent year. Set up a product of two matrices that gives the total number of pupils in the three states. (Source: National Center for Education Statistics, *Digest of Education Statistics.*)

Table 1 Teachers and Pupils

	Delaware	Maryland	Virginia
Teachers	9257	57,718	89,389
Pupils per Teacher	13.9	14.9	14.2

90. Population Table 2 gives the area and 2015 population density for three West Coast states. Set up a product of two matrices that gives total population of the three states.

Table 2 State Areas and Densities

	California	Oregon	Washington
Land Area (sq. mi.)	155,959	95,997	66,544
Pop. Density (per sq. mile)	250.0	42.0	107.8

In Exercises 91–96, calculate the given expression, where

$$A = \begin{bmatrix} .4 & 7 & -3 \\ 19 & .5 & 1.6 \\ -9 & 11 & 2 \end{bmatrix}, \qquad B = \begin{bmatrix} 6 & -9 & .3 \\ 1.5 & 22 & -4 \\ -5 & 6.6 & 14 \end{bmatrix}.$$

91. $A + B$ **92.** $B - A$ **93.** BA

94. AB **95.** $3A$ **96.** $I_3 + 2A$

97. Try adding two matrices of different sizes. How does your technology respond?

98. Try multiplying two matrices in which the number of columns of the first matrix differs from the number of rows of the second matrix. How does your technology respond?

Solutions to Check Your Understanding 2.3

1. Answer:

$$\begin{bmatrix} 3 & 1 & 2 \\ -1 & 0 & \frac{1}{2} \\ 0 & 4 & 1 \end{bmatrix}\begin{bmatrix} 7 & -1 & 0 \\ 5 & 4 & 2 \\ -6 & 0 & 4 \end{bmatrix} = \begin{bmatrix} 14 & 1 & 10 \\ -10 & 1 & 2 \\ 14 & 16 & 12 \end{bmatrix}.$$

The systematic steps to be taken are as follows:

(a) Determine the size of the product matrix. Since we have a

$$③ \times 3 \quad \text{times a} \quad 3 \times ③,$$
$$\underbrace{\qquad}_{\text{outer dimensions}}$$

the size of the product is given by the outer dimensions, or 3×3. Begin by drawing a 3×3 rectangular array.

(b) Find the entries one at a time. To find the entry in the first row, first column of the product, look at the first row of the left matrix and the first column of the right matrix, and form their product. In this case,

$$\begin{bmatrix} 3 & 1 & 2 \\ -1 & 0 & \frac{1}{2} \\ 0 & 4 & 1 \end{bmatrix}\begin{bmatrix} 7 & -1 & 0 \\ 5 & 4 & 2 \\ -6 & 0 & 4 \end{bmatrix} = \begin{bmatrix} 14 & & \\ & & \\ & & \end{bmatrix},$$

since $3 \cdot 7 + 1 \cdot 5 + 2(-6) = 14$. In general, to find the entry in the ith row, jth column of the product, put one finger on the ith row of the left matrix and another finger on the jth column of the right matrix. Then, multiply the row matrix times the column matrix to get the desired entry.

2. Denote the three matrices by A, X, and B, respectively. Since b_{11} (the entry of the first row, first column of B) is 5, this means that

$$[\text{first row of } A]\begin{bmatrix} \text{first} \\ \text{column} \\ \text{of } X \end{bmatrix} = [b_{11}].$$

That is,

$$[3 \quad -6]\begin{bmatrix} x \\ y \end{bmatrix} = [5] \quad \text{or} \quad 3x - 6y = 5.$$

Similarly, $b_{21} = 0$ says that $2x + y = 0$. Therefore, the corresponding system of linear equations is

$$\begin{cases} 3x - 6y = 5 \\ 2x + y = 0. \end{cases}$$

3. The coefficient matrix is

$$\begin{bmatrix} 8 & 3 \\ 9 & -2 \end{bmatrix}.$$

So the system is equivalent to the matrix equation

$$\begin{bmatrix} 8 & 3 \\ 9 & -2 \end{bmatrix}\begin{bmatrix} x \\ y \end{bmatrix} = \begin{bmatrix} 7 \\ -5 \end{bmatrix}.$$

2.4 The Inverse of a Square Matrix

In Section 2.3, we introduced the operations of addition, subtraction, and multiplication of matrices. In this section, let us pursue the algebra of matrices a bit further and consider equations involving matrices. Specifically, we consider equations of the form

$$AX = B, \tag{1}$$

where A and B are given matrices and X is an unknown matrix whose entries are to be determined. Such equations among matrices are intimately bound up with the theory of systems of linear equations. Indeed, we described the connection in a special case in Example 7 of Section 2.3. In that example, we wrote the system of linear equations

$$\begin{cases} -2x + 4y = 2 \\ -3x + 7y = 7 \end{cases}$$

as a matrix equation of the form (1), where

$$A = \begin{bmatrix} -2 & 4 \\ -3 & 7 \end{bmatrix}, \qquad B = \begin{bmatrix} 2 \\ 7 \end{bmatrix}, \qquad X = \begin{bmatrix} x \\ y \end{bmatrix}.$$

Note that, by determining the entries (x and y) of the unknown matrix X, we solve the system of linear equations. We will return to this example after we have made a complete study of the matrix equation (1).

As motivation for our solution of equation (1), let us consider the analogous equation among numbers:

$$ax = b,$$

where a and b are given numbers and x is to be determined. We may as well assume that $a \neq 0$. Otherwise, x does not occur. Let us examine its solution in great detail. Multiply both sides by $1/a$. (Note that $1/a$ makes sense, since $a \neq 0$.)

$$\left(\frac{1}{a}\right) \cdot (ax) = \frac{1}{a} \cdot b$$

$$\left(\frac{1}{a} \cdot a\right) \cdot x = \frac{1}{a} \cdot b$$

$$1 \cdot x = \frac{1}{a} \cdot b$$

$$x = \frac{1}{a} \cdot b$$

Let us model our solution of equation (1) on the preceding calculation. To do so, we need to multiply both sides of the equation by a matrix that plays the same role in matrix arithmetic as $1/a$ plays in ordinary arithmetic. Our first task, then, will be to introduce this matrix and study its properties.

The number $1/a$ has the following relationship to the number a:

$$\frac{1}{a} \cdot a = a \cdot \frac{1}{a} = 1. \tag{2}$$

The matrix analog of the number 1 is an identity matrix I. This prompts us to generalize equation (2) to matrices as follows.

DEFINITION Let A be an $n \times n$ matrix. The **inverse of A**, denoted A^{-1}, is the $n \times n$ matrix with the properties

$$A^{-1}A = I_n \quad \text{and} \quad AA^{-1} = I_n.$$

The matrix A^{-1} is the matrix analog of the number $1/a$. It can be shown that a matrix A has, at most, one inverse. However, A may not have an inverse at all.

If we are given a matrix A, then it is easy to determine whether or not a given matrix is its inverse. Merely check the two equations in the definition with the given matrix substituted for A^{-1}. For example, if

$$A = \begin{bmatrix} -2 & 4 \\ -3 & 7 \end{bmatrix}, \quad \text{then} \quad A^{-1} = \begin{bmatrix} -\frac{7}{2} & 2 \\ -\frac{3}{2} & 1 \end{bmatrix}.$$

Indeed, we have

$$\underbrace{\begin{bmatrix} -\frac{7}{2} & 2 \\ -\frac{3}{2} & 1 \end{bmatrix}}_{A^{-1}} \underbrace{\begin{bmatrix} -2 & 4 \\ -3 & 7 \end{bmatrix}}_{A} = \begin{bmatrix} 7-6 & -14+14 \\ 3-3 & -6+7 \end{bmatrix} = \underbrace{\begin{bmatrix} 1 & 0 \\ 0 & 1 \end{bmatrix}}_{I_2}$$

and

$$\underbrace{\begin{bmatrix} -2 & 4 \\ -3 & 7 \end{bmatrix}}_{A} \underbrace{\begin{bmatrix} -\frac{7}{2} & 2 \\ -\frac{3}{2} & 1 \end{bmatrix}}_{A^{-1}} = \begin{bmatrix} 7-6 & -4+4 \\ \frac{21}{2}-\frac{21}{2} & -6+7 \end{bmatrix} = \underbrace{\begin{bmatrix} 1 & 0 \\ 0 & 1 \end{bmatrix}}_{I_2}.$$

NOTE ▶ The definition of the inverse of a square matrix requires the verification of two equations. However, if one of the equations is confirmed, then the other equation will also be satisfied. Therefore, it is sufficient to confirm only one of the two equations. ◀◀

We provide a rather efficient computational method for calculating A^{-1} in the next section. For now, however, let us be content with the following general formula for A^{-1} in the case where A is a 2×2 matrix.

Inverse of a 2 × 2 Matrix To determine the inverse of a 2×2 matrix, let

$$A = \begin{bmatrix} a & b \\ c & d \end{bmatrix}.$$

Let $D = ad - bc$, and assume that $D \neq 0$. Then, A^{-1} is given by the formula

$$A^{-1} = \frac{1}{D} \begin{bmatrix} d & -b \\ -c & a \end{bmatrix}. \tag{3}$$

We will omit the derivation of this formula. Notice that formula (3) involves division by D. Since division by 0 is not permissible, it is necessary that $D \neq 0$ for formula (3) to be applied.

Obtaining equation (3) can be reduced to a simple step-by-step procedure.

Inverse of a 2 × 2 Matrix Method

To determine the inverse of $\begin{bmatrix} a & b \\ c & d \end{bmatrix}$ if $D = ad - bc \neq 0$,

1. Interchange a and d to get $\begin{bmatrix} d & b \\ c & a \end{bmatrix}$.

2. Change the signs of b and c to get $\begin{bmatrix} d & -b \\ -c & a \end{bmatrix}$.

3. Divide all entries by D to get $\begin{bmatrix} \dfrac{d}{D} & -\dfrac{b}{D} \\ -\dfrac{c}{D} & \dfrac{a}{D} \end{bmatrix}$.

EXAMPLE 1 **Using the Formula for the Inverse of a 2 × 2 Matrix** Calculate the inverse of

$$\begin{bmatrix} -2 & 4 \\ -3 & 7 \end{bmatrix}.$$

SOLUTION $D = (-2) \cdot 7 - 4 \cdot (-3) = -2$; so $D \neq 0$, and we may use the preceding computation.

1. Interchange a and d:

$$\begin{bmatrix} 7 & 4 \\ -3 & -2 \end{bmatrix}.$$

2. Change the signs of b and c:

$$\begin{bmatrix} 7 & -4 \\ 3 & -2 \end{bmatrix}.$$

3. Divide all entries by $D = -2$:

$$\begin{bmatrix} -\frac{7}{2} & 2 \\ -\frac{3}{2} & 1 \end{bmatrix}.$$

Thus,

$$\begin{bmatrix} -2 & 4 \\ -3 & 7 \end{bmatrix}^{-1} = \begin{bmatrix} -\frac{7}{2} & 2 \\ -\frac{3}{2} & 1 \end{bmatrix}.$$

≫ Now Try Exercise 5

NOTE ▶ Not every square matrix has an inverse. This phenomenon can even occur in the case of 2×2 matrices. One can show that *if $D = 0$, then the matrix does not have an inverse.* **≪**

We were led to introduce the inverse of a matrix from a discussion of the matrix equation $AX = B$. Let us now return to that discussion. Suppose that A and B are given matrices and that we wish to solve the matrix equation

$$AX = B$$

for the unknown matrix X. Suppose further that A has an inverse A^{-1}. Multiply both sides of the equation on the left by A^{-1} to obtain

$$A^{-1} \cdot AX = A^{-1}B.$$

Because $A^{-1} \cdot A = I$, we have

$$IX = A^{-1}B$$
$$X = A^{-1}B.$$

Thus, the matrix X is found by simply multiplying B on the left by A^{-1}, and we can summarize our findings as follows:

Solving a Matrix Equation   If the matrix A has an inverse, then the solution of the matrix equation

$$AX = B \quad \text{is given by} \quad X = A^{-1}B.$$

Solving Systems of Linear Equations with Inverses

Matrix equations can be used to solve systems of linear equations, as illustrated in the next example.

EXAMPLE 2   **Using a Matrix Inverse to Solve a System of Equations** Use a matrix equation to solve the system of linear equations

$$\begin{cases} -2x + 4y = 2 \\ -3x + 7y = 7. \end{cases}$$

SOLUTION   In Example 7 of Section 2.3, we saw that the system could be written as a matrix equation:

$$\underset{A}{\begin{bmatrix} -2 & 4 \\ -3 & 7 \end{bmatrix}} \underset{X}{\begin{bmatrix} x \\ y \end{bmatrix}} = \underset{B}{\begin{bmatrix} 2 \\ 7 \end{bmatrix}}.$$

We happen to know A^{-1} from Example 1—namely,

$$A^{-1} = \begin{bmatrix} -\frac{7}{2} & 2 \\ -\frac{3}{2} & 1 \end{bmatrix}.$$

So we may compute the matrix $X = A^{-1}B$:

$$X = \begin{bmatrix} x \\ y \end{bmatrix} = \begin{bmatrix} -\frac{7}{2} & 2 \\ -\frac{3}{2} & 1 \end{bmatrix} \begin{bmatrix} 2 \\ 7 \end{bmatrix} = \begin{bmatrix} 7 \\ 4 \end{bmatrix}.$$

Thus, the solution of the system is $x = 7$, $y = 4$.

≫ Now Try Exercise 11

EXAMPLE 3

Analyzing Marriage Trends Let x and y denote the number of married and single adults in a certain town as of January 1. Let m and s denote the corresponding numbers for the following year. A statistical survey shows that x, y, m, and s are related by the equations

$$.9x + .2y = m$$
$$.1x + .8y = s.$$

In a given year, there were found to be 490,000 married adults and 147,000 single adults.
(a) How many married adults were there in the preceding year?
(b) How many married adults were there two years ago?

SOLUTION

(a) The given equations can be written in the matrix form

$$AX = B,$$

where

$$A = \begin{bmatrix} .9 & .2 \\ .1 & .8 \end{bmatrix}, \quad X = \begin{bmatrix} x \\ y \end{bmatrix}, \quad B = \begin{bmatrix} m \\ s \end{bmatrix}.$$

Let's calculate A^{-1}. First, $D = (.9)(.8) - (.2)(.1) = .7 \neq 0$. Therefore,

$$A^{-1} = \frac{1}{.7} \begin{bmatrix} .8 & -.2 \\ -.1 & .9 \end{bmatrix} = \begin{bmatrix} \frac{8}{7} & -\frac{2}{7} \\ -\frac{1}{7} & \frac{9}{7} \end{bmatrix}.$$

We are given that $B = \begin{bmatrix} 490,000 \\ 147,000 \end{bmatrix}$. So, since $X = A^{-1}B$, we have

$$X = \begin{bmatrix} \frac{8}{7} & -\frac{2}{7} \\ -\frac{1}{7} & \frac{9}{7} \end{bmatrix} \begin{bmatrix} 490,000 \\ 147,000 \end{bmatrix} = \begin{bmatrix} 518,000 \\ 119,000 \end{bmatrix}.$$

Thus, last year there were 518,000 married adults and 119,000 single adults.
(b) We deduce x and y for two years ago from the values of m and s for last year—namely, $m = 518,000$, $s = 119,000$.

$$X = A^{-1}B = \begin{bmatrix} \frac{8}{7} & -\frac{2}{7} \\ -\frac{1}{7} & \frac{9}{7} \end{bmatrix} \begin{bmatrix} 518,000 \\ 119,000 \end{bmatrix} = \begin{bmatrix} 558,000 \\ 79,000 \end{bmatrix}.$$

That is, two years ago there were 558,000 married adults and 79,000 single adults.

>> Now Try Exercise 17

Using the method of matrix equations to solve a system of linear equations is especially efficient if one wishes to solve a number of systems all having the same left-hand sides, but different right-hand sides. For then, A^{-1} must be computed only once for all of the systems under consideration. (This point is useful in Exercises 19–26.)

EXAMPLE 4

Using a Matrix Inverse to Solve Systems of Equations In Section 2.5, we will show that if

$$A = \begin{bmatrix} 4 & -2 & 3 \\ 8 & -3 & 5 \\ 7 & -2 & 4 \end{bmatrix}, \quad \text{then} \quad A^{-1} = \begin{bmatrix} -2 & 2 & -1 \\ 3 & -5 & 4 \\ 5 & -6 & 4 \end{bmatrix}.$$

(a) Use this fact to solve the system of linear equations

$$\begin{cases} 4x - 2y + 3z = 1 \\ 8x - 3y + 5z = 4 \\ 7x - 2y + 4z = 5. \end{cases}$$

(b) Solve the system of equations

$$\begin{cases} 4x - 2y + 3z = 4 \\ 8x - 3y + 5z = 7 \\ 7x - 2y + 4z = 6. \end{cases}$$

SOLUTION   **(a)** The system can be written in the matrix form

$$\underbrace{\begin{bmatrix} 4 & -2 & 3 \\ 8 & -3 & 5 \\ 7 & -2 & 4 \end{bmatrix}}_{A} \underbrace{\begin{bmatrix} x \\ y \\ z \end{bmatrix}}_{X} = \underbrace{\begin{bmatrix} 1 \\ 4 \\ 5 \end{bmatrix}}_{B}.$$

The solution of this matrix equation is $X = A^{-1}B$, or

$$\begin{bmatrix} x \\ y \\ z \end{bmatrix} = \begin{bmatrix} -2 & 2 & -1 \\ 3 & -5 & 4 \\ 5 & -6 & 4 \end{bmatrix} \begin{bmatrix} 1 \\ 4 \\ 5 \end{bmatrix} = \begin{bmatrix} 1 \\ 3 \\ 1 \end{bmatrix}.$$

Thus, the solution of the system is $x = 1$, $y = 3$, $z = 1$.

(b) This system has the same left-hand side as the preceding system, so its solution is

$$\begin{bmatrix} x \\ y \\ z \end{bmatrix} = \begin{bmatrix} -2 & 2 & -1 \\ 3 & -5 & 4 \\ 5 & -6 & 4 \end{bmatrix} \begin{bmatrix} 4 \\ 7 \\ 6 \end{bmatrix} = \begin{bmatrix} 0 \\ 1 \\ 2 \end{bmatrix}.$$

That is, the solution of the system is $x = 0$, $y = 1$, $z = 2$.   **>> Now Try Exercise 19**

INCORPORATING TECHNOLOGY

The inverse of a square matrix can be obtained directly with the inverse key $\boxed{x^{-1}}$. See Fig. 1.

To obtain the inverse of the $n \times n$ matrix named A, select an n by n square of cells where you would like the inverse matrix to be displayed, type **=MINVERSE(A)**, and press Crtl+Shift+Enter.

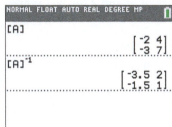

✹ WolframAlpha The inverse of a square matrix can be obtained by following the matrix with ^**–1** or by preceding the matrix with the word *inverse*. For instance, the instructions $\{\{-2, 4\}, \{-3, 7\}\}^\wedge -1$ and **inverse** $\{\{-2, 4\}, \{-3, 7\}\}$ each produce the result

$$\frac{1}{2}\begin{pmatrix} -7 & 4 \\ -3 & 2 \end{pmatrix}.$$

Figure 1

Check Your Understanding 2.4

Solutions can be found following the section exercises.

1. Show that the inverse of

$$\begin{bmatrix} -4 & 1 & 2 \\ 7 & -1 & -4 \\ -\frac{1}{2} & 0 & \frac{1}{2} \end{bmatrix} \text{ is } \begin{bmatrix} 1 & 1 & 4 \\ 3 & 2 & 4 \\ 1 & 1 & 6 \end{bmatrix}.$$

2. Use the method of this section to solve the system of linear equations

$$\begin{cases} .8x + .6y = 5 \\ .2x + .4y = 2. \end{cases}$$

EXERCISES 2.4

In Exercises 1 and 2, use the fact that

$$\begin{bmatrix} 2 & 2 \\ \frac{1}{2} & 1 \end{bmatrix}^{-1} = \begin{bmatrix} 1 & -2 \\ -\frac{1}{2} & 2 \end{bmatrix}.$$

1. Solve $\begin{cases} 2x + 2y = 4 \\ \frac{1}{2}x + y = 1. \end{cases}$ **2.** Solve $\begin{cases} 2x + 2y = 14 \\ \frac{1}{2}x + y = 4. \end{cases}$

In Exercises 3–10, find the inverse of the given matrix.

3. $\begin{bmatrix} 7 & 2 \\ 3 & 1 \end{bmatrix}$ **4.** $\begin{bmatrix} 2 & 3 \\ 5 & 7 \end{bmatrix}$

5. $\begin{bmatrix} 6 & 2 \\ 5 & 2 \end{bmatrix}$ **6.** $\begin{bmatrix} 1 & .5 \\ 0 & .5 \end{bmatrix}$

7. $\begin{bmatrix} .7 & .2 \\ .3 & .8 \end{bmatrix}$

8. $\begin{bmatrix} 0 & 1 \\ 1 & 0 \end{bmatrix}$

9. [3]

10. [.2]

In Exercises 11–14, use a matrix equation to solve the system of linear equations.

11. $\begin{cases} x + 2y = 3 \\ 2x + 6y = 5 \end{cases}$

12. $\begin{cases} 5x + 3y = 1 \\ 7x + 4y = 2 \end{cases}$

13. $\begin{cases} \frac{1}{2}x + 2y = 4 \\ 3x + 16y = 0 \end{cases}$

14. $\begin{cases} .8x + .6y = 2 \\ .2x + .4y = 1 \end{cases}$

15. Marriage Trends It is found that the number of married and single adults in a certain town are subject to the statistics that follow. Suppose that x and y denote the number of married and single adults, respectively, in a given year (say, as of January 1) and let m and s denote the corresponding numbers for the following year. Then,

$$.8x + .3y = m$$
$$.2x + .7y = s.$$

(a) Write this system of equations in matrix form.

(b) Solve the resulting matrix equation for $X = \begin{bmatrix} x \\ y \end{bmatrix}$.

(c) Suppose that, in a given year, there were found to be 100,000 married adults and 50,000 single adults. How many married and single adults were there the preceding year?

(d) How many married and single adults were there two years ago?

16. Epidemiology A flu epidemic is spreading through a town of 48,000 people. It is found that, if x and y denote the numbers of people sick and well in a given week, respectively, and if s and w denote the corresponding numbers for the following week, then

$$\tfrac{1}{3}x + \tfrac{1}{4}y = s$$
$$\tfrac{2}{3}x + \tfrac{3}{4}y = w.$$

(a) Write this system of equations in matrix form.

(b) Solve the resulting matrix equation for $X = \begin{bmatrix} x \\ y \end{bmatrix}$.

(c) Suppose that 13,000 people are sick in a given week. How many were sick the preceding week?

(d) Same question as part (c), except assume that 14,000 are sick.

17. Housing Trends Statistics show that, at a certain university, 70% of the students who live on campus during a given semester will remain on campus the following semester, and 90% of students living off campus during a given semester will remain off campus the following semester. Let x and y denote the number of students who live on and off campus this semester, and let u and v be the corresponding numbers for the next semester. Then,

$$.7x + .1y = u$$
$$.3x + .9y = v.$$

(a) Write this system of equations in matrix form.

(b) Solve the resulting matrix equation for $X = \begin{bmatrix} x \\ y \end{bmatrix}$.

(c) Suppose that, out of a group of 9000 students, 6000 currently live on campus and 3000 live off campus.

How many lived on campus last semester? How many will live off campus next semester?

18. Performance on Tests A teacher estimates that, of the students who pass a test, 80% will pass the next test, while of the students who fail a test, 50% will pass the next test. Let x and y denote the number of students who pass and fail a given test, and let u and v be the corresponding numbers for the following test.

(a) Write a matrix equation relating $\begin{bmatrix} x \\ y \end{bmatrix}$ to $\begin{bmatrix} u \\ v \end{bmatrix}$.

(b) Suppose that 25 of the teacher's students pass the third test and 8 fail the third test. How many students will pass the fourth test? Approximately how many passed the second test?

In Exercises 19–22, use the fact that the following two matrices are inverses of each other to solve the system of linear equations.

$$\begin{bmatrix} 1 & 2 & 2 \\ 1 & 3 & 2 \\ 1 & 2 & 3 \end{bmatrix} \text{ and } \begin{bmatrix} 5 & -2 & -2 \\ -1 & 1 & 0 \\ -1 & 0 & 1 \end{bmatrix}$$

19. $\begin{cases} x + 2y + 2z = 1 \\ x + 3y + 2z = -1 \\ x + 2y + 3z = -1 \end{cases}$

20. $\begin{cases} x + 2y + 2z = 1 \\ x + 3y + 2z = 0 \\ x + 2y + 3z = 0 \end{cases}$

21. $\begin{cases} 5x - 2y - 2z = 3 \\ -x + y \quad\;\; = 4 \\ -x \quad\;\;\; + z = 5 \end{cases}$

22. $\begin{cases} 5x - 2y - 2z = 0 \\ -x + y \quad\;\; = 1 \\ -x \quad\;\;\; + z = 2 \end{cases}$

In Exercises 23–26, use the fact that the following two matrices are inverses of each other to solve the system of linear equations.

$$\begin{bmatrix} 9 & 0 & 2 & 0 \\ -20 & -9 & -5 & 5 \\ 4 & 0 & 1 & 0 \\ -4 & -2 & -1 & 1 \end{bmatrix} \text{ and } \begin{bmatrix} 1 & 0 & -2 & 0 \\ 0 & 1 & 0 & -5 \\ -4 & 0 & 9 & 0 \\ 0 & 2 & 1 & -9 \end{bmatrix}$$

23. $\begin{cases} 9x \quad\;\; + 2z \quad\;\; = 1 \\ -20x - 9y - 5z + 5w = 0 \\ 4x \quad\;\; + z \quad\;\; = 0 \\ -4x - 2y - z + w = -1 \end{cases}$

24. $\begin{cases} 9x \quad\;\; + 2z \quad\;\; = 2 \\ -20x - 9y - 5z + 5w = 1 \\ 4x \quad\;\; + z \quad\;\; = 3 \\ -4x - 2y - z + w = 0 \end{cases}$

25. $\begin{cases} x \quad\;\; - 2z \quad\;\; = 0 \\ y \quad\;\; - 5w = 1 \\ -4x \quad\;\; + 9z \quad\;\; = 2 \\ 2y + z - 9w = 0 \end{cases}$

26. $\begin{cases} x \quad\;\; - 2z \quad\;\; = -1 \\ y \quad\;\; - 5w = 0 \\ -4x \quad\;\; + 9z \quad\;\; = 0 \\ 2y + z - 9w = 1 \end{cases}$

27. Show that if $a \neq 0$ and $b \neq 0$, then the inverse of $\begin{bmatrix} a & 0 \\ 0 & b \end{bmatrix}$ is $\begin{bmatrix} \frac{1}{a} & 0 \\ 0 & \frac{1}{b} \end{bmatrix}$.

28. (True or False) If B is the inverse of A, then A is the inverse of B.

29. Age Distribution There are two age groups for a particular species of organism. Group I consists of all organisms aged under 1 year, while group II consists of all organisms aged from 1 to 2 years. No organism survives more than 2 years. The average number of offspring per year born to each member of group I is 1, while the average number of offspring per year born to each member of group II is 2. Nine-tenths of group I survive to enter group II each year.

(a) Let x and y represent the initial number of organisms in groups I and II, respectively. Let a and b represent the number of organisms in groups I and II, respectively, after one year. Write a matrix equation relating $\begin{bmatrix} x \\ y \end{bmatrix}$ to $\begin{bmatrix} a \\ b \end{bmatrix}$.

(b) If there are initially 450,000 organisms in group I and 360,000 organisms in group II, calculate the number of organisms in each of the groups after 1 year and after 2 years.

(c) Suppose that, at a certain time, there were 810,000 organisms in group I and 630,000 organisms in group II. Determine the population of each group 1 year earlier.

30. If $A^2 = \begin{bmatrix} -2 & -1 \\ 2 & -1 \end{bmatrix}$ and $A^3 = \begin{bmatrix} -2 & 1 \\ -2 & -3 \end{bmatrix}$, what is A?

31. Show that, if AB is a matrix of all zeros and A has an inverse, then B is a matrix of all zeros.

32. Consider the matrices $A = \begin{bmatrix} 3 & 1 \\ 5 & 2 \end{bmatrix}$ and $B = \begin{bmatrix} 6 & 2 \\ 5 & 2 \end{bmatrix}$. Show that $(AB)^{-1} = B^{-1}A^{-1}$.

33. Find a 2×2 matrix A and a 2×1 column matrix B for which $AX = B$ has no solution.

34. Find a 2×2 matrix A and a 2×1 column matrix B for which $AX = B$ has infinitely many solutions.

TECHNOLOGY EXERCISES

In Exercises 35–38, use the inverse operation to find the inverse of the given matrix.

35. $\begin{bmatrix} .2 & 3 \\ 4 & 1.6 \end{bmatrix}$

36. $\begin{bmatrix} -12 & 3.3 \\ 6 & .4 \end{bmatrix}$

37. $\begin{bmatrix} .6 & 3 & -7 \\ 2.5 & -1 & 4 \\ -2 & .3 & 9 \end{bmatrix}$

38. $\begin{bmatrix} 5 & 2.3 & 6 \\ 1.2 & 5 & -7 \\ -3 & -4 & 6.5 \end{bmatrix}$

In Exercises 39–42, calculate the solution by using a matrix equation.

39. $\begin{cases} 2x - 4y + 7z = 11 \\ x + 3y - 5z = -9 \\ 3x - y + 3z = 7 \end{cases}$

40. $\begin{cases} 5x + 2y - 3z = 1 \\ 4x - y + z = 22 \\ -x + 5y - 6z = 4 \end{cases}$

41. $\begin{cases} 2x + 7z + 5w = 10 \\ 5x - y + 3z = -2 \\ x + 2y - 2w = 0 \\ 3x - 4y + 2z - 5w = -18 \end{cases}$

42. $\begin{cases} x + 4y - z + 2w = 9 \\ 3x - 8y + 2z + 4w = -2 \\ -x - 3y + 7z - 6w = 10 \\ 4x + 2y - 3z + w = -6 \end{cases}$

43. Try finding the inverse of a matrix that does not have an inverse. How does your technology respond?

Solutions to Check Your Understanding 2.4

1. To see whether this matrix is indeed the inverse, multiply it by the original matrix and find out if the product is an identity matrix.

$$\begin{bmatrix} 1 & 1 & 4 \\ 3 & 2 & 4 \\ 1 & 1 & 6 \end{bmatrix} \begin{bmatrix} -4 & 1 & 2 \\ 7 & -1 & -4 \\ -\frac{1}{2} & 0 & \frac{1}{2} \end{bmatrix} = \begin{bmatrix} 1 & 0 & 0 \\ 0 & 1 & 0 \\ 0 & 0 & 1 \end{bmatrix}.$$

2. The matrix form of this system is

$$\begin{bmatrix} .8 & .6 \\ .2 & .4 \end{bmatrix} \begin{bmatrix} x \\ y \end{bmatrix} = \begin{bmatrix} 5 \\ 2 \end{bmatrix}.$$

Therefore, the solution is

$$\begin{bmatrix} x \\ y \end{bmatrix} = \begin{bmatrix} .8 & .6 \\ .2 & .4 \end{bmatrix}^{-1} \begin{bmatrix} 5 \\ 2 \end{bmatrix}.$$

To compute the inverse of the 2×2 matrix, first compute D.

$$D = ad - bc = (.8)(.4) - (.6)(.2) = .32 - .12 = .2$$

Thus,

$$\begin{bmatrix} .8 & .6 \\ .2 & .4 \end{bmatrix}^{-1} = \begin{bmatrix} .4/.2 & -.6/.2 \\ -.2/.2 & .8/.2 \end{bmatrix} = \begin{bmatrix} 2 & -3 \\ -1 & 4 \end{bmatrix}.$$

Therefore,

$$\begin{bmatrix} x \\ y \end{bmatrix} = \begin{bmatrix} 2 & -3 \\ -1 & 4 \end{bmatrix} \begin{bmatrix} 5 \\ 2 \end{bmatrix} = \begin{bmatrix} 4 \\ 3 \end{bmatrix},$$

so the solution is $x = 4$, $y = 3$.

2.5 The Gauss–Jordan Method for Calculating Inverses

Of the several popular methods for finding the inverse of a matrix, the **Gauss–Jordan method** is probably the easiest to describe. It can be used on square matrices of any size. Also, the mechanical nature of the computations allows this method to be programmed for a computer, with relative ease. We shall illustrate the procedure with a

2×2 matrix, whose inverse can also be calculated by using the method of the previous section. Let

$$A = \begin{bmatrix} \frac{1}{2} & 1 \\ 1 & 3 \end{bmatrix}.$$

It is simple to check that

$$A^{-1} = \begin{bmatrix} 6 & -2 \\ -2 & 1 \end{bmatrix}.$$

Let us now derive this result, using the Gauss–Jordan method.

> **Step 1** Write down the matrix A, and on its right append an identity matrix of the same size.

This is most conveniently done by placing I_2 beside A in a single matrix.

$$\begin{bmatrix} \frac{1}{2} & 1 & 1 & 0 \\ 1 & 3 & 0 & 1 \end{bmatrix}$$
$$\underbrace{}_{A} \quad \underbrace{}_{I_2}$$

> **Step 2** Perform elementary row operations on the left-hand matrix so as to transform it into an identity matrix. Each operation performed on the left-hand matrix is also performed on the right-hand matrix.

This step proceeds exactly like the Gauss–Jordan elimination method and may be most conveniently expressed in terms of pivoting. Thus, we have

$$\begin{bmatrix} \textcircled{$\frac{1}{2}$} & 1 & 1 & 0 \\ 1 & 3 & 0 & 1 \end{bmatrix} \xrightarrow[R_2 + (-1)R_1]{2R_1} \begin{bmatrix} 1 & 2 & 2 & 0 \\ 0 & \textcircled{1} & -2 & 1 \end{bmatrix} \xrightarrow{R_1 + (-2)R_2} \begin{bmatrix} 1 & 0 & 6 & -2 \\ 0 & 1 & -2 & 1 \end{bmatrix}.$$

> **Step 3** When the matrix on the left becomes an identity matrix, the matrix on the right will be the desired inverse.

So, from the last matrix of our preceding calculation, we have

$$A^{-1} = \begin{bmatrix} 6 & -2 \\ -2 & 1 \end{bmatrix}.$$

This is the same result mentioned earlier.

EXAMPLE 1 **Finding the Inverse of a Matrix by the Gauss–Jordan Method** Find the inverse of the matrix

$$A = \begin{bmatrix} 4 & -2 & 3 \\ 8 & -3 & 5 \\ 7 & -2 & 4 \end{bmatrix}.$$

SOLUTION

$$\begin{bmatrix} \textcircled{4} & -2 & 3 & 1 & 0 & 0 \\ 8 & -3 & 5 & 0 & 1 & 0 \\ 7 & -2 & 4 & 0 & 0 & 1 \end{bmatrix}$$

$$\begin{bmatrix} 1 & -\frac{1}{2} & \frac{3}{4} & \frac{1}{4} & 0 & 0 \\ 0 & \textcircled{1} & -1 & -2 & 1 & 0 \\ 0 & \frac{3}{2} & -\frac{5}{4} & -\frac{7}{4} & 0 & 1 \end{bmatrix}$$

$$\begin{bmatrix} 1 & 0 & \frac{1}{4} & -\frac{3}{4} & \frac{1}{2} & 0 \\ 0 & 1 & -1 & -2 & 1 & 0 \\ 0 & 0 & \textcircled{$\frac{1}{4}$} & \frac{5}{4} & -\frac{3}{2} & 1 \end{bmatrix}$$

$$\left[\begin{array}{ccc|ccc} 1 & 0 & 0 & -2 & 2 & -1 \\ 0 & 1 & 0 & 3 & -5 & 4 \\ 0 & 0 & 1 & 5 & -6 & 4 \end{array}\right]$$

Therefore,

$$A^{-1} = \left[\begin{array}{ccc} -2 & 2 & -1 \\ 3 & -5 & 4 \\ 5 & -6 & 4 \end{array}\right].$$

» **Now Try Exercise 7**

Not all square matrices have inverses. If a matrix does not have an inverse, this will become apparent when applying the Gauss–Jordan method. At some point, there will be no way to continue transforming the left-hand matrix into an identity matrix. This is illustrated in the next example.

EXAMPLE 2   **Demonstrating That a Matrix Does Not Have an Inverse** Find the inverse of the matrix

$$A = \left[\begin{array}{ccc} 1 & 3 & 2 \\ 0 & 1 & 4 \\ 1 & 5 & 10 \end{array}\right].$$

SOLUTION

$$\left[\begin{array}{ccc|ccc} ① & 3 & 2 & 1 & 0 & 0 \\ 0 & 1 & 4 & 0 & 1 & 0 \\ 1 & 5 & 10 & 0 & 0 & 1 \end{array}\right]$$

$$\left[\begin{array}{ccc|ccc} 1 & 3 & 2 & 1 & 0 & 0 \\ 0 & ① & 4 & 0 & 1 & 0 \\ 0 & 2 & 8 & -1 & 0 & 1 \end{array}\right]$$

$$\left[\begin{array}{ccc|ccc} 1 & 0 & -10 & 1 & -3 & 0 \\ 0 & 1 & 4 & 0 & 1 & 0 \\ 0 & 0 & 0 & -1 & -2 & 1 \end{array}\right]$$

Since the third row of the left-hand matrix has only zero entries, it is impossible to complete the Gauss–Jordan method. Therefore, the matrix A has no inverse matrix.

» **Now Try Exercise 9**

Check Your Understanding 2.5

Solutions can be found following the section exercises.

1. Use the Gauss–Jordan method to calculate the inverse of the matrix

$$\left[\begin{array}{ccc} 1 & 0 & 2 \\ 0 & 1 & -4 \\ 0 & 0 & 2 \end{array}\right].$$

2. Solve the system of linear equations

$$\begin{cases} x & + 2z = 4 \\ & y - 4z = 6 \\ & 2z = 9. \end{cases}$$

ERCISES 2.5

es 1–12, use the Gauss–Jordan method to compute the
exists, of the matrix.

2. $\left[\begin{array}{cc} 5 & -2 \\ 6 & 2 \end{array}\right]$

4. $\left[\begin{array}{cc} 1 & -3 \\ 0 & 1 \end{array}\right]$

5. $\left[\begin{array}{cc} 2 & -4 \\ -1 & 2 \end{array}\right]$

7. $\left[\begin{array}{ccc} 1 & 2 & -2 \\ 1 & 1 & 1 \\ 0 & 0 & 1 \end{array}\right]$

6. $\left[\begin{array}{ccc} 1 & 3 & 1 \\ -1 & 2 & 0 \\ 2 & 11 & 3 \end{array}\right]$

8. $\left[\begin{array}{ccc} 2 & 2 & 0 \\ 0 & -2 & 0 \\ 3 & 0 & 1 \end{array}\right]$

9. $\begin{bmatrix} -2 & 5 & 2 \\ 1 & -3 & -1 \\ -1 & 2 & 1 \end{bmatrix}$ **10.** $\begin{bmatrix} 1 & 0 & 0 \\ 2 & 1 & -2 \\ -1 & 2 & 1 \end{bmatrix}$

11. $\begin{bmatrix} 1 & 6 & 0 & 0 \\ 1 & 5 & 0 & 0 \\ 0 & 0 & 4 & 2 \\ 0 & 0 & 50 & 2 \end{bmatrix}$ **12.** $\begin{bmatrix} 6 & 0 & 2 & 0 \\ -6 & 1 & 0 & 1 \\ 1 & 0 & 1 & 0 \\ -9 & 0 & -1 & 1 \end{bmatrix}$

In Exercises 13–18, use an inverse matrix to solve the system of linear equations.

13. $\begin{cases} x + y + 2z = 3 \\ 3x + 2y + 2z = 4 \\ x + y + 3z = 5 \end{cases}$ **14.** $\begin{cases} x + 2y + 3z = 4 \\ 3x + 5y + 5z = 3 \\ 2x + 4y + 2z = 4 \end{cases}$

15. $\begin{cases} x + 4y + 3z = 15 \\ x + 3y + 4z = 17 \\ 2x + 3y + 3z = 16 \end{cases}$ **16.** $\begin{cases} y + 2z = 1 \\ 2x + y + 3z = 2 \\ x + y + 2z = 3 \end{cases}$

17. $\begin{cases} x - 2z - 2w = 0 \\ y - 5w = 1 \\ -4x + 9z + 9w = 2 \\ 2y + z - 8w = 3 \end{cases}$

18. $\begin{cases} x + 2y - z + w = 15 \\ 2x - y + z + w = 12 \\ 2x + z + 2w = 18 \\ 3x + 2y - w = 21 \end{cases}$

19. Find a 2×2 matrix A for which

$$A \begin{bmatrix} 2 & 5 \\ 1 & 3 \end{bmatrix} = \begin{bmatrix} -1 & 0 \\ 4 & 2 \end{bmatrix}.$$

20. Find a 2×2 matrix A for which

$$\begin{bmatrix} 2 & 5 \\ 1 & 3 \end{bmatrix} A = \begin{bmatrix} -1 & 0 \\ 4 & 2 \end{bmatrix}.$$

21. College Degrees Figure 1 gives the responses of a group of 100 randomly selected college freshmen when asked for the highest academic degree that they intended to obtain. Twice as many students intended to obtain master's degrees as their highest degree than intended to obtain bachelor's degrees as their highest degree. The number of students intending to obtain bachelor's or master's degrees as their highest degree was 26 more than the number who intended to obtain other degrees as their highest degree. Let x, y, and z represent the three numbers shown in Fig. 1. Use the methods of this section to determine the values of x, y, and z.

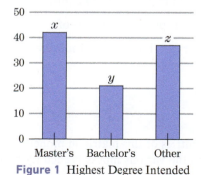

Figure 1 Highest Degree Intended

22. College Choices Figure 2 gives the responses of a group of 100 randomly selected college freshmen when asked whether the

college they were attending was their first choice or second choice. The number of students who attended their first-choice college was 16 more than the students who did not. The number of students who attended their second-choice college was 46 less than the number of students who did not. Let x, y, and z represent the three numbers shown in Fig. 2. Use the methods of this section to determine the values of x, y, and z.

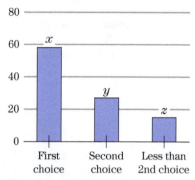

Figure 2 College Choice

23. High School Attended Figure 3 gives the responses of a group of 100 randomly selected college freshmen when asked for the type of high school attended. The number of students who attended public schools was 5 times the number who attended private schools minus 3 times the number who were home-schooled. The number of students who were homeschooled was 29 times the number who attended private schools minus 6 times the number who attended public schools. Use the methods of this section to determine the values of x, y, and z in Fig. 3.

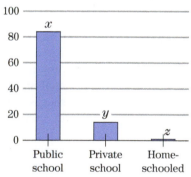

Figure 3 Type of High School Attended

24. Placement Tests Figure 4 gives the responses of a group of randomly selected college freshmen when asked for the types of placement tests taken. A total of 82 placement tests were taken. The number of mathematics placement tests taken was 2 more than twice the number of writing placement tests taken. Eight more writing placement tests were taken than reading placement tests. Determine the values of x, y, and z in Fig. 4.

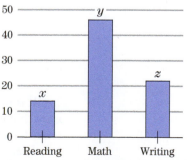

Figure 4 Type of Placement Tests Taken

Solutions to Check Your Understanding 2.5

1. First, write the given matrix beside an identity matrix of the same size

$$\left[\begin{array}{ccc|ccc} 1 & 0 & 2 & 1 & 0 & 0 \\ 0 & 1 & -4 & 0 & 1 & 0 \\ 0 & 0 & 2 & 0 & 0 & 1 \end{array}\right].$$

The object is to use elementary row operations to transform the 3×3 matrix on the left into the identity matrix. The first two columns are already in the correct form.

$$\left[\begin{array}{ccc|ccc} 1 & 0 & 2 & 1 & 0 & 0 \\ 0 & 1 & -4 & 0 & 1 & 0 \\ 0 & 0 & 2 & 0 & 0 & 1 \end{array}\right]$$

$$\xrightarrow{\frac{1}{2}R_3} \left[\begin{array}{ccc|ccc} 1 & 0 & 2 & 1 & 0 & 0 \\ 0 & 1 & -4 & 0 & 1 & 0 \\ 0 & 0 & 1 & 0 & 0 & \frac{1}{2} \end{array}\right]$$

$$\xrightarrow{R_1 + (-2)R_3} \left[\begin{array}{ccc|ccc} 1 & 0 & 0 & 1 & 0 & -1 \\ 0 & 1 & -4 & 0 & 1 & 0 \\ 0 & 0 & 1 & 0 & 0 & \frac{1}{2} \end{array}\right]$$

$$\xrightarrow{R_2 + (4)R_3} \left[\begin{array}{ccc|ccc} 1 & 0 & 0 & 1 & 0 & -1 \\ 0 & 1 & 0 & 0 & 1 & 2 \\ 0 & 0 & 1 & 0 & 0 & \frac{1}{2} \end{array}\right]$$

Thus, the inverse of the given matrix is

$$\left[\begin{array}{ccc} 1 & 0 & -1 \\ 0 & 1 & 2 \\ 0 & 0 & \frac{1}{2} \end{array}\right].$$

2. The matrix form of this system of equations is $AX = B$, where A is the matrix whose inverse was found in Problem 1, and

$$B = \left[\begin{array}{c} 4 \\ 6 \\ 9 \end{array}\right].$$

Therefore, $X = A^{-1}B$, so that

$$\left[\begin{array}{c} x \\ y \\ z \end{array}\right] = \left[\begin{array}{ccc} 1 & 0 & -1 \\ 0 & 1 & 2 \\ 0 & 0 & \frac{1}{2} \end{array}\right] \left[\begin{array}{c} 4 \\ 6 \\ 9 \end{array}\right] = \left[\begin{array}{c} -5 \\ 24 \\ \frac{9}{2} \end{array}\right].$$

So the solution of the system is $x = -5$, $y = 24$, $z = \frac{9}{2}$.

2.6 Input–Output Analysis

In recent years, matrix arithmetic has played an ever-increasing role in economics, especially in that branch of economics called **input–output analysis**. Pioneered by the Harvard economist Vassily Leontieff, input–output analysis is used to analyze an economy in order to meet given consumption and export demands. As we shall see, such analysis leads to matrix calculations and, in particular, to inverses. Input–output analysis has been of such great significance that Leontieff was awarded the 1973 Nobel Prize in economics for his fundamental work in the subject.

Suppose that we divide an economy into a number of industries—transportation, agriculture, steel, and so on. Each industry produces a certain output by using certain raw materials (or input). The input of each industry is made up in part by the outputs of other industries. For example, in order to produce food, agriculture uses as input the output of many industries, such as transportation (tractors and trucks) and oil (gasoline and fertilizers). This interdependence among the industries of the economy is summarized in a matrix—an **input–output matrix**. There is one column for each industry's input requirements. The entries in the column reflect the amount of input required from each of the industries. A typical input–output matrix looks like this:

Input requirements of:

Industry 1   Industry 2   Industry 3 . . .

$$\text{From} \quad \begin{array}{l} \text{Industry 1} \\ \text{Industry 2} \\ \text{Industry 3} \\ \vdots \end{array} \left[\begin{array}{ccccc} & & & & \\ & & & & \\ & & & & \\ & & & & \end{array}\right].$$

It is most convenient to express the entries of this matrix in monetary terms. That is, each column gives the dollar values of the various inputs needed by an industry in order to produce $1 worth of output.

There are consumers (other than the industries themselves) who want to purchase some of the output of these industries. The quantity of goods that these consumers want (or demand) is called the **consumer demand** on the economy. The consumer demand can be represented by a column matrix, with one entry for each industry, indicating the amount of consumable output demanded from the industry:

$$[\text{consumer-demand matrix}] = \begin{bmatrix} \text{amount from industry 1} \\ \text{amount from industry 2} \\ \vdots \end{bmatrix}.$$

We shall consider the situation in which the consumer-demand matrix is given, and it is necessary to determine how much output should be produced by each industry in order to provide the needed inputs of the various industries and also to satisfy the consumer demand. The proper level of output can be computed by using matrix calculations, as illustrated in the next example.

EXAMPLE 1   **Determining Industrial Production** Suppose that an economy is composed of only three industries—coal, steel, and electricity. Each of these industries depends on the others for some of its raw materials. Suppose that to make $1 of coal, it takes no coal, but $.02 of steel and $.01 of electricity; to make $1 of steel, it takes $.15 of coal, $.03 of steel, and $.08 of electricity; and to make $1 of electricity, it takes $.43 of coal, $.20 of steel, and $.05 of electricity. How much should each industry produce to allow for consumption (not used for production) at these levels: $2 billion of coal, $1 billion of steel, $3 billion of electricity?

SOLUTION   Put all the data indicating the interdependence of the industries in a matrix. In each industry's column, put the amount of input from each of the industries needed to produce $1 of output in that particular industry:

$$\begin{array}{c} \\ \text{Coal} \\ \text{Steel} \\ \text{Electricity} \end{array} \begin{array}{ccc} \text{Coal} & \text{Steel} & \text{Electricity} \\ \begin{bmatrix} 0 & .15 & .43 \\ .02 & .03 & .20 \\ .01 & .08 & .05 \end{bmatrix} \end{array} = A.$$

This matrix is the input–output matrix corresponding to the economy. Let D denote the consumer-demand matrix. Then, letting the numbers in D stand for billions of dollars, we have

$$D = \begin{bmatrix} 2 \\ 1 \\ 3 \end{bmatrix}.$$

Suppose that the coal industry produces x billion dollars of output, the steel industry y billion dollars, and the electrical industry z billion dollars. Our problem is to determine the values of x, y, and z that yield the desired amounts left over from the production process. As an example, consider coal. The amount of coal that can be consumed or exported is just

$$x - [\text{amount of coal used in production}].$$

To determine the amount of coal used in production, refer to the input–output matrix. Production of x billion dollars of coal takes $0 \cdot x$ billion dollars of coal; production of y billion dollars of steel takes $.15y$ billion dollars of coal; and production of z billion dollars of electricity takes $.43z$ billion dollars of coal. Thus,

$$[\text{amount of coal used in production}] = 0 \cdot x + .15y + .43z.$$

This quantity should be recognized as the first entry of a matrix product. Namely, if we let

$$X = \begin{bmatrix} x \\ y \\ z \end{bmatrix},$$

then

$$
\begin{bmatrix} \text{coal} \\ \text{steel} \\ \text{electricity} \end{bmatrix}_{\text{used in production}} = \begin{bmatrix} 0 & .15 & .43 \\ .02 & .03 & .20 \\ .01 & .08 & .05 \end{bmatrix} \begin{bmatrix} x \\ y \\ z \end{bmatrix} = AX.
$$

X is called the **production matrix** (or the **output matrix**) and AX is called the **internal consumption matrix**. The equations for the amounts of steel and electricity used in production are obtained in a manner similar to the equation for coal. But then, the amount of each output available for purposes other than production is $X - AX$. That is, we have the matrix equation

$$X - AX = D.$$

To solve this equation for X, proceed as follows. Since $IX = X$, write the equation in the form

$$IX - AX = D$$
$$(I - A)X = D$$

Solution to the Input–Output Problem   If the matrix $I - A$ has an inverse, then

$$X = (I - A)^{-1}D.$$

So, in other words, X may be found by multiplying D on the left by $(I - A)^{-1}$. Let us now do the arithmetic.

$$
I - A = \begin{bmatrix} 1 & 0 & 0 \\ 0 & 1 & 0 \\ 0 & 0 & 1 \end{bmatrix} - \begin{bmatrix} 0 & .15 & .43 \\ .02 & .03 & .20 \\ .01 & .08 & .05 \end{bmatrix} = \begin{bmatrix} 1 & -.15 & -.43 \\ -.02 & .97 & -.20 \\ -.01 & -.08 & .95 \end{bmatrix}
$$

Applying the Gauss–Jordan method (or using technology), we find that

$$
(I - A)^{-1} = \begin{bmatrix} 1.01 & .20 & .50 \\ .02 & 1.05 & .23 \\ .01 & .09 & 1.08 \end{bmatrix},
$$

where all figures are carried to two decimal places. Therefore,

$$
X = (I - A)^{-1}D = \begin{bmatrix} 1.01 & .20 & .50 \\ .02 & 1.05 & .23 \\ .01 & .09 & 1.08 \end{bmatrix} \begin{bmatrix} 2 \\ 1 \\ 3 \end{bmatrix} = \begin{bmatrix} 3.72 \\ 1.78 \\ 3.35 \end{bmatrix}.
$$

In other words, coal should produce $3.72 billion worth of output, steel $1.78 billion, and electricity $3.35 billion. This output will meet the required consumer demands from each industry.   **» Now Try Exercise 13**

The preceding analysis is useful in studying not only entire economies, but also segments of economies, and even individual companies.

EXAMPLE 2   **Determining Production for a Conglomerate** A conglomerate has three divisions, which produce computers, semiconductors, and business forms. For each $1 of output, the computer division needs $.02 worth of computers, $.20 worth of semiconductors, and $.10 worth of business forms. For each $1 of output, the semiconductor division needs $.02 worth of computers, $.01 worth of semiconductors, and $.02 worth of business forms. For each $1 of output, the business forms division requires $.10 worth of computers and $.01 worth of business forms. The conglomerate estimates the sales demand to be $300,000,000 for the computer division, $100,000,000 for the semiconductor division, and $200,000,000 for the business forms division. At what level should each division produce in order to satisfy this demand?

SOLUTION   The conglomerate can be viewed as a miniature economy and its sales as the consumer demand. The input–output matrix for this "economy" is

$$
\begin{array}{c}
 \\
\text{Computers} \\
\text{Semiconductors} \\
\text{Business forms}
\end{array}
\begin{array}{ccc}
\text{Computers} & \text{Semiconductors} & \begin{array}{c}\text{Business}\\\text{forms}\end{array}
\end{array}
\left[
\begin{array}{ccc}
.02 & .02 & .10 \\
.20 & .01 & 0 \\
.10 & .02 & .01
\end{array}
\right] = A.
$$

The consumer-demand matrix is

$$
D = \begin{bmatrix} 3 \\ 1 \\ 2 \end{bmatrix},
$$

where the demand is expressed in hundreds of millions of dollars. The matrix X, giving the desired levels of production for the various divisions, is given by

$$
X = (I - A)^{-1}D.
$$

But

$$
I - A = \begin{bmatrix} .98 & -.02 & -.10 \\ -.20 & .99 & 0 \\ -.10 & -.02 & .99 \end{bmatrix},
$$

and after computation,

$$
(I - A)^{-1} = \begin{bmatrix} 1.04 & .02 & .10 \\ .21 & 1.01 & .02 \\ .11 & .02 & 1.02 \end{bmatrix},
$$

rounded to two decimal places. Therefore,

$$
X = (I - A)^{-1}D = \begin{bmatrix} 3.34 \\ 1.68 \\ 2.39 \end{bmatrix}.
$$

That is, the computer division should produce $334,000,000, the semiconductor division $168,000,000, and the business forms division $239,000,000.   **» Now Try Exercise 15**

Input–output analysis is usually applied to the entire economy of a country having hundreds of industries. The resulting matrix equation $(I - A)X = D$ could be solved by the Gauss–Jordan elimination method. However, it is best to find the inverse of $I - A$ and solve for X as we have done in the examples of this section. Over a short period, D might change, but A is unlikely to change. Therefore, the proper outputs to satisfy the new consumer demand can easily be determined by using the already computed inverse of $I - A$.

INCORPORATING
TECHNOLOGY

The matrix $(I - A)^{-1}D$ can be calculated with a single press of the [ENTER] key. Figure 1 shows the calculation for the matrices of Example 2.

To display the matrix $(I - A)^{-1}D$ on an Excel spreadsheet, select a column of n cells, type the formula =MMULT(MINVERSE(I-A),D), and press Ctrl+Shift+Enter. (Here, D is an $n \times 1$ matrix, I is the $n \times n$ identity matrix, and A is an $n \times n$ square matrix.)

```
NORMAL FLOAT AUTO REAL DEGREE MP
(identity(3)-[A])⁻¹*[D]
            [3.339646051]
            [1.68477698 ]
            [2.391575904]
```

Figure 1

❋ WolframAlpha The solution to Example 2 can be obtained with either of the following two instructions:

solve (IdentityMatrix[3] − {{.02,.02,.1},{.2,.01,0}, {.1,.02,.01}}).{{x},{y},{z}} = {{3},{1},{2}}

(inverse of {IdentityMatrix[3] − {{.02,.02,.1},{.2,.01,0}, {.1,.02,.01}}}).{{3},{1},{2}}

Check Your Understanding 2.6

Solutions can be found following the section exercises.

1. Let

$$I = \begin{bmatrix} 1 & 0 & 0 \\ 0 & 1 & 0 \\ 0 & 0 & 1 \end{bmatrix}, \quad A = \begin{bmatrix} .1 & 0 & .1 \\ .2 & .1 & .1 \\ .1 & .2 & 0 \end{bmatrix},$$

$$X = \begin{bmatrix} x \\ y \\ z \end{bmatrix}, \quad D = \begin{bmatrix} 100 \\ 200 \\ 50 \end{bmatrix}.$$

Solve the matrix equation

$$(I - A)X = D.$$

2. Let I, A, and X be as in Problem 1, but let

$$D = \begin{bmatrix} 300 \\ 100 \\ 100 \end{bmatrix}.$$

Solve the matrix equation $(I - A)X = D$.

EXERCISES 2.6

Three-Sector Economy In Exercises 1–12, suppose that a simplified economy consisting of the three sectors Manufacturing, Energy, and Services has the input–output matrix

$$\begin{array}{c} \\ M \\ E \\ S \end{array} \begin{array}{ccc} M & E & S \\ \begin{bmatrix} .3 & .1 & .2 \\ .2 & .25 & .15 \\ .1 & .2 & .15 \end{bmatrix} \end{array} = A.$$

1. How many cents of energy are required to produce $1 worth of manufactured goods?

2. How many cents of energy are required to produce $1 worth of services?

3. Which sector of the economy requires the greatest amount of services in order to produce $1 worth of output?

4. Which sector of the economy requires the least amount of manufacturing in order to produce $1 worth of output?

5. What is the dollar amount of the energy costs needed to produce $10 million worth of goods from each sector?

6. What is the dollar amount of the costs for services needed to produce $10 million worth of goods from each sector?

7. On what sector are services most dependent?

8. On what sector is manufacturing least dependent?

9. Determine the internal consumption when the production matrix is

$$X = \begin{bmatrix} 10 \\ 20 \\ 30 \end{bmatrix}.$$

10. Show that

$$I - A = \begin{bmatrix} .7 & -.1 & -.2 \\ -.2 & .75 & -.15 \\ -.1 & -.2 & .85 \end{bmatrix}.$$

11. Given that

$$(I - A)^{-1} = \begin{bmatrix} 1.58 & .33 & .43 \\ .48 & 1.5 & .38 \\ .3 & .39 & 1.32 \end{bmatrix},$$

find the production matrix when the consumer demand matrix is

$$D = \begin{bmatrix} 5 \\ 6 \\ 7 \end{bmatrix}.$$

12. Repeat Exercise 11 for the consumer demand matrix

$$D = \begin{bmatrix} 10 \\ 5 \\ 20 \end{bmatrix}.$$

13. **Industrial Production** Suppose that, in the economy of Example 1, the demand for electricity triples and the demand for coal doubles, whereas the demand for steel increases by only 50%. At what levels should the various industries produce in order to satisfy the new demand?

14. **Conglomerate** Suppose that the conglomerate of Example 2 is faced with an increase of 50% in demand for computers, a doubling in demand for semiconductors, and a decrease of 50% in demand for business forms. At what levels should the various divisions produce in order to satisfy the new demand?

15. **Conglomerate** Suppose that the conglomerate of Example 2 experiences a doubling in the demand for business forms. At what levels should the computer and semiconductor divisions produce?

16. **Industrial Production** Suppose that the economy of Example 1 experiences a 20% increase in the demand for coal. At what levels should the three industries produce?

17. **Industrial Production** In the economy of Example 1, suppose that $4 billion worth of coal, $2 billion worth of steel, and $5 billion worth of electricity are produced. How much of each industry's output will be available for consumption?

18. **Conglomerate** In the conglomerate of Example 2, suppose that $400,000,000 worth of computers, $200,000,000 worth of semiconductors, and $300,000,000 worth of business forms are produced. How much of each division's output will be available for consumption?

19. **Two-Sector Economy** A simplified economy consists of the two sectors Transportation and Energy. For each $1 worth of output, the transportation sector requires $.25 worth of input from the transportation sector and $.20 of input from the energy sector. For each $1 worth of output, the energy sector requires $.30 from the transportation sector and $.15 from the energy sector.
    (a) Give the input–output matrix A for this economy.
    (b) Determine the matrix $(I - A)^{-1}$. (Round entries to two decimal places.)
    (c) What level of output should each sector produce to meet a consumer demand for $5 billion worth of transportation and $3 billion worth of energy?
    (d) How much of the output of each sector is used to meet internal consumption?

20. **Two-Sector Economy** Rework Exercise 19 under the condition that the consumer demand for transportation is $4 billion and the consumer demand for energy is $7 billion.

21. **Two-Product Corporation** A corporation has a plastics division and an industrial equipment division. For each $1 worth of output, the plastics division needs $.02 worth of plastics and $.10 worth of equipment. For each $1 worth of output, the industrial equipment division needs $.01 worth of plastics and $.05 worth of equipment. At what level should the divisions produce to meet a consumer demand for $930,000 worth of plastics and $465,000 worth of industrial equipment?

22. **Two-Product Corporation** Rework Exercise 21 under the condition that the consumer demand for plastics is $1,860,000 and the demand for industrial equipment is $2,790,000.

23. **Three-Sector Economy** In an economic system, each of three industries depends on the others for raw materials. To make $1 worth of processed wood requires 30 ¢ worth of wood, 20 ¢ steel, and 10 ¢ coal. To make $1 worth of steel requires no wood, 30 ¢ steel, and 20 ¢ coal. To make $1 worth of coal requires 10 ¢ wood, 20 ¢ steel, and 5 ¢ coal.
    (a) Give the input–output matrix A for this economy.
    (b) Determine the matrix $(I - A)^{-1}$. (Round entries to two decimal places.)
    (c) What level of output should each sector produce to meet a consumer demand for $1 worth of wood, $4 of steel, and $2 of coal?
    (d) How much of the output of each sector is used to meet internal consumption?

24. **Three-Sector Economy** Rework Exercise 23 under the condition that the consumer demand is $200 for wood, $300 for steel, and $800 for coal.

25. **Three-Sector Industry** An industrial system involves manufacturing, transportation, and agriculture. The interdependence of the three industries is given by the input–output matrix

$$\begin{array}{c} \\ M \\ T \\ A \end{array} \begin{array}{c} \begin{array}{ccc} M & T & A \end{array} \\ \begin{bmatrix} .4 & .3 & .1 \\ .2 & .2 & .2 \\ .1 & .1 & .4 \end{bmatrix} \end{array}$$

At what levels must the industries produce to satisfy a demand for $100 million worth of manufactured goods, $80 million of transportation, and $200 million worth of agricultural products?

26. **Three-Sector Economy** An economy consists of the three sectors agriculture, energy, and manufacturing. For each $1 worth of output, the agriculture sector requires $.08 worth of input from the agriculture sector, $.10 worth of input from the energy sector, and $.20 worth of input from the manufacturing sector. For each $1 worth of output, the energy sector requires $.15 worth of input from the agriculture sector, $.14 worth of input from the energy sector, and $.10 worth of input from the manufacturing sector. For each $1 worth of output, the manufacturing sector requires $.25 worth of input from the agriculture sector, $.12 worth of input from the energy sector, and $.05 worth of input from the manufacturing sector.
    (a) Give the input–output matrix A for this economy.
    (b) Determine the matrix $(I - A)^{-1}$. (Round entries to two decimal places.)
    (c) At what level of output should each sector produce to meet a demand for $4 billion worth of agriculture, $3 billion worth of energy, and $2 billion worth of manufacturing?
    (d) How much of the output of each sector is used to meet internal consumption?

27. **Localized Economy** A town has a merchant, a baker, and a farmer. To produce $1 worth of output, the merchant requires $.30 worth of baked goods and $.40 worth of the farmer's products. To produce $1 worth of output, the baker requires $.50 worth of the merchant's goods, $.10 worth of his own goods, and $.30 worth of the farmer's goods. To produce $1 worth of output, the farmer requires $.30 worth of the merchant's goods, $.20 worth of baked goods, and $.30 worth of his own products. How much should the merchant, baker, and farmer produce to meet a demand for $20,000 worth of output from the merchant, $15,000 worth of output from the baker, and $18,000 worth of output from the farmer?

28. **Multinational Corporation** A multinational corporation does business in the United States, Canada, and England. Its branches in one country purchase goods from the branches in other countries according to the matrix

		Branch in:		
		United States	Canada	England
Purchase from:	United States	.02	0	.02
	Canada	.01	.03	.01
	England	.03	0	.01

where the entries in the matrix represent proportions of total sales by the respective branch. The external sales by each of the offices are $800,000,000 for the U.S. branch, $300,000,000 for the Canadian branch, and $1,400,000,000 for the English branch. At what level should each of the branches produce in order to satisfy the consumer demand?

29. Consider the matrix $(I - A)^{-1}$ from Example 1. Show that if the consumer demand for coal is increased by \$1 billion, then the additional amounts (in billions of dollars) that must be produced by each of the three industries is given by the first column of $(I - A)^{-1}$.

$$\text{Hint: } (I - A)^{-1} \begin{bmatrix} 3 \\ 1 \\ 3 \end{bmatrix} = (I - A)^{-1} \left(\begin{bmatrix} 2 \\ 1 \\ 3 \end{bmatrix} + \begin{bmatrix} 1 \\ 0 \\ 0 \end{bmatrix} \right).$$

30. Refer to Exercise 29. Interpret the significance of the second and third columns of $(I - A)^{-1}$.

TECHNOLOGY EXERCISES

In Exercises 31 and 32, use the input–output matrix A and the consumer-demand matrix D to find the production matrix X for the Leontieff model. Round your answers to two decimal places.

31. $A = \begin{bmatrix} .1 & .2 & .4 & .1 \\ .05 & .3 & .25 & .15 \\ .15 & .1 & .2 & .04 \\ .25 & .1 & .05 & .03 \end{bmatrix}, D = \begin{bmatrix} 3 \\ 7 \\ 4 \\ 2 \end{bmatrix}$

32. $A = \begin{bmatrix} .2 & .35 & .15 & .05 \\ .1 & .1 & .3 & .2 \\ .075 & .2 & .05 & .1 \\ .3 & .04 & .1 & .15 \end{bmatrix}, D = \begin{bmatrix} 5 \\ 2 \\ 1 \\ 7 \end{bmatrix}$

Solutions to Check Your Understanding 2.6

1. The equation $(I - A)X = D$ has the form $CX = D$, where C is the matrix $I - A$. From Section 2.4 we know that $X = C^{-1}D$. That is, $X = (I - A)^{-1}D$. Now,

$$I - A = \begin{bmatrix} 1 & 0 & 0 \\ 0 & 1 & 0 \\ 0 & 0 & 1 \end{bmatrix} - \begin{bmatrix} .1 & 0 & .1 \\ .2 & .1 & .1 \\ .1 & .2 & 0 \end{bmatrix}$$

$$= \begin{bmatrix} .9 & 0 & -.1 \\ -.2 & .9 & -.1 \\ -.1 & -.2 & 1 \end{bmatrix}.$$

Using the Gauss–Jordan method to find the inverse of this matrix, we have (to two decimal places)

$$(I - A)^{-1} = \begin{bmatrix} 1.13 & .03 & .12 \\ .27 & 1.14 & .14 \\ .17 & .23 & 1.04 \end{bmatrix}.$$

Therefore, rounding to the nearest integer, we have

$$X = (I - A)^{-1}D = \begin{bmatrix} 1.13 & .03 & .12 \\ .27 & 1.14 & .14 \\ .17 & .23 & 1.04 \end{bmatrix} \begin{bmatrix} 100 \\ 200 \\ 50 \end{bmatrix}$$

$$= \begin{bmatrix} 125 \\ 262 \\ 115 \end{bmatrix}.$$

2. We have $X = (I - A)^{-1}D$, where $(I - A)^{-1}$ is as computed in Problem 1. So

$$X = (I - A)^{-1}D = \begin{bmatrix} 1.13 & .03 & .12 \\ .27 & 1.14 & .14 \\ .17 & .23 & 1.04 \end{bmatrix} \begin{bmatrix} 300 \\ 100 \\ 100 \end{bmatrix}$$

$$= \begin{bmatrix} 354 \\ 209 \\ 178 \end{bmatrix}.$$

CHAPTER 2   Summary

KEY TERMS AND CONCEPTS	EXAMPLES

2.1 Systems of Linear Equations with Unique Solutions

The three **elementary row operations** for a system of linear equations (or a matrix) are as follows:

(a) Interchange any two equations (rows).

(b) Multiply an equation (row) by a nonzero number.

(c) Change an equation (row) by adding to it a multiple of another equation (row).

The **Gauss–Jordan elimination method** is a systematic process that applies a sequence of elementary row operations to a system of linear equations (or a matrix) until the solutions can be easily obtained.

Solve the system

$$\begin{cases} x + 2y = 15 \\ -4x + 5y = -8 \end{cases}$$

$$\begin{cases} x + 2y = 15 \\ -4x + 5y = -8 \end{cases} \xrightarrow{R_2 + 4R_1} \begin{cases} x + 2y = 15 \\ 13y = 52 \end{cases} \xrightarrow{\frac{1}{13}R_2}$$

$$\begin{cases} x + 2y = 15 \\ y = 4 \end{cases} \xrightarrow{R_1 + (-2)R_2} \begin{cases} x = 7 \\ y = 4 \end{cases}$$

The solution is $x = 7$, $y = 4$.

In the matrix formulation, the augmented matrix

$$\begin{bmatrix} 1 & 2 & | & 15 \\ -4 & 5 & | & -8 \end{bmatrix} \text{ is transformed into the matrix } \begin{bmatrix} 1 & 0 & | & 7 \\ 0 & 1 & | & 4 \end{bmatrix}.$$

KEY TERMS AND CONCEPTS	EXAMPLES

2.2 General Systems of Linear Equations

The process of **pivoting** about a specific element of a matrix is to apply a sequence of elementary row operations so that the specific element becomes 1 and the other elements in its column become 0. To apply the Gauss–Jordan elimination method, proceed from left to right and perform pivots on as many columns to the left of the vertical line as possible, with the specific elements for the pivots coming from different rows.

Pivot about the circled element of the matrix.

$$\left[\begin{array}{cc|c} ② & 6 & 0 \\ 5 & 9 & 7 \end{array}\right] \xrightarrow[R_2 + (-5)R_1]{\frac{1}{2}R_1} \left[\begin{array}{cc|c} 1 & 3 & 0 \\ 0 & -6 & 7 \end{array}\right]$$

After a matrix corresponding to a system of linear equations has been completely reduced with the Gauss–Jordan elimination method, **all solutions** to the system of linear equations can be obtained. If the matrix on the left of the vertical line is a square matrix with ones on the main diagonal, then there is a unique solution. If one row of the matrix is of the form $0\,0\,0\ldots 0\,|\,a$, where a is a nonzero number, then there is no solution. Otherwise, there might be infinitely many solutions. In this case, variables corresponding to columns that have not been pivoted can assume any values, and the values of the other variables can be expressed in terms of those variables.

Completely reduced matrices:

$$\left[\begin{array}{ccc|c} 1 & 0 & 3 & 4 \\ 0 & 1 & 2 & 5 \\ 0 & 0 & 0 & 6 \end{array}\right]; \text{ no solution}$$

$$\left[\begin{array}{ccc|c} 1 & 0 & 2 & 6 \\ 0 & 1 & 8 & 5 \\ 0 & 0 & 0 & 0 \end{array}\right]; \text{ infinitely many solutions}$$

Since the z column has not been pivoted, z can assume any value. By the second row of the matrix,

$$y + 8z = 5 \quad \text{or} \quad y = 5 - 8z.$$

By the first row of the matrix,

$$x + 2z = 6 \quad \text{or} \quad x = 6 - 2z.$$

2.3 Arithmetic Operations on Matrices

To find the **sum** (or **difference**) of two matrices of the same size, add (or subtract) corresponding elements.

$$\begin{bmatrix} 3 & -2 & 4 \\ 0 & 6 & -1 \end{bmatrix} + \begin{bmatrix} 2 & 5 & 4 \\ 7 & 1 & 5 \end{bmatrix} = \begin{bmatrix} 5 & 3 & 8 \\ 7 & 7 & 4 \end{bmatrix}$$

$$\begin{bmatrix} 5 & 2 \\ 1 & 0 \end{bmatrix} - \begin{bmatrix} 3 & -5 \\ -6 & 5 \end{bmatrix} = \begin{bmatrix} 2 & 7 \\ 7 & -5 \end{bmatrix}$$

The **scalar product** of a number c and a matrix A is the matrix obtained by multiplying each element of A by c.

$$4\begin{bmatrix} 2 & 1 & -3 \\ -5 & 0 & 4 \end{bmatrix} = \begin{bmatrix} 8 & 4 & -12 \\ -20 & 0 & 16 \end{bmatrix}$$

The **product** of an $m \times n$ matrix and an $n \times r$ matrix is the $m \times r$ matrix whose ij^{th} element is obtained by multiplying the ith row of the first matrix by the jth column of the second matrix. (The product of each row and column is calculated as the sum of the products of successive entries.)

$$\begin{bmatrix} 1 & -3 & 2 \\ 4 & 5 & 0 \\ -2 & 6 & -6 \end{bmatrix} \begin{bmatrix} 2 \\ -4 \\ 1 \end{bmatrix} = \begin{bmatrix} 1(2) + (-3)(-4) + 2(1) \\ 4(2) + 5(-4) + 0(1) \\ (-2)(2) + 6(-4) + (-6)(1) \end{bmatrix}$$

$$\underset{3 \times 3}{\uparrow\,\uparrow} \underset{\text{same}}{\quad} \underset{3 \times 1}{\uparrow\,\uparrow}$$

size of product

$$= \begin{bmatrix} 16 \\ -12 \\ -34 \end{bmatrix}$$

$$3 \times 1$$

2.4 The Inverse of a Square Matrix

The **inverse** of an $n \times n$ matrix A is an $n \times n$ matrix A^{-1} with the property that $A^{-1}A = I_n$ and $AA^{-1} = I_n$.

The inverse of $\begin{bmatrix} 2 & 4 \\ 1 & 3 \end{bmatrix}$ is $\begin{bmatrix} \frac{3}{2} & -2 \\ -\frac{1}{2} & 1 \end{bmatrix}$, since

$$\begin{bmatrix} 2 & 4 \\ 1 & 3 \end{bmatrix} \begin{bmatrix} \frac{3}{2} & -2 \\ -\frac{1}{2} & 1 \end{bmatrix} = \begin{bmatrix} 1 & 0 \\ 0 & 1 \end{bmatrix} = \begin{bmatrix} \frac{3}{2} & -2 \\ -\frac{1}{2} & 1 \end{bmatrix} \begin{bmatrix} 2 & 4 \\ 1 & 3 \end{bmatrix}.$$

KEY TERMS AND CONCEPTS	EXAMPLES

A 2×2 matrix $\begin{bmatrix} a & b \\ c & d \end{bmatrix}$ has an inverse if $D = ad - bc \neq 0$. If so, the inverse matrix is

$$\frac{1}{D}\begin{bmatrix} d & -b \\ -c & a \end{bmatrix}.$$

A system of linear equations can be written as a **matrix equation** $AX = B$, where A is a rectangular matrix of coefficients of the variables, X is a column of variables, and B is a column matrix of the constants from the right side of the system. If the matrix A has an inverse, then the solution of the system is given by $X = A^{-1}B$.

Find the inverse of $\begin{bmatrix} 2 & 4 \\ 1 & 3 \end{bmatrix}$.

$D = 2 \cdot 3 - 4 \cdot 1 = 2$. Therefore,

$$\begin{bmatrix} 2 & 4 \\ 1 & 3 \end{bmatrix}^{-1} = \frac{1}{2}\begin{bmatrix} 3 & -4 \\ -1 & 2 \end{bmatrix} = \begin{bmatrix} \frac{3}{2} & -2 \\ -\frac{1}{2} & 1 \end{bmatrix}.$$

Solve $\begin{cases} 2x + 4y = 5 \\ x + 3y = 3 \end{cases}$ or $\begin{bmatrix} 2 & 4 \\ 1 & 3 \end{bmatrix}\begin{bmatrix} x \\ y \end{bmatrix} = \begin{bmatrix} 5 \\ 3 \end{bmatrix}$

$$\begin{bmatrix} x \\ y \end{bmatrix} = \begin{bmatrix} 2 & 4 \\ 1 & 3 \end{bmatrix}^{-1}\begin{bmatrix} 5 \\ 3 \end{bmatrix} = \begin{bmatrix} \frac{3}{2} & -2 \\ -\frac{1}{2} & 1 \end{bmatrix}\begin{bmatrix} 5 \\ 3 \end{bmatrix} = \begin{bmatrix} \frac{3}{2} \\ \frac{1}{2} \end{bmatrix}.$$

Therefore, $x = \frac{3}{2}, y = \frac{1}{2}$.

2.5 The Gauss–Jordan Method for Calculating Inverses

Follow these steps to obtain the inverse of an $n \times n$ matrix A by the **Gauss-Jordan method**.

1. Append I_n to the right of the original matrix to form $[A \mid I_n]$.

2. Perform pivots on $[A \mid I_n]$ to obtain a matrix of the form $[I_n \mid B]$. (If $[A \mid I_n]$ cannot be reduced to a matrix of the form $[I_n \mid B]$, then A does not have an inverse.)

3. The matrix B will then be A^{-1}.

Find A^{-1}, where $A = \begin{bmatrix} 2 & 1 \\ 5 & 3 \end{bmatrix}$.

$$\begin{bmatrix} ② & 1 & | & 1 & 0 \\ 5 & 3 & | & 0 & 1 \end{bmatrix}$$

$$\begin{bmatrix} 1 & \frac{1}{2} & | & \frac{1}{2} & 0 \\ 0 & ① ⁄② & | & -\frac{5}{2} & 1 \end{bmatrix}$$

$$\begin{bmatrix} 1 & 0 & | & 3 & -1 \\ 0 & 1 & | & -5 & 2 \end{bmatrix}.$$

Therefore, $A^{-1} = \begin{bmatrix} 3 & -1 \\ -5 & 2 \end{bmatrix}.$

2.6 Input–Output Analysis

An **input–output matrix** has rows and columns labeled with the different industries in an economy. The ij^{th} entry of the matrix gives the cost of the input from the industry in row i used in the production of $1 worth of the output of industry in column j.

The input–output matrix

$$\begin{array}{cc} & \text{industry 1} \quad \text{industry 2} \\ \begin{matrix} \text{industry 1} \\ \text{industry 2} \end{matrix} & \begin{bmatrix} .15 & .07 \\ .08 & .12 \end{bmatrix} \end{array}$$

states that industry 1 uses $.15 input from industry 1 and $.08 input from industry 2 in order to make $1 of output. Similarly for industry 2.

If A is an input–output matrix and D is a demand matrix giving the dollar values of the outputs from the various industries to be supplied to outside customers, then the proper **production amounts** for each sector are given by $X = (I - A)^{-1}D$.

The amount of input from each industry required to meet a demand of $2 billion of output from industry 1 and $3 billion of output from industry 2 is

$$\begin{bmatrix} \textit{input from industry 1} \\ \textit{input from industry 2} \end{bmatrix} = \left(\begin{bmatrix} 1 & 0 \\ 0 & 1 \end{bmatrix} - \begin{bmatrix} .15 & .07 \\ .08 & .12 \end{bmatrix}\right)^{-1}\begin{bmatrix} 2 \\ 3 \end{bmatrix}$$

$$= \begin{bmatrix} 2.65 \\ 3.65 \end{bmatrix}.$$

Therefore, industry 1 should produce $2.65 billion and industry 2 should produce $3.65 billion.

CHAPTER 2   Fundamental Concept Check Exercises

1. What is meant by a solution to a system of linear equations?

2. What is a matrix?

3. State the three elementary row operations on equations or matrices.

4. What does it mean for a system of equations or a matrix to be in *diagonal form*?

5. What is meant by *pivoting* a matrix about a nonzero entry?

6. State the Gauss–Jordan elimination method for solving a system of linear equations.

7. What is a row matrix? Column matrix? Square matrix? Identity matrix, I_n?

8. What is meant by a_{ij}, the ijth entry of a matrix?

9. Define the sum and difference of two matrices.

10. Define the product of two matrices.

11. Define scalar product.

12. Define the inverse of a matrix, A^{-1}.

13. Give the formula for the inverse of a 2×2 matrix.

14. Explain how to use the inverse of a matrix to solve a system of linear equations.

15. Describe the steps of the Gauss–Jordan method for calculating the inverse of a matrix.

16. What are an input–output matrix and a consumer-demand matrix?

17. Explain how to solve an input–output analysis problem.

CHAPTER 2   Review Exercises

In Exercises 1 and 2, pivot each matrix about the circled element.

1. $\begin{bmatrix} ③ & -6 & 1 \\ 2 & 4 & 6 \end{bmatrix}$

2. $\begin{bmatrix} -5 & -3 & 1 \\ 4 & ② & 0 \\ 0 & 6 & 7 \end{bmatrix}$

In Exercises 3–8, use the Gauss–Jordan elimination method to find all solutions of the system of linear equations.

3. $\begin{cases} \frac{1}{2}x - y = -3 \\ 4x - 5y = -9 \end{cases}$

4. $\begin{cases} 3x \quad\quad + 9z = \quad 42 \\ 2x + y + 6z = \quad 30 \\ -x + 3y - 2z = -20 \end{cases}$

5. $\begin{cases} 3x - 6y + 6z = -5 \\ -2x + 3y - 5z = \frac{7}{3} \\ x + y + 10z = 3 \end{cases}$

6. $\begin{cases} 3x + 6y - 9z = 1 \\ 2x + 4y - 6z = 1 \\ 3x + 4y + 5z = 0 \end{cases}$

7. $\begin{cases} x + 2y - 5z + 3w = \quad 16 \\ -5x - 7y + 13z - 9w = -50 \\ -x + y - 7z + 2w = \quad 9 \\ 3x + 4y - 7z + 6w = \quad 33 \end{cases}$

8. $\begin{cases} 5x - 10y = \quad 5 \\ 3x - 8y = -3 \\ -3x + 7y = \quad 0 \end{cases}$

In Exercises 9–12, perform the indicated matrix operation.

9. $\begin{bmatrix} 2 \\ -1 \\ 0 \end{bmatrix} + \begin{bmatrix} 3 \\ 4 \\ 7 \end{bmatrix}$

10. $\begin{bmatrix} 1 & 3 & -2 \\ 4 & 0 & -1 \end{bmatrix} \begin{bmatrix} 3 & 5 \\ 1 & 0 \\ 0 & -6 \end{bmatrix}$

11. $\frac{3}{4} \begin{bmatrix} 8 & -6 \\ \frac{2}{3} & 0 \end{bmatrix}$

12. $\begin{bmatrix} 1.4 & -3 \\ 8.2 & 0 \\ 4 & 5.5 \end{bmatrix} - \begin{bmatrix} .8 & 7 \\ 1.6 & -2 \\ 0 & -5.5 \end{bmatrix}$

13. Let $A = \begin{bmatrix} 2 & 5 \\ -3 & 1 \end{bmatrix}$ and $B = \begin{bmatrix} 4 & -5 \\ 3 & k \end{bmatrix}$. For what value(s) of k, if any, will $AB = BA$

14. Let $A = \begin{bmatrix} 5 & k \\ -6 & 5 \end{bmatrix}$ and $B = \begin{bmatrix} 1 & -2 \\ 3 & 1 \end{bmatrix}$. For what value(s) of k, if any, will $AB = BA$?

15. Find the inverse of the appropriate matrix, and use it to solve the system of equations

$$\begin{cases} 3x + 2y = 0 \\ 5x + 4y = 2. \end{cases}$$

16. The matrices

$$\begin{bmatrix} 4 & -2 & 3 \\ 8 & -3 & 5 \\ 7 & -2 & 4 \end{bmatrix} \text{ and } \begin{bmatrix} -2 & 2 & -1 \\ 3 & -5 & 4 \\ 5 & -6 & 4 \end{bmatrix}$$

are inverses of each other. Use these matrices to solve the following systems of linear equations:

(a) $\begin{cases} -2x + 2y - z = 1 \\ 3x - 5y + 4z = 0 \\ 5x - 6y + 4z = 3 \end{cases}$

(b) $\begin{cases} 4x - 2y + 3z = \quad 0 \\ 8x - 3y + 5z = -1 \\ 7x - 2y + 4z = \quad 2 \end{cases}$

In Exercises 17 and 18, use the Gauss–Jordan method to calculate the inverse of the matrix.

17. $\begin{bmatrix} 2 & 6 \\ 1 & 2 \end{bmatrix}$

18. $\begin{bmatrix} 1 & 1 & 1 \\ 3 & 4 & 3 \\ 1 & 1 & 2 \end{bmatrix}$

19. **Crop Allocation** Farmer Brown has 1000 acres of land on which he plans to grow corn, wheat, and soybeans. The cost of cultivating these crops is $357 per acre for corn, $127 per acre for wheat, and $181 per acre for soybeans. If Farmer Brown wishes to use all of his available land and his entire budget of $269,000, and if he wishes to plant the same number of acres of corn as wheat and soybeans combined, how many acres of each crop can he grow? (*Source:* USDA ERS Commodity Costs and Returns)

20. **Equipment Sales** A company makes backyard playground equipment such as swing sets, slides, and play sets. The cost (in dollars) to make a specific style of each piece is given in

matrix C. The sales price (in dollars) for each piece is given in matrix S. Two stores sell these specific pieces, and matrix A gives the quantities sold during one month.

$$C = \begin{bmatrix} 165 \\ 65 \\ 210 \end{bmatrix} \begin{matrix} \text{Swing set} \\ \text{Slide} \\ \text{Play set} \end{matrix}$$

$$S = \begin{bmatrix} 200 \\ 80 \\ 250 \end{bmatrix} \begin{matrix} \text{Swing set} \\ \text{Slide} \\ \text{Play set} \end{matrix}$$

$$A = \begin{bmatrix} \overset{\text{Swing set}}{15} & \overset{\text{Slide}}{20} & \overset{\text{Play set}}{8} \\ 10 & 17 & 12 \end{bmatrix}$$

Determine and interpret the following matrices:
(a) AC **(b)** AS **(c)** $S - C$ **(d)** $A(S - C)$

21. **Investment Earnings** A person wants to invest money in three different college savings plans. Matrix A contains the percentages (in decimal form) invested in bonds, stocks, and a conservative fixed income fund for each of the three different plans. Matrix B gives the total amount invested in each of the three savings plans, and matrix C gives the rates of return (in decimal form) for one year and for five years.

$$A = \begin{bmatrix} \overset{\text{Bonds}}{.50} & \overset{\text{Stocks}}{.43} & \overset{\text{Fixed Income}}{.07} \\ .45 & .26 & .29 \\ .40 & .40 & .20 \end{bmatrix} \begin{matrix} \text{Plan 1} \\ \text{Plan 2} \\ \text{Plan 3} \end{matrix}$$

$$B = \begin{bmatrix} \overset{\text{Plan 1}}{5000} & \overset{\text{Plan 2}}{8000} & \overset{\text{Plan 3}}{10,000} \end{bmatrix}$$

$$C = \begin{bmatrix} \overset{\text{One Year}}{.0032} & \overset{\text{Five Year}}{.1119} \\ .0233 & .0976 \\ .0320 & .0467 \end{bmatrix} \begin{matrix} \text{Plan 1} \\ \text{Plan 2} \\ \text{Plan 3} \end{matrix}$$

Calculate and interpret the following:
(a) BA **(b)** BC **(c)** $2B$
(d) the row 1, column 2 entry of BA
(e) the row 1, column 1 entry of BC

22. **Job Earnings** Sara, Quinn, Tamia, and Zack are working at the pool this summer. One week, they spend the following amounts of time at three different tasks:

$$A = \begin{bmatrix} \overset{\text{Concessions}}{11} & \overset{\text{Front Desk}}{7} & \overset{\text{Cleaning}}{12} \\ 9 & 5 & 16 \\ 13 & 8 & 9 \\ 13 & 7 & 10 \end{bmatrix} \begin{matrix} \text{Sara} \\ \text{Quinn} \\ \text{Tamia} \\ \text{Zack} \end{matrix}$$

The hourly pay (in dollars) for the three different tasks is given by

$$B = \begin{bmatrix} 11 \\ 9 \\ 12 \end{bmatrix} \begin{matrix} \text{Concessions} \\ \text{Front Desk} \\ \text{Cleaning} \end{matrix}$$

(a) Calculate and interpret the matrix AB.
(b) Who earned the most that week? Who earned the least?
(c) If the hourly pay for concessions is changed to $12 and the hourly pay for cleaning is changed to $11, who earns the most that week?
(d) How many hours did Sara work that week?

23. **Fruit Baskets** The produce department at a grocery store makes fruit baskets of apples, bananas, and oranges. The cost for each apple is $.85, for each banana is $.20, and for each orange is $.76. Suppose that each basket needs to have 18 pieces of fruit and costs $9 to make. How many apples, bananas, and oranges are in each basket if the number of bananas in each basket is the same as the number of apples and oranges together?

24. **Nuclear Disarmament** In an arms race between two superpowers, each nation takes stock of its own and its enemy's nuclear arsenal each year. Each nation has the policy of dismantling a certain percentage of its stockpile each year and adding that same percentage of its competitor's stockpile. Nation A dismantles 20%, and nation B dismantles 10%. Suppose that the current stockpiles of nations A and B are 10,000 and 7000 weapons, respectively.
(a) What will the stockpiles be in each of the next two years?
(b) What were the stockpiles in each of the preceding two years?
(c) Show that the "missile gap" between the superpowers decreases by 30% each year under these policies. Show that the total number of weapons decreases each year if nation A begins with the most weapons, and the total number of weapons increases if nation B begins with the most weapons.

25. **Two-Sector Economy** The economy of a small country can be regarded as consisting of two industries, I and II, whose input–output matrix is

$$A = \begin{bmatrix} .4 & .2 \\ .1 & .3 \end{bmatrix}.$$

How many units should be produced by each industry in order to meet a consumer demand for 8 units from industry I and 12 units from industry II?

26. **Coins** Joe has $3.30 in his pocket, made up of nickels, dimes, and quarters. There are 30 coins, and there are five times as many dimes as quarters. How many quarters does Joe have?

Conceptual Exercises

27. Identify each statement as true or false.
(a) If a system of linear equations has two different solutions, it must have infinitely many solutions.
(b) If a system of linear equations has more equations than variables, it cannot have a unique solution.
(c) If a system of linear equations has more variables than equations, then it must have infinitely many solutions.

28. Identify each statement as true or false.
(a) Every matrix can be added to itself.
(b) Every matrix can be multiplied by itself.

29. Make up a system of two linear equations, with two variables, that has infinitely many solutions.

30. Make up a system of two linear equations, with two variables, that has no solution.

31. If the product of two numbers is zero, then one of the numbers must be zero. Make up two 2×2 matrices A and B such that AB is a matrix of all zeros, but neither A nor B is a matrix of all zeros.

32. Suppose that we try to solve the matrix equation $AX = B$ by using an inverse matrix but find that even though the matrix A is a square matrix, it has no inverse. What can be said about the outcome from solving the associated system of linear equations by the Gauss–Jordan elimination method?

33. Why should the numbers in a single column of an input–output matrix have a sum that is less than 1?

CHAPTER 2 PROJECT

Population Dynamics

In 1991, the U.S. Fish and Wildlife Service proposed logging restrictions on nearly 12 million acres of Pacific Northwest forest to help save the endangered northern spotted owl. This decision caused considerable controversy between the logging industry and environmentalists.

Mathematical ecologists created a mathematical model to analyze the population dynamics of the spotted owl.[1] They divided the female owl population into three categories: juvenile (up to 1 year old), subadult (1 to 2 years old), and adult (over 2 years old). Suppose that in a certain region there are currently 2950 female spotted owls made up of 650 juveniles, 200 subadults, and 2100 adults. The ecologists used matrices to project the changes in the population from year to year. The original numbers can be displayed in the column matrix

$$X_0 = \begin{bmatrix} 650 \\ 200 \\ 2100 \end{bmatrix}.$$

The populations after one year are given by the column matrix

$$X_1 = \begin{bmatrix} 693 \\ 117 \\ 2116 \end{bmatrix}.$$

The subscript 1 tells us that the matrix gives the population after one year. The names of the matrices for subsequent years will have subscripts 2, 3, 4, etc.

1. How many subadult females are there after one year?

2. Did the total population of females increase or decrease during the year?

Let A denote the matrix

$$\begin{bmatrix} 0 & 0 & .33 \\ .18 & 0 & 0 \\ 0 & .71 & .94 \end{bmatrix}.$$

According to the mathematical model, subsequent population distributions are generated by multiplication on the left by A. That is,

$$A \cdot X_0 = X_1, \quad A \cdot X_1 = X_2, \quad A \cdot X_2 = X_3, \ldots .$$

If the population distribution at any time is given by $\begin{bmatrix} j \\ s \\ a \end{bmatrix}$, then the distribution one year later is $A \cdot \begin{bmatrix} j \\ s \\ a \end{bmatrix}$.

[1] Lamberson, R. H., R. McKelvey, B. R. Noon, and C. Voss, "A Dynamic Analysis of Northern Spotted Owl Viability in a Fragmented Forest Landscape," *Conservation Biology*, Vol. 6, No. 4, December 1992; 505–512.

3. Fill in the blanks in the following statements.
   (a) Each year, _____ juvenile females are born for each 100 adult females.
   (b) Each year, _____% of the juvenile females survive to become subadults.
   (c) Each year, _____% of the subadults survive to become adults and _____% of the adults survive.

4. Calculate the column matrices X_2, X_3, X_4, and X_5. *Note:* With a graphing calculator, you can specify matrix **[A]** to be the 3×3 matrix A and specify **[B]** to be the initial 3×1 population matrix X_0. Display **[B]** in the home screen and then enter the command **[A]*Ans**. Each press of the ENTER key will generate the next population distribution matrix. *Tip:* Prior to generating the matrices, invoke the MODE list and set Float to 0 so that all numbers in the population matrices will be rounded to whole numbers.

5. Refer to Exercise 4. Is the total female population increasing, decreasing, or neither during the first five years?

6. Explain why calculating **[A]^50*[B]** gives the column matrix for the population distribution after 50 years.

7. Find the projected population matrix after 50 years; after 100 years; after 150 years. On the basis of this mathematical model, what do you conclude about the prospects for the northern spotted owl?

   In this model, the main impediment to the survival of the owl is represented by the number .18 in the second row of matrix A. This number is low for two reasons: (1) The first year of life is precarious for most animals living in the wild. (2) Juvenile owls must eventually leave the nest and establish their own territory. If much of the forest near their original home has been cleared, then they are vulnerable to predators while searching for a new home.

8. Suppose that, through better forest management, the number .18 can be increased to .26. Find the total female population for the first five years under this new assumption. Repeat Exercise 7, and determine whether extinction will be avoided under the new assumption.

CHAPTER

3

Linear Programming, A Geometric Approach

Linear programming is a method for solving problems in which a linear function (representing cost, profit, distance, weight, or the like) is to be maximized or minimized. Such problems are called **optimization problems**. As we shall see, these problems, when translated into mathematical language, involve systems of linear inequalities, systems of linear equations, and eventually (in Chapter 4) matrices.

3.1   Linear Inequalities

In this section, we study the properties of inequalities. Start with a Cartesian coordinate system on the line (Fig. 1). Recall that it is possible to associate with each point of the line a number; and conversely, with each number (positive, negative, or zero) it is possible to associate a point on the line. Table 1 shows the four inequality signs that are used to denote relationships between numbers.

Figure 1

Table 1 Inequality Signs

Inequality	Meaning	Examples
$a < b$	a is less than b (a lies to the left of b on the number line)	$2 < 3, -1 < 2, -3 < -1$
$a > b$	a is greater than b (a lies to the right of b on the number line)	$3 > 2, 2 > -1, -1 > -3$
$a \leq b$	a is less than or equal to b	$2 \leq 3, 4 \leq 4$
$a \geq b$	a is greater than or equal to b	$2 \geq -1, 3 \geq 3$

An inequality expresses a relationship between the quantities on both of its sides. This relationship is very similar to the relationship expressed by an equation. And just as some problems require solving equations, others require solving inequalities. Our next task will be to state and illustrate the arithmetic operations permissible in dealing with inequalities.

> **Inequality Property 1** Suppose that $a < b$ and that c is any number. Then, $a + c < b + c$ and $a - c < b - c$. In other words, the same number can be added to or subtracted from both sides of the inequality.

For example, start with the inequality $2 < 3$ and add 4 to both sides to get

$$2 + 4 < 3 + 4.$$

That is,

$$6 < 7,$$

a correct inequality.

EXAMPLE 1 **Solving a Linear Inequality** Solve the inequality $x + 4 > 3$. That is, determine all values of x for which the inequality holds.

SOLUTION

$x + 4 > 3$	Given inequality
$(x + 4) - 4 > 3 - 4$	Subtract 4 from both sides (Inequality Property 1).
$x > -1$	Combine terms.

That is, the values of x for which the inequality holds are exactly those x greater than -1. «

In dealing with an equation, both sides may be multiplied or divided by a number. However, multiplying or dividing an inequality by a number requires some care. The result depends on whether the number is positive or negative. More precisely,

> **Inequality Property 2**
>
> **2A.** If $a < b$ and c is positive, then $ac < bc$.
> **2B.** If $a < b$ and c is negative, then $ac > bc$.

In other words, an inequality may be multiplied by a positive number, just as in the case of equations. But to multiply an inequality by a negative number, it is necessary to reverse the inequality sign. For example, the inequality $-1 < 2$ can be multiplied by 4 to get $-4 < 8$, a correct statement. But if we were to multiply by -4, it would be necessary to reverse the inequality sign, because -4 is negative. In this latter case, we would get $4 > -8$, a correct statement.

> **NOTE** ▶   Inequality Properties 1 and 2 are stated using only $<$. However, exactly the same properties hold if $<$ is replaced by $>$, $\leq$, or $\geq$. «

EXAMPLE 2   **Solving a Linear Inequality**   Solve the inequality $-3x + 2 \geq -1$.

SOLUTION

$$-3x + 2 \geq -1 \qquad \text{Given inequality}$$
$$(-3x + 2) - 2 \geq -1 - 2 \qquad \text{Subtract 2 from both sides (Inequality Property 1).}$$
$$-3x \geq -3 \qquad \text{Combine terms.}$$
$$-\tfrac{1}{3}(-3x) \leq -\tfrac{1}{3}(-3) \qquad \text{Multiply both sides by } -\tfrac{1}{3} \text{ and reverse the direction of the inequality sign (Inequality Property 2).}$$
$$x \leq 1 \qquad \text{Simplify.}$$

Therefore, the values of x satisfying the inequality are precisely those values that are ≤ 1.

» Now Try Exercise 7

The inequalities of greatest interest to us are those in two variables, x and y, and having the general form $cx + dy \leq e$ or $cx + dy \geq e$, where c, d, and e are given numbers, with c and d not both 0. (Such inequalities arise in our discussion of linear programming.) Such inequalities are called **linear inequalities**. When $d \neq 0$ (that is, when y actually appears), the inequality can be written in one of the **slope-intercept** forms $y \leq mx + b$ or $y \geq mx + b$. When $d = 0$, the inequality can be written in one of the **vertical** forms $x \leq a$ or $x \geq a$. The procedure for writing a linear inequality in one of these forms is analogous to that of writing a linear equation in slope-intercept or vertical form.

EXAMPLE 3   **Slope-Intercept Form of a Linear Inequality**   Write the linear inequality $2x - 3y \geq -9$ in slope-intercept form.

SOLUTION

$$2x - 3y \geq -9 \qquad \text{Given inequality}$$
$$-3y \geq -2x - 9 \qquad \text{Subtract } 2x \text{ from both sides (Inequality Property 1).}$$
$$y \leq -\tfrac{1}{3}(-2x - 9) \qquad \text{Multiply both sides by } -\tfrac{1}{3} \text{ and reverse the direction of the inequality sign (Inequality Property 2).}$$
$$y \leq \tfrac{2}{3}x + 3 \qquad \text{Simplify.} \qquad \text{**» Now Try Exercise 9**}$$

EXAMPLE 4   **Vertical Form of a Linear Inequality**   Find the vertical form of the inequality $\tfrac{1}{2}x \geq 4$.

SOLUTION   Since y does not appear in the inequality, we write the inequality with x on the left side.

$$\tfrac{1}{2}x \geq 4 \qquad \text{Given inequality}$$
$$2(\tfrac{1}{2}x) \geq 2 \cdot 4 \qquad \text{Multiply both sides by 2 (Inequality Property 2).}$$
$$x \geq 8 \qquad \text{Simplify.} \qquad \text{**» Now Try Exercise 13**}$$

Graphing Linear Inequalities

Associated with every linear inequality is a set of points in the plane, the set of all those points that satisfy the inequality. This set of points is called the **graph of the inequality**.

EXAMPLE 5   **Solution of a Linear Inequality**   Determine whether or not the given point satisfies the inequality $y \geq -\tfrac{2}{3}x + 4$.

(a) $(3, 4)$ **(b)** $(0, 0)$

SOLUTION   Substitute the x-coordinate of the point for x and the y-coordinate for y, and determine whether the resulting inequality is correct or not.

(a) $4 \geq -\frac{2}{3}(3) + 4$
$\quad\ 4 \geq -2 + 4$
$\quad\ 4 \geq 2 \quad \text{(correct)}$

(b) $0 \geq -\frac{2}{3}(0) + 4$
$\quad\ 0 \geq 0 + 4$
$\quad\ 0 \geq 4 \quad \text{(not correct)}$

Therefore, the point $(3, 4)$ satisfies the inequality and the point $(0, 0)$ does not.

>> *Now Try Exercises 15 and 17*

It is easiest to determine the graph of a given inequality after it has been written in slope-intercept or vertical form. Therefore, let us describe the graphs of each of these forms. The easiest to handle are the forms $x \geq a$ and $x \leq a$.

A point satisfies the inequality $x \geq a$ if, and only if, its x-coordinate is greater than or equal to a. The y-coordinate can be anything. Therefore, the graph of $x \geq a$ consists of all points to the right of and on the vertical line $x = a$. We will display the graph by crossing out the portion of the plane to the left of the line. (See Fig. 2.) Similarly, the graph of $x \leq a$ consists of the points to the left of and on the line $x = a$. This graph is shown in Fig. 3.

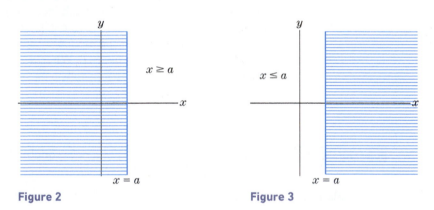

Figure 2   **Figure 3**

Here is a simple procedure for graphing the other two forms.

Graphing a Linear Inequality in Slope-Intercept Form   To graph $y \geq mx + b$ or $y \leq mx + b$,

1. Draw the graph of $y = mx + b$.

2. Throw away—that is, "cross out"—the portion of the plane not satisfying the inequality. The graph of $y \geq mx + b$ consists of all points above or on the line. The graph of $y \leq mx + b$ consists of all points below or on the line. See Fig. 4.

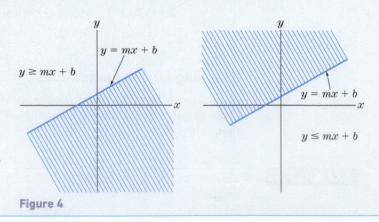

Figure 4

| **EXAMPLE 6** | **Graphing a Linear Inequality** Graph the inequality $2x + 3y \geq 15$. |

SOLUTION In order to apply the above procedure, the inequality must first be written in slope-intercept form.

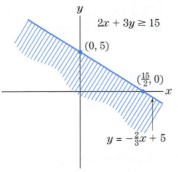

Figure 5

$$2x + 3y \geq 15 \qquad \text{Given inequality}$$
$$3y \geq -2x + 15 \qquad \text{Subtract } 2x \text{ from both sides.}$$
$$y \geq -\tfrac{2}{3}x + 5 \qquad \text{Divide both sides by 3.}$$

The last inequality is in slope-intercept form. Next, we graph the line $y = -\tfrac{2}{3}x + 5$. Its intercepts are $(0, 5)$ and $\left(\tfrac{15}{2}, 0\right)$. Since the inequality is "$y \geq$", we cross out the region below the line and label the region above with the inequality. The graph consists of all points above or on the line (Fig. 5).   **≫ Now Try Exercise 35**

| **EXAMPLE 7** | **Graphing a Linear Inequality** Graph the inequality $4x - 2y \geq 12$. |

SOLUTION First, write the inequality in slope-intercept form.

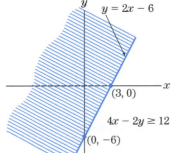

Figure 6

$$4x - 2y \geq 12 \qquad \text{Given inequality}$$
$$-2y \geq -4x + 12 \qquad \text{Subtract } 4x \text{ from both sides.}$$
$$y \leq 2x - 6 \qquad \begin{array}{l}\text{Divide both sides by } -2 \text{ and reverse the direction}\\ \text{of the inequality sign.}\end{array}$$

Next, graph $y = 2x - 6$. The intercepts are $(0, -6)$ and $(3, 0)$. Since the inequality is "$y \leq$", the graph consists of all points below or on the line (Fig. 6).   **≫ Now Try Exercise 31**

So far, we have been concerned only with graphing single inequalities. The next example concerns graphing a system of inequalities. That is, it asks us to determine all points of the plane that *simultaneously* satisfy all inequalities of a system.

| **EXAMPLE 8** | **Graphing a System of Linear Inequalities** Graph the system of inequalities |

$$\begin{cases} 2x + 3y \geq 15 \\ 4x - 2y \geq 12 \\ \qquad\quad y \geq 0. \end{cases}$$

SOLUTION The first two inequalities have already been graphed in Examples 6 and 7. The graph of $y \geq 0$ consists of all points above or on the x-axis. In Fig. 7, any point that is crossed out is *not* on the graph of at least one inequality. So the points that simultaneously satisfy all three inequalities are those in the remaining clear region and its border.

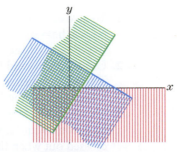

Figure 7   **≫ Now Try Exercise 45**

> **NOTE** At first, our convention of crossing out those points *not* on the graph of an inequality (instead of shading the points *on* the graph) may have seemed odd. However, the real advantage of this convention becomes apparent when graphing a system of inequalities. Imagine trying to find the graph of the system of Example 8 if the points *on* the graph of each inequality had been shaded. It would have been necessary to locate the points that had been shaded three times. This is hard to do. **«**

The graph of a system of inequalities is called a **feasible set**. The feasible set associated with the system of Example 8 is a three-sided unbounded region.

Given a specific point, we should be able to decide whether or not the point lies in the feasible set. The next example shows how this is done.

EXAMPLE 9

Feasible Set of a System of Linear Inequalities Determine whether the points $(5, 3)$ and $(4, 2)$ are in the feasible set of the system of inequalities of Example 8.

SOLUTION If we had very accurate measuring devices, we could plot the points in the graph of Fig. 8 and determine whether or not they lie in the feasible set. However, there is a simpler and more reliable algebraic method. Just substitute the coordinates of the points into each of the inequalities of the system and see whether or not *all* of the inequalities are satisfied. So doing, we find that $(5, 3)$ is in the feasible set and $(4, 2)$ is not.

$$(5, 3) \quad \begin{cases} 2(5) + 3(3) \geq 15 \\ 4(5) - 2(3) \geq 12 \\ (3) \geq 0 \end{cases} \quad \begin{cases} 19 \geq 15 \quad \text{true} \\ 14 \geq 12 \quad \text{true} \\ 3 \geq 0 \quad \text{true} \end{cases}$$

$$(4, 2) \quad \begin{cases} 2(4) + 3(2) \geq 15 \\ 4(4) - 2(2) \geq 12 \\ (2) \geq 0 \end{cases} \quad \begin{cases} 14 \geq 15 \quad \text{false} \\ 12 \geq 12 \quad \text{true} \\ 2 \geq 0 \quad \text{true} \end{cases}$$

» Now Try Exercises 49 and 51

INCORPORATING TECHNOLOGY

Graphing calculators use the Color/Line box settings and the **Shade** command to display graphs of linear inequalities. The colored squares and line style symbols on the left in Fig. 8(a) were specified with the Color/Line box shown in Fig. 8(b). The resulting graph in Fig. 8(c) shows the feasible set for Example 8.

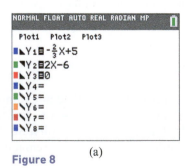
(a)

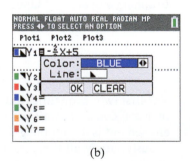
(b)

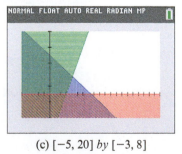
(c) $[-5, 20]$ *by* $[-3, 8]$

Figure 8

The following steps show how to specify the color and style for a line:

1. To invoke the Color/Line box for a function, press the left-arrow key until the cursor passes the $=$ sign, and then press the ENTER key. (The cursor will be on the small box to the right of the colored rectangle.)
2. Use the right- and left-arrow keys to scroll through the 15 possible colors. Press the ENTER key when you find the desired color. (The cursor will now be on the small box to the right of the Line style box.)
3. Use the right- and left-arrow keys to scroll through the 8 possible line style icons. The icons ◥ and ◣ specify that the region above and below the line, respectively, be crossed out when the line is graphed.
4. Press the ENTER key twice to exit the Color/Line box.

Regions to the left and right, respectively, of the vertical line $x = k$ can be crossed out with the command

$$\text{Shade}(\texttt{c-1,d+1,a,k,2,3}) \quad \text{and} \quad \text{Shade}(\texttt{c-1,d+1,k,b,2,3}),$$

where the window setting is $[a, b]$ by $[c, d]$. The **Shade** command is invoked from the home screen with 2nd [DRAW] **7**.

✹WolframAlpha An inequality of the form $ax + by \geq c$, where $b \neq 0$, can be converted to slope-intercept form with the instruction **solve $ax + by >= c$ for y**.

An instruction of the form

plot [first linear inequality],
[second linear inequality], . . . , [last linear inequality]

displays the feasible set of the system of linear inequalities as a shaded region. For instance, consider Example 8. The instruction

plot 2x + 3y >= 15, 4x − 2y >= 12, y >= 0

shades the feasible set shown in Figure 8.

Check Your Understanding 3.1

Solutions can be found following the section exercises.

1. Graph the inequality $3x - y \geq 3$.

2. Graph the feasible set for the system of inequalities

$$\begin{cases} x \geq 0, \ y \geq 0 \\ x + 2y \leq \quad 4 \\ 4x - 4y \geq -4. \end{cases}$$

EXERCISES 3.1

In Exercises 1–4, state whether the inequality is true or false.

1. $2 \leq -3$ **2.** $-2 \leq 0$

3. $7 \leq 7$ **4.** $0 \geq \frac{1}{2}$

In Exercises 5–7, solve for x.

5. $2x - 5 \geq 3$ **6.** $3x - 7 \leq 2$

7. $-5x + 13 \leq -2$

8. Which of the following results from solving $-x + 1 \leq 3$ for x?
    (a) $x \leq 4$ **(b)** $x \leq 2$
    (c) $x \geq -4$ **(d)** $x \geq -2$

In Exercises 9–14, write the linear inequality in slope-intercept or vertical form.

9. $2x + y \leq 5$ **10.** $-3x + y \geq 1$

11. $5x - \frac{1}{3}y \leq 6$ **12.** $\frac{1}{2}x - y \leq -1$

13. $4x \geq -3$ **14.** $-2x \leq 4$

In Exercises 15–22, determine whether or not the given point satisfies the given inequality.

15. $3x + 5y \leq 12, (2, 1)$ **16.** $-2x + y \geq 9, (3, 15)$

17. $y \geq -2x + 7, (3, 0)$ **18.** $y \leq \frac{1}{2}x + 3, (4, 6)$

19. $y \leq 3x - 4, (3, 5)$ **20.** $y \geq x, (-3, -2)$

21. $x \geq 5, (7, -2)$ **22.** $x \leq 7, (0, 0)$

In Exercises 23–26, graph the given inequality by crossing out (i.e., discarding) the points not satisfying the inequality.

23. $y \leq \frac{1}{3}x + 1$

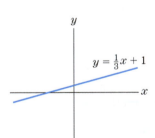

24. $y \geq -x + 1$

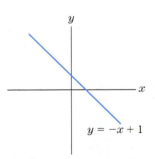

25. $x \geq 4$

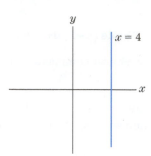

26. $y \leq 2$

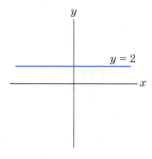

In Exercises 27–30, give the linear inequality corresponding to the graph. *Note:* The graph is the unshaded portion of the plane plus the boundary line.

27.

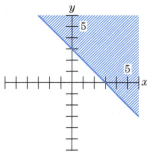

28.

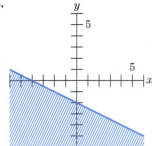

29.

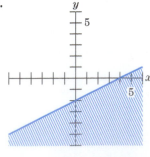

30.

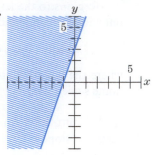

In Exercises 31–42, graph the given inequality.

31. $y \leq 2x + 1$ **32.** $y \geq -3x + 6$

33. $x \geq 2$ **34.** $x \geq 0$

35. $x + 4y \geq 12$ **36.** $4x - 4y \geq 8$

37. $4x - 5y + 25 \geq 0$ **38.** $.1y - x \geq .2$

39. $\frac{1}{2}x - \frac{1}{3}y \leq 1$ **40.** $3y + \frac{1}{2}x \leq 2y + x + 1$

41. $.5x + .4y \leq 2$ **42.** $y - 2x \geq \frac{1}{2}y - 2$

In Exercises 43–48, graph the feasible set for the system of inequalities.

43. $\begin{cases} y \leq 2x - 4 \\ y \geq 0 \end{cases}$ **44.** $\begin{cases} y \geq -\frac{1}{3}x + 1 \\ x \geq 0 \end{cases}$

45. $\begin{cases} x + 2y \geq 2 \\ 3x - y \geq 3 \end{cases}$ **46.** $\begin{cases} 3x + 6y \geq 24 \\ 3x + y \geq 6 \end{cases}$

47. $\begin{cases} x + 5y \leq 10 \\ x + y \leq 3 \\ x \geq 0, y \geq 0 \end{cases}$ **48.** $\begin{cases} x + 2y \geq 6 \\ x + y \geq 5 \\ x \geq 1 \end{cases}$

In Exercises 49–52, determine whether the given point is in the feasible set of this system of inequalities:

$$\begin{cases} 6x + 3y \leq 96 \\ x + y \leq 18 \\ 2x + 6y \leq 72 \\ x \geq 0, y \geq 0. \end{cases}$$

49. $(8, 7)$ **50.** $(14, 3)$ **51.** $(9, 10)$ **52.** $(16, 0)$

In Exercises 53–56, determine whether the given point is above or below the given line.

53. $y = 2x + 5, (3, 9)$ **54.** $3x - y = 4, (2, 3)$

55. $7 - 4x + 5y = 0, (0, 0)$ **56.** $x = 2y + 5, (6, 1)$

57. Give a system of inequalities for which the graph is the region between the pair of lines $8x - 4y - 4 = 0$ and $8x - 4y = 0$.

58. The shaded region in Fig. 9 is bounded by four straight lines. Which of the following is *not* an equation of one of the boundary lines?
(a) $y = 0$ (b) $y = 2$ (c) $x = 0$
(d) $2x + 3y = 12$ (e) $3x + 2y = 12$

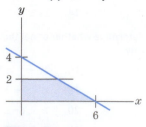

Figure 9

59. The shaded region in Fig. 10 is bounded by four straight lines. Which of the following is *not* an equation of one of the boundary lines?

(a) $x = 0$ (b) $y = x$ (c) $y = 5$

(d) $y = 0$ (e) $x + 2y = 6$

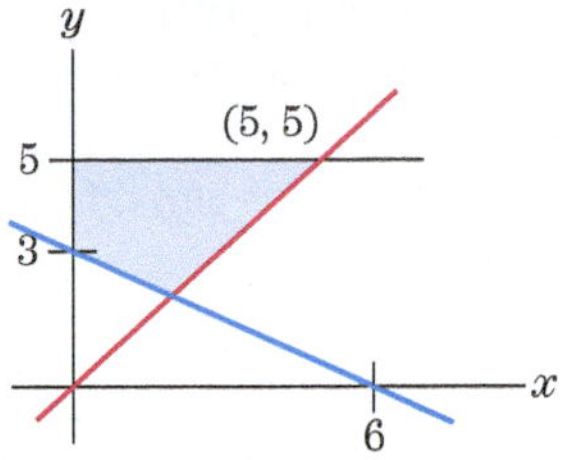

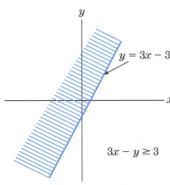

Figure 10

60. Which quadrant in Fig. 11 contains no points that satisfy the inequality $x + 2y \geq 6$?

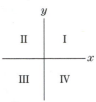

Figure 11

TECHNOLOGY EXERCISES

61. Graph the line $4x - 2y = 7$.

(a) Locate the point on the line with x-coordinate 3.6.

(b) Does the point (3.6, 3.5) lie above or below the line? Explain.

62. Graph the line $x + 2y = 11$.

(a) Locate the point on the line with x-coordinate 6.

(b) Does the point (6, 2.6) lie above or below the line? Explain.

63. Display the feasible set in Exercise 47.

64. Display the feasible set in Exercise 48.

Solutions to Check Your Understanding 3.1

1. Linear inequalities are easiest to graph if they are first written in slope-intercept or vertical form. Subtract $3x$ from both sides, and multiply by -1:

$$3x - y \geq 3$$
$$-y \geq -3x + 3$$
$$y \leq 3x - 3.$$

Now, graph the line $y = 3x - 3$ (Fig. 12). The graph of the inequality is the portion of the plane below and on the line ("$\leq$" corresponds to "below"), so throw away (that is, cross out) the portion above the line.

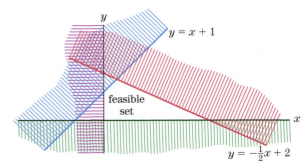

Figure 12

2. Begin by writing the linear inequalities in slope-intercept or vertical form and then graphing them all on the same coordinate system. Note that

$$\begin{cases} x \geq 0, y \geq 0 \\ x + 2y \leq 4 \\ 4x - 4y \geq -4 \end{cases}$$

becomes

$$\begin{cases} x \geq 0, y \geq 0 \\ y \leq -\frac{1}{2}x + 2 \\ y \leq x + 1. \end{cases}$$

A good procedure to follow is to graph all of the linear equations and then cross out the regions to be thrown away one at a time (Fig. 13). The inequalities $x \geq 0$ and $y \geq 0$ arise frequently in applications. The first has the form $x \geq a$, where $a = 0$, and the second has the form $y \geq mx + b$, where $m = 0$ and $b = 0$. To graph them, just cross out all points to the left of the y-axis and all points below the x-axis, respectively.

Figure 13

3.2

A Linear Programming Problem

Let us begin with a detailed discussion of a typical problem that can be solved by linear programming.

Furniture Manufacturing Problem

A furniture manufacturer makes two types of furniture—chairs and sofas. For simplicity, divide the production process into three distinct operations—carpentry, finishing, and upholstery. The amount of labor required for each operation varies. Manufacture of a chair requires 6 hours of carpentry, 1 hour of finishing, and 2 hours of upholstery. Manufacture of a sofa requires 3 hours of carpentry, 1 hour of finishing, and 6 hours of upholstery. Due to limited availability of skilled labor, as well as of tools and equipment, the factory has available each day 96 labor-hours for carpentry, 18 labor-hours for finishing, and 72 labor-hours for upholstery. The profit per chair is $80, and the profit per sofa is $70. How many chairs and how many sofas should be produced each day to maximize the profit?

It is often helpful to tabulate data given in verbal problems. Our first step, then, is to construct a chart.

	Chair	Sofa	Available Time
Carpentry	6 hours	3 hours	at most 96 labor-hours
Finishing	1 hour	1 hour	at most 18 labor-hours
Upholstery	2 hours	6 hours	at most 72 labor-hours
Profit	$80	$70	

The next step is to translate the problem into mathematical language. As you know, this is done by identifying what is unknown and denoting the unknown quantities by letters. Since the problem asks for the optimal number of chairs and sofas to be produced each day, there are two unknowns—the number of chairs produced each day and the number of sofas produced each day. Let x denote the number of chairs per day and y the number of sofas per day.

To achieve a large profit, one need only manufacture a large number of chairs and sofas. But due to restricted availability of tools and labor, the factory cannot manufacture an unlimited quantity of furniture. Let us translate the restrictions into mathematical language. Each of the first three rows of the chart gives one restriction. The first row says that the amount of carpentry required is 6 hours for each chair and 3 hours for each sofa. Also, there are available only 96 labor-hours of carpentry per day. We can compute the total number of labor-hours of carpentry required per day to produce x chairs and y sofas as follows:

[number of labor-hours per day of carpentry]

= [number of labor-hours of carpentry per chair] · [number of chairs per day]

+ [number of labor-hours of carpentry per sofa] · [number of sofas per day]

= $6 \cdot x + 3 \cdot y$.

The requirement that at most 96 labor-hours of carpentry be used per day means that x and y must satisfy the inequality

$$6x + 3y \leq 96. \tag{1}$$

The second row of the chart gives a restriction imposed by finishing. Since 1 hour of finishing is required for each chair and sofa, and since at most 18 labor-hours of finishing are available per day, the same reasoning as used to derive inequality (1) yields

$$x + y \leq 18. \tag{2}$$

Similarly, the third row of the chart gives the restriction due to upholstery:

$$2x + 6y \leq 72. \tag{3}$$

Further restrictions are given by the fact that the numbers of chairs and sofas must be nonnegative:

$$x \geq 0, \quad y \geq 0. \tag{4}$$

A restriction inequality on x and y is also called a **constraint**. Now, let us express the profit per day (which is to be maximized) in terms of x and y. The profit comes from two sources—chairs and sofas. Therefore,

$$[\text{profit}] = [\text{profit from chairs}] + [\text{profit from sofas}]$$
$$= [\text{profit per chair}] \cdot [\text{number of chairs per day}]$$
$$+ [\text{profit per sofa}] \cdot [\text{number of sofas per day}]$$
$$= 80x + 70y. \tag{5}$$

Since the objective of the problem is to optimize profit, the expression $80x + 70y$ is called the **objective function**. Combining (1) to (5), we arrive at the following:

Furniture Manufacturing Problem—Mathematical Formulation

Find numbers x and y for which the objective function $80x + 70y$ is as large as possible, and for which all the following inequalities hold simultaneously:

$$\begin{cases} 6x + 3y \le 96 \\ x + y \le 18 \\ 2x + 6y \le 72 \\ x \ge 0, \quad y \ge 0 \end{cases} \tag{6}$$

We are required to maximize an expression in a certain number of variables, where the variables are subject to restrictions in the form of one or more inequalities. Problems of this sort are called **mathematical programming problems**. Actually, general mathematical programming problems can be quite involved, and their solutions may require very sophisticated mathematical ideas. However, this is not the case with the furniture manufacturing problem. What makes it a rather simple mathematical programming problem is that both the expression to be maximized and the inequalities are linear. For this reason, the furniture manufacturing problem is called a **linear programming problem**.[1]

We will solve the furniture manufacturing problem in Section 3.3, where we develop a general technique for handling similar linear programming problems. At this point, it is worthwhile to attempt to gain some insights into the problem and possible methods for attacking it.

It seems clear that a factory will operate most efficiently when its labor is fully utilized. If no labor is to be wasted, then x and y must satisfy the system of equations

$$\begin{cases} 6x + 3y = 96 \\ x + y = 18 \\ 2x + 6y = 72 \end{cases} \tag{7}$$

Let us now graph the three equations of (7), which represent the conditions for full utilization of all forms of labor. See the following chart and Fig. 1:

Equation	Slope-Intercept Form	x-Intercept	y-Intercept
$6x + 3y = 96$	$y = -2x + 32$	$(16, 0)$	$(0, 32)$
$x + y = 18$	$y = -x + 18$	$(18, 0)$	$(0, 18)$
$2x + 6y = 72$	$y = -\frac{1}{3}x + 12$	$(36, 0)$	$(0, 12)$

[1] The theory of linear programming is a fairly recent advance in mathematics. It was developed over the last 78 years to deal with the increasingly complicated problems of our technological society. The 1975 Nobel Prize in economics was awarded to Kantorovich and Koopmans for their pioneering work in the field of linear programming.

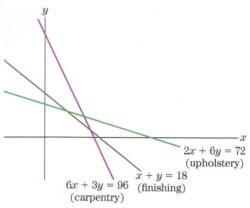

$$2x + 6y = 72$$
(upholstery)

$$x + y = 18$$
$$6x + 3y = 96 \quad \text{(finishing)}$$
(carpentry)

Figure 1

What does Fig. 1 say about the furniture manufacturing problem? Each particular pair of numbers (x, y) is called a **production schedule**. Each of the lines in Fig. 1 gives the production schedules that fully utilize one of the types of labor. Notice that the three lines do not have a common intersection point. This means that there is *no* production schedule that *simultaneously* makes full use of all three types of labor. In any production schedule, at least some of the labor-hours must be wasted. This is not a solution to the furniture manufacturing problem, but it is a valuable insight. It says that, in the inequalities of (6), not all of the corresponding equations can hold. This suggests that we take a closer look at the system of inequalities.

The simplified forms of the inequalities (6) are

$$\begin{cases} y \leq -2x + 32 \\ y \leq -x + 18 \\ y \leq -\frac{1}{3}x + 12 \\ x \geq 0, \quad y \geq 0 \end{cases}$$

By using the techniques of Section 3.1, we arrive at a feasible set for the preceding system of inequalities, as shown in Fig. 2.

The feasible set for the furniture manufacturing problem is a bounded five-sided region. The points on and inside the boundary of this feasible set give the production schedules that satisfy all of the restrictions. In Section 3.3, we show how to pick out a particular point of the feasible set that corresponds to a maximum profit.

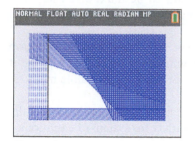

Figure 2

Check Your Understanding 3.2
Solutions can be found following the section exercises.

1. Determine whether the following points are in the feasible set of the furniture manufacturing problem:
   (a) $(10, 9)$ (b) $(14, 4)$

2. A physical fitness enthusiast decides to devote her exercise time to a combination of jogging and cycling. She wants to earn aerobic points (a measure of the benefit of the exercise to strengthening the heart and lungs) and also to achieve relaxation and enjoyment. She jogs at 6 miles per hour and cycles at 18 miles per hour. An hour of jogging earns 12 aerobic points, and an hour of cycling earns 9 aerobic points. Each week, she would like to earn at least 36 aerobic points, cover at least 54 miles, and cycle at least as many hours as she jogs.
   (a) Fill in the accompanying chart.
   (b) Let x be the number of hours of jogging and y the number of hours of cycling each week. Referring to the chart,

   give the inequalities that x and y must satisfy due to miles covered and aerobic points.
   (c) Give the inequalities that x and y must satisfy due to her preference for cycling and also due to the fact that x and y cannot be negative.
   (d) Express the time spent exercising each week as a linear function of x and y.
   (e) Graph the feasible set for the system of linear inequalities.

	One Hour of Jogging	One Hour of Cycling	Requirement
Miles covered			
Aerobic points			

EXERCISES 3.2

In Exercises 1–4, determine whether the given point is in the feasible set of the furniture manufacturing problem. The inequalities are as follows.

$$\begin{cases} 6x + 3y \le 96; & 2x + 6y \le 72 \\ x + y \le 18; & x \ge 0, y \ge 0 \end{cases}$$

1. $(8, 7)$ **2.** $(14, 3)$ **3.** $(9, 10)$ **4.** $(16, 0)$

5. Manufacturing Consider the furniture manufacturing problem discussed in this section. Suppose that the company manufactures only chairs. What is the maximum number of chairs that could be manufactured?

6. Manufacturing Consider the furniture manufacturing problem discussed in this section. Suppose that the company manufactures only sofas. What is the maximum number of sofas that could be manufactured?

7. Packaging Joe's Confectionary puts together two prepackaged assortments to be given to trick-or-treaters on Halloween. Assortment A contains 2 candy bars and 2 suckers and yields a profit of 40 cents. Assortment B contains 1 candy bar and 2 suckers and yields a profit of 30 cents. The store has available 500 candy bars and 600 suckers.
(a) Fill in the following chart:

	A	B	Available
Candy bars			
Suckers			
Profit			

(b) Let x be the number of A assortments and y be the number of B assortments. Referring to the chart, give the two inequalities that x and y must satisfy because of the availability of each confection.
(c) Give the inequalities that x and y must satisfy because x and y cannot be negative.
(d) Express the total earnings from producing x packages of assortment A and y packages of assortment B.
(e) Graph the feasible set for the packaging problem.

8. Nutrition—Animal Mr. Holloway decides to feed his pet Siberian husky two dog foods combined to create a nutritious low-sodium diet. Each can of brand A contains 3 units of protein, 1 unit of calories, and 5 units of sodium. Each can of brand B contains 1 unit of protein, 1 unit of calories, and 4 units of sodium. Mr. Holloway feels that each day, his dog should have at least 12 units of protein and 8 units of calories.
(a) Fill in the following chart:

	A	B	Requirements
Protein			
Calories			
Sodium			

(b) Let x be the number of cans of brand A and y be the number of cans of brand B. Referring to the chart, give the two inequalities that x and y must satisfy to meet the requirements for protein and calories.
(c) Give the inequalities that x and y must satisfy because x and y cannot be negative.
(d) Express the total amount of sodium in x cans of brand A and y cans of brand B.
(e) Graph the feasible set for the nutrition problem.

9. Shipping A truck traveling from New York to Baltimore is to be loaded with two types of cargo. Each crate of cargo A is 4 cubic feet in volume, weighs 100 pounds, and earns $13 for the driver. Each crate of cargo B is 3 cubic feet in volume, weighs 200 pounds, and earns $9 for the driver. The truck can carry no more than 300 cubic feet of crates and no more than 10,000 pounds. Also, the number of crates of cargo B must be less than or equal to twice the number of crates of cargo A.
(a) Fill in the following chart.

	A	B	Truck Capacity
Volume			
Weight			
Earnings			

(b) Let x be the number of crates of cargo A and y the number of crates of cargo B. Referring to the chart, give the two inequalities that x and y must satisfy because of the truck's capacity for volume and weight.
(c) Give the inequalities that x and y must satisfy because of the last sentence of the problem and also because x and y cannot be negative.
(d) Express the total earnings from carrying x crates of cargo A and y crates of cargo B.
(e) Graph the feasible set for the shipping problem.

10. Mining A coal company owns mines in two different locations. Each day, mine 1 produces 4 tons of anthracite (hard) coal, 4 tons of ordinary coal, and 7 tons of bituminous (soft) coal. Each day, mine 2 produces 10 tons of anthracite, 5 tons of ordinary coal, and 5 tons of bituminous coal. It costs the company $150 per day to operate mine 1 and $200 per day to operate mine 2. An order is received for 80 tons of anthracite, 60 tons of ordinary coal, and 75 tons of bituminous coal.
(a) Fill in the following chart:

	Mine 1	Mine 2	Ordered
Anthracite			
Ordinary			
Bituminous			
Daily cost			

(b) Let x be the number of days that mine 1 should be operated and y the number of days that mine 2 should be operated. Refer to the chart, and give three inequalities that x and y must satisfy to fill the order.
(c) Give other requirements that x and y must satisfy.
(d) Find the total cost of operating mine 1 for x days and mine 2 for y days.
(e) Graph the feasible set for the mining problem.

11. Exam Strategy A student is taking an exam consisting of 10 essay questions and 50 short-answer questions. They have 90 minutes to take the exam and know they cannot possibly answer every question. The essay questions are worth 20 points each, and the short-answer questions are worth

5 points each. An essay question takes 10 minutes to answer, and a short-answer question takes 2 minutes. The student must do at least 3 essay questions and at least 10 short-answer questions.

(a) Fill in the following chart. (*Note:* Fill in only the first entry of the last column.)

	Essay Questions	Short-Answer Questions	Available
Time to answer			
Quantity Required			
Worth			

(b) Let x be the number of essay questions to be answered and y be the number of short-answer questions to be answered. Refer to the chart, and give the inequality that x and y must satisfy due to the amount of time available.

(c) Give the inequalities that x and y must satisfy because of the numbers of each type of question and also because of the minimum number of each type of question that must be answered.

(d) Give an expression for the total score obtained from answering x essay questions and y short-answer questions.

(e) Graph the feasible set for the exam strategy problem.

12. **Political Campaign—Resource Allocation** A local politician has budgeted at most $80,000 for her media campaign. She plans to distribute these funds between TV ads and radio ads. Each one-minute TV ad is expected to be seen by 20,000 viewers, and each one-minute radio ad is expected to be heard by 4000 listeners. Each minute of TV time costs $8000, and each minute of radio time costs $2000. She has been advised to use at most 90% of her media campaign budget on television ads.

(a) Fill in the following chart.

	One-Minute TV Ads	One-Minute Radio Ads	Available
Cost			
Audience reached			

(b) Let x be the number of minutes of TV ads, and let y be the number of minutes of radio ads. Refer to the chart, and give an inequality that x and y must satisfy due to the amount of money available.

(c) Give the inequality that x must satisfy due to the limitation on the amount of money to be spent on TV ads. Also, give the inequalities that x and y must satisfy because x and y cannot be negative.

(d) Give an expression for the total audience reached by x minutes of TV ads and y minutes of radio ads.

(e) Graph the feasible set for the political campaign problem.

13. **Nutrition—Dairy Cows** A dairy farmer concludes that his small herd of cows will need at least 4550 pounds of protein in their winter feed, at least 26,880 pounds of total digestible nutrients (TDN), and at least 43,200 international units (IUs) of vitamin A. Each pound of alfalfa hay provides .13 pound of protein, .48 pound of TDN, and 2.16 IUs of vitamin A. Each pound of ground ears of corn supplies .065 pound of protein, .96 pound of TDN, and no vitamin A. Alfalfa hay costs $8 per 100-pound sack. Ground ears of corn costs $13 per 100-pound sack.

(a) Fill in the following chart:

	Alfalfa	Corn	Requirements
Protein			
TDN			
Vitamin A			
Cost/lb			

(b) Let x be the number of pounds of alfalfa hay and y be the number of pounds of ground ears of corn to be bought. Give the inequalities that x and y must satisfy.

(c) Graph the feasible set for the system of linear inequalities.

(d) Express the total cost of buying x pounds of alfalfa hay and y pounds of ground ears of corn.

14. **Manufacturing—Resource Allocation** A clothing manufacturer makes denim and hooded fleece jackets. Each denim jacket requires 2 labor-hours for cutting the pieces, 2 labor-hours for sewing, and 1 labor-hour for finishing. Each hooded fleece jacket requires 1 labor-hour for cutting, 4 labor-hours for sewing, and 1 labor-hour for finishing. There are 42 labor-hours for cutting, 90 labor-hours for sewing, and 27 labor-hours for finishing available each day. The profit is $9 per denim jacket and $5 per hooded fleece jacket.

(a) Fill in the following chart:

	Denim	Hooded Fleece	Available Hours
Cutting			
Sewing			
Finishing			
Profit			

(b) Let x be the number of denim jackets made each day. Let y be the number of hooded fleece jackets made each day. Refer to the chart, and give three inequalities that x and y must satisfy due to the available labor-hours.

(c) Give other requirements that x and y must satisfy.

(d) Find the total profit in making x denim jackets and y hooded fleece jackets.

(e) Graph the feasible set for the clothing problem.

Solutions to Check Your Understanding 3.2

1. A point is in the feasible set of a system of inequalities if it satisfies every inequality. Either the original forms or the changed forms of the inequalities can be used. The original form of the inequalities of the furniture manufacturing problem is on the right.

$$\begin{cases} 6x + 3y \le 96 \\ x + y \le 18 \\ 2x + 6y \le 72 \\ x \ge 0, \ y \ge 0 \end{cases}$$

(a) $(10, 9)$ $\begin{cases} 6(10) + 3(9) \leq 96 \\ 10 + 9 \leq 18 \\ 2(10) + 6(9) \leq 72 \\ 10 \geq 0, 9 \geq 0; \end{cases}$ $\begin{cases} 87 \leq 96 \\ 19 \leq 18 \\ 74 \leq 72 \\ 10 \geq 0, 9 \geq 0 \end{cases}$ true false false true

(b) $(14, 4)$ $\begin{cases} 6(14) + 3(4) \leq 96 \\ 14 + 4 \leq 18 \\ 2(14) + 6(4) \leq 72 \\ 14 \geq 0, 4 \geq 0; \end{cases}$ $\begin{cases} 96 \leq 96 \\ 18 \leq 18 \\ 52 \leq 72 \\ 14 \geq 0, 4 \geq 0 \end{cases}$ true true true true

Therefore, $(14, 4)$ is in the feasible set and $(10, 9)$ is not.

2. (a)

	One Hour of Jogging	One Hour of Cycling	Requirement
Miles covered	6	18	54
Aerobic points	12	9	36

(b) Miles covered: $6x + 18y \geq 54$
Aerobic points: $12x + 9y \geq 36$

(c) $y \geq x$, $x \geq 0$. It is not necessary to list $y \geq 0$, since this is automatically true if the other two inequalities hold.

(d) $x + y$. (An objective of the exercise program might be to minimize $x + y$.)

(e)

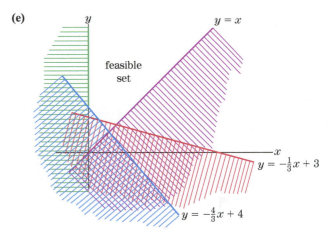

3.3 Fundamental Theorem of Linear Programming

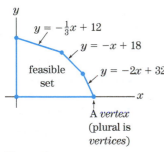

$y = -\frac{1}{3}x + 12$
$y = -x + 18$
feasible set
$y = -2x + 32$
x
A *vertex* (plural is *vertices*)

Figure 1

We have shown that the feasible set for the furniture manufacturing problem from Section 3.2 consists of the points in the interior and on the boundary of the five-sided region drawn in Fig. 1. For reference, we have labeled line segments with their equations. The line segments intersect in five points, each of which is a corner of the feasible set. Such a corner is called a **vertex**. Somehow, we must pick out of the feasible set an **optimal point**—that is, a point corresponding to a production schedule that yields a maximum profit. To assist us in this task, we have the following result:

> **Fundamental Theorem of Linear Programming** The maximum (or minimum) value of the objective function is achieved at one of the vertices of the feasible set.

This result is verified in Section 3.4. It does not completely solve the furniture manufacturing problem for us, but it comes close. It tells us that an optimal production schedule (a, b) corresponds to one of the five points labeled A–E in Fig. 2. So to complete the solution of the furniture manufacturing problem, it suffices to find the coordinates of the five points, evaluate the profit at each, and then choose the point corresponding to the maximum profit.

y
C $y = -\frac{1}{3}x + 12$
B $y = -x + 18$
feasible set A $y = -2x + 32$
D E x

Figure 2

Solution of the Furniture Manufacturing Problem

Let us begin by determining the coordinates of the points A–E in Fig. 2. Remembering that the x-axis has the equation $y = 0$ and the y-axis has the equation $x = 0$, we see from Fig. 2 that the coordinates of A–E can be found as intersections of the following lines:

$$A: \begin{cases} y = -x + 18 \\ y = -2x + 32 \end{cases}$$

$$B: \begin{cases} y = -x + 18 \\ y = -\frac{1}{3}x + 12 \end{cases}$$

$$C: \begin{cases} y = -\frac{1}{3}x + 12 \\ x = 0 \end{cases}$$

$$D: \begin{cases} y = 0 \\ x = 0 \end{cases}$$

$$E: \begin{cases} y = 0 \\ y = -2x + 32 \end{cases}$$

The point D is clearly $(0, 0)$, and C is clearly the point $(0, 12)$. We obtain A from

$-x + 18 = -2x + 32$	Equate two equations of lines through A.
$x + 18 = 32$	Add $2x$ to both sides.
$x = 14$	Subtract 18 from both sides.
$y = -x + 18 = -(14) + 18 = 4$	Substitute 14 for x in first equation.

Hence, $A = (14, 4)$. Similarly, we obtain B from

$-x + 18 = -\frac{1}{3}x + 12$	Equate two equations of lines through B.
$-\frac{2}{3}x + 18 = 12$	Add $\frac{1}{3}x$ to both sides.
$-\frac{2}{3}x = -6$	Subtract 18 from both sides.
$x = \left(-\frac{3}{2}\right)(-6) = 9$	Multiply both sides by $-\frac{3}{2}$.
$y = -x + 18 = -(9) + 18 = 9$	Substitute 9 for x in first equation.

So $B = (9, 9)$. Finally, E is obtained from

$0 = -2x + 32$	Equate two equations of lines through E.
$2x = 32$	Add $2x$ to both sides.
$x = 16$	Divide both sides by 2.
$y = 0$	$y = 0$ is a given equation.

Thus, $E = (16, 0)$. We have displayed the vertices in Fig. 3 and listed them in Table 1. In the second column, we have evaluated the profit, which is given by $80x + 70y$, at each of the vertices. Note that the largest profit occurs at the vertex $(14, 4)$, so the solution of the linear programming problem is $x = 14$, $y = 4$. In other words, the factory should produce 14 chairs and 4 sofas each day in order to achieve maximum profit, and the maximum profit is $1400 per day.

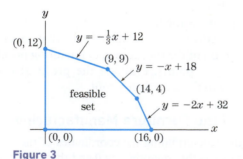

Figure 3

Table 1

Vertex	Profit $= 80x + 70y$
$(14, 4)$	$80(14) + 70(4) = 1400$
$(9, 9)$	$80(9) + 70(9) = 1350$
$(0, 12)$	$80(0) + 70(12) = 840$
$(0, 0)$	$80(0) + 70(0) = 0$
$(16, 0)$	$80(16) + 70(0) = 1280$

The furniture manufacturing problem is one particular example of a linear programming problem. Generally, such problems involve finding the values of x and y that maximize (or minimize) a particular linear expression in x and y (the objective function), where x and y are chosen so as to satisfy restrictions in the form of linear inequalities. On the basis of our experience with the furniture manufacturing problem, we can summarize the steps to be followed in approaching *any* linear programming problem.

Steps to Solve Linear Programming Problems

Step 1 Translate the problem into mathematical language.
   A. Organize the data into a chart.
   B. Identify the unknown quantities, and define corresponding variables.
   C. Translate the restrictions into linear inequalities.
   D. Form the objective function.

Step 2 Graph the feasible set.
   A. Simplify the inequalities.
   B. Graph the straight line corresponding to each inequality.
   C. Determine the side of the line belonging to the graph of each inequality. Cross out the other side. The remaining region is the feasible set.

Step 3 Determine the vertices of the feasible set.

Step 4 Evaluate the objective function at each vertex. Determine the optimal point.

Step 5 Interpret the result.

Linear programming can be used by dietitians in planning meals for large numbers of people. The object is to minimize the cost of the diet, and the restrictions reflect the minimum daily requirements of the various nutrients considered in the diet. The next example is representative of this type of problem. Whereas in actual practice, many nutritional factors are considered, we shall simplify the problem by considering only three: protein, calories, and riboflavin.

EXAMPLE 1   **Nutrition—People** Suppose that a person decides to make rice and soybeans part of their staple diet. The object is to design a lowest-cost diet that provides certain minimum levels of protein, calories, and vitamin B_2 (riboflavin). Suppose that one cup of uncooked rice costs 21 cents and contains 15 grams of protein, 810 calories, and $\frac{1}{9}$ milligram of riboflavin. On the other hand, one cup of uncooked soybeans costs 14 cents and contains 22.5 grams of protein, 270 calories, and $\frac{1}{3}$ milligram of riboflavin. Suppose that the minimum daily requirements are 90 grams of protein, 1620 calories, and 1 milligram of riboflavin. Design the lowest-cost diet meeting these specifications.

SOLUTION   **Step 1**   Translate the problem into mathematical language. The first part of this step is to organize the data, preferably into a chart (Table 2).

Table 2

	Rice	Soybeans	Required
Protein	15 grams/cup	22.5 grams/cup	90 grams per day
Calories	810 per cup	270 per cup	1620 per day
Riboflavin	$\frac{1}{9}$ milligram/cup	$\frac{1}{3}$ milligram/cup	1 milligram per day
Cost	21 cents/cup	14 cents/cup	

Now that we have organized the data, we ask for the unknowns. We wish to know how many cups each of rice and soybeans should comprise the diet, so we identify appropriate variables:

$$x = \text{number of cups of rice per day,}$$

$$y = \text{number of cups of soybeans per day.}$$

Next, we obtain the restrictions on the variables. There is one restriction corresponding to each nutrient. That is, there is one restriction for each of the first three rows of the chart. If x cups of rice and y cups of soybeans are consumed, the amount of protein is $15x + 22.5y$ grams. Thus, from the first row of the chart, $15x + 22.5y \geq 90$, a restriction expressing the fact that there must be at least 90 grams of protein per day. Similarly, the restrictions for calories and riboflavin lead to the inequalities $810x + 270y \geq 1620$ and $\frac{1}{9}x + \frac{1}{3}y \geq 1$, respectively. As in the furniture manufacturing problem, x and y cannot be negative, so there are two further restrictions: $x \geq 0$, $y \geq 0$. In all, there are five restrictions:

$$\begin{cases} 15x + 22.5y \geq 90 \\ 810x + 270y \geq 1620 \\ \frac{1}{9}x + \frac{1}{3}y \geq 1 \\ x \geq 0, \quad y \geq 0 \end{cases} \tag{1}$$

Now that we have the restrictions, we form the objective function, which tells us what we want to maximize or minimize. Since we wish to minimize cost, we express cost in terms of x and y. Now, x cups of rice cost $21x$ cents, and y cups of soybeans cost $14y$ cents, so the objective function is given by

$$[\text{cost}] = 21x + 14y. \tag{2}$$

The problem can finally be stated in mathematical form: Minimize the objective function (2) subject to the restrictions (1). This completes the first step of the solution process.

Step 2 Graph each of the inequalities (1). In Table 3 we have summarized all of the steps necessary to obtain the information from which to draw the graphs. We have sketched the graphs in Fig. 4. From Fig. 4(b), we see that the feasible set is an unbounded five-sided region. There are four vertices, two of which are known from Table 3, since they are intercepts of boundary lines. Label the remaining two vertices A and B (Fig. 5).

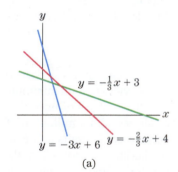

$y = -\frac{1}{3}x + 3$

$y = -3x + 6$ $y = -\frac{2}{3}x + 4$

(a)

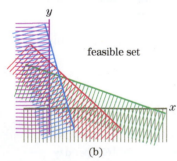

feasible set

(b)

Figure 4

Table 3

Inequality	Simplified Form	Line	Intercepts x	y	Graph
$15x + 22.5y \geq 90$	$y \geq -\frac{2}{3}x + 4$	$y = -\frac{2}{3}x + 4$	$(6,0)$	$(0,4)$	above
$810x + 270y \geq 1620$	$y \geq -3x + 6$	$y = -3x + 6$	$(2,0)$	$(0,6)$	above
$\frac{1}{9}x + \frac{1}{3}y \geq 1$	$y \geq -\frac{1}{3}x + 3$	$y = -\frac{1}{3}x + 3$	$(9,0)$	$(0,3)$	above
$x \geq 0$	$x \geq 0$	$x = 0$	$(0,0)$	—	right
$y \geq 0$	$y \geq 0$	$y = 0$	—	$(0,0)$	above

Step 3 Determine the coordinates of A and B. From Fig. 5, these coordinates can be found by solving the following systems of equations:

To find A, solve $\begin{cases} y = -3x + 6 \\ y = -\frac{2}{3}x + 4 \end{cases}$.   To find B, solve $\begin{cases} y = -\frac{2}{3}x + 4 \\ y = -\frac{1}{3}x + 3 \end{cases}$.

To solve the first system, equate the two expressions for y:

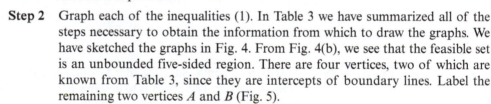

$-\frac{2}{3}x + 4 = -3x + 6$

$3x - \frac{2}{3}x = 6 - 4$ Add $3x$ and subtract 4 from both sides.

$\frac{7}{3}x = 2$ Combine terms.

$x = \frac{3}{7} \cdot 2 = \frac{6}{7}$ Multiply both sides by $\frac{3}{7}$.

$y = -3x + 6 = -3\left(\frac{6}{7}\right) + 6 = \frac{24}{7}$ Substitute $\frac{6}{7}$ for x in first equation.

y

$(0,6)$

$y = -3x + 6$

A

$y = -\frac{2}{3}x + 4$

B

$y = -\frac{1}{3}x + 3$

$(9,0)$

Figure 5

Therefore, $A = \left(\frac{6}{7}, \frac{24}{7}\right)$.

Similarly, we find B:

$-\frac{2}{3}x + 4 = -\frac{1}{3}x + 3$	Equate the two expressions for y.
$\frac{1}{3}x - \frac{2}{3}x = 3 - 4$	Add $\frac{1}{3}x$, and subtract 4 from both sides.
$-\frac{1}{3}x = -1$	Combine terms.
$x = (-3)\cdot(-1) = 3$	Multiply both sides by -3.
$y = -\frac{2}{3}x + 4 = -\frac{2}{3}(3) + 4 = 2$	Substitute 3 for x in first equation.

Therefore, $B = (3, 2)$.

Table 4

Vertex	Cost $= 21x + 14y$
$(0, 6)$	$21 \cdot 0 + 14 \cdot 6 = 84$
$\left(\frac{6}{7}, \frac{24}{7}\right)$	$21 \cdot \frac{6}{7} + 14 \cdot \frac{24}{7} = 66$
$(3, 2)$	$21 \cdot 3 + 14 \cdot 2 = 91$
$(9, 0)$	$21 \cdot 9 + 14 \cdot 0 = 189$

Step 4 Evaluate the objective function, in this case $21x + 14y$, at each vertex. From Table 4, we see that the minimum cost is achieved at the vertex $\left(\frac{6}{7}, \frac{24}{7}\right)$.

Step 5 The optimal diet—that is, the one that gives nutrients at the desired levels but at minimum cost—is the one that has $\frac{6}{7}$ cup of rice per day and $\frac{24}{7}$ cups of soybeans per day.

>> *Now Try Exercise 39*

NOTE ▷ We have assumed that all linear programming problems have solutions. Although every linear programming problem presented in this text has a solution, there are problems that have no optimal feasible solution. This can happen in two ways. First, there might be no points in the feasible set. Second, feasible solutions to the system of inequalities might exist, but the objective function might not have a maximum (or minimum) value within the feasible set. See Exercises 49 and 50. ◀

INCORPORATING
TECHNOLOGY

Graphing calculators can draw a feasible set, determine the vertices of the feasible set, and evaluate the objective function at the vertices. Figures 6 to 8 show how some of these tasks can be carried out for the furniture manufacturing problem.

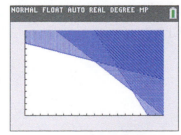

Figure 6 $[0, 18]$ *by* $[0, 14]$

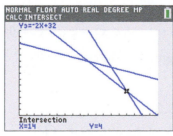

Figure 7

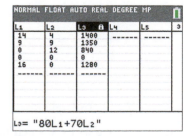

Figure 8

Solver can be used to solve linear programming problems. Each linear inequality should be written with the variables on the left and a constant on the right. For instance, the linear inequalities for the furniture manufacturing problem should be written as in system (6) on page 107. In Fig. 9 on the next page, the cells A1 and A2 have been named x and y, respectively. The figure shows the formulas that are typed into column B. The formulas entered into the first five cells of column B are the left sides of the inequalities, and the formula entered into cell B6 is the objective function.

	A	B
1		=6*x+3*y
2		=x+y
3		=2*x+6*y
4		=x
5		=y
6		=80*x+70*y

Figure 9

To obtain the solution of the linear programming problem, invoke Solver and fill in the Solver Parameters window, as shown in Fig. 10. The Max button is selected since the objective function is to be maximized. The rest of the window is completed in much the same way as when solving a system of linear equations. (For instance, to enter a constraint, click the Add button and fill in the Add Constraint dialog box that appears.) Figure 11 shows the answer generated after the Solve and then the OK buttons have been clicked.

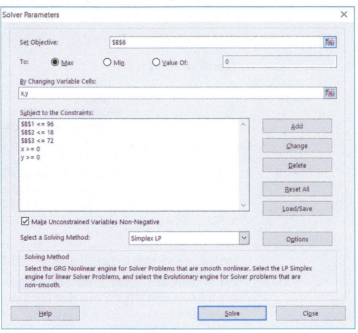

Figure 10 Solver Parameters window

	A	B
1	14	96
2	4	18
3		52
4		14
5		4
6		1400

Figure 11

✻WolframAlpha A linear programming problem can be solved directly with an instruction of the form

> **NMaximize [{objective function, 1st inequality &&**
> **2nd inequality && . . . && last inequality},{x,y}]**

or

> **NMinimize [{objective function, 1st inequality &&**
> **2nd inequality && . . . && last inequality},{x,y}]**

For instance, for the furniture manufacturing problem, the instruction

> **NMaximize [{80x + 70y,6x + 3y<= 96 && x + y<= 18 && 2x + 6y<= 72**
> **&& x>= 0 && y>= 0},{x,y}]**

produces the result "$\{1400., \{x \rightarrow 14., y \rightarrow 4.\}\}$".

Check Your Understanding 3.3

Solutions can be found following the section exercises.

1. The feasible set for the nutrition problem of Example 1 is shown in Fig. 12. The cost is $21x + 14y$. *Without* using the fundamental theorem of linear programming, explain why the cost could not possibly be minimized at the point $(4, 4)$.

2. Rework the nutrition problem, assuming that the cost of rice is changed to 7 cents per cup.

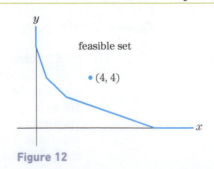

Figure 12

EXERCISES 3.3

For each of the feasible sets in Exercises 1–4, determine x and y so that the objective function $4x + 3y$ is maximized.

1.

2.

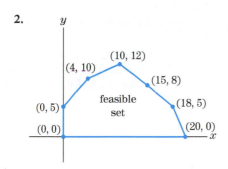

3.

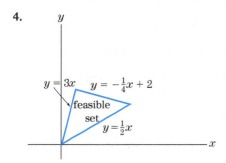

4.

In Exercises 5–8, find the values of x and y that maximize the given objective function for the feasible set in Fig. 13.

5. $x + 2y$

6. $x + y$

7. $2x + y$

8. $3 - x - y$

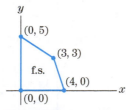

Figure 13

In Exercises 9–12, find the values of x and y that minimize the given objective function for the feasible set in Fig. 14.

9. $8x + y$

10. $3x + 2y$

11. $2x + 3y$

12. $x + 8y$

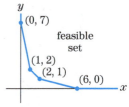

Figure 14

13. Nutrition—People Consider the nutrition problem of Example 1. Suppose that the only food available was rice. How many cups of rice would be required to meet the nutritional requirements?

14. Nutrition—People Consider the nutrition problem of Example 1. Suppose that the only food available was soybeans. How many cups of soybeans would be required to meet the nutritional requirements?

15. Packaging Refer to Exercises 3.2, Problem 7. How many of each assortment should be prepared in order to maximize profits? What is the maximum profit? (See the graph of the feasible set in Fig. 15.)

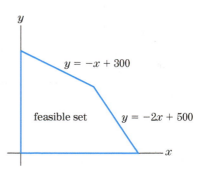

Figure 15 Packaging problem

16. Nutrition—Animal Refer to Exercises 3.2, Problem 8. How many cans of each dog food should he give to his dog each day to provide the minimum requirements with the least amount of sodium? What is the least amount of sodium? (See the graph of the feasible set in Fig. 16.)

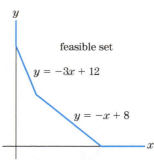

Figure 16 Animal nutrition problem

17. **Shipping** Refer to Exercises 3.2, Problem 9. How many crates of each cargo should be shipped in order to satisfy the shipping requirements and yield the greatest earnings? (See the graph of the feasible set in Fig. 17.)

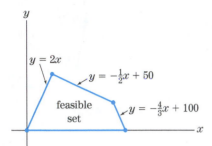

Figure 17 Shipping problem

18. **Mining** Refer to Exercises 3.2, Problem 10. Find the number of days that each mine should be operated in order to fill the order at the least cost. (See the graph of the feasible set in Fig. 18.)

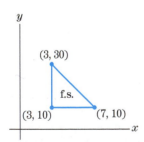

Figure 18 Mining problem

19. **Exam Strategy** Refer to Exercises 3.2, Problem 11. How many of each type of question should the student do to maximize the total score? (See the graph of the feasible set in Fig. 19.)

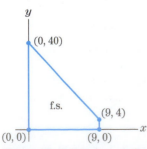

Figure 19 Exam strategy problem

20. **Political Campaign—Resource Allocation** Refer to Exercises 3.2, Problem 12. How should the media funds be allocated so as to maximize the total audience? (See the graph of the feasible set in Fig. 20.)

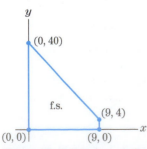

Figure 20 Political campaign problem

21. **Nutrition—Dairy Cows** Refer to Exercises 3.2, Problem 13. How many pounds of each food should be purchased in order to meet the nutritional requirements at the least cost? (See the graph of the feasible set in Fig. 21.)

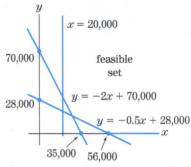

Figure 21 Animal nutrition problem

22. **Manufacturing—Resource Allocation** Refer to Exercises 3.2, Problem 14. How many of each type of jacket should be made to maximize the profit? (See the graph of the feasible set in Fig. 22.)

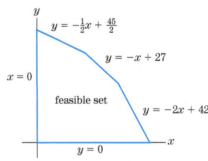

Figure 22 Manufacturing problem

Furniture Manufacturing In Exercises 23 and 24, rework the furniture manufacturing problem, where everything is the same except that the profit per chair is changed to the given value. (See Table 1 for vertices.)

23. $150   24. $60

In Exercises 25–32, find the optimal value for the linear programming problem.

25. Minimize the objective function $3x + 4y$ subject to the constraints
$$\begin{cases} 2x + y \geq 10 \\ x + 2y \geq 14 \\ x \geq 0, \quad y \geq 0 \end{cases}$$

26. Maximize the objective function $7x + 4y$ subject to the constraints
$$\begin{cases} 3x + 2y \leq 36 \\ x + 4y \leq 32 \\ x \geq 0, \quad y \geq 0 \end{cases}$$

27. Maximize the objective function $2x + 5y$ subject to the constraints
$$\begin{cases} x + 2y \leq 20 \\ 3x + 2y \geq 24 \\ x \leq 6 \\ x \geq 0, \quad y \geq 0 \end{cases}$$

28. Minimize the objective function $2x + 3y$ subject to the constraints

$$\begin{cases} x + y \geq 10 \\ y \leq x \\ x \leq 8 \end{cases}$$

29. Maximize the objective function $100x + 150y$ subject to the constraints

$$\begin{cases} x + 3y \leq 120 \\ 35x + 10y \leq 780 \\ x \leq 20 \\ x \geq 0, \quad y \geq 0 \end{cases}$$

30. Minimize the objective function $\frac{1}{2}x + \frac{3}{4}y$ subject to the constraints

$$\begin{cases} 2x + 2y \geq 8 \\ 3x + 5y \geq 16 \\ x \geq 0, \quad y \geq 0 \end{cases}$$

31. Minimize the objective function $7x + 4y$ subject to the constraints

$$\begin{cases} y \geq -2x + 11 \\ y \leq -x + 10 \\ y \leq -\frac{1}{3}x + 6 \\ y \geq -\frac{1}{4}x + 4 \end{cases}$$

32. Maximize the objective function $x + 2y$ subject to the constraints

$$\begin{cases} y \leq -x + 100 \\ y \geq \frac{1}{3}x + 20 \\ y \leq x \end{cases}$$

33. **Manufacturing—Resource Allocation** Infotron, Inc., makes electronic hockey and soccer games. Each hockey game requires 2 labor-hours of assembly and 2 labor-hours of testing. Each soccer game requires 3 labor-hours of assembly and 1 labor-hour of testing. Each day, there are 42 labor-hours available for assembly and 26 labor-hours available for testing. How many of each game should Infotron produce each day to maximize its total daily output?

34. **Manufacturing—Production Planning** An electronics company has factories in Cleveland and Toledo that manufacture Blu-ray and DVD players. Each day, the Cleveland factory produces 500 Blu-ray and 300 DVD players at a cost of $18,000. Each day, the Toledo factory produces 300 of each type of player at a cost of $15,000. An order is received for 25,000 Blu-ray and 21,000 DVD players. For how many days should each factory operate to fill the order at the least cost?

35. **Agriculture—Crop Planning** A farmer has 100 acres on which to plant oats or corn. Each acre of oats requires $18 capital and 2 hours of labor. Each acre of corn requires $36 capital and 6 hours of labor. Labor costs are $8 per hour. The farmer has $2100 available for capital and $2400 available for labor. If the revenue is $55 from each acre of oats and $125 from each acre of corn, what planting combination will produce the greatest total profit? (Profit here is revenue plus leftover capital and labor funds.) What is the maximum profit?

36. **Manufacturing—Resource Allocation** A company makes two items, I_1 and I_2, from three raw materials, M_1, M_2, and M_3.

Item I_1 uses 3 ounces of M_1, 2 ounces of M_2, and 2 ounces of M_3. Item I_2 uses 4 ounces of M_1, 1 ounce of M_2, and 3 ounces of M_3. The profit on item I_1 is $8 and on item I_2 is $6. The company has a daily supply of 40 ounces of M_1, 20 ounces of M_2, and 60 ounces of M_3.
   (a) How many of items I_1 and I_2 should be made each day to maximize profit?
   (b) What is the maximum profit?
   (c) How many ounces of each raw material are used?
   (d) If the profit on item I_1 increases to $13, how many of items I_1 and I_2 should be made each day to maximize profit?

37. **Manufacturing** The E-JEM Company produces two types of laptop computer bags. The regular version requires $32 in capital and 4 hours of labor and sells for $46. The deluxe version requires $38 in capital and 6 hours of labor and sells for $55. How many of each type of bag should the company produce in order to maximize their revenue if they have $2100 in capital and 280 labor-hours available?

38. **Refining** A refinery has two smelters that extract metallic iron from iron ore. Smelter A processes 1000 tons of iron ore per hour and uses 7 megawatts of energy per hour. Smelter B processes 2000 tons of iron ore per hour and uses 13 megawatts of energy per hour. Each refinery must be operated at least 8 hours per day and, of course, no more than 24 hours. If the refinery must process at least 30,000 tons of iron ore per day, how many hours should each smelter operate in order to expend as little energy as possible?

39. **Nutrition—People** A nutritionist, working for NASA, must meet certain nutritional requirements for astronauts and yet keep the weight of the food at a minimum. They are considering a combination of two foods, which are packaged in tubes. Each tube of food A contains 4 units of protein, 2 units of carbohydrates, and 2 units of fat, and weighs 3 pounds. Each tube of food B contains 3 units of protein, 6 units of carbohydrates, and 1 unit of fat, and weighs 2 pounds. The requirement calls for 42 units of protein, 30 units of carbohydrates, and 18 units of fat. How many tubes of each food should be supplied to the astronauts?

40. **Construction—Resource Allocation** A contractor builds two types of homes. The first type requires one lot, $12,000 capital, and 150 labor-days to build and is sold for a profit of $2400. The second type of home requires one lot, $32,000 capital, and 200 labor-days to build and is sold for a profit of $3400. The contractor owns 150 lots and has available for the job $2,880,000 capital and 24,000 labor-days. How many homes of each type will realize the greatest profit?

41. **Packaging—Product Mix** The Beautiful Day Fruit Juice Company makes two varieties of fruit drink. The given chart shows the composition and profit per can for each variety. Each week, the company has available 9000 ounces of pineapple juice, 2400 ounces of orange juice, and 1400 ounces of apricot juice. How many cans of Fruit Delight and of Heavenly Punch should be produced each week to maximize the total profit?

	Fruit Delight	Heavenly Punch
Pineapple juice	10 ounces	10 ounces
Orange juice	3 ounces	2 ounces
Apricot juice	1 ounce	2 ounces
Profit	$.40	$.60

42. Manufacturing—Resource Allocation The Bluejay Lacrosse Stick Company makes two kinds of lacrosse sticks. The accompanying chart shows the labor requirements and profits for each type of lacrosse stick. Each day, the company has available 120 labor-hours for cutting, 150 labor-hours for stringing, and 140 labor-hours for finishing. How many lacrosse sticks of each type should be manufactured each day to maximize the total profit?

	Attack	Defense
Cutting	2 labor-hours	1 labor-hour
Stringing	1 labor-hour	3 labor-hours
Finishing	2 labor-hours	2 labor-hours
Profit	$16	$20

43. Agriculture—Crop Planning Suppose that the farmer of Exercise 35 can allocate the $4500 available for capital and labor however he wants to.
   (a) Without solving the linear programming problem, explain why the optimal profit cannot be less than that found in Exercise 35.
   (b) Find the optimal solution in the new situation. Does it provide more profit than in Exercise 35?

44. Nutrition Pavan wants to add a sliced carrot and green pepper salad to his dinner each day to help him meet some of his nutritional needs. The nutritional information and costs per cup are contained in the given chart. He would like for his salad to contain at least 15 g of fiber, 100 mg of calcium, and 100 mg of Vitamin C. How many cups of sliced carrots and peppers should he use to make his salad for the least cost?

	Carrots	Green Peppers
Fiber	3.4 g	1.6 g
Calcium	40 mg	9 mg
Vitamin C	7.2 mg	74.0 mg
Cost	$.21	$.42

45. Packaging A small candy shop makes a special Cupid assortment, with 60 red pieces of candy and 40 white pieces of candy, that makes a profit of $8. A special Patriotic assortment has 30 pieces of red candy, 35 pieces of white candy, and 35 pieces of blue candy and makes a profit of $6. How many of each assortment should the shop produce to maximize profit if there are 2400 pieces of red candy, 1750 pieces of white candy, and 1470 pieces of blue candy available?

46. Packaging A portrait studio specializes in family portraits. They offer a Basic package that costs $25 to produce and an Heirloom package that costs $40 to produce. To have a successful week, the studio must sell at least 50 Basic packages at $30 each and at least 34 Heirloom packages at $75 each, with total revenue of at least $4500. How many of each package should they sell in order to minimize their costs?

47. Packaging A bath shop sells two different gift baskets. The Pamper Me basket contains 1 bottle of shower gel, 2 bottles of bubble bath, and 2 candles and makes a profit of $15. The Best Friends basket contains 2 bottles of shower gel and 2 bottles of bubble bath and makes a profit of $12. Each week, the shop has 400 bottles of shower gel, 550 bottles of bubble bath, and 400 candles available. How many of each type of basket should the shop prepare each week to maximize profits?

48. Packaging A florist offers two types of Thank You bouquets. The Thanks a Bunch bouquet consists of 3 roses, 4 carnations, and 2 stems of baby's breath and sells for $12. The Merci Beaucoup bouquet consists of 6 roses, 3 carnations, and 2 stems of baby's breath and sells for $16. The florist has 48 roses, 34 carnations, and 18 stems of baby's breath available each day for these bouquets. How many of each type of bouquet should be prepared in order to maximize the income from these bouquets?

49. Consider the following linear programming problem: Minimize $M = 10x + 6y$ subject to the constraints

$$\begin{cases} x + y \ge 6 \\ 4x + 3y \le 4 \\ x \ge 0, \quad y \ge 0 \end{cases}$$

Determine a point of the feasible set.

50. Consider the following linear programming problem: Maximize $M = 10x + 6y$ subject to the constraints

$$\begin{cases} x + y \ge 6 \\ x \ge 0, \quad y \ge 0 \end{cases}$$

   (a) Sketch the feasible set.
   (b) Determine three points in the feasible set, and calculate M at each of them.
   (c) Show that the objective function attains no maximum value for points in the feasible set.

TECHNOLOGY EXERCISES

51. Use Excel or Wolfram|Alpha to solve Exercise 25.

52. Use Excel or Wolfram|Alpha to solve Exercise 26.

Solutions to Check Your Understanding 3.3

1. The point P in Fig. 23 has a smaller value of x and a smaller value of y than $(4, 4)$ and is still in the feasible set. It therefore corresponds to a lower cost than $(4, 4)$ and still meets the requirements. We conclude that no interior point of the feasible set could possibly be an optimal point. This geometric argument indicates that an optimal point might be one that juts out far—that is, a vertex.

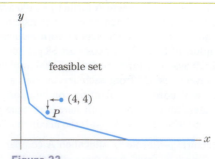

Figure 23

2. The system of linear inequalities, feasible set, and vertices will all be the same as before. Only the objective function changes. The new objective function is $7x + 14y$. The minimum cost occurs when using 3 cups of rice and 2 cups of soybeans.

Vertex	Cost $= 7x + 14y$
$(0, 6)$	84
$(\frac{6}{7}, \frac{24}{7})$	54
$(3, 2)$	49
$(9, 0)$	63

3.4 Linear Programming

In this section, we apply the technique of linear programming to the design of a portfolio for a retirement fund and to the transportation of goods from warehouses to retail outlets. The significant new feature of each of these problems is that, on the surface, they appear to involve more than two variables. However, they can be translated into mathematical language that requires only two variables.

EXAMPLE 1   **Investment Planning** A pension fund has $30 million to invest. The money is to be divided among Treasury notes, bonds, and stocks. The rules for administration of the fund require that at least $3 million be invested in each type of investment, that at least half the money be invested in Treasury notes and bonds, and that the amount invested in bonds not exceed twice the amount invested in Treasury notes. The annual yields for the various investments are 3 percent for Treasury notes, 4 percent for bonds, and 6 percent for stocks. How should the money be allocated among the various investments to produce the largest return?

SOLUTION   **Step 1**   Let all numbers stand for millions. That is, we write 30 to stand for 30 million dollars. This will save us from writing so many zeros. In examining the problem, we find that very little organization needs to be done. The rules for administration of the fund are written in a form from which inequalities can be read easily. Let us summarize the remaining data in the first row of a chart (Table 1).

Table 1

	Treasury Notes	Bonds	Stocks
Yield	.03	.04	.06
Variables	x	y	$30 - (x + y)$

There appear to be three variables—the amounts to be invested in each of the three categories. However, since the three investments must total 30, we need only two variables. Let $x =$ the amount to be invested in Treasury notes and $y =$ the amount to be invested in bonds. Then, the amount to be invested in stocks is $30 - (x + y)$. We have displayed the variables in Table 1.

Now for the restrictions. Since at least 3 (million dollars) must be invested in each category, we have the three inequalities

$$x \geq 3$$
$$y \geq 3$$
$$30 - (x + y) \geq 3.$$

Moreover, since at least half the money, or 15, must be invested in Treasury notes and bonds, we must have

$$x + y \geq 15.$$

Finally, since the amount invested in bonds must not exceed twice the amount invested in Treasury notes, we must have

$$y \leq 2x.$$

(In this example, we do not need to state that $x \geq 0$, $y \geq 0$, since we have already required that they be greater than or equal to 3.) Thus, there are five restriction inequalities:

$$\begin{cases} x \geq 3, \quad y \geq 3 \\ 30 - (x + y) \geq 3 \\ \qquad x + y \geq 15 \\ \qquad \quad y \leq 2x \end{cases} \tag{1}$$

Next, we form the objective function, which in this case equals the total return on the investment. Since x dollars are invested at 3 percent, y dollars at 4 percent, and $30 - (x + y)$ dollars at 6 percent, the total return is

$$[\text{return}] = .03x + .04y + .06[30 - (x + y)]$$
$$= .03x + .04y + 1.8 - .06x - .06y$$
$$= 1.8 - .03x - .02y. \tag{2}$$

So the mathematical statement of the problem is this: Maximize the objective function (2) subject to the restrictions (1).

Step 2   Graph the feasible set. The necessary information is compiled in Table 2.

Table 2

Inequality	Simplified Form	Line	Intercepts x	y	Graph
$x \geq 3$	$x \geq 3$	$x = 3$	$(3, 0)$	—	Right of line
$y \geq 3$	$y \geq 3$	$y = 3$	—	$(0, 3)$	Above line
$30 - (x + y) \geq 3$	$y \leq -x + 27$	$y = -x + 27$	$(27, 0)$	$(0, 27)$	Below line
$x + y \geq 15$	$y \geq -x + 15$	$y = -x + 15$	$(15, 0)$	$(0, 15)$	Above line
$y \leq 2x$	$y \leq 2x$	$y = 2x$	$(0, 0)$	$(0, 0)$	Below line

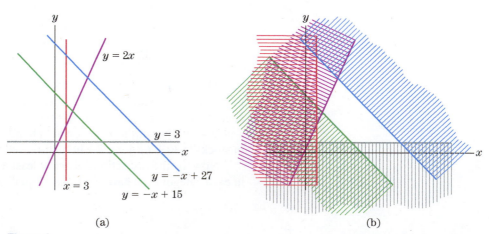

(a)   (b)

Figure 1

One point about the chart is worth noting: It contains enough data to graph each of the lines, with the exception of $y = 2x$. The reason is that the x- and y-intercepts of this line are the same, $(0, 0)$. So to graph $y = 2x$, we must find an additional point on the line. For example, if we set $x = 2$, then $y = 4$, so $(2, 4)$ is on the line. In Fig. 1(a), we have drawn the various lines, and in Fig. 1(b), we have crossed out the appropriate regions to produce the graph of the system. The feasible set, as well as the equations of the various lines that make up its boundary, are shown in Fig. 2.

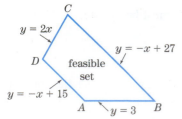

Figure 2

Step 3 From Fig. 2, we find the pairs of equations that determine each of the vertices A–D.

$$A: \begin{cases} y = 3 \\ y = -x + 15 \end{cases} \qquad B: \begin{cases} y = 3 \\ y = -x + 27 \end{cases}$$

$$C: \begin{cases} y = -x + 27 \\ y = 2x \end{cases} \qquad D: \begin{cases} y = 2x \\ y = -x + 15 \end{cases}$$

A and B are the easiest to determine. To find A, we must solve

$3 = -x + 15$	
$x = 12$	Add x and subtract 3 from both sides.
$y = 3$	Given by first equation.

Therefore, $A = (12, 3)$. Similarly, $B = (24, 3)$. To find C, we must solve

$2x = -x + 27$	
$3x = 27$	Add x to both sides.
$x = 9$	Divide both sides by 3.
$y = 2x = 2(9) = 18$	Substitute 9 for x in second equation.

Therefore, $C = (9, 18)$. Similarly, $D = (5, 10)$.

Step 4 List the four vertices and evaluate the objective function (2) at each one. The results are summarized in Table 3.

Table 3

Vertex	Return = $1.8 - .03x - .02y$
(12, 3)	$1.8 - .03(12) - .02(3) = \$1.38$ million
(24, 3)	$1.8 - .03(24) - .02(3) = \$1.02$ million
(9, 18)	$1.8 - .03(9) - .02(18) = \$1.17$ million
(5, 10)	$1.8 - .03(5) - .02(10) = \$1.45$ million

It is clear that the largest return occurs when $x = 5$ and $y = 10$.

Step 5 Five million dollars should be invested in Treasury notes, \$10 million in bonds, and $30 - (x + y) = 30 - (5 + 10) = \$15$ million in stocks.

>> Now Try Exercise 15

Linear programming is of use not only in analyzing investments but in the fields of transportation and shipping. It is often used to plan routes, determine locations of warehouses, and develop efficient procedures for getting goods to people. Many linear programming problems of this variety can be formulated as *transportation problems*. A typical transportation problem involves determining the least-cost scheme for delivering a commodity stocked in a number of different warehouses to a number of different locations—say, retail stores. Of course, in practical applications, it is necessary to consider problems involving perhaps dozens or even hundreds of warehouses and possibly just as many delivery locations. For problems on such a grand scale, the methods developed so far are inadequate. For one thing, the number of variables required is usually more than two. We must wait until Chapter 4 for methods that apply to such problems. However, the next example gives an instance of a transportation problem that does not involve too many warehouses or too many delivery points. It gives the flavor of general transportation problems.

EXAMPLE 2 **Transportation—Shipping** Suppose that a Maryland TV dealer has stores in Annapolis and Rockville and warehouses in College Park and Baltimore. The cost of shipping a TV set from College Park to Annapolis is \$6; from College Park to Rockville, \$3; from Baltimore to Annapolis, \$9; and from Baltimore to Rockville, \$5. Suppose that the Annapolis store orders 25 TV sets and the Rockville store 30. Suppose further that the College Park warehouse has a stock of 45 sets and the Baltimore warehouse has 40. What is the most economical way to supply the requested TV sets to the two stores?

SOLUTION   **Step 1**   Translate the problem into mathematical language. The first part of this step is to organize the information given, preferably in the form of a chart. In this case, since the problem is geographic, we draw a schematic diagram, as in Fig. 3, which shows the flow of goods between warehouses and retail stores. By each route, we have written the cost. Next to each warehouse, we have written its stock and below each retail store, the number of TV sets it ordered.

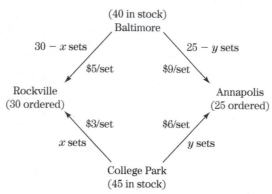

Figure 3

Next, let us determine the variables. It appears initially that four variables are required—namely, the number of TV sets to be shipped over each route. However, a closer look shows that only two variables are required. For, if x denotes the number of TV sets to be shipped from College Park to Rockville, then, since Rockville ordered 30 sets, the number shipped from Baltimore to Rockville is $30 - x$. Similarly, if y denotes the number of sets shipped from College Park to Annapolis, then the number shipped from Baltimore to Annapolis is $25 - y$. We have written the appropriate shipment sizes beside the various routes in Fig. 3.

As the third part of the translation process, let us write down the restrictions on the variables. Basically, there are two kinds of restrictions: None of x, y, $30 - x$, $25 - y$ can be negative, and a warehouse cannot ship more TV sets than it has in stock. Referring to Fig. 3, we see that College Park ships $x + y$ sets, so $x + y \leq 45$. Similarly, Baltimore ships $(30 - x) + (25 - y)$ sets, so $(30 - x) + (25 - y) \leq 40$. Simplifying this inequality, we get

$55 - x - y \leq 40$

$-x - y \leq -15$ Subtract 55 from both sides.

$x + y \geq 15$ Multiply both sides by -1. Reverse inequality sign.

The inequality $30 - x \geq 0$ can be simplified to $x \leq 30$, and the inequality $25 - y \geq 0$ can be written $y \leq 25$. So our restriction inequalities are these:

$$\begin{cases} x \geq 0, & y \geq 0 \\ x \leq 30, & y \leq 25 \\ x + y \leq 45 \\ x + y \geq 15. \end{cases} \tag{3}$$

The final step in the translation process is to form the objective function. In this problem, we are attempting to minimize cost, so the objective function must express the cost in terms of x and y. Refer again to Fig. 3. There are x sets going from College Park to Rockville, and each costs \$3 to transport, so the cost of delivering these x sets is $3x$. Similarly, the costs of making the other deliveries are $6y$, $5(30 - x)$, and $9(25 - y)$. Thus, the objective function is

$$\begin{aligned} \text{[cost]} &= 3x + 6y + 5(30 - x) + 9(25 - y) \\ &= 3x + 6y + 150 - 5x + 225 - 9y \\ &= 375 - 2x - 3y. \end{aligned} \tag{4}$$

So the mathematical problem we must solve is as follows: Find x and y that minimize the objective function (4) and satisfy the restrictions (3).

Step 2 Graph the feasible set. Four of the inequalities have graphs determined by horizontal or vertical lines. The only inequalities involving any work are $x + y \geq 15$ and $x + y \leq 45$. And even these are very easy to graph. The result is the graph in Fig. 4.

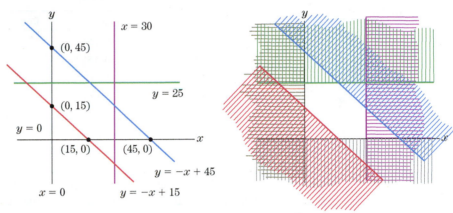

Figure 4

Step 3 In Fig. 5, we have drawn the feasible set and have labeled each boundary line with its equation. The vertices A–F are now simple to determine. First, A and F are the intercepts of the line $y = -x + 15$. Therefore, $A = (15, 0)$ and $F = (0, 15)$. Since B is the x-intercept of the line $x = 30$, we have $B = (30, 0)$. Similarly, $E = (0, 25)$. Since C is on the line $x = 30$, its x-coordinate is 30. Its y-coordinate is $y = -30 + 45 = 15$, so $C = (30, 15)$. Similarly, since D has y-coordinate 25, its x-coordinate is given by $25 = -x + 45$ or $x = 20$. Thus, $D = (20, 25)$.

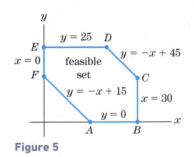

Figure 5

Table 4

Vertex	Cost = $375 - 2x - 3y$
$(0, 25)$	300
$(0, 15)$	330
$(15, 0)$	345
$(30, 0)$	315
$(30, 15)$	270
$(20, 25)$	260

Step 4 Table 4 lists the vertices A–F as well as the cost corresponding to each one. The minimum cost of \$260 occurs at the vertex $(20, 25)$. So $x = 20$, $y = 25$ yields the minimum of the objective function.

Step 5 Twenty TV sets should be shipped from College Park to Rockville, 25 from College Park to Annapolis, $30 - x = 10$ from Baltimore to Rockville, and $25 - y = 0$ from Baltimore to Annapolis. This solves our problem.

» Now Try Exercise 23

> **NOTE** ▶ The highest-cost route is the one from Baltimore to Annapolis. The solution that we have obtained eliminates any shipments over this route. One might infer from this that one should always avoid the most expensive route. But this is not correct reasoning. To see why, reconsider Example 2, except change the cost of transporting a TV set from Baltimore to Annapolis from $9 to $7. The Baltimore–Annapolis route is still the most expensive. However, in this case, the minimum cost is not obtained by eliminating the Baltimore–Annapolis route. For the revised problem, the linear inequalities stay the same. So the feasible set and the vertices remain the same. The only change is in the objective function, which now is given by

$$[\text{cost}] = 3x + 6y + 5(30 - x) + 7(25 - y) = 325 - 2x - y.$$

The costs at the various vertices are given in Table 5. So the minimum cost of $250 is achieved when $x = 30$, $y = 15$, $30 - x = 0$, and $25 - y = 10$. Note that 10 sets are being shipped from Baltimore to Annapolis even though this is the most expensive route.

Table 5

Vertex	Cost = $325 - 2x - y$
$(0, 25)$	300
$(0, 15)$	310
$(15, 0)$	295
$(30, 0)$	265
$(30, 15)$	250
$(20, 25)$	260

It is even possible for the cost function to be optimized simultaneously at two different vertices. For example, if the cost from Baltimore to Annapolis is $8 and all other data are the same as in Example 2, then the optimal cost is $260 and is achieved at both vertices $(30, 15)$ and $(20, 25)$. ◀◀

Verification of the Fundamental Theorem

The fundamental theorem of linear programming asserts that the objective function assumes its optimal value at a vertex of the feasible set. Let us verify this fact. For simplicity, we give the argument only in a special case, namely, for the furniture manufacturing problem. However, this is for convenience of exposition only. The same argument given in the example that follows may be used to prove the fundamental theorem in general. Our argument relies on the parallel property for straight lines, which asserts that parallel lines have the same slope.

The profit derived from producing x chairs and y sofas is $80x + 70y$ dollars. Let us examine all production schedules having a given profit. As an example, consider a profit of $2800. Then, x and y must satisfy $80x + 70y = 2800$. That is, (x, y) must lie on the line whose equation is $80x + 70y = 2800$, or in slope-intercept form, $y = -\frac{8}{7}x + 40$. The slope of this line is $-\frac{8}{7}$, and its y-intercept is $(0, 40)$. We have drawn this line in Fig. 6(a),

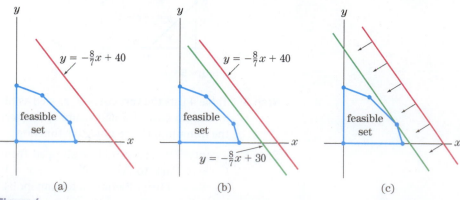

(a)   (b)   (c)

Figure 6

in which we have also drawn the feasible set for the furniture manufacturing problem. Note two fundamental facts: (1) Every production schedule on the line corresponds to a profit of $2800. (2) The line lies above the feasible set. In particular, no production schedule on the line satisfies all of the restrictions of the problem. The difficulty is that $2800 is too high a profit for us to obtain.

So now, lower the profit, say, to $2100. In this case, the production schedule (x, y) lies on the line $80x + 70y = 2100$, or in slope-intercept form, $y = -\frac{8}{7}x + 30$. This line is drawn in Fig. 6(b). Note that, since both lines have slope $-\frac{8}{7}$, they are parallel, by the parallel property. Actually, if we look at the production schedules yielding any fixed profit p, then they will lie along a line of slope $-\frac{8}{7}$, which is parallel to the two lines already drawn. For, if the production schedule (x, y) yields a profit p, then

$$80x + 70y = p \quad \text{or} \quad y = -\frac{8}{7}x + \frac{p}{70}.$$

In other words, (x, y) lies on a line of slope $-\frac{8}{7}$ and y-intercept $\left(0, \dfrac{p}{70}\right)$. In particular, all of the *lines of constant profit* are parallel to one another. So let us go back to the line of $2800 profit. It does not touch the feasible set. So now, lower the profit and therefore translate the line downward parallel to itself. Next, lower the profit until we first touch the feasible set. This line now touches the feasible set at a vertex [Fig. 6(c)]. And this vertex corresponds to the optimal production schedule, since any other point of the feasible set lies on a *line of constant profit* corresponding to an even lower profit. This shows that the fundamental theorem of linear programming is true.

Range of Optimality

The objective function for the furniture manufacturing problem is $80x + 70y$, and the optimal production schedule is $(14, 4)$. The numbers 80 and 70 in the objective function were the profits per chair and sofa, respectively. If the profit per chair is increased to $90 and all other numbers in the problem remain the same, then the optimal production schedule will still be $(14, 4)$. See Table 6. Actually, if

$$70 \leq [\text{profit per chair}] \leq 140,$$

Table 6

Vertex	Profit $= 90x + 70y$
$(14, 4)$	$90(14) + 70(4) = 1540$
$(9, 9)$	$90(9) + 70(9) = 1440$
$(0, 12)$	$90(0) + 70(12) = 840$
$(0, 0)$	$90(0) + 70(0) = 0$
$(16, 0)$	$90(16) + 70(0) = 1440$

then the production schedule $(14, 4)$ will still provide an optimal solution. The interval $[70, 140]$ is called the **range of optimality** for the profit per chair. (Exercise 27 shows how to use the concept developed above to calculate ranges of optimality.) Since $80 - 70 = 10$ and $140 - 80 = 60$, the profit per chair can be decreased by as much as $10 or increased by as much as $60 while still keeping $(14, 4)$ as an optimal production schedule. We say that the **allowable decrease** for the profit per chair is 10 and the **allowable increase** is 60. The range of optimality for the profit per sofa is $[40, 80]$. Therefore, the profit per sofa has an allowable decrease of 30 and an allowable increase of 10.

If the objective function is parallel to one of the boundary lines of the feasible set, there will be infinitely many solutions—all points on that boundary line segment provide optimal values for the objective function. At least one such point is a vertex of the feasible set.

EXAMPLE 3   **Transportation—Shipping** Reconsider the TV shipping problem discussed in Example 2 with the cost of shipping from Baltimore to Annapolis now $8 and all other data the same as in Example 2. Minimize the shipping costs.

SOLUTION   The cost function to be minimized is now

$$[\text{cost}] = 350 - 2x - 2y.$$

Note that, for any fixed value of the cost, the slope of the objective function is -1. The feasible set is identical to that of Example 2 (Fig. 5). The slope of the boundary line $y = -x + 45$ is also -1. A check of the vertices of the feasible set (Table 7) shows that the minimal cost of $260 is achieved at the two vertices $C = (30, 15)$ and $D = (20, 25)$. Figure 7 shows the feasible set and the line of constant cost $260. Every production schedule on the line segment from $(30, 15)$ to $(20, 25)$ has the minimum cost $260, and therefore is a solution to the problem.

» Now Try Exercise 17

Table 7

Vertex	Cost = $350 - 2x - 2y$
(0, 25)	300
(0, 15)	320
(15, 0)	320
(30, 0)	290
(30, 15)	260
(20, 25)	260

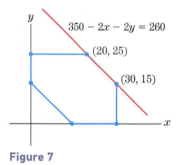

Figure 7

INCORPORATING TECHNOLOGY

Graphing calculators can draw accurate feasible sets and accurate lines of constant profit (or cost). These precise drawings often permit us to select the optimal vertex by inspection. Figure 8 shows a TI-84 Plus screen for the shipping problem of Example 2. The yellow line is the line of constant cost, $253. If the yellow line were to be lowered, the point $(20, 25)$ would be selected.

Solver can be asked to provide allowable increases and decreases for the coefficients of the objective function along with the solution. Fill in the Solver Parameter window for the furniture manufacturing problem as in Section 3.3. However, make sure that the Select a Solving Method box is set to Simplex LP. As before, click on the Solve button to invoke the Solver Results window. In this window, click on Sensitivity in the Reports list, and then click the OK button. The requested information will be contained in a table in a sheet named "Sensitivity Report 1." See Table 8. The sheet contains a second table that is discussed at the end of Section 4.4.

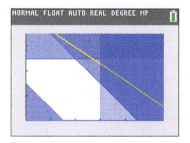

Figure 8

Table 8 **The First Table of the Sensitivity Sheet**

Cell	Name	Final Value	Reduced Cost	Objective Coefficient	Allowable Increase	Allowable Decrease
$A$1	x	14	0	80	60	10
$A$2	y	4	0	70	10	30

Check Your Understanding 3.4

Solutions can be found following the section exercises.

Problems 1–3 refer to Example 1. Translate the statement into an inequality.

1. The amount to be invested in bonds is at most $5 million more than the amount to be invested in Treasury notes.

2. No more than $25 million should be invested in stocks and bonds.

3. Rework Example 1, assuming that the yield for Treasury notes goes up to 4%.

4. A linear programming problem has objective function [cost] $= 5x + 10y$, which is to be minimized. Figure 9 shows the feasible set and the straight line of all combinations of x and y for which [cost] $= \$20$.

  (a) Give the linear equation (in slope-intercept form) of the line of constant cost c.

  (b) As c increases, does the line of constant cost c move up or down?

  (c) By inspection, find the vertex of the feasible set that gives the optimal solution.

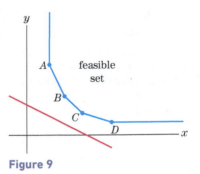

Figure 9

EXERCISES 3.4

1. Figure 10(a) shows the feasible set of the nutrition problem in Example 1 of Section 3.3 and the straight line of all combinations of rice and soybeans for which the cost is 42 cents.

  (a) The objective function is $21x + 14y$. Give the linear equation (in slope-intercept form) of the line of constant cost c.

  (b) As c increases, does the line of constant cost move up or down?

  (c) By inspection, find the vertex of the feasible set that gives the optimal solution.

2. Figure 10(b) shows the feasible set of the transportation problem of Example 2 and the straight line of all combinations of shipments for which the transportation cost is $\$240$.

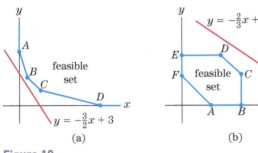

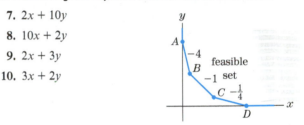

Figure 10

  (a) The objective function is [cost] $= 375 - 2x - 3y$. Give the linear equation (in slope-intercept form) of the line of constant cost c.

  (b) As c increases, does the line of constant cost move up or down?

  (c) By inspection, find the vertex of the feasible set that gives the optimal solution.

Consider the feasible set in Fig. 11, where three of the boundary lines are labeled with their slopes. In Exercises 3–6, find the point at which the given objective function has its greatest value.

3. $3x + 2y$

4. $2x + 10y$

5. $10x + 2y$

6. $2x + 3y$

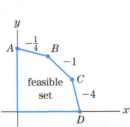

Figure 11

Consider the feasible set in Fig. 12, where three of the boundary lines are labeled with their slopes. In Exercises 7–10, find the point at which the given objective function has its least value.

7. $2x + 10y$

8. $10x + 2y$

9. $2x + 3y$

10. $3x + 2y$

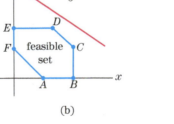

Figure 12

11. Consider the feasible set in Fig. 13. For what values of k will the objective function $x + ky$ be maximized at the vertex $(3, 4)$?

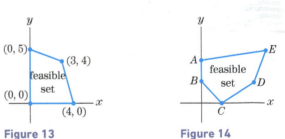

Figure 13   **Figure 14**

12. Consider the feasible set in Fig. 14. Explain why the objective function $ax + by$, with a and b positive, must have its maximum value at point E.

13. **Nutrition—Animal** Mr. Smith decides to feed his pet Doberman pinscher a combination of two dog foods. Each can of brand A contains 3 units of protein, 1 unit of carbohydrates, and 2 units of fat and costs 80 cents. Each can of brand B contains 1 unit of protein, 1 unit of carbohydrates, and 6 units of fat and costs 50 cents. Mr. Smith feels that, each day, his dog should have at least 6 units of protein, 4 units of carbohydrates, and 12 units of fat. How many cans of each dog food should he give to his dog each day to provide the minimum requirements at the least cost?

14. **Oil Production** An oil company owns two refineries. Refinery I produces each day 100 barrels of high-grade oil, 200 barrels of medium-grade oil, and 300 barrels of low-grade oil and costs $\$10,000$ to operate. Refinery II produces each day 200 barrels of high-grade, 100 barrels of medium-grade, and 200 barrels of low-grade oil and costs $\$9000$ to operate. An order is received for 1000 barrels of high-grade oil, 1000 barrels of medium-gr

oil, and 1800 barrels of low-grade oil. How many days should each refinery be operated to fill the order at the least cost?

15. **Investment Planning** Mr. Jones has $9000 to invest in three types of stocks: low-risk, medium-risk, and high-risk. He invests according to three principles. The amount invested in low-risk stocks will be at most $1000 more than the amount invested in medium-risk stocks. At least $5000 will be invested in low- and medium-risk stocks. No more than $7000 will be invested in medium- and high-risk stocks. The expected yields are 6% for low-risk stocks, 7% for medium-risk stocks, and 8% for high-risk stocks. How much money should Mr. Jones invest in each type of stock to maximize his total expected yield?

16. **Shipping—Product Mix** A produce dealer in Florida ships oranges, grapefruits, and avocados to New York by truck. Each truckload consists of 100 crates, of which at least 20 crates must be oranges, at least 10 crates must be grapefruits, at least 30 crates must be avocados, and there must be at least as many crates of oranges as grapefruits. The profit per crate is $5 for oranges, $6 for grapefruits, and $4 for avocados. How many crates of each type should be shipped to maximize the profit? (*Hint:* Let $x =$ number of crates of oranges, $y =$ number of crates of grapefruit. Then, $100 - x - y =$ number of crates of avocados.)

17. **Transportation—Shipping** A foreign-car wholesaler with warehouses in New York and Baltimore receives orders from dealers in Philadelphia and Trenton. The dealer in Philadelphia needs 4 cars, and the dealer in Trenton needs 7. The New York warehouse has 6 cars, and the Baltimore warehouse has 8. The cost of shipping cars from Baltimore to Philadelphia is $120 per car, from Baltimore to Trenton is $90 per car, from New York to Philadelphia is $100 per car, and from New York to Trenton is $70 per car. Find the number of cars to be shipped from each warehouse to each dealer to minimize the shipping cost. (*Hint:* Let $x =$ number of cars to be shipped from Baltimore to Philadelphia, $y =$ number of cars to be shipped from Baltimore to Trenton, $4 - x =$ number of cars to be shipped from New York to Philadelphia, and $7 - y =$ number of cars to be shipped from New York to Trenton.)

18. **Transportation—Shipping** Consider the foreign-car wholesaler discussed in Exercise 17. Suppose that the cost of shipping a car from New York to Philadelphia is increased to $110 and all other costs remain the same. How many cars should be shipped from each warehouse to each dealer in order to minimize the cost?

19. **Manufacturing—Production Planning** An oil refinery produces gasoline, jet fuel, and diesel fuel. The profits per gallon from the sale of these fuels are $.15, $.12, and $.10, respectively. The refinery has a contract with an airline to deliver a minimum of 20,000 gallons per day of jet fuel and/or gasoline. It has a contract with a trucking firm to deliver a minimum of 50,000 gallons per day of diesel fuel and/or gasoline. The refinery can produce 100,000 gallons of fuel per day, distributed among the fuels in any fashion. It wishes to produce at least 5000 gallons per day of each fuel. How many gallons of each should be produced to maximize the daily profit?

20. **Manufacturing—Production Planning** Suppose that a price war reduces the profits of gasoline in Exercise 19 to $.05 per gallon and that the profits on jet fuel and diesel fuel are unchanged. How many gallons of each fuel should now be produced to maximize the profit?

21. **Shipping—Resource Allocation** A shipping company is buying new trucks. The high-capacity trucks cost $50,000 and hold 320 cases of merchandise. The low-capacity trucks cost $30,000 and hold 200 cases of merchandise. The company has budgeted $1,080,000 for the new trucks and has a maximum of 30 people qualified to drive the trucks. Due to availability limitations, the company can purchase at most 15 high-capacity trucks. How many of each type of truck should the company purchase to maximize the number of cases of merchandise that can be shipped simultaneously?

22. **Shipping—Resource Allocation** Suppose that the shipping company of Exercise 21 needs to buy enough new trucks to be able to ship 11,200 cases of merchandise. Of course, the company is willing to increase its budget.
    (a) How many of each type of truck should the company purchase to minimize cost?
    (b) What if the company hires 23 additional qualified drivers?

23. **Transportation—Shipping** A major coffee supplier has warehouses in Seattle and San José. The coffee supplier receives orders from coffee retailers in Salt Lake City and Reno. The retailer in Salt Lake City needs 400 pounds of coffee, and the retailer in Reno needs 350 pounds of coffee. The Seattle warehouse has 700 pounds available, and the warehouse in San José has 500 pounds available. The cost of shipping from Seattle to Salt Lake City is $2.50 per pound, from Seattle to Reno $3.50 per pound, from San José to Salt Lake City $2 per pound, and from San José to Reno $2.50 per pound. Find the number of pounds to be shipped from each warehouse to each retailer to minimize the cost.

24. **Transportation—Shipping** Consider the coffee supplier discussed in Exercise 23. Suppose that the cost of shipping from San José to Reno is increased to $3.00 per pound and all other costs remain the same. How many pounds of coffee should be shipped from each warehouse to each retailer to minimize the cost?

25. **Packaging—Product Mix** A pet store sells three different starter kits for 10-gallon aquariums. The accompanying chart shows the contents of each kit. The store has 54 filters, 100 pounds of gravel, and 53 packages of fish food available. If the store makes as many of kit I as kits II and III together, how many of each kit should be created to maximize profit?

	Kit I	Kit II	Kit III
Filters	1	2	1
Gravel (pounds)	2	2	3
Fish food (packages)	1	0	2
Profit	$7	$10	$13

26. **Manufacturing—Production Planning** An automobile manufacturer has assembly plants in Detroit and Cleveland, each of which can assemble cars and trucks. The Detroit plant can assemble at most 800 vehicles in one day at a cost of $1200 per car and $2100 per truck. The Cleveland plant can assemble at most 500 vehicles in one day at a cost of $1000 per car and $2000 per truck. A rush order is received for 600 cars and 300 trucks. How many vehicles of each type should each plant produce to fill the order at the least cost? (*Hint:* Let $x =$ number of cars to be produced in Detroit, $y =$ number of trucks to be produced in Detroit, $600 - x =$ number of cars to be produced in Cleveland, and $300 - y =$ number of trucks to be produced in Cleveland.)

27. Refer to Fig. 6. As the lines of constant profit were lowered, the final line had slope $-\frac{8}{7}$ and contained the optimal vertex of the feasible set. Figure 15 shows that, as long as the slope of the final line is between -2 and -1, the optimal solution would still be at the same vertex.

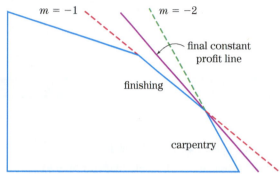

Figure 15

(a) Suppose that the profit per chair is changed from \$80 to \$ A, but that the profit per sofa remains at \$70. Show that the lines of constant profit will each have slope $-\frac{A}{70}$.
(b) Show that, when $-2 \leq -\frac{A}{70} \leq -1$, then $70 \leq A \leq 140$. Conclude that the interval $[70, 140]$ is the range of optimality for the profit per chair.
(c) Use similar reasoning to show that the range of optimality for the profit per sofa is $[40, 80]$.

28. Figure 16 shows the feasible set for the nutrition problem discussed in Example 1 of Section 3.3, along with the purple line of final cost passing through the optimal point. The figure shows that, as long as the slope of the final cost line is between -3 and $-\frac{2}{3}$, the optimal solution will still be at the same vertex.
(a) Suppose that the cost of a cup of rice is changed from 21¢ to A¢ but that the cost of a cup of soybeans remains at 14¢. Show that the lines of constant cost will each have slope $-\frac{A}{14}$.
(b) Show that, when $-3 \leq -\frac{A}{14} \leq -\frac{2}{3}$, then $\frac{28}{3} \leq A \leq 42$. Conclude that the interval $\left[\frac{28}{3}, 42\right]$ is the range of optimality for the cost of a cup of rice.
(c) Use similar reasoning to show that the range of optimality for the cost of a cup of soybeans is $[7, 31.5]$.

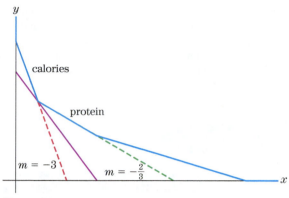

Figure 16

Consider the feasible set in Fig. 17(a). In Exercises 29–32, find an objective function of the form $ax + by$ that has its greatest value at the given point.

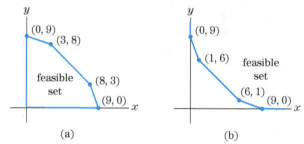

Figure 17

29. $(9, 0)$ **30.** $(3, 8)$ **31.** $(8, 3)$ **32.** $(0, 9)$

Consider the feasible set in Fig. 17(b). In Exercises 33–36, find an objective function of the form $ax + by$ that has its least value at the given point.

33. $(9, 0)$ **34.** $(1, 6)$ **35.** $(6, 1)$ **36.** $(0, 9)$

TECHNOLOGY EXERCISES

37. Create a sensitivity report for the nutrition problem of Example 1 of Section 3.3. Use the report to determine the range of optimality for the cost of rice and for the cost of soybeans.

Solutions to Check Your Understanding 3.4

1. Amount invested in bonds is y. Five million dollars more than the amount invested in Treasury notes is $x + 5$. Therefore, $y \leq x + 5$.

2. Amount invested in stocks $= 30 - (x + y)$.
   Amount invested in bonds $= y$.
   Therefore,
   $$30 - (x + y) + y \leq 25$$
   $$30 - x \leq 25$$
   $$x \geq 5.$$

3. The feasible set stays the same, but the return becomes
   $$[\text{return}] = .04x + .04y + .06[30 - (x + y)]$$
   $$= .04x + .04y + 1.8 - .06x - .06y$$
   $$= 1.8 - .02x - .02y.$$

When the return is evaluated at each of the vertices of the feasible set, the greatest return is achieved at two vertices. Either of these vertices yields an optimal solution.

Vertex	Return $= 1.8 - .02x - .02y$
$(12, 3)$	1.50
$(24, 3)$	1.26
$(9, 18)$	1.26
$(5, 10)$	1.50

4. (a) The values of x and y for which the cost is c dollars satisfy $5x + 10y = c$. The slope-intercept form of this linear equation is $y = -\frac{1}{2}x + \frac{c}{10}$.

(b) The line $y = -\frac{1}{2}x + \frac{c}{10}$ has slope $-\frac{1}{2}$ and y-intercept $(0, \frac{c}{10})$. As c increases, the slope stays the same, but the y-intercept moves up. Therefore, the line moves up.

(c) The line of constant cost \$20 does not contain any points of the feasible set, so such a low cost cannot be achieved.

Increase the cost until the line of constant cost just touches the feasible set. As c increases, the line moves up (keeping the same slope) and first touches the feasible set at vertex C. Therefore, taking x and y to be the coordinates of C yields the minimum cost.

CHAPTER 3   Summary

KEY TERMS AND CONCEPTS	EXAMPLES

3.1 Linear Inequalities

When solving a linear inequality, the direction of the **inequality sign** is unchanged when a number is added to or subtracted from both sides of an inequality, or when both sides are multiplied or divided by the same positive number. The direction is reversed when both sides are multiplied or divided by the same negative number.

The graph of the solution of a **linear inequality** is a half-plane.

The graph of the solution of a system of linear equations is called the **feasible set** of the solution.

$2 \le 3$

$8 \le 12$ (**multiply by 4**); $6 \le 7$ (**add 4**)

$-2 \le -1$ (**subtract 4**); $-8 \ge -12$ (**multiply by -4**)

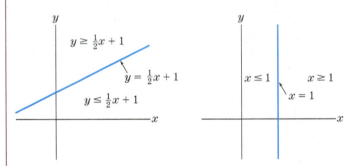

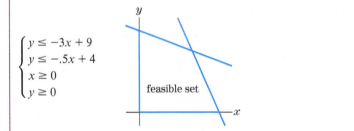

$$\begin{cases} y \le -3x + 9 \\ y \le -.5x + 4 \\ x \ge 0 \\ y \ge 0 \end{cases}$$

3.2 A Linear Programming Problem

A linear programming problem asks us to maximize or minimize the value of a linear function subject to restrictions given by linear inequalities.

First Steps for Solving a Linear Programming Problem:

1. Translate the problem into mathematical language.
2. Graph the feasible set.

Problem: Each unit of food A contains 120 milligrams of sodium, 1 gram of fat, and 5 grams of protein. Each unit of food B contains 60 milligrams of sodium, 1 gram of fat, and 4 grams of protein. Suppose that a meal consisting of these two foods is required to have at most 480 milligrams of sodium and at most 6 grams of fat. Find the combination of these two foods that meets the requirements and has the greatest amount of protein.

Solution:

1. Let x be the number of units of food A and y the number of units of food B. The restrictions for x and y are

$$\begin{cases} 120x + 60y \le 480 \text{ (sodium requirement)} \\ x + y \le 6 \qquad \text{(fat requirement)} \\ x \ge 0, y \ge 0 \end{cases}$$

KEY TERMS AND CONCEPTS	EXAMPLES

EXAMPLES

2. The amount of protein is $5x + 4y$ grams.

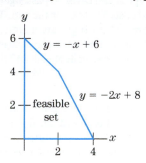

3.3 Fundamental Theorem of Linear Programming

Fundamental Theorem of Linear Programming: The optimal value of the objective function for a linear programming problem occurs at a vertex of the feasible set.

Final Steps for Solving a Linear Programming Problem:
3. Determine the vertices of the feasible set.
4. Evaluate the objective function at each vertex. Determine the optimal point.
5. Interpret the result.

Complete the solution to the previous problem.

3.

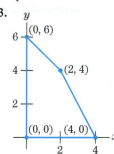

4.

Vertex	$5x + 4y$
(0, 6)	24
(2, 4)	26
(4, 0)	20
(0, 0)	0

5. Use 2 units of food A and 4 units of food B to obtain 26 grams of protein.

3.4 Linear Programming

Some linear programming problems have infinitely many solutions.

Consider the previous linear programming problem, but assume that both food A and food B have 4 grams of protein each. The new table for Step (4) is shown. The coordinates of the vertices (0, 6) and (2, 4) are both optimal solutions, as well as the coordinates of any point on the line connecting them.

Vertex	Protein = $4x + 4y$
(0, 6)	24
(2, 4)	24
(4, 0)	16
(0, 0)	0

CHAPTER 3   Fundamental Concept Check Exercises

1. State the inequality properties for addition, subtraction, and multiplication.

2. What are the general forms of a linear inequality in x and y?

3. Explain how to obtain the graph of a linear inequality.

4. What is meant by the *feasible set of a system of linear inequalities*?

5. What is the nature of a linear programming problem?

6. What is the role of the objective function in a linear programming problem?

7. What is the feasible set of a linear programming problem?

8. State the fundamental theorem of linear programming.

9. Give a procedure for solving a linear programming problem.

CHAPTER 3   Review Exercises

1. Does the point $(1, 2)$ satisfy the linear inequality $3x + 4y \geq 11$?

2. Graph the linear inequality $x - 3y \geq 12$.

3. Write the inequality whose graph is the half-plane above and on the line through $(2, -1)$ and $(6, 8.6)$.

4. **Travel—Resource Allocation** Terrapin Airlines wants to fly 1400 members of a ski club to Colorado. The airline owns two types of planes. Type A can carry 50 passengers, requires 3 flight attendants, and costs $14,000 for the trip. Type B can carry 300 passengers, requires 4 flight attendants, and costs $90,000 for the trip. If the airline must use at least as many type A planes as type B and has available only 42 flight attendants, how many planes of each type should be used to minimize the cost for the trip?

5. **Nutrition—People** A nutritionist is designing a new breakfast cereal, using wheat germ and enriched oat flour as the basic ingredients. The given chart shows the nutritional information and costs per ounce. The nutritionist wants each serving of the cereal to have at least 7 milligrams of niacin, 9 milligrams of iron, and 1 milligram of thiamin. How many ounces of wheat germ and how many ounces of enriched oat flour should be used in each serving to meet the nutritional requirements at the least cost?

	Wheat Germ	Oat Flour
Niacin	2 mg	3 mg
Iron	3 mg	3 mg
Thiamin	.5 mg	.25 mg
Cost	$.06	$.08

6. **Manufacturing—Resource Allocation** An automobile manufacturer makes hardtops and sports cars. The given chart shows the labor requirements and profits per automobile. During each day, 360 labor-hours are available to assemble, 50 labor-hours to paint, and 40 labor-hours to upholster automobiles. How many hardtops and sports cars should be produced each day to maximize the total profit?

	Hardtops	Sports Cars
Assemble	8 labor-hours	18 labor-hours
Paint	2 labor-hours	2 labor-hours
Upholster	2 labor-hours	1 labor-hour
Profit	$90	$100

7. **Packaging—Product Mix** A confectioner makes two raisin–nut mixtures. A box of mixture A contains 6 ounces of peanuts, 1 ounce of raisins, and 4 ounces of cashews and sells for $4.25. A box of mixture B contains 12 ounces of peanuts, 3 ounces of raisins, and 2 ounces of cashews and sells for $6.55. They have available 5400 ounces of peanuts, 1200 ounces of raisins, and 2400 ounces of cashews. How many boxes of each mixture should they make to maximize revenue?

8. **Publishing—Product Mix** A textbook publisher puts out 72 new books each year, which are classified as elementary, intermediate, and advanced. The company's policy for new books is to publish at least four advanced books, at least three times as many elementary books as intermediate books, and at least twice as many intermediate books as advanced books. On the average, the annual profits are $8000 for each elementary book, $7000 for each intermediate book, and $1000 for each advanced book. How many new books of each type should be published to maximize the annual profit while conforming to company policy?

9. **Packaging—Resource Allocation** A computer company has two manufacturing plants, one in Rochester and one in Queens. Packaging a computer in Rochester takes 1.5 hours and costs $15, while packaging a computer in Queens takes 2 hours and costs $30. The profit on each computer manufactured in Rochester is $40, and the profit on each computer manufactured in Queens is $30. The Rochester plant has 80 computers available, and the Queens plant has 120 computers available. If there are 210 hours and $3000 allotted for packaging the computers, how many computers should be packaged at each of the two plants to maximize the company's profits?

10. **Transportation—Shipping** An appliance company has two warehouses and two retail outlets. Warehouse A has 400 refrigerators, and warehouse B has 300 refrigerators. Outlet I needs 200 refrigerators, and outlet II needs 300 refrigerators. It costs $36 to ship a refrigerator from warehouse A to outlet I and $30 to ship a refrigerator from warehouse A to outlet II. It costs $30 to ship a refrigerator from warehouse B to outlet I and $25 to ship a refrigerator from warehouse B to outlet II. How should the company ship the refrigerators to minimize the cost?

11. **Investment Planning** Portia has $10,000 to invest. She is considering a certificate of deposit (CD) that is expected to yield 5%, a mutual fund expected to yield 7%, and stocks expected to yield 9%. The amount invested in the mutual fund can be no more than the amount invested in the CD and stocks together. The amount in the mutual fund and stocks must be no more than $8000. How much should Portia invest in each investment vehicle to maximize her total expected yield?

Conceptual Exercises

12. Suppose that a constraint is added to a cost minimization problem. Is it possible for the new optimal cost to be greater than the original optimal cost? Is it possible for the new optimal cost to be less than the original optimal cost?

13. Suppose that a constraint is removed from a profit maximization problem. Is it possible for the new optimal profit to be greater than the original optimal profit? Is it possible for the new optimal profit to be less than the original optimal profit?

14. Make up a linear programming problem for which the objective function assumes neither a minimum nor a maximum value.

15. Explain why a linear programming problem will always have a solution if the feasible set is bounded.

16. Suppose that the maximum value of an objective function occurs at two vertices. Explain why every point on the line segment between the two vertices yields a maximum value of the objective function.

Shadow Prices

When mathematicians are presented with a linear programming problem, they will not only determine the optimal solution but will also supply what are called **shadow prices** for each resource. This chapter project develops the concept of a shadow price.

Consider the furniture manufacturing problem. The constraint for finishing is $x + y \leq 18$. The number 18 came from the fact that 18 hours are available for finishing each day. Suppose that you could increase the number of hours available for finishing by one hour. The shadow price for the finishing resource is the maximum price that you would be willing to pay for that additional hour.

1. What is the new constraint for the finishing resource?

2. The figure shows the graph of the original feasible set for the furniture manufacturing problem drawn with a red boundary. The blue line segments show the change in the feasible set when the new finishing constraint is used. Find the coordinates of points *A* and *B*.

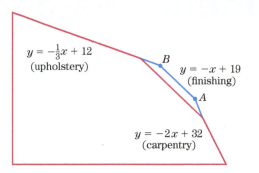

$y = -\frac{1}{3}x + 12$
(upholstery)

B

$y = -x + 19$
(finishing)

A

$y = -2x + 32$
(carpentry)

3. Determine the optimal solution for the revised linear programming problem. What is the new maximum profit? By how much was the profit increased due to the additional hour for finishing? This amount is the shadow price.

4. Return to the original furniture manufacturing problem, and assume that one additional hour is available for carpentry. Solve the altered problem, and determine the shadow price for the carpentry resource.

5. Use your knowledge of the solution of the original furniture manufacturing problem and the definition of shadow price to explain why the shadow price for upholstery is 0. (*Hint:* No computation is necessary.)

6. Fill in the blanks in the following sentence: The shadow price associated with a resource can be interpreted as the change in value of the _____ _____ per unit change of the availability of the resource.

The Simplex Method

In Chapter 3, we introduced a graphical method for solving linear programming problems. This method, although very simple, applies only to problems that involve (or can be reduced to) two variables. On the other hand, linear programming applications in business and economics can involve dozens or even hundreds of variables. In this chapter, we describe a method for handling such applications. This method, called the **simplex method**, was developed by the mathematician George B. Dantzig in the late 1940s and today is the principal method used in solving complex linear programming problems. The simplex method can be used for problems in any number of variables and is easily adapted to computer calculations.

4.1   Slack Variables and the Simplex Tableau

In this section and the next, we explain how the simplex method can be used to solve linear programming problems. Let us reconsider the furniture manufacturing problem of Chapter 3. You may recall that the problem is to determine the number of chairs and the number of sofas that should be produced each day in order to maximize the profit. The requirements and availability of resources for carpentry, finishing, and upholstery determine the constraints on the production schedule. Thus, we try to find numbers x and y for which

$$80x + 70y$$

is as large as possible subject to the constraints

$$\begin{cases} 6x + 3y \leq 96 \\ x + y \leq 18 \\ 2x + 6y \leq 72 \\ x \geq 0, \quad y \geq 0. \end{cases}$$

Here, x is the number of chairs to be produced each day and y is the number of sofas to be produced each day.

This problem exhibits certain features that make it particularly convenient to work with

1. The objective function is to be maximized.
2. Each variable is constrained to be greater than or equal to 0.
3. All other constraints are of the form

$$[\text{linear expression}] \leq [\text{nonnegative constant}].$$

A linear programming problem satisfying these conditions is said to be in **standard maximization form**. Our initial discussion of the simplex method will involve only such problems. Then, in Section 4.3, we will consider problems in nonstandard form.

The essential feature of the simplex method is that it provides a systematic method of testing selected vertices of the feasible set until an optimal vertex is reached. The method usually begins at the origin, if it is in the feasible set, and then considers the adjacent vertex that most improves the value of the objective function. This process continues until the optimal vertex is found.

The first step of the simplex method is to convert the given linear programming problem into a system of linear equations. To see how this is done, consider the furniture manufacturing problem. It specifies that the variables x and y are subject to the constraint

$$6x + 3y \leq 96.$$

Let us introduce another variable, u, which turns the inequality into an equation:

$$6x + 3y + u = 96.$$

The variable u "takes up the slack" between $6x + 3y$ and 96 and is therefore called a **slack variable**. Moreover, since $6x + 3y$ is at most 96, the variable u must be greater than or equal to 0. In a similar way, the constraint

$$x + y \leq 18$$

can be turned into the equation

$$x + y + v = 18,$$

where v is a slack variable and $v \geq 0$. The third constraint,

$$2x + 6y \leq 72,$$

becomes the equation

$$2x + 6y + w = 72,$$

where w is also a slack variable and $w \geq 0$. Let us even turn our objective function $80x + 70y$ into an equation by introducing the new variable M defined by $M = 80x + 70y$. Then M is the variable that we want to maximize. Moreover, it satisfies the equation

$$-80x - 70y + M = 0.$$

Thus, the furniture manufacturing problem can be restated in terms of a system of linear equations.

Furniture Manufacturing Problem

Among all of the solutions of the system of linear equations

$$
\begin{cases}
6x + 3y + u & = 96 \\
x + y + v & = 18 \\
2x + 6y + w & = 72 \\
-80x - 70y + M = & 0,
\end{cases}
$$

find one for which $x \geq 0$, $y \geq 0$, $u \geq 0$, $v \geq 0$, $w \geq 0$ and for which M is as large as possible.

In a similar way, any linear programming problem in standard maximization form can be reduced to that of determining a certain type of solution of a system of linear equations.

EXAMPLE 1 **Using Linear Equations in a Linear Programming Problem** Formulate the following linear programming problem in terms of a system of linear equations:

Maximize the objective function $3x + 4y$ subject to the constraints

$$
\begin{cases}
x + y \leq 20 \\
x + 2y \leq 25 \\
x \geq 0 \\
y \geq 0.
\end{cases}
$$

SOLUTION The two constraints $x + y \leq 20$ and $x + 2y \leq 25$ yield the equations

$$
\begin{aligned}
x + y + u & = 20 \\
x + 2y + v & = 25,
\end{aligned}
$$

where u and v are slack variables and $u \geq 0$ and $v \geq 0$. The objective function gives the equation $M = 3x + 4y$, or

$$-3x - 4y + M = 0.$$

So the problem can be reformulated: Among all of the solutions of the system of linear equations

$$
\begin{cases}
x + y + u & = 20 \\
x + 2y + v & = 25 \\
-3x - 4y + M = & 0,
\end{cases}
$$

find one for which $x \geq 0$, $y \geq 0$, $u \geq 0$, $v \geq 0$, and M is as large as possible.

>> *Now Try Exercise 1*

EXAMPLE 2 **Using Linear Equations in a Linear Programming Problem** Formulate the following linear programming problem in terms of a system of linear equations:

Maximize the objective function $x + 2y + z$ subject to the constraints

$$
\begin{cases}
x - y + 2z \leq 10 \\
2x + y + 3z \leq 12 \\
x \geq 0, \quad y \geq 0, \quad z \geq 0.
\end{cases}
$$

SOLUTION The two constraints $x - y + 2z \leq 10$ and $2x + y + 3z \leq 12$ yield the equations

$$
\begin{aligned}
x - y + 2z + u & = 10 \\
2x + y + 3z + v & = 12.
\end{aligned}
$$

The objective function yields the equation $M = x + 2y + z$; that is,

$$-x - 2y - z + M = 0.$$

So the problem can be reformulated: Among all solutions of the system of linear equations

$$\begin{cases} x - y + 2z + u & = 10 \\ 2x + y + 3z & + v & = 12 \\ -x - 2y - z & + M = 0, \end{cases}$$

find one for which $x \geq 0$, $y \geq 0$, $z \geq 0$, $u \geq 0$, $v \geq 0$, and M is as large as possible.

> **>> Now Try Exercise 3**

We shall now discuss a scheme for solving systems of equations like those just encountered. For the moment, we will not worry about maximizing M or keeping the variables ≥ 0. Rather, let us concentrate on a particular method for determining solutions. In order to be concrete, consider the system of linear equations from the furniture manufacturing problem:

$$\begin{cases} 6x + 3y + u & = 96 \\ x + y & + v & = 18 \\ 2x + 6y & + w & = 72 \\ -80x - 70y & + M = 0 \end{cases} \tag{1}$$

This system of equations has an infinite number of solutions. We can rewrite the equations as

$$\begin{aligned} u &= 96 - 6x - 3y \\ v &= 18 - x - y \\ w &= 72 - 2x - 6y \\ M &= 80x + 70y. \end{aligned}$$

Given any values of x and y, we can determine corresponding values for u, v, w, and M. For example, if $x = 0$ and $y = 0$, then $u = 96$, $v = 18$, $w = 72$, and $M = 0$. These values for u, v, w, and M are precisely the numbers that appear to the right of the equality signs in our original system of linear equations. Therefore, this particular solution could have been read directly from system (1) without any computation. This method of generating solutions is used in the simplex method, so let us further explore the special properties of the system that allowed us to read off a specific solution so easily.

Note that the system of linear equations has six variables: x, y, u, v, w, and M. These variables can be divided into two groups. **Group I** consists of those variables that were set equal to 0, namely, x and y. **Group II** consists of those variables whose particular values were read from the right-hand sides of the equations, namely, u, v, w, and M. Note also that the system has a special form that allows the particular values of the group II variables to be read off: Each of the equations involves exactly one of the group II variables, and these variables always appear with coefficient 1. Thus, for example, the first equation involves the group II variable u:

$$6x + 3y + u = 96.$$

Therefore, when all group I variables (x and y) are set equal to 0, only the u-term remains on the left and the particular value of u can be read from the right-hand side.

The special form of the system can best be described in matrix form. Write the system in the usual way as a matrix, but add column headings corresponding to the variables:

x	y	u	v	w	M	
6	3	1	0	0	0	96
1	1	0	1	0	0	18
2	6	0	0	1	0	72
−80	−70	0	0	0	1	0

Note closely the columns corresponding to the group II variables u, v, w, and M:

$$\begin{array}{ccccccc} x & y & u & v & w & M & \\ \begin{bmatrix} 6 & 3 & 1 & 0 & 0 & 0 & 96 \\ 1 & 1 & 0 & 1 & 0 & 0 & 18 \\ 2 & 6 & 0 & 0 & 1 & 0 & 72 \\ -80 & -70 & 0 & 0 & 0 & 1 & 0 \end{bmatrix} \end{array}$$

The presence of these columns gives the system the special form discussed previously. Indeed, the u column asserts that u appears only in the first equation and its coefficient there is 1, and similarly for the v, w, and M columns.

The property of allowing us to read a particular solution from the right-hand column is shared by all linear systems whose matrices contain the columns

$$\begin{matrix} 1 & 0 & 0 & \cdots & 0 \\ 0 & 1 & 0 & \cdots & 0 \\ 0 & 0 & 1 & \cdots & 0 \\ \vdots & \vdots & \vdots & & \vdots \\ 0 & 0 & 0 & \cdots & 1. \end{matrix}$$

(These columns need not appear in exactly the order shown.) The variables corresponding to these columns are called the group II variables. The group I variables consist of all of the other variables. To get one particular solution to the system, set all group I variables equal to zero and read the values of the group II variables from the right-hand side of the system. This procedure is illustrated in the following example.

EXAMPLE 3 **Finding Group II Variables from the Matrix Form** Determine by inspection one set of solutions to each of these systems of linear equations:

(a) $\begin{cases} x - 5y + u & = 3 \\ -2x + 8y & + v & = 11 \\ -\frac{1}{2}x & + M = 0 \end{cases}$ **(b)** $\begin{cases} -y + 2u + v & = 12 \\ x + \frac{1}{2}y - 6u & = -1 \\ 3y + 8u & + M = 4 \end{cases}$

SOLUTION **(a)** The matrix of the system is

$$\begin{array}{ccccc} x & y & u & v & M & \\ \begin{bmatrix} 1 & -5 & 1 & 0 & 0 & 3 \\ -2 & 8 & 0 & 1 & 0 & 11 \\ -\frac{1}{2} & 0 & 0 & 0 & 1 & 0 \end{bmatrix} \end{array}.$$

We look for each variable whose column contains one entry of 1 and all other entries 0.

$$\begin{array}{ccccc} x & y & u & v & M & \\ \begin{bmatrix} 1 & -5 & 1 & 0 & 0 & 3 \\ -2 & 8 & 0 & 1 & 0 & 11 \\ -\frac{1}{2} & 0 & 0 & 0 & 1 & 0 \end{bmatrix} \end{array}$$

The group II variables should be u, v, and M, with x, y as the group I variables. Set all group I variables equal to 0. The corresponding values of the group II variables may then be read from the last column: $u = 3$, $v = 11$, $M = 0$. So one solution of the system is

$$x = 0, \quad y = 0, \quad u = 3, \quad v = 11, \quad M = 0.$$

(b) The matrix of the system is

$$\begin{array}{ccccc} x & y & u & v & M & \\ \begin{bmatrix} 0 & -1 & 2 & 1 & 0 & 12 \\ 1 & \frac{1}{2} & -6 & 0 & 0 & -1 \\ 0 & 3 & 8 & 0 & 1 & 4 \end{bmatrix} \end{array}.$$

The shaded columns show that the group II variables should be v, x, and M, with y and u as the group I variables. So the corresponding solution is

$$x = -1, \quad y = 0, \quad u = 0, \quad v = 12, \quad M = 4.$$ **» Now Try Exercise 9**

> **DEFINITION**   A **simplex tableau** (plural: tableaux) is a matrix corresponding to a linear system in which each of the columns
>
> $$\begin{matrix} 1 & 0 & \cdots & 0 \\ 0 & 1 & \cdots & 0 \\ \vdots & \vdots & & \vdots \\ 0 & 0 & \cdots & 1 \end{matrix}$$
>
> is present (in some order) to the left of the vertical line.

We have seen how to construct a simplex tableau corresponding to a linear programming problem in standard maximization form. From this initial tableau, we can identify one particular solution of the linear system by using the method described previously. This particular solution may or may not correspond to the solution of the original optimization problem. If it does not, we replace the initial tableau with another one whose corresponding solution is "closer" to the optimum. How do we replace the initial simplex tableau with another? Just pivot it about a nonzero entry! Indeed, one of the key reasons that the simplex method works is that pivoting transforms one simplex tableau into another. Note also that, since pivoting consists of elementary row operations, a solution corresponding to a transformed tableau is a solution of the original linear system. The next example illustrates how pivoting transforms a tableau into another one.

EXAMPLE 4

Finding a Feasible Solution by Pivoting   Consider the simplex tableau obtained from the furniture manufacturing problem:

$$\begin{array}{cccccc} x & y & u & v & w & M \\ \end{array}$$
$$\left[\begin{array}{cccccc|c} ⑥ & 3 & 1 & 0 & 0 & 0 & 96 \\ 1 & 1 & 0 & 1 & 0 & 0 & 18 \\ 2 & 6 & 0 & 0 & 1 & 0 & 72 \\ -80 & -70 & 0 & 0 & 0 & 1 & 0 \end{array}\right]$$

(a) Pivot this tableau around the circled entry, 6.
(b) Calculate the particular solution corresponding to the transformed tableau that results from setting the new group I variables equal to 0.

SOLUTION

(a) The first step in pivoting is to replace the pivot element 6 by a 1. To do this, multiply the first row of the tableau by $\frac{1}{6}$ to get

$$\begin{array}{cccccc} x & y & u & v & w & M \\ \end{array}$$
$$\left[\begin{array}{cccccc|c} 1 & \frac{1}{2} & \frac{1}{6} & 0 & 0 & 0 & 16 \\ 1 & 1 & 0 & 1 & 0 & 0 & 18 \\ 2 & 6 & 0 & 0 & 1 & 0 & 72 \\ -80 & -70 & 0 & 0 & 0 & 1 & 0 \end{array}\right].$$

Next, we must replace all nonpivot elements in the first column by zeros.

$$\begin{array}{cccccc} x & y & u & v & w & M \\ \end{array}$$
$$\left[\begin{array}{cccccc|c} 1 & \frac{1}{2} & \frac{1}{6} & 0 & 0 & 0 & 16 \\ 1 & 1 & 0 & 1 & 0 & 0 & 18 \\ 2 & 6 & 0 & 0 & 1 & 0 & 72 \\ -80 & -70 & 0 & 0 & 0 & 1 & 0 \end{array}\right]$$

$$\xrightarrow[\substack{R_2 + (-1)R_1 \\ R_3 + (-2)R_1 \\ R_4 + 80R_1}]{}$$

$$\begin{array}{cccccc} x & y & u & v & w & M \\ \end{array}$$
$$\left[\begin{array}{cccccc|c} 1 & \frac{1}{2} & \frac{1}{6} & 0 & 0 & 0 & 16 \\ 0 & \frac{1}{2} & -\frac{1}{6} & 1 & 0 & 0 & 2 \\ 0 & 5 & -\frac{1}{3} & 0 & 1 & 0 & 40 \\ 0 & -30 & \frac{40}{3} & 0 & 0 & 1 & 1280 \end{array}\right]$$

Note that we indeed get a new simplex tableau. The new group II variables are x, v, w, and M. The group I variables are y and u.

$$
\begin{array}{cccccc}
x & y & u & v & w & M \\
\end{array}
$$

$$
\left[
\begin{array}{cccccc|c}
1 & \frac{1}{2} & \frac{1}{6} & 0 & 0 & 0 & 16 \\
0 & \frac{1}{2} & -\frac{1}{6} & 1 & 0 & 0 & 2 \\
0 & 5 & -\frac{1}{3} & 0 & 1 & 0 & 40 \\
0 & -30 & \frac{40}{3} & 0 & 0 & 1 & 1280
\end{array}
\right]
$$

(b) Set the group I variables equal to 0:

$$
y = 0, \qquad u = 0.
$$

Identify the particular values of the group II variables from the right-hand column:

$$
x = 16, \qquad v = 2, \qquad w = 40, \qquad M = 1280.
$$

So the particular solution corresponding to the transformed tableau is

$$
x = 16, \qquad y = 0, \qquad u = 0, \qquad v = 2, \qquad w = 40, \qquad M = 1280.
$$

We see that the simplex tableau leads to the vertex $(16, 0)$ that was listed and tested in the graphical solution of the problem presented in Chapter 3.

» Now Try Exercise 21(a)

Check Your Understanding 4.1

Solutions can be found following the section exercises.

1. Determine by inspection a particular solution of the following system of linear equations:

$$
\begin{cases}
x + 2y + 3u & = 6 \\
y & + v & = 4 \\
5y + 2u & + M = 0
\end{cases}
$$

2. Pivot the simplex tableau about the circled element.

$$
\left[
\begin{array}{ccccc|c}
2 & 4 & 1 & 0 & 0 & 6 \\
3 & ① & 0 & 1 & 0 & 0 \\
1 & 1 & 0 & 0 & 1 & 1
\end{array}
\right]
$$

EXERCISES 4.1

For each of the following linear programming problems, determine the corresponding linear system and restate the linear programming problem in terms of the linear system.

1. Maximize $8x + 13y$ subject to the constraints

$$
\begin{cases}
20x + 30y \le 3500 \\
50x + 10y \le 5000 \\
x \ge 0 \\
y \ge 0.
\end{cases}
$$

2. Maximize $x + 15y$ subject to the constraints

$$
\begin{cases}
3x + 2y \le 10 \\
x \qquad \le 15 \\
\qquad y \le 3 \\
x + y \le 5 \\
x \ge 0 \\
y \ge 0.
\end{cases}
$$

3. Maximize $x + 2y - 3z$ subject to the constraints

$$
\begin{cases}
x + y + z \le 100 \\
3x \qquad + z \le 200 \\
5x + 10y \qquad \le 100 \\
x \ge 0 \\
y \ge 0 \\
z \ge 0.
\end{cases}
$$

4. Maximize $2x + y + 50$ subject to the constraints

$$
\begin{cases}
x + 3y \le 24 \\
y \le 5 \\
x + 7y \le 10 \\
x \ge 0 \\
y \ge 0.
\end{cases}
$$

5. Maximize $3x + 5y + 12z$ subject to the constraints

$$\begin{cases} 4x + 6y - 7z \le 16 \\ 3x + 2y \qquad \le 11 \\ \qquad 9y + 3z \le 21 \\ x \ge 0 \\ y \ge 0 \\ z \ge 0. \end{cases}$$

6. Maximize $3x + 7y + 5z$ subject to the constraints

$$\begin{cases} x + 2y + 6z \le 60 \\ 4x \quad + \ z \le 50 \\ 2x + \ y + 5z \le 47 \\ x \ge 0 \\ y \ge 0 \\ z \ge 0. \end{cases}$$

7–12. For each of the linear programming problems in Exercises 1–6,

   (a) Set up the initial simplex tableau.

   (b) Determine the particular solution corresponding to the initial tableau.

In Exercises 13–20, find the particular solution corresponding to the tableau.

13.

x	y	u	v	M	
0	2	1	0	0	10
1	3	0	12	0	15
0	−1	0	17	1	20

14.

x	y	u	v	M	
1	0	3	11	0	6
0	1	10	17	0	16
0	0	5	−1	1	3

15.

x	y	z	u	v	M	
0	−2	1	6	4	0	17
1	1	0	3	5	0	14
0	8	0	12	9	1	56

16.

x	y	z	u	v	M	
9	1	0	5	6	0	32
7	0	1	−2	4	0	25
5	0	0	9	10	1	431

17.

x	y	z	u	v	w	M	
0	1	0	7	2	4	0	54
0	0	1	5	6	5	0	61
1	0	0	6	9	8	0	42
0	0	0	11	10	9	1	604

18.

x	y	z	u	v	w	M	
4	0	0	1	2	2	0	25
8	0	1	0	3	4	0	34
5	1	0	0	9	6	0	19
7	0	0	0	10	8	1	178

19.

x	y	z	u	v	w	M	
0	3	1	0	1	15	0	15
1	−1	0	0	2	−5	0	10
0	2	0	1	−5	4	0	23
0	11	0	0	11	6	1	−11

20.

x	y	z	u	v	w	M	
6	0	1	0	5	−1	0	$\frac{1}{4}$
5	1	0	0	3	$\frac{1}{3}$	0	100
4	0	0	1	8	$\frac{1}{2}$	0	11
2	0	0	0	6	$\frac{1}{7}$	1	$-\frac{1}{2}$

21. Pivot the simplex tableau

x	y	u	v	M	
2	3	1	0	0	12
1	1	0	1	0	10
−10	−20	0	0	1	0

about each indicated element, and compute the particular solution corresponding to the new tableau.

  (a) 2　　　　**(b)** 3

  (c) 1 (second row, first column)

  (d) 1 (second row, second column)

  (e) Determine which of the pivot operations increases M the most.

22. Pivot the simplex tableau

x	y	u	v	M	
5	4	1	0	0	100
10	6	0	1	0	1200
−1	2	0	0	1	0

about each indicated element, and compute the solution corresponding to the new tableau.

  (a) 5　　**(b)** 4　　**(c)** 10　　**(d)** 6

  (e) Determine which of the pivot operations increases M the most.

23. (a) Name the group I and group II variables in the tableau as given.

x	y	u	v	M	
2	5	1	0	0	100
3	1	0	1	0	300
−10	−7	0	0	1	0

  (b) Pivot the simplex tableau about each indicated element, and compute the solution corresponding to the new tableau. Which solutions are feasible (that is, have all values ≥ 0)? Which variables are now in group I; which are in group II?

    (i) 2　(ii) 5　(iii) 3　(iv) 1 (row 2, column 2)

  (c) Which of the feasible solutions increases the value of M the most?

24. (a) Name the group I and group II variables in the tableau as given.

x	y	u	v	w	M	
1	2	1	0	0	0	15
1	1	0	1	0	0	27
2	1	0	0	1	0	42
−9	−5	0	0	0	1	0

(b) Pivot the simplex tableau about each indicated element, and compute the solution corresponding to the new tableau. Which solutions are feasible (that is, have all values ≥ 0)? Which variables are now in group I; which are in group II?

(i) 1 (row 1, column 1)   (ii) 2 (row 1)
(iii) 2 (row 3)   (iv) 1 (row 3, column 2)

(c) Which of the feasible solutions increases the value of M the most?

Solutions to Check Your Understanding 4.1

1. The matrix of the system is

$$
\begin{array}{ccccc}
x & y & u & v & M \\
\end{array}
$$
$$
\left[\begin{array}{ccccc|c}
1 & 2 & 3 & 0 & 0 & 6 \\
0 & 1 & 0 & 1 & 0 & 4 \\
0 & 5 & 2 & 0 & 1 & 0
\end{array}\right],
$$

from which we see that the group II variables are x, v, and M, and the group I variables are y and u. To obtain a solution, we set the group I variables equal to 0. We obtain from the first row that $x = 6$, from the second that $v = 4$, and from the third that $M = 0$. Thus, a solution of the system is $x = 6$, $y = 0$, $u = 0$, $v = 4$, $M = 0$.

2. We must use elementary row operations to transform the second column into $\begin{bmatrix} 0 \\ 1 \\ 0 \end{bmatrix}$.

$$
\begin{bmatrix}
2 & 4 & 1 & 0 & 0 & 6 \\
3 & ① & 0 & 1 & 0 & 0 \\
1 & 1 & 0 & 0 & 1 & 1
\end{bmatrix}
$$

$$\xrightarrow{R_1 + (-4)R_2}
\begin{bmatrix}
-10 & 0 & 1 & -4 & 0 & 6 \\
3 & 1 & 0 & 1 & 0 & 0 \\
1 & 1 & 0 & 0 & 1 & 1
\end{bmatrix}
$$

$$\xrightarrow{R_3 + (-1)R_2}
\begin{bmatrix}
-10 & 0 & 1 & -4 & 0 & 6 \\
3 & 1 & 0 & 1 & 0 & 0 \\
-2 & 0 & 0 & -1 & 1 & 1
\end{bmatrix}
$$

4.2 The Simplex Method I: Maximum Problems

We can now describe the simplex method for solving linear programming problems in standard maximization form. The procedure will be illustrated as we solve the furniture manufacturing problem of Section 4.1. Recall that we must maximize the objective function $80x + 70y$ subject to the constraints

$$
\begin{cases}
6x + 3y \leq 96 \\
x + y \leq 18 \\
2x + 6y \leq 72 \\
x \geq 0, \quad y \geq 0.
\end{cases}
$$

First, introduce slack variables, and state the problem in terms of a system of linear equations. We carried out this step in Section 4.1. The result was the following restatement of the problem.

Mathematical Model for the Furniture Manufacturing Problem

Among all of the solutions of the system of linear equations

$$
\begin{cases}
6x + 3y + u & = 96 \\
x + y + v & = 18 \\
2x + 6y + w & = 72 \\
-80x - 70y + M = 0,
\end{cases}
$$

find one for which $x \geq 0$, $y \geq 0$, $u \geq 0$, $v \geq 0$, $w \geq 0$ and for which M is as large as possible.

Next, construct the simplex tableau corresponding to the linear system. This step was also carried out in Section 4.1. The tableau is

$$
\begin{array}{c}
 \\
u \\
v \\
w \\
M
\end{array}
\begin{array}{c}
\begin{array}{cccccc}
x & y & u & v & w & M
\end{array} \\
\left[\begin{array}{cccccc|c}
6 & 3 & 1 & 0 & 0 & 0 & 96 \\
1 & 1 & 0 & 1 & 0 & 0 & 18 \\
2 & 6 & 0 & 0 & 1 & 0 & 72 \\
\hline
-80 & -70 & 0 & 0 & 0 & 1 & 0
\end{array}\right].
\end{array}
$$

Note that we have made two additions to the previously found tableau. First, we have separated the last row from the others by means of a horizontal line. This is because the last row, which corresponds to the objective function in the original problem, will play a special role in what follows. The second addition is that we have labeled each row with one of the group II variables—namely, the variable whose value is determined by the row. Thus, for example, the first row gives the particular value of u, which is 96, so the row is labeled with a u. We will find these labels convenient.

Corresponding to this tableau, there is a particular solution to the linear system, namely, the one obtained by setting all group I variables equal to 0. Reading the values of the group II variables from the last column, we obtain

$$x = 0, \quad y = 0, \quad u = 96, \quad v = 18, \quad w = 72, \quad M = 0.$$

Our objective is to make M as large as possible. How can the value of M be increased? Look at the equation corresponding to the last row of the tableau. It reads

$$-80x - 70y + M = 0.$$

Note that two of the coefficients, -80 and -70, are negative. Or, what amounts to the same thing, if we solve for M and get

$$M = 80x + 70y,$$

then the coefficients on the right-hand side are *positive*. This fact is significant. It says that M can be increased by increasing either the value of x or the value of y. A unit change in x will increase M by 80 units, whereas a unit change in y will increase M by only 70 units. And since we wish to increase M by as much as possible, it is reasonable to attempt to increase the value of x. Let us indicate this by drawing an arrow pointing to the x column of the tableau:

$$
\begin{array}{c}
\begin{array}{cccccc}
x & y & u & v & w & M
\end{array} \\
\begin{array}{c} u \\ v \\ w \\ M \end{array}
\left[
\begin{array}{cccccc|c}
6 & 3 & 1 & 0 & 0 & 0 & 96 \\
1 & 1 & 0 & 1 & 0 & 0 & 18 \\
2 & 6 & 0 & 0 & 1 & 0 & 72 \\
-80 & -70 & 0 & 0 & 0 & 1 & 0
\end{array}
\right] \\
\uparrow
\end{array}
\tag{1}
$$

To increase x (from its present value, zero), we will pivot about one of the entries (above the horizontal line) in the x column. In this way, x will become a group II variable and hence will not necessarily be zero in our next particular solution. But around which entry should we pivot? To find out, let us experiment. The results from pivoting about the 6, the 1, and the 2 in the x column are, respectively,

$$
\begin{array}{c}
\begin{array}{cccccc}
x & y & u & v & w & M
\end{array} \\
\begin{array}{c} x \\ v \\ w \\ M \end{array}
\left[
\begin{array}{cccccc|c}
1 & \frac{1}{2} & \frac{1}{6} & 0 & 0 & 0 & 16 \\
0 & \frac{1}{2} & -\frac{1}{6} & 1 & 0 & 0 & 2 \\
0 & 5 & -\frac{1}{3} & 0 & 1 & 0 & 40 \\
0 & -30 & \frac{40}{3} & 0 & 0 & 1 & 1280
\end{array}
\right],
\end{array}
$$

Pivot about 6

$$
\begin{array}{c}
\begin{array}{cccccc}
x & y & u & v & w & M
\end{array} \\
\begin{array}{c} u \\ x \\ w \\ M \end{array}
\left[
\begin{array}{cccccc|c}
0 & -3 & 1 & -6 & 0 & 0 & -12 \\
1 & 1 & 0 & 1 & 0 & 0 & 18 \\
0 & 4 & 0 & -2 & 1 & 0 & 36 \\
0 & 10 & 0 & 80 & 0 & 1 & 1440
\end{array}
\right], \text{ and}
\end{array}
$$

Pivot about 1

$$
\begin{array}{c}
\begin{array}{cccccc}
x & y & u & v & w & M
\end{array} \\
\begin{array}{c} u \\ v \\ x \\ M \end{array}
\left[
\begin{array}{cccccc|c}
0 & -15 & 1 & 0 & -3 & 0 & -120 \\
0 & -2 & 0 & 1 & -\frac{1}{2} & 0 & -18 \\
1 & 3 & 0 & 0 & \frac{1}{2} & 0 & 36 \\
0 & 170 & 0 & 0 & 40 & 1 & 2880
\end{array}
\right].
\end{array}
$$

Pivot about 2

Note that the labels on the rows have *changed* because the group II variables are now *different*. The solutions corresponding to these tableaux are, respectively,

$$x = 16, \quad y = 0, \quad u = 0, \quad v = 2, \quad w = 40, \quad M = 1280,$$
$$x = 18, \quad y = 0, \quad u = -12, \quad v = 0, \quad w = 36, \quad M = 1440,$$
$$x = 36, \quad y = 0, \quad u = -120, \quad v = -18, \quad w = 0, \quad M = 2880.$$

The second and third solutions violate the requirement that all variables be ≥ 0. Thus, we use the first solution, in which we pivoted about 6. Using this solution, we have increased the value of M to 1280 and have replaced our original tableau by

$$
\begin{array}{c}
x \\ v \\ w \\ M
\end{array}
\left[
\begin{array}{cccccc|c}
x & y & u & v & w & M & \\
1 & \frac{1}{2} & \frac{1}{6} & 0 & 0 & 0 & 16 \\
0 & \frac{1}{2} & -\frac{1}{6} & 1 & 0 & 0 & 2 \\
0 & 5 & -\frac{1}{3} & 0 & 1 & 0 & 40 \\
\hline
0 & -30 & \frac{40}{3} & 0 & 0 & 1 & 1280
\end{array}
\right].
$$

Can M be increased further? To answer this question, look at the last row of the tableau, which corresponds to the equation

$$-30y + \tfrac{40}{3}u + M = 1280.$$

There is a negative coefficient for the variable y in this equation. Correspondingly, when the equation is solved for M, there is a positive coefficient for y:

$$M = 1280 + 30y - \tfrac{40}{3}u.$$

Now it is clear that we should try to increase y. So we pivot about one of the entries in the y column. A calculation for each of the possible pivots shows that pivoting about the first or the third entries leads to solutions having some negative values. Therefore, we pivot about the second entry in the y column. The result is

$$
\begin{array}{c}
x \\ y \\ w \\ M
\end{array}
\left[
\begin{array}{cccccc|c}
x & y & u & v & w & M & \\
1 & 0 & \frac{1}{3} & -1 & 0 & 0 & 14 \\
0 & 1 & -\frac{1}{3} & 2 & 0 & 0 & 4 \\
0 & 0 & \frac{4}{3} & -10 & 1 & 0 & 20 \\
\hline
0 & 0 & \frac{10}{3} & 60 & 0 & 1 & 1400
\end{array}
\right].
$$

The corresponding solution is

$$x = 14, \quad y = 4, \quad u = 0, \quad v = 0, \quad w = 20, \quad M = 1400.$$

Note that with this pivot operation, we have increased M from 1280 to 1400.

Can we increase M any further? Let us reason as before. Use the last row of the current tableau to write M in terms of the other variables:

$$\tfrac{10}{3}u + 60v + M = 1400, \qquad M = 1400 - \tfrac{10}{3}u - 60v.$$

Note, however, that in contrast to the previous expressions for M, this one has *no positive coefficients*. And since u and v are ≥ 0, M can be *at most* 1400. But M is already 1400. So M cannot be increased further. Thus, we have shown that the maximum value of M is 1400, and this occurs when $x = 14$ and $y = 4$. So, to maximize profits, the furniture manufacturer should make 14 chairs and 4 sofas each day. The maximum profit is \$1400. From the tableau, we can identify the values of the slack variables: $u = 0$, $v = 0$, and $w = 20$. This shows that we have no slack resulting from the first inequality, so we have used all of the labor-hours available for carpentry. Similarly, since $v = 0$, we have used all of the labor-hours available for finishing. But since $w = 20$, we have 20 labor-hours of upholstery remaining when we manufacture the optimal number of chairs and sofas.

Let us compare the simplex method solution of the furniture manufacturing problem with the geometric solution carried out in Chapter 3. Both solutions yield the same optimal production schedule. In the geometric solution, we found *all* of the vertices of the feasible set and then evaluated the objective function at every one of these vertices. The following table was obtained:

Vertex	Profit $= 80x + 70y$
$(14, 4)$	$80(14) + 70(4) = 1400$
$(9, 9)$	$80(9) + 70(9) = 1350$
$(0, 12)$	$80(0) + 70(12) = 840$
$(0, 0)$	$80(0) + 70(0) = 0$
$(16, 0)$	$80(16) + 70(0) = 1280$

We selected the optimal solution ($x = 14$, $y = 4$) because it produced the greatest profit.

With the simplex method, we had to consider only *some* of the vertices. In the initial tableau, we first considered the vertex $(0, 0)$ —that is, both x and y were 0. M was also 0. In the second tableau, we looked at the vertex $(16, 0)$ —that is, $x = 16$ and $y = 0$, and the tableau showed that $M = 1280$. Finally, as a result of the last pivot operation, we came to the vertex $(14, 4)$. This meant that $x = 14$ and $y = 4$. The value of the objective function was read from the tableau: $M = 1400$. Since we could not increase M any more, we did not have to consider any other vertices. In large linear programming problems, the time saved from looking at just *some* of the vertices, rather than *all* of the vertices, can be substantial.

On the basis of the preceding discussion, we can state several general principles. First of all, the following criterion determines when a simplex tableau yields a maximum:

> **Condition for a Maximum**   The particular solution derived from a simplex tableau is a maximum if and only if the bottom row contains no negative entries except perhaps the entry in the last column. (In Section 4.3, we will encounter maximum problems whose final tableau has a negative number in the lower right-hand corner.)

We saw this condition illustrated in the previous example. Each of the first two tableaux had negative entries in the last row, and as we showed, their corresponding solutions were not maxima. However, the third tableau, with no negative entries in the last row, did yield a maximum.

The crucial point of the simplex method is the correct choice of a pivot element. In the preceding example, we decided to choose a pivot element from the column corresponding to the most-negative entry in the last row. It can be proved that this is the proper choice in general; that is, we have the following rule:

> **Choosing the Pivot Column**   The pivot element should be chosen from that column to the left of the vertical line that has the most-negative entry in the last row. In case two or more columns are tied for the honor of being the pivot column, an arbitrary choice among them may be made.

Choosing the correct pivot element from the designated column is somewhat more complicated. Our approach before was to calculate the tableau associated with each element and observe that only one corresponded to a solution with nonnegative elements. However, there is a simpler way to make the choice. As an illustration, let us reconsider tableau (1). We have already decided to pivot around some entry in the first column. For each *positive* entry in the pivot column, we compute a ratio: the corresponding entry in the right-hand column divided by the entry in the pivot column. So, for example, for the first entry the ratio is $\frac{96}{6}$; for the second entry, the ratio is $\frac{18}{1}$;

and for the third entry, the ratio is $\frac{72}{2}$. We write these ratios to the right of the matrix as follows:

$$
\begin{array}{c}
\\ u \\ v \\ w \\ M
\end{array}
\begin{array}{c}
\begin{array}{cccccc}
x & y & u & v & w & M
\end{array}\\
\left[\begin{array}{cccccc|c}
6 & 3 & 1 & 0 & 0 & 0 & 96 \\
1 & 1 & 0 & 1 & 0 & 0 & 18 \\
2 & 6 & 0 & 0 & 1 & 0 & 72 \\
\hline
-80 & -70 & 0 & 0 & 0 & 1 & 0
\end{array}\right]
\end{array}
\begin{array}{l}
\frac{96}{6} = 16 \\
\frac{18}{1} = 18 \\
\frac{72}{2} = 36 \\

\end{array}
$$

It is possible to prove the following rule, which allows us to determine the pivot element from the preceding display:

> **Choosing the Pivot Element** For each positive entry of the pivot column, compute the appropriate ratio. Choose as pivot element the one corresponding to the smallest nonnegative ratio.

For instance, consider the choice of pivot element in the preceding example. The least of the ratios is 16. So we choose 6 as the pivot element.

Rationale for Method of Choosing Pivot Element

At first, this method for choosing the pivot element might seem odd. However, it is just a way of guaranteeing that the last column of the new tableau will have entries ≥ 0. And that is just the basis on which we chose the pivot element earlier. To obtain further insight, let us analyze the preceding example yet further.

Suppose that we pivot our tableau about the 6 in column 1. The first step in pivoting is to divide the pivot row by the pivot element (in this case, 6). This gives the array

$$
\begin{array}{c}
\begin{array}{cccccc}
x & y & u & v & w & M
\end{array}\\
\left[\begin{array}{cccccc|c}
1 & \frac{1}{2} & \frac{1}{6} & 0 & 0 & 0 & \frac{96}{6} \\
1 & 1 & 0 & 1 & 0 & 0 & 18 \\
2 & 6 & 0 & 0 & 1 & 0 & 72 \\
\hline
-80 & -70 & 0 & 0 & 0 & 1 & 0
\end{array}\right]
\end{array},
$$

where we have written $\frac{96}{6}$ rather than 16 to emphasize that we have divided by the pivot element. The result of completing the pivot is as follows.

$$
\begin{array}{c}
\begin{array}{cccccc}
x & y & u & v & w & M
\end{array}\\
\left[\begin{array}{cccccc|c}
1 & \frac{1}{2} & \frac{1}{6} & 0 & 0 & 0 & \frac{96}{6} \\
1 & 1 & 0 & 1 & 0 & 0 & 18 \\
2 & 6 & 0 & 0 & 1 & 0 & 72 \\
\hline
-80 & -70 & 0 & 0 & 0 & 1 & 0
\end{array}\right]
\end{array}
$$

$$
\begin{array}{c}
\begin{array}{r}
R_2 + (-1)R_1 \\
R_3 + (-2)R_1 \\
\hline
R_4 + 80R_1
\end{array}\\

\end{array}
\xrightarrow{}
\begin{array}{c}
\begin{array}{cccccc}
x & y & u & v & w & M
\end{array}\\
\left[\begin{array}{cccccc|c}
1 & \frac{1}{2} & \frac{1}{6} & 0 & 0 & 0 & \frac{96}{6} \\
0 & \frac{1}{2} & -\frac{1}{6} & 1 & 0 & 0 & 18 - \frac{96}{6} \\
0 & 5 & -\frac{1}{3} & 0 & 1 & 0 & 72 - 2(\frac{96}{6}) \\
\hline
0 & -30 & \frac{40}{3} & 0 & 0 & 1 & 1280
\end{array}\right]
\end{array}
$$

The entries in the upper part of the right-hand column may be written

$$
\frac{96}{6}, \quad \frac{18}{1} - \frac{96}{6}, \quad 2\left(\frac{72}{2} - \frac{96}{6}\right).
$$

If we had pivoted about the 1 or 2 in the first column of the original tableau, the upper entries in the last column of the tableau would have been

$$6\left(\frac{96}{6} - \frac{18}{1}\right), \qquad \frac{18}{1}, \qquad 2\left(\frac{72}{2} - \frac{18}{1}\right)$$

or

$$6\left(\frac{96}{6} - \frac{72}{2}\right), \qquad \frac{18}{1} - \frac{72}{2}, \qquad \frac{72}{2},$$

respectively. Notice that all of the combinations of the differences of the pairs of ratios appear in these triples. In the first case, the ratio $\frac{96}{6}$ is subtracted from each of the other two ratios, whereas in the next two cases, the ratios $\frac{18}{1}$ and $\frac{72}{2}$ are subtracted. In order for the upper entries in the last column to be nonnegative, we must subtract off the smallest of the ratios. That is, we should pivot about the entry corresponding to the smallest ratio. This is the rationale governing our choice of pivot element!

Now that we have assembled all of the components of the simplex method, we can summarize it as follows:

> ### The Simplex Method for Problems in Standard Maximization Form
>
> 1. Introduce slack variables, and state the problem in terms of a system of linear equations.
>
> 2. Construct the simplex tableau corresponding to the system.
>
> 3. Determine whether the left part of the bottom row contains negative entries. If none is present, the solution corresponding to the tableau yields a maximum and the problem is solved.
>
> 4. If the left part of the bottom row contains negative entries, construct a new simplex tableau.
>   (a) Choose the pivot column by inspecting the entries of the bottom row of the current tableau, excluding the right-hand entry. The pivot column is the one containing the most negative of these entries.
>   (b) Choose the pivot element by computing ratios associated with the positive entries of the pivot column and the right column. The pivot element is the one corresponding to the smallest nonnegative ratio.
>   (c) Construct the new simplex tableau by pivoting around the selected element.
>
> 5. Return to step 3. Repeat steps 3 and 4 as many times as necessary to find a maximum.
>
> 6. Interpret the result.

Let us now work some problems to see how this method is applied.

EXAMPLE 1 **Using the Simplex Method** Maximize the objective function $10x + y$, subject to the constraints

$$\begin{cases} x + 2y \le 10 \\ 3x + 4y \le 6 \\ x \ge 0, \quad y \ge 0. \end{cases}$$

SOLUTION **Step 1** The corresponding system of linear equations with slack variables is

$$\begin{cases} x + 2y + u & = 10 \\ 3x + 4y & + v & = 6 \\ -10x - y & + M = 0, \end{cases}$$

and we must find that solution of the system for which $x \ge 0$, $y \ge 0$, $u \ge 0$, $v \ge 0$, and M is as large as possible.

Step 2 Here is the initial simplex tableau:

$$
\begin{array}{c}
\begin{array}{ccccc} x & y & u & v & M \end{array} \\
\begin{array}{c} u \\ v \\ M \end{array}
\left[\begin{array}{ccccc|c}
1 & 2 & 1 & 0 & 0 & 10 \\
3 & 4 & 0 & 1 & 0 & 6 \\
\hline
-10 & -1 & 0 & 0 & 1 & 0
\end{array} \right]
\end{array}
$$

Step 3 Note that this tableau does not correspond to a maximum, since the left part of the bottom row has negative entries. So we pivot to create a new tableau.

Step 4 Since -10 is the most-negative entry in the last row, we choose the first column as the pivot column. To determine the pivot element, we compute ratios as follows:

$$
\begin{array}{c}
\begin{array}{ccccc} x & y & u & v & M \end{array} \qquad\quad \text{Ratios}\\
\begin{array}{c} u \\ v \\ M \end{array}
\left[\begin{array}{ccccc|c}
1 & 2 & 1 & 0 & 0 & 10 \\
③ & 4 & 0 & 1 & 0 & 6 \\
\hline
-10 & -1 & 0 & 0 & 1 & 0
\end{array} \right]
\begin{array}{c} 10/1 = 10 \\ 6/3 = 2 \\ \\ \end{array}
\end{array}
$$

The smallest ratio is 2, so we pivot about 3, which we have circled. The new tableau is therefore

$$
\begin{array}{c}
\begin{array}{ccccc} x & y & u & v & M \end{array} \\
\begin{array}{c} u \\ x \\ M \end{array}
\left[\begin{array}{ccccc|c}
0 & \frac{2}{3} & 1 & -\frac{1}{3} & 0 & 8 \\
1 & \frac{4}{3} & 0 & \frac{1}{3} & 0 & 2 \\
\hline
0 & \frac{37}{3} & 0 & \frac{10}{3} & 1 & 20
\end{array} \right]
\end{array}.
$$

Step 5 Note that this tableau corresponds to a maximum, since there are no negative entries in the left part of the last row. The solution corresponding to the tableau is

$$ x = 2, \qquad y = 0, \qquad u = 8, \qquad v = 0, \qquad M = 20. $$

Step 6 Therefore, the objective function assumes its maximum value of 20 when $x = 2$ and $y = 0$. **» Now Try Exercise 7**

The simplex method can be used to solve problems in any number of variables. Let us illustrate the method for three variables.

EXAMPLE 2 **Using the Simplex Method** Maximize the objective function $x + 2y + z$ subject to the constraints

$$
\begin{cases}
x - y + 2z \le 10 \\
2x + y + 3z \le 12 \\
x \ge 0, \quad y \ge 0, \quad z \ge 0.
\end{cases}
$$

SOLUTION We determined the corresponding linear system in Example 2 of Section 4.1:

$$
\begin{cases}
x - y + 2z + u & = 10 \\
2x + y + 3z & + v & = 12 \\
-x - 2y - z & + M = 0
\end{cases}
$$

So the simplex method works as follows:

$$
\begin{array}{c}
 \\
u \\
v \\
M
\end{array}
\begin{array}{cccccc}
\;\;x & \;\;y & \;\;z & \;\;u & \;\;v & \;\;M \\
\end{array}
$$

	x	y	z	u	v	M		
u	1	-1	2	1	0	0	10	(negative ratio)
v	2	①	3	0	1	0	12	$12/1 = 12$
M	-1	-2	-1	0	0	1	0	

	x	y	z	u	v	M	
u	3	0	5	1	1	0	22
y	2	1	3	0	1	0	12
M	3	0	5	0	2	1	24

Thus, the solution of the original problem ($x = 0$, $y = 12$, $z = 0$) yields the maximum value of the objective function $x + 2y + z$. The maximum value is 24.

» Now Try Exercise 15

Check Your Understanding 4.2

Solutions can be found following the section exercises.

1. Which of these simplex tableaux has a solution that corresponds to a maximum for the associated linear programming problem?

(a)
	x	y	u	v	M	
	3	1	0	1	0	5
	2	0	0	0	1	0
	-1	-2	1	0	0	3

(b)
	x	y	u	v	M	
	2	1	0	11	0	10
	1	0	1	7	0	1
	1	0	0	4	1	-2

2. Suppose that, in the solution of a linear programming problem by the simplex method, we encounter the simplex tableau shown below. What is the next step in the solution?

	x	y	u	v	M	
	0	4	1	2	0	4
	1	5	0	1	0	9
	0	2	0	-3	1	6

EXERCISES 4.2

In Exercises 1–6, determine the next pivot element for the tableau.

1.
	x	y	u	v	M	
	1	1	1	0	0	30
	2	1	0	1	0	50
	-4	-3	0	0	1	0

2.
	x	y	u	v	M	
	0	$\frac{1}{2}$	1	$-\frac{1}{2}$	0	5
	1	$\frac{1}{2}$	0	$\frac{1}{2}$	0	25
	0	-10	0	20	1	100

3.
	x	y	z	u	v	M	
	1	$\frac{1}{4}$	$\frac{1}{4}$	0	0	0	8
	0	$\frac{1}{4}$	$-\frac{5}{4}$	0	1	0	0
	0	-4	2	0	0	1	9

4.
	x	y	z	u	v	M	
	2	3	4	1	0	0	5
	5	6	8	0	1	0	8
	-2	-3	-4	0	0	1	0

5.
	x	y	u	v	w	M	
	1	2	1	0	0	0	4
	4	3	0	1	0	0	9
	-2	1	0	0	1	0	0
	-3	-7	0	0	0	1	0

6.
	x	y	u	v	w	M	
	$\frac{1}{2}$	0	$\frac{1}{2}$	0	0	0	7
	$\frac{1}{2}$	0	$-\frac{1}{2}$	1	0	0	2
	2	0	-1	0	1	0	10
	$-\frac{1}{2}$	0	$\frac{3}{2}$	0	0	1	21

For each of the simplex tableaux in Exercises 7–10,
(a) Determine the next pivot element.
(b) Determine the next tableau.
(c) Determine the particular solution corresponding to the tableau of part (b).

7.
	x	y	u	v	M	
	6	2	1	0	0	10
	1	3	0	1	0	6
	-4	-12	0	0	1	0

8.
	x	y	u	v	M	
	1	0	3	1	0	5
	0	1	2	0	0	12
	-6	0	5	0	1	10

9.
	x	y	u	v	M	
	5	12	1	0	0	12
	15	10	0	1	0	5
	4	-2	0	0	1	0

10.
$$\begin{bmatrix} x & y & u & v & M & \\ 0 & 6 & 3 & 1 & 0 & 5 \\ 1 & -5 & 2 & 0 & 0 & 8 \\ \hline 0 & 20 & -10 & 0 & 1 & 22 \end{bmatrix}$$

In Exercises 11–20, solve the linear programming problem by the simplex method.

11. Maximize $x + 3y$ subject to the constraints
$$\begin{cases} x + y \le 7 \\ x + 2y \le 10 \\ x \ge 0, \quad y \ge 0. \end{cases}$$

12. Maximize $x + 2y$ subject to the constraints
$$\begin{cases} -x + y \le 100 \\ 6x + 6y \le 1200 \\ x \ge 0, \quad y \ge 0. \end{cases}$$

13. Maximize $4x + 2y$ subject to the constraints
$$\begin{cases} 5x + y \le 80 \\ 3x + 2y \le 76 \\ x \ge 0, \quad y \ge 0. \end{cases}$$

14. Maximize $2x + 6y$ subject to the constraints
$$\begin{cases} -x + 8y \le 160 \\ 3x - y \le 3 \\ x \ge 0, \quad y \ge 0. \end{cases}$$

15. Maximize $x + 3y + 5z$ subject to the constraints
$$\begin{cases} x + 2z \le 10 \\ 3y + z \le 24 \\ x \ge 0, \quad y \ge 0, \quad z \ge 0. \end{cases}$$

16. Maximize $-x + 8y + z$ subject to the constraints
$$\begin{cases} x - 2y + 9z \le 10 \\ y + 4z \le 12 \\ x \ge 0, \quad y \ge 0, \quad z \ge 0. \end{cases}$$

17. Maximize $2x + 3y$ subject to the constraints
$$\begin{cases} 5x + y \le 30 \\ 3x + 2y \le 60 \\ x + y \le 50 \\ x \ge 0, \quad y \ge 0. \end{cases}$$

18. Maximize $10x + 12y + 10z$ subject to the constraints
$$\begin{cases} x - 2y \le 6 \\ 3x + z \le 9 \\ y + 3z \le 12 \\ x \ge 0, \quad y \ge 0, \quad z \ge 0. \end{cases}$$

19. Maximize $6x + 7y + 300$ subject to the constraints
$$\begin{cases} 2x + 3y \le 400 \\ x + y \le 150 \\ x \ge 0, \quad y \ge 0. \end{cases}$$

20. Maximize $10x + 20y + 50$ subject to the constraints
$$\begin{cases} x + y \le 10 \\ 5x + 2y \le 20 \\ x \ge 0, \quad y \ge 0. \end{cases}$$

21. Toy Factory A toy manufacturer makes lightweight balls for indoor play. The large basketball uses 4 ounces of foam and

20 minutes of labor and brings a profit of $2.50. The football uses 3 ounces of foam and 30 minutes of labor and brings a profit of $2. The manufacturer has available 48 pounds of foam and 120 labor-hours a week. Use the simplex method to determine the optimal production schedule so as to maximize profits. Show that the geometric method gives the same solution.

22. Agriculture A large agricultural firm has 250 acres and $8000 available for cultivating three crops: barley, oats, and wheat. Barley requires $10 per acre for cultivation, oats require $15 per acre for cultivation, and wheat requires $12 per acre for cultivation. Barley requires 7 hours of labor per acre, oats require 9 hours of labor per acre, and wheat requires 8 hours of labor per acre. The firm has 2100 hours of labor available. The profits per acre of each crop are barley $60, oats $75, and wheat $70. How many acres of each crop should be planted to maximize profit?

23. Furniture Factory Suppose that a furniture manufacturer makes chairs, sofas, and tables. The amounts of labor of various types, as well as the relative availability of each type, are summarized by the following chart:

	Chair	Sofa	Table	Daily Labor Available (labor-hours)
Carpentry	6	3	8	768
Finishing	1	1	2	144
Upholstery	2	5	0	216

The profit per chair is $80, per sofa $70, and per table $120. How many pieces of each type of furniture should be manufactured each day to maximize the profit?

24. Stereo Store A stereo store sells three brands of stereo systems, brands A, B, and C. It can sell a total of 100 stereo systems per month. Brands A, B, and C take up, respectively, 5, 4, and 4 cubic feet of warehouse space, and a maximum of 480 cubic feet of warehouse space is available. Brands A, B, and C generate sales commissions of $40, $20, and $30, respectively, and $3200 is available to pay the sales commissions. The profit generated from the sale of each brand is $70, $210, and $140, respectively. How many of each brand of stereo system should be sold to maximize the profit?

25. Weight Loss and Exercise As part of a weight-reduction program, a person designs a monthly exercise program consisting of bicycling, jogging, and swimming. They would like to exercise at most 30 hours, devote at most 4 hours to swimming, and jog for no more than the total number of hours of bicycling and swimming. The calories burned per hour by bicycling, jogging, and swimming are 200, 475, and 275, respectively. How many hours should be allotted to each activity to maximize the number of calories burned?

26. Furniture Factory A furniture manufacturer produces small sofas, large sofas, and chairs. The profits per item are, respectively, $60, $60, and $50. The pieces of furniture require the following numbers of labor-hours for their manufacture:

	Carpentry	Upholstery	Finishing
Small sofas	10	30	20
Large sofas	10	30	0
Chairs	10	10	10

The following amounts of labor are available per month: carpentry, at most 1200 hours; upholstery, at most 3000 hours; and finishing, at most 1800 hours. How many each of small sofas, large sofas, and chairs should be manufactured to maximize the profit?

27. **Fast-Food Restaurants** The XYZ Corporation plans to open three different types of fast-food restaurants. Type A restaurants require an initial cash outlay of $600,000, need 15 employees, and are expected to make an annual profit of $40,000. Type B restaurants require an initial cash outlay of $400,000, need 9 employees, and are expected to make an annual profit of $30,000. Type C restaurants require an initial cash outlay of $300,000, need 5 employees, and are expected to make an annual profit of $25,000. The XYZ Corporation has $48,000,000 available for initial outlays, does not want to hire more than 1000 new employees, and would like to open at most 70 restaurants. How many restaurants of each type should be opened to maximize the expected annual profit?

28. **Baby Products** A baby products company makes car seats, strollers, and travel yards. A particular model of car seat requires 5 labor-hours for parts creation, 4 labor-hours for assembly, and 2 labor-hours for finishing and packaging. A particular model of stroller requires 3 labor-hours for parts creation, 3 labor-hours for assembly, and 1 labor-hour for finishing and packaging. A particular model of travel yard requires 2 labor-hours for parts creation, 3 labor-hours for assembly, and 1 labor-hour for finishing and packaging. There are 450 labor-hours available for parts creation, 380 for assembly, and 140 for finishing and packaging. If the profit per car seat is $30, per stroller is $10, and per travel yard is $18, how many of each should be produced to maximize profit? What advice would you give this company?

29. **Potting Soil Mixes** A lawn and garden store creates three different potting mixes sold in 20-pound bags. Mix A contains 12 pounds of peat, 5 pounds of perlite, and 3 pounds of organic material and earns a profit of $3 per 20-pound bag. Mix B contains 10 pounds of peat, 6 pounds of perlite, and 4 pounds of organic material and earns a profit of $5 per 20-pound bag. Mix C contains 8 pounds of peat, 8 pounds of perlite, and 4 pounds of organic material and earns a profit of $6 per 20-pound bag. There are 1200 pounds of peat, 800 pounds of perlite, and 600 pounds of organic material available every week to make these potting mixes. How many 20-pound bags of each mix should be made in order to maximize weekly profit?

30. **Construction** A builder has $6 million and 12 acres of land on which to build cottages, ranch houses, and McMansions. Each cottage requires $1/4$ acre of land and $100,000 to build, and yields a profit of $6000. Each ranch house requires $1/2$ acre of land and $400,000 to build, and yields a profit of $13,000. Each McMansion requires 1 acre of land and $200,000 to

build, and yields a profit of $14,000. How many houses of each type should be built in order to maximize the profit?

31. **Manufacturing-Resources Allocation** A widget manufacturer has the capability of making three types of widgets. Type A widgets require 12 minutes on a lathe, 2 minutes on a grinder, and 2 minutes on a drill press, and produce a profit of $10. Type B widgets require 8 minutes on a lathe, 6 minutes on a grinder, and 2 minutes on a drill press, and produce a profit of $20. Type C widgets require 4 minutes on a lathe, 4 minutes on a grinder, and 8 minutes on a drill press, and produce a profit of $30. Each day, 800 minutes of lathe time, 600 minutes of grinder time, and 300 minutes of drill press time are available. How many of each type of widget should be produced each day to maximize the profit? What is the maximum profit?

32. Maximize $200x + 500y$ subject to the constraints
$$\begin{cases} x + 4y \le 300 \\ x + 2y \le 200 \\ x \ge 0, \quad y \ge 0. \end{cases}$$

33. Maximize $60x + 90y + 300z$ subject to the constraints
$$\begin{cases} x + y + z \le 600 \\ x + 3y \quad\quad \le 600 \\ 2x \quad\quad + z \le 900 \\ x \ge 0, \quad y \ge 0, \quad z \ge 0. \end{cases}$$

TECHNOLOGY EXERCISES

34. Maximize $4x + 6y$ subject to the constraints
$$\begin{cases} x + 4y \le 4 \\ 3x + 2y \le 6 \\ x \ge 0, \quad y \ge 0. \end{cases}$$

35. Maximize $2x + 4y$ subject to the constraints
$$\begin{cases} 5x + y \le 8 \\ x + 2y \le 10 \\ x \ge 0, \quad y \ge 0. \end{cases}$$

36. Maximize $16x + 4y - 20z$ subject to the constraints
$$\begin{cases} 4x + y + 5z \le 20 \\ x + 2y + 4z \le 50 \\ 4x + 10y + z \le 32 \\ x \ge 0, \quad y \ge 0, \quad z \ge 0. \end{cases}$$

37. Maximize $3x + 2y + 2z$ subject to the constraints
$$\begin{cases} x + y + 2z \le 6 \\ x + 5y + 2z \le 20 \\ 2x + y + z \le 4 \\ x \ge 0, \quad y \ge 0, \quad z \ge 0. \end{cases}$$

Solutions to Check Your Understanding 4.2

1. (a) The tableau shown here does not correspond to a maximum, since among the entries $-1, -2, 1, 0, 0$ in the last row, at least one is negative.

   (b) The tableau corresponds to a maximum, since none of the entries $1, 0, 0, 4, 1$ of the last row is negative. Note that it does not matter that the entry -2 in the lower

right-hand corner of the matrix is negative. This number gives the value of M. In this example, -2 is as large as M can become.

2. First, choose the column corresponding to the most-negative entry of the final row—that is, the fourth column. For each

positive entry in the fourth column that is above the horizontal line, compute the ratio with the sixth column. The smallest ratio is 2 and appears in the first row, so the next operation is to pivot around the 2 in the first row of the fourth column.

$$\begin{bmatrix} 0 & 4 & 1 & \fbox{2} & 0 & 4 \\ 1 & 5 & 0 & 1 & 0 & 9 \\ 0 & 2 & 0 & -3 & 1 & 6 \end{bmatrix} \begin{matrix} \frac{4}{2}=2 \\ \frac{9}{1}=9 \\ \\ \end{matrix}$$

$\uparrow$

4.3   The Simplex Method II: Nonstandard and Minimum Problems

In the preceding section, we developed the simplex method and applied it to a number of problems. However, throughout, we restricted ourselves to linear programming problems in standard maximization form. Recall that such problems satisfied three properties: (1) The objective function is to be maximized; (2) each variable must be ≥ 0; and (3) all constraints other than those implied by (2) must be of the form

[linear expression] $\leq$ [nonnegative constant].

In this section, we do what we can to relax these restrictions.

Nonstandard Problems

Let us begin with restriction (3). This could be violated in two ways. First, the constant on the right-hand side of one or more constraints could be negative. Thus, for example, one constraint might be

$$x - y \leq -2.$$

A second way in which restriction (3) could be violated is for some constraints to involve $\geq$ rather than $\leq$. An example of such a constraint is

$$2x + 3y \geq 5.$$

However, we can convert such a constraint into one involving $\leq$ by multiplying both sides of the inequality by -1:

$$-2x - 3y \leq -5.$$

Of course, the right-hand constant is no longer nonnegative. Thus, if we allow negative constants on the right, we can write such constraints in the form

[linear expression] $\leq$ [constant].

Henceforth, the first step in solving a linear programming problem will be to write the constraints in this form. Let us now see how to deal with the phenomenon of negative constants.

EXAMPLE 1   **Working with Nonstandard Inequalities** Maximize the objective function $5x + 10y$ subject to the constraints

$$\begin{cases} x + y \leq 20 \\ 2x - y \geq 10 \\ x \geq 0, \quad y \geq 0. \end{cases}$$

SOLUTION   The first step is to put the second constraint into $\leq$ form. Multiply the second inequality by -1 to obtain

$$\begin{cases} x + y \leq 20 \\ -2x + y \leq -10 \\ x \geq 0, \quad y \geq 0. \end{cases}$$

Just as before, write the linear programming problem as a linear system:

$$\begin{cases} x + y + u & = 20 \\ -2x + y + v & = -10 \\ -5x - 10y + M = & 0 \end{cases}$$

From the linear system, construct the simplex tableau:

$$\begin{array}{c} \\ u \\ v \\ M \end{array} \begin{array}{ccccc} x & y & u & v & M \\ \left[\begin{array}{ccccc|c} 1 & 1 & 1 & 0 & 0 & 20 \\ -2 & 1 & 0 & 1 & 0 & -10 \\ \hline -5 & -10 & 0 & 0 & 1 & 0 \end{array}\right] \end{array}$$

Everything would proceed exactly as before, except that the right-hand column has a −10 in it. This means that the initial value for v is −10, which violates the condition that all variables be ≥ 0. Before we can apply the simplex method of Section 4.2, we must first put the tableau into standard maximization form. This can be done by pivoting to remove the negative entry in the right column.

We choose the pivot element as follows. Look along the left side of the −10 row of the tableau, and locate any negative entry. There is only one: −2. Use the column containing the −2 as the pivot column (that is, use column 1). Now, compute ratios as before. Note, however, that in this circumstance, we compute ratios corresponding to both positive *and* negative entries (except the last) in the pivot column, considering further only positive ratios.

$$\begin{array}{c} \\ u \\ v \\ M \end{array} \begin{array}{ccccc} x & y & u & v & M \\ \left[\begin{array}{ccccc|c} 1 & 1 & 1 & 0 & 0 & 20 \\ \boxed{-2} & 1 & 0 & 1 & 0 & -10 \\ \hline -5 & -10 & 0 & 0 & 1 & 0 \end{array}\right] \end{array} \begin{array}{l} \frac{20}{1} = 20 \\ \frac{-10}{-2} = 5 \end{array}$$

The smallest positive ratio is 5, so we choose −2 as the pivot element. The new tableau is

$$\begin{array}{c} \\ u \\ x \\ M \end{array} \begin{array}{ccccc} x & y & u & v & M \\ \left[\begin{array}{ccccc|c} 0 & \frac{3}{2} & 1 & \frac{1}{2} & 0 & 15 \\ 1 & -\frac{1}{2} & 0 & -\frac{1}{2} & 0 & 5 \\ \hline 0 & -\frac{25}{2} & 0 & -\frac{5}{2} & 1 & 25 \end{array}\right] \end{array}.$$

Note that all entries in the right-hand column are now nonnegative; that is, the corresponding solution has all variables ≥ 0. From here on, we follow the simplex method for tableaux in standard maximization form.

$$\begin{array}{c} \\ u \\ x \\ M \end{array} \begin{array}{ccccc} x & y & u & v & M \\ \left[\begin{array}{ccccc|c} 0 & \boxed{\frac{3}{2}} & 1 & \frac{1}{2} & 0 & 15 \\ 1 & -\frac{1}{2} & 0 & -\frac{1}{2} & 0 & 5 \\ \hline 0 & -\frac{25}{2} & 0 & -\frac{5}{2} & 1 & 25 \end{array}\right] \end{array} \begin{array}{l} 15/\frac{3}{2} = 10 \\ \text{(negative ratio)} \end{array}$$

$$\begin{array}{c} \\ y \\ x \\ M \end{array} \begin{array}{ccccc} x & y & u & v & M \\ \left[\begin{array}{ccccc|c} 0 & 1 & \frac{2}{3} & \frac{1}{3} & 0 & 10 \\ 1 & 0 & \frac{1}{3} & -\frac{1}{3} & 0 & 10 \\ \hline 0 & 0 & \frac{25}{3} & \frac{5}{3} & 1 & 150 \end{array}\right] \end{array}$$

So the maximum value of M is 150, which is attained for $x = 10$, $y = 10$.

>> Now Try Exercise 9

In summary:

The Simplex Method for Problems in Nonstandard Form

1. If necessary, convert all inequalities (except $x \geq 0$, $y \geq 0$, . . .) into the form

$$[\text{linear expression}] \leq [\text{constant}].$$

2. If a negative number appears in the upper part of the last column of the simplex tableau, remove it by pivoting.
   (a) Select one of the negative entries in its row. The column containing the entry will be the pivot column.
   (b) Select the pivot element by determining the least of the positive ratios associated with entries in the pivot column (except the bottom entry).
   (c) Pivot.

3. Repeat step 2 until there are no negative entries in the upper part of the right-hand column of the simplex tableau. (If this process will not remove all negative entries from the upper part of the right-hand column, then the linear programming problem does not have an optimal solution.)

4. Proceed to apply the simplex method for tableaux in standard maximization form.

Minimum Problems

The method we have just developed can be used to solve *minimum* problems as well as maximum problems. Minimizing the objective function f is the same as maximizing $(-1) \cdot f$. This is so, since multiplying an inequality by -1 reverses the direction of the inequality sign. Thus, to apply our method to a minimum problem, we merely multiply the objective function by -1 and turn the problem into a maximum problem.

EXAMPLE 2   **Minimizing the Objective Function** Minimize the objective function $3x + 2y$ subject to the constraints

$$\begin{cases} x + y \geq 10 \\ x - y \leq 15 \\ x \geq 0, \quad y \geq 0. \end{cases}$$

SOLUTION   First, transform the problem so that the first two constraints are in $\leq$ form:

$$\begin{cases} -x - y \leq -10 \\ x - y \leq 15 \\ x \geq 0, \quad y \geq 0. \end{cases}$$

Instead of minimizing $3x + 2y$, let us maximize $(-1)(3x + 2y) = -3x - 2y$. Let $M = -3x - 2y$. Then, our initial simplex tableau reads

$$\begin{array}{c|ccccc|c}
 & x & y & u & v & M & \\
\hline
u & -1 & \boxed{-1} & 1 & 0 & 0 & -10 \\
v & 1 & -1 & 0 & 1 & 0 & 15 \\
\hline
M & 3 & 2 & 0 & 0 & 1 & 0
\end{array}$$

$\frac{-10}{-1} = 10$

(negative ratio)

We first eliminate the -10 in the right-hand column. We have a choice of two negative entries in the -10 row. Let us choose the one in the y column. The ratios are then calculated as before, and we pivot around the circled element. The new tableau is

$$\begin{array}{c c} & \begin{array}{c c c c c} x & y & u & v & M \end{array} \\ \begin{array}{c} u \\ v \\ M \end{array} & \left[\begin{array}{c c c c c|c} 1 & 1 & -1 & 0 & 0 & 10 \\ 2 & 0 & -1 & 1 & 0 & 25 \\ 1 & 0 & 2 & 0 & 1 & -20 \end{array}\right]. \end{array}$$

Since all entries in the bottom row, except the last, are positive, this tableau corresponds to a maximum. Thus, the maximum value of $-3x - 2y$ (subject to the constraints) is -20, and this value occurs for $x = 0$, $y = 10$. Therefore, the *minimum* value of $3x + 2y$ subject to the constraints is 20.   **» Now Try Exercise 11**

Let us now rework an applied problem previously treated (see Example 2 in Section 3.4), this time by using the simplex method. For easy reference, we restate the problem.

EXAMPLE 3   **Transportation Problem** Suppose that a TV dealer has stores in Annapolis and Rockville and warehouses in College Park and Baltimore. The cost of shipping sets from College Park to Annapolis is $6 per set; from College Park to Rockville, $3; from Baltimore to Annapolis, $9; and from Baltimore to Rockville, $5. Suppose that the Annapolis store orders 25 TV sets and the Rockville store 30. Further suppose that the College Park warehouse has a stock of 45 sets, and the Baltimore warehouse 40. What is the most economical way to supply the requested TV sets to the two stores?

SOLUTION   As in the geometric solution, let x be the number of sets shipped from College Park to Rockville, and y the number shipped from College Park to Annapolis. The flow of sets is depicted in Fig. 1.

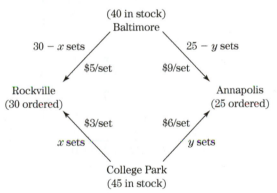

Figure 1

Exactly as in the geometric solution, we reduce the problem to the following algebraic form: Minimize $375 - 2x - 3y$ subject to the constraints

$$\begin{cases} x \le 30, \quad y \le 25 \\ x + y \ge 15 \\ x + y \le 45 \\ x \ge 0, \quad y \ge 0. \end{cases}$$

Two changes are needed. First, instead of minimizing $375 - 2x - 3y$, we maximize $-(375 - 2x - 3y) = 2x + 3y - 375$. Second, we write the constraint $x + y \ge 15$ in the form

$$-x - y \le -15.$$

With these changes made, we can write the linear system.

$$\begin{cases} x & + t & & & & = & 30 \\ & y & + u & & & = & 25 \\ -x - & y & & + v & & = & -15 \\ x + & y & & & + w & = & 45 \\ -2x - & 3y & & & & + M = & -375. \end{cases}$$

From here on, we follow our routine procedure in a mechanical way.

	x	y	t	u	v	w	M		
t	1	0	1	0	0	0	0	30	$\frac{30}{1} = 30$
u	0	1	0	1	0	0	0	25	(ratio undefined)
v	⊝1	−1	0	0	1	0	0	−15	$\frac{-15}{-1} = 15$
w	1	1	0	0	0	1	0	45	$\frac{45}{1} = 45$
M	−2	−3	0	0	0	0	1	−375	

$\uparrow$ (under x)

	x	y	t	u	v	w	M		
t	0	−1	1	0	①1	0	0	15	$\frac{15}{1} = 15$
u	0	1	0	1	0	0	0	25	(ratio undefined)
x	1	1	0	0	−1	0	0	15	(negative ratio)
w	0	0	0	0	1	1	0	30	$\frac{30}{1} = 30$
M	0	−1	0	0	−2	0	1	−345	

$\uparrow$ (under v)

	x	y	t	u	v	w	M		
v	0	−1	1	0	1	0	0	15	(negative ratio)
u	0	1	0	1	0	0	0	25	$\frac{25}{1} = 25$
x	1	0	1	0	0	0	0	30	(ratio undefined)
w	0	①1	−1	0	0	1	0	15	$\frac{15}{1} = 15$
M	0	−3	2	0	0	0	1	−315	

$\uparrow$ (under y)

	x	y	t	u	v	w	M		
v	0	0	0	0	1	1	0	30	(ratio undefined)
u	0	0	①1	1	0	−1	0	10	$\frac{10}{1} = 10$
x	1	0	1	0	0	0	0	30	$\frac{30}{1} = 30$
y	0	1	−1	0	0	1	0	15	(negative ratio)
M	0	0	−1	0	0	3	1	−270	

$\uparrow$ (under t)

	x	y	t	u	v	w	M	
v	0	0	0	0	1	1	0	30
t	0	0	1	1	0	−1	0	10
x	1	0	0	−1	0	1	0	20
y	0	1	0	1	0	0	0	25
M	0	0	0	1	0	2	1	−260

The last tableau corresponds to a maximum. So $2x + 3y - 375$ has a maximum value -260, and therefore, $375 - 2x - 3y$ has a minimum value 260. This value occurs when $x = 20$ and $y = 25$. This is in agreement with our previous graphical solution of the problem. **» Now Try Exercise 21**

The calculations used in Example 3 are not much simpler than those in the original solution. Why, then, should we concern ourselves with the simplex method? For one thing, the simplex method is so mechanical in its execution that it is much easier to program for a computer. For another, our previous method was restricted to problems in two variables. However, suppose that the two warehouses were to deliver their TV sets to three, four, or perhaps even 100 stores. Our previous method could not be applied. However, the simplex method, although yielding very large matrices and very tedious calculations, is applicable. Indeed, this is the method that many industries use to optimize distribution of their products.

Some Further Comments on the Simplex Method

Our discussion has omitted some of the technical complications arising in the simplex method. A complete discussion of these is beyond the scope of this book. However, let us mention three. First, it is possible that a given linear programming problem has more than one solution. This can occur, for example, if there are ties for the choice of pivot column. For instance, if the bottom row of the simplex tableau is

$$\begin{bmatrix} -3 & -7 & 4 & -7 & 1 & 3 \end{bmatrix},$$

then -7 is the most-negative entry and we may choose as pivot column either the second or fourth. In such a circumstance, the pivot column may be chosen arbitrarily. Different choices, however, may lead to different solutions of the problem.

A second difficulty is that a given linear programming problem may have no solution at all. In this case, the method will break down at some point. For example, among the ratios at a given stage, there may be no nonnegative ones to consider. Then we cannot choose a pivot element. Such a breakdown of the method indicates that the associated linear programming problem has no solution.

Finally, whenever there is a tie for the choice of a pivot element, we choose one of the candidates arbitrarily. Occasionally, this may lead to a loop in which the simplex method leads back to a previously encountered tableau. To prevent the loop from recurring, we should then make a different selection from the tied pivot possibilities.

Check Your Understanding 4.3

Solutions can be found following the section exercises.

1. Convert the following minimum problem into a problem in standard maximization form: Minimize $3x + 4y$ subject to the constraints

$$\begin{cases} x - y \geq 0 \\ 3x - 4y \geq 0 \\ x \geq 0, \quad y \geq 0. \end{cases}$$

2. Suppose that the solution of a minimum problem yields the final simplex tableau

x	y	u	v	M	
1	6	-1	0	0	11
0	5	3	1	0	16
0	2	4	0	1	-40

What is the minimum value sought in the original problem?

EXERCISES 4.3

In Exercises 1–4, write each linear programming problem as a maximization problem with all inequalities (except $x \geq 0, y \geq 0, \ldots$) written in the form [linear inequality] $\leq$ [constant].

1. Maximize $3x + 4y$ subject to the constraints

$$\begin{cases} 5x + 3y \geq 6 \\ 2x - 3y \leq 7 \\ x \geq 0, \quad y \geq 0. \end{cases}$$

2. Maximize $7x + 6y + 5z$ subject to the constraints

$$\begin{cases} 3x - 4y - z \geq 1 \\ 2x - y - 3z \geq 7 \\ x \geq 0, \quad y \geq 0, \quad z \geq 0. \end{cases}$$

3. Minimize $x + y + z$ subject to the constraints

$$\begin{cases} 2x + 3y + z \geq 7 \\ 5x + 6y + 7z \geq 8 \\ x \geq 0, \quad y \geq 0, \quad z \geq 0. \end{cases}$$

4. Minimize $375 - 2x - 3y$ subject to the constraints

$$\begin{cases} x \geq 0, y \geq 0 \\ x \leq 30, y \leq 25 \\ x + y \leq 45 \\ x + y \geq 15. \end{cases}$$

In Exercises 5–8, determine the next pivot element for the tableau.

5.

x	y	u	v	M	
1	1	-2	0	0	4
0	-3	0	1	0	-6
0	3	1	0	1	-2

6.

x	y	u	v	M	
-2	2	1	0	0	-6
4	0	0	1	0	5
7	4	0	0	1	0

7.
$$\begin{array}{cccccc|c}
x & y & z & u & v & M & \\
1 & \frac{1}{2} & 0 & 0 & 0 & 0 & 5 \\
0 & 1 & -\frac{1}{2} & 1 & 0 & 0 & 7 \\
0 & -\frac{1}{2} & 3 & 0 & 1 & 0 & -3 \\
\hline
0 & 4 & 6 & 0 & 0 & 1 & -5
\end{array}$$

8.
$$\begin{array}{cccccc|c}
x & y & z & u & v & M & \\
1 & 3 & 0 & -3 & 0 & 0 & 2 \\
0 & 4 & 0 & -2 & 1 & 0 & -6 \\
0 & 0 & 1 & 8 & 0 & 0 & 7 \\
\hline
0 & -5 & 0 & -4 & 0 & 1 & -8
\end{array}$$

In Exercises 9–16, solve the linear programming problem by the simplex method.

9. Maximize $40x + 30y$ subject to the constraints
$$\begin{cases} x + y \le 5 \\ -2x + 3y \ge 12 \\ x \ge 0, \quad y \ge 0. \end{cases}$$

10. Maximize $3x - y$ subject to the constraints
$$\begin{cases} 2x + 5y \le 100 \\ x \quad\quad \ge 10 \\ \quad\quad y \ge 0. \end{cases}$$

11. Minimize $3x + y$ subject to the constraints
$$\begin{cases} x + y \ge 3 \\ 2x \quad\quad \ge 5 \\ x \ge 0, \quad y \ge 0. \end{cases}$$

12. Minimize $3x + 5y + z$ subject to the constraints
$$\begin{cases} x + y + z \ge 20 \\ y + 2z \ge 10 \\ x \ge 0, \quad y \ge 0, \quad z \ge 0. \end{cases}$$

13. Minimize $13x + 4y$ subject to the constraints
$$\begin{cases} y \ge -2x + 11 \\ y \le -x + 10 \\ y \le -\frac{1}{3}x + 6 \\ y \ge -\frac{1}{4}x + 4 \\ x \ge 0, \quad y \ge 0. \end{cases}$$

14. Minimize $500 - 10x - 3y$ subject to the constraints
$$\begin{cases} x + y \le 20 \\ 3x + 2y \ge 50 \\ x \ge 0, \quad y \ge 0. \end{cases}$$

15. Minimize $2x + 7y$ subject to the constraints
$$\begin{cases} 2x + 5y \ge 30 \\ -3x + 5y \ge 5 \\ 8x + 3y \le 101 \\ -9x + 7y \le 42 \\ x \ge 0, \quad y \ge 0. \end{cases}$$

16. Minimize $10x + y$ subject to the constraints
$$\begin{cases} 3x + y \ge 16 \\ x + 2y \ge 12 \\ x \quad\quad \ge 2 \\ x \ge 0, \quad y \ge 0. \end{cases}$$

17. **Nutrition** A dietitian is designing a daily diet that is to contain at least 60 units of protein, 40 units of carbohydrates, and 120 units of fat. The diet is to consist of two types of foods. One serving of food A contains 30 units of protein, 10 units of carbohydrates, and 20 units of fat and costs $3. One serving of food B contains 10 units of protein, 10 units of carbohydrates, and 60 units of fat and costs $1.50. Design the diet that provides the daily requirements at the least cost.

18. **Electronics Manufacture** A manufacturing company has two plants, each capable of producing smartphones, tablets, and Bluetooth headphones. The daily production capacities of each plant are as follows.

	Plant I	Plant II
Smartphones	1000	2000
Tablets	3000	2000
Bluetooth headphones	2000	1000

Plant I costs $15,000 per day to operate, whereas plant II costs $12,000. How many days should each plant be operated to fill an order for 100,000 smartphones, 180,000 tablets, and 100,000 Bluetooth headphones at the minimum cost?

19. **Supply and Demand** An appliance store sells three brands of TV sets, brands A, B, and C. The profit per set is $30 for brand A, $50 for brand B, and $60 for brand C. The total warehouse space allotted to all brands is sufficient for 600 sets, and the inventory is delivered only once per month. At least 100 customers per month will demand brand A, at least 50 will demand brand B, and at least 200 will demand either brand B or brand C. How can the appliance store satisfy all of these constraints and earn maximum profit?

20. **Political Campaign** A citizen decides to campaign for the election of a candidate for city council. Her goal is to generate at least 210 votes by a combination of door-to-door canvassing, letter writing, and phone calls. She figures that each hour of door-to-door canvassing will generate four votes, each hour of letter writing will generate two votes, and each hour on the phone will generate three votes. She would like to devote at least seven hours to phone calls and spend at most half her time at door-to-door canvassing. How much time should she allocate to each task in order to achieve her goal in the least amount of time?

21. **Inventory** A manufacturer of computers must fill orders from two dealers. The computers are stored in two warehouses located at two airports, one in Boston (BOS) and one in Chicago (MDW). The dealers are located in Detroit, Michigan, and Fletcher, North Carolina. There are 50 computers in stock in Boston and 80 in stock in Chicago. The dealer in Detroit orders 40 computers, and the dealer in Fletcher orders 30 computers. The table that follows shows the costs of shipping one computer from each warehouse to each dealer. Find the shipping schedule with the minimum cost. What is the minimum cost?

	Detroit	Fletcher
Boston	$125	$180
Chicago	$100	$160

22. **Inventory** The manufacturer of computers in Exercise 21 gets a revised order from the dealer in Fletcher, now requiring 50 computers. How should the manufacturer adjust the schedule? What is the shipping schedule with the minimum cost? What is the cost?

23. Maximize $x - 2y$ subject to the constraints

$$\begin{cases} 4x + y \le 5 \\ x + 3y \ge 4 \\ x \ge 0, \quad y \ge 0. \end{cases}$$

24. Minimize $30x + 20y$ subject to the constraints

$$\begin{cases} 5x + 10y \ge 3 \\ 3x + 2y \ge 2 \\ x \ge 0, \quad y \ge 0. \end{cases}$$

Solutions to Check Your Understanding 4.3

1. To minimize $3x + 4y$, we maximize $-(3x + 4y) = -3x - 4y$. So the associated maximum problem is as follows: Maximize $-3x - 4y$ subject to the constraints

$$\begin{cases} -x + y \le 0 \\ -3x + 4y \le 0 \\ x \ge 0, \quad y \ge 0. \end{cases}$$

2. The value -40 in the lower right corner gives the solution of the associated *maximum* problem. The minimum value originally sought is the negative of the maximum value—that is, $-(-40) = 40$.

4.4 Sensitivity Analysis and Matrix Formulations of Linear Programming Problems

Sensitivity Analysis of Linear Programming Problems

The simplex method not only provides the optimal solutions to linear programming problems but also gives other useful information. As a matter of fact, each number appearing in the final simplex tableau has an interpretation that not only sheds light on the current situation but also can be used to analyze the benefits of small changes in the available resources.

Consider the final simplex tableau of the furniture manufacturing problem. Since the variable u was introduced to take up the slack in the carpentry inequality, the u column has been labeled *Carpentry*. Similarly, the v column and the w column have been labeled *Finishing* and *Upholstery*, respectively:

	x	y	(Carpentry) u	(Finishing) v	(Upholstery) w	M	
x	1	0	$\frac{1}{3}$	-1	0	0	14
y	0	1	$-\frac{1}{3}$	2	0	0	4
w	0	0	$\frac{4}{3}$	-10	1	0	20
M	0	0	$\frac{10}{3}$	60	0	1	1400

The optimum profit is $1400, which occurs when $x = 14$ chairs and $y = 4$ sofas.

We now consider the following question: If additional labor becomes available, how will this change the production level and the profit? To be specific, suppose that we had 3 more labor-hours available for carpentry. The initial tableau of the furniture manufacturing problem would become

	x	y	(Carpentry) u	(Finishing) v	(Upholstery) w	M	
u	6	3	1	0	0	0	$96 + 3$
v	1	1	0	1	0	0	$18 + 0$
w	2	6	0	0	1	0	$72 + 0$
M	-80	-70	0	0	0	1	$0 + 0$

Note that only the first entry of the right-hand column has been changed. The increment to the right-hand column can be written as the column

$$
\begin{array}{c}
3 \\
0 \\
0 \\
\hline
0
\end{array}
= 3 \cdot
\begin{bmatrix}
1 \\
0 \\
0 \\
0
\end{bmatrix},
$$

which is three times the u column and is referred to as the **increment column**. Now, when the simplex method is performed on the new initial tableau, all of the row operations will affect the increment column exactly as they affected the u column in the original initial tableau. Therefore, the final increment column will be three times the final u column. Hence, the new final tableau will be

	x	y	(Carpentry) u	(Finishing) v	(Upholstery) w	M	
x	1	0	$\frac{1}{3}$	-1	0	0	$14 + 3(\frac{1}{3})$
y	0	1	$-\frac{1}{3}$	2	0	0	$4 + 3(-\frac{1}{3})$
w	0	0	$\frac{4}{3}$	-10	1	0	$20 + 3(\frac{4}{3})$
M	0	0	$\frac{10}{3}$	60	0	1	$1400 + 3(\frac{10}{3})$

Thus, when 3 additional labor-hours of carpentry are available,

$$x = 14 + 3(\tfrac{1}{3}) = 15 \text{ chairs} \quad \text{and} \quad y = 4 + 3(-\tfrac{1}{3}) = 3 \text{ sofas}$$

should be produced. The maximum profit will increase to the new value of

$$M = 1400 + 3(\tfrac{10}{3}) = 1410 \text{ dollars.}$$

The number 3 was arbitrary. If h is a suitable number, then adding h hours of labor for carpentry to the original problem results in the final tableau:

	x	y	(Carpentry) u	(Finishing) v	(Upholstery) w	M	
x	1	0	$\frac{1}{3}$	-1	0	0	$14 + h(\frac{1}{3})$
y	0	1	$-\frac{1}{3}$	2	0	0	$4 + h(-\frac{1}{3})$
w	0	0	$\frac{4}{3}$	-10	1	0	$20 + h(\frac{4}{3})$
M	0	0	$\frac{10}{3}$	60	0	1	$1400 + h(\frac{10}{3})$

The number h can be positive or negative. For instance, if three fewer labor-hours are available for carpentry, then setting $h = -3$ yields the optimal production schedule $x = 13$, $y = 5$ and a profit of \$1390. The only restriction on h is that the three numbers in the upper part of the right-hand column of the final tableau must all be nonnegative. In order to determine the restriction on h, we must determine the values of h for which each of the three numbers in the column will be nonnegative.

$14 + h(\frac{1}{3}) \geq 0$	First number in column
$42 + h \geq 0$	Multiply both sides by 3.
$h \geq -42$	Subtract 42 from both sides.

$4 + h(-\frac{1}{3}) \geq 0$	Second number in column
$-12 + h \leq 0$	Multiply both sides by -3. Reverse inequality sign.
$h \leq 12$	Add 12 to both sides.

$20 + h(\frac{4}{3}) \geq 0$	Third number in column
$15 + h \geq 0$	Multiply both sides by $\frac{3}{4}$.
$h \geq -15$	Subtract 15 from both sides.

All three inequalities will be satisfied provided that $-15 \leq h \leq 12$. These values give the amount of decrease or increase that can occur on the right-hand side of the constraint. We say that the **allowable decrease** in the labor-hours available for carpentry is 15, and the **allowable increase** in the labor-hours available for carpentry is 12.

Similarly, if in the original problem, the amount of labor available for finishing is increased by h labor-hours, the initial tableau becomes

			(Carpentry)	(Finishing)	(Upholstery)		
	x	y	u	v	w	M	
u	6	3	1	0	0	0	$96 + 0$
v	1	1	0	1	0	0	$18 + h$
w	2	6	0	0	1	0	$72 + 0$
M	-80	-70	0	0	0	1	$0 + 0$

The right-hand column of the original tableau was changed by adding an increment column that is h times the v column, and therefore the right-hand column of the new final tableau will be the original final right-hand column plus h times the final v column:

			(Carpentry)	(Finishing)	(Upholstery)		
	x	y	u	v	w	M	
x	1	0	$\frac{1}{3}$	-1	0	0	$14 + h(-1)$
y	0	1	$-\frac{1}{3}$	2	0	0	$4 + h(2)$
w	0	0	$\frac{4}{3}$	-10	1	0	$20 + h(-10)$
M	0	0	$\frac{10}{3}$	60	0	1	$1400 + h(60)$

Hence, if one additional labor-hour were available for finishing, the optimal production schedule would be $x = 13$ chairs, $y = 6$ sofas, and the profit would be $1460.

Finally, if in the original problem, the amount of labor available for upholstery is increased by h labor-hours, the initial and final tableaux become

			(Carpentry)	(Finishing)	(Upholstery)		
	x	y	u	v	w	M	
u	6	3	1	0	0	0	$96 + 0$
v	1	1	0	1	0	0	$18 + 0$
w	2	6	0	0	1	0	$72 + h$
M	-80	-70	0	0	0	1	$0 + 0$

and

			(Carpentry)	(Finishing)	(Upholstery)		
	x	y	u	v	w	M	
x	1	0	$\frac{1}{3}$	-1	0	0	$14 + h(0)$
y	0	1	$-\frac{1}{3}$	2	0	0	$4 + h(0)$
w	0	0	$\frac{4}{3}$	-10	1	0	$20 + h(1)$
M	0	0	$\frac{10}{3}$	60	0	1	$1400 + h(0)$

, respectively.

Therefore, a small change in the amount of labor available for upholstery has no effect on the production schedule or the profit. This makes sense, since we had excess labor available for upholstery in the solution to the original problem. The slack in carpentry and finishing was used up (u and v were 0), but there was slack in the labor available for upholstery (w was 20).

In summary, each of the slack variable columns in the final tableau of the original furniture manufacturing problem gives the sensitivity to change in the production schedule and in the profit due to a suitable change in one of the factors of production. The final values in each of these columns ($u = \frac{10}{3}$, $v = 60$, and $w = 0$) are called the **shadow prices** or *marginal values* of the three factors of production—carpentry, finishing, and upholstery. For example, the shadow price for carpentry is $3.33 ($= \frac{10}{3}$), and represents the increase (or decrease) in profit caused by increasing (or decreasing) the labor-hours available for carpentry by 1 hour.

The following example shows a complete analysis of a new linear programming problem.

EXAMPLE 1　**A Production Problem—Adjusting to Changing Resources** The exclusive Cutting Edge Knife Company manufactures chef's knives and pocket knives. Each chef's knife requires 3 labor-hours, 7 units of steel, and 4 units of wood. Each pocket knife requires 6 labor-hours, 5 units of steel, and 3 units of wood. The profit on each chef's knife is $30, and the profit on each pocket knife is $50. Each day, the company has available 90 labor-hours, 138 units of steel, and 120 units of wood.

(a) How many of each type of knife should the Cutting Edge Knife Company manufacture daily to maximize its profits?

(b) Suppose that an additional 18 units of steel were available each day. What effect would this have on the optimal solution?

(c) Generalize the result in part (b) to the case in which the increase in the number of units of steel available each day is h. (The value of h can be positive or negative.) For what range of values will the result be valid?

SOLUTION　We need to find the number of chef's knives, x, and pocket knives, y, that will maximize the profit, $M = 30x + 50y$, subject to the constraints

$$\begin{cases} 3x + 6y \leq 90 & \text{(labor-hours constraint)} \\ 7x + 5y \leq 138 & \text{(steel constraint)} \\ 4x + 3y \leq 120 & \text{(wood constraint)} \\ x \geq 0, \quad y \geq 0. \end{cases}$$

The initial tableau with slack variables u, v, and w added for labor, steel, and wood, respectively, is

	x	y	u (Labor)	v (Steel)	w (Wood)	M		
u	3	⑥	1	0	0	0	90	$\frac{90}{6} = 15$
v	7	5	0	1	0	0	138	$\frac{138}{5} = 27.6$
w	4	3	0	0	1	0	120	$\frac{120}{3} = 40$
M	−30	−50	0	0	0	1	0	

The proper pivot element is the entry 6 in the y column. The next tableau is

	x	y	u (Labor)	v (Steel)	w (Wood)	M		
y	$\frac{1}{2}$	1	$\frac{1}{6}$	0	0	0	15	$15/\frac{1}{2} = 30$
v	$\frac{9}{2}$	0	$-\frac{5}{6}$	1	0	0	63	$63/\frac{9}{2} = 14$
w	$\frac{5}{2}$	0	$-\frac{1}{2}$	0	1	0	75	$75/\frac{5}{2} = 30$
M	−5	0	$\frac{25}{3}$	0	0	1	750	

The proper pivot element is the entry $\frac{9}{2}$ in the x column. The next tableau is

	x	y	u (Labor)	v (Steel)	w (Wood)	M	
y	0	1	$\frac{7}{27}$	$-\frac{1}{9}$	0	0	8
x	1	0	$-\frac{5}{27}$	$\frac{2}{9}$	0	0	14
w	0	0	$-\frac{1}{27}$	$-\frac{5}{9}$	1	0	40
M	0	0	$\frac{200}{27}$	$\frac{10}{9}$	0	1	820

.

Since there are no negative entries in the last row of this tableau, the simplex method is complete.

(a) The Cutting Edge Knife Company should produce 14 chef's knives and 8 pocket knives each day for a profit of $820. (Since the slack variable w has the value 40, there will be 40 excess units of wood each day.)

(b) Since 18 additional units of steel are available, the final tableau of the revised problem can be obtained from the final tableau of the original problem by adding 18 times the v column to the right-hand column:

$$\begin{array}{c} & & & (Labor) & (Steel) & (Wood) \\ & x & y & u & v & w & M \\ \begin{matrix} y \\ x \\ w \\ M \end{matrix} & \left[\begin{matrix} 0 & 1 & \frac{7}{27} & -\frac{1}{9} & 0 & 0 \\ 1 & 0 & -\frac{5}{27} & \frac{2}{9} & 0 & 0 \\ 0 & 0 & -\frac{1}{27} & -\frac{5}{9} & 1 & 0 \\ 0 & 0 & \frac{200}{27} & \frac{10}{9} & 0 & 1 \end{matrix}\right. & \left.\begin{matrix} 8 + 18(-\frac{1}{9}) \\ 14 + 18(\frac{2}{9}) \\ 40 + 18(-\frac{5}{9}) \\ 820 + 18(\frac{10}{9}) \end{matrix}\right] \end{array}$$

The company should make 4 more chef's knives $[18(\frac{2}{9}) = 4]$ and 2 fewer pocket knives $[18(-\frac{1}{9}) = -2]$. Doing so will increase the profits by $20 $[18(\frac{10}{9}) = 20]$.

(c) With h additional units of steel available, the right-hand column of the final tableau will be similar to the preceding tableau but with 18 replaced by h:

$$\begin{array}{c} & & & (Labor) & (Steel) & (Wood) \\ & x & y & u & v & w & M \\ \begin{matrix} y \\ x \\ w \\ M \end{matrix} & \left[\begin{matrix} 0 & 1 & \frac{7}{27} & -\frac{1}{9} & 0 & 0 \\ 1 & 0 & -\frac{5}{27} & \frac{2}{9} & 0 & 0 \\ 0 & 0 & -\frac{1}{27} & -\frac{5}{9} & 1 & 0 \\ 0 & 0 & \frac{200}{27} & \frac{10}{9} & 0 & 1 \end{matrix}\right. & \left.\begin{matrix} 8 + h(-\frac{1}{9}) \\ 14 + h(\frac{2}{9}) \\ 40 + h(-\frac{5}{9}) \\ 820 + h(\frac{10}{9}) \end{matrix}\right] \end{array}$$

Therefore, the number of chef's knives made should be $14 + h(\frac{2}{9})$, and the number of pocket knives made should be $8 + h(-\frac{1}{9})$. The new profit will be $820 + h(\frac{10}{9})$ dollars. This analysis is valid, provided that each of the entries in the upper part of the right-hand column of the tableau is nonnegative. The restrictions on h given by each of these entries are as follows:

Entry	Restriction
$8 + h(-\frac{1}{9})$	$h \le 72$
$14 + h(\frac{2}{9})$	$h \ge -63$
$40 + h(-\frac{5}{9})$	$h \le 72$

All of these restrictions will be satisfied if h is between -63 and 72. Since $138 - 63 = 75$ and $138 + 72 = 210$, the production problem will have a solution provided there are between 75 and 210 units of steel available.

» Now Try Exercise 1

Matrix Formulations of Linear Programming Problems

Linear programming problems can be neatly stated in terms of matrices. Such formulations provide a convenient way to define the *dual* of a linear programming problem, an important concept that is studied in Section 4.5. To introduce the matrix formulation of a linear programming problem, we first need the concept of inequality for matrices.

Let A and B be two matrices of the same size. We say that A is less than or equal to B (denoted $A \leq B$) if each entry of A is less than or equal to the corresponding entry of B. For instance, we have the following matrix inequalities:

$$\begin{bmatrix} 2 & -3 \\ \frac{1}{2} & 0 \end{bmatrix} \leq \begin{bmatrix} 5 & -1 \\ 1 & 0 \end{bmatrix} \quad \text{and} \quad \begin{bmatrix} 5 \\ 6 \end{bmatrix} \leq \begin{bmatrix} 8 \\ 9 \end{bmatrix}.$$

The symbol $\geq$ has an analogous meaning for matrices.

EXAMPLE 2   **Matrix Formulation of a Linear Programming Problem**   Let

$$A = \begin{bmatrix} 6 & 3 \\ 1 & 1 \\ 2 & 6 \end{bmatrix}, \quad B = \begin{bmatrix} 96 \\ 18 \\ 72 \end{bmatrix}, \quad C = \begin{bmatrix} 80 & 70 \end{bmatrix}, \quad X = \begin{bmatrix} x \\ y \end{bmatrix}.$$

Carry out the indicated matrix multiplications in the following statement: Maximize CX, subject to the constraints $AX \leq B$, $X \geq \mathbf{0}$. ($\mathbf{0}$ is a matrix of all zeros.)

SOLUTION $CX = \begin{bmatrix} 80 & 70 \end{bmatrix} \begin{bmatrix} x \\ y \end{bmatrix} = \begin{bmatrix} 80x + 70y \end{bmatrix}$

$$AX = \begin{bmatrix} 6 & 3 \\ 1 & 1 \\ 2 & 6 \end{bmatrix} \begin{bmatrix} x \\ y \end{bmatrix} = \begin{bmatrix} 6x + 3y \\ x + y \\ 2x + 6y \end{bmatrix}$$

$$AX \leq B \text{ means } \begin{bmatrix} 6x + 3y \\ x + y \\ 2x + 6y \end{bmatrix} \leq \begin{bmatrix} 96 \\ 18 \\ 72 \end{bmatrix} \quad \text{or} \quad \begin{cases} 6x + 3y \leq 96 \\ x + y \leq 18 \\ 2x + 6y \leq 72 \end{cases}$$

$$X \geq \mathbf{0} \text{ means } \begin{bmatrix} x \\ y \end{bmatrix} \geq \begin{bmatrix} 0 \\ 0 \end{bmatrix} \quad \text{or} \quad \begin{cases} x \geq 0 \\ y \geq 0 \end{cases}$$

Hence, the statement "Maximize CX, subject to the constraints $AX \leq B$, $X \geq \mathbf{0}$" is a matrix formulation of the furniture manufacturing problem.   **» Now Try Exercise 13**

Another concept that is needed for the definition of the dual of a linear programming problem is the *transpose of a matrix*.

DEFINITION Transpose of a Matrix   If A is an $m \times n$ matrix, then the matrix A^T (read "A transpose") is the $n \times m$ matrix whose ij^{th} entry is the ji^{th} entry of A. The rows of A^T are the columns of A, and vice versa.

EXAMPLE 3   **Finding the Transpose of a Matrix**   Find the transpose of the following:

(a) $\begin{bmatrix} 3 & -2 & 4 \\ 6 & 5 & 0 \end{bmatrix}$ 　　　　　　　(b) $\begin{bmatrix} 5 \\ 2 \\ 1 \end{bmatrix}$

SOLUTION   (a) Since the given matrix has two rows and three columns, its transpose will have three rows and two columns. The entries of the first row of the transpose will be the entries in the first column of the original matrix:

$$\begin{bmatrix} 3 & 6 \\ & \\ & \end{bmatrix}$$

The entries of the second and third rows are obtained in a similar manner from the second and third columns of the original matrix. Therefore,

$$\begin{bmatrix} 3 & -2 & 4 \\ 6 & 5 & 0 \end{bmatrix}^T = \begin{bmatrix} 3 & 6 \\ -2 & 5 \\ 4 & 0 \end{bmatrix}.$$

(b) Since the given matrix has three rows and one column, its transpose has one row and three columns. That row consists of the single column of the original matrix:

$$\begin{bmatrix} 5 \\ 2 \\ 1 \end{bmatrix}^T = \begin{bmatrix} 5 & 2 & 1 \end{bmatrix}.$$

» Now Try Exercise 7

INCORPORATING

TECHNOLOGY

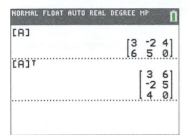

Figure 1

The transpose of a matrix can be obtained with the TI-84 Plus calculator. See Fig. 1. The **T** command is found in the MATRIX/MATH menu.

In Section 3.4, we showed how to generate a sensitivity report with Solver. Table 1 shows the second table in the sensitivity sheet report for the furniture manufacturing problem. The cells $B$1, $B$2, and $B$3 represent carpentry, finishing, and upholstery, respectively. For each of the three factors of production, the table gives its shadow price along with the amounts by which the right side of its constraint can be increased or decreased and still have a feasible solution. *Note:* The spreadsheet uses the expression 1E+30 to represent infinity.

TABLE 1 The Second Table of the Sensitivity Sheet

Constraints

Cell	Name	Final Value	Shadow Price	Constraint R.H. Side	Allowable Increase	Allowable Decrease
$B$1		96	3.333333333	96	12	15
$B$2		18	60	18	2	2
$B$3		52	0	72	1E+30	20

Check Your Understanding 4.4

Solutions can be found following the section exercises.

Consider the furniture manufacturing problem, whose final simplex tableau appears in Section 4.2.

1. Suppose that the number of labor-hours for finishing that are available each day is decreased by 2. What will be the effect on the optimal number of chairs and sofas produced and on the profit?

2. For what range of values of h will a sensitivity analysis on the effect of a change of h labor-hours for finishing be valid?

EXERCISES 4.4

Exercises 1 and 2 refer to the Cutting Edge Knife Company problem of Example 1.

1. Suppose that the number of labor-hours that are available each day is increased by 54. Use sensitivity analysis to determine the effect on the optimal number of knives produced and on the profit.

2. For what values of h will a change of h labor-hours not change the shadow price of labor?

Exercises 3 and 4 refer to the transportation problem of Example 3 in Section 4.3.

3. Suppose that the number of TV sets stocked in the College Park warehouse is increased to 50. What will be the effect on the optimal numbers of TV sets shipped from each warehouse to each store, and what will be the change in the cost?

4. For what range of values of h will a sensitivity analysis on the effect of a change of h in the number of TV sets stocked in the College Park store be valid?

5. Refer to Exercise 29 of Section 4.2. For what values of h will a sensitivity analysis on the effect of a change of h pounds of peat be valid?

6. Consider the nutrition problem in Example 1 of Section 3.3. Solve the problem by the simplex method, and then determine the optimal quantities of soybeans and rice in the diet, and the new cost, if the daily requirement for calories is increased to 1700. For what range of values of h will a sensitivity analysis on the effect of a change of h calories be valid?

In Exercises 7–10, find the transpose of each given matrix.

7. $\begin{bmatrix} 9 & 4 \\ 1 & 8 \\ 1 & -3 \end{bmatrix}$

8. $\begin{bmatrix} 4 \\ 0 \\ 6 \end{bmatrix}$

9. $\begin{bmatrix} 7 & 6 & 5 & 1 \end{bmatrix}$

10. $\begin{bmatrix} 5 & 2 \\ 3 & -1 \end{bmatrix}$

11. Is it true that the transpose of the transpose of a matrix is the original matrix?

12. Give an example of a matrix that is its own transpose.

In Exercises 13 and 14, give the matrix formulation of the linear programming problem.

13. Minimize $7x + 5y + 4z$ subject to

$$\begin{cases} 3x + 8y + 9z \geq 75 \\ x + 2y + 5z \geq 80 \\ 4x + y + 7z \geq 67 \\ x \geq 0, \quad y \geq 0, \quad z \geq 0. \end{cases}$$

14. Maximize $20x + 30y$ subject to

$$\begin{cases} 7x + 8y \leq 55 \\ x + 2y \leq 78 \\ x \quad\quad \leq 25 \\ x \geq 0, \quad y \geq 0. \end{cases}$$

15. Give a matrix formulation of the Cutting Edge Knife Company problem of Example 1.

16. Does every linear programming problem have a matrix formulation? If not, under what conditions will a linear programming problem have a matrix formulation?

In Exercises 17 and 18, let

$$C = \begin{bmatrix} 2 & 3 \end{bmatrix}, \quad x = \begin{bmatrix} x \\ y \end{bmatrix}, \quad A = \begin{bmatrix} 7 & 4 \\ 5 & 8 \\ 1 & 3 \end{bmatrix},$$

$$B = \begin{bmatrix} 33 \\ 44 \\ 55 \end{bmatrix}, \quad \text{and} \quad U = \begin{bmatrix} u \\ v \\ w \end{bmatrix}.$$

17. Give the linear programming problem whose matrix formulation is "Minimize CX, subject to the constraints $AX \geq B$, $X \geq \mathbf{0}$."

18. Give the linear programming problem whose matrix formulation is "Maximize $B^T U$, subject to the constraints $A^T U \leq C^T$, $U \geq \mathbf{0}$."

19. Create a sensitivity report for the transportation problem of Example 2 of Section 3.4. Use the report to determine the shadow price and the allowable increase and decrease for the numbers of TV sets shipped from College Park.

20. Create a sensitivity report for the nutrition problem of Example 1 of Section 3.3. Use the report to determine the shadow prices and the allowable increase and decrease for each of the three nutritional factors.

Solutions to Check Your Understanding 4.4

1. Since the finishing column in the original final tableau is

$$\begin{matrix} -1 \\ 2 \\ -10 \\ \underline{} \\ 60, \end{matrix}$$

the right-hand column in the new tableau is

$$\begin{matrix} 14 + (-2)(-1) \\ 4 + (-2)(2) \\ 20 + (-2)(-10) \\ \overline{1400 + (-2)(60).} \end{matrix}$$

Therefore, the new values of x, y, and M are 16, 0, and 1280.

2. Using h instead of -2 in the marginal analysis, we find that the right-hand column of the new final tableau is

$$\begin{matrix} 14 + h(-1) \\ 4 + h(2) \\ 20 + h(-10) \\ \overline{1400 + h(60).} \end{matrix}$$

Of course, this analysis is valid only if the three numbers above the line are not negative. That is, $14 + h(-1) \geq 0$, $4 + h(2) \geq 0$, and $20 + h(-10) \geq 0$. These three inequalities can be simplified to $h \leq 14$, $h \geq -2$, and $h \leq 2$. Therefore, in order to satisfy all three inequalities, h must be in the range $-2 \leq h \leq 2$.

4.5 Duality

Each linear programming problem may be converted into a related linear programming problem called its *dual*. The dual problem is sometimes easier to solve than the original problem, and moreover, it has the same optimum value. Furthermore, the solution of the dual problem often can provide valuable insights into the original problem. To understand the relationship between a linear programming problem and its dual, it is best to begin with a concrete example.

Problem A

Maximize the objective function $6x + 5y$ subject to the constraints

$$\begin{cases} 4x + 8y \le 32 \\ 3x + 2y \le 12 \\ x \ge 0, \quad y \ge 0. \end{cases}$$

The dual of Problem A is the following problem.

Problem B

Minimize the objective function $32u + 12v$ subject to the constraints

$$\begin{cases} 4u + 3v \ge 6 \\ 8u + 2v \ge 5 \\ u \ge 0, \quad v \ge 0. \end{cases}$$

The relationship between the two problems is easiest to see if we write them in their matrix formulations. Problem A is

$$\text{Maximize } \begin{bmatrix} 6 & 5 \end{bmatrix} \begin{bmatrix} x \\ y \end{bmatrix} \text{ subject to the constraints}$$

$$\begin{bmatrix} 4 & 8 \\ 3 & 2 \end{bmatrix} \begin{bmatrix} x \\ y \end{bmatrix} \le \begin{bmatrix} 32 \\ 12 \end{bmatrix} \quad \text{and} \quad \begin{bmatrix} x \\ y \end{bmatrix} \ge \begin{bmatrix} 0 \\ 0 \end{bmatrix}.$$

Problem B is

$$\text{Minimize } \begin{bmatrix} 32 & 12 \end{bmatrix} \begin{bmatrix} u \\ v \end{bmatrix} \text{ subject to the constraints}$$

$$\begin{bmatrix} 4 & 3 \\ 8 & 2 \end{bmatrix} \begin{bmatrix} u \\ v \end{bmatrix} \ge \begin{bmatrix} 6 \\ 5 \end{bmatrix} \quad \text{and} \quad \begin{bmatrix} u \\ v \end{bmatrix} \ge \begin{bmatrix} 0 \\ 0 \end{bmatrix}.$$

Each of the numeric matrices in Problem B is the transpose of one of the matrices in Problem A. Let

$$C = \begin{bmatrix} 6 & 5 \end{bmatrix}, \quad X = \begin{bmatrix} x \\ y \end{bmatrix}, \quad A = \begin{bmatrix} 4 & 8 \\ 3 & 2 \end{bmatrix}, \quad B = \begin{bmatrix} 32 \\ 12 \end{bmatrix}, \quad U = \begin{bmatrix} u \\ v \end{bmatrix}, \quad \mathbf{0} = \begin{bmatrix} 0 \\ 0 \end{bmatrix}.$$

Problem A is

"Maximize CX subject to the constraints $AX \le B, X \ge \mathbf{0}$."

Problem B is

"Minimize $B^T U$ subject to the constraints $A^T U \ge C^T, U \ge \mathbf{0}$."

Problem A is referred to as the **primal problem** and Problem B as its **dual problem**.

Note that Problem A is a **standard maximization problem**. That is, all of the inequalities involve $\le$ except for $x \ge 0$ and $y \ge 0$. Problem B is a **standard minimization problem** in that all of its inequalities are $\ge$. The coefficients of the objective function for Problem A are the numbers on the right-hand side of the inequalities of Problem B and vice versa. The coefficient matrices for the left-hand sides of the inequalities are transposes of one another. In an analogous way, we can start with a standard minimization problem and define its dual to be a standard maximization problem. Any linear programming problem can be put into one of these two standard forms. (If an inequality points in the wrong direction, we need only multiply it by -1.) Therefore, every linear programming problem has a dual.

> **The Dual of a Linear Programming Problem**
>
> **1.** If the original (primal) problem has the form
>
> $$\text{Maximize } CX \text{ subject to the constraints } AX \leq B, X \geq \mathbf{0},$$
>
> then the dual problem is
>
> $$\text{Minimize } B^T U \text{ subject to the constraints } A^T U \geq C^T, U \geq \mathbf{0}.$$
>
> **2.** If the original (primal) problem has the form
>
> $$\text{Minimize } CX \text{ subject to the constraints } AX \geq B, X \geq \mathbf{0},$$
>
> then the dual problem is
>
> $$\text{Maximize } B^T U \text{ subject to the constraints } A^T U \leq C^T, U \geq \mathbf{0}.$$

EXAMPLE 1   **Finding the Dual in Standard Form** Determine the dual of the following linear programming problem: Minimize $18x + 20y + 2z$ subject to the constraints

$$\begin{cases} 3x - 5y - 2z \leq 4 \\ 6x \quad\quad - 8z \geq 9 \\ x \geq 0, \quad y \geq 0, \quad z \geq 0. \end{cases}$$

SOLUTION   We first put the problem into standard form. Since the primal problem is a minimization problem, we must write all constraints with the inequality sign $\geq$. To put the first inequality in this form, we multiply by -1 to obtain

$$-3x + 5y + 2z \geq -4.$$

We now write the problem in matrix form:

$$\text{Minimize } \begin{bmatrix} 18 & 20 & 2 \end{bmatrix} \begin{bmatrix} x \\ y \\ z \end{bmatrix} \text{ subject to the constraints}$$

$$\begin{bmatrix} -3 & 5 & 2 \\ 6 & 0 & -8 \end{bmatrix} \begin{bmatrix} x \\ y \\ z \end{bmatrix} \geq \begin{bmatrix} -4 \\ 9 \end{bmatrix} \quad \text{and} \quad \begin{bmatrix} x \\ y \\ z \end{bmatrix} \geq \begin{bmatrix} 0 \\ 0 \\ 0 \end{bmatrix}.$$

The dual problem is

$$\text{Maximize } \begin{bmatrix} -4 & 9 \end{bmatrix} \begin{bmatrix} u \\ v \end{bmatrix} \text{ subject to the constraints}$$

$$\begin{bmatrix} -3 & 6 \\ 5 & 0 \\ 2 & -8 \end{bmatrix} \begin{bmatrix} u \\ v \end{bmatrix} \leq \begin{bmatrix} 18 \\ 20 \\ 2 \end{bmatrix} \quad \text{and} \quad \begin{bmatrix} u \\ v \end{bmatrix} \geq \begin{bmatrix} 0 \\ 0 \end{bmatrix}.$$

Multiplying the matrices, we obtain the following:
Maximize $-4u + 9v$ subject to

$$\begin{cases} -3u + 6v \leq 18 \\ 5u \quad\quad \leq 20 \\ 2u - 8v \leq 2 \\ u \geq 0, \quad v \geq 0. \end{cases}$$

     » Now Try Exercise 1

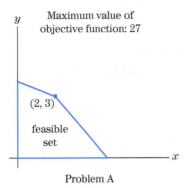

Problem A

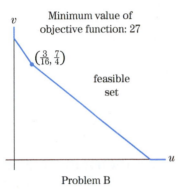

Problem B

Figure 1

Let us now return to Problems A and B to examine the connection between the solutions of a linear programming problem and its dual problem. Problems A and B both involve two variables and hence can be solved by the geometric method of Chapter 3. Figure 1 shows their respective feasible sets and the vertices that yield the optimum values of the objective functions. The feasible sets do not look alike, and the optimal vertices are different. However, both problems have the same optimum value, 27. The relationship between the two problems is brought into even sharper focus by looking at the final tableaux that arise when the two problems are solved by the simplex method. *Note:* In Problem B, the original variables are u and v and the slack variables have been named x and y.

FINAL TABLEAUX

$$
\text{Problem A}\quad
\begin{array}{c}
y \\ x \\ M
\end{array}
\begin{array}{c}
x \\ \hline
\left[\begin{array}{ccccc|c}
0 & 1 & \frac{3}{16} & -\frac{1}{4} & 0 & 3 \\
1 & 0 & -\frac{1}{8} & \frac{1}{2} & 0 & 2 \\
0 & 0 & \frac{3}{16} & \frac{7}{4} & 1 & 27
\end{array}\right]
\end{array}
$$

$$
\begin{array}{ccccc}
x & y & u & v & M
\end{array}
$$

$$
\text{Problem B}\quad
\begin{array}{c}
v \\ u \\ M
\end{array}
\left[\begin{array}{ccccc|c}
0 & 1 & -\frac{1}{2} & \frac{1}{4} & 0 & \frac{7}{4} \\
1 & 0 & \frac{1}{8} & -\frac{3}{16} & 0 & \frac{3}{16} \\
0 & 0 & 2 & 3 & 1 & -27
\end{array}\right]
$$

$$
\begin{array}{ccccc}
u & v & x & y & M
\end{array}
$$

The final tableau for Problem A contains the solution to Problem B ($u = \frac{3}{16}$, $v = \frac{7}{4}$) in the final entries of the u and v columns. Similarly, the final tableau for Problem B gives the solution to Problem A ($x = 2$, $y = 3$) in the final entries of its x and y columns. This situation always occurs. The solutions to a linear programming problem and its dual problem may be obtained simultaneously by solving just one of the problems by the simplex method and applying the following theorem:

Fundamental Theorem of Duality

1. If either the primal problem or the dual problem has an optimal solution, then they both have an optimal solution and their objective functions have the same value at these optimal points.
2. The solution of one of these problems by the simplex method yields the solution of the other problem as the final entries in the columns associated with the slack variables.

EXAMPLE 2 **Solving a Minimization Problem by Using its Dual** Solve the linear programming problem of Example 1 by applying the simplex method to its dual problem.

SOLUTION In the solution to Example 1, the dual problem is as follows:
Maximize $-4u + 9v$ subject to the constraints

$$
\begin{cases}
-3u + 6v \le 18 \\
5u \le 20 \\
2u - 8v \le 2 \\
u \ge 0, \quad v \ge 0.
\end{cases}
$$

Since there are three nontrivial inequalities, the simplex method calls for three slack variables. Denote the slack variables by x, y, and z. Let $M = -4u + 9v$, and apply the simplex method.

$$
\begin{array}{c}
\begin{array}{ccccccc} & u & v & x & y & z & M \end{array} \\
\begin{array}{c} x \\ y \\ z \\ M \end{array}
\left[
\begin{array}{cccccc|c}
-3 & ⑥ & 1 & 0 & 0 & 0 & 18 \\
5 & 0 & 0 & 1 & 0 & 0 & 20 \\
2 & -8 & 0 & 0 & 1 & 0 & 2 \\
4 & -9 & 0 & 0 & 0 & 1 & 0
\end{array}
\right]
\begin{array}{l}
\frac{18}{6}=3 \\
\text{(ratio undefined)} \\
\text{(negative ratio)} \\

\end{array}
\end{array}
$$

$$
\begin{array}{c}
\begin{array}{ccccccc} & u & v & x & y & z & M \end{array} \\
\begin{array}{c} v \\ y \\ z \\ M \end{array}
\left[
\begin{array}{cccccc|c}
-\frac{1}{2} & 1 & \frac{1}{6} & 0 & 0 & 0 & 3 \\
⑤ & 0 & 0 & 1 & 0 & 0 & 20 \\
-2 & 0 & \frac{4}{3} & 0 & 1 & 0 & 26 \\
-\frac{1}{2} & 0 & \frac{3}{2} & 0 & 0 & 1 & 27
\end{array}
\right]
\begin{array}{l}
\text{(negative ratio)} \\
\frac{20}{5}=4 \\
\text{(negative ratio)} \\

\end{array}
\end{array}
$$

$$
\begin{array}{c}
\begin{array}{ccccccc} & u & v & x & y & z & M \end{array} \\
\begin{array}{c} v \\ u \\ z \\ M \end{array}
\left[
\begin{array}{cccccc|c}
0 & 1 & \frac{1}{6} & \frac{1}{10} & 0 & 0 & 5 \\
1 & 0 & 0 & \frac{1}{5} & 0 & 0 & 4 \\
0 & 0 & \frac{4}{3} & \frac{2}{5} & 1 & 0 & 34 \\
0 & 0 & \frac{3}{2} & \frac{1}{10} & 0 & 1 & 29
\end{array}
\right]
\end{array}
$$

Since the maximum value of the dual problem is 29, we know that the minimum value of the original problem is also 29. Looking at the last row of the final tableau, we conclude that this minimum value is assumed when $x = \frac{3}{2}$, $y = \frac{1}{10}$, and $z = 0$.

≫ Now Try Exercise 11

In Example 2, the dual problem was easier to solve than the original problem given in Example 1. Thus, we see how consideration of the dual problem may simplify the solution of linear programming problems in some cases.

An Economic Interpretation of the Dual Problem

To illustrate the interpretation of the dual problem, let us reconsider the furniture manufacturing problem. Recall that this problem asked us to maximize the profit from the sale of x chairs and y sofas, subject to limitations on the amount of labor available for carpentry, finishing, and upholstery. In mathematical terms, the problem required us to maximize $80x + 70y$ subject to the constraints

$$
\begin{cases}
6x + 3y \le 96 \\
x + y \le 18 \\
2x + 6y \le 72 \\
x \ge 0, \quad y \ge 0.
\end{cases}
$$

Its dual problem is to minimize $96u + 18v + 72w$ subject to the constraints

$$
\begin{cases}
6u + v + 2w \ge 80 \\
3u + v + 6w \ge 70 \\
u \ge 0, \quad v \ge 0, \quad w \ge 0.
\end{cases}
$$

The variables u, v, and w can be assigned a meaning so that the dual problem has a significant interpretation in terms of the original problem.

First, recall the following table of data (labor-hours except as noted):

	Chair	Sofa	Available Labor-Hours
Carpentry	6	3	96
Finishing	1	1	18
Upholstery	2	6	72
Profit	$80	$70	

Suppose that we have an opportunity to hire out all of our workers. Suppose that hiring out carpenters will yield a profit of u dollars per hour, the finishers v dollars per hour, and the upholsterers w dollars per hour. Of course, u, v, and w all must be ≥ 0. However, there are other constraints that, reasonably, we should impose. Any scheme for hiring out the workers should generate at least as much profit as is currently being generated in the construction of chairs and sofas. In terms of the potential profits from hiring the workers out, the labor involved in constructing a chair will generate

$$6u + v + 2w$$

dollars of profit. And this amount should be at least equal to the \$80 profit that could be earned by using the labor to construct a chair. That is, we have the constraint

$$6u + v + 2w \geq 80.$$

Similarly, considering the labor involved in building a sofa, we derive the constraint

$$3u + v + 6w \geq 70.$$

Since there are available 96 hours of carpentry, 18 hours of finishing, and 72 hours of upholstery, the total profit from hiring out the workers would be

$$96u + 18v + 72w.$$

Thus, the problem of determining the smallest acceptable profit from hiring out the workers is equivalent to the following: Minimize $96u + 18v + 72w$ subject to the constraints

$$\begin{cases} 6u + v + 2w \geq 80 \\ 3u + v + 6w \geq 70 \\ u \geq 0, \quad v \geq 0, \quad w \geq 0. \end{cases}$$

This is just the dual of the furniture manufacturing problem. The values u, v, and w are measures of the value of an hour's labor by each type of worker. Earlier, we referred to them as *shadow prices*. The fundamental theorem of duality asserts that the minimum acceptable profit that can be achieved by hiring the workers out is equal to the maximum profit that can be generated if they make furniture.

As we saw in the furniture manufacturing problem, the maximum profit, $M = 1400$, is achieved when

$$x = 14, \quad y = 4, \quad u = 0, \quad v = 0, \quad w = 20.$$

For the dual problem, we can read from the final tableau of the primal problem that the minimum acceptable profit, $M = 1400$, is achieved when

$$u = \tfrac{10}{3}, \quad v = 60, \quad w = 0, \quad x = 0, \quad y = 0.$$

The solution of the primal problem gives the activity level that meets the constraints imposed by the resources and provides the maximum profit. Such a model is called an **allocation problem**. On the other hand, the solution of the dual problem assigns values to each of the resources of production. The solution of the dual problem might be used by an insurance salesperson or by an accountant to impute a value to each resource. It is referred to as a **valuation problem**.

We summarize the situation briefly as follows:

If the original problem is a maximization problem,

$$\text{Maximize } CX \text{ subject to } AX \leq B \text{ and } X \geq \mathbf{0},$$

then we interpret the solution matrix X as the **activity matrix** in which each entry gives the optimal level of each activity. The matrix B is the **capacity matrix**, or **resources matrix**, where each entry represents the available amount of a (scarce) resource. The matrix C is the **profit matrix**, whose entries are the unit profits for each activity represented in X.

The dual solution matrix U is the **imputed value matrix**, which gives the imputed value of each of the resources in the production process.

If the original problem is a minimization problem,

$$\text{Minimize } CX \text{ subject to } AX \geq B \text{ and } X \geq \mathbf{0},$$

then we interpret X as the activity matrix, in which each entry gives the optimal level of an activity. B is the **requirements matrix**, in which each entry is a minimum required level of production for some commodity. C is the **cost matrix**, where each entry is the unit cost of the corresponding activity in X. The dual solution matrix U is the **imputed cost matrix**, whose entries are the costs imputed to the required commodities.

We see that $x = 14 > 0$ in the solution to the primal furniture manufacturing problem, and $x = 0$ in the dual. Also, $u = \frac{10}{3} > 0$ in the dual and $u = 0$ in the primal problem. In general, this is the complementary nature of the solutions to the primal and dual problems. Each variable that has a positive value in the solution of the primal problem has the value 0 in the solution of the dual problem. Similarly, if a variable has a positive value in the solution to the dual problem, then it has the value 0 in the solution of the primal problem. This result is called the **principle of complementary slackness**.

The economic interpretation of complementary slackness is this: If a slack variable from the primal problem is positive, then having available more of the corresponding resource cannot improve the value of the objective function. For instance, since $w > 0$ in the solution of the primal problem and $2x + 6y + w = 72$, not all of the labor available for upholstery will be used in the optimal production schedule. Therefore, the shadow price of the resource is zero. As we saw in Section 4.4, the shadow price of a resource is the final value in the column of its slack variable in the primal problem. But this number is the value of the main variable in the solution of the dual problem.

When we gave an economic interpretation to the dual of the furniture manufacturing problem, we first had to assign units to each of the variables of the dual problem. For instance, u was in units of profit per labor-hour of carpentry, or

$$\frac{\text{profit}}{\text{labor-hours of carpentry}}.$$

The systematic procedure that follows can often be used to obtain the units for the variables of the dual problem from units appearing in the primal problem. Assume that the primal problem is stated in one of the two forms given previously.

1. Replace each entry of the matrix A by its units, written in fraction form. Label each column and row with the corresponding variable.
2. Replace each entry of the matrix C by its units, written in fraction form.
3. To find the units for a variable of the dual problem, select any entry in its row in A, divide the corresponding entry in C by the entry chosen in A, and simplify the fraction.

For the furniture manufacturing problem, the matrices are

$$
\begin{array}{c}
u \\ v \\ w
\end{array}
\left[
\begin{array}{cc}
\dfrac{\text{labor-hours of carpentry}}{\text{chair}} & \dfrac{\text{labor-hours of carpentry}}{\text{sofa}} \\[2ex]
\dfrac{\text{labor-hours of finishing}}{\text{chair}} & \dfrac{\text{labor-hours of finishing}}{\text{sofa}} \\[2ex]
\dfrac{\text{labor-hours of upholstery}}{\text{chair}} & \dfrac{\text{labor-hours of upholstery}}{\text{sofa}}
\end{array}
\right] = A
$$

$$
\left[
\begin{array}{cc}
\dfrac{\text{profit}}{\text{chair}} & \dfrac{\text{profit}}{\text{sofa}}
\end{array}
\right] = C.
$$

Using the first entry in the row labeled u, we find that the units for u in the dual problem are

$$\frac{\text{profit}}{\text{chair}} \div \frac{\text{labor-hours of carpentry}}{\text{chair}} = \frac{\text{profit}}{\text{chair}} \cdot \frac{\text{chair}}{\text{labor-hours of carpentry}}$$

$$= \frac{\text{profit}}{\text{labor-hours of carpentry}}.$$

Here, u is measured in dollars per labor-hour of carpentry.

EXAMPLE 3   **Optimizing Dietary Allotments** A rancher needs to provide a daily dietary supplement to the minks on his ranch. He needs 6 units of protein and 5 units of carbohydrates to add to their regular feed each day. Bran X costs 32 cents per ounce and supplies 4 units of protein and 8 units of carbohydrates per ounce. Wheatchips cost 12 cents per ounce, and each ounce supplies 3 units of protein and 2 units of carbohydrates.

(a) Determine the mixture of Bran X and Wheatchips that will meet the daily requirements at minimum cost.

(b) Solve and interpret the dual problem.

SOLUTION   (a) Let x be the number of ounces of Bran X and y the number of ounces of Wheatchips to be added to the feed. Then we must find the values of x and y that minimize $32x + 12y$ subject to the constraints

$$\begin{cases} 4x + 3y \geq 6 \\ 8x + 2y \geq 5 \\ x \geq 0, \quad y \geq 0. \end{cases}$$

For simplicity, we solve the dual problem. Let u and v be the dual variables. We must find the values of u and v that maximize $6u + 5v$ subject to the constraints

$$\begin{cases} 4u + 8v \leq 32 \\ 3u + 2v \leq 12 \\ u \geq 0, \quad v \geq 0. \end{cases}$$

The final tableau for the dual problem (with slack variables x and y) is

	u	v	x	y	M	
v	0	1	$\frac{3}{16}$	$-\frac{1}{4}$	0	3
u	1	0	$-\frac{1}{8}$	$\frac{1}{2}$	0	2
M	0	0	$\frac{3}{16}$	$\frac{7}{4}$	1	27

The solution to the rancher's problem is to add $x = \frac{3}{16}$ ounce of Bran X and $y = \frac{7}{4} = 1\frac{3}{4}$ ounces of Wheatchips to the daily feed at a minimum cost of 27 cents per day.

(b) The solution of the dual problem can be read from the preceding tableau: $u = 2$ and $v = 3$. To find the units of u and v, we construct a chart from the original problem:

	x	y
u	$\dfrac{\text{units of protein}}{\text{ounce of Bran X}}$	$\dfrac{\text{units of protein}}{\text{ounce of Wheatchips}}$
v	$\dfrac{\text{units of carbohydrates}}{\text{ounce of Bran X}}$	$\dfrac{\text{units of carbohydrates}}{\text{ounce of Wheatchips}}$
	$\dfrac{\text{cents}}{\text{ounce of Bran X}}$	$\dfrac{\text{cents}}{\text{ounce of Wheatchips}}$

The units for u in the dual problem are

$$\frac{\text{cents}}{\text{ounce of Bran X}} \div \frac{\text{units of protein}}{\text{ounce of Bran X}} = \frac{\text{cents}}{\text{ounce of Bran X}} \cdot \frac{\text{ounce of Bran X}}{\text{units of protein}}$$

$$= \frac{\text{cents}}{\text{units of protein}}.$$

We can interpret the dual problem this way: If someone were to provide the perfect daily supplement for the minks with 6 units of protein and 5 units of carbohydrates, the rancher would expect to pay 27 cents. He certainly would pay no more, since he could mix his own supplement for that price per day. The value of the protein is 2 cents per unit, and the value of the carbohydrates is 3 cents per unit.   ≪

A Useful Application of the Dual Problem

One type of decision that businesses must make is whether or not to introduce new products. The following example uses a linear programming problem and its dual to determine the proper course of action.

EXAMPLE 4

Introducing New Products Consider the furniture manufacturing problem once again. Suppose that the manufacturer has the same resources but is considering adding a new product to his line, love seats. The manufacture of a love seat requires 3 hours of carpentry, 2 hours of finishing, and 4 hours of upholstery. What profit must the manufacturer gain per love seat in order to justify adding love seats to his product line?

SOLUTION   Let z denote the number of love seats to be produced each day, and let p be the profit per love seat. The new furniture manufacturing problem is as follows:
  Maximize $80x + 70y + pz$ subject to

$$\begin{cases} 6x + 3y + 3z \le 96 \\ x + y + 2z \le 18 \\ 2x + 6y + 4z \le 72 \\ x \ge 0, \quad y \ge 0, \quad z \ge 0. \end{cases}$$

What is the dual of this problem? All inequalities are $\le$ in the original maximization problem, so we may proceed directly to write the dual problem.
  Minimize $96u + 18v + 72w$ subject to

$$\begin{cases} 6u + v + 2w \ge 80 \\ 3u + v + 6w \ge 70 \\ 3u + 2v + 4w \ge p \\ u \ge 0, \quad v \ge 0, \quad w \ge 0. \end{cases}$$

Whereas the new furniture manufacturing problem has one more variable than the original problem, its dual has one more constraint than the dual of the original problem,

$$3u + 2v + 4w \ge p.$$

If the optimal solution to the original problem (just making chairs and sofas) were to remain optimal in the new problem, the variable z would not enter the set of group II variables. If that were the case, the solution to the dual problem would also remain optimal. But that solution was

$$u = \tfrac{10}{3}, \qquad v = 60, \qquad w = 0.$$

And the new dual problem would require that $3u + 2v + 4w \ge p$. Since

$$3\left(\tfrac{10}{3}\right) + 2(60) + 4(0) = 130,$$

if the profit per love seat is at most $130, the previous solution will remain optimal. That is, the manufacturer should make 14 chairs, 4 sofas, and 0 love seats. However, if the profit per love seat exceeds $130, the variable z will enter the set of group II variables, and we will find an optimal production schedule in which $z > 0$.

» Now Try Exercise 19

Check Your Understanding 4.5

Solutions can be found following the section exercises.

A linear programming problem involving three variables and four nontrivial inequalities has the number 52 as the maximum value of its objective function.

1. How many variables and nontrivial inequalities will the dual problem have?

2. What is the optimum value for the objective function of the dual problem?

EXERCISES 4.5

In Exercises 1–6, determine the dual problem of the given linear programming problem.

1. Maximize $4x + 2y$ subject to the constraints
$$\begin{cases} 5x + y \le 80 \\ 3x + 2y \le 76 \\ x \ge 0, \quad y \ge 0. \end{cases}$$

2. Minimize $30x + 60y + 50z$ subject to the constraints
$$\begin{cases} 5x + 3y + z \ge 2 \\ x + 2y + z \ge 3 \\ x \ge 0, \quad y \ge 0, \quad z \ge 0. \end{cases}$$

3. Minimize $10x + 12y$ subject to the constraints
$$\begin{cases} x + 2y \ge 1 \\ -x + y \ge 2 \\ 2x + 3y \ge 1 \\ x \ge 0, \quad y \ge 0. \end{cases}$$

4. Maximize $80x + 70y + 120z$ subject to the constraints
$$\begin{cases} 6x + 3y + 8z \le 768 \\ x + y + 2z \le 144 \\ 2x + 5y \qquad \le 216 \\ x \ge 0, \quad y \ge 0, \quad z \ge 0. \end{cases}$$

5. Minimize $3x + 5y + z$ subject to the constraints
$$\begin{cases} 2x - 4y - 6z \le 7 \\ y \ge 10 - 8x - 9z \\ x \ge 0, \quad y \ge 0, \quad z \ge 0. \end{cases}$$

6. Maximize $2x - 3y + 4z - 5w$ subject to the constraints
$$\begin{cases} x + y + z + w - 6 \le 10 \\ 7x + 9y - 4z - 3w \qquad \ge 5 \\ x \ge 0, \quad y \ge 0, \quad z \ge 0, \quad w \ge 0. \end{cases}$$

7. The final simplex tableau for the linear programming problem of Exercise 1 is as follows. Give the solution to the problem and to its dual.

	x	y	u	v	M	
x	1	0	$\frac{2}{7}$	$-\frac{1}{7}$	0	12
y	0	1	$-\frac{3}{7}$	$\frac{5}{7}$	0	20
M	0	0	$\frac{2}{7}$	$\frac{6}{7}$	1	88

8. The final simplex tableau for the *dual* of the linear programming problem of Exercise 2 is as follows. Give the solution to the problem and to its dual.

	u	v	x	y	z	M	
v	5	1	1	0	0	0	30
y	-7	0	-2	1	0	0	0
z	-4	0	-1	0	1	0	20
M	13	0	3	0	0	1	90

9. The final simplex tableau for the *dual* of the linear programming problem of Exercise 3 is as follows. Give the solution to the problem and to its dual.

	u	v	w	x	y	M	
x	3	0	5	1	1	0	22
v	2	1	3	0	1	0	12
M	3	0	5	0	2	1	24

10. The final simplex tableau for the linear programming problem of Exercise 4 is as follows. Give the solution to the problem and to its dual.

	x	y	z	u	v	w	M	
x	1	0	0	$\frac{5}{12}$	$-\frac{5}{3}$	$\frac{1}{12}$	0	98
z	0	0	1	$-\frac{1}{8}$	1	$-\frac{1}{8}$	0	21
y	0	1	0	$-\frac{1}{6}$	$\frac{2}{3}$	$\frac{1}{6}$	0	4
M	0	0	0	$\frac{20}{3}$	$\frac{100}{3}$	$\frac{10}{3}$	1	10,640

In Exercises 11–14, determine the dual problem. Solve either the original problem or its dual by the simplex method, and then give the solutions to both.

11. Minimize $3x + y$ subject to the constraints
$$\begin{cases} x + y \ge 3 \\ 2x \qquad \ge 5 \\ x \ge 0, \quad y \ge 0. \end{cases}$$

12. Minimize $3x + 5y + z$ subject to the constraints
$$\begin{cases} x + y + z \ge 20 \\ y + 2z \ge 0 \\ x \ge 0, \quad y \ge 0, \quad z \ge 0. \end{cases}$$

13. Maximize $10x + 12y + 10z$ subject to the constraints

$$\begin{cases} x - 2y & \leq 6 \\ 3x & + z \leq 9 \\ & y + 3z \leq 12 \\ x \geq 0, & y \geq 0, \quad z \geq 0. \end{cases}$$

14. Maximize $x + 3y$ subject to the constraints

$$\begin{cases} x + y \leq 7 \\ x + 2y \leq 10 \\ x \geq 0, \quad y \geq 0. \end{cases}$$

15. Cutting Edge Knife Co. Give an economic interpretation to the dual of the Cutting Edge Knife Company problem of Example 1 of Section 4.4.

16. Electronics Manufacture Give an economic interpretation to the dual problem of Exercise 18 of Section 4.3.

17. Mining Give an economic interpretation to the dual of the mining problem of Exercise 10 of Section 3.2 and Exercise 18 of Section 3.3.

18. Nutrition Give an economic interpretation to the dual of the nutrition problem of Example 1 of Section 3.3.

19. Cutting Edge Knife Co. Consider the Cutting Edge Knife Company problem of Example 1 of Section 4.4. Suppose that the company is thinking of also making Bowie knives. If each Bowie knife requires 4 labor-hours, 6 units of steel, and 2 units of wood, what profit must be realized per knife to justify adding this product?

TECHNOLOGY EXERCISES

Use the dual to solve Exercises 20 and 21.

20. Minimize $3x + y$ subject to the constraints

$$\begin{cases} x + y \geq 3 \\ 2x \quad \geq 5 \\ x \geq 0, \quad y \geq 0. \end{cases}$$

21. Minimize $16x + 42y$ subject to the constraints

$$\begin{cases} x + 3y \geq 5 \\ 2x + 4y \geq 8 \\ x \geq 0, \quad y \geq 0. \end{cases}$$

Solutions to Check Your Understanding 4.5

1. The dual problem will have four variables and three nontrivial inequalities. The number of variables in the dual problem is always the same as the number of nontrivial inequalities in the original problem. The number of nontrivial inequalities in the dual problem is the same as the number of variables in the original problem.

2. The optimum is a minimum value of 52. The original problem and the dual problem always have the same optimum values. However, if this value is a maximum for one of the problems, it will be a minimum for the other.

CHAPTER 4   Summary

KEY TERMS AND CONCEPTS	EXAMPLES
4.1 Slack Variables and the Simplex Tableau	

A linear programming problem is said to be in **standard maximization form** if it is a maximization problem, each variable is greater than or equal to zero, and all other (nontrivial) inequalities are of the form

[linear expression] $\leq$ [nonnegative number].

Maximize $7x + 4y$ subject to the constraints

$$\begin{cases} 3x + 2y \leq 36 \\ x + 4y \leq 32 \\ x \geq 0, \quad y \geq 0 \end{cases}$$

is in standard maximization form.

A **slack variable** is a variable that, when added to the left-hand side of a constraint, makes the inequality into an equation. It is a nonnegative quantity and so "picks up the slack." To write the objective function as an equation, let M equal the objective function.

Use u as the first slack variable so that the first constraint becomes $3x + 2y + u = 36$.

Let $M = 7x + 4y$, so $-7x - 4y + M = 0$.

Reformulate the problem as a system of linear equations.

Rewrite the preceding linear programming problem as a system of linear equations,

$$\begin{cases} 3x + 2y + u & = 36 \\ x + 4y \quad + v & = 32 \\ -7x - 4y \quad + M = & 0, \end{cases}$$

where $x \geq 0, y \geq 0, u \geq 0, v \geq 0$, and M is as big as possible.

KEY TERMS AND CONCEPTS	EXAMPLES

A simplex tableau is a matrix (corresponding to a linear programming problem) in which each of the columns of the appropriately sized identity matrix are present (in some order) to the left of the right-hand column.

The initial tableau is

$$\begin{array}{ccccc} x & y & u & v & M \\ \end{array}$$
$$\left[\begin{array}{ccccc|c} 3 & 2 & 1 & 0 & 0 & 36 \\ 1 & 4 & 0 & 1 & 0 & 32 \\ \hline -7 & -4 & 0 & 0 & 1 & 0 \end{array}\right].$$

Group II variables correspond to the columns that are columns from the identity matrix. **Group I variables** correspond to the remaining columns to the left of the vertical line. A particular solution can be found by setting the Group I variables equal to 0 and the Group II variables equal to the right-hand sides.

The particular solution for the following matrix is Group I: $y = 0$, $u = 0$, $v = 0$; Group II: $x = 21$, $z = 34$, $w = 45$, $M = 897$:

$$\begin{array}{ccccccc} x & y & z & u & v & w & M \\ \end{array}$$
$$\left[\begin{array}{ccccccc|c} 0 & 9 & 1 & 3 & -5 & 0 & 0 & 34 \\ 1 & -7 & 0 & 8 & 14 & 0 & 0 & 21 \\ 0 & 6 & 0 & -1 & 8 & 1 & 0 & 45 \\ \hline 0 & 11 & 0 & 0 & 4 & 0 & 1 & 897 \end{array}\right]$$

4.2 The Simplex Method I: Maximum Problems

Condition for a Maximum The particular solution derived from a simplex tableau is a maximum if and only if the bottom row contains no negative numbers to the left of the vertical bar.

Consider the simplex tableau

$$\begin{array}{ccccc} x & y & u & v & M \\ \end{array}$$
$$\left[\begin{array}{ccccc|c} 3 & 2 & 1 & 0 & 0 & 36 \\ 1 & 4 & 0 & 1 & 0 & 32 \\ \hline -7 & -4 & 0 & 0 & 1 & 0 \end{array}\right].$$

Since the bottom row of the simplex tableau has a negative number to the left of the vertical bar, the maximum has not yet been found.

Choosing a Pivot Column Consider the left side of the bottom row. Select the column with the most negative entry. Ties are broken arbitrarily.

The pivot column is the x-column, since -7 is the most negative entry of the bottom row.

Choosing a Pivot Element Find the ratios of the right-most column to the entries in the pivot column. Determine the smallest nonnegative ratio, and select the corresponding entry in the pivot column as the pivot element.

$$\begin{array}{ccccc} x & y & u & v & M \\ \end{array}$$
$$\left[\begin{array}{ccccc|c} ③ & 2 & 1 & 0 & 0 & 36 \\ 1 & 4 & 0 & 1 & 0 & 32 \\ \hline -7 & -4 & 0 & 0 & 1 & 0 \end{array}\right] \begin{array}{l} 36/3 = 12 \\ 32/1 = 32 \end{array}$$

Since 12 is the smallest nonnegative ratio, the 3 in the x-column is the first pivot element.

The Simplex Method for Problems in Standard Maximization Form

See the box on page 149.

Pivot to get

$$\begin{array}{ccccc} x & y & u & v & M \\ \end{array}$$
$$\begin{array}{l} \frac{1}{3}R_1 \\ R_2 + (-1)R_1 \\ R_3 + 7R_1 \end{array} \left[\begin{array}{ccccc|c} 1 & \frac{2}{3} & \frac{1}{3} & 0 & 0 & 12 \\ 0 & \frac{10}{3} & -\frac{1}{3} & 1 & 0 & 20 \\ \hline 0 & \frac{2}{3} & \frac{7}{3} & 0 & 1 & 84 \end{array}\right].$$

Since the bottom row contains no negative entries, a maximum has been found. The solution is $x = 12$, $y = 0$, $u = 0$, $v = 20$, $M = 84$.

4.3 The Simplex Method II: Nonstandard and Minimum Problems

The Simplex Method for Problems in Nonstandard Form

1. If necessary, convert all (nontrivial) inequalities into the form

   [linear expression] $\leq$ [constant]

   by multiplying by -1.

Maximize $10x + 13y$ subject to the constraints

$$\begin{cases} 5x + 6y \geq 60 \\ 8x + 4y \leq 64 \\ x \geq 0, \quad y \geq 0. \end{cases}$$

KEY TERMS AND CONCEPTS	EXAMPLES

2. If a negative number occurs in the upper part of the right-hand column, remove it by pivoting.
   (a) Pick any negative entry in that row. That is the pivot column.
   (b) Find the ratios of the last column entries to the pivot column entries, and select as pivot element the one with the smallest positive ratio.
   (c) Pivot.

3. Repeat Step 2 until there are no negative numbers in the upper portion of the right-hand column.

4. If necessary, apply the simplex method for tableaux in standard maximization form.

Multiply the first inequality by -1 to get the problem:

Maximize $10x + 13y$ subject to the constraints

$$\begin{cases} -5x - 6y \le -60 \\ 8x + 4y \le 64 \\ x \ge 0, \quad y \ge 0. \end{cases}$$

The initial simplex tableau is

$$\begin{array}{ccccc} x & y & u & v & M \\ \end{array}$$
$$\left[\begin{array}{ccccc|c} -5 & \boxed{-6} & 1 & 0 & 0 & -60 \\ 8 & 4 & 0 & 1 & 0 & 64 \\ \hline -10 & -13 & 0 & 0 & 1 & 0 \end{array}\right] \begin{array}{l} -60/-6 = 10 \\ 64/4 = 16 \end{array}$$

We need to eliminate the -60. Choose the y-column, since it has -6 in it. The smallest positive ratio is 10, so the -6 in the y-column is the pivot element. The new tableau is

$$\begin{array}{ccccc} x & y & u & v & M \\ \end{array}$$
$$\left[\begin{array}{ccccc|c} \frac{5}{6} & 1 & \frac{1}{6} & 0 & 0 & 10 \\ \frac{14}{3} & 0 & \frac{2}{3} & 1 & 0 & 24 \\ \hline \frac{5}{6} & 0 & -\frac{13}{6} & 0 & 1 & 130 \end{array}\right].$$

To apply the simplex method to a **minimum problem**, convert it to a maximum problem by multiplying the objective function by -1. Apply the appropriate simplex method, and multiply the "maximum" obtained by -1 to find the minimum value.

Minimize $8x + 9y$ subject to the constraints

$$\begin{cases} 2x + 4y \ge 6 \\ 5x + 5y \ge 10 \\ x \ge 0, \quad y \ge 0. \end{cases}$$

Rewrite the problem:

Maximize $-8x - 9y$ subject to the constraints

$$\begin{cases} -2x - 4y \le -6 \\ -5x - 5y \le -10 \\ x \ge 0, \quad y \ge 0. \end{cases}$$

The initial simplex tableau is

$$\begin{array}{ccccc} x & y & u & v & M \\ \end{array}$$
$$\left[\begin{array}{ccccc|c} -2 & -4 & 1 & 0 & 0 & -6 \\ -5 & -5 & 0 & 1 & 0 & -10 \\ \hline 8 & 9 & 0 & 0 & 1 & 0 \end{array}\right].$$

The final tableau is

$$\begin{array}{ccccc} x & y & u & v & M \\ \end{array}$$
$$\left[\begin{array}{ccccc|c} 0 & 1 & -\frac{1}{2} & \frac{1}{5} & 0 & 1 \\ 1 & 0 & \frac{1}{2} & -\frac{2}{5} & 0 & 1 \\ \hline 0 & 0 & \frac{1}{2} & \frac{7}{5} & 1 & -17 \end{array}\right].$$

Since $M = -17$, the minimum value is $-1(-17) = 17$.

4.4 Sensitivity Analysis and Matrix Formulations of Linear Programming Problems

Sensitivity analysis provides an indication of how changes in the coefficients of the objective function and/or the constraint equations will change the optimal solution of the problem.

KEY TERMS AND CONCEPTS	EXAMPLES
The **allowable increase** and **allowable decrease** for a constraint give the amount of increase and decrease that can occur on the right-hand side of the constraint before there is no feasible solution to the linear programming problem.	In the furniture manufacturing problem, the allowable increase for the carpentry constraint is 12 and the allowable decrease is 15.
The **shadow price** of a constraint is the amount by which the value of the objective function could be improved (increased for maximization problems and decreased for minimization problems) if you had one more unit on the right-hand side of the constraint.	In the furniture manufacturing problem, the shadow price for the carpentry constraint is $\frac{10}{3}$.
Certain linear programming problems have a **matrix formulation**.	The matrix formulation of the furniture manufacturing problem is Maximize CX subject to the constraints $AX \leq B$, $X \geq \mathbf{0}$, where $$A = \begin{bmatrix} 6 & 3 \\ 1 & 1 \\ 2 & 6 \end{bmatrix}, \quad B = \begin{bmatrix} 96 \\ 18 \\ 72 \end{bmatrix}, \quad C = [80 \quad 70], \quad X = \begin{bmatrix} x \\ y \end{bmatrix}.$$
The **transpose** of an $m \times n$ matrix A is the $n \times m$ matrix, denoted A^T, that results from interchanging the rows and columns of the matrix.	The transpose of $\begin{bmatrix} 1 & 2 & 3 \\ 4 & 5 & 6 \end{bmatrix}$ is $\begin{bmatrix} 1 & 4 \\ 2 & 5 \\ 3 & 6 \end{bmatrix}$.

4.5 Duality

The Dual of a Linear Programming Problem

1. If the original (primal) problem has the form Maximize CX, subject to the constraints $AX \leq B$, $X \geq \mathbf{0}$, then the dual problem is Minimize $B^T U$, subject to the constraints $A^T U \geq C^T$, $U \geq \mathbf{0}$.

2. If the original (primal) problem has the form Minimize CX, subject to the constraints $AX \geq B$, $X \geq \mathbf{0}$, then the dual problem is Maximize $B^T U$, subject to the constraints $A^T U \leq C^T$, $U \geq \mathbf{0}$,

where C is the row matrix of the coefficients of the objective function, X is the column matrix of the variables, A is the coefficient matrix of the constraints, B is the column matrix of the right-hand sides of the constraints, U is the column matrix of the slack variables, and $\mathbf{0}$ is a column matrix of zeros.

Fundamental Theorem of Duality

See the box on page 171.

If the original (primal) problem is

Maximize $7x + 4y$ subject to the constraints

$$\begin{cases} 3x + 2y \leq 36 \\ x + 4y \leq 32 \\ x \geq 0, \quad y \geq 0, \end{cases}$$

then the dual problem is

Minimize $36u + 32v$ subject to the constraints

$$\begin{cases} 3u + v \leq 7 \\ 2u + 4v \leq 4 \\ u \geq 0, \quad v \geq 0. \end{cases}$$

CHAPTER 4 Fundamental Concept Check Exercises

1. What is the standard maximization form of a linear programming problem?

2. What is a slack variable? A group I variable? A group II variable?

3. Explain how to construct a simplex tableau corresponding to a linear programming problem in standard maximization form.

4. Give the steps for carrying out the simplex method for linear programming problems in standard maximization form.

5. Explain how to convert a minimization problem to a maximization problem.

6. Give the steps for carrying out the simplex method for problems in nonstandard form.

7. Describe how to obtain the dual of a linear programming problem.

8. State the fundamental theorem of duality.

9. Explain how to obtain the matrix formulation of a linear programming problem.

10. What is meant by "sensitivity analysis"?

11. Explain how the dual problem can be used to decide whether to introduce a new product.

CHAPTER 4 Review Exercises

In Exercises 1–10, use the simplex method to solve the linear programming problem.

1. Maximize $3x + 4y$ subject to the constraints

$$\begin{cases} 2x + y \le 7 \\ -x + y \le 1 \\ x \ge 0, \quad y \ge 0. \end{cases}$$

2. Maximize $2x + 5y$ subject to the constraints

$$\begin{cases} x + y \le 7 \\ 4x + 3y \le 24 \\ x \ge 0, \quad y \ge 0. \end{cases}$$

3. Maximize $2x + 3y$ subject to the constraints

$$\begin{cases} x + 2y \le 14 \\ x + y \le 9 \\ 3x + 2y \le 24 \\ x \ge 0, \quad y \ge 0. \end{cases}$$

4. Maximize $3x + 7y$ subject to the constraints

$$\begin{cases} x + 2y \le 10 \\ -4x + 3y \le 30 \\ -2x + y \le 0 \\ x \ge 0, \quad y \ge 0. \end{cases}$$

5. Minimize $x + y$ subject to the constraints

$$\begin{cases} 7x + 5y \ge 40 \\ x + 4y \ge 9 \\ x \ge 0, \quad y \ge 0. \end{cases}$$

6. Minimize $3x + 2y$ subject to the constraints

$$\begin{cases} x + y \ge 6 \\ x + 2y \ge 0 \\ x \ge 0, \quad y \ge 0. \end{cases}$$

7. Minimize $20x + 30y$ subject to the constraints

$$\begin{cases} x + 4y \ge 8 \\ x + y \ge 5 \\ 2x + y \ge 7 \\ x \ge 0, \quad y \ge 0. \end{cases}$$

8. Minimize $5x + 7y$ subject to the constraints

$$\begin{cases} 2x + y \ge 10 \\ 3x + 2y \ge 18 \\ x + 2y \ge 10 \\ x \ge 0, \quad y \ge 0. \end{cases}$$

9. Maximize $36x + 48y + 70z$ subject to the constraints

$$\begin{cases} x \le 4 \\ y \le 6 \\ x \le 8 \\ 4x + 3y + 2z \le 38 \\ x \ge 0, \quad y \ge 0, \quad z \ge 0. \end{cases}$$

10. Maximize $3x + 4y + 5z + 4w$ subject to the constraints

$$\begin{cases} 6x + 9y + 12z + 15w \le 672 \\ x - y + 2z + 2w \le 92 \\ 5x + 10y - 5z + 4w \le 280 \\ x \ge 0, \quad y \ge 0, \quad z \ge 0, \quad w \ge 0. \end{cases}$$

11. Determine the dual problem of the linear programming problem in Exercise 3.

12. Determine the dual problem of the linear programming problem in Exercise 7.

13. The final simplex tableau for the linear programming problem of Exercise 3 is as follows. Give the solution to the problem and to its dual.

	x	y	u	v	w	M	
y	0	1	1	-1	0	0	5
x	1	0	-1	2	0	0	4
w	0	0	1	-4	1	0	2
M	0	0	1	1	0	1	23

14. The final simplex tableau for the *dual* of the linear programming problem of Exercise 7 is as follows. Give the solution to the problem and to its dual.

	u	v	w	x	y	M	
v	0	1	$\frac{7}{3}$	$\frac{4}{3}$	$-\frac{1}{3}$	0	$\frac{50}{3}$
u	1	0	$-\frac{1}{3}$	$-\frac{1}{3}$	$\frac{1}{3}$	0	$\frac{10}{3}$
M	0	0	2	4	1	1	110

15. Consider the linear programming problem in Exercise 3. Identify the matrices A, B, C, X, and U and state the problem and its dual in terms of matrices.

16. Consider the linear programming problems in Exercise 7. Identify the matrices A, B, C, X, and U and state the problem and its dual in terms of matrices.

17. **Manufacturing-Resources Allocation** A lens manufacturer has the capability of making three types of lenses. Type A lenses require 4 minutes of grinding, 2 minutes of polishing, and 4 minutes of coating, and produce a profit of $12. Type B lenses require 2 minutes of grinding, 6 minutes of polishing, and

2 minutes of coating, and produce a profit of $10. Type C lenses require 2 minutes of grinding, 4 minutes of polishing, and 4 minutes of coating, and produce a profit of $8. Each day, 360 labor-hours for grinding, 600 labor hours for polishing, and 480 labor-hours for coating are available. How many of each type of lens should be produced each day to maximize the profit? What will be the profit?

18. **Nutrition** A camp counselor wants to make a smoothie for a group of children using at most 75 pounds of fruit consisting of oranges, cherries, and blueberries. Each pound of oranges contains 230 calories, 3 mg of sodium, and 4 g of protein. Each pound of cherries contains 260 calories, 7 mg of sodium, and 5 g of protein. Each pound of blueberries contains 250 calories, 5 mg of sodium, and 3 g of protein. How many pounds of each fruit should she use in order to minimize the number of calories, keep the amount of sodium less than 251 mg, and have the smoothie contain at least 300 g of protein? How many calories will the smoothie contain?

19. **Manufacturing—Resource Allocation** Consider Exercise 42 of Section 3.3.
    (a) Solve the problem by the simplex method.
    (b) The Bluejay Lacrosse Stick Company is considering diversifying by also making tennis rackets. A tennis racket requires 1 labor-hour for cutting, 4 labor-hours for stringing, and 2 labor-hours for finishing. How much profit must the company be able to make on each tennis racket in order to justify the diversification?

20. **Stereo Store** Consider the stereo store of Exercise 24 in Section 4.2. A fourth brand of stereo system has appeared on the market. Brand D takes up 3 cubic feet of storage space and generates a sales commission of $30. What profit would the store have to realize on the sale of each brand D stereo set in order to justify carrying it?

CHAPTER 4 PROJECT

Shadow Prices Revisited

Jason's House of Cheese offers two cheese assortments for holiday gift giving. In his supply refrigerator, Jason has 3600 ounces of cheddar, 1498 ounces of Brie, and 2396 ounces of Stilton. The St. Nick assortment contains 10 ounces of cheddar, 5 ounces of Brie, and 6 ounces of Stilton. The Holly assortment contains 8 ounces of cheddar, 3 ounces of Brie, and 8 ounces of Stilton. Each St. Nick assortment sells for $16, and each Holly assortment sells for $14. How many of each assortment should be produced and sold in order to maximize Jason's revenue?

1. Solve the problem geometrically.

2. By looking at your graph from part 1, can you determine the shadow price of cheddar?

3. Solve the problem by the simplex method. The solution should be the same as in part 1. Verify your answer to part 2 by looking at your final tableau.

4. What are the shadow prices for Brie and Stilton?

5. What would the maximum revenue be if there were 3620 ounces of cheddar, 1500 ounces of Brie, and 2400 ounces of Stilton?

6. Go back to the original problem, and state its dual problem. What information do the original slack variables u, v, and w give us about the dual problem? Determine the solution to the dual problem from your final tableau in part 3, and give an economic interpretation.

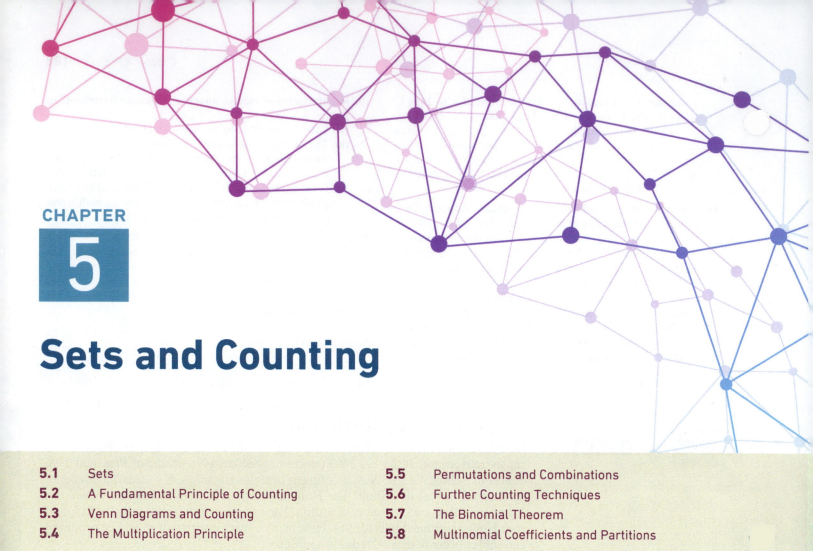

CHAPTER 5

Sets and Counting

In this chapter, we introduce some ideas useful in the study of probability (Chapter 6). Our first topic, the theory of sets, will provide a convenient language and notation in which to discuss probability. Using set theory, we develop a number of counting principles that can also be applied to computing probabilities.

5.1 Sets

In many applied problems, one must consider collections of various sorts of objects. For example, a survey of unemployment trends might consider the collection of all U.S. cities with current unemployment greater than 9 percent. A study of birthrates might consider the collection of countries with a current birthrate less than 20 per 1000 population. Such collections are examples of sets. A **set** is any collection of objects. The objects, which may be countries, cities, years, numbers, letters, or anything else, are called the **elements** of the set. A set is often specified by listing its elements inside a pair of braces. For example, the set whose elements are the first six letters of the alphabet is written

$$\{a, b, c, d, e, f\}.$$

Similarly, the set whose elements are the even numbers between 1 and 11 is written

$$\{2, 4, 6, 8, 10\}.$$

We can also specify a set by giving a description of its elements (without actually listing the elements). For example, the set $\{a, b, c, d, e, f\}$ can also be written

$$\{\text{the first six letters of the alphabet}\},$$

and the set $\{2, 4, 6, 8, 10\}$ can be written

$$\{\text{all even numbers between 1 and 11}\}.$$

For convenience, we usually denote sets by capital letters, A, B, C, and so on.

Two sets A and B are said to be **equal** if every element of A is also in B, and every element of B is also in A. For example,

$$\{a, b, c, d, e, f\} = \{f, e, d, b, c, a\}$$

and

$$\{2, 4, 6, 8\} = \{2, 2, 4, 4, 4, 6, 6, 8\}.$$

The great diversity of sets is illustrated by the following examples:

1. Let $C = \{\text{possible sequences of outcomes of tossing a coin three times}\}$. If we let H denote "heads" and T denote "tails," the various sequences can be easily described:

$$C = \{\text{HHH, THH, HTH, HHT, TTH, THT, HTT, TTT}\},$$

where, for instance, THH means "first toss tails, second toss heads, third toss heads."

2. Let $B = \{\text{license plate numbers consisting of three letters followed by three digits}\}$. Some elements of B are

$$\text{SBG 602,} \qquad \text{GXZ 179,} \qquad \text{YHJ 006.}$$

The number of elements in B is sufficiently large so that listing all of them is impractical. However, in this chapter, we develop a technique that allows us to calculate the number of elements of B.

3. The graph of the equation $y = x^2$ is the set of all points (a, b) in the plane for which $b = a^2$. This set has infinitely many elements.

Sets arise in many practical contexts, as the next example shows.

EXAMPLE 1   **Listing the Elements of a Set** Table 1 gives the rate of inflation, as measured by the percentage change in the consumer price index, for the years from 1996 to 2015. Let

$$A = \{\text{years from 1996 to 2015 in which inflation was above 3\%}\}$$

$$B = \{\text{years from 1996 to 2015 in which inflation was below 2\%}\}.$$

Determine the elements of A and B.

Table 1 U.S. Inflation Rates

Year	Inflation (%)	Year	Inflation (%)
1996	3.0	2006	3.2
1997	2.3	2007	2.8
1998	1.6	2008	3.8
1999	2.2	2009	−0.4
2000	3.4	2010	1.6
2001	2.8	2011	3.2
2002	1.6	2012	2.1
2003	2.3	2013	1.5
2004	2.7	2014	1.6
2005	3.4	2015	0.1

(Source: www.bls.gov)

SOLUTION   By reading Table 1, we see that

$$A = \{2000, 2005, 2006, 2008, 2011\}$$
$$B = \{1998, 2002, 2009, 2010, 2013, 2014, 2015\}.$$

>> *Now Try Exercises 9(a) and (b)*

Suppose that we are given two sets, A and B. Then it is possible to form new sets from A and B.

DEFINITIONS   Union and Intersection of Two Sets   The **union** of A and B, written $A \cup B$ and pronounced "A union B," is defined as follows:

$A \cup B$ is the set of all elements that belong to either A or B (or both).

The **intersection** of A and B, written $A \cap B$ and pronounced "A intersection B," is defined as follows:

$A \cap B$ is the set of all elements that belong to both A and B.

For example, let $A = \{1, 2, 3, 4\}$ and $B = \{1, 3, 5, 7, 11\}$. Then,

$$A \cup B = \{1, 2, 3, 4, 5, 7, 11\}$$
$$A \cap B = \{1, 3\}.$$

EXAMPLE 2   **The Intersection and Union of Sets**   Table 2 gives the rates of unemployment and inflation for the years from 2001 to 2015. Let

$$A = \{\text{years from 2001 to 2015 in which unemployment is at least 5\%}\}$$
$$B = \{\text{years from 2001 to 2015 in which the inflation rate is at least 3\%}\}.$$

(a) Describe the sets $A \cap B$ and $A \cup B$.
(b) Determine the elements of A, B, $A \cap B$, and $A \cup B$.

Table 2 U.S. Unemployment and Inflation Rates

Year	Unemployment (%)	Inflation (%)
2001	4.7	2.8
2002	5.8	1.6
2003	6.0	2.3
2004	5.5	2.7
2005	5.1	3.4
2006	4.6	3.2
2007	4.6	2.8
2008	5.8	3.8
2009	9.3	−0.4
2010	9.6	1.6
2011	8.9	3.2
2012	8.1	2.1
2013	7.4	1.5
2014	6.2	1.6
2015	5.3	0.1

(Source: www.bls.gov)

SOLUTION   (a) From the descriptions of A and B, we have

$$A \cap B = \{\text{years from 2001 to 2015 in which unemployment is}$$
$$\text{at least 5\% and inflation is at least 3\%}\}$$

$$A \cup B = \{\text{years from 2001 to 2015 in which either unemployment is}$$
$$\text{at least 5\% or inflation is at least 3\% (or both)}\}.$$

(b) From the table, we see that

$$A = \{2002, 2003, 2004, 2005, 2008, 2009, 2010, 2011, 2012, 2013, 2014, 2015\}$$
$$B = \{2005, 2006, 2008, 2011\}$$
$$A \cap B = \{2005, 2008, 2011\}$$
$$A \cup B = \{2002, 2003, 2004, 2005, 2006, 2008, 2009, 2010, 2011, 2012, 2013, 2014, 2015\}.$$

>> *Now Try Exercises 9(c) and (d)*

We have defined the union and the intersection of two sets. In a similar manner, we can define the union and intersection of any number of sets. For example, if A, B, and C are three sets, then their union, denoted $A \cup B \cup C$, is the set whose elements are precisely those that belong to at least one of the sets A, B, and C. Similarly, the intersection of A, B, and C, denoted $A \cap B \cap C$, is the set consisting of those elements that belong to all of the sets A, B, and C. In a similar way, we may define the union and intersection of more than three sets.

Suppose that we are given a set A. We may form new sets by selecting elements from A. Sets formed in this way are called *subsets* of A.

> **DEFINITION Subset of a Set**   The set B is a **subset** of the set A, written $B \subseteq A$ and pronounced "B is a subset of A," provided that every element of B is an element of A.

For example, $\{1, 3\} \subseteq \{1, 2, 3\}$.

One set that is considered very often is the set that contains no elements at all. This set is called the **empty set** (or *null set*) and is written $\varnothing$ or $\{\ \}$. The empty set is a subset of every set. (Here is why: Let A be any set. Every element of $\varnothing$ also belongs to A. If you do not agree, then you must produce an element of $\varnothing$ that does not belong to A. But you cannot, since $\varnothing$ has no elements. So $\varnothing \subseteq A$.)

> **DEFINITION Disjoint Sets**   Two sets A and B are **disjoint** if they have no elements in common, that is, if $A \cap B = \varnothing$.

EXAMPLE 3   **Listing the Subsets of a Set**   Let $A = \{a, b, c\}$. Find all subsets of A.

SOLUTION   Since A contains three elements, every subset of A has at most three elements. We look for subsets according to the number of elements:

Number of Elements in Subset	Possible Subsets
0	$\varnothing$
1	$\{a\}, \{b\}, \{c\}$
2	$\{a, b\}, \{a, c\}, \{b, c\}$
3	$\{a, b, c\}$

Thus, we see that A has eight subsets, namely, those listed on the right. (Note that we count A as a subset of itself.)

>> *Now Try Exercise 5*

In general, if a set A contains n elements, then A will have 2^n subsets. (In Section 5.7, we will prove this fact.)

It is usually convenient to regard all sets involved in a particular discussion as subsets of a single larger set. Thus, for example, if a problem involves the sets $\{a, b, c\}, \{e, f\}, \{g\}, \{b, x, y\}$, then we can regard all of these as subsets of the set

$$U = \{\text{all letters of the alphabet}\}.$$

Since U contains all elements being discussed, it is called a **universal set** (for the particular problem). In this book, we shall specify the particular universal set that we have in mind or it will be clearly defined by the context.

The set A contained in the universal set U has a counterpart, called its *complement*.

DEFINITION Complement of a Set   The **complement** of A, written A' and pronounced "A complement," is defined as follows:

A' is the set of all elements in the universal set U that do not belong to A.

For example, let $U = \{1, 2, 3, 4, 5, 6, 7, 8, 9\}$ and $A = \{2, 4, 6, 8\}$. Then,

$$A' = \{1, 3, 5, 7, 9\}.$$

EXAMPLE 4   **Finding the Complement of a Set**   Let $U = \{a, b, c, d, e, f, g\}, S = \{a, b, c\}$, and $T = \{a, c, d\}$. List the elements of the following sets:

(a) S' (b) T' (c) $(S \cap T)'$ (d) $S' \cap T'$ (e) $S' \cup T'$

SOLUTION   (a) S' consists of those elements of U that are not in S, so $S' = \{d, e, f, g\}$.

(b) Similarly, $T' = \{b, e, f, g\}$.

(c) As is the case in arithmetic, we perform the operation in parentheses first. To determine $(S \cap T)'$, we must first determine $S \cap T$:

$$S \cap T = \{a, c\}.$$

Then, we determine the complement of this set:

$$(S \cap T)' = \{b, d, e, f, g\}.$$

(d) We determined S' and T' in parts (a) and (b). The set $S' \cap T'$ consists of the elements that belong to both S' and T'. Therefore, referring to parts (a) and (b), we have

$$S' \cap T' = \{e, f, g\}.$$

(e) Since $S' \cup T'$ consists of the elements that belong to S' or T' (or both),

$$S' \cup T' = \{b, d, e, f, g\}.$$

» Now Try Exercises 9(e) and (f)

NOTE▶   The results of parts (c) and (d) show that in general $(S \cap T)'$ *is not* the same as $S' \cap T'$. In the next section, we show that $(S \cap T)'$ *is* the same as $S' \cup T'$. **«**

EXAMPLE 5   **College Students**   Let $U = \{\text{students at Gotham College}\}, E = \{\text{students at Gotham College who are at most 18 years old}\}$, and $S = \{\text{STEM majors at Gotham College}\}$.

(a) Use set-theoretic notation (that is, union, intersection, and complement symbols) to describe {students at Gotham College who are at most 18 years old or are not STEM majors}.

(b) Use set-theoretic notation to describe {students at Gotham College who are older than 18 and are STEM majors}.

(c) Describe in words the set $E' \cap S'$.

(d) Describe in words the set $E' \cup S$.

SOLUTION   (a) $E \cup S'$
(b) $E' \cap S$
(c) {students at Gotham College who are older than 18 and are not STEM majors}
(d) {students at Gotham College who are older than 18 or are STEM majors}

>> *Now Try Exercises 21 and 33*

The symbol $\in$ is commonly used with sets as a shorthand for "is an element of." For instance, if $S = \{H, T\}$, then $H \in S$. The symbol $\in$ should not be confused with the symbol $\subseteq$. For instance, $H \in S$, but $\{H\} \subseteq S$. The symbol $\notin$ is shorthand for "is not an element of." The symbol $\in$ can be used when defining union, intersection, complement, and subset.

DEFINITIONS

Union	$A \cup B$ is the set of all x such that $x \in A$ or $x \in B$.
Intersection	$A \cap B$ is the set of all x such that $x \in A$ and $x \in B$.
Complement	A' is the set of all $x \in U$ such that $x \notin A$.
Subset	$B \subseteq A$ if $x \in A$ whenever $x \in B$.

Check Your Understanding 5.1

Solutions can be found following the section exercises.

1. Let $U = \{a, b, c, d, e, f, g\}$, $R = \{a, b, c, d\}$, $S = \{c, d, e\}$, and $T = \{c, e, g\}$. List the elements of the following sets:
   (a) R' (b) $R \cap S$
   (c) $(R \cap S) \cap T$ (d) $R \cap (S \cap T)$

2. Let $U = $ {all Nobel winners}, $W = $ {women who have received Nobel Prizes}, $A = $ {Americans who have received

Nobel Prizes}, $L = $ {Nobel winners in literature}. Describe the following sets:
   (a) W' (b) $A \cap L'$ (c) $W \cap A \cap L'$

3. Refer to Problem 2. Use set-theoretic notation to describe {Nobel winners who are American men or recipients of the Nobel Prize in literature}.

EXERCISES 5.1

1. Let $U = \{1, 2, 3, 4, 5, 6, 7\}$, $S = \{1, 2, 3, 4\}$, and $T = \{1, 3, 5, 7\}$. List the elements of the following sets:
   (a) S' (b) $S \cup T$ (c) $S \cap T$ (d) $S' \cap T$

2. Let $U = \{1, 2, 3, 4, 5\}$, $S = \{1, 2, 3\}$, and $T = \{5\}$. List the elements of the following sets:
   (a) S' (b) $S \cup T$ (c) $S \cap T$ (d) $S' \cap T$

3. Let $U = $ {all letters of the alphabet}, $R = \{a, b, c\}$, $S = \{a, e, i, o, u\}$, and $T = \{x, y, z\}$. List the elements of the following sets:
   (a) $R \cup S$ (b) $R \cap S$ (c) $S \cap T$ (d) $S' \cap R$

4. Let $U = \{a, b, c, d, e, f, g\}$, $R = \{a\}$, $S = \{a, b\}$, and $T = \{b, d, e, f, g\}$. List the elements of the following sets:
   (a) $R \cup S$ (b) $R \cap S$ (c) T' (d) $T' \cup S$

5. List all subsets of the set $\{1, 2\}$.

6. List all subsets of the set $\{1, 2, 3, 4\}$.

7. **College Students** Let $U = $ {all college students}, $F = $ {all freshman college students}, and $B = $ {all college students who like basketball}. Describe the elements of the following sets:
   (a) $F \cap B$ (b) B' (c) $F' \cap B'$ (d) $F \cup B$

8. **Corporations** Let $U = $ {all corporations}, $S = $ {all corporations with headquarters in New York City}, and $T = $ {all privately owned corporations}. Describe the elements of the following sets:
   (a) S' (b) T' (c) $S \cap T$ (d) $S \cap T'$

9. **S&P Index** The Standard and Poor's Index measures the price of a certain collection of 500 stocks. Table 3 on the next page

compares the percentage change in the index during the first 5 business days of certain years with the percentage change for the entire year. Let $U = $ {all years from 1996 to 2015}, $S = $ {all years during which the index increased by 2% or more during the first 5 business days}, and $T = $ {all years for which the index increased by 16% or more during the entire year}. List the elements of the following sets:
   (a) S (b) T (c) $S \cap T$
   (d) $S \cup T$ (e) $S' \cap T$ (f) $S \cap T'$

10. **S&P Index** Refer to Table 3 on the next page. Let $U = $ {all years from 1996 to 2015}, $A = $ {all years during which the index declined during the first 5 business days}, and $B = $ {all years during which the index declined for the entire year}. List the elements of the following sets:
   (a) A (b) B (c) $A \cap B$
   (d) $A' \cap B$ (e) $A \cap B'$

11. **S&P Index** Refer to Exercise 9. Describe in words the fact that $S \cap T'$ has two elements.

12. **S&P Index** Refer to Exercise 10. Describe in words the fact that $A' \cap B$ has two elements.

13. Let $U = \{a, b, c, d, e, f\}$, $R = \{a, b, c\}$, $S = \{a, b, d\}$, and $T = \{e, f\}$. List the elements of the following sets:
   (a) $(R \cup S)'$ (b) $R \cup S \cup T$
   (c) $R \cap S \cap T$ (d) $R \cap S \cap T'$
   (e) $R' \cap S \cap T$ (f) $S \cup T$
   (g) $(R \cup S) \cap (R \cup T)$ (h) $(R \cap S) \cup (R \cap T)$
   (i) $R' \cap T'$

Table 3 Percentage Change in the Standard and Poor's Index

Year	Percent Change for First 5 Days	Percent Change for Year	Year	Percent Change for First 5 Days	Percent Change for Year
2015	0.0	−0.7	2005	−2.1	3.0
2014	−0.5	11.4	2004	1.8	9.0
2013	2.2	29.6	2003	3.4	26.4
2012	1.7	13.4	2002	1.1	−23.4
2011	1.1	0.0	2001	−1.9	−13.0
2010	2.7	12.8	2000	−1.9	−10.1
2009	0.7	23.5	1999	3.7	19.5
2008	−5.3	−38.5	1998	−1.5	26.7
2007	−0.4	4.8	1997	1.0	31.0
2006	3.0	13.6	1996	0.4	20.3

(Source: www.forecast-chart.com, www.investing.com)

14. Let $U = \{1, 2, 3, 4, 5\}$, $R = \{1, 3, 5\}$, $S = \{3, 4, 5\}$, and $T = \{2, 4\}$. List the elements of the following sets:
(a) $R \cap S \cap T$ **(b)** $R \cap S \cap T'$ **(c)** $R \cap S' \cap T$
(d) $R' \cap T$ **(e)** $R \cup S$ **(f)** $R' \cup R$
(g) $(S \cap T)'$ **(h)** $S' \cup T'$

In Exercises 15–20, simplify each given expression.

15. $(S')'$ **16.** $S \cap S'$ **17.** $S \cup S'$

18. $S \cap \varnothing$ **19.** $T \cap S \cap T'$ **20.** $S \cup \varnothing$

Corporation A large corporation classifies its many divisions by their performance in the preceding year. Let $P = \{\text{divisions that made a profit}\}$, $L = \{\text{divisions that had an increase in labor costs}\}$, and $T = \{\text{divisions whose total revenue increased}\}$. Describe the sets in Exercises 21–26 by using set-theoretic notation.

21. {divisions that had increases in labor costs or total revenue}

22. {divisions that did not make a profit}

23. {divisions that made a profit despite an increase in labor costs}

24. {divisions that had an increase in labor costs and either were unprofitable or did not increase their total revenue}

25. {profitable divisions with increases in labor costs and total revenue}

26. {divisions that were unprofitable or did not have increases in either labor costs or total revenue}

Automobile Insurance An automobile insurance company classifies applicants by their driving records for the previous three years. Let $S = \{\text{applicants who have received speeding tickets}\}$, $A = \{\text{applicants who have caused accidents}\}$, and $D = \{\text{applicants who have been arrested for driving while intoxicated}\}$. Describe the sets in Exercises 27–32 by using set-theoretic notation.

27. {applicants who have not received speeding tickets}

28. {applicants who have caused accidents and been arrested for drunk driving}

29. {applicants who have received speeding tickets, caused accidents, or been arrested for drunk driving}

30. {applicants who have not been arrested for drunk driving, but have received speeding tickets or have caused accidents}

31. {applicants who have not both caused accidents and received speeding tickets but who have been arrested for drunk driving}

32. {applicants who have not caused accidents or have not been arrested for drunk driving}

College Teachers and Students Let $U = \{\text{people at Mount College}\}$, $A = \{\text{students at Mount College}\}$, $B = \{\text{teachers at Mount College}\}$, $C = \{\text{people at Mount College who are older than 35}\}$, and $D = \{\text{people at Mount College who are younger than 35}\}$. Describe verbally the sets in Exercises 33–40.

33. $A \cap D$ **34.** $B \cap C$ **35.** $A \cap B$ **36.** $B \cup C$

37. $A \cup C'$ **38.** $(A \cap D)'$ **39.** D' **40.** $D \cap U$

Ice Cream Preferences Let $U = \{\text{all people}\}$, $S = \{\text{people who like strawberry ice cream}\}$, $V = \{\text{people who like vanilla ice cream}\}$, and $C = \{\text{people who like chocolate ice cream}\}$. Describe the sets in Exercises 41–46 by using set-theoretic notation.

41. {people who don't like vanilla ice cream}

42. {people who like vanilla but not chocolate ice cream}

43. {people who like vanilla but not chocolate or strawberry ice cream}

44. {people who don't like any of the three flavors of ice cream}

45. {people who like neither chocolate nor vanilla ice cream}

46. {people who like only strawberry and chocolate ice cream}

47. Let U be the set of vertices in Fig. 1. Let $R = \{\text{vertices } (x, y) \text{ with } x > 0\}$, $S = \{\text{vertices } (x, y) \text{ with } y > 0\}$, and $T = \{\text{vertices } (x, y) \text{ with } x \leq y\}$. List the elements of the following sets:
(a) R **(b)** S **(c)** T
(d) $R' \cup S$ **(e)** $R' \cap T$ **(f)** $R \cap S \cap T$

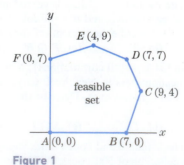

Figure 1

48. Let U be the set of vertices in Fig. 2. Let $R = \{$vertices (x, y) with $x \geq 150\}$, $S = \{$vertices (x, y) with $y \leq 100\}$, and $T = \{$vertices (x, y) with $x + y \leq 400\}$. List the elements of the following sets.

(a) R **(b)** S **(c)** T

(d) $R \cap S'$ **(e)** $R' \cup T$ **(f)** $R' \cap S' \cap T'$

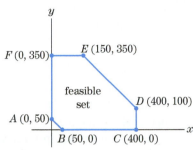

Figure 2

49. Sandwich Toppings Ed's Cheesesteaks offers any combination of three toppings on his sandwiches: peppers, onions, and mushrooms. How many different ways can you order a sandwich from Ed? List them.

50. Toppings Choices Amy ordered a baked potato at a restaurant. The server offered her butter, cheese, chives, and bacon as toppings. How many different ways could she have her potato? List them.

51. Let $S = \{1, 3, 5, 7\}$ and $T = \{2, 5, 7\}$. Give an example of a subset of T that is not a subset of S.

52. Suppose that S and T are subsets of the set U. Under what circumstance will $S \cap T = T$?

53. Suppose that S and T are subsets of the set U. Under what circumstance will $S \cup T = T$?

54. Find three subsets of the set of integers from 1 through 10, R, S, and T, such that $R \cup (S \cap T)$ is different from $(R \cup S) \cap T$.

In Exercises 55–62, determine whether the statement is true or false.

55. $5 \in \{3, 5, 7\}$ **56.** $\{1, 3\} \subseteq \{1, 2, 3\}$

57. $\{b\} \subseteq \{b, c\}$ **58.** $0 \in \{1, 2, 3\}$

59. $0 \in \varnothing$ **60.** $\varnothing \subseteq \{a, b, c\}$

61. $\{b, c\} \subseteq \{b, c\}$ **62.** $1 \notin \{1\}$

Solutions to Check Your Understanding 5.1

1. (a) $\{e, f, g\}$ **(b)** $\{c, d\}$

(c) $\{c\}$. This problem asks for the intersection of two sets. The first set is $R \cap S = \{c, d\}$, and the second set is $T = \{c, e, g\}$. The intersection of these sets is $\{c\}$.

(d) $\{c\}$. Here again, the problem asks for the intersection of two sets. However, now the first set is $R = \{a, b, c, d\}$ and the second set is $S \cap T = \{c, e\}$. The intersection of these sets is $\{c\}$.

Note: It should be expected that the set $(R \cap S) \cap T$ is the same as the set $R \cap (S \cap T)$, for each set consists of those elements that are in all three sets. Therefore, each of these sets equals the set $R \cap S \cap T$.

2. (a) $W' = \{$men who have received Nobel Prizes$\}$. This is so, since W' consists of those elements of U that are not in W—that is, those Nobel winners who are not women.

(b) $A \cap L' = \{$Americans who have received Nobel Prizes in fields other than literature$\}$

(c) $W \cap A \cap L' = \{$American women who have received Nobel Prizes in fields other than literature$\}$. This is so, since to qualify for $W \cap A \cap L'$, a Nobel winner must simultaneously be in W, in A, and in L'—that is, a woman, an American, and not a Nobel winner in literature.

3. $(A \cap W') \cup L$

5.2   A Fundamental Principle of Counting

A counting problem is one that requires us to determine the number of elements in a set S. Counting problems arise in many applications of mathematics and comprise the mathematical field of combinatorics. We shall study a number of different sorts of counting problems in the remainder of this chapter.

If S is any set, we will denote the number of elements in S by $n(S)$. For example, if $S = \{1, 7, 11\}$, then $n(S) = 3$, and if $S = \{a, b, c, d, e, f, g, h, i\}$, then $n(S) = 9$. Of course, if $S = \varnothing$, the empty set, then $n(S) = 0$.

Let us begin by stating one of the fundamental principles of counting, the **inclusion–exclusion principle**.

> **Inclusion–Exclusion Principle**   Let S and T be sets. Then,
>
> $$n(S \cup T) = n(S) + n(T) - n(S \cap T). \tag{1}$$

Notice that formula (1) connects the four quantities $n(S \cup T)$, $n(S)$, $n(T)$, and $n(S \cap T)$. Given any three, the remaining quantity can be determined by using this formula.

To test the plausibility of the inclusion–exclusion principle, consider this example. Let $S = \{a, b, c, d, e\}$ and $T = \{a, c, g, h\}$. Then,

$$S \cup T = \{a, b, c, d, e, g, h\} \qquad n(S \cup T) = 7$$
$$S \cap T = \{a, c\} \qquad n(S \cap T) = 2.$$

In this case, the inclusion–exclusion principle reads

$$n(S \cup T) = n(S) + n(T) - n(S \cap T)$$
$$7 \quad = \quad 5 \; + \; 4 \; - \quad 2,$$

which is correct.

Here is the reason for the validity of the inclusion–exclusion principle: The left side of formula (1) is $n(S \cup T)$, the number of elements in either S or T (or both). As a first approximation to this number, add the number of elements in S to the number of elements in T, obtaining $n(S) + n(T)$. However, if an element lies in both S and T, it is counted twice—once in $n(S)$ and again in $n(T)$. To make up for this double counting, we must subtract the number of elements counted twice, namely, $n(S \cap T)$. So doing gives us $n(S) + n(T) - n(S \cap T)$ as the number of elements in $S \cup T$.

When S and T are disjoint, the inclusion–exclusion principle reduces to a simple sum.

Special Case of the Inclusion–Exclusion Principle If $S \cap T = \varnothing$, then

$$n(S \cup T) = n(S) + n(T).$$

The next example illustrates a typical use of the inclusion–exclusion principle in an applied problem.

EXAMPLE 1 **Using the Inclusion–Exclusion Principle** In the year 2016, *Executive* magazine surveyed the presidents of the 500 largest corporations in the United States. Of these 500 people, 310 had degrees (of any sort) in business, 238 had undergraduate degrees in business, and 184 had graduate degrees in business. How many presidents had both undergraduate and graduate degrees in business?

SOLUTION Let

$$S = \{\text{presidents with an undergraduate degree in business}\}$$
$$T = \{\text{presidents with a graduate degree in business}\}.$$

Then,

$$S \cup T = \{\text{presidents with at least one degree in business}\}$$
$$S \cap T = \{\text{presidents with both undergraduate and graduate degrees in business}\}.$$

From the data given, we have

$$n(S) = 238 \qquad n(T) = 184 \qquad n(S \cup T) = 310.$$

The problem asks for $n(S \cap T)$. By the inclusion–exclusion principle, we have

$$n(S \cup T) = n(S) + n(T) - n(S \cap T)$$
$$310 = 238 + 184 - n(S \cap T)$$
$$n(S \cap T) = 112.$$

That is, exactly 112 of the presidents had both undergraduate and graduate degrees in business.

>> *Now Try Exercise 9*

Venn Diagrams

It is possible to visualize sets geometrically by means of drawings known as **Venn diagrams**. Such graphical representations of sets are very useful tools in solving counting problems. In order to describe Venn diagrams, let us begin with a single set S contained in a universal set U. Draw a rectangle, and view its points as the elements of U [Fig. 1(a)]. To show that S is a subset of U, we draw a circle inside the rectangle and view S as the set of points in the circle [Fig. 1(b)]. The resulting diagram is called a Venn diagram of S. It illustrates the proper relationship between S and U. Since S' consists of those elements of U that are not in S, we may view the portion of the rectangle that is outside of the circle as representing S' [Fig. 1(c)].

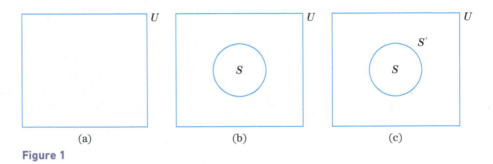

(a)　　　　　　　　(b)　　　　　　　　(c)

Figure 1

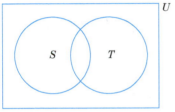

Figure 2

Venn diagrams are particularly useful for visualizing the relationship between two or three sets. Suppose that we are given two sets S and T in a universal set U. As before, we represent each of the sets by means of a circle inside the rectangle (Fig. 2).

We can now illustrate a number of sets by shading in appropriate regions of the rectangle. For instance, in Fig. 3(a), (b), and (c), we have shaded the regions corresponding to T, $S \cup T$, and $S \cap T$, respectively.

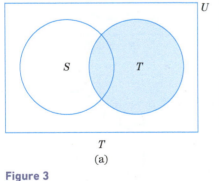

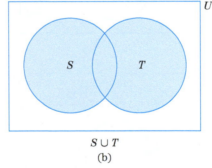
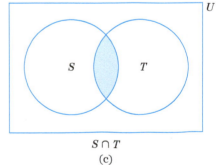

T　　　　　　　　$S \cup T$　　　　　　　　$S \cap T$

(a)　　　　　　　　(b)　　　　　　　　(c)

Figure 3

EXAMPLE 2　　**Shading Portions of a Venn Diagram**　Shade the portions of the rectangle corresponding to the sets

(a) $S \cap T'$　　　**(b)** $(S \cap T')'$.

SOLUTION　**(a)** $S \cap T'$ consists of the points in S and in T'—that is, the points in S and not in T [Figs. 4(a), 4(b) on the next page]. So we shade the points that are in the circle S but are not in the circle T [Fig. 4(c)].

(b) $(S \cap T')'$ is the complement of the set $S \cap T'$. Therefore, it consists of exactly those points not shaded in Fig. 4(c). [See Fig. 4(d).]

» *Now Try Exercises 15 and 19*

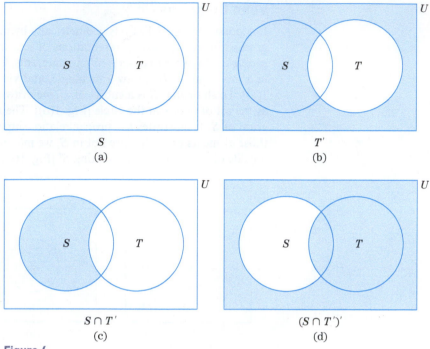

Figure 4

In a similar manner, Venn diagrams can illustrate intersections and unions of three sets. Some representative regions are shaded in Fig. 5.

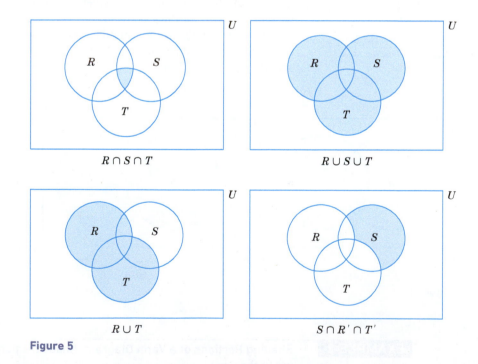

Figure 5

There are many formulas expressing relationships between intersections and unions of sets. Possibly the most fundamental are the two formulas known as *De Morgan's laws*.

De Morgan's Laws   Let S and T be sets. Then,

$$(S \cup T)' = S' \cap T' \quad \text{and} \quad (S \cap T)' = S' \cup T'.$$

In other words, De Morgan's laws state that, to form the complement of a union (or intersection), form the complements of the individual sets and change unions to intersections (or intersections to unions).

Verification of De Morgan's Laws

Let us use Venn diagrams to describe $(S \cup T)'$. In Fig. 6(a), we have shaded the region corresponding to $S \cup T$. In Fig. 6(b), we have shaded the region corresponding to $(S \cup T)'$. In Figs. 6(c) and 4(d), we have shaded the regions corresponding to S' and T'. By considering the common shaded regions of Figs. 6(c) and (d), we arrive at the shaded region corresponding to $S' \cap T'$ [Fig. 6(e)]. Note that this is the same region as shaded in Fig. 6(b). Therefore,

$$(S \cup T)' = S' \cap T'.$$

This verifies the first of De Morgan's laws. The proof of the second law is similar.

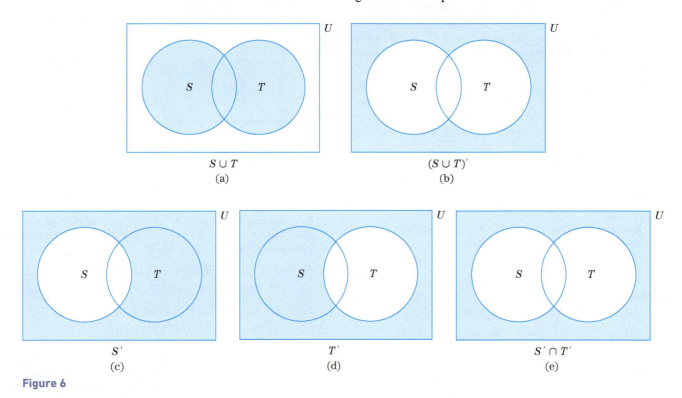

Figure 6

INCORPORATING TECHNOLOGY

✹WolframAlpha Venn diagrams can be displayed with instructions containing the words **intersect**, **union**, and **complement**. For instance, the instruction

S intersect (complement R) intersect (complement T)

produces the fourth Venn diagram of Fig. 5.

Check Your Understanding 5.2

Solutions can be found following the section exercises.

1. Draw a two-circle Venn diagram, and shade the portion corresponding to the set $(S \cap T') \cup (S \cap T)$.

2. What does the inclusion–exclusion principle conclude when T is a subset of S?

EXERCISES 5.2

1. Find $n(S \cup T)$, given that $n(S) = 4, n(T) = 4$, and $n(S \cap T) = 2$.

2. Find $n(S \cup T)$, given that $n(S) = 17, n(T) = 13$, and $n(S \cap T) = 0$.

3. Find $n(S \cap T)$, given that $n(S) = 6, n(T) = 9$, and $n(S \cup T) = 15$.

4. Find $n(S \cap T)$, given that $n(S) = 4, n(T) = 12$, and $n(S \cup T) = 15$.

5. Find $n(S)$, given that $n(T) = 7, n(S \cap T) = 5$, and $n(S \cup T) = 10$.

6. Find $n(T)$, given that $n(S) = 14, n(S \cap T) = 6$, and $n(S \cup T) = 14$.

7. If $n(S) = n(S \cap T)$, what can you conclude about S and T?

8. If $n(T) = n(S \cup T)$, what can you conclude about S and T?

9. **Languages** Suppose that each of the 314 million adults in South America is fluent in Portuguese or Spanish. If 170 million are fluent in Portuguese and 155 million are fluent in Spanish, how many are fluent in both languages?

10. **Course Enrollments** Suppose that all of the 1000 first-year students at a certain college are enrolled in a math or an English course. Suppose that 400 are taking both math and English and 600 are taking English. How many are taking a math course?

11. **Symmetry of Letters** Of the 26 capital letters of the alphabet, 11 have vertical symmetry (for instance, A, M, and T), 9 have horizontal symmetry (such as B, C, and D), and 4 have both (H, I, O, X). How many letters have neither horizontal nor vertical symmetry?

12. **Streaming Subscriptions** A survey of employees in a certain company revealed that 250 people subscribe to a streaming video service, 75 subscribe to a streaming music service, and 25 subscribe to both. How many people subscribe to at least one of these services?

13. **Automobile Options** Motors Inc. manufactured 325 cars with navigation systems, 216 with push-button start, and 89 with both of these options. How many cars were manufactured with at least one of the two options?

14. **Investments** A survey of 120 investors in stocks and bonds revealed that 90 investors owned stocks and 70 owned bonds. How many investors owned both stocks and bonds?

In Exercises 15–26, draw a two-circle Venn diagram and shade the portion corresponding to the set.

15. $S \cap T'$ 16. $S' \cap T'$

17. $S' \cup T$ 18. $S' \cup T'$

19. $(S \cap T')'$ 20. $(S \cap T)'$

21. $(S \cap T') \cup (S' \cap T)$ 22. $(S \cap T) \cup (S' \cap T')$

23. $S \cup (S \cap T)$ 24. $S \cup (T' \cup S)$

25. $S \cup S'$ 26. $S \cap S'$

In Exercises 27–38, draw a three-circle Venn diagram and shade the portion corresponding to the set.

27. $R \cap S \cap T'$ 28. $R' \cap S' \cap T$

29. $R \cup (S \cap T)$ 30. $R \cap (S \cup T)$

31. $R \cap (S' \cup T)$ 32. $R' \cup (S \cap T')$

33. $R \cap T$ 34. $S \cap T'$

35. $R' \cup S' \cup T'$ 36. $(R \cap S \cap T)'$

37. $(R \cap T) \cup (S \cap T')$ 38. $(R \cup S') \cap (R \cup T')$

In Exercises 39–44, use De Morgan's laws to simplify each given expression.

39. $S' \cup (S \cap T)'$ 40. $S \cap (S \cup T)'$

41. $(S' \cup T)'$ 42. $(S' \cap T')'$

43. $T \cup (S \cap T)'$ 44. $(S' \cap T)' \cup S$

In Exercises 45–50, give a set-theoretic expression that describes the shaded portion of each Venn diagram.

45.   46.

47.   48.

49.   50.

By drawing a Venn diagram, simplify each of the expressions in Exercises 51–54 to involve at most one union and the complement symbol applied only to R, S, and T.

51. $(T \cap S) \cup (T \cap R) \cup (R \cap S') \cup (T \cap R' \cap S')$

52. $(R \cap S) \cup (S \cap T) \cup (R \cap S' \cap T')$

53. $((R \cap S') \cup (S \cap T') \cup (T \cap R'))'$

54. $(R \cap T) \cup (R \cap S) \cup (S \cap T') \cup (R \cap S' \cap T')$

Citizenship Assume that the universal set U is the set of all people living in the United States. Let A be the set of all U.S. citizens, let B be the set of all children under 5 years of age, let C be the set of children from 5 to 18 years of age, let D be the set of everyone over the age of 18, and let E be the set of all people who are employed. Describe in words each set in Exercises 55–60.

55. $A' \cup (D \cap E)$ 56. $A \cap C \cap E$

57. $D \cap E'$ 58. $A \cap (D \cup E)$

59. $A' \cap B'$ 60. $B \cap E$

Solutions to Check Your Understanding 5.2

1. $(S \cap T') \cup (S \cap T)$ is given as a union of two sets, $S \cap T'$ and $S \cap T$. The Venn diagrams for these two sets are given in Figs. 7(a) and (b). The desired set consists of the elements that are in one or the other (or both) of the two sets. Therefore, its Venn diagram is obtained by shading everything that is shaded in either Fig. 7(a) or (b). [See Fig. 7(c).] *Note:* Looking at Fig. 7(c) reveals that $(S \cap T') \cup (S \cap T)$ and S are the same set. Often, Venn diagrams can be used to simplify complicated set-theoretic expressions.

2. When $T \subseteq S$, $S \cup T = S$ and $S \cap T = T$; the inclusion–exclusion principle becomes

$$n(S \cup T) = n(S) + n(T) - n(S \cap T)$$
$$n(S) = n(S) + n(T) - n(T)$$
$$n(S) = n(S).$$

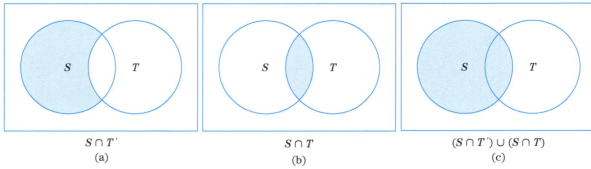

$S \cap T'$
(a)

$S \cap T$
(b)

$(S \cap T') \cup (S \cap T)$
(c)

Figure 7

5.3   Venn Diagrams and Counting

In this section, we discuss the use of Venn diagrams in solving counting problems. The techniques developed are especially useful in analyzing survey data.

Each Venn diagram divides the universal set U into a certain number of regions. For example, the Venn diagram for a single set divides U into two regions—the inside and outside of the circle [Fig. 1(a)]. The Venn diagram for two sets divides U into four regions [Fig. 1(b)]. The Venn diagram for three sets divides U into eight regions [Fig. 1(c)]. Each of the regions is called a **basic region** for the Venn diagram. Knowing the number of elements in each basic region is of great use in many applied problems. As an illustration, consider the next example.

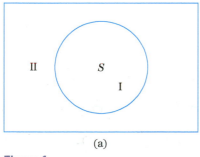

(a)

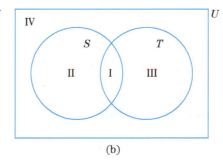

(b)

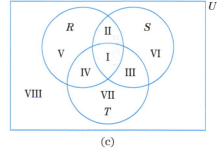

(c)

Figure 1

EXAMPLE 1   **Nobel Prize Winners** Let

$U = \{$Nobel winners during the period 1901–2015$\}$

$A = \{$American Nobel winners during the period 1901–2015$\}$

$C = \{$Chemistry Nobel winners during the period 1901–2015$\}$

$P = \{$Nobel Peace Prize winners during the period 1901–2015$\}$.

These sets are illustrated in the Venn diagram of Fig. 2, in which each basic region has been labeled with the number of elements in it.

(a) How many Americans received a Nobel Prize during this period 1901–2015?

(b) How many Americans received Nobel Prizes in fields other than chemistry and peace during this period?

(c) How many Americans received the Nobel Peace Prize during this period?

(d) How many Nobel Prize winners were there during this period?

Figure 2

SOLUTION   **(a)** The number of Americans who received a Nobel Prize is the total contained in the circle A [Fig. 3(a) on the next page], which is

$$249 + 26 + 1 + 69 = 345.$$

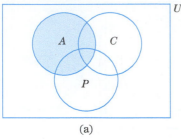
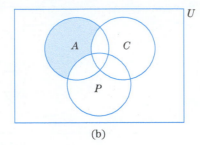
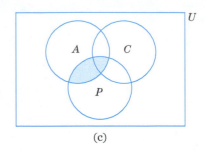

(a)   (b)   (c)

Figure 3

(b) The question asks for the number of Nobel winners in A but not in C and not in P. So start with the A circle, and eliminate those basic regions belonging to C or P [Fig. 3(b)]. There remains a single basic region with 249. Nobel winners. Note that this region corresponds to $A \cap C' \cap P'$.

(c) The question asks for the number of elements in both A and P—that is, $n(A \cap P)$. But $A \cap P$ comprises two basic regions [Fig. 3(c)]. Thus, to compute $n(A \cap P)$, we add the numbers in these basic regions to obtain $26 + 1 = 27$ Americans who have received the Nobel Peace Prize.

(d) The number of recipients is just $n(U)$, and we obtain it by adding together the numbers corresponding to the basic regions. We obtain

$$355 + 249 + 69 + 1 + 26 + 101 + 0 + 99 = 900.$$

>> *Now Try Exercises 1–4*

One need not always be given the number of elements in each of the basic regions of a Venn diagram. Very often, this data can be deduced from given information.

EXAMPLE 2   **Corporate Presidents** Consider the set of 500 corporate presidents of Example 1, Section 5.2.
(a) Draw a Venn diagram displaying the given data, and determine the number of elements in each basic region.
(b) Determine the number of presidents having exactly one degree (graduate or undergraduate) in business.

SOLUTION   (a) Recall that we defined the following sets:

$$S = \{\text{presidents with an undergraduate degree in business}\}.$$

$$T = \{\text{presidents with a graduate degree in business}\}.$$

We were given the following data:

$$n(U) = 500 \qquad n(S) = 238 \qquad n(T) = 184 \qquad n(S \cup T) = 310.$$

We draw a Venn diagram corresponding to S and T (Fig. 4). Notice that none of the given information corresponds to a basic region of the Venn diagram. So we must use our wits to determine the number of presidents in each of the regions I–IV. Region IV is the complement of $S \cup T$, so it contains

$$n(U) - n(S \cup T) = 500 - 310 = 190$$

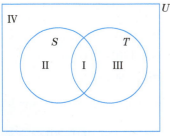

Figure 4

presidents. Region I is just $S \cap T$. By using the inclusion–exclusion principle, in Example 1, Section 5.2, we determined that $n(S \cap T) = 112$. Now, the total number of presidents in I and II combined equals $n(S)$, or 238. Therefore, the number of presidents in II is

$$n(S) - n(S \cap T) = 238 - 112 = 126.$$

Similarly, the number of presidents in III is

$$n(T) - n(S \cap T) = 184 - 112 = 72.$$

Thus, we may fill in the data to obtain a completed Venn diagram (Fig. 5).

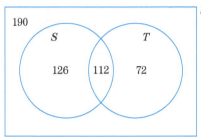

Figure 5

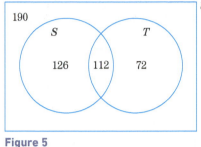

Figure 6

(b) The number of people with exactly one business degree corresponds to the shaded region in Fig. 6. Adding together the number of presidents in each of these regions gives $126 + 72 = 198$ presidents with exactly one business degree.

》 Now Try Exercise 23

Here is another example illustrating the procedure for determining the number of elements in each of the basic regions of a Venn diagram.

EXAMPLE 3　**Advertising Media** An advertising agency finds that the media use of its 170 clients is as follows:

　　115 use television (T)　　　　　　95 use the Internet and mobile apps
　　100 use the Internet (I)　　　　　　85 use television and mobile apps
　　130 use mobile apps (M)　　　　　70 use all three.
　　75 use television and the Internet

Use this data to complete the Venn diagram in Fig. 7 to display the clients' use of mass media.

SOLUTION　Of the various data given, only the last item corresponds to one of the eight basic regions of the Venn diagram—namely, the "70" corresponding to the use of all three media. So we begin by entering this number in the diagram [Fig. 8(a)]. We can fill in the rest of the Venn diagram by working with the remaining information one piece at a time in the reverse order that it is given. Since 85 clients advertise in television and mobile apps, $85 - 70 = 15$ advertise in television and mobile apps but not on the Internet. The appropriate region is labeled in Fig. 8(b). In Fig. 8(c), the next two pieces of information have been used in the same way to fill in two more basic regions. In Fig. 8(c), we observe that three of the four basic regions comprising M have been filled in. Since $n(M) = 130$, we deduce that the number of clients advertising only in

$n(U) = 170$

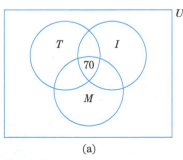

Figure 7

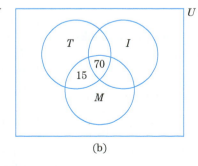

(a)

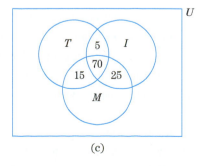

(b)　　　　　　　　　　(c)

Figure 8

mobile apps is $130 - (15 + 70 + 25) = 130 - 110 = 20$ [Fig. 9(a)]. By similar reasoning, the number of clients using only Internet advertising and the number using only television advertising can be determined [Fig. 9(b)]. Adding together the numbers in the three circles gives the number of clients utilizing television, Internet, or mobile apps as $25 + 5 + 0 + 15 + 70 + 25 + 20 = 160$. Since there were 170 clients in total, the remainder—or $170 - 160 = 10$ clients—use none of these media. Figure 9(c) gives a complete display of the data.

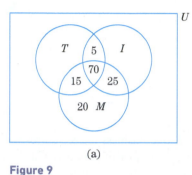
(a)

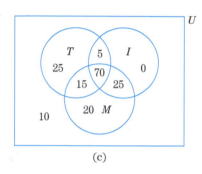
(b)

(c)

Figure 9

>> *Now Try Exercise 31(a)*

Check Your Understanding 5.3

Solutions can be found following the section exercises.

1. Of the 1000 first-year students at a certain college, 700 take mathematics courses, 300 take mathematics and economics courses, and 200 do not take any mathematics or economics courses. Represent this data in a Venn diagram.

2. Refer to the Venn diagram from Problem 1.
   (a) How many of the first-year students take an economics course?
   (b) How many take an economics course but not a mathematics course?

EXERCISES 5.3

Family Library The Venn diagram in Fig. 10 classifies the 100 books in a family's library as hardback (H), fiction (F), and children's (C). Exercises 1–10 refer to this Venn diagram.

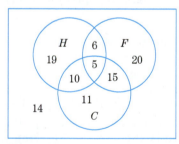

Figure 10

1. How many books are hardback fiction?

2. How many books are paperback fiction?

3. How many books are fiction?

4. How many books are nonfiction?

5. How many books are paperback nonfiction children's books?

6. How many books are adult hardback nonfiction?

7. How many books are either hardback or fiction?

8. How many hardback books are either fiction or children's books?

9. How many children's books are either hardback or fiction?

10. How many books are either hardback, fiction, or children's books?

In Exercises 11–22, let R, S, and T be subsets of the universal set U. Draw an appropriate Venn diagram, and use the given data to determine the number of elements in each basic region.

11. $n(U) = 17, n(S) = 12, n(T) = 7, n(S \cap T) = 5$

12. $n(U) = 20, n(S) = 11, n(T) = 7, n(S \cap T) = 7$

13. $n(U) = 20, n(S) = 12, n(T) = 14, n(S \cup T) = 18$

14. $n(S') = 4, n(S \cup T) = 12, n(S \cap T) = 5, n(T) = 9$

15. $n(U) = 75, n(S) = 15, n(T) = 25, n(S' \cap T') = 40$

16. $n(S) = 10, n(T) = 10, n(S \cap T) = 5, n(S') = 13$

17. $n(S) = 3, n(S \cup T) = 6, n(T) = 4, n(S' \cup T') = 9$

18. $n(U) = 15, n(S) = 8, n(T) = 9, n(S \cup T) = 14$

19. $n(U) = 28, n(R) = 12, n(S) = 12, n(T) = 9, n(R \cap S) = 5,$
    $n(S \cap T) = 3, n(R \cap T) = 7, n(R \cap S \cap T) = 2$

20. $n(U) = 29, n(R) = 10, n(S) = 12, n(T) = 10, n(R \cap S) = 1,$
    $n(R \cap T) = 5, n(S \cap T) = 4, n(R \cap S \cap T) = 1$

21. $n(R') = 22, n(R \cup S) = 21, n(S) = 14, n(T) = 22,$
    $n(R \cap S) = 7, n(S \cap T) = 9, n(R \cap T) = 11,$
    $n(R \cap S \cap T) = 5$

22. $n(U) = 64, n(R \cup S \cup T) = 45, n(R) = 22, n(T) = 26,$
    $n(R \cap S) = 4, n(S \cap T) = 6, n(R \cap T) = 8,$
    $n(R \cap S \cap T) = 1$

23. **Music Preferences** A survey of 70 high school students revealed that 35 like rock music, 15 like hip-hop music, and 5 like both. How many of the students surveyed do not like either rock or hip-hop music?

24. **Nobel Winners** A total of 900 Nobel Prizes had been awarded by 2015. Fourteen of the 112 prizes in literature were awarded to Scandinavians. Scandinavians received a total of 57 awards. How many Nobel Prizes outside of literature have been awarded to non-Scandinavians?

25. **Analysis of Sonnet** One of Shakespeare's sonnets has a verb in 11 of its 14 lines, an adjective in 9 lines, and both in 7 lines. How many lines have a verb but no adjective? An adjective but no verb? Neither an adjective nor a verb?

Exam Performance The results from an exam taken by 150 students were as follows:

> 90 students correctly answered the first question,
> 71 students correctly answered the second question,
> 66 students correctly answered both questions.

Exercises 26–30 refer to these students.

26. How many students correctly answered either the first or second question?

27. How many students did not answer either of the two questions correctly?

28. How many students answered either the first or the second question correctly, but not both?

29. How many students answered the second question correctly, but not the first?

30. How many students missed the second question?

31. **Class Enrollment** Out of 35 students in a finite math class, 22 are male, 19 are business majors, 27 are first-year students, 14 are male business majors, 17 are male first-year students, 15 are first-year business majors, and 11 are male first-year business majors.
    (a) Use this data to complete a Venn diagram displaying the characteristics of the students.
    (b) How many students in the class are neither first-year, nor male, nor business majors?
    (c) How many non-male business majors are in the class?

32. **Exercise Preferences** A survey of 100 college faculty who exercise regularly found that 45 jog, 30 swim, 20 cycle, 6 jog and swim, 1 jogs and cycles, 5 swim and cycle, and 1 does all three. How many of the faculty members do not do any of these activities? How many just jog?

Flag Colors The three most common colors in the 193 flags of the member nations of the United Nations are red, white, and blue:

> 52 flags contain all three colors
> 103 flags contain both red and white
> 66 flags contain both red and blue
> 73 flags contain both white and blue
> 145 flags contain red
> 132 flags contain white
> 104 flags contain blue.

Exercises 33–38 refer to the 193 flags.

33. How many flags contain red, but not white or blue?

34. How many flags contain exactly one of the three colors?

35. How many flags contain none of the three colors?

36. How many flags contain exactly two of the three colors?

37. How many flags contain red and white, but not blue?

38. How many flags contain red or white, but not both?

News Dissemination A merchant surveyed 400 people to determine from what source they found out about an upcoming sale. The results of the survey follow:

> 180 from the Internet
> 190 from television
> 190 from newspapers
> 80 from the Internet and television
> 90 from the Internet and newspapers
> 50 from television and newspapers
> 30 from all three sources.

Exercises 39–44 refer to the people in this survey.

39. How many people learned of the sale from newspapers or the Internet but not from both?

40. How many people learned of the sale only from newspapers?

41. How many people learned of the sale from the Internet or television but not from newspapers?

42. How many people learned of the sale from at least two of the three media?

43. How many people learned of the sale from exactly one of the three media?

44. How many people learned of the sale from the Internet and television but not from newspapers?

45. **Course Enrollments** Table 1 shows the number of students enrolled in each of three science courses at Gotham College. Although no students are enrolled in all three courses, 15 are enrolled in both chemistry and physics, 10 are enrolled in both physics and biology, and 5 are enrolled in both biology and chemistry. How many students are enrolled in at least one of these science courses?

Table 1

Course	Enrollment
Chemistry	60
Physics	40
Biology	30

Foreign Language Courses A survey in a local high school shows that, of the 4000 students in the school,

> 2000 take French (F)
> 3000 take Spanish (S)
> 500 take Latin (L)
> 1500 take both French and Spanish
> 300 take both French and Latin
> 200 take Spanish and Latin
> 50 take all three languages.

Use a Venn diagram to find the number of people in the sets given in Exercises 46–50.

46. $L \cap (F \cup S)$ 47. $(L \cup F \cup S)'$ 48. L'

49. $L \cup S \cup F'$ 50. $F \cap S' \cap L'$

51. **Voting Preferences** One hundred college students were surveyed after voting in an election involving a Democrat and a Republican. There were 50 first-year students, 55 voted Democratic, and 25 were non-first-year students who voted Republican. How many first-year students voted Democratic?

52. Union Membership and Education Status A group of 100 workers were asked whether they were college graduates and whether they belonged to a union. According to their responses, 60 were not college graduates, 20 were nonunion college graduates, and 30 were union members. How many of the workers were neither college graduates nor union members?

53. Diagnostic Test Results A class of 30 students was given a diagnostic test on the first day of a mathematics course. At the end of the semester, only 2 of the 21 students who had passed the diagnostic test failed the course. A total of 23 students passed the course. How many students managed to pass the course even though they failed the diagnostic test?

54. Air-Traffic Controllers A group of applicants for training as air-traffic controllers consists of 35 pilots, 20 veterans, 30 pilots who were not veterans, and 50 people who were neither veterans nor pilots. How large was the group?

College Majors A group of 61 students has the following characteristics:

6 are biology majors and seniors
17 are biology majors and not seniors
12 are not seniors and are majoring in a field other than biology.

Exercises 55–60 refer to these students.

55. How many of the students are either seniors or biology majors?

56. How many of the students are seniors?

57. How many of the students are not seniors?

58. How many of the students are biology majors?

59. How many of the seniors are not biology majors?

60. How many of the students are not biology majors?

Music Preferences A campus radio station surveyed 190 students to determine the genres of music they liked. The survey results follow:

114 like rock
50 like country
15 like rock and rap
11 like rap and country
20 like rap only
10 like rock and rap, but not country
9 like rock and country, but not rap
20 don't like any of the three types of music.

Exercises 61–68 refer to the students in this survey.

61. How many students like rock only?

62. How many students like country but not rock?

63. How many students like rap and country but not rock?

64. How many students like rap or country but not rock?

65. How many students like exactly one of the genres?

66. How many students like all three genres?

67. How many students like at least two of the three genres?

68. How many students do not like either rock or country?

69. Website Preferences One hundred and sixty business executives were surveyed to determine whether they regularly visit the *CNN Money*, *Bloomberg*, or *The Wall Street Journal* websites. The survey showed that 70 visit *CNN Money*, 60 visit *Bloomberg*, 55 visit *The Wall Street Journal*, 45 visit exactly two of the three websites, 20 visit *CNN Money* and *Bloomberg*, 20 visit *Bloomberg* and *The Wall Street Journal*, and 5 visit all three websites. How many do not visit any of the three websites?

70. Small Businesses A survey of the characteristics of 100 small businesses that had failed revealed that 95 of them either were undercapitalized, had inexperienced management, or had a poor location. Four of the businesses had all three of these characteristics. Forty businesses were undercapitalized but had experienced management and good location. Fifteen businesses had inexperienced management but sufficient capitalization and good location. Seven were undercapitalized and had inexperienced management. Nine were undercapitalized and had poor location. Ten had inexperienced management and poor location. How many of the businesses had poor location? Which of the three characteristics was most prevalent in the failed businesses?

71. Music Each of the 100 students attending a conservatory of music plays at least one of three instruments: piano, violin, and clarinet. Of the students, 65 play the piano, 42 play the violin, 28 play the clarinet, 20 play the piano and the violin, 10 play the violin and the clarinet, and 8 play the piano and the clarinet. How many play all three instruments? *Hint:* Let x represent the number of students who play all three instruments.

72. Courses Students living in a certain dormitory were asked about their enrollment in mathematics and history courses. Ten percent were taking both types of courses, and twenty percent were taking neither type of course. One hundred sixty students were taking a mathematics course but not a history course, and one hundred twenty students were taking a history course but not a mathematics course. How many students were taking a mathematics course? *Hint:* Let x be the total number of students living in the dormitory.

Solutions to Check Your Understanding 5.3

1. Draw a Venn diagram with two circles, one for mathematics (M) and one for economics (E) [Fig. 11(a)]. This Venn diagram has four basic regions, and our goal is to label each basic region with the proper number of students. The numbers for two of the basic regions are given directly. Since "300 take mathematics and economics," $n(M \cap E) = 300$. Since "200 do not take any mathematics or economics courses," $n((M \cup E)') = 200$ [Fig. 11(b)]. Now, "700 take mathematics courses." Since M is made up of two basic regions and one region has 300 elements, the other basic region of M must contain 400 elements [Fig. 11(c)]. At this point, all but one of the basic regions have been labeled and $400 + 300 + 200 = 900$ students

have been accounted for. Since there is a total of 1000 students, the remaining basic region has 100 students [Fig. 11(d)].

2. (a) 400. "Economics" refers to the entire circle E, which is made up of two basic regions, one having 300 elements and the other 100. (A common error is to interpret the question as asking for the number of first-year students who take economics exclusively and therefore give the answer 100. To say that a person takes an economics course does not imply anything about the person's enrollment in mathematics courses.)

(b) 100

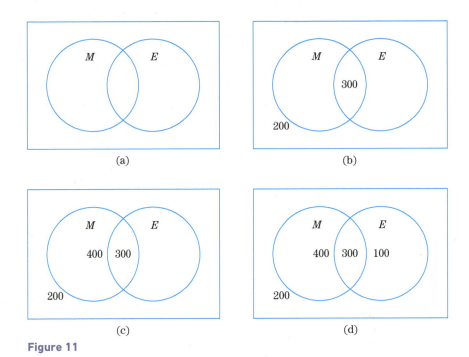

Figure 11

5.4 The Multiplication Principle

In this section, we introduce a second fundamental principle of counting, the *multiplication principle*. By way of motivation, consider the following example:

EXAMPLE 1   **Counting Paths through a Maze** A medical researcher wishes to test the effect of a drug on a rat's perception by studying the rat's ability to run a maze while under the influence of the drug. The maze is constructed so that, to arrive at the exit point C, the rat must pass through a central point B. There are five paths from the entry point A to B, and three paths from B to C. In how many different ways can the rat run the maze from A to C? (See Fig. 1.)

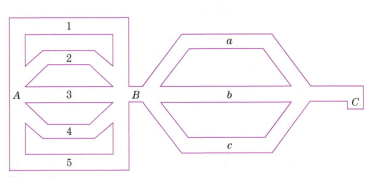

Figure 1

SOLUTION   The paths from A to B have been labeled 1 through 5, and the paths from B to C have been labeled a through c. The various paths through the maze can be schematically represented as in Fig. 2 on the next page. The diagram shows that there are five ways to go from A to B. For each of these five ways, there are three ways to go from B to C. So there are five groups of three paths each and therefore $5 \cdot 3 = 15$ possible paths from A to C. (A diagram such as Fig. 2, called a **tree diagram**, is useful in enumerating the various possibilities in counting problems.)

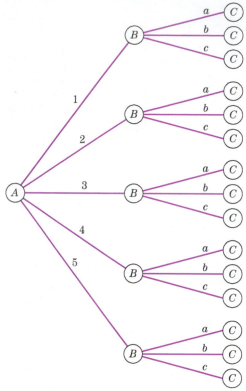

Figure 2   **» Now Try Exercise 1**

In the preceding problem, selecting a path is a task that can be broken up into two consecutive choices.

Choose path from A to B
Choice 1

Choose path from B to C
Choice 2

The first choice can be performed in five ways, and after the first choice has been carried out, the second can be performed in three ways. And we determined that the entire task can be performed in $5 \cdot 3 = 15$ ways. The same reasoning as just used yields the following useful counting principle.

> **Multiplication Principle**   Suppose that a task is composed of two consecutive choices. If choice 1 can be performed in m ways and, for each of these, choice 2 can be performed in n ways, then the complete task can be performed in $m \cdot n$ ways.

EXAMPLE 2   **Counting Routes for a Trip**   An airline passenger must fly from New York to Frankfurt via London. There are 8 flights leaving New York for London. All of these provide connections on any one of 19 flights from London to Frankfurt. In how many different ways can the passenger book reservations?

SOLUTION   The task *Fly from New York to Frankfurt* is composed of two consecutive choices:

Select a flight from New York to London
Choice 1

Select a flight from London to Frankfurt
Choice 2

From the data given, the multiplication principle implies that the task can be accomplished in $8 \cdot 19 = 152$ ways.   **» Now Try Exercise 3**

It is possible to generalize the multiplication principle to tasks consisting of more than two choices.

> **Generalized Multiplication Principle**   Suppose that a task consists of t choices performed consecutively. Suppose that choice 1 can be performed in m_1 ways; for each of these, choice 2 in m_2 ways; for each of these, choice 3 in m_3 ways; and so forth. Then the task can be performed in
>
> $$m_1 \cdot m_2 \cdot m_3 \cdot \cdots \cdot m_t \quad \text{ways.}$$

EXAMPLE 3   **Officers for a Board of Directors**   A corporation has a board of directors consisting of 10 members. The board must select from among its members a chairperson, vice chairperson, and secretary. In how many ways can this be done?

SOLUTION   The task *Select the three officers* can be divided into three consecutive choices:

Select chairperson	Select vice chairperson	Select secretary

Since there are 10 directors, choice 1 can be performed in 10 ways. After the chairperson has been selected, there are 9 directors left as possible candidates for vice chairperson so that for each way of performing choice 1, choice 2 can be performed in 9 ways. After this has been done, there are 8 directors who are possible candidates for secretary, so choice 3 can be performed in 8 ways. By the generalized multiplication principle, the number of possible ways to perform the sequence of three choices equals $10 \cdot 9 \cdot 8$, or 720. So the officers of the board can be selected in 720 ways.

» Now Try Exercise 7

In Example 3, we made important use of the phrase "for each of these" in the generalized multiplication principle. The choice *Select a vice chairperson* can be performed in 10 ways, since any member of the board is eligible. However, when we view the selection process as a sequence of choices of which *Select a vice chairperson* is the second choice, the situation has changed. *For each way* that the first choice is performed, one person will have been used up; hence, there will be only 9 possibilities for choosing the vice chairperson.

Note that the order of the choices doesn't matter. For example, we could choose the vice chairman first, then the secretary, then the chairperson, and we would arrive at the same result.

EXAMPLE 4   **Posing for a Group Picture**   In how many ways can a baseball team of nine players arrange themselves in a line for a group picture?

SOLUTION   Choose the players by their place in the picture—say, from left to right. The first can be chosen in nine ways; for each of these outcomes, the second can be chosen in eight ways; for each of these outcomes, the third can be chosen in seven ways; and so forth. So the number of possible arrangements is

$$9 \cdot 8 \cdot 7 \cdot 6 \cdot 5 \cdot 4 \cdot 3 \cdot 2 \cdot 1 = 362{,}880.$$ **» Now Try Exercise 11**

We can write the product $9 \cdot 8 \cdot 7 \cdot 6 \cdot 5 \cdot 4 \cdot 3 \cdot 2 \cdot 1$ from the previous example in a condensed way by using **factorial** notation:

> **DEFINITION Factorial**   If n is a positive integer, the number n **factorial**, denoted $n!$, is defined to be the product
>
> $$n! = n(n-1)(n-2) \cdots (2)(1).$$
>
> In addition, we define $0! = 1$.

EXAMPLE 5   **License Plates** A certain state uses automobile license plates that consist of three letters followed by three digits. How many such license plates are there?

SOLUTION   The task in this case, *Form a license plate*, consists of a sequence of six choices: three for choosing letters and three for choosing digits. Each letter can be chosen in 26 ways and each digit in 10 ways. So the number of license plates is

$$26 \cdot 26 \cdot 26 \cdot 10 \cdot 10 \cdot 10 = 17{,}576{,}000.$$ **≫ Now Try Exercise 17**

Check Your Understanding 5.4

Solutions can be found following the section exercises.

1. There are five seats available in a sedan. In how many ways can five people be seated if only three can drive?

2. A multiple-choice exam contains 10 questions, each having 3 possible answers. Assuming you answer each question, how many different ways are there of completing the exam?

EXERCISES 5.4

1. Jolene wants to drive from her house to the grocery store and then to the library. If her GPS suggests four routes from her house to the grocery store, and two routes from the grocery store to the library, how many total ways are there for Jolene to do this? (See Fig. 3.)

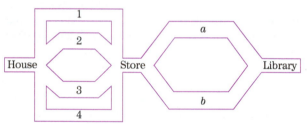

Figure 3

2. There are three bridges from the west shore of a river to an island in the center, and three bridges from the island to the east shore. How many different ways are there to cross from the west shore to the east shore? (See Fig. 4.)

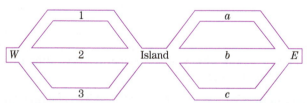

Figure 4

3. **Travel Options** If you can travel from Frederick, Maryland, to Baltimore, Maryland, by car, bus, or train and from Baltimore to London by airplane or ship, how many different ways are there to go from Frederick to London via Baltimore?

4. **Travel Options** Suppose that Maria wants to go from Florida to Maine via New York and can travel each leg of the journey by bus, car, train, or airplane. How many different ways can Maria make the trip?

5. **Daytona 500** Forty-four race cars competed in the 2016 Daytona 500. Assuming no ties, how many possibilities were there for the first-, second-, and third-place finishers?

6. **Kentucky Derby** Twenty horses competed in the 2016 Kentucky Derby. Assuming no ties, how many possibilities were there for the first, second, and third place finishers?

7. **Winners** Twenty athletes enter an Olympic event. Assuming no ties, how many different possibilities are there for winning the Gold Medal, Silver Medal, and Bronze Medal?

8. **Ranking Teams** A sportswriter is asked to rank six teams. How many different orderings are possible?

9. **Electing Captains** A football squad can elect a captain and an assistant captain in 870 possible ways. How many members does the squad have?

10. **Club Officers** A club can elect a president and a treasurer in 600 different ways. How many members does the club have?

11. **Group Picture** A group of five boys and three girls is to be photographed.
    (a) How many ways can they be arranged in one row?
    (b) How many ways can they be arranged with the girls in the front row and the boys in the back row?

12. **Arranging Books** Three history books and six novels are to be arranged on a bookshelf.
    (a) How many ways can they be arranged?
    (b) How many ways can they be arranged with the history books to the left of the novels?

13. **Rearranging Letters** How many different words (including nonsense words) can be formed by using the four letters of the word "MATH"?

14. **Three-Letter Words** How many different three-letter words (including nonsense words) are there in which successive letters are different?

15. **Selecting an Outfit** How many different outfits consisting of a coat and a hat can be selected from two coats and three hats?

16. **Selecting an Outfit** How many different outfits can be selected from two coats, four hats, and two scarves?

17. **Serial Numbers** A computer manufacturer assigns serial numbers to its computers. The first symbol of a serial number is either A, B, or C, indicating the manufacturing plant. The second and third symbols taken together are one of the numbers 01, 02, . . . , 12, indicating the month of manufacture. The final four symbols are digits. How many possible serial numbers are there?

18. **License Plates** Suppose that a license plate consists of a nonzero digit followed by three letters and then three nonzero digits. How many such license plates are there?

19. **Social Security Numbers** How many Social Security numbers are available if the only restriction is that the number 000-00-0000 cannot be assigned?

20. **Call Letters** In 1923, the Federal Communications Commission directed that all new radio stations east of the Mississippi River have call letters beginning with the letter W. How many different three- or four-letter call letters are possible?

21. **Area Codes** Before 1995, three-digit area codes for the United States had the following restrictions:
    (i)   Neither 0 nor 1 could be used as the first digit.
    (ii) 0 or 1 had to be used for the second digit.
    (iii) There were no restrictions on the third digit.
    How many different area codes were possible?

22. **Area Codes** Refer to Exercise 21. Beginning in 1995, restriction (ii) was lifted and any digit could be used in the second position. How many different area codes were then possible?

Palindromes A number or word is said to be a *palindrome* if it reads the same backward as forward (e.g., 58485 or radar).

23. How many 5-digit numbers are palindromes?

24. How many 6-digit numbers are palindromes?

25. How many 4-letter words (including nonsense words) are palindromes?

26. How many 3-letter words (including nonsense words) are palindromes?

27. **World Series** The World Series of Baseball is played between the American League and National League champions, in which each league consists of 15 teams. How many different possible matchups are there for the World Series?

28. **Super Bowl** The Super Bowl is a game played between the National Football Conference and American Football Conference champions. Each conference consists of 16 teams. How many different possible matchups are there for the Super Bowl?

29. **Bridge Games** There are 3200 duplicate bridge clubs sanctioned by the American Contract Bridge Association. If each club holds two games per week and each duplicate game consists of 24 deals, how many deals are played in club games each year?

30. **Ice Cream Selections** An ice cream parlor offers 25 flavors of ice cream. Assuming that order matters, how many different two-scoop cones can be made? What if no flavor can be repeated?

31. **Internet Accounts** A college of 20,000 students provides each student with an Internet account. Explain why letting each student have his or her initials as the username cannot possibly work. Assume that each person has a first name, middle name, and last name and therefore that each person's initials consist of three letters.

32. **Pairs of Initials** A company has 700 employees. Explain why there must be two people with the same first and last initials.

33. **Game Outcomes** The final score in a soccer game is 6 to 4. How many different halftime scores are possible?

34. **Selecting an Outfit** Each day, Gloria dresses in a blouse, a skirt, and shoes. She wants to wear a different combination on every day of the year. If she has the same number of blouses, skirts, and pairs of shoes, how many of each article would she need to have a different combination every day?

35. **Mismatched Gloves** A man has five different pairs of gloves. In how many ways can he select a right-hand glove and a left-hand glove that do not match?

36. **Mismatched Shoes** Fred has 11 different pairs of shoes. In how many ways can he put on a pair of shoes that do not match?

37. **Coin Tosses** Toss a coin six times, and observe the sequence of heads and tails that results. How many different sequences are possible?

38. **Coin Tosses** Refer to Exercise 37. In how many of the sequences are the first and last tosses identical?

39. **Exam Questions** An exam contains five true-or-false questions. In how many different ways can the exam be completed? Assume that every question must be answered.

40. **Exam Questions** An exam contains five true-or-false questions. In how many ways can the exam be completed if leaving the answer blank is also an option?

41. **Exam Questions** Each of the 10 questions on a multiple-choice exam has four possible answers. How many different ways are there for a student to answer the questions? Assume that every question must be answered.

42. **Exam Questions** Rework Exercise 41 under the assumption that not every question must be answered.

ZIP Codes ZIP (Zone Improvement Plan) codes, sequences of five digits, were introduced by the United States Post Office Department in 1963.

43. How many ZIP codes are possible?

44. ZIP codes for Delaware, New York, and Pennsylvania begin with the digit 1. How many such ZIP codes are possible?

45. **Group Pictures** How many ways can eight people stand in a line for a group picture? If you took a picture every 15 seconds (day and night with no breaks), how long would it take to photograph every possible arrangement?

46. **License Plates** A company is manufacturing license plates with the pattern LL#-##LL, where L represents a letter and # represents a digit from 1 through 9. If a letter can be any letter from A to Z except O, how many different license plates are possible? If the company produces 500,000 license plates per week, how many years will be required to make every possible license plate?

47. **Menu Selections** A college student eats all of their meals at a restaurant offering six breakfast specials, seven lunch specials, and four dinner specials. How many days can they go without repeating an entire day's menu selections?

48. **Menu Selections** A restaurant menu lists 7 appetizers, 10 entrées, and 4 desserts. How many ways can a diner select a three-course meal?

49. **Gift Wrapping** The gift-wrap desk at a large department store offers 5 box sizes, 10 wrapping papers, 7 colors of ribbon in 2 widths, and 9 special items to be added on the box. How many different ways are there to package a gift, assuming that the customer must choose at least a box but need not choose any of the other offerings?

50. **Selecting Fruit** José was told to create a gift basket containing one dozen oranges, eight apples, and a half-pound of grapes. When he gets to the store, he finds five varieties of oranges, five varieties of apples, and two varieties of grapes. Assuming

that he selects only one variety of each type of fruit, how many different gift baskets of fruit could he bring home?

51. **Shading of Venn Diagrams** How many different ways can a Venn diagram with two circles be shaded?

52. **Shading of Venn Diagrams** How many different ways can a Venn diagram with three circles be shaded?

Roulette An American roulette wheel consists of 38 numbered pockets. Two of them (numbered 0 and 00) are colored green, 18 are colored red, and 18 are colored black. Gamblers bet on which numbered pocket a ball will fall into when the wheel is spun. For Exercises 53 and 54, assume the wheel is spun three times.

53. How many outcomes are possible if the first number is green?

54. How many outcomes are possible if all three numbers are red and no number repeats?

55. **Batting Orders** The manager of a Little League baseball team has picked the nine starting players for a game. How many different batting orders are possible under each of the following conditions?
   (a) There are no restrictions.
   (b) The pitcher must bat last.
   (c) The pitcher must bat last, the catcher eighth, and the shortstop first.

56. **Test Volunteers** A physiologist wants to test the effects of exercise and meditation on blood pressure. She devises four different exercise programs and three different meditation programs. If she wants 10 subjects for each combination of exercise and meditation program, how many volunteers must she recruit?

57. **Handshakes** Two 10-member basketball teams play a game. After the game, each of the members of the winning team shakes hands once with each member of both teams. How many handshakes take place?

58. **Band Selections** In how many ways can a band play a set of three waltzes and three tangos without repeating any song, such that the first, third, and fifth songs are waltzes?

59. **Colored Houses** Six houses in a row are each to be painted with one of the colors red, blue, green, and yellow. In how many different ways can the houses be painted so that no two adjacent houses are of the same color?

60. **Chair Varieties** A furniture manufacturer makes three types of upholstered chairs and offers 20 fabrics. How many different chairs are available?

61. **Ballots** Seven candidates for mayor, four candidates for city council president, and six propositions are being put before the electorate. How many different ballots could be cast, assuming that every voter votes on each of the items? If voters can choose to leave any item blank, how many different ballots are possible?

62. **College Applications** Allison is preparing her applications for college. She will apply to three community colleges and has to fill out six parts in each of those applications. She will apply to three four-year schools, each of which has a seven-part application. How many application segments must she complete?

63. **Paths to Texas** Consider the triangular display of letters below. Start with the letter T at the top and move down the triangle to a letter S at the bottom. From any given letter, move only to one of the letters directly below it on the left or right. How many different paths spell *TEXAS*?

$$T$$
$$E \quad E$$
$$X \quad X \quad X$$
$$A \quad A \quad A \quad A$$
$$S \quad S \quad S \quad S$$

64. **Railroad Tickets** A railway has 20 stations. If the names of the point of departure and the destination are printed on each ticket, how many different kinds of single tickets must be printed? How many different kinds of tickets are needed if each ticket may be used in either direction between two stations?

Solutions to Check Your Understanding 5.4

1. 72. Pretend that you are given the task of seating the five people. This task consists of five choices performed consecutively, as shown in Table 1. After you have performed choice 1, four people will remain, and any one of these four can be seated in the right front seat. After choice 2, three people remain, and so on. By the generalized multiplication principle, the task can be performed in $3 \cdot 4 \cdot 3 \cdot 2 \cdot 1 = 72$ ways.

2. 3^{10}. The task of answering the questions consists of 10 consecutive choices, each of which can be performed in three ways. Therefore, by the generalized multiplication principle, the task can be performed in

$$\underbrace{3 \cdot 3 \cdot 3 \cdot \cdots \cdot 3}_{10 \text{ terms}} \text{ ways.}$$

Note: The answer can be left as 3^{10} or can be multiplied out to 59,049.

Table 1

Choice	Number of Ways Choice Can Be Performed
1: Select person to drive.	3
2: Select person for right front seat.	4
3: Select person for left rear seat.	3
4: Select person for middle rear seat.	2
5: Select person for right rear seat.	1

5.5 Permutations and Combinations

In preceding sections, we have solved a variety of counting problems by using Venn diagrams and the generalized multiplication principle. Let us now turn our attention to two types of counting problems that occur very frequently and that can be solved by using formulas derived from the generalized multiplication principle. These problems involve what are called *permutations* and *combinations*, which are particular types of arrangements of elements of a set. The sorts of arrangements that we have in mind are illustrated in two problems:

Problem A   How many words (by which we mean *strings of letters*) of two distinct letters can be formed from the letters $\{a, b, c\}$?

Problem B   A construction crew has three members. A team of two must be chosen for a particular job. In how many ways can the team be chosen?

Each of the two problems can be solved by enumerating all possibilities.

Solution of Problem A   There are six possible words, namely,

$$ab \quad ac \quad ba \quad bc \quad ca \quad cb.$$

Solution of Problem B   Designate the three crew members by a, b, and c. Then there are three possible two-person teams, namely,

$$\{a, b\} \quad \{a, c\} \quad \{b, c\}.$$

Note that $\{b, a\}$, the team consisting of b and a, is the same as the team $\{a, b\}$.

We deliberately set up both problems with the same letters in order to facilitate comparison. Both problems are concerned with counting the numbers of arrangements of the elements of the set $\{a, b, c\}$, taken two at a time, without allowing repetition. (For example, *aa* was not allowed.) However, in Problem A, the order of the arrangement mattered, whereas in Problem B it did not. Arrangements of the sort considered in Problem A are called *permutations*, whereas those in Problem B are called *combinations*.

More precisely, suppose that we are given a set of n distinguishable objects.

> **Permutations**   A **permutation of n objects taken r at a time** is an arrangement of r of the n objects in a specific order.

So, for example, Problem A was concerned with permutations of the three objects a, b, c ($n = 3$) taken two at a time ($r = 2$).

> **Combinations**   A **combination of n objects taken r at a time** is a selection of r objects from among the n, with order disregarded.

Thus, for example, in Problem B, we considered combinations of the three objects a, b, c ($n = 3$) taken two at a time ($r = 2$).

It is convenient to introduce the notation that follows for counting permutations and combinations. Let

$$P(n, r) = \text{the number of permutations of } n \text{ objects taken } r \text{ at a time}$$

$$C(n, r) = \text{the number of combinations of } n \text{ objects taken } r \text{ at a time.}$$

Thus, for example, from our solutions to Problems A and B, we have

$$P(3, 2) = 6 \quad\quad C(3, 2) = 3.$$

Very simple formulas for $P(n, r)$ and $C(n, r)$ allow us to calculate these quantities for any n and r. Let us begin by stating the formula for $P(n, r)$. For $r = 1, 2, 3$, respectively,

$$P(n, 1) = n$$
$$P(n, 2) = n(n - 1) \qquad \text{(two factors)},$$
$$P(n, 3) = n(n - 1)(n - 2) \qquad \text{(three factors)}.$$

Continuing, we obtain the following formula:

Permutation Formula The number of permutations of n objects taken r at a time, $P(n, r)$, is the product of the r whole numbers counting down by 1 from n. That is,

$$P(n, r) = n(n - 1)(n - 2) \cdots (n - r + 1) \qquad (r \text{ factors}), \qquad (1)$$

or

$$P(n, r) = \frac{n!}{(n - r)!} \qquad (2)$$

This formula is verified at the end of the section.

EXAMPLE 1   **Applying the Permutation Formula** Compute the following numbers:
(a) $P(100, 2)$ (b) $P(6, 4)$ (c) $P(5, 5)$

SOLUTION   (a) Here, $n = 100$, $r = 2$. So we take the product of two factors, beginning with 100:

$$P(100, 2) = 100 \cdot 99 = 9900.$$

(b) $P(6, 4) = 6 \cdot 5 \cdot 4 \cdot 3 = 360$
(c) $P(5, 5) = 5 \cdot 4 \cdot 3 \cdot 2 \cdot 1 = 120$ **≫ Now Try Exercise 1**

Combination Formula The number of combinations of n objects taken r at a time, $C(n, r)$, is

$$C(n, r) = \frac{P(n, r)}{r!} = \frac{n(n - 1)(n - 2) \cdots (n - r + 1)}{r!} \qquad (r \text{ factors}), \qquad (3)$$

or

$$C(n, r) = \frac{n!}{r! \, (n - r)!}$$

This formula is verified at the end of this section.

EXAMPLE 2   **Applying the Combination Formula** Compute the following numbers:
(a) $C(100, 2)$ (b) $C(6, 4)$ (c) $C(5, 5)$

SOLUTION   (a) $C(100, 2) = \dfrac{P(100, 2)}{2!} = \dfrac{100 \cdot 99}{2 \cdot 1} = 4950$

(b) $C(6, 4) = \dfrac{P(6, 4)}{4!} = \dfrac{6 \cdot 5 \cdot 4 \cdot 3}{4 \cdot 3 \cdot 2 \cdot 1} = 15$

(c) $C(5, 5) = \dfrac{P(5, 5)}{5!} = \dfrac{5 \cdot 4 \cdot 3 \cdot 2 \cdot 1}{5 \cdot 4 \cdot 3 \cdot 2 \cdot 1} = 1$

≫ Now Try Exercise 5

EXAMPLE 3	**Applying the Permutation and Combination Formulas** Solve Problems A and B, using formulas (1) and (3).

SOLUTION   The number of two-letter words that can be formed from the three letters a, b, and c is equal to $P(3, 2) = 3 \cdot 2 = 6$, in agreement with our previous solution.

The number of two-worker teams that can be formed from three individuals is equal to $C(3, 2)$, and

$$C(3, 2) = \frac{P(3, 2)}{2!} = \frac{3 \cdot 2}{2 \cdot 1} = 3,$$

in agreement with our previous result.   **» Now Try Exercises 27 and 29**

EXAMPLE 4	**Selecting a Committee** The board of directors of a corporation has 10 members. In how many ways can they choose a committee of three board members to negotiate a merger?

SOLUTION   Since the committee of three involves no ordering of its members, we are concerned here with combinations. The number of combinations of 10 people taken 3 at a time is $C(10, 3)$, which is

$$C(10, 3) = \frac{10 \cdot 9 \cdot 8}{3 \cdot 2 \cdot 1} = 120.$$

Thus, there are 120 possibilities for the committee.   **» Now Try Exercise 31**

EXAMPLE 5	**Selecting Club Officers** A club has 10 members. In how many ways can they choose a slate of four officers, consisting of a president, vice president, secretary, and treasurer?

SOLUTION   In this problem, we are dealing with an ordering of four members. (The first is the president, the second the vice president, and so on.) So we are dealing with permutations, and the number of ways of choosing the officers is

$$P(10, 4) = 10 \cdot 9 \cdot 8 \cdot 7 = 5040.$$ **» Now Try Exercise 33**

EXAMPLE 6	**Outcomes of a Horse Race** Eight horses are entered in a race in which a first, second, and third prize will be awarded. Assuming no ties, how many different outcomes are possible?

SOLUTION   In this example, we are considering ordered arrangements of three horses, so we are dealing with permutations. The number of permutations of eight horses taken three at a time is

$$P(8, 3) = 8 \cdot 7 \cdot 6 = 336,$$

so the number of possible outcomes of the race is 336.   **» Now Try Exercise 39**

EXAMPLE 7	**Polling Sample** A political pollster wishes to survey 1500 individuals chosen from a sample of 5,000,000 adults. In how many ways can the 1500 individuals be chosen?

SOLUTION   No ordering of the 1500 individuals is involved, so we are dealing with combinations. So the number in question is $C(5{,}000{,}000, 1500)$, a number too large to be written down in digit form. (It has several thousand digits!) But it could be calculated with the aid of a computer.   **» Now Try Exercise 29**

EXAMPLE 8	**Seating Arrangements** Three couples go on a movie date. In how many ways can they be seated in a row of six seats so that each couple is seated together?

SOLUTION   A seating arrangement can be composed of four consecutive choices:

Select an order in which to seat the couples	Seat the left-most couple	Seat the middle couple	Seat the right-most couple
Choice 1	Choice 2	Choice 3	Choice 4

Since there are three couples, choice 1 can be performed in 3! ways. After the order of the three couples has been chosen, there are 2! ways to seat the two members of the leftmost couple. Similarly, there are 2! ways to perform choice 3 and 2! ways to perform choice 4.

By the generalized multiplication principle, the number of possible arrangements is $3! \cdot 2! \cdot 2! \cdot 2! = 48$.   **≫ Now Try Exercise 59**

EXAMPLE 9   **Arranging Books** If you have six books, in how many ways can you select four books and arrange them on a shelf?

SOLUTION   Arrange four of the six books in order. This can be done in $P(6, 4) = 360$ ways.

As is often the case in counting problems, there are multiple ways to arrive at a solution.

ALTERNATIVE SOLUTION   First, select the four books from the group of six. Since order does not matter at this point, this can be done in $C(6, 4)$ ways. Next, arrange the selection of four books on the shelf, which can be done in 4! ways. By the multiplication principle, the number of possible arrangements is $C(6, 4) \cdot 4! = 15 \cdot 24 = 360$.   **≫ Now Try Exercise 58**

Example 9 shows that $P(n, r) = C(n, r) \cdot r!$ when $n = 6$ and $r = 4$. The discussion that follows shows that this formula holds for all values of n and r.

Verification of the Formulas for $P(n, r)$ and $C(n, r)$

Let us first derive the formula for $P(n, r)$, the number of permutations of n objects taken r at a time. The task of choosing r objects (in a given order) consists of r consecutive choices (Fig. 1). The first choice can be performed in n ways. For each way that the first choice is performed, one object will have been used up so that we can perform the second choice in $n - 1$ ways, and so on. For each way of performing the sequence of choices $1, 2, 3, \ldots, r - 1$, the rth choice can be performed in $n - (r - 1) = n - r + 1$ ways. By the generalized multiplication principle, the task of choosing the r objects from among the n can be performed in

$$n(n - 1) \cdot \cdots \cdot (n - r + 1) \quad \text{ways.}$$

That is,

$$P(n, r) = n(n - 1) \cdot \cdots \cdot (n - r + 1),$$

which is formula (1).

Choose 1st object	Choose 2nd object	$\cdots$	Choose rth object

Figure 1

Let us now verify the formula for $C(n, r)$, the number of combinations of n objects taken r at a time. Each such combination is a set of r objects and, therefore, can be ordered in

$$P(r, r) = r(r - 1) \cdot \cdots \cdot 2 \cdot 1 = r!$$

ways by formula (1). In other words, each different combination of r objects gives rise to $r!$ permutations of the same r objects. On the other hand, each permutation of n objects taken r at a time gives rise to a combination of n objects taken r at a time, by simply ignoring the order of the permutation. Thus, if we start with the $P(n, r)$ permutations, we will have all of the combinations of n objects taken r at a time, with each combination repeated $r!$ times. Thus,

$$P(n, r) = r! \, C(n, r).$$

On dividing both sides of the equation by $r!$, we obtain formula (3).

INCORPORATING	
TECHNOLOGY	

Computing Permutations, Combinations, and Factorials Most graphing calculators have commands to compute $P(n, r)$, $C(n, r)$, and $n!$. For instance, the MATH key on the TI-84 Plus leads to the PROB menu of Fig. 2, which contains the commands **nPr**, **nCr**, and **!**. Figure 3 shows how these commands are used. *Note:* The number **8.799226775E15** represents $8.799226775 \times 10^{15}$.

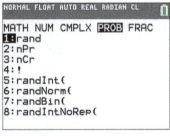

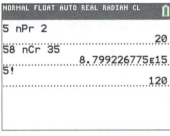

Figure 2   **Figure 3**

The values of $n!$, $P(n, r)$, and $C(n, r)$ are calculated in an Excel spreadsheet with the functions FACT(n), PERMUT(n, r), and COMBIN(n, r).

WolframAlpha Factorials, permutations, and combinations can be calculated in Wolfram|Alpha with the instructions $n!$, **P(n, r)**, and **C(n, r)**. Some alternative instructions are **factorial n**, **permutations (n, r)**, and **combinations (n, r)** or **n choose r**.

Check Your Understanding 5.5

Solutions can be found following the section exercises.

1. Calculate the following values:
   (a) $5!$ (b) $P(5, 5)$ (c) $P(7, 3)$ (d) $C(7, 3)$

2. A newborn child is to be given a first name and a middle name from a selection of 10 names. How many different possibilities are there?

EXERCISES 5.5

For Exercises 1–20, calculate the values.

1. $P(4, 2)$
2. $P(5, 1)$
3. $P(6, 3)$
4. $P(5, 4)$
5. $C(10, 3)$
6. $C(12, 2)$
7. $C(5, 4)$
8. $C(6, 3)$
9. $P(7, 1)$
10. $P(5, 5)$
11. $P(n, 1)$
12. $P(n, 2)$
13. $C(4, 4)$
14. $C(n, 2)$
15. $C(n, n - 2)$
16. $C(n, 1)$
17. $6!$
18. $\dfrac{10!}{4!}$
19. $\dfrac{9!}{7!}$
20. $7!$

In Exercises 21–26, determine whether the computation involves a permutation, a combination, or neither.

21. **Stock Abbreviations** The number of different stock abbreviations for which each abbreviation consists of four letters, none repeated.

22. **Airport Codes** The number of different airport codes in which each code consists of three letters, none repeated.

23. **Ice Cream Flavors** The selection of three different flavors of ice cream (out of 29 flavors) for the three scoops of ice cream in a sundae.

24. **Ice Cream Flavors** The selection of three different flavors of ice cream (out of 29 flavors) for the three scoops of ice cream on an ice cream cone, where order matters.

25. **Rolling Dice** The number of possible sums when two dice are rolled and the numbers displayed are added.

26. **Meal Choices** The number of possible meals consisting of an appetizer, a main course, and a dessert from a restaurant that offers 5 different appetizers, 10 main courses, and 4 desserts.

27. **Group Picture** In how many ways can four people line up in a row for a group picture?

28. **Waiting in Line** In how many ways can six people line up at a single counter to order food at a fast-food restaurant?

29. **Book Selection** How many different selections of seven books can be made from nine books?

30. **Pizza Varieties** A pizzeria offers five toppings for the plain cheese base of the pizzas. How many pizzas can be made that use three different toppings?

31. **Selecting Colleges** A high school student decides to apply to four of the eight Ivy League colleges. In how many possible ways can the four colleges be selected?

32. **Senate Committees** In how many different ways can a committee of 5 senators be selected from the 100 members of the U.S. Senate?

33. **Ranking Teams** A sportswriter makes a preseason guess of the top five football teams (in order) from among the 65 Power Five conference teams. How many different possibilities are there?

34. **CD Changer** Suppose that you have 36 CDs and your CD player has five slots numbered 1 through 5. How many ways can you fill your CD player?

35. **Guest Lists** How many ways can you choose five out of 10 friends to invite to a dinner party?

36. **Choosing Exam Questions** A student is required to work exactly four problems from an eight-problem exam. In how many ways can the problems be chosen?

37. **DVDs** In a batch of 100 DVDs, seven are defective. A sample of three DVDs is to be selected from the batch. How many samples are possible? How many of the samples consist of all defective DVDs?

38. **Debates** There are 17 candidates for an elected office. If 10 candidates are selected to participate in a debate, determine the total number of possible debate groups.

39. **Race Winners** Theoretically, assuming no ties, how many possibilities are there for first, second, and third places in a marathon race with 150 entries?

40. **Player Introductions** The five starting players of a basketball team are introduced one at a time. In how many different ways can they be introduced?

Poker Hands Exercises 41–44 refer to poker hands. A poker hand consists of 5 cards selected from a standard deck of 52 cards.

41. How many different poker hands are there?

42. How many different poker hands consist entirely of aces and kings?

43. How many different poker hands consist entirely of clubs?

44. How many different poker hands consist entirely of red cards?

45. **Distributing Sandwiches** Five students order different sandwiches at a campus eatery. The waiter forgets who ordered what and gives out the sandwiches at random. In how many different ways can the sandwiches be distributed?

46. **Nautical Signals** A nautical signal consists of three flags arranged vertically on a flagpole. If a sailor has six flags, each of a different color, how many different signals are possible?

47. **Selecting Sweaters** Suppose that you own 10 sweaters.
    (a) How many ways can you select four of them to take on a trip?
    (b) How many ways can you select six of the sweaters to leave at home?
    (c) Explain why the answers to parts (a) and (b) are the same.

48. **Distributing Books** Fred has 12 different books.
    (a) Suppose that Fred first gives three books to Jill and then gives four of the remaining books to Jack. How many different outcomes are possible?
    (b) Suppose that Fred first gives four books to Jack and then gives three of the remaining books to Jill. How many different outcomes are possible?
    (c) Explain why the answers to parts (a) and (b) are the same.

49. **Conference Games** In an eight-team football conference, each team plays every other team exactly once. How many games must be played?

50. **League Games** In a six-team softball league, each team plays every other team three times during the season. How many games must be scheduled?

51. **Powerball** In the Powerball lottery, five white balls are drawn out of a drum with 69 numbered white balls and then one red ball is drawn out of a drum with 26 numbered red balls. The jackpot is won by guessing all five white balls in any order and the red Powerball. Determine the number of possible outcomes.

52. **Baseball Lineup** On a children's baseball team, there are four players who can play at any of the following infield positions: catcher, first base, third base, and shortstop. There are five possible pitchers, none of whom plays any other position. And there are four players who can play any of the three outfield positions (right, left, and center) or second base. In how many ways can the coach assign players to positions?

Lotto Exercises 53 and 54 refer to the New York State lottery (Lotto). When it was first established, a contestant had to select six numbers from 1 to 49. A few years later, the numbers 50 through 54 were added. In March 2007, the numbers 55 through 59 were added.

53. The number of possible combinations of six numbers selected from 1 to 59 is approximately _____ times the number of combinations selected from 1 through 49.
    (a) 2   (b) 3   (c) 10   (d) 100

54. Drawings for Lotto are held twice per week. Suppose that you decide to purchase 110 tickets for each drawing and never use the same combination twice. Approximately how many years would be required before you would have bet on every possible combination?
    (a) 100   (b) 1000   (c) 2000   (d) 4000

55. **Choosing Candy** Two children, Moe and Joe, are allowed to select candy from a plate of nine pieces of candy. Moe, being younger, is allowed to choose first but can take only two candies. Joe is then allowed to take four of the remaining candies. Joe complains that he has fewer options than Moe. Is Joe correct? How many options will each child have?

56. **Committee Selection** The 12 members of the Gotham City Council consists of four members from each of the city's three wards. In how many ways can a committee of six council

members be selected if the committee must contain at least one council member from each ward?

57. Group Picture The student council at Gotham College is made up of four freshmen, five sophomores, six juniors, and seven seniors. A yearbook photographer would like to line up three council members from each class for a picture. How many different pictures are possible if each group of classmates stands together?

58. Arranging Books George has three books by each of his five favorite authors. In how many ways can the books be placed on a shelf if books by the same author must be together?

59. Seating Arrangements In the quiz show *It's Academic*, three-person teams from three high schools are seated in a row, with each team seated together. How many different seating arrangements are possible?

60. Displaying Paintings An art gallery has four paintings, by each of three artists, hanging in a row, with paintings by the same artist grouped together. How many different arrangements are possible?

61. Handshakes At a party, everyone shakes hands with everyone else. If 45 handshakes take place, how many people are at the party?

62. Football Games In a football league, each team plays one game against each other team in the league. If 55 games are played, how many teams are in the league?

63. Side Dishes A restaurant offers its customers a choice of three side dishes with each meal. The side dishes can be chosen from a list of 15 possibilities, with duplications allowed. For instance, a customer can order two sides of mashed potatoes and one side of string beans. Show that there are 680 possible options for the three side dishes.

64. Ice Cream Specials An ice cream parlor offers a special consisting of three scoops of ice cream chosen from 16 different flavors. Duplication of flavors is allowed. For instance, one possibility is two scoops of chocolate and one scoop of vanilla. Show that there are 816 different possible options for the special.

65. Determine Position There are 6! = 720 six-letter words (that is, strings of letters) that can be made from the letters C, N, O, S, T, and U. If these 720 words are listed in alphabetical order, what position in the list will be occupied by TUCSON?

Hint: Count the number of words that follow TUCSON in the list. They must each be of the form TUNxxx or Uxxxxx.

66. Detour-Prone ZIP Codes A five-digit ZIP code is said to be detour prone if it looks like a valid and different ZIP code when read upside down (Fig. 4). For instance, 68901 and 88111 are detour prone, whereas 32145 and 10801 are not. How many of the 10^5 possible ZIP code numbers are detour prone?

Figure 4

TECHNOLOGY EXERCISES

67. Lottery
(a) Calculate the number of possible lottery tickets if the player must choose five distinct numbers from 0 to 44, inclusive, where the order does not matter. The winner must match all five.
(b) Calculate the number of lottery tickets if the player must choose four distinct numbers from 0 to 99, inclusive, where the order does not matter. The winner must match all four.
(c) In which lottery does the player have a better chance of choosing the randomly selected winning numbers?

68. Bridge A bridge hand contains 13 cards.
(a) What percent of bridge hands contain all four aces?
(b) What percent of bridge hands contain the two red kings, the two red queens, and no other kings or queens?
(c) Which is more likely—a bridge hand with four aces or one with the two red kings, the two red queens, and no other kings or queens?

69. Cards versus Atoms Are there more ways to order a deck of 52 cards than there are atoms on Earth? *Note:* There are about 10^{50} atoms on Earth.

70. Alphabet versus Atoms Are there more ways to rearrange the 26 letters of the alphabet than there are atoms on Earth?

Solutions to Check Your Understanding 5.5

1. (a) $5! = 5 \cdot 4 \cdot 3 \cdot 2 \cdot 1 = 120$

(b) $P(5, 5) = 5 \cdot 4 \cdot 3 \cdot 2 \cdot 1 = 120$
[$P(n, n)$ is the same as $n!$.]

(c) $P(7, 3) = \underbrace{7 \cdot 6 \cdot 5}_{3 \text{ factors}} = 210$

[$P(7, 3)$ is the product of the 3 whole numbers, counting down by 1 from 7.]

(d) $C(7, 3) = \dfrac{7 \cdot 6 \cdot 5}{3 \cdot 2 \cdot 1} = \dfrac{7 \cdot 6 \cdot 5}{3 \cdot 2 \cdot 1} = 35$

[A convenient procedure to follow when calculating $C(n, r)$ is first to write the product expansion of $r!$ in the denominator and then to write in the numerator a whole number above each integer in the denominator. The whole numbers should begin with n and successively decrease by 1.]

2. 90. The first question to be asked here is whether permutations or combinations are involved. Two names are to be selected, and the order of the names is important. (The name Amanda Beth is different from the name Beth Amanda.) Since the problem asks for arrangements of 10 names taken 2 at a time in a *specific order*, the number of arrangements is $P(10, 2) = 10 \cdot 9 = 90$. In general, order is important if a different outcome results when two items in the selection are interchanged.

5.6 Further Counting Techniques

In Section 5.5, we introduced permutations and combinations and developed formulas for counting all permutations (or combinations) of a given type. Many counting problems can be formulated in terms of permutations or combinations. But to use the formulas of Section 5.5 successfully, we must be able to recognize these problems when they occur and to translate them into a form in which the formulas may be applied. In this section, we practice doing that. We consider five typical applications giving rise to permutations or combinations. At first glance, the first two applications may seem to have little practical significance. However, they suggest a common way to "model" outcomes of real-life situations having two equally likely results.

As our first application, consider a coin-tossing experiment in which we toss a coin a fixed number of times. We can describe the outcome of the experiment as a sequence of "heads" and "tails." For instance, if a coin is tossed three times, then one possible outcome is "heads on the first toss, tails on the second toss, and tails on the third toss." This outcome can be abbreviated as HTT. We can use the methods of the preceding sections to count the number of possible outcomes having various prescribed properties.

EXAMPLE 1 **Tossing a Coin Ten Times** Suppose that an experiment consists of tossing a coin 10 times and observing the sequence of heads and tails.
(a) How many different outcomes are possible?
(b) How many different outcomes have exactly four heads?

SOLUTION (a) Visualize each outcome of the experiment as a sequence of 10 boxes, where each box contains one letter, H or T, with the first box recording the result of the first toss, the second box recording the result of the second toss, and so forth.

$$\boxed{H}\,\boxed{T}\,\boxed{H}\,\boxed{T}\,\boxed{T}\,\boxed{T}\,\boxed{H}\,\boxed{T}\,\boxed{H}\,\boxed{T}$$
$$1\quad 2\quad 3\quad 4\quad 5\quad 6\quad 7\quad 8\quad 9\quad 10$$

Each box can be filled in two ways. So by the generalized multiplication principle, the sequence of 10 boxes can be filled in

$$\underbrace{2\cdot 2\cdot\,\cdots\,\cdot 2}_{\text{10 factors}} = 2^{10}$$

ways. So there are $2^{10} = 1024$ different possible outcomes.
(b) An outcome with 4 heads corresponds to filling the boxes with 4 H's and 6 T's. A particular outcome is determined as soon as we decide where to place the H's. The 4 boxes to receive H's can be selected from the 10 boxes in $C(10, 4)$ ways. So the number of outcomes with 4 heads is

$$C(10, 4) = \frac{10\cdot 9\cdot 8\cdot 7}{4\cdot 3\cdot 2\cdot 1} = 210.$$

» Now Try Exercise 1

Ideas similar to those applied in Example 1 are useful in counting even more complicated sets of outcomes of coin-tossing experiments. The second part of our next example highlights a technique that can often save time and effort.

EXAMPLE 2 **Tossing a Coin Ten Times** Consider the coin-tossing experiment of Example 1.
(a) How many different outcomes have at most two heads?
(b) How many different outcomes have at least three heads?

SOLUTION (a) The outcomes with at most two heads are those having 0, 1, or 2 heads. Let us count the number of these outcomes separately:

0 heads: There is 1 outcome, namely, T T T T T T T T T T.
1 head: To determine such an outcome, we just select the box in which to put the single H. And this can be done in $C(10, 1) = 10$ ways.

2 heads: To determine such an outcome, we just select the boxes in which to put the two H's. And this can be done in $C(10, 2) = (10 \cdot 9)/(2 \cdot 1) = 45$ ways.

Adding up all the possible outcomes, we see that the number of outcomes with at most two heads is $1 + 10 + 45 = 56$.

(b) "At least three heads" refers to an outcome with either 3, 4, 5, 6, 7, 8, 9, or 10 heads. The total number of such outcomes is

$$C(10, 3) + C(10, 4) + \cdots + C(10, 10).$$

This sum can, of course, be calculated, but there is a less tedious way to solve the problem. Just start with all outcomes [1024 of them by Example 1(a)], and subtract those with at most two heads [56 of them by part (a)]. So the number of outcomes with at least three heads is $1024 - 56 = 968$. **≫ Now Try Exercise 3**

The solution to part (b) of Example 2 employs a useful counting technique.

Complement Rule for Counting If U is the set of all possible outcomes and S is a subset of U, then S' is the set of all outcomes for which S does *not* occur, and

$$n(S') = n(U) - n(S).$$

Let us now turn to a different sort of counting problem, namely, one that involves counting the number of paths between two points.

EXAMPLE 3   **Routes through a City** In Fig. 1, we have drawn a partial map of the streets in New York City. A tourist wishes to walk from Times Square to Grand Central Station. We have drawn two possible routes. What is the total number of routes (with no backtracking) from Times Square to Grand Central Station?

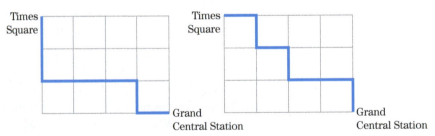

Figure 1

SOLUTION   Any particular route can be described by giving the directions of each block walked in the appropriate order. For instance, the route on the left of Fig. 1 is described as "a block south, a block south, a block east, a block east, a block east, a block south, a block east." Using S for south and E for east, this route can be designated by the string of letters SSEEESE. Similarly, the route on the right is ESESEES. Note that each route is then described by a string of seven letters, of which three are S's (we must go three blocks south) and four are E's (we must go four blocks east). Selecting a route is thus the same as placing three S's in a string of seven boxes:

The three boxes to receive S's can be selected in $C(7, 3) = 35$ ways. So the number of paths from Times Square to Grand Central Station is 35. **≫ Now Try Exercise 11**

EXAMPLE 4   **Routes through a City** Refer to the street map of Example 3. In how many of the routes does the tourist never walk south for two consecutive blocks?

SOLUTION   Each route is a string of letters containing three S's and four E's. We are asked to count the number of such strings in which two S's are never adjacent to each other. One way to construct such a string is to write down four E's and then decide where to insert the three S's. The arrows below show the five places that the S's can be inserted:

$$\textbf{E E E E}$$
$$\uparrow \uparrow \uparrow \uparrow \uparrow$$

The three insertion points can be selected in $C(5, 3) = 10$ ways. Therefore, there are 10 routes without consecutive souths.   **» Now Try Exercise 15**

Let us now move on to a third type of counting problem. Suppose that we have an urn in which there are a certain number of red balls and a certain number of white balls. We perform an experiment that consists of selecting a number of balls from the urn and observing the color distribution of the sample selected. (This model may be used, for example, to describe the process of selecting people to be polled in a survey. The different colors would correspond to different opinions.) By using familiar counting techniques, we can calculate the number of possible samples having a given color distribution. The next example illustrates a typical computation.

EXAMPLE 5   **Selecting Balls from an Urn** An urn contains 25 numbered balls, of which 15 are red and 10 are white. A sample of 5 balls is to be selected.
(a) How many different samples are possible?
(b) How many samples contain all red balls?
(c) How many samples contain 3 red balls and 2 white balls?
(d) How many samples contain at least 4 red balls?

SOLUTION   (a) A sample is just an unordered selection of 5 balls out of 25. There are $C(25, 5)$ such samples. Numerically, we have

$$C(25, 5) = \frac{25 \cdot 24 \cdot 23 \cdot 22 \cdot 21}{5 \cdot 4 \cdot 3 \cdot 2 \cdot 1} = 53{,}130$$

samples.

(b) To form a sample of all red balls, we must select 5 balls from the 15 red ones. This can be done in $C(15, 5)$ ways—that is, in

$$C(15, 5) = \frac{15 \cdot 14 \cdot 13 \cdot 12 \cdot 11}{5 \cdot 4 \cdot 3 \cdot 2 \cdot 1} = 3003$$

ways.

(c) To answer this question, we use both the multiplication principle and the formula for $C(n, r)$. We form a sample of 3 red balls and 2 white balls, using a sequence of two choices:

Select 3 red balls	Select 2 white balls
Choice 1	Choice 2

The first choice can be performed in $C(15, 3)$ ways and the second in $C(10, 2)$ ways. Thus, the total number of samples having 3 red and 2 white balls is $C(15, 3) \cdot C(10, 2)$. That is,

$$C(15, 3) = \frac{15 \cdot 14 \cdot 13}{3 \cdot 2 \cdot 1} = 455$$

$$C(10, 2) = \frac{10 \cdot 9}{2 \cdot 1} = 45$$

$$C(15, 3) \cdot C(10, 2) = 455 \cdot 45 = 20{,}475.$$

So the number of possible samples is 20,475.

(d) A sample with at least 4 red balls has either 4 or 5 red balls. By part (b), the number of samples with 5 red balls is 3003. Using the same reasoning as in part (c), the number of samples with 4 red balls is $C(15, 4) \cdot C(10, 1) = 1365 \cdot 10 = 13,650$. Thus, the total number of samples having at least 4 red balls is $13,650 + 3003 = 16,653$.

>> *Now Try Exercise 25*

EXAMPLE 6 **Photo Session** The nine justices of the United States Supreme Court can be seated in a row for a photo session in 9! different ways. In how many of the choices will Justice Elena Kagan be seated to the left of, but not necessarily beside, Justice John Roberts?

FIRST SOLUTION A possible seating for the justices can be obtained as follows:

1. Select the set of two chairs for Justices Kagan and Roberts. $C(9, 2)$ possibilities
2. Seat Justice Kagan in the left chair and Justice Roberts in the right chair. 1 possibility
3. Seat the remaining justices in the unoccupied seven chairs. 7! possibilities

Therefore, by the multiplication principle, the number of such seatings are

$$C(9, 2) \cdot 1 \cdot 7! = 36 \cdot 1 \cdot 5040 = 181,440.$$

SECOND SOLUTION There are a total of $9! = 362,880$ possible seatings of the nine justices. In half of them, Justice Kagan will be seated to the left of Justice Roberts. Therefore, the answer to the question is $\frac{1}{2} \cdot 362,880 = 181,440$. >> *Now Try Exercise 37*

EXAMPLE 7 **Numbers** How many six-digit numbers are there in which the digits strictly decrease when read from left to right? (Two examples are 865,421 and 976,532.)

SOLUTION One way to obtain a six-digit number is to start with the ten-digit number 9876543210 and remove four digits. Since the four digits can be selected in $C(10, 4) = 210$ different ways, there are 210 six-digit numbers whose digits are strictly decreasing.

>> *Now Try Exercise 39*

NOTE▶ In counting problems, we often compute permutations or combinations as an intermediate step, then add, subtract, or multiply to solve the problem. A useful guideline to determine when to perform which arithmetic operation is:

- **Addition** If a set of outcomes can be expressed as the union of disjoint subsets of outcomes, then we may calculate the number of outcomes in each subset and add [see Example 2(a)].
- **Subtraction** If a set of outcomes can be expressed as the complement of another set, then we may use the complement rule [see Example 2(b)].
- **Multiplication** If a set of outcomes can be expressed as a sequence of choices, then we may use the multiplication principle (see Example 5). ◀◀

Check Your Understanding 5.6

Solutions can be found following the section exercises.

1. **School Board** A newspaper reporter wants an indication of how the 15 members of the school board feel about a certain proposal. She decides to question a sample of 6 of the board members.
   (a) How many different samples are possible?
   (b) Suppose that 10 of the board members support the proposal, and 5 oppose it. How many of the samples reflect the distribution of the board? That is, in how many of the samples do 4 people support the proposal and 2 oppose it?

2. **Free Throws** A basketball player shoots eight free throws and lists the sequence of results of each trial in order. Let S represent *success* and F represent *failure*. Then, for instance, FFSSSSSS represents the outcome of missing the first two shots and hitting the rest.
   (a) How many different outcomes are possible?
   (b) How many of the outcomes have six successes?

EXERCISES 5.6

1. **Tossing a Coin** An experiment consists of tossing a coin eight times and observing the sequence of heads and tails.
   (a) How many different outcomes are possible?
   (b) How many different outcomes have exactly four heads?

2. **Tossing a Coin** An experiment consists of tossing a coin nine times and observing the sequence of heads and tails.
   (a) How many different outcomes are possible?
   (b) How many different outcomes have exactly two tails?

3. **Tossing a Coin** An experiment consists of tossing a coin seven times and observing the sequence of heads and tails.
   (a) How many different outcomes have at least five heads?
   (b) How many different outcomes have at most four heads?

4. **Tossing a Coin** An experiment consists of tossing a coin six times and observing the sequence of heads and tails.
   (a) How many different outcomes have at most three heads?
   (b) How many different outcomes have four or more heads?

5. **World Series** In the World Series, the American League team ("A") and the National League team ("N") play until one team wins four games. Each sequence of winners can be designated by a sequence of As and Ns. For instance, NAAAA means the National League won the first game and lost the next four games. In how many ways can a series end in seven games? Six games?

6. **World Series** Refer to Exercise 5. How many different sequences are possible?

7. **Game Outcomes** A football team plays 11 games. In how many ways can these games result in five wins, five losses, and one tie?

8. **Game Outcomes** A chess master plays 15 games. In how many ways can these games result in 10 wins, 2 losses, and 3 draws?

Bytes Exercises 9 and 10 refer to computer bytes. A computer *byte* is a string of eight digits, where each digit is either a zero or a one. Two examples are 01001001 and 11001101.

9. How many bytes have exactly five ones?

10. In how many of the bytes with exactly five ones are no two zeroes next to each other?

11. **Routes through City Streets** Refer to the map in Fig. 2. How many shortest routes are there from A to B?

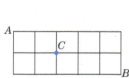
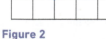
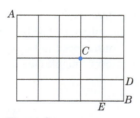

Figure 2   **Figure 3**

12. **Routes through City Streets** Refer to the map in Fig. 3. How many shortest routes are there from A to B?

13. **Routes through City Streets** Refer to the map in Fig. 3. How many shortest routes are there from A to B that pass through the point C?

14. **Routes through City Streets** Refer to the map in Fig. 2. How many shortest routes are there from A to B that pass through the point C?

15. **Routes through City Streets** Refer to the map in Fig. 2. How many shortest routes are there from A to B that do not have two consecutive souths?

16. **Routes through City Streets** Refer to the map in Fig. 3. How many shortest routes are there from A to B that do not have two consecutive souths?

17. **Routes through City Streets** Refer to the map in Fig. 3. The number of shortest routes from A to B is $C(9, 4)$.
   (a) Observe that the number of shortest routes from A to D is $C(8, 3)$.
   (b) Observe that the number of shortest routes from A to E is $C(8, 4)$.
   (c) By looking at Fig. 3, explain why $C(8, 3) + C(8, 4)$ should equal $C(9, 4)$.
   (d) Calculate the values in part (c) to verify the equality.

18. **Routes through City Streets** Imagine a street map similar to the map in Fig. 3 but having r streets vertically and $n - r$ street horizontally. Then, the number of shortest routes from A to B would be $C(n, r)$.
   (a) Explain why the number of shortest routes A to D, the intersection directly north of B, would be $C(n - 1, r - 1)$.
   (b) Explain why the number of shortest routes A to E, the intersection directly west of B, would be $C(n - 1, r)$.
   (c) Explain why $C(n - 1, r - 1) + C(n + 1, r) = C(n, r)$.

19. **Tossing a Coin** A coin is tossed 10 times, and the sequence of heads and tails is observed. How many of the possible outcomes contain three heads, with no two heads adjacent to each other?

20. **Arranging Books** Four mathematics books and seven history books are arranged on a bookshelf. In how many of the possible arrangements are no two mathematics books next to each other?

21. **Seating Arrangements** In how many ways can six people be seated in a row of ten chairs so that at least two adjacent chairs are vacant? (*Hint:* First use the reasoning in Example 4 to count the number of ways they can be seated so that no two adjacent chairs are empty.)

22. **Photo Session** In 2015, there were three women and six men on the United States Supreme Court. In how many ways could the justices be seated in a row for a group picture where no women sat next to each other?

23. **Selecting Balls from an Urn** An urn contains 12 numbered balls, of which 7 are red and 5 are white. A sample of 5 balls is to be selected.
   (a) How many different samples are possible?
   (b) How many samples contain all red balls?
   (c) How many samples contain two red balls and three white balls?
   (d) How many samples contain at least four red balls?

24. **Selecting Balls from an Urn** An urn contains 15 numbered balls, of which 6 are red and 9 are white. A sample of six balls is to be selected.
   (a) How many different samples are possible?
   (b) How many samples contain all white balls?
   (c) How many samples contain two red balls and four white balls?
   (d) How many samples contain at least two red balls?

25. **Selecting Apples** A bag of 10 apples contains 2 rotten apples and 8 good apples. A shopper selects a sample of three apples from the bag.
    (a) How many different samples are possible?
    (b) How many samples contain all good apples?
    (c) How many samples contain at least one rotten apple?

26. **Selecting Light Bulbs** A package contains 100 LED light bulbs, of which 10 are defective. A sample of five bulbs is selected at random.
    (a) How many different samples are there?
    (b) How many of the samples contain two defective bulbs?
    (c) How many of the samples contain at least one defective bulb?

27. **Subcommittee Selection** A committee has four male and six female members. In how many ways can a subcommittee consisting of two males and two females be selected?

28. **Investment Portfolio** In how many ways can an investor put together a portfolio of five stocks and six bonds selected from her favorite nine stocks and seven bonds?

Poker Hands Exercises 29–32 refer to poker hands. A poker hand consists of 5 cards selected from a standard deck of 52 cards.

29. How many poker hands consist of three aces and two kings?

30. How many poker hands consist of two aces, two cards of another rank, and one card of a third rank?

31. How many poker hands consist of three cards of one rank and two cards of another rank? (Such a poker hand is called a "full house.")

32. How many poker hands consist of two cards of one rank, two cards of another (different) rank, and one card of a third rank? (Such a poker hand is called "two pairs.")

In Exercises 33–36, a "word" is interpreted to be a sequence of letters.

33. **SEQUOIA** How many seven-letter words with no repeated letters contain all five vowels?

34. **FACETIOUS** How many nine-letter words with no repeated letters contain the five vowels in alphabetical order?

35. **ABSTEMIOUS** How many 10-letter words with no repeated letters contain the five vowels in alphabetical order?

36. **DIALOGUE** How many eight-letter words with no repeated letters contain all five vowels?

37. **Photo Session** In 2015, there were three women and six men on the United States Supreme Court. In how many ways could the justices be seated in a row for a group picture in which the three women sat next to each other?

38. **Seating Arrangements** You, a friend, and four other people are to be seated in a row of six chairs. How many arrangements are there in which you and your friend are seated next to each other?

39. **Numbers** In how many five-digit numbers (without zeros) are the digits strictly increasing when read from left to right?

40. **Alphabetical Order** In how many four-letter words (including nonsense words) using four different letters from A through J are the letters in alphabetical order?

41. **License Plates** Suppose that license plates from a certain state consist of four different letters followed by three different digits. In how many license plates are the letters in alphabetical order and the digits in increasing order?

42. **License Plates** Suppose that license plates from a certain state consist of two different letters followed by four different digits. In how many license plates are the letters in alphabetical order and the digits in increasing order?

43. **Seating Arrangements** A family has 12 members. In how many ways can six family members be seated in a row so that their ages increase from left to right?

44. **Arranging Books** In how many ways can five books out of eight be selected and lined up on a bookshelf so that their page counts increase from left to right? *Note:* Assume that no two books have the same page count.

TECHNOLOGY EXERCISES

45. **Tossing a Coin** What percent of the possible outcomes resulting from tossing a coin 100 times contain exactly 50 heads?

46. **Tossing a Coin** What percent of the possible outcomes resulting from tossing a coin 200 times contain exactly 100 heads?

47. **Selecting Balls from an Urn** Suppose that a sample of 20 balls is selected from an urn containing 50 white balls and 50 red balls. What percent of the possible outcomes contains 10 white balls and 10 red balls?

48. **Selecting Balls from an Urn** Suppose that a sample of 20 balls is selected from an urn containing 100 white balls and 100 red balls. What percent of the possible outcomes contains 10 white balls and 10 red balls?

Solutions to Check Your Understanding 5.6

1. (a) $C(15, 6)$. Each sample is an unordered selection of 15 objects taken 6 at a time.

   (b) $C(10, 4) \cdot C(5, 2)$. Asking for the number of samples of a certain type is the same as asking for the number of ways that the task of forming such a sample can be performed. This task is composed of two consecutive choices. Choice 1, selecting 4 people from among the 10 who support the proposal, can be performed in $C(10, 4)$ ways. Choice 2, selecting two people from among the five people who oppose the proposal, can be performed in $C(5, 2)$ ways. Therefore, by the multiplication principle, the complete task can be performed in $C(10, 4) \cdot C(5, 2)$ ways.

   Note: $C(15, 6) = 5005$ and $C(10, 4) \cdot C(5, 2) = 2100$. Therefore, less than half of the possible samples reflect the true distribution of the school board.

2. (a) 2^8, or 256. Apply the generalized multiplication principle.

   (b) $C(8, 6)$, or 28. Each outcome having six successes corresponds to a sequence of eight letters of which six are S's and two are F's. Such an outcome is specified by selecting the six locations for the S's from among the eight locations, and this has $C(8, 6)$ possibilities.

5.7 The Binomial Theorem

In Sections 5.5 and 5.6, we dealt with permutations and combinations and, in particular, derived a formula for $C(n, r)$, the number of combinations of n objects taken r at a time. Namely, we have

$$C(n, r) = \frac{P(n, r)}{r!} = \frac{n(n - 1) \cdot \cdots \cdot (n - r + 1)}{r!}. \tag{1}$$

Actually, formula (1) was verified in case both n and r are positive integers. But it is useful to consider $C(n, r)$ also in case $r = 0$. In this case, we are considering the number of combinations of n things taken 0 at a time. There is clearly only one such combination: the one containing no elements. Therefore,

$$C(n, 0) = 1.$$

In Section 5.5, we encountered another formula for $C(n, r)$:

$$C(n, r) = \frac{n!}{r!(n - r)!}. \tag{2}$$

Note that, for $r = 0$, formula (2) reads

$$C(n, 0) = \frac{n!}{0! \, (n - 0)!} = \frac{n!}{0! \, n!} = \frac{1}{0!}.$$

As stated in Section 5.4, the value of 0! is 1. Then the right-hand side of the preceding equation is 1, so formula (2) also holds for $r = 0$.

Formula (2) can be used to prove many facts about $C(n, r)$. For example, the following formula is useful in calculating $C(n, r)$ for large values of r:

$$C(n, r) = C(n, n - r). \tag{3}$$

EXAMPLE 1 **Applying Formula (3)** Calculate $C(100, 98)$.

SOLUTION If we apply formula (3), we have

$$C(100, 98) = C(100, 100 - 98) = C(100, 2) = \frac{100 \cdot 99}{2 \cdot 1} = 4950.$$

>> *Now Try Exercise 1*

Verification of Formula (3) Apply formula (2) to evaluate $C(n, n - r)$:

$$C(n, n - r) = \frac{n!}{(n - r)! \, (n - (n - r))!} = \frac{n!}{(n - r)! \, r!}$$

$$= C(n, r) \quad \text{[by formula (2) again].}$$

The formula is intuitively reasonable, since each time we select a subset of r elements, we are selecting a subset of $n - r$ elements to be excluded. Thus, there are as many subsets of $n - r$ elements as there are subsets of r elements. 《

An alternative notation for $C(n, r)$ is $\binom{n}{r}$. Thus, for example,

$$\binom{5}{2} = C(5, 2) = \frac{5 \cdot 4}{2 \cdot 1} = 10.$$

The symbol $\binom{n}{r}$ is called a **binomial coefficient**. To discover why, let us tabulate the values of $\binom{n}{r}$ for some small values of n and r.

$$n = 2: \quad \binom{2}{0} = 1 \quad \binom{2}{1} = 2 \quad \binom{2}{2} = 1$$

$$n = 3: \quad \binom{3}{0} = 1 \quad \binom{3}{1} = 3 \quad \binom{3}{2} = 3 \quad \binom{3}{3} = 1$$

$$n = 4: \quad \binom{4}{0} = 1 \quad \binom{4}{1} = 4 \quad \binom{4}{2} = 6 \quad \binom{4}{3} = 4 \quad \binom{4}{4} = 1$$

$$n = 5: \quad \binom{5}{0} = 1 \quad \binom{5}{1} = 5 \quad \binom{5}{2} = 10 \quad \binom{5}{3} = 10 \quad \binom{5}{4} = 5 \quad \binom{5}{5} = 1$$

Each row consists of the coefficients that arise in expanding $(x + y)^n$. To see this, inspect the results of expanding $(x + y)^n$ for $n = 2, 3, 4$, and 5:

$$(x + y)^2 = x^2 + 2xy + y^2$$
$$(x + y)^3 = x^3 + 3x^2y + 3xy^2 + y^3$$
$$(x + y)^4 = x^4 + 4x^3y + 6x^2y^2 + 4xy^3 + y^4$$
$$(x + y)^5 = x^5 + 5x^4y + 10x^3y^2 + 10x^2y^3 + 5xy^4 + y^5.$$

Compare the coefficients in any row with the values in the corresponding row of binomial coefficients. Note that they are the same. Thus, we see that the binomial coefficients arise as coefficients in multiplying out powers of the binomial $x + y$—hence the name *binomial coefficient*.

What we observed for the exponents $n = 2, 3, 4$, and 5 holds true for any positive integer n. We have the following result, a proof of which is given at the end of this section:

Binomial Theorem

$$(x + y)^n = \binom{n}{0}x^n + \binom{n}{1}x^{n-1}y + \binom{n}{2}x^{n-2}y^2 + \cdots + \binom{n}{n-1}xy^{n-1} + \binom{n}{n}y^n.$$

EXAMPLE 2 **Applying the Binomial Theorem** Expand $(x + y)^6$.

SOLUTION By the binomial theorem,

$$(x + y)^6 = \binom{6}{0}x^6 + \binom{6}{1}x^5y + \binom{6}{2}x^4y^2 + \binom{6}{3}x^3y^3$$
$$+ \binom{6}{4}x^2y^4 + \binom{6}{5}xy^5 + \binom{6}{6}y^6.$$

Furthermore,

$$\binom{6}{0} = 1 \qquad \binom{6}{1} = \frac{6}{1} = 6 \qquad \binom{6}{2} = \frac{6 \cdot 5}{2 \cdot 1} = 15$$

$$\binom{6}{3} = \frac{6 \cdot 5 \cdot 4}{3 \cdot 2 \cdot 1} = 20 \qquad \binom{6}{4} = \binom{6}{2} = 15$$

$$\binom{6}{5} = \binom{6}{1} = 6 \qquad \binom{6}{6} = \binom{6}{0} = 1.$$

Thus,

$$(x + y)^6 = x^6 + 6x^5y + 15x^4y^2 + 20x^3y^3 + 15x^2y^4 + 6xy^5 + y^6.$$

>> *Now Try Exercise 21*

The binomial theorem can be used to count the number of subsets of a set, as shown in the next example.

EXAMPLE 3   **Counting the Number of Subsets of a Set** Determine the number of subsets of a set with five elements.

SOLUTION   Let us count the number of subsets of each possible size. A subset of r elements can be chosen in $\binom{5}{r}$ ways, since $C(5, r) = \binom{5}{r}$. So the set has $\binom{5}{0}$ subsets with 0 elements, $\binom{5}{1}$ subsets with 1 element, $\binom{5}{2}$ subsets with 2 elements, and so on. Therefore, the total number of subsets is

$$\binom{5}{0} + \binom{5}{1} + \binom{5}{2} + \binom{5}{3} + \binom{5}{4} + \binom{5}{5}. \tag{4}$$

Rather than calculate this sum directly, we can take advantage of the binomial theorem. Setting $n = 5$ gives

$$(x + y)^5 = \binom{5}{0}x^5 + \binom{5}{1}x^4y + \binom{5}{2}x^3y^2 + \binom{5}{3}x^2y^3 + \binom{5}{4}xy^4 + \binom{5}{5}y^5.$$

We want the right-hand side of this formula to equal the sum (5). This can be done by setting $x = 1$ and $y = 1$.

$$(1 + 1)^5 = \binom{5}{0}1^5 + \binom{5}{1}1^4 \cdot 1 + \binom{5}{2}1^3 \cdot 1^2 + \binom{5}{3}1^2 \cdot 1^3 + \binom{5}{4}1 \cdot 1^4 + \binom{5}{5}1^5$$

$$2^5 = \binom{5}{0} + \binom{5}{1} + \binom{5}{2} + \binom{5}{3} + \binom{5}{4} + \binom{5}{5}$$

Thus, the total number of subsets of a set with five elements (the right side) equals $2^5 = 32$.   **» Now Try Exercise 37**

There is nothing special about the number 5 in the preceding example. An analogous argument gives the following result:

> A set of n elements has 2^n subsets.

EXAMPLE 4   **Counting Pizza Options** A pizza parlor offers a plain cheese pizza to which any number of six possible toppings can be added. How many different pizzas can be ordered?

SOLUTION   Ordering a pizza requires selecting a subset of the six possible toppings. Since the set of six toppings has 2^6 different subsets, there are 2^6, or 64, different pizzas. Note that the plain cheese pizza corresponds to selecting the empty subset of toppings.
      » Now Try Exercise 41

EXAMPLE 5   **Book Selection** In how many ways can a selection of at least two books be made from a set of six books?

SOLUTION   There are $2^6 = 64$ different subsets of the six books. However, one subset contains no books and six subsets contain just one book. Therefore, the number of possibilities is $64 - 1 - 6 = 57$.   **» Now Try Exercise 43**

Proof of the Binomial Theorem   Note that

$$(x + y)^n = \underbrace{(x + y)(x + y) \cdot \cdots \cdot (x + y)}_{n \text{ factors}}.$$

Multiplying out these factors involves forming all products, where one term is selected from each factor, and then combining like products. For instance,

$$(x + y)(x + y)(x + y) = x \cdot x \cdot x + x \cdot x \cdot y + x \cdot y \cdot x + y \cdot x \cdot x$$
$$+ x \cdot y \cdot y + y \cdot x \cdot y + y \cdot y \cdot x + y \cdot y \cdot y.$$

The first product on the right, $x \cdot x \cdot x$, is obtained by selecting the x-term from each of the three factors. The next term, $x \cdot x \cdot y$, is obtained by selecting the x-terms from the first two factors and the y-term from the third. The next product, $x \cdot y \cdot x$, is obtained by selecting the x-terms from the first and third factors and the y-term from the second, and so on. There are as many products containing two x's and one y as there are ways of selecting the factor from which to pick the y-term—namely, $\binom{3}{1}$.

In general, when multiplying the n factors $(x + y)(x + y) \cdots (x + y)$, the number of products having k y's (and therefore, $(n - k)$ x's) is equal to the number of different ways of selecting the k factors from which to take the y-term—that is, $\binom{n}{k}$. Therefore, the coefficient of $x^{n-k}y^k$ is $\binom{n}{k}$. This proves the binomial theorem. «

Check Your Understanding 5.7

Solutions can be found following the section exercises.

1. Calculate $\binom{12}{8}$.

2. An ice cream parlor offers 10 flavors of ice cream and 5 toppings. How many different servings are possible if each serving consists of one flavor of ice cream and as many toppings as desired?

EXERCISES 5.7

Calculate the value for each of Exercises 1–18.

1. $C(18, 16)$ 2. $C(25, 24)$ 3. $\binom{6}{2}$ 4. $\binom{7}{3}$

5. $\binom{8}{1}$ 6. $\binom{9}{9}$ 7. $\binom{7}{0}$ 8. $\binom{6}{1}$

9. $\binom{8}{8}$ 10. $\binom{9}{0}$ 11. $\binom{n}{n-1}$ 12. $\binom{n}{n}$

13. $0!$ 14. $1!$ 15. $n \cdot (n-1)!$ 16. $\dfrac{n!}{n}$

17. $\binom{6}{0} + \binom{6}{1} + \binom{6}{2} + \binom{6}{3} + \binom{6}{4} + \binom{6}{5} + \binom{6}{6}$

18. $\binom{7}{0} + \binom{7}{1} + \binom{7}{2} + \binom{7}{3} + \binom{7}{4} + \binom{7}{5} + \binom{7}{6} + \binom{7}{7}$

19. How many terms are there in the binomial expansion of $(x + y)^{19}$?

20. How many terms are there in the binomial expansion of $(x + y)^{25}$?

21. Determine the first three terms in the binomial expansion of $(x + y)^{10}$.

22. Determine the first three terms in the binomial expansion of $(x + y)^{20}$.

23. Determine the last three terms in the binomial expansion of $(x + y)^{15}$.

24. Determine the last three terms in the binomial expansion of $(x + y)^{12}$.

25. Determine the middle term in the binomial expansion of $(x + y)^{20}$.

26. Determine the middle term in the binomial expansion of $(x + y)^{10}$.

27. Determine the coefficient of x^2 in the expansion of $(1 + x)^4$.

28. Determine the coefficient of x^3 in the expansion of $(2 + x)^6$.

29. Determine the coefficient of x^4y^7 in the binomial expansion of $(x + y)^{11}$.

30. Determine the coefficient of x^9y^4 in the binomial expansion of $(x + y)^{13}$.

31. Determine the first three terms in the binomial expansion of $(x + 2y)^9$.

32. Determine the last three terms in the binomial expansion of $(x - y)^8$.

33. Determine the middle term in the binomial expansion of $(x - 3y)^{12}$.

34. Determine the coefficient of x^4y^2 in the binomial expansion of $(x + 3y)^6$.

35. Determine the term containing y^3 in the expansion of $(x - 3y)^7$.

36. Determine the term containing y^5 in the expansion of $(x - 3y)^8$.

37. How many different subsets can be chosen from a set of eight elements?

38. How many different subsets can be chosen from a set of nine elements?

39. **Restaurant Tip** How many different tips could you leave in a tip jar if you had a nickel, a dime, a quarter, and a half-dollar?

40. **Pizza Options** A pizza parlor offers mushrooms, green peppers, onions, and sausage as toppings for the plain cheese base. How many different types of pizzas with no duplicate toppings can be made?

41. **Cable TV Options** A cable TV franchise offers 20 basic channels plus a selection (at an extra cost per channel) from a

collection of 5 premium channels. How many different options are available to the subscriber?

42. **Salad Options** A salad bar offers a base of lettuce to which tomatoes, chickpeas, beets, pinto beans, olives, and green peppers can be added. Five salad dressings are available. How many different salads are possible? (Assume that each salad contains at least lettuce and at most one salad dressing.)

43. **Tie Selection** In how many ways can a selection of at least one tie be made from a set of eight ties?

44. **Dessert Choices** In how many ways can a selection of at most five desserts be made from a dessert trolley containing six desserts?

45. **Pizza Options** Armand's Chicago Pizzeria offers thin-crust and deep-dish pizzas in 9-, 12-, and 14-inch sizes, with 13 possible toppings. How many different types of pizzas with no duplicate toppings can be ordered?

46. **Ice Cream Sundaes** An ice cream parlor offers four flavors of ice cream, three sauces, and two types of nuts. How many different sundaes consisting of a single flavor of ice cream plus one or more toppings are possible?

47. **Appetizer Selection** In how many ways can a selection of at most five appetizers be made from a menu containing seven appetizers?

48. **CD Selection** In how many ways can a selection of at least two CDs be made from a set of seven CDs?

49. **Lab Projects** Students in a physics class are required to complete at least two out of a collection of eight lab projects. In how many ways can a student satisfy the requirement?

50. **Election** A voter is asked to vote for at most six of the eight candidates for school board. In how many ways can the voter cast their ballot?

51. **CD Selection** James has nine jazz CDs and ten top 40/pop CDs. How many possible combinations of CDs can he select if he decides to take at least two CDs of each type to a party?

52. **Book Selection** Sarah has five nonfiction and six fiction books on her reading list. When packing for her summer vacation, she decides to pack at least two nonfiction books and at least one fiction book. How many different book selections are possible?

53. Can $\binom{8}{5} x^3 y^4$ be a term of a binomial expansion?

54. Can $\binom{8}{5} x^3$ be a term of a binomial expansion?

55. Show that half of the subsets of a set of five elements have an odd number of elements. *Hint:* Show that $\binom{5}{0} + \binom{5}{2} + \binom{5}{4} = \binom{5}{1} + \binom{5}{3} + \binom{5}{5}$.

56. **(a)** Use equation (3) to show that
$$\binom{5}{0} - \binom{5}{1} + \binom{5}{2} - \binom{5}{3} + \binom{5}{4} - \binom{5}{5} = 0.$$

(b) For what values of n can equation (3) be used to show that
$$\binom{n}{0} - \binom{n}{1} + \binom{n}{2} - \binom{n}{3} + \cdots \pm \binom{n}{n} = 0?$$

(c) Use the binomial theorem to prove the result in part (a). *Hint:* Apply the binomial theorem to $(x + y)^5$ with $x = 1$ and $y = -1$.

(d) For what values of n can the binomial theorem be used to prove the result in part (b)?

57. Find a simple expression for the value of
$$\binom{10}{1} + \binom{10}{2} + \binom{10}{3} + \binom{10}{4} + \binom{10}{5}$$
$$+ \binom{10}{6} + \binom{10}{7} + \binom{10}{8} + \binom{10}{9}.$$

58. **Seating Passengers** A blue van and a red van, each having nine passenger seats, have arrived to take ten people to the airport. In how many different ways can the passengers be placed into the vans?

Solutions to Check Your Understanding 5.7

1. 495. $\binom{12}{8}$ is the same as $C(12, 8)$, which equals $C(12, 12 - 8)$ or $C(12, 4)$.

$$C(12, 4) = \frac{12 \cdot 11 \cdot 10 \cdot 9}{4 \cdot 3 \cdot 2 \cdot 1} = \frac{12 \cdot 11 \cdot \overset{5}{\cancel{10}} \cdot 9}{\cancel{4} \cdot \cancel{3} \cdot \cancel{2} \cdot 1} = 495$$

2. 320. The task of deciding what sort of serving to have consists of two choices. The first choice, selecting the flavor of

ice cream, can be performed in 10 ways. The second choice, selecting the toppings, can be performed in 2^5, or 32, ways, since selecting the toppings amounts to selecting a subset from the set of 5 toppings and a set of 5 elements has 2^5 subsets. (Notice that selecting the empty subset corresponds to ordering a plain dish of ice cream.) By the multiplication principle, the task can be performed in $10 \cdot 32 = 320$ ways.

5.8 Multinomial Coefficients and Partitions

Permutation and combination problems are only two of the many types of counting problems. One such type, problems involving *partitions*, arise as a generalization of a problem we've already studied. To introduce the notion of a partition, let us return to combinations and look at them from another viewpoint.

First, consider the number of ways to select two elements from the set $\{a, b, c\}$:

$$\{a, b\} \qquad \{a, c\} \qquad \{b, c\}$$

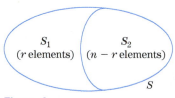

Figure 1

Alternatively, we can write the selection $\{a, b\}$ as $(\{a, b\}, \{c\})$, separating the elements of the set $\{a, b, c\}$ that we selected from the element we did not select. We call $(\{a, b\}, \{c\})$ an **ordered partition** of type $(2, 1)$. There are three such partitions of the set:

$$(\{a, b\}, \{c\}), \qquad (\{a, c\}, \{b\}) \qquad (\{b, c\}, \{a\})$$

In general, suppose that we consider combinations of n objects taken r at a time. View the n objects as the elements of a set S. Then each combination determines an ordered division of S into two subsets, S_1 and S_2, the first containing the r elements selected and the second containing the $n - r$ elements remaining (Fig. 1). We see that

$$S = S_1 \cup S_2 \quad \text{and} \quad n(S_1) + n(S_2) = n.$$

This ordered division is called an **ordered partition of type** $(r, n - r)$. We know that the number of such partitions is just the number of ways of selecting the first subset—that is, $n!/[r!(n - r)!]$. If we let $n_1 = n(S_1) = r$ and $n_2 = n(S_2) = n - r$, then we find that the number of ordered partitions of type (n_1, n_2) is $n!/[n_1! \, n_2!]$.

We may generalize the aforementioned situation as follows: Let S be a set of n elements. An **ordered partition of S of type $(n_1, n_2, \ldots, n_m)$** is a decomposition of S into m subsets (given in a specific order) $S_1, S_2, \ldots, S_m$, where no two of these intersect and where

$$n(S_1) = n_1, \qquad n(S_2) = n_2, \qquad \ldots \qquad , n(S_m) = n_m$$

(Fig. 2). Since S has n elements, we clearly must have $n = n_1 + n_2 + \cdots + n_m$.

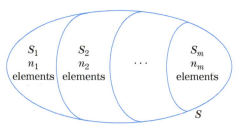

Figure 2

EXAMPLE 1   **Ordered Partitions of a Set**   List all ordered partitions of $S = \{a, b, c, d\}$ of type $(1, 1, 2)$.

SOLUTION

$(\{a\}, \{b\}, \{c, d\})$	$(\{c\}, \{a\}, \{b, d\})$
$(\{a\}, \{c\}, \{b, d\})$	$(\{c\}, \{b\}, \{a, d\})$
$(\{a\}, \{d\}, \{b, c\})$	$(\{c\}, \{d\}, \{a, b\})$
$(\{b\}, \{a\}, \{c, d\})$	$(\{d\}, \{a\}, \{b, c\})$
$(\{b\}, \{c\}, \{a, d\})$	$(\{d\}, \{b\}, \{a, c\})$
$(\{b\}, \{d\}, \{a, c\})$	$(\{d\}, \{c\}, \{a, b\})$

Note that the ordered partition $(\{a\}, \{b\}, \{c, d\})$ is different from the ordered partition $(\{b\}, \{a\}, \{c, d\})$, since in the first, S_1 is $\{a\}$, whereas in the second, S_1 is $\{b\}$. The order in which the subsets are given is significant. $\ll$

We saw earlier that the number of ordered partitions of type (n_1, n_2) for a set of n elements is $n!/[n_1! \, n_2!]$. This result generalizes.

> **Number of Ordered Partitions of Type $(n_1, n_2, \ldots, n_m)$**   Let S be a set of n elements. Then the number of ordered partitions of S of type $(n_1, n_2, \ldots, n_m)$, where $n_1 + n_2 + \cdots + n_m = n$ is
>
> $$\frac{n!}{n_1! \, n_2! \cdots n_m!}. \tag{1}$$

The number of ordered partitions of type $(n_1, n_2, \ldots, n_m)$ for a set of n elements is often denoted

$$\binom{n}{n_1, n_2, \ldots, n_m}.$$

Using the preceding notation, result (1) says that

$$\binom{n}{n_1, n_2, \ldots, n_m} = \frac{n!}{n_1! \, n_2! \cdots n_m!}.$$

The binomial coefficient $\binom{n}{r}$ can also be written $\binom{n}{r, n-r}$. The number

$$\binom{n}{n_1, n_2, \ldots, n_m}$$

is known as a **multinomial coefficient**, since it appears as the coefficient of $x_1^{n_1} x_2^{n_2} \cdots x_m^{n_m}$ in the expansion of $(x_1 + x_2 + \cdots + x_m)^n$.

EXAMPLE 2   **Ordered Partitions of a Set** Let S be a set of four elements. Use the formula in (1) to determine the number of ordered partitions of S of type $(1, 1, 2)$.

SOLUTION   Here, $n = 4$, $n_1 = 1$, $n_2 = 1$, and $n_3 = 2$. Therefore, the number of ordered partitions of type $(1, 1, 2)$ is

$$\binom{4}{1, 1, 2} = \frac{4!}{1! \, 1! \, 2!} = \frac{4 \cdot 3 \cdot 2 \cdot 1}{1 \cdot 1 \cdot 2 \cdot 1} = 12.$$

This result is the same as obtained in Example 1 by enumeration.

>> *Now Try Exercise 1*

EXAMPLE 3   **Assigning Tasks to Construction Workers** A work crew consists of 12 construction workers, all having the same skills. A construction job requires four welders, three concrete workers, three heavy equipment operators, and two bricklayers. In how many ways can the 12 workers be assigned to the required tasks?

SOLUTION   Each assignment of jobs corresponds to an ordered partition of the type $(4, 3, 3, 2)$. The number of such ordered partitions is

$$\binom{12}{4, 3, 3, 2} = \frac{12!}{4! \, 3! \, 3! \, 2!} = 277{,}200.$$

>> *Now Try Exercise 15*

Sometimes, the m subsets of an ordered partition are required to have the same number of elements. If the set has n elements and each of the m subsets has r elements, then the number of ordered partitions of type

$$\underbrace{(r, r, \ldots, r)}_{m}$$

is

$$\binom{n}{r, r, \ldots, r} = \frac{n!}{r! \, r! \cdots r!} = \frac{n!}{(r!)^m}. \tag{2}$$

EXAMPLE 4   **Counting the Number of Bridge Hands** In the game of bridge, four players seated in a specific order are each dealt 13 cards. How many different possibilities are there for the hands dealt to the players?

SOLUTION   Each deal results in an ordered partition of the 52 cards of type (13, 13, 13, 13). The number of such partitions is

$$\binom{52}{13, 13, 13, 13} = \frac{52!}{(13!)^4}.$$

This number is approximately 5.36×10^{28}.   «

EXAMPLE 5   **MISSISSIPPI** How many 11-letter words consist of one M, four Is, four Ss, and two Ps?

SOLUTION   Think of forming a word as filling slots numbered 1, 2, . . . , 11 with the letters. Partition the set of 11 slots into four subsets where the first subset gives the location for the M, the second subset gives the locations for the four Is, the third subset gives the locations for the four Ss, and the fourth subset gives the locations for the two Ps. The number of partitions is

$$\binom{11}{1, 4, 4, 2} = \frac{11!}{1!\,4!\,4!\,2!} = \frac{39{,}916{,}800}{1 \cdot 24 \cdot 24 \cdot 2} = \frac{39{,}916{,}800}{1152} = 34{,}650.$$

»» Now Try Exercise 17

Unordered Partitions

Determining the number of unordered partitions of a certain type is a complex matter. We will restrict our attention to the special case in which each subset is of the same size.

EXAMPLE 6   **Unordered Partitions of a Set** List all unordered partitions of $S = \{a, b, c, d\}$ of type (2, 2).

SOLUTION

$$(\{a, b\}, \{c, d\})$$
$$(\{a, c\}, \{b, d\})$$
$$(\{a, d\}, \{b, c\})$$ «

Note that the partition $(\{c, d\}, \{a, b\})$ is the same as the partition $(\{a, b\}, \{c, d\})$ when order is not taken into account.

Number of Unordered Partitions of Type $(r, r, \ldots, r)$ Let S be a set of n elements where $n = m \cdot r$. Then the number of unordered partitions of S of type $(r, r, \ldots, r)$ is

$$\frac{1}{m!} \cdot \frac{n!}{(r!)^m}. \tag{3}$$

Formula (3) follows from the fact that each unordered partition of the m subsets gives rise to $m!$ ordered partitions. Therefore,

$$(m!)\,[\text{number of unordered partitions}] = [\text{number of ordered partitions}]$$

or

$$[\text{number of unordered partitions}] = \frac{1}{m!} \cdot [\text{number of ordered partitions}]$$

$$= \frac{1}{m!} \cdot \frac{n!}{(r!)^m} \quad [\text{by formula (2)}].$$

EXAMPLE 7   **Unordered Partitions of a Set** Let S be a set of four elements. Use formula (3) to determine the number of unordered partitions of S of type $(2, 2)$.

SOLUTION   Here, $n = 4$, $r = 2$, and $m = 2$. Therefore, the number of unordered partitions of type $(2, 2)$ is

$$\frac{1}{2!} \cdot \frac{4!}{(2!)^2} = \frac{1}{2} \cdot \frac{4 \cdot 3 \cdot 2 \cdot 1}{(2 \cdot 1)^2} = 3.$$

This result is the same as that obtained in Example 6 by enumeration.

≫ Now Try Exercise 11

EXAMPLE 8   **Grouping Construction Workers** A construction crew contains 12 workers, all having similar skills. In how many ways can the workers be divided into four groups of three?

SOLUTION   The order of the four groups is not relevant. (It does not matter which is labeled S_1 and which S_2, and so on. Only the composition of the groups is important.) Applying formula (3) with $n = 12$, $r = 3$, and $m = 4$, we see that the number of ways is

$$\frac{1}{m!} \cdot \frac{n!}{(r!)^m} = \frac{1}{4!} \cdot \frac{12!}{(3!)^4}$$

$$= 15{,}400.$$

≫ Now Try Exercise 21

INCORPORATING TECHNOLOGY

✳ **WolframAlpha** The number of ordered partitions of type $(n_1, n_2, \ldots, n_m)$ is given by the instruction

multinomial $(n_1, n_2, \ldots, n_m)$.

For instance, the answer to Example 3 is given by **multinomial (4, 3, 3, 2)** and the answer to Example 8 is given by **(1/4!)*multinomial (3, 3, 3, 3)**.

Check Your Understanding 5.8

Solutions can be found following the section exercises.

1. A foundation wishes to award one grant of \$100,000, two grants of \$10,000, five grants of \$5000, and five grants of \$2000. Its list of potential grant recipients has been narrowed to 13 possibilities. In how many ways can the awards be made?

2. In how many different ways can six medical interns be put into three groups of two and assigned to
   (a) the radiology, neurology, and surgery departments?
   (b) share offices?

EXERCISES 5.8

Let S be a set of n elements. Determine the number of ordered partitions of the types in Exercises 1–10.

1. $n = 5$; $(3, 1, 1)$
2. $n = 5$; $(2, 1, 2)$
3. $n = 6$; $(2, 1, 2, 1)$
4. $n = 6$; $(3, 3)$
5. $n = 7$; $(3, 2, 2)$
6. $n = 7$; $(4, 1, 2)$
7. $n = 12$; $(4, 4, 4)$
8. $n = 8$; $(3, 3, 2)$
9. $n = 12$; $(5, 3, 2, 2)$
10. $n = 8$; $(2, 2, 2, 2)$

Let S be a set of n elements. Determine the number of unordered partitions of the types in Exercises 11–14.

11. $n = 15$; $(3, 3, 3, 3, 3)$
12. $n = 10$; $(5, 5)$
13. $n = 18$; $(6, 6, 6)$
14. $n = 12$; $(4, 4, 4)$

15. **Stock Reports** A brokerage house regularly reports the behavior of a group of 20 stocks, each stock being reported as "up," "down," or "unchanged." How many different reports can show seven stocks up, five stocks down, and eight stocks unchanged?

16. **Investment Ratings** An investment advisory service rates investments as A, AA, and AAA. On a certain week, it rates 15 investments. In how many ways can it rate five investments in each of the categories?

17. **RIFFRAFF** How many eight-letter words consist of two Rs, one I, four Fs, and one A?

18. **RAZZMATAZZ** How many 10-letter words consist of one R, three As, four Zs, one M, and one T?

19. **Postage Stamps** In how many ways can three Forever stamps, two 20¢ stamps, and four 1¢ stamps be pasted in a row on an envelope?

20. **Nautical Signals** A nautical signal consists of six flags arranged vertically on a flagpole. If a sailor has three red flags, two blue flags, and one white flag, how many different signals are possible?

21. **Observation Groups** A psychology experiment observes groups of four individuals. In how many ways can an experimenter choose 5 groups of 4 from among 20 subjects?

22. **Orientation Groups** During orientation, new students are divided into groups of five people. In how many ways can 4 groups be chosen from among 20 people?

23. **Weather** In a certain month (of 30 days) it rains 10 days, snows 2 days, and is clear 18 days. In how many ways can such weather be distributed over the month?

24. **Awarding Prizes** Of the nine contestants in a contest, three will receive cars, three will receive TV sets, and three will receive radios. In how many different ways can the prizes be awarded?

25. **Job Promotions** A corporation has four employees that it wants to place in high executive positions. One will become president, one will become vice president, and two will be appointed to the board of directors. In how many different ways can this be accomplished?

26. **Forming Committees** The 10 members of a city council decide to form two committees of six to study zoning ordinances and street-repair schedules, with an overlap of two committee members. In how many ways can the committees be formed? (*Hint:* Specify three groups, not two.)

27. **Field Trip** In how many ways can the 14 children in a third-grade class be paired up for a trip to a museum?

28. **Work Schedule** A sales representative must travel to three cities, twice each, in the next 10 days. Her nontravel days are spent in the office. In how many different ways can she schedule her travel?

29. **Basketball Teams** Ten students in a physical education class are to be divided into five-member teams for a basketball game. In how many ways can the two teams be selected?

30. **Sorting Sweaters** In how many ways can 12 sweaters be stored into three boxes of different sizes if 6 sweaters are to be stored in the large box, 4 in the medium box, and 2 in the small box?

31. What is the value of $\binom{n}{1,1,\ldots,1}$ where there are n 1s?

32. Derive formula (1), using the generalized multiplication principle and the formula for $\binom{n}{r}$. (*Hint:* First select the elements of S_1, then the elements of S_2, and so on.)

TECHNOLOGY EXERCISES

33. **Assignments to Seminars** Calculate the number of ways that 38 students can be assigned to four seminars of size 10, 12, 10, and 6, respectively.

34. **Campaign Tasks** Calculate the number of ways that 65 phone numbers can be distributed to 5 campaign workers if each worker gets the same number of names.

35. **Bridge Deals** One octillion is 10^{28}, or 10 billion billion billion. Is the number of possible deals in bridge greater than or less than one octillion?

Solutions to Check Your Understanding 5.8

1. Each choice of recipients is an ordered partition of the 13 finalists into a first subset of one ($100,000 award), a second subset of two ($10,000 award), a third subset of five ($5000 award), and a fourth subset of five ($2000 award). The number of ways to choose the recipients is thus

$$\binom{13}{1,2,5,5} = \frac{13!}{1!\,2!\,5!\,5!}$$

$$= \frac{13 \cdot 12 \cdot 11 \cdot \overset{3}{\cancel{10}} \cdot \cancel{9} \cdot \cancel{8} \cdot 7 \cdot \cancel{6} \cdot \cancel{5} \cdot \cancel{4} \cdot \cancel{3} \cdot \cancel{2} \cdot \cancel{1}}{1 \cdot \cancel{2} \cdot 1 \cdot \cancel{5} \cdot \cancel{4} \cdot \cancel{3} \cdot \cancel{2} \cdot 1 \cdot \cancel{5} \cdot \cancel{4} \cdot \cancel{3} \cdot \cancel{2} \cdot \cancel{1}}$$

$$= 13 \cdot 12 \cdot 11 \cdot 3 \cdot 7 \cdot 6 = 216{,}216.$$

2. Each partition is of the type (2, 2, 2). In part (a), the order of the subsets is important, whereas in part (b), the order is irrelevant. Consider the partitions

$(\{\text{Dr. A, Dr. B}\}, \quad \{\text{Dr. C, Dr. D}\}, \quad \{\text{Dr. E, Dr. F}\})$

and

$(\{\text{Dr. C, Dr. D}\}, \quad \{\text{Dr. A, Dr. B}\}, \quad \{\text{Dr. E, Dr. F}\}).$

With respect to part (a), these two partitions are different, since in one, Drs. A and B are assigned to the radiology department and in the other, they are assigned to the neurology department. With respect to part (b), these two partitions are the same, since, for instance, Drs. A and B are officemates in both partitions. Therefore, the answers are as follows:

(a) $\binom{6}{2,2,2} = \frac{6!}{(2!)^3} = \frac{6 \cdot 5 \cdot \cancel{4} \cdot 3 \cdot \cancel{2} \cdot 1}{2 \cdot 2 \cdot 2} = 90.$

(b) $\frac{1}{3!} \cdot \frac{6!}{(2!)^3} = \frac{1}{\cancel{6}} \cdot \frac{\cancel{6} \cdot 5 \cdot \cancel{4} \cdot 3 \cdot \cancel{2} \cdot 1}{2 \cdot 2 \cdot 2} = 15.$

CHAPTER 5   Summary

KEY TERMS AND CONCEPTS	EXAMPLES

5.1 Sets

Let A and B be sets with universal set U.

The **intersection** of A and B, written $A \cap B$, is the set of all elements that belong to both A and B.

The **union** of A and B, written $A \cup B$, is the set of all elements that belong to either A or B or both.

The **complement** of A, written A', is the set of all elements in U that do not belong to A.

The set S is a **subset** of the set T, written $S \subseteq T$, if every element of S is an element of T.

The **empty set**, denoted $\varnothing$, is the set containing no elements.

Let $U = \{a, b, c, d, e, f\}$, $A = \{a, b, c, d\}$, and $B = \{c, d, f\}$. Then,
$$A \cap B = \{c, d\}$$

$$A \cup B = \{a, b, c, d, f\}$$

$$A' = \{e, f\}.$$

$$\{a, c, f\} \subseteq \{a, b, c, d, e, f\}$$

5.2 A Fundamental Principle of Counting

Number of elements in set S is denoted $n(S)$.

Inclusion–Exclusion Principle:
$n(S \cup T) = n(S) + n(T) - n(S \cap T)$.

$n(\{a, b, c\}) = 3$

Let $S = \{a, b, c, d\}$ and $T = \{b, d, e\}$. Then,
$n(S) = 4$, $n(T) = 3$, $n(S \cap T) = 2$, and $S \cup T = \{a, b, c, d, e\}$.
By the inclusion–exclusion principle, $n(S \cup T) = 4 + 3 - 2 = 5$.

Venn diagrams are used to illustrate set operations.

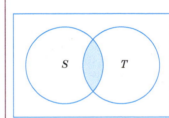

$S \cap T$ $(S \cup T)'$

De Morgan's Laws:
$(S \cup T)' = S' \cap T'$ and $(S \cap T)' = S' \cup T'$.

Let $U = \{a, b, c, d, e, f\}$, $S = \{a, b, c, d\}$, and $T = \{c, d, f\}$. Then,
$$(S \cup T)' = \{a, b, c, d, f\}' = \{e\}$$
$$S' \cap T' = \{e, f\} \cap \{a, b, e\} = \{e\}$$
$$(S \cap T)' = \{c, d\}' = \{a, b, e, f\}$$
$$S' \cup T' = \{e, f\} \cup \{a, b, e\} = \{a, b, e, f\}.$$

5.3 Venn Diagrams and Counting

Venn diagrams can be used to solve certain types of counting problems.

Suppose that $n(U) = 47$, $n(S \cap T') = 10$, $n(S) = 15$, $n(T) = 25$. Find $n((S \cup T)')$.

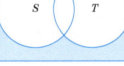

$n(U) = 47$
$n(S \cap T') = 10$

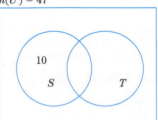

KEY TERMS AND CONCEPTS	EXAMPLES

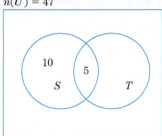

Since $n(S) = 15$,
$n(S \cap T) = 5$

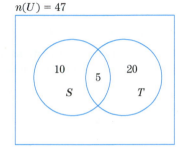

Since $n(T) = 25$,
$n(S' \cap T) = 20$

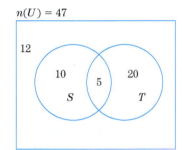

$n((S \cup T)') =$
$47 - 10 - 5 - 20 = 12$

5.4 The Multiplication Principle

The **multiplication principle** states that if there are m ways to make a selection and, for each of these, there are n ways to make a second selection, then there are $m \times n$ ways to make the two selections.

If you have three coats and four hats, then there are $3 \times 4 = 12$ ways to select a coat and hat.

The multiplication principle generalizes to any number of selections.

If you also have two scarves, there are $3 \times 4 \times 2 = 24$ ways to select a coat, hat, and scarf.

5.5 Permutations and Combinations

The number of **permutations** of n elements taken r at a time (ordered selections) is

$$P(n, r) = \frac{n!}{(n-r)!} = \underbrace{n(n-1)(n-2) \cdots (n-r+1)}_{r \text{ factors}}.$$

How many ways can you select three of ten photos and hang them in a row on the wall?

Answer: $P(10, 3) = 10 \cdot 9 \cdot 8 = 720$

The number of **combinations** of n elements taken r at a time (unordered selections) is

$$C(n, r) = \frac{n!}{r!(n-r)!} = \frac{n(n-1)(n-2) \cdots (n-r+1)}{r(r-1)(r-2) \cdots 1}.$$

How many ways can you select three of ten photos?

Answer: $C(10, 3) = \frac{10 \cdot 9 \cdot 8}{3 \cdot 2 \cdot 1} = 120$

5.6 Further Counting Techniques

The **complement rule of counting** says that
$n(S) = n(U) - n(S')$.

When a coin is tossed five times, how many possible outcomes contain two or more heads?

Answer: Let S be the set of outcomes containing two or more heads. Then S' is the number of outcomes containing zero or one head. Since $n(S') = 1 + 5 = 6$, $n(S) = 2^5 - 6 = 32 - 6 = 26$.

KEY TERMS AND CONCEPTS	EXAMPLES

5.7 The Binomial Theorem

$C(n, r) = C(n, n - r)$

The **binomial theorem** states that

$$(x + y)^n = \binom{n}{0}x^n + \binom{n}{1}x^{n-1}y + \binom{n}{2}x^{n-2}y^2 +$$

$$\cdots + \binom{n}{n-1}xy^{n-1} + \binom{n}{n}y^n.$$

A set of n elements has 2^n subsets.

$$C(5, 3) = \frac{5 \cdot 4 \cdot 3}{3 \cdot 2 \cdot 1} = \frac{5 \cdot 4 \cdot \cancel{3}}{\cancel{3} \cdot 2 \cdot 1} = C(5, 2)$$

$$(x + y)^3 = \binom{3}{0}x^3 + \binom{3}{1}x^{3-1}y + \binom{3}{2}x^{3-2}y^2 + \binom{3}{3}x^{3-3}y^3$$

$$= x^3 + 3x^2y + 3xy^2 + y^3$$

The set $\{a, b, c\}$ has 3 elements and $2^3 = 8$ subsets:
$\{a, b, c\}, \{a, b\}, \{a, c\}, \{b, c\}, \{a\}, \{b\}, \{c\}, \varnothing.$

5.8 Multinomial Coefficients and Partitions

The number of **ordered partitions** of type $(n_1, n_2, \ldots, n_m)$ for a set of n elements is $\frac{n!}{n_1!n_2!\ldots n_m!}$, denoted $\binom{n}{n_1, n_2, \ldots, n_m}$.

How many different signals can be created by lining up 10 flags in a vertical column if 2 flags are green, 3 are red, and 5 are blue?

A signal is an ordered partition of the numbers 1 through 10. For instance, one example of an ordered partition is ({3, 7}, {1, 4, 9}, {2, 5, 6, 8, 10}) corresponding to the green flags being in the 3rd and 7th positions, the red flags being in the 1st, 4th, and 9th positions, and the blue flags being in the remaining positions. Therefore, the total number of signals is $\frac{10!}{2!3!5!} = 2520$.

The number of **unordered partitions** of type $(r, r, \ldots, r)$ for a set of n elements where $n = m \cdot r$ is $\frac{1}{m!} \cdot \frac{n!}{(r!)^m}$.

In how many ways can six students be paired up to work on a group project? Answer: Since $n = 6$, $r = 2$, $m = 3$, there are $\frac{1}{3!} \cdot \frac{6!}{(2!)^3} = 15$ different ways.

CHAPTER 5   Fundamental Concept Check Exercises

1. What is a set?

2. What is a subset of a set?

3. What is an element of a set?

4. Define a universal set.

5. Define the empty set.

6. Define the complement of the set A.

7. Define the intersection of the two sets A and B.

8. Define the union of the two sets A and B.

9. State the generalized multiplication principle for counting.

10. What is meant by a permutation of n items taken r at a time?

11. How would you calculate the number of permutations of n items taken r at a time?

12. What is the difference between a permutation and a combination?

13. How would you calculate the number of combinations of n items taken r at a time?

14. Give a formula that can be used to calculate each of the following:

$$n! \qquad \binom{n}{r} \qquad C(n, r) \qquad P(n, r)$$

15. State the binomial theorem.

16. If a set contains n elements, how many subsets does it have?

17. Explain what is meant by an ordered partition of a set.

18. Explain how to calculate the number of ordered partitions of a set.

CHAPTER 5   Review Exercises

1. List all subsets of the set $\{a, b\}$.

2. Draw a two-circle Venn diagram, and shade the portion corresponding to the set $(S \cup T')'$.

3. **Tennis Finalists** There are 16 contestants in a tennis tournament. How many different possibilities are there for the two people who will play in the final round?

4. **Team Picture** In how many ways can a coach and five basketball players line up in a row for a picture if the coach insists on standing at one of the ends of the row?

5. Draw a three-circle Venn diagram, and shade the portion corresponding to the set $R' \cap (S \cup T)$.

6. Determine the first three terms in the binomial expansion of $(x - 2y)^{12}$.

7. **Balls in an Urn** An urn contains 14 numbered balls, of which 8 are red and 6 are green. How many different possibilities are there for selecting a sample of 5 balls in which 3 are red and 2 are green?

8. **Testing a Drug** Sixty people with a certain medical condition were given pills. Fifteen of these people received placebos. Forty people showed improvement, and 30 of these had received an actual drug. How many of the people who received the drug showed no improvement?

9. **Appliance Purchase** An appliance store carries seven different types of washing machines and five different types of dryers. How many different combinations are possible for a customer who wants to purchase a washing machine and a dryer?

10. **Contest Prizes** There are 12 contestants in a contest. Two will receive trips around the world, four will receive cars, and six will receive TV sets. In how many different ways can the prizes be awarded?

11. **Languages** Out of a group of 115 applicants for jobs at the World Bank, 70 speak French, 65 speak Spanish, 65 speak German, 45 speak French and Spanish, 35 speak Spanish and German, 40 speak French and German, and 35 speak all three languages. How many of the people speak none of the three languages?

12. Calculate $\binom{17}{15}$.

Environmental Poll The 100 members of the Earth Club were asked what they felt the club's priorities should be in the coming year: clean water, clean air, or recycling. The responses were 45 for clean water, 30 for clean air, 42 for recycling, 13 for both clean air and clean water, 20 for clean air and recycling, 16 for clean water and recycling, and 9 for all three. Exercises 13–20 refer to this poll.

13. How many members thought that the priority should be clean air only?

14. How many members thought that the priority should be clean water or clean air, but not both?

15. How many members thought that the priority should be clean water or recycling but not clean air?

16. How many members thought that the priority should be clean air and recycling but not clean water?

17. How many members thought that the priority should be exactly one of the three issues?

18. How many members thought that recycling should not be a priority?

19. How many members thought that the priority should be recycling but not clean air?

20. How many members thought that the priority should be something other than one of these three issues?

21. **Nine-Letter Words** How many different nine-letter words (i.e., sequences of letters) can be made by using four Ss and five Ts?

22. **Passing an Exam** Twenty people take an exam. How many different possibilities are there for the set of people who pass the exam?

23. **Winter Sports** A survey at a small New England college showed that 400 students skied, 300 played ice hockey, and 150 did both. How many students participated in at least one of these sports?

24. **Meal Choices** How many different meals can be chosen if there are 6 appetizers, 10 main dishes, and 8 desserts, assuming that a meal consists of one item from each category?

25. **Test Scoring** On an essay test, there are five questions worth 20 points each. In how many ways can a student get 10 points on one question, 15 points on each of three questions, and 20 points on another question?

26. **Seven-Digit Numbers** How many seven-digit numbers are even and have a 3 in the hundreds place?

27. **Telephone Numbers** How many telephone numbers are theoretically possible if all numbers are of the form *abc-def-ghij* and neither of the first two leading digits (*a* and *d*) is zero?

28. **Three-Digit Numbers** How many three-digit numbers can be formed from the digits 1, 2, 3, 4, 5, 6, and 7 if no digit is repeated?

29. **Eight-Letter Words** How many strings of length 8 can be formed from the letters *A*, *B*, *C*, *D*, and *E* ? How many of the strings have at least one *E*?

30. **Basketball Teams** How many different five-person basketball teams can be formed from a pool of 12 players?

31. **Selecting Students** Fourteen students in the 30-student eighth grade are to be chosen to tour the United Nations. How many different groups of 14 are possible?

32. **Journal Subscriptions** In one ZIP code, there are 40,000 households. Of them, 4000 households get *Fancy Diet Magazine*, 10,000 households get *Clean Living Journal*, and 1500 households get both publications. How many households get neither?

33. **Computer Program** At each stage in a decision process, a computer program has three branches. There are 10 stages at which these branches appear. How many different paths could the process follow?

34. **Filling Jobs** Sixty people apply for 10 job openings. In how many ways can all of the jobs be filled?

35. **Generating Tests** A computerized test generator can generate any one of five problems for each of the 10 areas being tested. How many different tests can be generated?

36. **Six-Letter Words** If a string of six letters cannot contain any vowels (A, E, I, O, U), how many strings are possible?

37. **Choosing Delegations** How many different four-person delegations can be chosen from 10 ambassadors?

38. **Subdividing a Class** In how many ways can a teacher divide a class of 21 students into groups of 7 students each?

39. **Distributing Candy** In how many ways can 14 different candies be distributed to 14 scouts?

40. **Subdividing People** In how many ways can 20 people be divided into groups of 5 each?

41. **Executive Positions** An Internet company is considering three candidates for CEO, five candidates for CFO, and four candidates for marketing director. In how many different ways can these positions be filled?

42. **Softball League** If every team in a 10-team softball league plays every other team three times, how many games are played?

43. **Diagonals** How many diagonals does an *n*-sided polygon have?

44. **Racetrack Betting** Racetracks have a compound bet called the *daily double*, in which the bettor tries to select the winners of the first two races. If eight horses compete in the first race and six horses compete in the second race, determine the possible number of daily double bets.

45. **Hat Displays** A designer of a window display wants to form a pyramid with 15 hats. She wants to place the five men's hats in the bottom row, the four women's hats in the next row, next the three baseball caps, then the two berets, and a clown's hat at the top. All of the hats are different. How many displays are possible?

46. **Seat Assignments** In how many ways can five people be assigned to seats in a 12-seat room?

47. **Poker** A poker hand consists of five cards. How many different poker hands contain all cards of the same suit? (Such a hand is called a "flush.")

48. **Poker** How many hands of five cards contain exactly three aces?

49. **Three-Digit Numbers** How many three-digit numbers are there in which exactly two digits are alike?

50. **Three-Digit Numbers** How many three-digit numbers are there in which no two digits are alike?

Baseball In Exercises 51 and 52, suppose there are three pitchers and three catchers at a baseball training camp.

51. How many different pitcher–catcher pairs can be formed?

52. In how many ways can the six players be seated in a row if no one sits next to someone who plays the same position?

53. **Greek-Letter Societies** Fraternity and sorority names consist of two or three letters from the Greek alphabet. How many different names are there in which no letter appears more than once? (The Greek alphabet contains 24 letters.)

54. **Family Picture** A family consisting of two parents and four children is to be seated in a row for a picture. How many different arrangements are possible in which the children are seated together?

55. **Feasible Set** If 10 lines are drawn in the plane so that no two of them are parallel and no three lines intersect at the same point, how many points of intersection are there?

56. **Splitting Classes** Two elementary school teachers have 24 students each. The first teacher splits his students into four groups of six. The second teacher splits her students into six groups of four. Which teacher has more options?

57. **Consultation Schedule** A consulting engineer agrees to spend three days at Widgets International, four days at Gadgets Unlimited, and three days at Doodads Incorporated in the next two workweeks. In how many different ways can she schedule her consultations?

58. **Arranging Books** A set of books can be arranged on a bookshelf in 120 different ways. How many books are in the set?

59. **Batting Order** How many batting orders are possible in a nine-member baseball team if the catcher must bat fourth and the pitcher last?

60. **Choosing Committees** In the United States Senate, each state is represented by one junior senator and one senior senator. In how many ways can a committee of five senators from the 12 Midwest states be represented if
   (a) the committee must consist of two senior senators and three junior senators?
   (b) no two senators from the same state can serve on the committee?

61. **Voting** Suppose that you are voting in an election for state delegate. Two state delegates are to be elected from among seven candidates. In how many different ways can you cast your ballot? *Note:* You may vote for two candidates. However, some people "single-shoot," and others don't pull any levers.

62. **Spelling** *Algebra* The object is to start with the letter *A* on top and to move down the diagram to an *A* at the bottom. From any given letter, move only to one of the letters directly below it on the left or right. If these rules are followed, how many different paths spell *ALGEBRA*?

$$A$$
$$L \quad L$$
$$G \quad G \quad G$$
$$E \quad E \quad E \quad E$$
$$B \quad B \quad B \quad B \quad B$$
$$R \quad R \quad R \quad R \quad R \quad R$$
$$A \quad A \quad A \quad A \quad A \quad A \quad A$$

63. **Call Letters** The call letters of radio stations in the United States consist of either three or four letters, where the first letter is K or W. How many different call letters are possible?

TECHNOLOGY EXERCISES

In Exercises 64–66, use technology to calculate the answer.

64. **Arranging Students** A group of students can be arranged in a row of seats in 479,001,600 ways. How many students are there?

65. **Assigning Jobs** There are 25 people in a department who must be deployed to work on three projects requiring 10, 9, and 6 people. All of the people are eligible for all jobs. Calculate the number of ways this can be done.

66. **License Plates** Calculate the number of license plates that can be formed by using three distinct letters and three distinct digits in any order.

Conceptual Exercises

67. What is the relationship between the two sets A and B if $A \cap B = \varnothing$?

68. What is the relationship between the two sets A and B if $n(A \cup B) = n(A) + n(B)$?

69. Suppose that A and B are subsets of the set U. Under what circumstance will $A \cap B = B$?

70. Suppose that A and B are subsets of the set U. Under what circumstance will $A \cup B = B$?

71. (True or False) $n(A) + n(B) = n(A \cap B) + n(A \cup B)$.

72. (True or False) The empty set is a subset of every set.

73. Explain why $n \cdot (n - 1)! = n!$.

74. Use the result in Exercise 73 to explain why 0! is defined to be 1.

75. Express in your own words the difference between a permutation and a combination.

76. Consider a group of 10 people. Without doing any computation, explain why the number of committees of six people is equal to the number of committees of four people.

77. Without doing any computation, explain why $C(10, 3) = C(10, 7)$.

78. Without doing any computation, explain why $C(10, 4) + C(10, 5) = C(11, 5)$. *Hint:* Suppose that a committee of size five is to be chosen from a pool of 11 people, and John Doe is one of the people. How many committees are there that include John? How many committees are there that don't include John?

CHAPTER

5 PROJECT

Pascal's Triangle

In the following triangular table, known as **Pascal's triangle**, the entries in the nth row are the binomial coefficients $\binom{n}{0}, \binom{n}{1}, \binom{n}{2}, \ldots, \binom{n}{n}$.

							1							0th row
						1		1						1st row
					1		2		1					2nd row
				1		3		3		1				3rd row
			1		4		6		4		1			4th row
		1		5		10		10		5		1		5th row
	1		6		15		20		15		6		1	6th row
1		7		21		35		35		21		7		1

7th row

Observe that each number (other than the ones) is the sum of the two numbers directly above it. For example, in the 5th row, the number 5 is the sum of the numbers 1 and 4 from the 4th row, and the number 10 is the sum of the numbers 4 and 6 from the 4th row. This fact is known as **Pascal's formula**. Namely, the formula says that

$$\binom{n}{r} = \binom{n-1}{r-1} + \binom{n-1}{r}.$$

1. For what values of n and r does Pascal's formula say that the number 10 in the triangle is the sum of the numbers 4 and 6?

2. Derive Pascal's formula from the fact that $C(n, r) = \dfrac{n!}{r!\,(n-r)!}$.

3. Derive Pascal's formula from the fact that $C(n, r)$ is the number of ways of selecting r objects from a set of n objects. (*Hint:* Let x denote the nth object of the set. Count the number of ways that a subset of r objects containing x can be selected, and then count the number of ways that a subset of r objects not containing x can be selected.)

4. Use Pascal's formula to extend Pascal's triangle to the 12th row. Determine the values of $\binom{12}{5}$ and $\binom{12}{6}$ from the extended triangle.

5. (a) Show that, for any positive integer n,

$$\binom{n}{0} + \binom{n}{1} + \binom{n}{2} + \binom{n}{3} + \cdots + \binom{n}{n} = 2^n.$$

Hint: Apply the binomial theorem to $(x + y)^n$ with $x = 1, y = 1$.

(b) Show that, for any positive integer n,

$$\binom{n}{0} - \binom{n}{1} + \binom{n}{2} - \binom{n}{3} + \cdots \pm \binom{n}{n} = 0.$$

Hint: Apply the binomial theorem to $(x + y)^n$ with $x = 1, y = -1$.

(c) Show that, for the 7th row of Pascal's triangle, the sum of the even-numbered elements equals the sum of the odd-numbered elements; that is,

$$\binom{7}{0} + \binom{7}{2} + \binom{7}{4} + \binom{7}{6} = \binom{7}{1} + \binom{7}{3} + \binom{7}{5} + \binom{7}{7}.$$

(d) Use the result of parts (a) and (b) to show that, for any row of Pascal's triangle, the sum of the even-numbered elements equals the sum of the odd-numbered elements, and give that common sum for the nth row in terms of n. [*Hint:* Add the two equations in parts (a) and (b).]

(e) Suppose that S is a set of n elements. Use the result of part (d) to determine the number of subsets of S that have an even number of elements.

6. (a) Show that, for any positive integer n,

$$1 + 2 + 4 + 8 + \cdots + 2^n = 2^{n+1} - 1.$$

(*Hint:* Let $S = 1 + 2 + 4 + 8 + \cdots + 2^n$, multiply both sides of the equation by 2, and subtract the first equation from the second.)

(b) Show that the sum of the elements of any row of Pascal's triangle equals one more than the sum of the elements of all previous rows.

7. (a) Consider the 7th row of Pascal's triangle. Observe that each interior number (that is, a number other than 1) is divisible by 7. For what values of n, for $1 \le n \le 12$, are the interior numbers of the nth row divisible by n?

(b) Confirm that each of the values of n from part (a) is a prime number. (*Note:* A number p is a *prime* number if the only positive integers that divide it are p and 1.) Prove that, if p is a prime number, then each interior number of the pth row of Pascal's triangle is divisible by p. $\left[\text{\textit{Hint:} Use the fact that } \binom{p}{r} = \frac{p(p-1)(p-2)\cdots(p-r+1)}{1 \cdot 2 \cdot 3 \cdot \cdots \cdot r}.\right]$

(c) Show that, for any prime number p, the sum of the interior numbers of the pth row is $2^p - 2$.

(d) Calculate $2^p - 2$ for $p = 7$, and show that it is a multiple of 7.

(e) Use the results of parts (b) and (c) to show that, for any prime number p, $2^p - 2$ is a multiple of p. *Note:* This is a special case of Fermat's theorem, which states that, for any prime number p and any integer a, $a^p - a$ is a multiple of p.

8. There are four odd numbers in the 6th row of Pascal's triangle (1, 15, 15, 1), and $4 = 2^2$ is a power of 2. For the 0th through 12th rows of Pascal's triangle, show that the number of odd numbers in each row is a power of 2.

In Fig. 1(a), each odd number in the first eight rows of Pascal's triangle has been replaced by a dot and each even number has been replaced by a space. In Fig. 1(b), the pattern for the first four rows is shown in blue. Notice that this pattern appears twice in the next four rows, with the two appearances separated by an inverted triangle. Fig. 1(c) shows the locations of the odd numbers in the first 16 rows of Pascal's triangle, and Fig. 1(d) shows that the pattern for the first eight rows appears twice in the next eight rows, separated by an inverted triangle. Figure 2 demonstrates that the first 32 rows of Pascal's triangle have the same property.

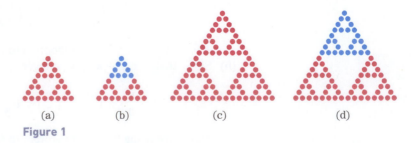

(a)　　　(b)　　　(c)　　　(d)

Figure 1

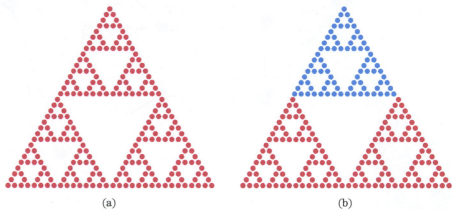

(a)   (b)

Figure 2

9. Assume that the property shown in Figs. 1 and 2 continues to hold for subsequent rows of Pascal's triangle. Use this result to explain why the number of odd numbers in each row of Pascal's triangle is a power of 2.

Probability is the mathematics of chance. A probability is a numerical measure of the likelihood that a specific event will occur. In this chapter, we consider the basic concepts that allow us to associate realistic probabilities to random events in many diverse fields.

6.1 Experiments, Outcomes, Sample Spaces, and Events

The events whose probabilities we wish to compute all arise as outcomes of experiments. So as our first step in developing probability theory, let us define *experiments*, *outcomes*, *events*, and the associated concept *sample spaces*.

> **DEFINITIONS**   An **experiment** is an activity with an observable result. Each possible result is called an **outcome** of the experiment. The set of all possible outcomes is called the **sample space** of the experiment. An **event** is a subset of the sample space. (We say that the event E has occurred when the outcome of the experiment is an element of E.)

EXAMPLE 1	**Rolling a Die** Illustrate the preceding definitions for the experiment of rolling a die and observing the number on the uppermost face.

SOLUTION
(a) The possible outcomes are the numbers 1, 2, 3, 4, 5, and 6.
(b) The sample space is $\{1, 2, 3, 4, 5, 6\}$.
(c) One possible event is $\{2, 4, 6\}$. This event also can be described as "the event that the outcome is an even number." ◄◄

EXAMPLE 2	**Rolling a Pair of Dice** Illustrate the preceding definitions for the experiment of rolling two dice, one red and one green, and observing the number on the uppermost face of each.

SOLUTION
(a) Each outcome of the experiment can be regarded as an ordered pair of numbers, the first representing the number on the red die and the second the number on the green die. Thus, for example, the pair of numbers $(3, 5)$ represents the outcome "3 on the red die, 5 on the green die." That is, each outcome is an ordered pair of numbers (r, g), where r and g are each one of the numbers 1, 2, 3, 4, 5, 6.
(b) The sample space, which has 36 elements, is as follows:

$$\{(1, 1), (1, 2), (1, 3), (1, 4), (1, 5), (1, 6),$$
$$(2, 1), (2, 2), (2, 3), (2, 4), (2, 5), (2, 6),$$
$$(3, 1), (3, 2), (3, 3), (3, 4), (3, 5), (3, 6),$$
$$(4, 1), (4, 2), (4, 3), (4, 4), (4, 5), (4, 6),$$
$$(5, 1), (5, 2), (5, 3), (5, 4), (5, 5), (5, 6),$$
$$(6, 1), (6, 2), (6, 3), (6, 4), (6, 5), (6, 6)\}.$$

(c) One possible event is $\{(1, 4), (2, 3), (3, 2), (4, 1)\}$. This event also can be described as "the event that the sum of the two numbers is 5." **»Now Try Exercise 3(a)**

EXAMPLE 3	**Number of People in a Queue** Once every hour, a supermarket manager observes the number of people standing in a checkout line. The store has space for at most 30 customers to wait in line. Illustrate the preceding definitions for this situation.

SOLUTION
(a) The possible outcomes are the numbers 0, 1, 2, 3, . . . , 30.
(b) The sample space is $\{0, 1, 2, 3, . . . , 30\}$.
(c) One possible event is $\{0, 1, 2, 3, 4, 5\}$. This event also can be described as "the event that there are at most five people standing in the line." ◄◄

EXAMPLE 4	**Sample Space for Pollutant Levels** The Environmental Protection Agency ordered Middle States Edison Corporation to install "scrubbers" to remove the pollutants from its smokestacks. To monitor the effectiveness of the scrubbers, the corporation installed monitoring devices to record the levels of sulfur dioxide, particulate matter, and oxides of nitrogen in the smokestack emissions. Consider the monitoring operation as an experiment. Describe the associated sample space.

SOLUTION
Each reading of the instruments consists of an ordered triple of numbers (x, y, z), where $x =$ level of sulfur dioxide, $y =$ level of particulate matter, and $z =$ level of oxides of nitrogen. The sample space thus consists of all possible triples (x, y, z), where $x \geq 0, y \geq 0$, and $z \geq 0$. **»Now Try Exercise 7(a)**

The sample spaces in Examples 1, 2, and 3 are **finite**. That is, the associated experiments have only a finite number of possible outcomes. However, the sample space of Example 4 is **infinite**, since there are infinitely many triples (x, y, z), where $x \geq 0, y \geq 0$, and $z \geq 0$.

EXAMPLE 5   **Tossing a Coin Three Times** Suppose that an experiment consists of tossing a coin three times and observing the sequence of heads and tails. (Order counts.)
(a) Determine the sample space S.
(b) Determine the event $E =$ "exactly two heads."

SOLUTION   (a) Denote "heads" by H and "tails" by T. Then a typical outcome of the experiment is a sequence of Hs and Ts. So, for instance, the sequence HTT would stand for a head followed by two tails. We exhibit all such sequences and arrive at the sample space S:

$$S = \{HHH, HHT, HTH, THH, HTT, THT, TTH, TTT\}.$$

(b) Here are the outcomes in which exactly two heads occur: HHT, HTH, THH. Therefore, event E is

$$E = \{HHT, HTH, THH\}. \qquad \text{\textbf{» Now Try Exercise 3(b)}}$$

EXAMPLE 6   **Political Poll** A political poll surveys a group of people to determine their income levels and political affiliations. People are classified as either low-, middle-, or upper-level income and as either Democrat, Republican, or Independent.
(a) Find the sample space corresponding to the poll.
(b) Determine the event $E_1 =$ "Independent."
(c) Determine the event $E_2 =$ "low income and not Independent."
(d) Determine the event $E_3 =$ "neither upper income nor Independent."

SOLUTION   (a) Let us abbreviate low, middle, and upper income by the letters L, M, and U, respectively. And let us abbreviate Democrat, Republican, and Independent by the letters D, R, and I, respectively. Then a response to the poll can be represented as a pair of letters. For example, the pair (L, D) refers to a low-income-level Democrat. The sample space S is then given by

$$S = \{(L, D), (L, R), (L, I), (M, D), (M, R), (M, I), (U, D), (U, R), (U, I)\}.$$

(b) For event E_1, the income level may be anything, but the political affiliation is Independent. Thus,

$$E_1 = \{(L, I), (M, I), (U, I)\}.$$

(c) For event E_2, the income level is low and the political affiliation may be either Democrat or Republican. In this case,

$$E_2 = \{(L, D), (L, R)\}.$$

(d) For event E_3, the income level may be either low or middle and the political affiliation may be Democrat or Republican. Thus,

$$E_3 = \{(L, D), (L, R), (M, D), (M, R)\}. \qquad \text{\textbf{» Now Try Exercise 9}}$$

As we have seen, an event is a subset of a sample space. Two events are worthy of special mention. The first is the event corresponding to the empty set, $\varnothing$. This is called the **impossible event**, since it can never occur. The second special event is the set S, the sample space itself. Every outcome is an element of S, so S always occurs. For this reason, S is called the **certain event**.

One particular advantage of defining experiments and events in terms of sets is that it allows us to define new events from given ones by applying the operations of set theory. When so doing, we always let the sample space S play the role of universal set. (All outcomes belong to the universal set.)

If E and F are events, then so are $E \cup F$, $E \cap F$, and E'. For example, consider the die-rolling experiment of Example 1. Then

$$S = \{1, 2, 3, 4, 5, 6\}.$$

Let E and F be the events given by

$$E = \{3, 4, 5, 6\} \qquad F = \{1, 4, 6\}.$$

Then we have

$$E \cup F = \{1, 3, 4, 5, 6\}$$
$$E \cap F = \{4, 6\}$$
$$E' = \{1, 2\}.$$

Let us interpret the events $E \cup F$, $E \cap F$, and E' by using Venn diagrams. In Fig. 1, we have drawn a Venn diagram for $E \cup F$. Note that $E \cup F$ occurs precisely when the experimental outcome belongs to the shaded region—that is, to E or F. Thus, we have the following result:

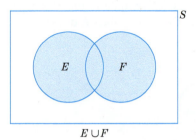

$E \cup F$

Figure 1

> **DEFINITION Union of Two Events**   The event $E \cup F$ occurs precisely when either E or F (or both) occurs.

Similarly, we can interpret the event $E \cap F$. This event occurs when the experimental outcome belongs to the shaded region of Fig. 2—that is, to both E and F. Thus, we have an interpretation for $E \cap F$:

> **DEFINITION Intersection of Two Events**   The event $E \cap F$ occurs precisely when both E and F occur.

Finally, the event E' consists of all of those outcomes not in E (Fig. 3). Therefore, we have the following:

> **DEFINITION Complement of an Event**   The event E' occurs precisely when E does not occur.

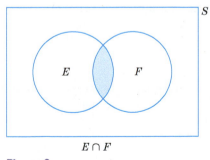

$E \cap F$

Figure 2

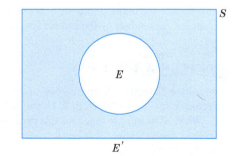

E'

Figure 3

EXAMPLE 7   **Events Related to Pollution Control** Consider the pollution monitoring described in Example 4. Let E, F, and G be the events

$$E = \text{"level of sulfur dioxide} \geq 100\text{"}$$

$$F = \text{"level of particulate matter} \leq 50\text{"}$$

$$G = \text{"level of oxides of nitrogen} \leq 30\text{."}$$

Describe the following events:
(a) $E \cap F$ **(b)** E' **(c)** $E \cup G$ **(d)** $E' \cap F \cap G$

SOLUTION   (a) $E \cap F =$ "level of sulfur dioxide ≥ 100 *and* level of particulate matter ≤ 50."
(b) $E' =$ "level of sulfur dioxide < 100."
(c) $E \cup G =$ "level of sulfur dioxide ≥ 100 *or* level of oxides of nitrogen ≤ 30."
(d) $E' \cap F \cap G =$ "level of sulfur dioxide < 100 *and* level of particulate matter ≤ 50 *and* level of oxides of nitrogen ≤ 30."   **» Now Try Exercise 7(b)**

Suppose that E and F are events in a sample space S. We say that E and F are **mutually exclusive** (or *disjoint*) provided that $E \cap F = \varnothing$. In terms of Venn diagrams, we may represent a pair of mutually exclusive events as a pair of circles with no points in common (Fig. 4). If the events E and F are mutually exclusive, then E and F cannot simultaneously occur; if E occurs, then F does not; and if F occurs, then E does not.

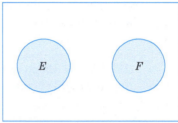

Figure 4 E and F are mutually exclusive.

EXAMPLE 8   **Determining Whether Events Are Mutually Exclusive** Let $S = \{a, b, c, d, e, f, g\}$ be a sample space, and let $E = \{a, b, c\}$, $F = \{e, f, g\}$, and $G = \{c, d, f\}$.
(a) Are E and F mutually exclusive?
(b) Are F and G mutually exclusive?

SOLUTION   (a) $E \cap F = \varnothing$, so E and F are mutually exclusive.
(b) $F \cap G = \{f\}$, so F and G are *not* mutually exclusive.   **» Now Try Exercise 11**

Check Your Understanding 6.1

Solutions can be found following the section exercises.

1. **Lightbulbs** A machine produces lightbulbs. As part of a quality control procedure, a sample of five lightbulbs is collected each hour and the number of defective lightbulbs among these is observed.
   (a) What is the sample space for this experiment?
   (b) Describe the event "there are at most two defective lightbulbs" as a subset of the sample space.

2. **Citrus Fruit** Suppose that there are two crates of citrus fruit and each crate contains oranges, grapefruit, and tangelos. An experiment consists of selecting a crate and then selecting a piece of fruit from that crate. Both the crate and the type of fruit are noted. Refer to the crates as crate I and crate II.
   (a) What is the sample space for this experiment?
   (b) Describe the event "a tangelo is selected" as a subset of the sample space.

EXERCISES 6.1

1. **Committee Selection** A committee of two people is to be selected from five people, R, S, T, U, and V.
   (a) What is the sample space for this experiment?
   (b) Describe the event "R is on the committee" as a subset of the sample space.
   (c) Describe the event "neither R nor S is on the committee" as a subset of the sample space.

2. **Selecting Letters** A letter is selected at random from the word "MISSISSIPPI."
   (a) What is the sample space for this experiment?
   (b) Describe the event "the letter chosen is a vowel" as a subset of the sample space.

3. **Heads and Tails** An experiment consists of tossing a coin two times and observing the sequence of heads and tails.
   (a) What is the sample space of this experiment?
   (b) Describe the event "the first toss is a head" as a subset of the sample space.

4. **Four-Sided Dice** A pair of four-sided dice—each with the numbers from 1 to 4 on their sides—are rolled, and the numbers facing down are observed. (*Note:* A four-sided die is similar to a pyramid with a triangular base.)
   (a) List the sample space.
   (b) Describe each of the following events as a subset of the sample space:
      (i) Both numbers are even.
      (ii) At least one number is odd.
      (iii) Neither number is less than or equal to 2.
      (iv) The sum of the numbers is 7.
      (v) The sum of the numbers is greater than or equal to 6.
      (vi) The numbers are the same.
      (vii) A 2 or 3 occurs, but not both 2 and 3.
      (viii) No 4 appears.

5. **Selecting from Urns** Suppose that we have two urns—call them urn I and urn II—each containing red balls and white

balls. An experiment consists of selecting an urn and then selecting a ball from that urn and noting its color.
(a) What is a suitable sample space for this experiment?
(b) Describe the event "urn I is selected" as a subset of the sample space.

6. **Coin Tosses** An experiment consists of tossing a coin four times and observing the sequence of heads and tails.
(a) What is the sample space of this experiment?
(b) Determine the event $E_1 = $ "more heads than tails occur."
(c) Determine the event $E_2 = $ "the first toss is a head."
(d) Determine the event $E_1 \cap E_2$.

7. **Efficiency Studies** An efficiency expert records the time it takes an assembly line worker to perform a particular task. Let E be the event "more than 5 minutes," F the event "less than 8 minutes," and G the event "less than 4 minutes."
(a) Describe the sample space for this experiment.
(b) Describe the events $E \cap F$, $E \cap G$, E', F', $E' \cap F$, $E' \cap F \cap G$, and $E \cup F$.

8. **Census Data** A census taker records the annual income of each household they visit. Let E be the event "more than \$50,000" and F be the event "less than \$75,000." Describe the events $E \cap F$, E', and F'.

9. **Student Poll** A campus survey is taken to correlate the number of years that students have been on campus with their political leanings. Students are classified as first-year, sophomore, junior, or senior and as conservative or liberal.
(a) Find the sample space corresponding to the poll.
(b) Determine the event $E_1 = $ "conservative."
(c) Determine the event $E_2 = $ "junior and liberal."
(d) Determine the event $E_3 = $ "neither first-year nor conservative."

10. **Automobiles** An experiment consists of selecting a car at random from a college parking lot and observing the color and make. Let E be the event "the car is red," F be the event "the car is a Chevrolet," G be the event "the car is a green Ford," and H be the event "the car is black or a Chrysler."
(a) Which of the following pairs of events are mutually exclusive?
　(i) E and F　　　　(ii) E and G
　(iii) F and G　　　(iv) E and H
　(v) F and H　　　 (vi) G and H
　(vii) E' and G　　 (viii) F' and H'
(b) Describe each of the following events:
　(i) $E \cap F$　　　　 (ii) $E \cup F$
　(iii) E'　　　　　　(iv) F'
　(v) G'　　　　　　 (vi) H'
　(vii) $E \cup G$　　　 (viii) $E \cap G$
　(ix) $E \cap H$　　　　(x) $E \cup H$
　(xi) $G \cap H$　　　　(xii) $E' \cap F'$
　(xiii) $E' \cup G'$

11. Let $S = \{1, 2, 3, 4, 5, 6\}$ be a sample space,

$$E = \{1, 2\} \qquad F = \{2, 3\} \qquad G = \{1, 5, 6\}.$$

(a) Are E and F mutually exclusive?
(b) Are F and G mutually exclusive?

12. Draw the events E and E' on two separate Venn diagrams. Are E and E' mutually exclusive?

13. Let $S = \{a, b, c\}$ be a sample space. Determine all possible events associated with S.

14. Let S be a sample space with n outcomes. How many events are associated with S?

15. Let $S = \{1, 2, 3, 4\}$ be a sample space, $E = \{1\}$, and $F = \{2, 3\}$. Are the events $E \cup F$ and $E' \cap F'$ mutually exclusive?

16. Let S be any sample space, and E, F any events associated with S. Are the events $E \cup F$ and $E' \cap F'$ mutually exclusive? (*Hint:* Apply De Morgan's laws.)

17. **Coin Tosses** Suppose that 10 coins are tossed and the number of heads observed.
(a) Describe the sample space for this experiment.
(b) Describe the event "more heads than tails" in terms of the sample space.

18. **Three-Digit Numbers** An experiment consists of forming a three-digit number by using the digits 1, 7, and 8 without repetition.
(a) Describe the sample space for this experiment.
(b) Describe the event "the number is greater than 300" as a subset of the sample space.

19. **Genetic Traits** An experiment consists of observing the eye color and age of all United States citizens. Let E be the event "blue eyes," F the event "at least 18 years old," and G the event "brown eyes and younger than 18."
(a) Are E and F mutually exclusive?
(b) Are E and G mutually exclusive?
(c) Are F and G mutually exclusive?

20. **Genetic Traits** Consider the experiment and events of Exercise 19. Describe the following events:
(a) $E \cup F$　　　　　　(b) $E \cap G$
(c) E'　　　　　　　　(d) F'
(e) $(G \cup F) \cap E$　　(f) $G' \cap E$

21. **Shuttle Bus** Suppose that you observe the number of passengers arriving at a metro station on a shuttle bus that holds up to eight passengers. Describe the sample space.

22. **Dice** A pair of dice is rolled, and the sum of the numbers on the two uppermost faces is observed. What is the sample space of this experiment?

23. **Selecting Balls from an Urn** An urn contains balls numbered 1 through 9. Suppose that you draw a ball from the urn, observe its number, replace the ball, draw a ball again, and observe its number. Give an example of an outcome of the experiment. How large is the sample space?

24. **Selecting Balls from an Urn** Repeat Exercise 23 in the case that the first ball is not replaced.

25. **NBA Draft Lottery** In the NBA, the 14 basketball teams that did not make the playoffs participate in the draft lottery. Ping-pong balls numbered 1 through 14 are placed in a lottery machine, and a sample of four balls is drawn randomly to determine which team will have the first overall draft pick. The order in which the numbers are drawn does not matter. The combination $\{11, 12, 13, 14\}$ is ignored if it is drawn. (*Note:* Each team is assigned between 5 and 250 combinations prior to the drawing in the order of their regular season record.) Give an example of an outcome of the draw. In 2015, the Minnesota Timberwolves, who had the worst record in the NBA and were assigned 250 combinations, won the first overall pick. What percentage of the combinations were assigned to the Minnesota Timberwolves?

26. **Coin & Die** Suppose that a coin is tossed and a die is rolled, and the face and number appearing are observed. How many outcomes are in the sample space?

27. **The Game of Clue** *Clue* is a board game in which players are given the opportunity to solve a murder that has six suspects, six possible weapons, and nine possible rooms where the murder may have occurred. The six suspects are Colonel Mustard, Miss Scarlet, Professor Plum, Mrs. White, Mr. Green, and Mrs. Peacock. Determine a sample space

for the choice of murderer. Discuss how to form a sample space with the entire solution to the murder, giving murderer, weapon, and site.

(a) How many outcomes would the sample space for the entire solution have?

Let E be the event that the murder occurred in the library. Let F be the event that the weapon was a gun.

(b) Describe $E \cap F$.
(c) Describe $E \cup F$.

Solutions to Check Your Understanding 6.1

1. (a) $\{0, 1, 2, 3, 4, 5\}$. The sample space is the set of all outcomes of the experiment. At first glance, it might seem that each outcome is a set of five lightbulbs. What is observed, however, is not the specific sample but rather the number of defective bulbs in the sample. Therefore, the outcome must be a number.

   (b) $\{0, 1, 2\}$. "At most 2" means "2 or less."

2. (a) {(crate I, orange), (crate I, grapefruit), (crate I, tangelo), (crate II, orange), (crate II, grapefruit), (crate II, tangelo)}. Two selections are being made, and both should be recorded.

   (b) {(crate I, tangelo), (crate II, tangelo)}. This set consists of those outcomes in which a tangelo is selected.

6.2   Assignment of Probabilities

A **probability** is a numerical measure of the likelihood that a specific event will occur. The number is between 0 and 1 and is called the *probability of the event*. It is written Pr(*event*). The larger the number, the more confident we are that the event will occur.

There are three types of probabilities.

1. A **logical probability** is obtained by mathematical reasoning—often, by the use of the counting techniques of the previous chapter. Some examples are as follows.

   - Pr(coin toss showing heads)
   - Pr(obtaining a full house in poker)
   - Pr(winning the Powerball Lottery)

2. An **empirical probability** is obtained by sampling or observation and is calculated by dividing the number of times the event occurs by the total number of observations. This value is also called the **relative frequency**. Some examples are as follows.

   - Pr(a certain brand of lightbulb will last more than 1000 hours)
   - Pr(a person selected at random will be a smoker)
   - Pr(taking small amounts of aspirin daily will lower your risk of having a stroke)

3. A **judgmental probability** is obtained by an educated guess. Also known as a **subjective probability**, it describes an individual's personal judgment about how likely a particular event is to occur. Some examples are as follows.

   - Pr(New England Patriots will win next year's Super Bowl)
   - Pr(a Democrat will be elected president in 2024)
   - Pr(you will get an A on your next mathematics exam)

EXAMPLE 1   **Forecasting Rain** Classify the following probability as *logical*, *empirical*, or *judgmental*:

$$\text{Pr(it will rain tomorrow)} = 30\%.$$

SOLUTION   Empirical. Weather probabilities are widely misunderstood. Some people think that the statement means that it will rain in 30% of the area. Others think that it will rain 30% of the time. Actually, what the weather person means is that, based on past observations when weather conditions were similar to today's weather conditions, 30% of the time it rained the next day.

» *Now Try Exercise 1*

In this chapter, we develop techniques for calculating logical probabilities and use tables acquired by sampling to obtain empirical probabilities. We will not explore judgmental probabilities.

In the previous section, we introduced the sample space of an experiment and used it to describe events. We complete our description of experiments by introducing probabilities associated with events. For the remainder of this chapter, we limit our discussion to experiments with only a finite number of outcomes. This restriction will remain in effect until our discussion of the normal distribution in the next chapter.

Suppose that an experiment has a sample space S consisting of a finite number of outcomes $s_1, s_2, \ldots, s_N$. To each outcome we associate a number, called the *probability* of the outcome, which represents the likelihood that the outcome will occur. Suppose that to the outcome s_1 we associate the probability p_1, to the outcome s_2 the probability p_2, and so forth. We can summarize this data in a chart of the following sort:

Outcome	Probability
s_1	p_1
s_2	p_2
$\vdots$	$\vdots$
s_N	p_N

Such a chart is called the **probability distribution** for the experiment.

We require that probabilities obey two fundamental properties:

Fundamental Property 1   Each of the numbers $p_1, p_2, \ldots, p_N$ is between 0 and 1.

Fundamental Property 2 $p_1 + p_2 + \cdots + p_n = 1$.

Fundamental Property 1 says that the likelihood of each outcome lies between 0% and 100%, whereas Fundamental Property 2 says that there is a 100% chance that one of the outcomes $s_1, s_2, \ldots, s_N$ will occur.

In the experiments associated with many common applications, all outcomes are **equally likely**—that is, they all have the same probability. This is the case, for example, if we toss an unbiased coin or select a person at random from the population. If a sample space has N equally likely outcomes, then the probability of each outcome is $1/N$ (since the probabilities must add up to 1).

The next three examples illustrate some methods for determining probability distributions.

EXAMPLE 2   **Probability Distributions** Determine the probability distributions for the following experiments.
(a) Toss an unbiased coin and observe the side that faces upward.
(b) Roll a die and observe the side that faces upward.

SOLUTION   (a) Since the coin is unbiased, we expect each of the outcomes "heads" and "tails" to be equally likely. We assign the two outcomes equal probabilities, namely, $\frac{1}{2}$. The probability distribution is as follows:

Outcome	Probability
Heads	$\frac{1}{2}$
Tails	$\frac{1}{2}$

(b) There are six possible outcomes, namely, 1, 2, 3, 4, 5, and 6. Assuming that the die is unbiased, these outcomes are equally likely. So we assign to each outcome the probability $\frac{1}{6}$. Here is the probability distribution for the experiment:

Outcome	Probability	Outcome	Probability
1	$\frac{1}{6}$	4	$\frac{1}{6}$
2	$\frac{1}{6}$	5	$\frac{1}{6}$
3	$\frac{1}{6}$	6	$\frac{1}{6}$

>> *Now Try Exercise 5*

Probabilities may be assigned to the elements of a sample space by using common sense about the physical nature of the experiment. The fair coin has two sides, both equally likely to be face up. The unbiased die has six equally probable faces. However, in general, it may not be possible to use intuition alone to decide on a realistic probability to assign to individual sample elements. Sometimes it is necessary to use experimental data to determine the relative frequency with which events occur. The following example demonstrates this technique:

EXAMPLE 3   **College Majors** A group of 141,000 college freshmen were asked the question, "What is your intended major?" Table 1 shows the results of this question. (The data was obtained from *The American Freshman: National Norms Fall 2015*.) Consider the experiment of selecting a student at random from the group surveyed and observing their answer.
(a) Determine the probability distribution for this experiment.
(b) Verify that the probabilities satisfy Fundamental Properties 1 and 2.

Table 1 Intended Majors

Intended Major	Number of Freshmen
Arts and humanities	14,241
Biological science	21,009
Business	18,894
Engineering	18,471
Health professions	15,933
Math, computer science, and physical science	11,421
Social science	15,228
Other and undecided	25,803

(Source: www.heri.ucla.edu)

SOLUTION   **(a)** For each of the eight possible outcomes, we use the data to compute its relative frequency. For example, of the 141,000 freshmen sampled, 18,471 intended to

Table 2 Probability Distribution of Intended Majors

Outcome	Probability
Arts and humanities	.101
Biological science	.149
Business	.134
Engineering	.131
Health professions	.113
Math, computer science, and physical science	.081
Social science	.108
Other and undecided	.183

major in engineering. So the outcome "Engineering" occurred in $\frac{18,471}{141,000} = 13.1\%$ of the answers. Therefore, we assign to the outcome "Engineering" the probability .131. Similarly, we assign probabilities to the other outcomes on the basis of the fraction of times that they occurred.

(b) The probabilities are .101, .149, .134, .131, .113, .081, .108, and .183. Clearly, each is between 0 and 1, so Fundamental Property 1 is satisfied. Adding the probabilities shows that their sum is 1, satisfying Fundamental Property 2.

>> *Now Try Exercise 15*

Unlike in Example 2, we cannot always rely on intuition to assign a value to a logical probability. Instead, we must use our knowledge of sets and counting to construct a theoretical model of the experiment along with associated probabilities. This allows us to determine the probability for *events* rather than just single *outcomes*, and we can construct probability distributions for events that satisfy Fundamental Properties 1 and 2. We can observe this in the following probability distribution for the number of heads in four tosses of a fair coin:

Events	Probability
0 heads	$\frac{1}{16}$
1 heads	$\frac{4}{16} = \frac{1}{4}$
2 heads	$\frac{6}{16} = \frac{3}{8}$
3 heads	$\frac{4}{16} = \frac{1}{4}$
4 heads	$\frac{1}{16}$
Total	1

Note that the column labeled "Events" contains all possible outcomes in the sample space. Also, all of the entries in the probability column are nonnegative numbers between 0 and 1. Furthermore, the sum of the probabilities is 1.

The Addition and Inclusion–Exclusion Principles

Suppose that we are given an experiment with a finite number of outcomes. Let us now assign to each event E a probability, which we denote by $\Pr(E)$. If E consists of a single outcome, say, $E = \{s\}$, then E is called an **elementary event**. In this case, we associate with E the probability of the outcome s. If E consists of more than one outcome, we may compute $\Pr(E)$ via the **addition principle**.

Addition Principle   Suppose that an event E consists of the finite number of outcomes $s, t, u, \ldots, z$. That is,

$$E = \{s, t, u, \ldots, z\}.$$

Then

$$\Pr(E) = \Pr(s) + \Pr(t) + \Pr(u) + \cdots + \Pr(z).$$

We supplement the addition principle with the convention that the probability of the impossible event $\varnothing$ is 0. This is certainly reasonable, since the impossible event never occurs.

EXAMPLE 4   **Probability Associated with a Die**   Suppose that we roll a die and observe the side that faces upward. What is the probability that an odd number will occur?

SOLUTION   The event "odd number occurs" corresponds to the subset of the sample space given by

$$E = \{1, 3, 5\}.$$

That is, the event occurs if a 1, 3, or 5 appears on the side that faces upward. By the addition principle,

$$\Pr(E) = \Pr(1) + \Pr(3) + \Pr(5).$$

As we observed in Example 2(b), each of the outcomes in the die-rolling experiment has probability $\frac{1}{6}$. Therefore,

$$\Pr(E) = \tfrac{1}{6} + \tfrac{1}{6} + \tfrac{1}{6} = \tfrac{1}{2}.$$

So we expect an odd number to occur approximately half of the time.

>> *Now Try Exercise 7*

EXAMPLE 5   **Probability Distribution for the Number of Boys in Two-Child Families** Observe the genders and birth orders in two-child families, and calculate the probabilities of 0, 1, and 2 boys.

SOLUTION   The sample space $S = \{GG, GB, BG, BB\}$ describes the gender and birth order in two-child families. (Here, for instance, GB denotes the birth sequence "first child is a girl, second child is a boy.") If we assume that each of the four outcomes in S is equally likely to occur, we should assign probability $\frac{1}{4}$ to each outcome. Then

$$\Pr(\text{no boys}) = \Pr(GG) = \tfrac{1}{4}$$
$$\Pr(\text{one boy}) = \Pr(GB) + \Pr(BG) = \tfrac{1}{4} + \tfrac{1}{4} = \tfrac{1}{2}$$
$$\Pr(\text{two boys}) = \Pr(BB) = \tfrac{1}{4}.$$

EXAMPLE 6   **Probability Associated with College Majors** Consider the college major survey of Example 3. What is the probability that a student selected at random from the group surveyed intends to major in one of the three most popular fields outside the "other and undecided" choice?

SOLUTION   The event "top three most popular fields" is the same as

$$\{\text{biological science, business, engineering}\}.$$

Thus, the probability of the event is

$$\Pr(\text{biological science}) + \Pr(\text{business}) + \Pr(\text{engineering}) = .149 + .134 + .131 = .414.$$

>> *Now Try Exercise 17*

EXAMPLE 7   **Probability Associated with the Roll of a Pair of Dice** Suppose that we roll a red die and a green die and observe the numbers on the sides that face upward.
(a) Calculate the probabilities of the elementary events.
(b) Calculate the probability that the two dice show the same number.

SOLUTION   **(a)** As shown in Example 2 of the previous section, the sample space consists of 36 pairs of numbers:

$$S = \{(1, 1), (1, 2), \ldots, (6, 5), (6, 6)\}.$$

Each of these pairs is equally likely to occur. (How could the dice show favoritism to a particular pair?) Therefore, each outcome is expected to occur about $\frac{1}{36}$ of the time, and the probability of each elementary event is $\frac{1}{36}$.
(b) The event

$$E = \text{"both dice show the same number"}$$

consists of six outcomes:

$$E = \{(1, 1), (2, 2), (3, 3), (4, 4), (5, 5), (6, 6)\}.$$

Thus, by the addition principle,

$$\Pr(E) = \tfrac{1}{36} + \tfrac{1}{36} + \tfrac{1}{36} + \tfrac{1}{36} + \tfrac{1}{36} + \tfrac{1}{36} = \tfrac{6}{36} = \tfrac{1}{6}.$$ **» Now Try Exercise 13**

EXAMPLE 8   **Probability Associated with a Scratch-Off Ticket** A person playing a certain scratch-off ticket can win $100, $10, or $1; can break even; or can lose $10. These five outcomes with their corresponding probabilities are given by the probability distribution in Table 3.

(a) Which outcome has the greatest probability?
(b) Which outcome has the least probability?
(c) What is the probability that the person will win some money?

SOLUTION

(a) Table 3 reveals that the outcome -10 has the greatest probability, .50. (A person playing the lottery repeatedly can expect to lose $10 about 50% of the time.) This outcome is just as likely to occur as not.

(b) The outcome 100 has the least probability, .02. A person playing the lottery can expect to win $100 about 2% of the time. (This outcome is quite unlikely to occur.)

(c) We are asked to determine the probability that the event E occurs, where $E = \{100, 10, 1\}$. By the addition principle,

$$\begin{aligned}
\Pr(E) &= \Pr(100) + \Pr(10) + \Pr(1) \\
&= \ \ .02 \ \ + \ \ .05 \ \ + \ \ .40 \\
&= \ \ .47.
\end{aligned}$$ **«**

Table 3

Winnings	Probability
100	.02
10	.05
1	.40
0	.03
−10	.50

Here is a useful formula that relates $\Pr(E \cup F)$ to $\Pr(E \cap F)$:

Inclusion–Exclusion Principle   Let E and F be any events. Then

$$\Pr(E \cup F) = \Pr(E) + \Pr(F) - \Pr(E \cap F).$$

In particular, if E and F are mutually exclusive, then

$$\Pr(E \cup F) = \Pr(E) + \Pr(F).$$

Note the similarity of this principle to the principle of the same name that was used in Section 5.2 to count the elements in a set.

EXAMPLE 9   **Resource Availability** A factory needs two raw materials in order to operate. The probability of not having an adequate supply of material A is .05, whereas the probability of not having an adequate supply of material B is .03. A study determines that the probability of a shortage of both A and B is .01. What proportion of the time can the factory operate?

SOLUTION   Let E be the event "shortage of A" and F the event "shortage of B." We are given that

$$\Pr(E) = .05 \quad \Pr(F) = .03 \quad \Pr(E \cap F) = .01.$$

The factory can operate only if it has both raw materials. Therefore, we must calculate the proportion of the time in which there is no shortage of material A or material B. A shortage of A or B is the event $E \cup F$. By the inclusion–exclusion principle,

$$\begin{aligned}
\Pr(E \cup F) &= \Pr(E) + \Pr(F) - \Pr(E \cap F) \\
&= \ \ .05 \ \ + \ \ .03 \ \ - \ \ .01 \\
&= \ \ .07.
\end{aligned}$$

Thus, the factory is likely to be short of one raw material or the other 7% of the time. Therefore, the factory can expect to operate 93% of the time.

» Now Try Exercise 37

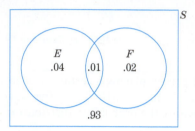

Figure 1

Probabilities involving unions and intersections of events are often conveniently displayed in a Venn diagram, similar to counting problems. Figure 1 displays the probabilities from Example 9.

Odds

Odds are similar to probability, but relate the likelihood of an event occurring to the likelihood of the event not occurring rather than to the total possible outcomes. As an example, we say that the odds in favor of obtaining a *three* when rolling a die are 1 to 5. (There is one way for the event to be achieved and five ways for the event to fail.)

As another example, if a meteorologist says that the odds for rain tomorrow are 3 to 4, they mean that on days when weather conditions are similar to today's, we can expect three tomorrows with rain for every four without rain. In general, when we say that the odds in favor of event E occurring are a to b, we expect that in $a + b$ trials the event would occur a times and fail to occur b times.

Converting between Odds and Probabilities If the odds in favor of the event E occurring are a to b, then

$$\Pr(E) = \frac{a}{a + b}.$$

If $\Pr(E) = p$, then the odds in favor of E are found by reducing the fraction $\frac{p}{1 - p}$ to the form a/b, where a and b are integers having no common divisor. Then the odds in favor of E are

$$a \text{ to } b.$$

EXAMPLE 10 **Chance of Rain** Suppose that the odds of rain tomorrow are 5 to 3. What is the probability that rain will occur tomorrow?

SOLUTION The probability that rain will occur tomorrow is

$$\frac{5}{5 + 3} = \frac{5}{8} = .625.$$

  » Now Try Exercise 45

EXAMPLE 11 **Rolling a Pair of Dice** The probability of obtaining a sum of eight or more when rolling a pair of dice is $15/36$. What are the odds of obtaining a sum of eight or more?

SOLUTION By the explanation in the preceding box with $p = 15/36$,

$$\frac{p}{1 - p} = \frac{\frac{15}{36}}{1 - \frac{15}{36}} = \frac{\frac{15}{36}}{\frac{21}{36}} = \frac{15}{21} = \frac{5}{7}.$$

Therefore, the odds in favor of obtaining a sum of eight or more are 5 to 7.

  » Now Try Exercise 43

Odds are often stated as *odds against*. This is always the case with gambling odds. If the odds in favor of an event are a to b, then the odds against the event are b to a. For instance, in Las Vegas roulette, the odds against winning when *red* is bet are 10 to 9.

Casinos also have the concept of *house odds*, which are different from true odds. House odds are always in the form m to 1, where m is the amount of money that a player wins on a one-dollar bet. For instance, the house odds on a *red* bet are 1 to 1. If the ball lands on red, the player wins \$1 (in addition to getting his \$1 bet back).

Big numbers are easier to get a handle on than small numbers. For instance, the odds of winning the Powerball Lottery jackpot with a single ticket are 1 to 292,201,338 and the probability of winning the lottery is about .000000003422298. We can easily pronounce the large number in the odds, and we have a reasonable idea of how large 292 million dollars is. But, how does one even pronounce the small probability? One way to pronounce it is 3,422,298 quadrillionths. But very few people have a good feel for how small that number is. Therefore, odds are usually preferable to probabilities in the discussion of highly unlikely events.

Paradoxes

Counterintuitive results abound in probability.

EXAMPLE 12   **A Paradox** In the December 1, 1996, issue of *Parade* magazine, Marilyn vos Savant presented the following question in her column *Ask Marilyn:* "A woman and a man (who are unrelated) each has two children. At least one of the woman's children is a boy, and the man's older child is a boy. Does the chance that the woman has two boys equal the chance that the man has two boys?"

SOLUTION   The answer is no. The sample space for the man is {BB, BG}, and each outcome has the same probability, $\frac{1}{2}$. Therefore, his likelihood of having two boys is $\Pr(BB) = \frac{1}{2}$. The sample space for the woman is {BB, GB, BG}, and each outcome has the same probability, $\frac{1}{3}$. Therefore, her likelihood of having two boys is $\Pr(BB) = \frac{1}{3}$.

A *paradox* is a statement that is counter to many people's intuition and yet is actually true. After the column was published, Marilyn vos Savant received many letters from irate readers criticizing her answer. They reasoned that among families with at least one boy, the other child is just as likely to be a boy as a girl. Therefore, they concluded incorrectly that the probability is $\frac{1}{2}$ in each case.   **«**

Some other examples of counterintuitive results discussed in this book are as follows.

- *The Famous Birthday Problem* In a group as small as 23 people, the probability that at least two people have the same birthday is greater than $\frac{1}{2}$. See the discussion following Example 6 in Section 6.3.

- *Suit Distribution in Bridge* The most likely distribution of suits in a bridge hand is not 4-3-3-3; that is, four cards of one suit and three cards of each of the other suits. See Exercise 47 in Section 6.3.

- *Medical Screening* Even with highly reliable tests, a person who tests positive for a certain disease might have a very low probability of actually having the disease. See Example 3 in Section 6.5.

- *Matches Consisting of Multiple Games* You can have a higher probability than your opponent of winning each game, yet your opponent can have a higher probability of winning the match. See Exercise 36 in Section 6.5.

- *Tennis Tournament* Under certain circumstances, you have your best chance of winning a tennis tournament if you play most of your games against the best possible opponent. See the first project at the end of Chapter 6.

- *Gender Bias in College Admissions* In 1973, the University of California at Berkeley was sued for discrimination against women in graduate school admissions. That year, 44% of male applicants were admitted and only 30% of female applicants were admitted. However, admission is decided by departments, and nearly every department favored the female applicants. The second project at the end of Chapter 6 shows how the probability of admission for women can be higher than for men in *every* department and yet the probability of admission for men will be higher in the university as a whole.

- *Family Composition* A family of four children is more likely to consist of three children of one gender and one of the other, than to consist of two boys and two girls. See Exercise 27 in Section 7.3.

Check Your Understanding 6.2

Solutions can be found following the section exercises.

1. **T-Maze** A mouse is put into a T-maze (a maze shaped like a "T") (Fig. 2). If it turns to the left, it receives cheese, and if it turns to the right, it receives a mild shock. This trial is done twice with the same mouse and the directions of the turns recorded.
   (a) What is the sample space for this experiment?
   (b) Why would it not be reasonable to assign each outcome the same probability?

2. What are the odds in favor of an event that is just as likely to occur as not?

3. The New York State lottery (Lotto) requires a contestant to select six numbers from 1 to 59. The probability of the same combination of six numbers occurring twice in one year is exceptionally low. Lotto system books advise people to avoid number combinations that have been drawn before. What is your assessment of this advice?

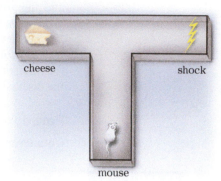

Figure 2

EXERCISES 6.2

In Exercises 1–4, classify the type of probability as logical, empirical, or judgmental.

1. Pr(China will win the most gold medals in the 2020 Olympics)

2. Pr(a person who smokes two packs of cigarettes each day has an increased risk of lung cancer)

3. Pr(obtaining a sum of 7 when rolling a pair of dice)

4. Pr(NASDAQ will increase by more than 10% in the next 12 months)

In Exercises 5 and 6, determine the probability distribution for the given experiment.

5. **Coin Tossing** Toss an unbiased coin twice, and count the number of heads.

6. **Selecting a Letter** A box contains seven slips of paper: one with a letter A printed on it, one with a B, three with a C, and two with a D. Draw a slip of paper and observe which letter is selected.

7. **Roulette** The modern American roulette wheel has 38 slots, which are labeled with 36 numbers evenly divided between red and black, plus two green numbers 0 and 00. What is the probability that the ball will land on a green number?

8. **U.S. States** A state is selected at random from the 50 states of the United States. What is the probability that it is one of the six New England states?

9. **Word Frequencies** There are 4487 words in the U.S. Constitution. The word "shall" occurs 191 times, and the word "States" occurs 81 times. Suppose that a word is selected at random from the U.S. Constitution.
   (a) What is the probability that the word is "shall"?
   (b) What is the probability that the word is "shall" or "States"?
   (c) What is the probability that the word is neither "shall" nor "States"?

10. **United Nations** Of the 193 member countries of the United Nations, 54 are in the African Group and 23 are in the Eastern European Group. Suppose that a country is selected at random from the members of the United Nations.
    (a) What is the probability that the country is in the African Group?
    (b) What is the probability that the country is in the African Group or the Eastern European Group?
    (c) What is the probability that the country is in neither the African Group nor the Eastern European Group?

11. **Selecting a Letter** An experiment consists of selecting a letter at random from the alphabet. Find the probability that the letter selected
    (a) precedes G alphabetically.
    (b) is a vowel (A, E, I, O, or U).
    (c) precedes G alphabetically or is a vowel.

12. **Selecting a Number** An experiment consists of selecting a number at random from the set of numbers {1, 2, 3, 4, 5, 6, 7, 8, 9}. Find the probability that the number selected is
    (a) less than 4.   (b) odd.   (c) less than 4 or odd.

13. **Dice** Suppose that a red die and a green die are rolled and the numbers on the sides that face upward are observed. (See Example 7 of this section and Example 2 of the first section.)
    (a) What is the probability that the numbers add up to 9?
    (b) What is the probability that the sum of the numbers is less than 5?

14. **Children** An experiment consists of observing the genders and birth orders in three-child families. Assume all eight outcomes have the same probability of occurring.
    (a) What is the probability that a family has at least two boys?
    (b) What is the probability that the oldest child is a girl?

15. **Kind of High School** The given table shows the results from *The American Freshman: National Norms Fall 2015* when the question "From what kind of high school did you graduate?" was posed to 141,000 college freshmen. Consider the experiment of selecting a student at random from the group surveyed and observing his or her answer. Determine the probability distribution for this experiment.

Kind of High School	Number of Freshmen
Public	115,620
Private	24,252
Home school	1128

Source: www.heri.ucla.edu

16. Highest Degree Planned The next table shows the results from *The American Freshman: National Norms Fall 2015* of a question posed to 141,000 college freshmen. Consider the experiment of selecting a student at random from the group surveyed and observing his or her answer. Determine the probability distribution for this experiment.

Highest Academic Degree Planned	Number of Freshmen
Master's	59,361
Bachelor's	29,751
Ph.D. or Ed.D.	26,931
M.D., D.O., D.D.S., D.V.M.	15,792
Other	9,165

Source: www.heri.ucla.edu

17. Grade Distributions The following table shows the probability distributions of letter grades from a mathematics class. What is the probability that a randomly chosen student received a letter grade higher than F but lower than A?

Letter Grade	Probability
A	.29
B	.34
C	.21
D	.09
F	.07

18. Candy Colors The colors in a bag of candy-coated milk chocolate candies have the probability distribution in the table that follows. What is the probability of randomly selecting a brown, orange, or red candy?

Color	Probability
Brown	.13
Yellow	.14
Red	.13
Orange	.20
Blue	.24
Green	.16

19. An experiment with outcomes s_1, s_2, s_3, s_4 has the following probability distribution:

Outcome	Probability
s_1	.1
s_2	.5
s_3	.2
s_4	.2

Let $E = \{s_1, s_2\}$ and $F = \{s_2, s_4\}$.
(a) Determine $\Pr(E)$ and determine $\Pr(F)$.
(b) Determine $\Pr(E')$.
(c) Determine $\Pr(E \cap F)$.
(d) Determine $\Pr(E \cup F)$.

20. An experiment with outcomes $s_1, s_2, s_3, s_4, s_5, s_6$ has the following probability distribution:

Outcome	Probability
s_1	.05
s_2	.25
s_3	.05
s_4	.01
s_5	.63
s_6	.01

Let $E = \{s_1, s_2\}$ and $F = \{s_3, s_5, s_6\}$.
(a) Determine $\Pr(E)$ and determine $\Pr(F)$.
(b) Determine $\Pr(E')$.
(c) Determine $\Pr(E \cap F)$.
(d) Determine $\Pr(E \cup F)$.

21. College Applications The table that follows was derived from a survey of college freshmen in 2015. Each probability is the likelihood that a randomly selected freshman applied to the specified number of colleges. For instance, 10% of the freshmen applied to just one college, and therefore, the probability that a student selected at random applied to just one college is .10.
(a) Convert these data into a probability distribution with outcomes 1, 2, 3, 4, and $\geq$ 5.
(b) What is the probability that a student applied to three or more colleges?

Number of Colleges Applied To	Probability
1	.10
2 or fewer	.17
3 or fewer	.27
4 or fewer	.40
20 or fewer	1

Source: www.heri.ucla.edu

22. Employees' Ages The next table summarizes the age distribution for a company's employees. Each probability is the likelihood that a randomly selected employee is in the specified age group.
(a) Convert this data into a probability distribution with outcomes 20–34 years, 35–49 years, 50–64 years, and 65–79 years.
(b) What is the probability that an employee selected at random is at least 50 years old?

Age (years)	Probability
20–34	.15
20–49	.70
20–64	.90
20–79	1

23. Which of the following probabilities are feasible for an experiment having sample space $\{s_1, s_2, s_3\}$? Explain your answer.
(a) $\Pr(s_1) = .4, \Pr(s_2) = .4, \Pr(s_3) = .4$
(b) $\Pr(s_1) = .5, \Pr(s_2) = .7, \Pr(s_3) = -.2$
(c) $\Pr(s_1) = \frac{1}{5}, \Pr(s_2) = \frac{2}{5}, \Pr(s_3) = \frac{1}{5}$

24. Which of the following probabilities are feasible for an experiment having sample space $\{s_1, s_2, s_3\}$? Explain your answer.
    (a) $\Pr(s_1) = .25, \Pr(s_2) = .25, \Pr(s_3) = .4$
    (b) $\Pr(s_1) = .7, \Pr(s_2) = .8, \Pr(s_3) = -.5$
    (c) $\Pr(s_1) = .2, \Pr(s_2) = .3, \Pr(s_3) = \frac{1}{2}$

25. **Car Race** Three cars, a Mazda, a Honda, and a Ford, are in a quarter-mile race. The probability that the Mazda will win the race is 2/3, and the probability that the Honda will win is 1/4. Assuming no ties are possible, what is the probability that the Ford will win the race?

26. **Hair Color** In a study, the residents of Edinburgh, Scotland, were classified as having either black hair, brown hair, blonde hair, or red hair. The probabilities of a randomly selected resident having black, brown, or blonde hair are .17, .47, and .20 respectively. Assuming each resident has one of these four hair colors,
    (a) what is the probability that a randomly selected resident has red hair?
    (b) what is the probability that a randomly selected resident has brown or black hair?
    (c) what is the probability that a randomly selected resident does not have blonde hair?

27. **Political Views** On a certain campus, the probability that a student selected at random has liberal political views is .28. The probability of having middle-of-the-road political views is twice the probability of having conservative political views. What is the probability that a student selected at random has conservative political views?

28. **Tennis** The probability that Alice beats Ben in a game of tennis is twice the probability that Ben beats Alice. Determine the two probabilities.

29. **Pair of Dice** Suppose that a pair of dice is rolled. Find $\Pr(\text{sum of the two numbers is odd}) + \Pr(\text{sum of the two numbers is even})$.

30. **Coin Tossing** An experiment consists of tossing a coin five times and observing the sequence of heads and tails. Find $\Pr(\text{an even number of heads occurs}) + \Pr(\text{an odd number of heads occurs})$.

31. Suppose that $\Pr(E) = .4$ and $\Pr(F) = .5$, where E and F are mutually exclusive. Find $\Pr(E \cup F)$.

32. Suppose that $\Pr(E) = .3$ and $\Pr(E \cup F) = .7$, where E and F are mutually exclusive. Find $\Pr(F)$.

In Exercises 33–36, consider the probabilities shown in the Venn diagram in Figure 3.

33. Determine the probability that the event E occurs.

34. Determine the probability that exactly one of the events E or F occurs.

35. Determine the probability that F occurs, but E does not occur.

36. Determine the probability that E does not occur.

In Exercises 37–40, use a Venn diagram similar to the one in Fig. 1 to solve the problem.

37. Suppose that $\Pr(E) = .6, \Pr(F) = .5$, and $\Pr(E \cap F) = .4$. Find (a) $\Pr(E \cup F)$ (b) $\Pr(E \cap F')$.

38. Suppose that $\Pr(E) = .6, \Pr(F) = .5$, and $\Pr(E' \cup F') = .6$. Find (a) $\Pr(E \cap F)$ (b) $\Pr(E \cup F')$.

39. **Grades** Joe feels that the probability of his getting an A in history is .7, the probability of getting an A in psychology is .8, and the probability of getting an A in history or psychology is .9. What is the probability that he will get an A in both subjects?

40. **Courses** At a certain engineering college, the probability that a student selected at random is taking a mathematics course is 1/2, the probability that they are taking a computer science course is 3/8, and the probability they are taking either a mathematics course or a computer science course is 3/4. What is the probability that a student selected at random is taking both types of courses?

41. Convert the odds of "10 to 1" to a probability.

42. Convert the odds of "4 to 5" to a probability.

43. Convert the probability .2 to odds.

44. Convert the probability $\frac{3}{7}$ to odds.

45. **Coin Tosses** The probability of getting three heads in five tosses of a coin is .3125. What are the odds of getting three heads?

46. **Advanced Degree** The probability that a graduate of a Big Ten school will eventually earn a Ph.D. degree is .05. What are the odds of a Big Ten graduate eventually earning a Ph.D. degree?

47. **Demographic** The odds of a person in the continental United States living within 50 miles of the Atlantic Ocean are 2 to 9. What is the probability of such a person living within 50 miles of the Atlantic Ocean?

48. **Election Odds** In March 2016, a betting website listed the odds of Hillary Clinton winning the 2016 U.S. presidential election as 2 to 5. According to the website, what was the probability that she would win the election?

49. **Bookies** Gamblers usually give odds *against* an event happening. For instance, if a bookie gives the odds 4 to 1 that the Yankees will win the next World Series, he is stating that the probability that the Yankees will win is 1/5, or .2. Also, if a bettor bets $1 that the Yankees will win and the Yankees do win, then the bettor will receive $5 (his original bet plus a profit of $4). The following are typical odds set by a bookie for the eventual winner in a fictitious four-team league: Sparks (5 to 3), Meteors (3 to 1), Asteroids (3 to 2), Suns (4 to 1).
    (a) Convert the odds to probabilities of winning for each team.
    (b) Add the four probabilities.
    (c) Explain why the answer to (b) makes sense.

50. **Odds of an Earthquake** The probability that there will be a major earthquake in the San Francisco area during the next 30 years is .63. What are the corresponding odds?

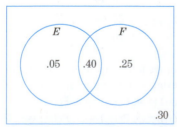

Figure 3

Exercises 51 and 52 can be answered without any computations.

51. Zodiac Signs A high school astrology club has 13 members. What is the probability that two or more members have the same zodiac sign? *Note:* There are 12 zodiac signs.

52. Absent-Minded Attendant Six people check their coats at a restaurant's coat-check counter. If the attendant returns the coats to the people at random, what is the probability that exactly five of the people receive the correct coat?

Solutions to Check Your Understanding 6.2

1. (a) {LL, LR, RR, RL}. Here, LL means that the mouse turned left both times, LR means that the mouse turned left the first time and right the second, and so on.

   (b) The mouse will learn something from the first trial. If it turned left the first time and got rewarded, then it is more likely to turn left again on the second trial. Hence, LL should have a greater probability than LR. Similarly, RL should be more likely than RR.

2. 1 to 1. An event that is just as likely to occur as not has probability $\frac{1}{2}$ and $\frac{.5}{1-.5} = \frac{.5}{.5} = \frac{1}{1}$.

3. The ping-pong balls don't remember what combinations they produced in the past. Any combination of six numbers is just as likely to occur as any other combination.

6.3 Calculating Probabilities of Events

As mentioned in Section 6.2, there are several ways to assign probabilities to the events of a sample space. One way is to perform the experiment many times and assign probabilities based on empirical data. Sometimes, our intuition about situations suffices, as in simple coin-tossing experiments. But we frequently are faced with forming a model of an experiment to assign probabilities consistent with Fundamental Properties 1 and 2. This section shows how the counting techniques of Chapter 5 may be used to extend these ideas to more complex situations. In addition, the exceptionally wide range of applications that make use of probability theory is illustrated.

Experiments with Equally Likely Outcomes

In Section 6.2, we saw that, if a sample space has N equally likely outcomes, then the probability of each outcome is $1/N$. If we use this fact, the probability of any event is easy to compute. For instance, suppose that E is an event consisting of M outcomes. Then, by the addition principle,

$$\Pr(E) = \underbrace{\frac{1}{N} + \frac{1}{N} + \cdots + \frac{1}{N}}_{M \text{ terms}} = \frac{M}{N}.$$

We can restate this fundamental result as follows:

> **Probability with Equally Likely Outcomes**   Let S be a sample space consisting of N equally likely outcomes. Let E be any event. Then, if $n(E)$ is the number of outcomes in E,
>
> $$\Pr(E) = \frac{n(E)}{N} = \frac{[\text{number of outcomes in } E]}{[\text{number of outcomes in } S]}. \tag{1}$$

In order to apply formula (1) in particular examples, it is necessary to compute N, the number of equally likely outcomes in the sample space, and $n(E)$. Often, these quantities can be determined by using the counting techniques of Chapter 5. Some illustrative computations are provided in Examples 1 through 6 here.

We should mention that, although the urn and dice problems considered in this section and the next might seem artificial and removed from applications, many applied problems can be described in mathematical terms as urn or dice experiments. We begin our discussion with two examples involving abstract urn problems. Then in two more examples, we show the utility of urn models by applying them to quality control and medical screening problems (Examples 2 and 4, respectively).

EXAMPLE 1   **Selecting Balls from an Urn**   An urn contains eight white balls and two green balls. A sample of three balls is selected at random.
  (a) What is the probability that the sample contains only white balls?
  (b) What is the probability that the sample contains at least one green ball?

SOLUTION   **(a)** The experiment consists of selecting 3 balls from the 10. Since the order in which the 3 balls are selected does not matter, the samples are combinations of 10 balls taken 3 at a time. The total number of samples is therefore $C(10, 3)$, and this is N, the number of elements in the sample space. Since the selection of the sample is random, all samples are equally likely, and thus we can use formula (1) to compute the probability of any event. The problem asks us to compute the probability of the event $E =$ "all three balls selected are white." Since there are 8 white balls, the number of different samples in which all are white is $C(8, 3)$. Thus,

$$\Pr(E) = \frac{n(E)}{N} = \frac{C(8, 3)}{C(10, 3)} = \frac{56}{120} = \frac{7}{15}.$$

(b) As in Part (a), there are $N = C(10, 3)$ equally likely outcomes. Let F be the event "at least one green ball is selected." Let us determine the number of different outcomes in F. These outcomes contain either one or two green balls. There are $C(2, 1)$ ways to select one green ball from two; and for each of these, there are $C(8, 2)$ ways to select two white balls from eight. By the multiplication principle, the number of samples containing one green ball equals $C(2, 1) \cdot C(8, 2)$. Similarly, the number of samples containing two green balls equals $C(2, 2) \cdot C(8, 1)$. Note that, although the sample size is three, there are only two green balls in the urn. Since the sampling is done without replacement, no sample can have more than two greens. Therefore, the number of outcomes in F—namely, the number of samples having at least one green ball—equals

$$\underbrace{C(2, 1) \cdot C(8, 2)}_{\substack{\text{number of samples} \\ \text{containing} \\ \text{one green ball}}} + \underbrace{C(2, 2) \cdot C(8, 1)}_{\substack{\text{number of samples} \\ \text{containing} \\ \text{two green balls}}} = 2 \cdot 28 + 1 \cdot 8 = 64,$$

so

$$\Pr(F) = \frac{n(F)}{N} = \frac{64}{C(10, 3)} = \frac{64}{120} = \frac{8}{15}.$$

» Now Try Exercise 3

EXAMPLE 2   **Quality Control**   A toy manufacturer inspects boxes of toys before shipment. Each box contains 10 toys. The inspection procedure consists of randomly selecting three toys from the box. If any are defective, the box is not shipped. Suppose that a given box has two defective toys. What is the probability that it will be shipped?

SOLUTION   This problem is not really new! We solved it in disguise as Example 1(a). The urn can be regarded as a box of toys, and the balls as individual toys. The white balls are nondefective toys and the green balls defective toys. The random selection of three balls from the urn is just the inspection procedure. And the event "all three balls selected are white" corresponds to the box being shipped. As we calculated previously, the probability of this event is $\frac{7}{15}$. (Since $\frac{7}{15} \approx .47$, there is approximately a 47% chance of shipping a box with two defective toys. This inspection procedure is not particularly effective!)

» Now Try Exercise 11

EXAMPLE 3   **Selecting Students**   A professor is randomly choosing a group of three students to do an oral presentation. In her class of 10 students, 2 are on the debate team. What is the chance that the professor chooses at least one of the debaters for the group?

SOLUTION   This is the same as Example 1(b). Just think of the debate team members as the green balls and the others in the class as white balls. The chance of getting at least one debate team member in the group is $\frac{8}{15} \approx .53$.

» Now Try Exercise 9

| EXAMPLE 4 | **Medical Screening** Suppose that a cruise ship returns to the United States from the Far East. Unknown to anyone, 4 of its 600 passengers have contracted a rare disease. Suppose that the Public Health Service screens 20 passengers, selected at random, to see whether the disease is present aboard ship. What is the probability that the presence of the disease will escape detection? |

SOLUTION   The sample space consists of samples of 20 drawn from among the 600 passengers. There are $C(600, 20)$ such samples. The number of samples containing none of the sick passengers is $C(596, 20)$. Therefore, the probability of not detecting the disease is

$$\frac{C(596, 20)}{C(600, 20)} \approx .87.$$

So there is approximately an 87% chance that the disease will escape detection.

≫ Now Try Exercise 13

| EXAMPLE 5 | **Rolling a Die Five Times** A die is rolled five times. What is the probability of obtaining exactly three 4s? |

SOLUTION   Think of each of the 6^5 possible outcomes as a sequence of five digits from 1 through 6. To obtain an outcome containing three 4s, select the three positions to fill with 4s [$C(5, 3)$ possibilities], and then select digits other than 4 for the remaining two positions ($5 \cdot 5$ possibilities). Therefore, the probability of obtaining three 4s is

$$\frac{C(5, 3) \cdot 5 \cdot 5}{6^5} = \frac{10 \cdot 5 \cdot 5}{7776} \approx .03.$$

Thus, there is about a 3% chance of obtaining three 4s in five rolls of a die.

≫ Now Try Exercise 25

The Complement Rule

The **complement rule** relates the probability of an event E to the probability of its complement E'. When applied together with counting techniques, it often simplifies computation of probabilities.

> **Complement Rule**   Let E be any event and E' its complement. Then
>
> $$\Pr(E) = 1 - \Pr(E').$$

For example, recall Example 1(a). We determined the probability of the event

$$E = \text{"all three balls selected are white"}$$

associated with the experiment of selecting three balls from an urn containing eight white balls and two green balls. We found that $\Pr(E) = \frac{7}{15}$. On the other hand, in Example 1(b), we determined the probability of the event

$$F = \text{"at least one green ball is selected."}$$

The event E is the complement of F:

$$E = F'.$$

So, by the complement rule,

$$\Pr(F) = 1 - \Pr(F') = 1 - \Pr(E) = 1 - \tfrac{7}{15} = \tfrac{8}{15},$$

in agreement with the calculations of Example 1(b).

The complement rule is especially useful in situations where $\Pr(E')$ is easier to compute than $\Pr(E)$. One of these situations arises in the celebrated *birthday problem*.

> **EXAMPLE 6**   **The Famous "Birthday Problem"** A group of five people is to be selected at random. What is the probability that two or more of them have the same birthday?

> **SOLUTION**   For simplicity, we ignore February 29. Furthermore, we assume that each of the 365 days in a year is an equally likely birthday (not an unreasonable assumption). The experiment we have in mind is this. Pick out five people, and observe their birthdays. The outcomes of this experiment are strings of five dates, corresponding to the birthdays. For example, one outcome of the experiment is

$$(\text{June 2, April 6, Dec. 20, Feb.12, Aug. 5}).$$

Each birth date has 365 different possibilities. So, by the generalized multiplication principle, the total number N of possible outcomes of the experiment is

$$N = 365 \cdot 365 \cdot 365 \cdot 365 \cdot 365 = 365^5.$$

Let E be the event "at least two people have the same birthday." It is very difficult to calculate directly the number of outcomes in E. However, it is comparatively simple to compute the number of outcomes in E' and hence to compute $\Pr(E')$. This is because E' is the event "all five birthdays are different." An outcome in E' can be selected in a sequence of five steps:

① Select a day	② Select a different day	③ Select yet a different day	④ Select yet a different day	⑤ Select yet a different day

These five steps will result in a sequence of five different birthdays. The first step can be performed in 365 ways; for each of these, the next step in 364; for each of these, the next step in 363; for each of these, the next step in 362; and for each of these, the last step in 361 ways. Therefore, E' contains $365 \cdot 364 \cdot 363 \cdot 362 \cdot 361$ [or $P(365, 5)$] outcomes, and

$$\Pr(E') = \frac{365 \cdot 364 \cdot 363 \cdot 362 \cdot 361}{365^5} \approx .973.$$

By the complement rule,

$$\Pr(E) = 1 - \Pr(E') \approx 1 - .973 = .027.$$

So the likelihood is about 2.7% that two or more of the five people will have the same birthday.   **» Now Try Exercise 17**

The experiment of Example 6 can be repeated by using samples of 8, 10, 20, or any number of people. As before, let E be the event "at least two people have the same birthday," so that $E' = $ "all the birthdays are different." If a sample of r people is used, then the same reasoning as used previously yields

$$\Pr(E') = \frac{365 \cdot 364 \cdot \cdots \cdot (365 - r + 1)}{365^r}.$$

Table 1 gives the values of $\Pr(E) = 1 - \Pr(E')$ for various values of r. You may be surprised by the numbers in the table. Even with as few as 23 people, it is more likely than not that at least two people have the same birthday. With a sample of 50 people, we are almost certain to have two with the same birthday. (Try this experiment in your class.)

Table 1 Probability That, in a Randomly Selected Group of *r* People, at Least Two People Will Have the Same Birthday

r	5	10	15	20	22	23	25	30	40	50
$\Pr(E)$	.027	.117	.253	.411	.476	.507	.569	.706	.891	.970

Health Statistics

In health statistics, the likelihood that something will happen is typically called the *risk* rather than the *probability* or *chance*.

Suppose that a study produced the following results: Out of a group of 1000 people who ate 1.75 ounces of chocolate per week, 39 of them had a stroke during a certain time period. In a control group of 1000 who did not eat chocolate regularly, 50 of them had a stroke during the same time period. There are three ways to report the benefit of eating chocolate.

The method of *absolute risk reduction* would report that

$$\text{risk reduction} = \frac{50 - 39}{1000} = \frac{11}{1000} = 1.1\%.$$

The method of *relative risk reduction* would report that

$$\text{risk reduction} = \frac{50 - 39}{50} = \frac{11}{50} = 22\%.$$

The *number needed to treat* method would report the number of people who must participate in the treatment in order to prevent one stroke. Eleven people out of 1000 were helped; that is, 1 out of $\frac{1000}{11} \approx 91$. *Note:* Since $\frac{1000}{11} = \frac{1}{\text{absolute risk reduction}}$, the number can be calculated as $\frac{1}{.011} \approx 91$. Therefore,

$$\text{number needed to treat} = 91.$$

The researcher is most likely to report that eating chocolate reduced the risk of stroke by 22% since that number is more impressive than the other two numbers. Drug advertisements on television usually report the relative risk reduction.

Verification of the Complement Rule   If S is the sample space, then $\Pr(S) = 1$, $E \cup E' = S$, and $E \cap E' = \varnothing$. Therefore, by the inclusion–exclusion principle,

$$\Pr(S) = \Pr(E \cup E') = \Pr(E) + \Pr(E').$$

So we have

$$1 = \Pr(E) + \Pr(E') \quad \text{and} \quad \Pr(E) = 1 - \Pr(E'). \qquad \ll$$

Check Your Understanding 6.3

Solutions can be found following the section exercises.

1. **Children** A couple decides to have four children. What is the probability that among the children, there will be at least one boy and at least one girl?

2. **Roulette** Find the probability that all of the numbers are different in three spins of an American roulette wheel. [*Note:* An American roulette wheel has 38 numbers.]

EXERCISES 6.3

1. A number is chosen at random from the whole numbers between 1 and 17, inclusive.
   (a) What is the probability that the number is odd?
   (b) What is the probability that the number is even?
   (c) What is the probability that the number is a multiple of 3?
   (d) What is the probability that the number is odd or a multiple of 3?

2. A number is chosen at random from the whole numbers between 1 and 100, inclusive.
   (a) What is the probability that the number ends in a zero?
   (b) What is the probability that the number is odd?
   (c) What is the probability that the number is odd or ends in a zero?

3. **Balls in an Urn** An urn contains five red balls and six white balls. A sample of two balls is selected at random from the urn. Find the probability that
   (a) only red balls are selected.
   (b) at least one white ball is selected.

4. **Balls in an Urn** An urn contains seven green balls and five white balls. A sample of three balls is selected at random from the urn. Find the probability that
   (a) only green balls are selected.
   (b) at least one white ball is selected.

5. **Balls in an Urn** An urn contains six green balls and seven white balls. A sample of four balls is selected at random from the urn. Find the probability that
   (a) the four balls have the same color.
   (b) the sample contains more green balls than white balls.

6. **Balls in an Urn** An urn contains eight red balls and six white balls. A sample of three balls is selected at random from the urn. Find the probability that
   (a) the three balls have the same color.
   (b) the sample contains more white balls than red balls.

7. **Opinion Polling** Two out of the seven members of a school board feel that all high school students should be required to take a course in coding. A pollster selects three members of the board at random and asks them for their opinion on requiring a coding course. What is the probability that at least one of the members polled favors requiring the course?

8. **Opinion Polling** Of the 15 members on a Senate committee, 10 plan to vote "yes" and 5 plan to vote "no" on an important issue. A reporter attempts to predict the outcome of the vote by questioning six of the senators. Find the probability that this sample is precisely representative of the final vote. That is, find the probability that four of the six senators questioned plan to vote "yes."

9. **Committee Selection** In the 114th United States Congress, the House Committee on Rules consisted of nine Republicans and four Democrats. If a subcommittee of three committee members was selected at random, what is the probability that at least one Democrat was on the committee?

10. **Committee Selection** The U.S. Senate consists of two senators from each of the 50 states. Five senators are to be selected at random to form a committee. What is the probability that no two members of the committee are from the same state?

11. **Quality Control** A factory produces LCD panels, which are packaged in boxes of 10. Three panels are selected at random from each box for inspection. The box is rejected if at least one of these three panels is defective. What is the probability that a box containing six defective panels will be rejected?

12. **Rotten Tomato** A bag contains nine tomatoes, of which one is rotten. A sample of three tomatoes is selected at random. What is the probability that the sample contains the rotten tomato?

Selecting Students Exercises 13–16 refer to a classroom of children (12 boys and 10 girls) in which seven students are chosen to go to the blackboard.

13. What is the probability that no boys are chosen?

14. What is the probability that the first three children chosen are boys?

15. What is the probability that at least two girls are chosen?

16. What is the probability that at least three boys are chosen?

17. **Birthday** Three people are chosen at random. What is the probability that at least two of them were born on the same day of the week?

18. **Birthday** Four people are chosen at random. What is the probability that at least two of them were born in the same month? Assume that each month is as likely as any other.

19. **Date Conflict** Without consultation with each other, each of four organizations announces a one-day convention to be held during June. Find the probability that at least two organizations specify the same day for their convention.

20. **Presidential Choices** There were 16 presidents of the Continental Congress from 1774 to 1788. Each of the five students in a seminar in American history chooses one of these presidents on whom to do a report. If all presidents are equally likely to be chosen, calculate the probability that at least two students choose the same president.

21. **Name Badges** Eight workers need an employee number to be printed on a name badge. The company's security office uses a random-number generator to assign each employee a number between 81 and 100, inclusive. What is the probability that at least two workers receive the same number?

22. **Random Selection** Each person in a group of 10 people randomly selects a number from 1 to 100, inclusive. What is the probability that at least two people select the same number?

23. **Birthday Problem** What is the probability that, in a group of 25 people, at least one person has a birthday on June 13? Why is your answer different from the probability displayed in Table 1 for $r = 25$?

24. **Birthday Problem** Johnny Carson, host of *The Tonight Show* from 1962–1992, discussed the birthday problem during one of his monologues. To test the hypothesis, he asked the audience whether any of them were born on his birthday, October 23. Carson was surprised that no one in the audience of 100 people shared his birthday. What was wrong with Carson's reasoning? What is the probability that one or more people in the audience were born on October 23?

25. **Dice** A die is rolled twice. What is the probability that the two numbers are different?

26. **Dice** A die is rolled three times. What is the probability of obtaining three different numbers?

27. **Dice** A die is rolled four times. What is the probability of obtaining only even numbers?

28. **Dice** A die is rolled three times. What is the probability that the number 1 does not appear?

29. **Coin Tosses** A coin is tossed 10 times. What is the probability of obtaining four heads and six tails?

30. **Coin Tosses** A coin is tossed seven times. What is the probability of obtaining five heads and two tails?

31. **Orientation Teams** A university admissions office randomly assigns each student from a group of four incoming freshmen to an orientation team. If there are seven orientation teams, what is the probability that two or more of these students are assigned to the same team?

32. **Elevator** An elevator has six buttons: *L, 1, 2, 3, 4,* and *5.* Suppose that five people get on the elevator at the Lobby. What is the probability that they are each going to a different floor? Assume that each of the floors 1 through 5 is equally likely.

33. **Street Routs** Figure 1 shows a partial map of the streets in New York City. (Such maps are discussed in Chapter 5.) A tourist starts at point *A* and selects at random a shortest path to point *B*. That is, they walk only south and east. Find the probability that
    (a) they pass through point *C*.
    (b) they pass through point *D*.
    (c) they pass through point *C* and point *D*.
    (d) they pass through point *C* or point *D*.

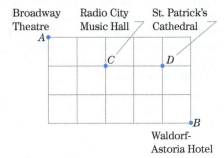

Figure 1

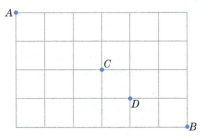

Figure 2

34. **Street Routes** Repeat Exercise 33 for Fig. 2.

35. **Baseball Predictions** In the American League, the East, Central, and West divisions each consists of five teams. A sportswriter predicts the winner of each of the three divisions by choosing a team completely at random in each division. What is the probability that the sportswriter will predict at least one winner correctly?

36. **Baseball Predictions** Suppose that the sportswriter in Exercise 35 eliminates from each division one team that clearly has no chance of winning and predicts a winner at random from the remaining teams. Assuming that the eliminated teams don't end up surprising anyone, what is the writer's chance of predicting at least one winner?

37. **Baseball Predictions** Suppose that the sportswriter in Exercise 35 simply puts the 15 team names in a hat and draws 3 completely at random. Does this increase or decrease the writer's chance of picking at least one winner?

38. **Baseball Predictions** Suppose that the sportswriter in Exercise 36 simply puts the 12 team names in a hat and draws 3 completely at random. Does this increase or decrease the writer's chance of picking at least one winner?

39. **Place Settings** Fred has five place settings consisting of a dinner plate, a salad plate, and a bowl. Each setting is a different color. If Fred randomly selects a dinner plate, a salad plate, and a bowl, what is the probability that they will all have different colors?

40. **Track Positions** Michael and Christopher are among seven contestants in a race to be run on a seven-lane track. If the runners are assigned to the lanes at random, what is the probability that Michael will be assigned to the inside lane and Christopher will be assigned to the outside lane?

41. **Group Picture** A man, a woman, and their three children randomly stand in a row for a family picture. What is the probability that the parents will be standing next to each other?

42. **Letter Positions** What is the probability that a random arrangement of the letters in the word GEESE has all the E's adjacent to one another?

Poker A poker hand consists of five cards drawn from a deck of 52 cards. Each card has one of 13 ranks (2, 3, 4, . . . , 10, jack, queen, king, ace) and one of four suits (spades, hearts, diamonds, clubs). In Exercises 43–46, determine the probability of the specified type of poker hand.

43. Full house (three cards of one rank and two cards of another rank)

44. Three of a kind (three cards of one rank and two cards of distinct rank, both different from the rank of the triple)

45. Two pairs (two cards of one rank, two cards of a different rank, and one card of a rank other than those two ranks)

46. One pair (two cards of one rank and three cards of distinct ranks, where each of the three cards has a different rank from the rank of the pair)

47. **Bridge** A bridge hand consists of thirteen cards drawn from a deck of 52 cards. Each card has one of 13 ranks (2, 3, 4, . . . , 10, jack, queen, king, ace) and one of four suits (spades, hearts, diamonds, clubs). What is the probability of each of the following suit distributions in a bridge hand?
    (a) 4-3-3-3. That is, four cards of one suit and three cards of each of the other three suits. *Hint:* The number of hands having this distribution is $4 \cdot C(13, 4) \cdot C(13, 3) \cdot C(13, 3) \cdot C(13, 3)$.
    (b) 4-4-3-2. That is, four cards of each of two suits, three cards of another suit, and two cards of the remaining suit.

48. **Powerball Lottery** The winner of the Powerball lottery must correctly pick a set of 5 numbers from 1 through 69 and then correctly pick one number (called the Powerball) from 1 through 26.
    (a) What is the probability of winning the Powerball lottery?
    (b) What are the odds of winning the Powerball lottery?

Illinois Lotto Exercises 49 and 50 refer to the Illinois Lottery Lotto game. (The data for these exercises were taken from Allan J. Gottlieb's "Puzzle Corner" in *Technology Review*, February/March 1985.) In this game, the player chooses six different integers from 1 to 40. If the six match (in any order) the six different integers drawn by the lottery, the player wins the grand prize jackpot, which starts at $1 million and grows weekly until won. Multiple winners split the pot equally. For each $1 bet, the player must pick two (presumably different) sets of six integers.

49. What is the probability of winning the Illinois Lottery Lotto with a $1 bet?

50. In the game week ending June 18, 1983, a total of 2 million people bought $1 tickets, and 78 people matched all six winning integers and split the jackpot. If all numbers were selected randomly, the likelihood of having so many joint winners would be about 10^{-115}. Can you think of any reason that such an unlikely event would have occurred? (*Note:* The winning numbers were 7, 13, 14, 21, 28, and 35.) What would be the best strategy in selecting the numbers to ensure that in the event that you won, you would probably not have to share the jackpot with too many people?

51. **California Lottery** In the California Fantasy 5 lottery, a player pays $1 for a ticket and selects 5 numbers from the numbers 1 through 39. If they match exactly three of the five numbers drawn, they receive $15. What is the probability of selecting exactly three of the five numbers drawn?

52. **Florida Lottery** The winning combination in the Florida lottery consists of six numbers drawn from the numbers 1 through 53. What is the probability that a single ticket has no matches?

53. **Health Statistics** Suppose that a study produced the following results: Out of a group of 80 people who took Sleep Helper before going to bed, only 20 of them had difficulty falling asleep. Out of a group of 80 people who took a placebo before going to bed, 25 of them had difficulty falling asleep. Calculate the absolute and the relative risk reduction due to taking Sleep Helper. Also, calculate the number of people who must take Sleep Helper in order for one person to be helped.

54. **Health Statistics** Suppose that a study produced the following results: Out of a group of 60 people who took Math Helper before their math exam, only 8 of them failed the exam. Out of a group of 60 people who took a placebo before the exam, 14 of them failed the exam. Calculate the absolute and relative risk reduction due to taking Math Helper. Also, calculate the number of people who must take Math Helper in order for one person to be helped.

55. **Health Statistics** Table 2 shows the experiences of 100 people who took a medication designed to prevent a certain condition. Calculate the absolute and the relative risk reduction due to taking the medication. Also, calculate the number of people who must take the medication in order for one person to be helped.

Table 2

	Took Medication	Took Placebo
Developed condition	15	20
Did not develop condition	85	80
Total	100	100

56. **Health Statistics** Table 3 shows the experiences of 200 people who took a medication designed to prevent a certain condition. Calculate the absolute and the relative risk reduction

Table 3

	Took Medication	Took Placebo
Developed condition	40	50
Did not develop condition	160	150
Total	200	200

due to taking the medication. Also, calculate the number of people who must take the medication in order for one person to be helped.

57. **Health Statistics** Suppose that a certain drug intervention has a relative risk reduction of 25% and 12 people taking the drug developed the condition. How many people in the control group developed the condition?

58. **Health Statistics** Suppose that 80 people must take a certain medication in order for one person to be helped. What is the absolute risk reduction of the intervention?

59. **License Plate Game** Johnny and Doyle are driving on a lightly traveled road. Johnny proposes the following game: They will look at the license plates of oncoming cars and focus on the last two digits. For instance, the license plates ABC512 and 7412BG would yield 12, and the license plate XY406T would yield the number 6. Johnny bets Doyle that at least 2 of the next 15 cars will yield the same number. What is the probability that Johnny wins? Assume that each of the 100 possible numbers is equally likely to occur.

60. **Parking Spaces** Ten people randomly park their cars in a row of fourteen parking spaces. Then Mr. Jones arrives in a wide camper that requires two adjacent parking spaces. What is the probability that Mr. Jones will be able to park his camper? *Hint*: Use the complement rule and the counting principle from Example 4 of Section 5.6.

TECHNOLOGY EXERCISES

61. **Pick a Card** Find the probability that at least two people in a group of size $n = 5$ select the same card when drawing from a 52-card deck with replacement. Determine the group size n for which the probability of such a match first exceeds 50%.

62. **Term Papers** A political science class has 20 students, each of whom chooses a topic from a list for a term paper. How big a pool of topics is necessary for the probability of at least one duplicate to drop below 50%?

63. **Birthday Problem** Refer to Exercise 24. How large would the audience have to have been so that the probability that someone in the audience had the same birthday as Carson was at least 50%?

64. **Birthday Problem** A year on planet Ork has 100 days. Find the smallest number of Orkians for which the probability that at least two of them have the same birthday is 50% or more.

65. **Lottery** In many state lotteries, six numbers are selected from a set of numbers. Quite often, the winning selection contains two consecutive numbers. When six numbers are selected from the numbers 1 through n, the probability that there will be two consecutive integers in the selection is .5771 when n is 40 and the probability is .4209 when n is 60. Determine the largest value of n for which the probability of having two consecutive numbers is greater than .5. *Note:* There are $C(n - k + 1, k)$ ways that k numbers selected from the numbers 1 through n will have no two consecutive numbers.

66. **The de Méré Problem** How many times must you roll two dice so that the probability of rolling a pair of 6s at least once is greater than or equal to .5? *Note:* This question was posed in 1654 by the French writer and gambler Chevalier de Méré.

Solutions to Check Your Understanding 6.3

1. Each possible outcome is a string of four letters composed of Bs and Gs. By the generalized multiplication principle, there are 2^4, or 16, possible outcomes. Let E be the event "children of both genders." Then $E' = \{BBBB, GGGG\}$, and

$$\Pr(E') = \frac{n(E')}{N} = \frac{2}{16} = \frac{1}{8}.$$

Therefore,

$$\Pr(E) = 1 - \Pr(E') = 1 - \frac{1}{8} = \frac{7}{8}.$$

So the probability is 87.5% that they will have children of both genders.

2. Each sequence of three numbers is just as likely to occur as any other. Therefore,

$$\Pr(\text{numbers different})$$
$$= \frac{[\text{number of outcomes with numbers different}]}{[\text{number of possible outcomes}]}$$
$$= \frac{38 \cdot 37 \cdot 36}{38^3} \approx .92.$$

6.4 Conditional Probability and Independence

The probability of an event depends, often in a critical way, on the sample space in question. In this section, we explore this dependence in some detail by introducing what are called *conditional probabilities*.

To illustrate the dependence of probabilities on the sample space, consider the following example:

EXAMPLE 1 **College Students** Suppose that a certain mathematics class contains 26 students. Of these, 14 are economics majors, 15 are first-year students, and 7 are neither. Suppose that a person is selected at random from the class.
(a) What is the probability that the person is both an economics major and a first-year student?
(b) Suppose that we are given the additional information that the person selected is a first-year student. What is the probability that they are also an economics major?

SOLUTION Let E denote the set of economics majors and F the set of first-year students. A complete Venn diagram of the class can be obtained with the techniques of Section 5.3. See Fig. 1.

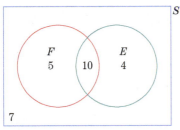

Figure 1

(a) In selecting a student from the class, the sample space consists of all 26 students. Since the choice is random, all students are equally likely to be selected. The event "economics major and first-year student" corresponds to the set $E \cap F$ of the Venn diagram. Therefore,

$$\Pr(E \cap F) = \frac{n(E \cap F)}{N} = \frac{10}{26} = \frac{5}{13}.$$

So the probability of selecting a first-year economics major is $\frac{5}{13}$.

(b) If we know that the student selected is a first-year student, then the possible outcomes of the experiment are restricted. They must belong to F. In other words, given the additional information, we must alter the sample space from "all students" to "first-year students." Since each of the 15 first-year students is equally likely to be selected, and since 10 of the 15 first-year students are economics majors, the probability of choosing an economics major under these circumstances is equal to $\frac{10}{15} = \frac{2}{3}$.   **«**

Let us consider Example 1 more carefully. In part (a), the sample space is the set of all students in the mathematics class, E is the event "student is an economics major," and F is the event "student is a first-year student." On the other hand, part (b) poses a condition, "student is a first-year student." The condition is satisfied by every element of F. We are being asked to find the **conditional probability** of E given F, written $\Pr(E \mid F)$; this is the probability of E, assuming that F has occurred. To do this, we shall restrict our attention to the new (restricted) sample space, now just the elements of F.

Thus, we consider only first-year students and ask, of these, what is the probability of choosing an economics major? We assign a value to $\Pr(E|F)$ via the following formula:

DEFINITION Conditional Probability

$$\Pr(E|F) = \frac{\Pr(E \cap F)}{\Pr(F)}, \quad \text{provided that } \Pr(F) \neq 0. \tag{1}$$

We will provide an intuitive justification of this formula shortly. However, we first give an application.

EXAMPLE 2   **Earnings and Education**   Twenty percent of the employees of Acme Steel Company are college graduates. Of all of its employees, 25% earn more than $50,000 per year and 15% are college graduates earning more than $50,000. What is the probability that an employee selected at random earns more than $50,000 per year, given that they are a college graduate?

SOLUTION   Let H and C be the events

$$H = \text{"earns more than \$50,000 per year"}$$

$$C = \text{"college graduate."}$$

We are asked to calculate $\Pr(H|C)$. The given data is

$$\Pr(H) = .25 \qquad \Pr(C) = .20 \qquad \Pr(H \cap C) = .15.$$

By formula (1), we have

$$\Pr(H|C) = \frac{\Pr(H \cap C)}{\Pr(C)} = \frac{.15}{.20} = \frac{3}{4}.$$

Thus $\frac{3}{4}$ of all college graduates at Acme Steel earn more than $50,000 per year.

» Now Try Exercise 15

Suppose that an experiment has N equally likely outcomes. Then we may apply the following formula to calculate $\Pr(E|F)$:

Conditional Probability in Case of Equally Likely Outcomes

$$\Pr(E|F) = \frac{n(E \cap F)}{n(F)}, \tag{2}$$

provided that $n(F) \neq 0$.

This formula was actually used in Example 1(b) to compute $\Pr(E|F)$. In that example, each student had the same likelihood of being selected.

Let us now justify formulas (1) and (2).

Justification of Formula (1)   Formula (1) is a definition of conditional probability and as such does not really need any justification. (We can make whatever definitions we choose!) However, let us proceed intuitively and show that the definition is reasonable, in the sense that formula (1) gives the expected long-run proportion of occurrences of E, given that F occurs. Assume that our experiment is performed repeatedly, say, for 10,000 trials. We would expect F to occur in approximately $10{,}000 \Pr(F)$ trials. Among these, the trials for which E also occurs are exactly those for which both E and F occur. In other words, the trials for which E also occurs are exactly those for which the event $E \cap F$ occurs; and this event has probability $\Pr(E \cap F)$. Therefore, out of the original

10,000 trials, there should be approximately $10{,}000\,\Pr(E \cap F)$ in which E and F both occur. Thus, considering only those trials in which F occurs, the proportion in which E also occurs is

$$\frac{10{,}000\,\Pr(E \cap F)}{10{,}000\,\Pr(F)} = \frac{\Pr(E \cap F)}{\Pr(F)}.$$

Thus, at least intuitively, it seems reasonable to define $\Pr(E \mid F)$ by formula (1).   «

Justification of Formula (2)   Suppose that the number of outcomes of the experiment is N. Then

$$\Pr(F) = \frac{n(F)}{N}$$

$$\Pr(E \cap F) = \frac{n(E \cap F)}{N}$$

Therefore, using formula (1), we have

$$\Pr(E \mid F) = \frac{\Pr(E \cap F)}{\Pr(F)}$$

$$= \frac{\dfrac{n(E \cap F)}{N}}{\dfrac{n(F)}{N}}$$

$$= \frac{n(E \cap F)}{n(F)}.$$

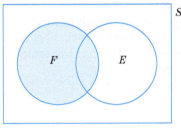

Figure 2

Alternatively, consider the Venn Diagram in Fig. 2. If we assume that the event F occurs, we are restricted to only the outcomes in the shaded region; that is, we consider F to be the sample space. Then, E occurs exactly when $E \cap F$ occurs, so the conditional probability is

$$\Pr(E \mid F) = \frac{n(E \cap F)}{n(F)}.$$ «

From formula (1), multiplying both sides of the equation by $\Pr(F)$, we can deduce the following useful fact:

> **Product Rule** If $\Pr(F) \neq 0$,
>
> $$\Pr(E \cap F) = \Pr(F) \cdot \Pr(E \mid F).$$

The next example illustrates the use of this rule.

EXAMPLE 3   **Color-Blind Males** Assume that a certain school contains an equal number of female and male students and that 5% of the male population is color blind. Find the probability that a randomly selected student is a color-blind male.

SOLUTION   Let $M =$ "male" and $B =$ "color blind." We wish to calculate $\Pr(B \cap M)$. From the given data,

$$\Pr(M) = .5 \quad \text{and} \quad \Pr(B \mid M) = .05.$$

Therefore, by the product rule,

$$\Pr(B \cap M) = \Pr(M) \cdot \Pr(B \mid M) = (.5)(.05) = .025.$$

» Now Try Exercise 29

Often, an event G can be described as a sequence of two other events E and F. That is, G occurs if F occurs and then E occurs. The product rule allows us to compute the probability of G as the probability of F times the conditional probability $\Pr(E\,|\,F)$. The next example illustrates this point.

EXAMPLE 4   **Cards** A sequence of two playing cards is drawn at random (without replacement) from a standard deck of 52 cards. What is the probability that the first card is red and the second is black?

SOLUTION   The event in question is a sequence of two events, namely,

$$F = \text{``the first card is red''}$$

$$E = \text{``the second card is black.''}$$

Since half of the deck consists of red cards, $\Pr(F) = 1/2$. If we are given that F occurs, then there are only 51 cards left in the deck, of which 26 are black, so

$$\Pr(E\,|\,F) = \tfrac{26}{51}.$$

By the product rule,

$$\Pr(E \cap F) = \Pr(F) \cdot \Pr(E\,|\,F) = \tfrac{1}{2} \cdot \tfrac{26}{51} = \tfrac{13}{51}.$$

»» Now Try Exercise 25

The product rule may be generalized to sequences of three events, E_1, E_2, and E_3.

$$\Pr(E_1 \cap E_2 \cap E_3) = \Pr(E_1) \cdot \Pr(E_2\,|\,E_1) \cdot \Pr(E_3\,|\,E_1 \cap E_2).$$

Similar formulas hold for sequences of four or more events.

One of the most important applications of conditional probability is in the discussion of independent events. Intuitively, two events are **independent** of each other if the occurrence of one has no effect on the likelihood that the other will occur. For example, suppose that we roll a die twice. Let the events E and F be

$$F = \text{``first roll is a 6''}$$

$$E = \text{``second roll is a 3.''}$$

Then intuitively, we conclude that these events are independent of one another. Rolling a 6 on the first roll has no effect whatsoever on the outcome of the second roll. On the other hand, suppose that we draw a sequence of two cards at random (without replacement) from a deck. Then we know that the events

$$F = \text{``first card is red''}$$

$$E = \text{``second card is black''}$$

are not independent of one another, at least intuitively. Indeed, whether or not we draw a red on the first card affects the likelihood of drawing a black on the second.

The notion of independence of events is easily formulated. If E and F are events in a sample space and $\Pr(F) \neq 0$, then the product rule states that $\Pr(E \cap F) = \Pr(E\,|\,F) \cdot \Pr(F)$. However, if the occurrence of event F does not affect the likelihood of the occurrence of event E, we would expect that $\Pr(E\,|\,F) = \Pr(E)$. Substitution then shows that $\Pr(E \cap F) = \Pr(E) \cdot \Pr(F)$.

DEFINITION Let E and F be events. We say that E and F are **independent**, provided that

$$\Pr(E \cap F) = \Pr(E) \cdot \Pr(F).$$

If $\Pr(E) \neq 0$ and $\Pr(F) \neq 0$, then our definition is equivalent to the intuitive statement of independence stated in terms of conditional probability. The two may be used interchangeably.

> **DEFINITION** Let E and F be events with nonzero probability. E and F are **independent**, provided that
>
> $$\Pr(E|F) = \Pr(E) \quad \text{and} \quad \Pr(F|E) = \Pr(F).$$

EXAMPLE 5 **Determining Whether Probabilities Are Independent** For each pair of probabilities, determine whether E and F are independent.
(a) An experiment consists of observing the outcome of two consecutive rolls of a die. Let E and F be the events

$$E = \text{“the first roll is a 3”}$$
$$F = \text{“the second roll is a 6”}$$

(b) An experiment consists of observing the results of drawing two consecutive cards from a 52-card deck. Let E and F be the events

$$E = \text{“the second card is black”}$$
$$F = \text{“the first card is red”}$$

SOLUTION (a) Clearly, $\Pr(E) = \Pr(F) = \frac{1}{6}$. To compute $\Pr(E|F)$, assume that F occurs. Then there are six possible outcomes:

$$F = \{(1, 6), (2, 6), (3, 6), (4, 6), (5, 6), (6, 6)\},$$

and all outcomes are equally likely. Moreover.

$$E \cap F = \{(3, 6)\},$$

so that

$$\Pr(E|F) = \frac{n(E \cap F)}{n(F)} = \frac{1}{6} = \Pr(E).$$

Similarly, $\Pr(E|F) = \Pr(F)$. So E and F are independent events, in agreement with our intuition.

(b) There are the same number of outcomes with the second card red as with the second card black, so $\Pr(E) = \frac{1}{2}$. To compute $\Pr(E|F)$, note that if F occurs, then there are 51 equally likely choices for the second card, of which 26 are black, so that $\Pr(E|F) = \frac{26}{51}$. Note that $\Pr(E|F) \neq \Pr(E)$, so E and F are not independent, in agreement with our intuition. **» Now Try Exercise 45**

EXAMPLE 6 **Determining Whether Probabilities Are Independent** For each pair of probabilities, determine whether E and F are independent.
(a) We toss a coin three times and record the sequence of heads and tails. Let E be the event "at most one head occurs" and F the event "both heads and tails occur."
(b) A family has four children. Let E be the event "at most one boy" and F the event "at least one child of each gender."

SOLUTION (a) Using the abbreviations H for "heads" and T for "tails," we have

$$E = \{TTT, HTT, THT, TTH\}$$
$$F = \{HTT, HTH, HHT, THH, THT, TTH\}$$
$$E \cap F = \{HTT, THT, TTH\}.$$

The sample space contains eight equally likely outcomes so that

$$\Pr(E) = \tfrac{1}{2} \qquad \Pr(F) = \tfrac{3}{4} \qquad \Pr(E \cap F) = \tfrac{3}{8}.$$

Moreover,

$$\Pr(E) \cdot \Pr(F) = \tfrac{1}{2} \cdot \tfrac{3}{4} = \tfrac{3}{8},$$

which equals $\Pr(E \cap F)$. So E and F are independent.

(b) Let B stand for "boy" and G for "girl." Then

$$E = \{\text{GGGG, GGGB, GGBG, GBGG, BGGG}\}$$
$$F = \{\text{GGGB, GGBG, GBGG, BGGG, BBBG, BBGB, BGBB,}$$
$$\text{GBBB, BBGG, BGBG, BGGB, GBBG, GBGB, GGBB}\},$$

and the sample space consists of 16 equally likely outcomes. Furthermore,

$$E \cap F = \{\text{GGGB, GGBG, GBGG, BGGG}\}.$$

Therefore,

$$\Pr(E) = \tfrac{5}{16} \qquad \Pr(F) = \tfrac{7}{8} \qquad \Pr(E \cap F) = \tfrac{1}{4}.$$

In this example,

$$\Pr(E) \cdot \Pr(F) = \tfrac{5}{16} \cdot \tfrac{7}{8} \neq \Pr(E \cap F).$$

So E and F are *not* independent events.　　**» Now Try Exercise 43**

Examples 6(a) and 6(b) are similar, yet the events they describe are independent in one case and not the other. Although intuition is frequently a big help, in complex problems we shall need to use the definition of independence to verify that our intuition is correct.

EXAMPLE 7　**Reliability of a Calculator** A new calculator is designed to be extra reliable by having two independent calculating units. The probability that a given calculating unit fails within the first 1000 hours of operation is .001. What is the probability that at least one calculating unit will operate without failure for the first 1000 hours of operation?

SOLUTION　Let

$$E = \text{"calculating unit 1 fails in first 1000 hours"}$$
$$F = \text{"calculating unit 2 fails in first 1000 hours."}$$

Then E and F are independent events, since the calculating units are independent of one another. Therefore,

$$\Pr(E \cap F) = \Pr(E) \cdot \Pr(F) = (.001)^2 = .000001$$
$$\Pr[(E \cap F)'] = 1 - .000001 = .999999.$$

Since $(E \cap F)' = $ "not both calculating units fail in first 1000 hours," the desired probability is .999999.　　**«**

The concept of independent events can be extended to more than two events:

DEFINITION A set of events is said to be **independent** if, for each collection of events chosen from them, say, $E_1, E_2, \ldots, E_n$, we have

$$\Pr(E_1 \cap E_2 \cap \cdots \cap E_n) = \Pr(E_1) \cdot \Pr(E_2) \cdot \cdots \cdot \Pr(E_n).$$

EXAMPLE 8 **Probabilities Associated with Independent Events** Three events A, B, and C are independent: $\Pr(A) = .5$, $\Pr(B) = .3$, and $\Pr(C) = .2$.
(a) Calculate $\Pr(A \cap B \cap C)$. (b) Calculate $\Pr(A \cap C)$.

SOLUTION (a) $\Pr(A \cap B \cap C) = \Pr(A) \cdot \Pr(B) \cdot \Pr(C) = (.5)(.3)(.2) = .03$.
(b) $\Pr(A \cap C) = \Pr(A) \cdot \Pr(C) = (.5)(.2) = .1$.   **» Now Try Exercise 41**

We shall leave as an exercise the intuitively reasonable result that if E and F are independent events, so are E and F', E' and F, and E' and F'. This result also generalizes to any collection of independent events.

EXAMPLE 9 **Quality Control** A company manufactures stereo components. Experience shows that defects in manufacture are independent of one another. Quality-control studies reveal that

2% of CD players are defective,
3% of amplifiers are defective,
7% of speakers are defective.

A system consists of a CD player, an amplifier, and two speakers. What is the probability that the system is not defective?

SOLUTION Let C, A, S_1, and S_2 be events corresponding to defective CD player, amplifier, speaker 1, and speaker 2, respectively. Then

$$\Pr(C) = .02 \qquad \Pr(A) = .03 \qquad \Pr(S_1) = \Pr(S_2) = .07.$$

We wish to calculate $\Pr(C' \cap A' \cap S_1' \cap S_2')$. By the complement rule, we have

$$\Pr(C') = .98 \qquad \Pr(A') = .97 \qquad \Pr(S_1') = \Pr(S_2') = .93.$$

Since we have assumed that C, A, S_1, and S_2 are independent, so are C', A', S_1', and S_2'. Therefore,

$$\Pr(C' \cap A' \cap S_1' \cap S_2') = \Pr(C') \cdot \Pr(A') \cdot \Pr(S_1') \cdot \Pr(S_2')$$
$$= (.98)(.97)(.93)^2 \approx .822.$$

Thus, there is an 82.2% chance that the system is not defective.

» Now Try Exercise 53

Check Your Understanding 6.4

Solutions can be found following the section exercises.

1. **Cards** Suppose that there are three cards: one red on both sides; one white on both sides; and one having a side of each color. A card is selected at random and placed on a table. If the up side is red, what is the probability that the down side is red? (Try guessing at the answer before working it by using the formula for conditional probability.)

2. Show that if events E and F are independent of each other, then so are E and F'. [*Hint:* Since $E \cap F$ and $E \cap F'$ are mutually exclusive, we have

$$\Pr(E) = \Pr(E \cap F) + \Pr(E \cap F').]$$

EXERCISES 6.4

1. The Venn diagram in Fig. 3 shows the probabilities for its four basic regions. Find
   (a) $\Pr(E)$ (b) $\Pr(F)$ (c) $\Pr(E|F)$ (d) $\Pr(F|E)$.

2. The Venn diagram in Fig. 4 shows the probabilities for its four basic regions. Find
   (a) $\Pr(E)$ (b) $\Pr(F)$ (c) $\Pr(E|F)$ (d) $\Pr(F|E)$.

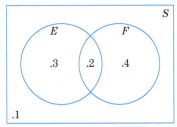

Figure 3

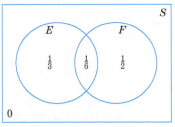

Figure 4

3. Let S be a sample space and E and F be events associated with S. Suppose that $\Pr(E) = .5, \Pr(F) = .4$, and $\Pr(E \cap F) = .1$. Calculate
   (a) $\Pr(E|F)$ (b) $\Pr(F|E)$
   (c) $\Pr(E|F')$ (d) $\Pr(E'|F')$.

4. Let S be a sample space and E and F be events associated with S. Suppose that $\Pr(E) = .6, \Pr(F) = .3$, and $\Pr(E \cap F) = .2$. Calculate
   (a) $\Pr(E|F)$ (b) $\Pr(F|E)$
   (c) $\Pr(E|F')$ (d) $\Pr(E'|F')$.

5. Let S be a sample space and E and F be events associated with S. Suppose that $\Pr(E) = \frac{1}{3}, \Pr(F) = \frac{5}{12}$, and $\Pr(E \cup F) = \frac{2}{3}$. Calculate
   (a) $\Pr(E \cap F)$ (b) $\Pr(E|F)$ (c) $\Pr(F|E)$.

6. Let S be a sample space and E and F be events associated with S. Suppose that $\Pr(E) = \frac{1}{2}, \Pr(F) = \frac{1}{3}$, and $\Pr(E \cup F) = \frac{7}{12}$. Calculate
   (a) $\Pr(E \cap F)$ (b) $\Pr(E|F)$ (c) $\Pr(F|E)$.

7. Let S be a sample space and E and F be events associated with S. Suppose that $\Pr(E) = .4, \Pr(F|E) = .25$, and $\Pr(F) = .3$. Calculate
   (a) $\Pr(E \cap F)$ (b) $\Pr(E \cup F)$
   (c) $\Pr(E|F)$ (d) $\Pr(E' \cap F)$.

8. Let S be a sample space and E and F be events associated with S. Suppose that $\Pr(E) = .5, \Pr(F|E) = .4$, and $\Pr(F) = .3$. Calculate
   (a) $\Pr(E \cap F)$ (b) $\Pr(E \cup F)$
   (c) $\Pr(E|F)$ (d) $\Pr(E \cap F')$.

9. **Dice** When a pair of dice is rolled, what is the probability that the sum of the dice is 8, given that the outcome is not 7?

10. **Dice** When a pair of dice is rolled, what is the probability that the sum of the dice is 5, given that exactly one of the dice shows a 3?

11. **Coins** A coin is tossed three times. What is the probability that the outcome is three heads, given that exactly one of the coins shows a head?

12. **Coins** A coin is tossed three times. What is the probability that the outcome contains no heads, given that exactly one of the coins shows a tail?

13. **Bag of Marbles** A bag contains five red marbles and seven white marbles. If a sample of four marbles contains at least one white marble, what is the probability that all the marbles in the sample are white?

14. **Balls in an Urn** Two balls are selected at random from an urn containing two white balls and three red balls. What is the conditional probability that both balls are white, given that at least one of them is white?

15. **Children** Suppose a family has two children and the youngest is a girl. What is the probability that both children are girls?

16. **Children** Suppose a family has two children and at least one is a girl. What is the probability that both children are girls?

17. **Value of College** Twenty-five percent of individuals in a certain city earn more than $45,000 per year. The percentage of individuals earning more than $45,000 and having a college degree is 10. Suppose that a person is randomly chosen and turns out to be earning more than $45,000. What is the probability that they are a college graduate?

18. **Advanced Degrees** Sixty percent of the teachers at a certain high school are female. Forty percent of the teachers are females with a master's degree. What is the probability that a randomly selected teacher has a master's degree, given that the teacher is female?

19. **Advanced Degrees** Table 1 shows the projected number of advanced degrees (in thousands) earned in the United States in 2016 by gender and type of degree. Find the probability that a person selected at random who received an advanced degree
    (a) received a master's degree.
    (b) is male.
    (c) is female, given that the person received a master's degree.
    (d) received a doctor's degree, given that the person is female.

Table 1

	Bachelor's	Master's	Doctor's	Total
Male	779	329	93	1201
Female	1082	522	93	1697
Total	1861	851	186	2898

Source: www.nces.ed.gov

20. **Voting** Table 2 shows the number of registered voting-age U.S. citizens (in millions) by gender and their reported participation in the 2014 congressional election. Find the probability that a voting-age citizen selected at random
    (a) voted.
    (b) is male.
    (c) is female, given that the citizen voted.
    (d) voted, given that the citizen is male.

Table 2

	Voted	Did Not Vote	Total
Male	43.0	45.5	88.5
Female	49.2	47.6	96.8
Total	92.2	93.1	185.3

Source: www.nces.ed.gov

21. **Military Personnel** Table 3 shows the numbers (in thousands) of officers and enlisted persons on active military duty on December 31, 2015. Find the probability that a person in the military selected at random is
    (a) an officer.
    (b) a Marine.
    (c) an officer in the Marines.
    (d) an officer, given that they are a Marine.
    (e) a Marine, given that they are an officer.

Table 3

	Army	Navy	Marine Corps	Air Force	Total
Officer	93.6	54.0	20.7	60.3	228.6
Enlisted	384.3	269.7	162.5	246.7	1063.2
Total	477.9	323.7	183.2	307.0	1291.8

Source: www.dmdc.osd.mil

22. **College Majors** Table 4 shows the probable field of study for 1500 freshman males and 1000 freshman females. Find the probability that a freshman selected at random
    (a) intends to major in business.
    (b) is female.
    (c) is a female intending to major in business.
    (d) is male, given that the freshman intends to major in social science.
    (e) intends to major in social science, given that the freshman is female.

Table 4

	Business	Social Science	Other	Total
Male	260	122	1118	1500
Female	102	130	768	1000
Total	362	252	1886	2500

Source: www.heri.ucla.edu

23. **Bills in Envelopes** Each of three sealed opaque envelopes contains two bills. One envelope contains two $1 bills, another contains two $5 bills, and the third contains a $1 bill and a $5 bill. An envelope is selected at random, and a bill is taken from the envelope at random. If it is a $5 bill, what is the probability that the other bill in the envelope is also a $5 bill?

24. **Gold and Silver Coins** Consider three boxes. One box contains two gold coins, one box contains two silver coins, and one box contains a gold coin and a silver coin. Suppose that you select a box at random and then select a coin at random from that box. If the coin is gold, what is the probability that the other coin in the box is gold?

25. **Cards** A sequence of two playing cards is drawn at random (without replacement) from a standard deck of 52 cards. What is the probability that both cards are kings?

26. **Cards** A sequence of two playing cards is drawn at random (without replacement) from a standard deck of 52 cards. What is the probability that both cards are diamonds?

27. **Coin Tosses** A coin is tossed five times. What is the probability that heads appears on every toss, given that heads appears on the first four tosses?

28. **Coin Tosses** A coin is tossed twice. What is the probability that heads appeared on the first toss, given that tails appeared on the second toss?

29. **Exit Polling** According to exit polling for the 2016 Missouri Republican primary election, 48% of the primary voters were women. Nine percent of the women polled voted for John Kasich. What is the probability that a randomly selected voter from the poll is a woman who voted for John Kasich?

30. **Population** Twenty percent of the world's population in China. The residents of Shanghai constitute China's population. If a person is selected few the entire world, what is the probabil (prob-ability of Shanghai?

31. **Basketball** Suppose that you and you have the ball on y seconds left in the game. Y ability of success is .48) o

success is .29). Which choice gives your team the greater probability of winning the game? Assume that your shot will be taken just before the buzzer sounds and that each team has the same chance of winning in overtime.

32. **Password** Fred remembers all but the last character of the password for his e-mail account. However, he knows that it is a digit. He tries to log into his account by guessing the digit. What is the probability that Fred will be successful within two attempts?

33. Let E and F be events with $P(E) = .4$, $\Pr(F) = .5$, and $\Pr(E \cup F) = .7$. Are E and F independent events?

34. Let E and F be events with $P(E) = .2$, $\Pr(F) = .5$, and $\Pr(E \cup F) = .6$. Are E and F independent events?

35. Let E and F be independent events with $P(E) = .5$ and $\Pr(F) = .6$. Find $\Pr(E \cup F)$.

36. Let E and F be independent events with $P(E) = .25$ and $\Pr(F) = .4$. Find $\Pr(E \cup F)$.

In Exercises 37–40, assume that E and F are independent events. Use the given information to find $\Pr(F)$.

37. $\Pr(E) = .7$ and $\Pr(F \mid E) = .6$.

38. $\Pr(E) = .4$ and $\Pr(F' \mid E') = .3$.

39. $\Pr(E') = .6$ and $\Pr(E \cap F) = .1$.

40. $\Pr(E) = .8$ and $\Pr(E \cap F) = .4$.

41. Let A, B, and C be independent events with $\Pr(A) = .4$, $\Pr(B) = .1$, and $\Pr(C) = .2$. Calculate $\Pr[(A \cap B \cap C)']$.

42. Let A, B, and C be independent events with $\Pr(A) = .2$, $\Pr(A \cap B) = .12$, and $\Pr(A \cap C) = .06$. Calculate $\Pr(B \cap C)$.

43. **Balls in an Urn** A sample of two balls is drawn from an urn containing two white balls and three red balls. Are the events "the sample contains at least one white" and "the sample contains balls of both colors" independent?

44. **Balls in an Urn** An urn contains white balls and three red balls. A ball is withdrawn (without being replaced), and then a second ball is drawn. Are the events "the first ball is red" and "the second ball is red" independent?

45. **Roll a Die** Roll a die. Consider the following two events: $E = \{2, 4, 6\}$. Are the events E and F independent?

46. **Roll a Die** Roll a die, and consider the following two events: $E = \{3, 4, 6\}$. Are the events E and F independent?

47. **Rolling Dice** Roll a pair of dice, and consider the sum of the two numbers. Are the events "the sum is an odd number" and "the sum is 5, or 7" independent?

Epidemiology A doctor studies the known cancer patients in a certain town. The probability that a randomly chosen resident has cancer is found to be .001. It is found that 30% of the town works for Ajax Chemical Company. The probability that an employee of Ajax has cancer is equal to .003. Are the events "has cancer" and "works for Ajax" independent of one another?

22. **College Majors** Table 4 shows the probable field of study for 1500 freshman males and 1000 freshman females. Find the probability that a freshman selected at random
    (a) intends to major in business.
    (b) is female.
    (c) is a female intending to major in business.
    (d) is male, given that the freshman intends to major in social science.
    (e) intends to major in social science, given that the freshman is female.

Table 4

	Business	Social Science	Other	Total
Male	260	122	1118	1500
Female	102	130	768	1000
Total	362	252	1886	2500

Source: www.heri.ucla.edu

23. **Bills in Envelopes** Each of three sealed opaque envelopes contains two bills. One envelope contains two $1 bills, another contains two $5 bills, and the third contains a $1 bill and a $5 bill. An envelope is selected at random, and a bill is taken from the envelope at random. If it is a $5 bill, what is the probability that the other bill in the envelope is also a $5 bill?

24. **Gold and Silver Coins** Consider three boxes. One box contains two gold coins, one box contains two silver coins, and one box contains a gold coin and a silver coin. Suppose that you select a box at random and then select a coin at random from that box. If the coin is gold, what is the probability that the other coin in the box is gold?

25. **Cards** A sequence of two playing cards is drawn at random (without replacement) from a standard deck of 52 cards. What is the probability that both cards are kings?

26. **Cards** A sequence of two playing cards is drawn at random (without replacement) from a standard deck of 52 cards. What is the probability that both cards are diamonds?

27. **Coin Tosses** A coin is tossed five times. What is the probability that heads appears on every toss, given that heads appears on the first four tosses?

28. **Coin Tosses** A coin is tossed twice. What is the probability that heads appeared on the first toss, given that tails appeared on the second toss?

29. **Exit Polling** According to exit polling for the 2016 Missouri Republican primary election, 48% of the primary voters were women. Nine percent of the women polled voted for John Kasich. What is the probability that a randomly selected voter from the poll is a woman who voted for John Kasich?

30. **Population** Twenty percent of the world's population live in China. The residents of Shanghai constitute 1.6% of China's population. If a person is selected at random from the entire world, what is the probability that they live in Shanghai?

31. **Basketball** Suppose that your team is behind by two points and you have the ball on your opponent's court with a few seconds left in the game. You can try a two-point shot (probability of success is .48) or a three-point shot (probability of success is .29). Which choice gives your team the greater probability of winning the game? Assume that your shot will be taken just before the buzzer sounds and that each team has the same chance of winning in overtime.

32. **Password** Fred remembers all but the last character of the password for his e-mail account. However, he knows that it is a digit. He tries to log into his account by guessing the digit. What is the probability that Fred will be successful within two attempts?

33. Let E and F be events with $P(E) = .4, \Pr(F) = .5$, and $\Pr(E \cup F) = .7$. Are E and F independent events?

34. Let E and F be events with $P(E) = .2, \Pr(F) = .5$, and $\Pr(E \cup F) = .6$. Are E and F independent events?

35. Let E and F be independent events with $P(E) = .5$ and $\Pr(F) = .6$. Find $\Pr(E \cup F)$.

36. Let E and F be independent events with $P(E) = .25$ and $\Pr(F) = .4$. Find $\Pr(E \cup F)$.

In Exercises 37–40, assume that E and F are independent events. Use the given information to find $\Pr(F)$.

37. $\Pr(E) = .7$ and $\Pr(F \mid E) = .6$.

38. $\Pr(E) = .4$ and $\Pr(F' \mid E') = .3$.

39. $\Pr(E') = .6$ and $\Pr(E \cap F) = .1$.

40. $\Pr(E) = .8$ and $\Pr(E \cap F) = .4$.

41. Let A, B, and C be independent events with $\Pr(A) = .4$, $\Pr(B) = .1$, and $\Pr(C) = .2$. Calculate $\Pr[(A \cap B \cap C)']$.

42. Let A, B, and C be independent events with $\Pr(A) = .2$, $\Pr(A \cap B) = .12$, and $\Pr(A \cap C) = .06$. Calculate $\Pr(B \cap C)$.

43. **Balls in an Urn** A sample of two balls is drawn from an urn containing two white balls and three red balls. Are the events "the sample contains at least one white ball" and "the sample contains balls of both colors" independent?

44. **Balls in an Urn** An urn contains two white balls and three red balls. A ball is withdrawn at random (without being replaced), and then a second ball is drawn. Are the events "the first ball is red" and "the second ball is white" independent?

45. **Roll a Die** Roll a die, and consider the following two events: $E = \{2, 4, 6\}, F = \{3, 6\}$. Are the events E and F independent?

46. **Roll a Die** Roll a die, and consider the following two events: $E = \{2, 4, 6\}, F = \{3, 4, 6\}$. Are the events E and F independent?

47. **Rolling Dice** Roll a pair of dice, and consider the sum of the two numbers. Are the events "the sum is an odd number" and "the sum is 4, 5, or 6" independent?

48. **Rolling Dice** Roll a pair of dice, and consider the sum of the two numbers. Are the events "the sum is an odd number" and "the sum is 5, 6, or 7" independent?

49. **Epidemiology** A doctor studies the known cancer patients in a certain town. The probability that a randomly chosen resident has cancer is found to be .001. It is found that 30% of the town works for Ajax Chemical Company. The probability that an employee of Ajax has cancer is equal to .003. Are the events "has cancer" and "works for Ajax" independent of one another?

50. Blood Tests A hospital uses two tests to classify blood. Every blood sample is subjected to both tests. The first test correctly identifies blood type with probability .7, and the second test correctly identifies blood type with probability .8. The probability that at least one of the tests correctly identifies the blood type is .9.

(a) Find the probability that the second test is correct, given that the first test is correct.

(b) Are the events "test I correctly identifies the blood type" and "test II correctly identifies the blood type" independent?

51. Medical Screening A medical screening program administers three independent tests. Of the persons taking the tests, 80% pass test I, 75% pass test II, and 60% pass test III. A participant is chosen at random.

(a) What is the probability that they will pass all three tests?

(b) What is the probability that they will pass at least two of the three tests?

52. Guessing on an Exam A "true–false" exam has 10 questions. Assuming that the questions are independent and that a student is guessing, find the probability that they get 100%.

53. System Reliability A TV set contains five circuit boards of type A, five of type B, and three of type C. The probability of failing in its first 5000 hours of use is .01 for a type A circuit board, .02 for a type B circuit board, and .025 for a type C circuit board. Assuming that the failures of the various circuit boards are independent of one another, compute the probability that no circuit board fails in the first 5000 hours of use.

54. System Reliability In November 2015, Intel announced the release of a business desktop computer with 72 processor cores. Suppose the probability that a given processor core will crash in 120 hours of use is .003, and that the failures of the various processor cores are independent of one another. What is the probability that no processor core will crash during the first 120 hours of use?

55. Smartphones Suppose that in Sleepy Valley, 70% of those over 50 years old own smartphones. Find the probability that among four randomly chosen people in that age group, none owns a smartphone.

56. Fishing The probability that a fisherman catches a tuna in any one excursion is .15. What is the probability that he catches a tuna on each of three excursions? On at least one of three excursions?

57. Baseball A baseball player's batting average changes every time he goes to bat and therefore should not be used as the probability of his getting a hit. However, we can still make a subjective assessment of his ability.

(a) If a player with .3 probability of getting a hit bats four times in a game and each at-bat is an independent event, what is the probability of the player getting at least one hit in the game?

(b) What is the probability of the player in part (a) starting off the season with at least one hit in each of the first 10 games?

(c) If there are 20 players with .3 probability of getting a hit, what is the probability that at least one of them will start the season with a 10-game hitting streak?

58. Roulette If you bet on the number 7 in roulette, the probability of winning on a single spin of the wheel is $\frac{1}{38}$. Suppose that you bet on 7 for 38 consecutive spins.

(a) Which of the following numbers do you think is closest to the probability of winning at least once: 1, .64, or .5?

(b) Calculate the probability of winning at least once.

59. Free-Throws A basketball player makes each free-throw with a probability of .6 and is on the line for a one-and-one free throw. (That is, a second throw is allowed only if the first is successful.) Assume that the two throws are independent. What do you think is the most likely result: scoring 0 points, 1 point, or 2 points? After making a guess, calculate the three probabilities.

60. Free-Throws Rework Exercise 59 with a probability of .7.

61. Free-Throws Consider Exercise 59, but let the probability of success be p, where $0 < p < 1$. Explain why the probability of scoring 0 points is always greater than the probability of scoring 1 point.

62. Free-Throws Consider Exercise 59, but let the probability of success be p, where $0 < p < 1$. For what value of p will the probability of scoring 1 point be the same as the probability of scoring 2 points?

63. Coin Toss A biased coin shows heads with probability .6. What is the probability of obtaining the sequence HT in two tosses of the coin? TH? Explain how this coin can be used at the start of a football game to fairly determine which team is to kick off.

64. Coin Toss A coin is tossed five times. Is the outcome HTHHT more likely to occur than the outcome HHHHH?

65. Show that, if events E and F are independent of each other, then so are E' and F'.

66. Show that, if E and F are independent events, then $\Pr(E \cup F) = 1 - \Pr(E') \cdot \Pr(F')$.

67. Let $\Pr(F) > 0$.

(a) Show that $\Pr(E' \mid F) = 1 - \Pr(E \mid F)$.

(b) Find an example for which $\Pr(E \mid F') \neq 1 - \Pr(E \mid F)$.

68. Use the inclusion–exclusion principle for (nonconditional) probabilities to show that, if E, F, and G are events in S, then $\Pr(E \cup F \mid G) = \Pr(E \mid G) + \Pr(F \mid G) - \Pr(E \cap F \mid G)$.

TECHNOLOGY EXERCISES

69. Roulette Find that value of N for which the probability of winning in Exercise 58 at least once in N successive plays is about .5.

70. Roulette If you bet "even" in roulette, the probability of winning is $\frac{9}{19}$. Find the smallest value of N for which the probability of winning at least once on the "even" bet in N successive plays is greater than .99.

Solutions to Check Your Understanding 6.4

1. $\frac{2}{3}$. Let F be the event that the up side is red and E be the event that the down side is red. $\Pr(F) = \frac{1}{2}$, since half of the faces are red. $F \cap E$ is the event that both sides of the card are red—that is, that the card that is red on both sides was selected, an event with probability $\frac{1}{3}$. By formula (1),

$$\Pr(E|F) = \frac{\Pr(E \cap F)}{\Pr(F)} = \frac{\frac{1}{3}}{\frac{1}{2}} = \frac{2}{3}.$$

(A common error is to conclude that the answer is $\frac{1}{2}$, since the card must be either the red/red card or the red/white card, and each of them is equally likely to have been selected. The correct probability is intuitively evident when you realize that two-thirds of the time, the card will have the same color on the bottom as on the top.)

2. By the hint,

$$\Pr(E \cap F') = \Pr(E) - \Pr(E \cap F)$$
$$= \Pr(E) - \Pr(E) \cdot \Pr(F)$$

(since E and F are independent)

$$= \Pr(E)[1 - \Pr(F)]$$
$$= \Pr(E) \cdot \Pr(F')$$

(by the complement rule).

Therefore, E and F' are independent events.

6.5 Tree Diagrams

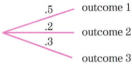

Figure 1

In solving many probability problems, it is helpful to represent the various events and their associated probabilities by a **tree diagram**. To explain this useful tool, suppose that we wish to compute the probability of an event that results from performing a sequence of experiments. The various outcomes of each experiment are represented as branches emanating from a point. For example, Fig. 1 represents an experiment with three outcomes. Notice that each branch has been labeled with the probability of the associated outcome. For example, the probability of outcome 1 is .5.

We represent experiments performed one after another by stringing together diagrams of the sort shown in Fig. 1, proceeding from left to right. For example, the diagram in Fig. 2 indicates that first we perform experiment A, having three outcomes, labeled 1–3. If the outcome is 1 or 2, we perform experiment B. If the outcome is 3, we perform experiment C. The probabilities on the right are conditional probabilities. For example, the top probability is the probability of outcome a (of B), given outcome 1 (of A). The probability of a sequence of outcomes may then be computed by multiplying the probabilities along a path. For example, to calculate the probability of outcome 2 followed by outcome b, we must calculate $\Pr(2 \text{ and } b) = \Pr(2) \cdot \Pr(b|2)$. To carry out this calculation, trace out the sequence of outcomes. Multiplying the probabilities along the path gives $(.2)(.6) = .12$ —the probability of outcome 2 followed by outcome b.

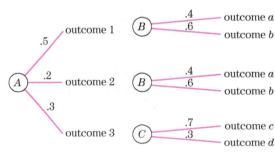

Figure 2

Tree diagrams come in all shapes and sizes. Some trees may not have the symmetry of the tree in Fig. 2. Tree diagrams arise whenever an activity can be thought of as a sequence of simpler activities.

Quality Control A box contains five good lightbulbs and two defective ones. Bulbs are selected one at a time (without replacement) until a good bulb is found. Find the probability that the number of bulbs selected is **(i)** one, **(ii)** two, **(iii)** three.

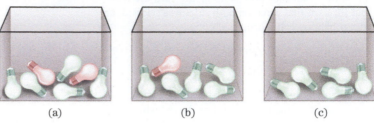

Figure 3   (a)   (b)   (c)

SOLUTION   The initial situation in the box is shown in Fig. 3(a). A bulb selected at random will be good (G) with probability $\frac{5}{7}$ and defective (D) with probability $\frac{2}{7}$. If a good bulb is selected, the activity stops. Otherwise, the situation is as shown in Fig. 3(b), and a bulb selected at random has probability $\frac{5}{6}$ of being good and probability $\frac{1}{6}$ of being defective. If the second bulb is good, the activity stops. If the second bulb is defective, then the situation is as shown in Fig. 3(c). At this point, a bulb has probability 1 of being good.

The tree diagram corresponding to the sequence of activities is given in Fig. 4. Each of the three paths leading to a G has a different length. The probability associated with the length of each path has been computed by multiplying the probabilities for its branches. The first path corresponds to the situation in which only one bulb is selected, the second path corresponds to two bulbs, and the third path to three bulbs. Therefore,

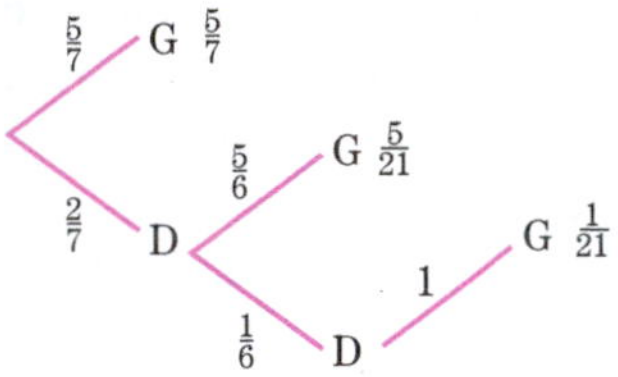

$$\textbf{(i) } \Pr(1) = \tfrac{5}{7} \qquad \textbf{(ii) } \Pr(2) = \tfrac{5}{21} \qquad \textbf{(iii) } \Pr(3) = \tfrac{1}{21}.$$

Figure 4

$\tfrac{5}{7}$ G $\tfrac{5}{7}$

$\tfrac{5}{6}$ G $\tfrac{5}{21}$

$\tfrac{2}{7}$ D

$\tfrac{1}{6}$ D 1 G $\tfrac{1}{21}$

>> *Now Try Exercise 11*

Political Polling A presidential candidate uses a phone bank to determine their support among the voters of Pennsylvania's two big cities: Philadelphia and Pittsburgh. Each phone bank worker has an auto-dialer that selects one of the cities at random and calls a random voter from that city. Suppose that, in Philadelphia, two-fifths of the voters favor the Republican candidate and three-fifths favor the Democratic candidate. Suppose that, in Pittsburgh, two-thirds of the voters favor the Republican candidate and one-third favor the Democratic candidate.

(a) Draw a tree diagram describing the survey.

(b) Find the probability that the voter polled is from Philadelphia and favors the Republican candidate.

(c) Find the probability that the voter favors the Republican candidate.

(d) Find the probability that the voter is from Philadelphia, given that they favor the Republican candidate.

SOLUTION   **(a)** The survey proceeds in two steps: First, select a city, and second, select and poll a voter. Figure 5(a) shows the possible outcomes of the first step and the associated probabilities. For each outcome of the first step, there are two possibilities for the

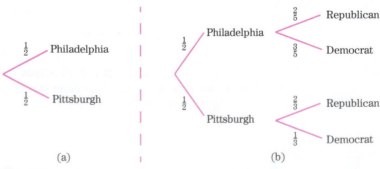

Figure 5

second step: The person selected could favor the Republican or the Democrat. In Fig. 5(b), we have represented these possibilities by drawing branches emanating from each of the outcomes of the first step. The probabilities on the new branches are actually conditional probabilities. For instance,

$$\tfrac{2}{5} = \Pr(\text{Rep} \,|\, \text{Phila}),$$

the probability that the voter favors the Republican candidate, given that the voter is from Philadelphia.

(b) $\Pr(\text{Phila} \cap \text{Rep}) = \Pr(\text{Phila}) \cdot \Pr(\text{Rep} \,|\, \text{Phila}) = \tfrac{1}{2} \cdot \tfrac{2}{5} = \tfrac{1}{5}.$

That is, the probability is $\tfrac{1}{5}$ that the combined outcome corresponds to the blue path in Fig. 6(a). We have written the probability $\tfrac{1}{5}$ at the end of the path to which it corresponds.

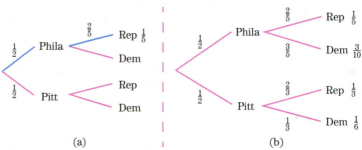

(a)   (b)

Figure 6

(c) In Fig. 6(b) we have computed the probabilities for each path of the tree as in part (b). Namely, the probability for a given path is the product of the probabilities for each of its segments. We are asked for $\Pr(\text{Rep})$. There are two paths through the tree leading to Republican, namely,

$$\text{Philadelphia} \cap \text{Republican} \quad \text{or} \quad \text{Pittsburgh} \cap \text{Republican}.$$

The probabilities of these two paths are $\tfrac{1}{5}$ and $\tfrac{1}{3}$, respectively. So the probability that the Republican is favored equals $\tfrac{1}{5} + \tfrac{1}{3} = \tfrac{8}{15}$.

(d) Here, we are asked for $\Pr(\text{Phila} \,|\, \text{Rep})$. By the definition of conditional probability,

$$\Pr(\text{Phila} \,|\, \text{Rep}) = \frac{\Pr(\text{Phila} \cap \text{Rep})}{\Pr(\text{Rep})} = \frac{\tfrac{1}{5}}{\tfrac{8}{15}} = \frac{3}{8}.$$

>> *Now Try Exercise 13*

Note that, from part (c), we might be led to conclude that the Republican candidate is leading with $\tfrac{8}{15}$ of the vote. However, we must always be careful when interpreting surveys. The results depend heavily on the auto-dialer design. For example, the phone bank drew half of its sample from each of the cities. However, Philadelphia is a much larger city and is leaning toward the Democratic candidate—so much so, in fact, that in terms of popular vote, the Democratic candidate would win, contrary to our expectations drawn from (c).

EXAMPLE 3   **Medical Screening** Suppose that the reliability of a skin test for active pulmonary tuberculosis (TB) is specified as follows: Of people with TB, 98% have a positive reaction and 2% have a negative reaction; of people free of TB, 99% have a negative reaction and 1% have a positive reaction. From a large population of which 2 per 10,000 persons have TB, a person is selected at random and given a skin test, which turns out to be positive. What is the probability that the person has active pulmonary tuberculosis?

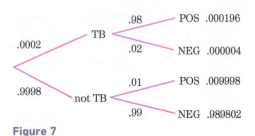

Figure 7

SOLUTION   The given data is organized in Fig. 7. The procedure called for is as follows: First select a person at random from the population. There are two possible outcomes: The person has TB,

$$\Pr(\text{TB}) = \frac{2}{10{,}000} = .0002,$$

or the person does not have TB,

$$\Pr(\text{not TB}) = 1 - .0002 = .9998.$$

For each of these two possibilities, the possible test results and conditional probabilities are given. Multiplying the probabilities along each of the paths through the tree gives the probabilities of the different outcomes. The resulting probabilities are written on the right in Fig. 7. The problem asks for the conditional probability that a person has TB, given that the test is positive. By definition,

$$\Pr(\text{TB}\,|\,\text{POS}) = \frac{\Pr(\text{TB} \cap \text{POS})}{\Pr(\text{POS})} = \frac{.000196}{.000196 + .009998} = \frac{.000196}{.010194} \approx .02.$$

>> The numerical data presented in Example 3 is only approximate. Variations in air quality for different localities within the United States cause variations in the incidence of TB and the reliability of skin tests.

Therefore, the probability is .02 that a person with a positive skin test has TB. In other words, although the skin test is quite reliable, only about 2% of those with a positive test turn out to have active TB. This result must be taken into account when large-scale medical diagnostic tests are planned. Because the group of people without TB is so much larger than the group with TB, the small error in the former group is magnified to the point where it dominates the calculation.   **>> Now Try Exercise 45**

DEFINITION   The **sensitivity** of a medical test is the probability of a positive test in the presence of the condition. The **specificity** is the probability of a negative test in the absence of the condition. The **positive predictive value** (PPV) is the probability of having the condition, given a positive test result. The **negative predictive value** (NPV) is the probability of not having the condition, given a negative test result.

From Example 3, the sensitivity, specificity, and positive predictive values of the TB skin test are .98, .99, and .02, respectively.

The positive and negative predictive values of a medical test can also be used to determine the probability of a **false positive** result, when a patient does not have the condition but has a positive test result, or a **false negative,** when a patient has the condition but tests negative.

EXAMPLE 4   **Medical Screening**   Consider the TB test of Example 3. Find
(a) the probability of a false negative.
(b) the probability of a false positive.
(c) the negative predictive value.

SOLUTION The probabilities can be read directly from Fig. 7.

(a) .02. A false negative occurs when the patient has the condition, but the test result is negative.

(b) .01. A false positive occurs when the patient does not have the condition, but the test result is positive.

(c) The negative predictive value is

$$\Pr(\text{not TB}\,|\,\text{NEG}) = \frac{\Pr(\text{not TB and NEG})}{\Pr(\text{NEG})} = \frac{.989802}{.000004 + .989802} = \frac{.989802}{.989806} = .999996.$$

>> *Now Try Exercise 47*

NOTE From the preceding definitions and example, we see that

$$\Pr(\text{false positive}) = 1 - \text{ sensitivity and } \Pr(\text{false negative}) = 1 - \text{ specificity. } \ll$$

Check Your Understanding 6.5

Solutions can be found following the section exercises.

Fifty percent of the students enrolled in a business statistics course had previously taken a finite mathematics course. Thirty percent of these students received an A for the statistics course, whereas 20 percent of the other students received an A for the statistics course.

1. Draw a tree diagram, and label it with the appropriate probabilities.

2. What is the probability that a student selected at random previously took a finite mathematics course and did not receive an A in the statistics course?

3. What is the probability that a student selected at random received an A in the statistics course?

4. What is the conditional probability that a student previously took a finite mathematics course, given that they received an A in the statistics course?

EXERCISES 6.5

In Exercises 1–4, draw trees representing the sequence of experiments.

1. Experiment I is performed. Outcome *a* occurs with probability .4, and outcome *b* occurs with probability .6. Then experiment II is performed. Its outcome *c* occurs with probability .8, and its outcome *d* occurs with probability .2.

2. Experiment I is performed twice. The three outcomes of experiment I are equally likely.

3. **Personnel Categories** A training program is used by a corporation to direct hirees to appropriate jobs. The program consists of two steps. Step I identifies 30% as management trainees, 60% as nonmanagerial workers, and 10% to be transferred to a different department. In step II, 75% of the management trainees are assigned to managerial positions, 20% are assigned to nonmanagerial positions, and 5% are transferred. In step II, 60% of the nonmanagerial workers are kept in the same category, 10% are assigned to management positions, and 30% are transferred.

4. **Tax Returns** An accounting firm uses a two-step auditing procedure to find problems in erroneous tax returns. Step I identifies the problem in the tax return with probability .7. Step II (which is executed only if step I fails to locate the problem) identifies the problem with probability .6.

5. **Personnel Categories** Refer to Exercise 3. What is the probability that a randomly chosen hiree will be assigned to a management position at the end of the training period?

6. **Tax Returns** Refer to Exercise 4. What is the probability that the procedure will fail to locate the problem?

7. **Personnel Categories** Refer to Exercise 3. What is the probability that a randomly chosen hiree will be transferred by the end of the training period?

8. **Personnel Categories** Refer to Exercise 3. What is the probability that a randomly chosen hiree will be designated a management trainee but *not* be appointed to a management position?

9. **Selecting from Urns** Suppose that there is a white urn containing two white balls and one red ball and there is a red urn containing one white ball and three red balls. An experiment consists of selecting at random a ball from the white urn and then (without replacing the first ball) selecting at random a ball from the urn having the color of the first ball. Find the probability that the second ball is red.

10. **Cards, Coins, Dice** A card is drawn from a 52-card deck. If the card is a face card (jack, queen, or king), we toss a coin. If the card is not a face card, we roll a die. Find the probability that we end the sequence with a "6" on the die. Find the probability that we end the sequence with a "head" on the coin.

11. **Cards** A card is drawn from a 52-card deck. We continue to draw until we have drawn a king or until we have drawn five cards, whichever comes first. Draw a tree diagram that illustrates the experiment. Put the appropriate probabilities on the tree. Find the probability that the drawing ends before the fourth draw.

12. **Balls in an Urn** An urn contains six white balls and two red balls. Balls are selected one at a time (without replacement) until a white ball is selected. Find the probability that the number of balls selected is (a) one, (b) two, (c) three.

13. **Quality Control** Twenty percent of the library books in the fiction section are worn and need replacement. Ten percent of the nonfiction holdings are worn and need replacement. The library's holdings are 40% fiction and 60% nonfiction. Use a tree diagram to find the probability that a book chosen at random from this library is worn and needs replacement.

14. **Water Testing** In a recent environmental study of 440 copper and galvanized steel water lines in Flint, Michigan, 77% of the lines were copper and 23% were galvanized steel. The study demonstrated that water in 6% of the copper lines had elevated levels of lead (above 15 parts per billion), while water in 11% of the galvanized lines had elevated lead levels. Use a tree diagram to find the probability that a water line chosen at random has elevated water lead levels. (*Source:* www.mlive.com)

15. **Color Blindness** Color blindness is a gender-linked inherited condition that is much more common among men than women. Suppose that 8% of all men and .5% of all women are color-blind. A person is chosen at random and found to be color-blind. What is the probability that the person is male? (You may assume that 50% of the population are men and 50% are women.)

16. **Manufacturing** A factory has two machines that produce bolts. Machine I produces 60% of the daily output of bolts, and 3% of its bolts are defective. Machine II produces 40% of the daily output, and 2% of its bolts are defective.
    (a) What is the probability that a bolt selected at random will be defective?
    (b) If a bolt is selected at random and found to be defective, what is the probability that it was produced by machine I?

17. **T-maze** A mouse is put into a T-maze (a maze shaped like a T). In this maze, it has the choice of turning to the left and being rewarded with cheese or going to the right and receiving a mild shock. Before any conditioning takes place (i.e., on trial 1), the mouse is equally likely to go to the left or to the right. After the first trial, its decision is influenced by what happened on the previous trial. If it receives cheese on any trial, the probabilities of going to the left or right become .9 and .1, respectively, on the following trial. If it receives the electric shock on any trial, the probabilities of going to the left or right on the next trial become .7 and .3, respectively. What is the probability that the mouse will turn left on the second trial?

18. **T-maze** Refer to Exercise 17. What is the probability that the mouse will turn left on the third trial?

19. **Heads or Tails** Three ordinary quarters and a fake quarter with two heads are placed in a hat. One quarter is selected at random and tossed twice. If the outcome is "HH," what is the probability that the fake quarter was selected?

20. **Selecting from a Bag** A bag is equally likely to contain either one white ball or one red ball. A white ball is added to the bag, and then a ball is selected at random from the bag. If the selected ball is white, what is the probability that the bag originally contained a white ball? (*Note:* This problem has been attributed to the famous English author Lewis Carroll.)

21. **Tennis** Kim has a strong first serve; whenever it is good (that is, in), she wins the point 75% of the time. Whenever her second serve is good, she wins the point 50% of the time. Sixty percent of her first serves and 75% of her second serves are good.
    (a) What is the probability that Kim wins the point when she serves?

    (b) If Kim wins a service point, what is the probability that her first serve was good?

22. **Tennis** When a tennis player hits his first serve as hard as possible (called a *blast*), he gets the ball *in* (that is, within bounds) 60% of the time. When the blast first serve is *in*, he wins the point 80% of the time. When the first serve is *out*, his gentler second serve wins the point 45% of the time. Draw a tree diagram representing the probabilities of winning the point for the first two serves. Use the tree diagram to determine the probability that the server eventually wins the point when his first serve is a blast.

23. **Accidental Nuclear War** Suppose that, during any year, the probability of an accidental nuclear war is .0001 (provided, of course, that there hasn't been one in a previous year). Draw a tree diagram representing the possibilities for the next three years. What is the probability that there will be an accidental nuclear war during the next three years?

24. **Accidental Nuclear War** Refer to Exercise 23. What is the probability that there will be an accidental nuclear war during the next n years?

25. **Coin Tosses** A coin is to be tossed at most five times. The tosser wins as soon as the number of heads exceeds the number of tails and loses as soon as three tails have been tossed. Use a tree diagram for this game to calculate the probability of winning.

26. **Cards** Suppose that, instead of tossing a coin, the player in Exercise 25 draws up to five cards from a deck consisting only of three red and three black cards. The player wins as soon as the number of red cards exceeds the number of black cards and loses as soon as three black cards have been drawn. Does the tree diagram for the card game have the same shape as the tree diagram for the coin game? Is there any difference in the probability of winning? If so, which game has the greater probability of winning?

27. **Genetics** Traits passed from generation to generation are carried by genes. For a certain type of pea plant, the color of the flower produced by the plant (either red or white) is determined by a pair of genes. Each gene is of one of the types C (dominant gene) or c (recessive gene). Plants for which both genes are of type c (said to have genotype cc) produce white flowers. All other plants—that is, plants of genotypes CC and Cc—produce red flowers. When two plants are crossed, the offspring receives one gene from each parent.

Genotype	Color
cc	white
Cc	red
CC	red

(a) Suppose that you cross two pea plants of genotype Cc. What is the probability that the offspring produces white flowers? Red flowers?

(b) Suppose that you have a batch of red-flowering pea plants, of which 60% have genotype Cc and 40% have genotype CC. If you select one of these plants at random and cross it with a white-flowering pea plant, what is the probability that the offspring will produce red flowers?

28. **Genetics** Refer to Exercise 27. Suppose that a batch of 99 pea plants contains 33 plants of each of the three genotypes.
    (a) If you select one of these plants at random and cross it with a white-flowering pea plant, what is the probability that the offspring will produce white flowers?
    (b) If you select one of the 99 pea plants at random and cross it with a white-flowering pea plant, and the offspring produces red flowers, what is the probability that the selected plant had genotype *Cc*?

29. **College Faculty** At a local college, five sections of economics are taught during the day and two sections are taught at night. Sixty percent of the day sections are taught by full-time faculty. Forty percent of the evening sections are taught by full-time faculty. If Jane has a part-time teacher for her economics course, what is the probability that she is taking a night class?

30. **Quality Control** A lightbulb manufacturer knows that .05% of all bulbs manufactured are defective. A testing machine is 99% effective; that is, 99% of good bulbs will be declared fine and 99% of flawed bulbs will be declared defective. If a randomly selected lightbulb is tested and found to be defective, what is the probability that it actually is defective?

31. **Balls in an Urn** Urn I contains 5 red balls and 5 white balls. Urn II contains 12 white balls. A ball is selected at random from urn I and placed in urn II. Then a ball is selected at random from urn II. What is the probability that the second ball is white?

32. **Balls in an Urn** An urn contains five red balls and three green balls. One ball is selected at random and then replaced by a ball of the other color. Then a second ball is selected at random. What is the probability that the second ball is green?

33. **Coin Tosses** Two people toss two coins each. What is the probability that they get the same number of heads?

34. **Selecting from Urns** An urn contains four red marbles and three green marbles. One marble is removed, its color noted, and the marble is not replaced. A second marble is removed and its color noted.
    (a) What is the probability that both marbles are red? Green?
    (b) What is the probability that exactly one marble is red?

35. **Industrial Production** A factory that produces three-dimensional models has two 3D printers. Printer A is very reliable and produces 200 models every week. Printer B is a little less reliable and produces 201 models on 99% of the weeks, but breaks down and produces 0 models the rest of the weeks.
    (a) In a random week, what is the probability that Printer B produces more models that Printer A?
    (b) After 200 weeks have elapsed, what it the probability that Printer B has produced more total models than Printer A? *Hint:* Printer A will have produced 40,000 models. If Printer B breaks down at least once, it will have produced fewer models.

36. **Golf** Bud is a very consistent golfer. On par-three holes, he always scores a 4. Lou, on the other hand, is quite erratic. On par-three holes, Lou scores a 3 seventy percent of the time and scores a 6 thirty percent of the time.
    (a) If Bud and Lou play a single par-three hole together, who is more likely to win—that is, to have the lowest score?
    (b) If Bud and Lou play two consecutive par-three holes, who is more likely to have the lowest total score?

37. **Nontransitive Dice** Consider three dice: one red, one blue, and one green. The sides of the red die contain the numbers 3 3 3 3 3 6, the sides of the blue die contain the numbers 2 2 2 5 5 5, and the sides of the green die contain the numbers 1 4 4 4 4 4.
    (a) Determine the probability that the red die will show a higher number than the blue die when both are tossed.
    (b) Determine the probability that the blue die will show a higher number than the green die when both are tossed.
    (c) Determine the probability that the green die will show a higher number than the red die when both are tossed.
    (d) What is surprising about the results in parts (a)–(c)?

38. **U.S. Car Production** Car production in North America in January 2016 was distributed among car manufacturers as follows.

North American Car Production	Type	Percentage of Type by Brand	
60%	Domestic	Chrysler	33%
		Ford	39%
		General Motors	28%
40%	Foreign	Honda	34%
		Toyota	33%
		Other	33%

Source: www.wardsauto.com

This means that 60% of the cars produced in North America were manufactured by domestic companies; of them, 33% were Chryslers, 39% were Fords, and 28% were General Motors products.
    (a) A January 2016 automobile is chosen at random. What is the probability that it is a General Motors car?
    (b) What is the probability that a randomly selected January 2016 automobile is a Ford or a Toyota?

Exercises 39–44 apply to medical diagnostic tests.

39. (True or False) *Sensitivity* also can be called the *true positive rate*.

40. (True or False) *Specificity* also can be called the *true negative rate*.

41. (True or False) Specificity = 1 − false negative rate.

42. (True or False) Sensitivity = 1 − false positive rate.

43. (True or False) Sensitivity is related to a test's ability to rule *in* a condition.

44. (True or False) Specificity is related to a test's ability to rule *out* a condition.

45. **Medical Screening** Suppose that a test for hepatitis has a sensitivity of 95% and a specificity of 90%. A person is selected at random from a large population, of which .05% of the people have hepatitis, and given the test. What is the positive predictive value of the test?

46. **Medical Screening** The probability .0002 (or .02%) in Fig. 7 on page 278 is called the *prevalence* of the disease. It states that in the population being considered, 2 people per 10,000 have TB. Find the positive predictive value for a population in which 2 people per 100 have TB.

47. Medical Screening The results of a trial used to determine the capabilities of a new diagnostic test are shown in Table 1. Use the empirical probabilities obtained from the table to calculate the positive predictive value for the diagnostic test.

Table 1

	Has Condition	Does Not Have Condition	Total
Test positive	9	10	19
Test negative	1	980	981
Total	10	990	1000

48. Medical Screening The results of a trial used to determine the capabilities of a new diagnostic test are shown in Table 2. Use the empirical probabilities obtained from the table to calculate the positive predictive value for the diagnostic test.

Table 2

	Has Condition	Does Not Have Condition	Total
Test positive	13	7	20
Test negative	3	77	80
Total	16	84	100

49. Drug Testing Suppose that 500 athletes are tested for a drug, 1 in 20 has used the drug, and the test has a 99% specificity and a 100% sensitivity. If an athlete in the group tests positive, what is the probability that they have used the drug?

50. Polygraph Test Recent studies have indicated that polygraph tests have a sensitivity of .88 and a specificity of .86. (Currently, polygraph tests are admissible as evidence in court in 19 U.S. states.) Suppose that a robbery is committed and polygraph tests are given to 10 suspects, 1 of whom committed the crime. If the polygraph test for a person indicates that they are guilty, what is the probability they are actually guilty?

Solutions to Check Your Understanding 6.5

1.

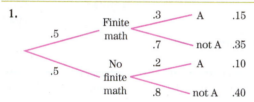

2. The event "finite math and not A" corresponds to the second path of the tree diagram, which has probability .35.

3. This event is satisfied by the first or third paths and therefore has probability .15 + .10 = .25.

4. $\Pr(\text{finite math} \,|\, A) = \dfrac{\Pr(\text{finite math and A})}{\Pr(A)} = \dfrac{.15}{.25} = .6.$

6.6 Bayes' Theorem, Natural Frequencies

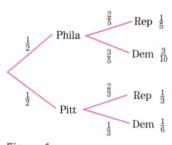

Figure 1

In the preceding section, we explored how to calculate conditional probabilities using tree diagrams. *Bayes' theorem*, named after English statistician Thomas Bayes, gives us a second method for computing these probabilities. The advantages of Bayes' theorem over the use of tree diagrams are that (1) we do not need to draw the tree diagram to calculate the desired probability, and (2) we need not compute extraneous probabilities. These advantages become significant in dealing with experiments having many outcomes.

Let us reconsider the phone survey in Example 2(d) of Section 6.5. (See Fig. 1.) Given that the person chosen at random favors the Republican candidate, what is the probability that the respondent is from Philadelphia? We found this probability by calculating

$$\Pr(\text{Phila} \,|\, \text{Rep}) = \frac{\Pr(\text{Phila} \cap \text{Rep})}{\Pr(\text{Rep})}.$$

Let us analyze the components of this calculation. First, recall that

$$\Pr(\text{Phila} \cap \text{Rep}) = \Pr(\text{Phila}) \cdot \Pr(\text{Rep} \,|\, \text{Phila}).$$

Second,

$$\Pr(\text{Rep}) = \Pr(\text{Phila} \cap \text{Rep}) + \Pr(\text{Pitt} \cap \text{Rep})$$

$$= \Pr(\text{Phila}) \cdot \Pr(\text{Rep} \,|\, \text{Phila}) + \Pr(\text{Pitt}) \cdot \Pr(\text{Rep} \,|\, \text{Pitt}),$$

by using the tree diagram. Denote the events "Phila," "Pitt," "Rep," and "Dem" by the letters P_1, P_2, R, and D, respectively. Then

$$\Pr(\text{Phila} \,|\, \text{Rep}) = \Pr(P_1 \,|\, R)$$

$$= \frac{\Pr(P_1 \cap R)}{\Pr(R)}$$

$$= \frac{\Pr(P_1) \cdot \Pr(R \,|\, P_1)}{\Pr(P_1) \cdot \Pr(R \,|\, P_1) + \Pr(P_2) \cdot \Pr(R \,|\, P_2)}.$$

This is a special case of Bayes' theorem.

We summarize a simple form of Bayes' theorem.

> **Bayes' Theorem ($n = 2$)**　If B_1 and B_2 are mutually exclusive events, and $B_1 \cup B_2 = S$ (that is, $B_2 = B_1'$), then
>
> $$\Pr(B_1 \,|\, A) = \frac{\Pr(B_1) \cdot \Pr(A \,|\, B_1)}{\Pr(B_1) \cdot \Pr(A \,|\, B_1) + \Pr(B_2) \cdot \Pr(A \,|\, B_2)}$$
>
> for any event A in S with $\Pr(A) \neq 0$.

We have the same type of result for the situation in which we have three mutually exclusive sets B_1, B_2, and B_3, whose union is all of S. We state Bayes' theorem for that case and leave the general case for n mutually exclusive sets for the end of the section.

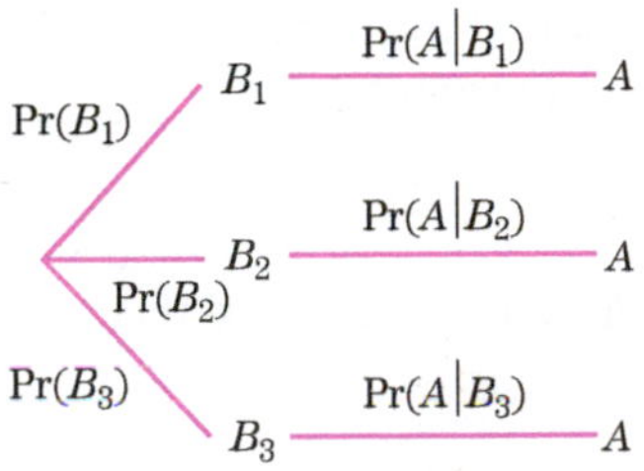

Figure 2

> **Bayes' Theorem ($n = 3$)**　If B_1, B_2, and B_3 are mutually exclusive events, and $B_1 \cup B_2 \cup B_3 = S$, then for any event A in S with $\Pr(A) \neq 0$,
>
> $$\Pr(B_1 \,|\, A) = \frac{\Pr(B_1) \cdot \Pr(A \,|\, B_1)}{\Pr(B_1) \cdot \Pr(A \,|\, B_1) + \Pr(B_2) \cdot \Pr(A \,|\, B_2) + \Pr(B_3) \cdot \Pr(A \,|\, B_3)}.$$
>
> See Fig. 2.

EXAMPLE 1　**Medical Screening**　Solve the tuberculosis skin test problem of Example 3 of Section 6.5 by using Bayes' theorem.

SOLUTION　The observed event A is "positive skin test result." There are two possible events leading to A—namely,

$$B_1 = \text{"person has tuberculosis"}$$

$$B_2 = \text{"person does not have tuberculosis."}$$

We wish to calculate $\Pr(B_1 \,|\, A)$. From the data given, we have

$$\Pr(B_1) = \frac{2}{10{,}000} = .0002$$

$$\Pr(B_2) = .9998$$

$$\Pr(A \,|\, B_1) = \Pr(\text{POS} \,|\, \text{TB}) = .98$$

$$\Pr(A \,|\, B_2) = \Pr(\text{POS} \,|\, \text{not TB}) = .01.$$

Therefore, by Bayes' theorem,

$$\Pr(B_1 \,|\, A) = \frac{\Pr(B_1)\Pr(A \,|\, B_1)}{\Pr(B_1)\Pr(A \,|\, B_1) + \Pr(B_2)\Pr(A \,|\, B_2)}$$

$$= \frac{(.0002)(.98)}{(.0002)(.98) + (.9998)(.01)} \approx .02,$$

in agreement with our calculation of Example 3 in Section 6.5.　**≫ Now Try Exercise 11**

EXAMPLE 2   **Quality Control** A printer has five book-binding machines. For each machine, Table 1 gives the proportion of the total book production that it binds and the probability that the machine produces a defective binding. For instance, machine 1 binds 10% of the books and produces a defective binding with probability .03. Suppose that a book is selected at random and found to have a defective binding. What is the probability that it was bound by machine 1?

Table 1

Machine	Proportion of Books Bound	Probability of Defective Binding
1	.10	.03
2	.10	.02
3	.40	.02
4	.15	.03
5	.25	.01

SOLUTION   In this example, we have five mutually exclusive events whose union is the entire sample space (the book was bound by one, and only one, of the five machines). Bayes' theorem can be extended to any finite number of mutually exclusive events.

Let B_i ($i = 1, 2, 3, 4, 5$) be the event that the book was bound by machine i, and let A be the event that the book has a defective binding. Then, for example,

$$\Pr(B_1) = .10 \quad \text{and} \quad \Pr(A \mid B_1) = .03.$$

The problem asks for the reversed conditional probability, $\Pr(B_1 \mid A)$. By Bayes' theorem,

$$\Pr(B_1 \mid A) = \frac{\Pr(B_1)\Pr(A \mid B_1)}{\Pr(B_1)\Pr(A \mid B_1) + \Pr(B_2)\Pr(A \mid B_2) + \cdots + \Pr(B_5)\Pr(A \mid B_5)}$$

$$= \frac{(.10)(.03)}{(.10)(.03) + (.10)(.02) + (.40)(.02) + (.15)(.03) + (.25)(.01)}$$

$$= \frac{.003}{.02} = .15.$$

>> *Now Try Exercise 1*

Natural Frequencies

In Example 3 of the previous section, we used a tree diagram to solve the TB medical screening problem. In Example 1 of this section, we solved the same problem with Bayes' theorem. The method of natural frequencies uses simple counts of occurrences of events and provides a third way to obtain the probability. Although the method of natural frequencies is not as rigorous as Bayes' theorem, many people find it easier to use and understand. We will illustrate the method by using it to solve the TB medical screening problem.

EXAMPLE 3   **Medical Screening** Use the method of natural frequencies to solve the TB medical screening problem.

SOLUTION   Let us begin with a group of 1 million people. Out of this group, $.0002 \cdot 1,000,000 = 200$ can be expected to have TB and $.9998 \cdot 1,000,000 = 999,800$ to not have TB. Of the 200 people with TB, $.98 \cdot 200 = 196$ should test positive and $.02 \cdot 200 = 4$ should test negative. Of the 999,800 without TB, $.01 \cdot 999,800 = 9998$ should test positive and $.99 \cdot 999,800 = 989,802$ should test negative. These numbers are displayed in Fig. 3.

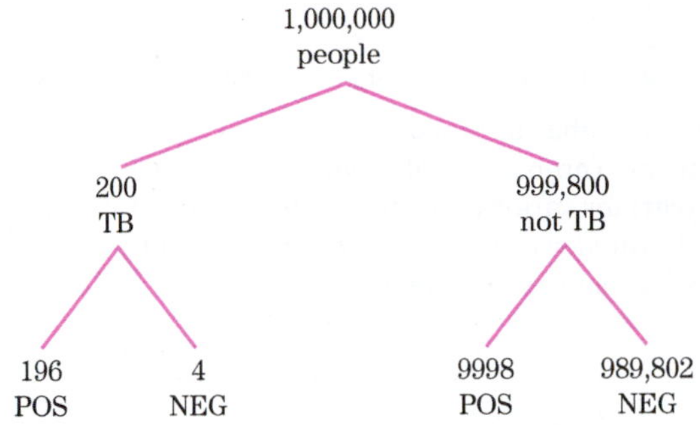

1,000,000
people
200
TB
999,800
not TB
196
POS
4
NEG
9998
POS
989,802
NEG

Figure 3

The tree diagram shows that 196 + 9998 = 10,194 of the people tested positive and that 196 of those people actually had TB. Therefore, the probability that a person who tested positive actually has TB is $\frac{196}{10,194} = .019227 \approx .02$. **» Now Try Exercise 25**

Pr(B1)
B1
Pr(A|B1)
A
B2
Pr(B2)
Pr(A|B2)
A
Pr(Bn)
Bn
Pr(A|Bn)
A

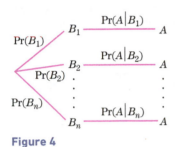

Figure 4

Derivation of Bayes' Theorem To derive Bayes' theorem in general, we consider a two-stage tree. Suppose that, at the first stage, there are the events $B_1, B_2, \ldots, B_n$, which are mutually exclusive and exhaust all possibilities. Let us examine only the paths of the tree leading to event A at the second stage of the experiment. (See Fig. 4.) Suppose we are given that the event A occurs. What is $\Pr(B_1 \mid A)$? First, consider $\Pr(B_1 \cap A)$. This can be seen from Fig. 4 to be $\Pr(B_1) \cdot \Pr(A \mid B_1)$. Next, we calculate $\Pr(A)$. Recall that A occurs at stage 2, preceded at stage 1 by either event $B_1, B_2, \ldots,$ or B_n. Since $B_1, B_2, \ldots, B_n$ are mutually exclusive,

$$\Pr(A) = \Pr(B_1 \cap A) + \Pr(B_2 \cap A) + \cdots + \Pr(B_n \cap A).$$

Each of the elements in the sum can be calculated by the product rule, or directly from Fig. 4:

$$\begin{aligned}\Pr(B_1 \cap A) &= \Pr(B_1) \cdot \Pr(A \mid B_1)\\ \Pr(B_2 \cap A) &= \Pr(B_2) \cdot \Pr(A \mid B_2)\\ &\vdots\\ \Pr(B_n \cap A) &= \Pr(B_n) \cdot \Pr(A \mid B_n).\end{aligned}$$

«

The result is the following:

Bayes' Theorem If $B_1, B_2, \ldots, B_n$ are mutually exclusive events, and if $B_1 \cup B_2 \cup \cdots \cup B_n = S$, then for any event A in S with $\Pr(A) \neq 0$,

$$\Pr(B_1 \mid A) = \frac{\Pr(B_1) \cdot \Pr(A \mid B_1)}{\Pr(B_1) \cdot \Pr(A \mid B_1) + \Pr(B_2) \cdot \Pr(A \mid B_2) + \cdots + \Pr(B_n) \cdot \Pr(A \mid B_n)}$$

$$\Pr(B_2 \mid A) = \frac{\Pr(B_2) \cdot \Pr(A \mid B_2)}{\Pr(B_1) \cdot \Pr(A \mid B_1) + \Pr(B_2) \cdot \Pr(A \mid B_2) + \cdots + \Pr(B_n) \cdot \Pr(A \mid B_n)},$$

and so forth.

Check Your Understanding 6.6

Solutions can be found following the section exercises.

1. **Quality Control** Refer to Example 2. Suppose that a book is selected at random and found to have a defective binding. What is the probability that the book was bound by machine 2?

2. **Political Polling** Use the method of natural frequencies with a group of 30 people to solve the political polling problem presented in Example 2 of the previous section and discussed at the beginning of this section.

EXERCISES 6.6

In Exercises 1–22, use Bayes' theorem to calculate the probabilities.

1. **Accident Rates** An automobile insurance company has determined the accident rate (probability of having at least one accident during a year) for various age groups. (See Table 2.) Suppose that a policyholder calls in to report an accident. What is the probability that they are over 60?

Table 2

Age Group	Proportion of Total Insured	Accident Rate
Under 21	.05	.06
21–30	.10	.04
31–40	.25	.02
41–50	.20	.015
51–60	.30	.025
Over 60	.10	.04

2. **Quality Control** A scoreboard has six different types of LEDs. For each type of LED, Table 3 gives the proportion of the total number of LEDs of that type and the failure rate (probability of failing within one year). If an LED fails, what is the probability that it is type 1?

Table 3

Type	Proportion of Total	Failure Rate
1	.30	.0002
2	.25	.0004
3	.20	.0005
4	.10	.0010
5	.05	.0020
6	.10	.0040

3. **Student Performance** The enrollment in a certain course is 10% first-year students, 30% sophomores, 40% juniors, and 20% seniors. Experience has shown that the likelihood of receiving an A in the course is .2 for first-year students, .4 for sophomores, .3 for juniors, and .1 for seniors. Find the probability that a student who receives an A is a sophomore.

4. **Larceny Rates** A metropolitan police department maintains statistics of larcenies reported in the various precincts of the city. It records the proportion of the city population in each precinct and the precinct larceny rate (= the proportion of the precinct population reporting a larceny within the past year). These statistics are summarized in Table 4. A larceny victim is randomly chosen from the city population. What is the probability that they come from Precinct 3?

Table 4

Precinct	Proportion of Population	Larceny Rate
1	.20	.01
2	.10	.02
3	.40	.05
4	.30	.04

5. **Cars and Income** Table 5 gives the distribution of incomes and shows the proportion of two-car families by income level for a certain suburban county. Suppose that a randomly chosen family has two or more cars. What is the probability that its income is at least $75,000 per year?

Table 5

Annual Family Income	Proportion of People	Proportion Having Two or More Cars
< $30,000	.10	.20
$30,000–$44,999	.20	.50
$45,000–$59,999	.35	.60
$60,000–$74,999	.30	.75
≥ $75,000	.05	.90

6. **Voter Turnout** Table 6 gives the distribution of voter registration and voter turnouts for a certain city. A randomly chosen person is questioned at the polls. What is the probability that the person is an Independent?

Table 6

	Proportion Registered	Proportion of Turnout
Democrat	.50	.4
Republican	.20	.5
Independent	.30	.7

7. **Mathematics Exam** In a calculus course, the instructor gave an algebra exam on the first day of class to help students determine whether or not they had enrolled in the appropriate course. Eighty percent of the students in the class passed the exam. Forty percent of those who passed the exam on the first day of class earned an A in the course, whereas only twenty percent of those who failed the exam earned an A in the course. What is the probability that a student selected at random passed the exam on the first day of class, given that they earned an A in the course?

8. **Demographics** Table 7 shows the percentages of various portions of the U.S. population in 2012, based on age and gender. Suppose that a person is chosen at random from the entire population.

Table 7

| Age Group | U.S. Population | |
	% of Population	% Male
Under 5 yrs	7	51
5–19 yrs	20	51
20–44 yrs	33	50
45–64 yrs	27	48
Over 64 yrs	13	44

Source: www.census.gov

(a) What is the probability that the person chosen is male?
(b) Given that the person chosen is male, find the probability that he is between 5 and 19 years old.

9. **Bilingual Employees** A multinational company has five divisions: A, B, C, D, and E. The percentage of employees from each division who speak at least two languages fluently is shown in Table 8.

Table 8

Division	Number of Employees	Percentage of Employees Who Are Bilingual
A	20,000	20
B	15,000	15
C	25,000	12
D	30,000	10
E	10,000	10
Total	100,000	

(a) Find the probability that an employee selected at random is bilingual.

(b) Find the probability that a bilingual employee selected at random works for division C.

10. **Customized Dice** A specially made pair of dice has only one- and two-spots on the faces. One of the dice has three faces with a one-spot and three faces with a two-spot. The other die has two faces with a one-spot and four faces with a two-spot. One of the dice is selected at random and then rolled six times. If a two-spot shows up only once, what is the probability that it is the die with four two-spots?

Exercises 11–15 refer to diagnostic tests. A *false negative* in a diagnostic test is a test result that is negative even though the patient has the condition. A *false positive*, on the other hand, is a test result that is positive although the patient does not have the condition.

11. **Mammogram Accuracy** The *New York Times* of January 24, 1997, discussed the recommendation of a special panel concerning mammograms for women in their 40s. About 2% of women aged 40 to 49 years old develop breast cancer in their 40s. But the mammogram used for women in that age group has a high rate of false positives and false negatives; the false positive rate is .30, and the false negative rate is .25. If a woman in her 40s has a positive mammogram, what is the probability that she actually has breast cancer?

12. **Drug Screening** A drug-testing laboratory produces false negative results 2% of the time and false positive results 5% of the time. Suppose that the laboratory has been hired by a company in which 10% of the employees use drugs.

(a) If an employee tests positive for drug use, what is the probability that they actually use drugs?

(b) What is the probability that a nondrug user will test positive for drug use twice in a row?

(c) What is the probability that someone who tests positive twice in a row is not a drug user?

13. **Pregnancy Test** An over-the-counter pregnancy test claims to be 99% accurate. Actually, what the insert says is that if the test is performed properly, it is 99% sure to detect a pregnancy.

(a) What is the probability of a false negative?

(b) Let us assume that the probability is 98% that the test result is negative for a woman who is not pregnant. If the

woman estimates that her chances of being pregnant are about 40% and the test result is positive, what is the probability that she is actually pregnant?

14. **Medical Screening** A test for a condition has a high probability of false positives, 20%. Its rate of false negatives is 10%. The condition is estimated to exist in 65% of all patients sent for screening. If the test is positive, what is the chance the patient has the condition? Suppose that the condition is much more rare in the population—say, Pr(condition) = .30. Given the same testing situation, what is Pr(condition | pos)?

15. **Steroid Testing** It is estimated that 10% of Olympic athletes use steroids. The test currently being used to detect steroids is said to be 93% effective in correctly detecting steroids in users. It yields false positives in only 2% of the tests. A country's best weightlifter tests positive. What is the probability that he actually takes steroids?

16. **Medical Screening** The results of a trial used to determine the capabilities of a new diagnostic test are shown in Table 9. Use the empirical probabilities obtained from the table to calculate the probability that a person who tests positive actually has the condition.

Table 9

	Has Condition	Does Not Have Condition	Total
Test positive	9	11	20
Test negative	1	179	180
Total	10	190	200

17. **Exit Polling** According to exit polling from the 2016 Virginia Democratic primary, 43% of primary voters were men and 57% were women. Fifty-seven percent of Democratic men voting in the primary supported Hillary Clinton, while 70% of Democratic women supported Hillary Clinton. If a Hillary Clinton supporter from the Virginia primary exit poll is chosen at random, what is the probability that they are male?

18. **Exit Polling** According to exit polling from the 2014 U.S. midterm elections, 36% of voters had a household income less than $50,000, while 64% had a household income of at least $50,000. Forty-three percent of voters from households making less than $50,000 voted for the Republican party in the election, while 55% percent of voters from households making at least $50,000 voted Republican. What is the probability that a randomly selected Republican voter from the exit poll is from a household that makes at least $50,000?

19. **Cards** Thirteen cards are dealt from a deck of 52 cards.

(a) What is the probability that the ace of spades is one of the 13 cards?

(b) Suppose that one of the 13 cards is chosen at random and found *not* to be the ace of spades. What is the probability that *none* of the 13 cards is the ace of spades?

(c) Suppose that the experiment in part (b) is repeated a total of 10 times (replacing the card looked at each time) and the ace of spades is not seen. What is the probability that the ace of spades actually *is* one of the 13 cards?

20. College Majors There are three sections of English 101. In Section I, there are 25 students, of whom 5 are mathematics majors. In Section II, there are 20 students, of whom 6 are mathematics majors. In Section III, there are 35 students, of whom 5 are mathematics majors. A student in English 101 is chosen at random. Find the probability that the student is from Section I, given that they are a mathematics major.

21. Scholarship Winners Twenty percent of the contestants in a scholarship competition come from Pylesville High School, 40% come from Millerville High School, and the remainder come from Lakeside High School. Two percent of the Pylesville students are among the scholarship winners; 3% of the Millerville contestants and 5% of the Lakeside contestants win.

(a) If a winner is chosen at random, what is the probability that they are from Lakeside?

(b) What percentage of the winners are from Pylesville?

22. Manufacturing Reliability Ten percent of the pens made by Apex are defective. Only 5% of the pens made by its competitor, B-ink, are defective. Since Apex pens are cheaper than B-ink pens, an office orders 70% of its stock from Apex and 30% from B-ink. A pen is chosen at random and found to be defective. What is the probability that it was produced by Apex?

In Exercises 23–30, use the method of natural frequencies to calculate the probabilities.

23. Rework Exercise 3, starting with a class size of 50 people.

24. Rework Exercise 6, starting with 100 people.

25. Rework Exercise 7, starting with a class size of 25 people.

26. Rework Exercise 12(a), starting with 1000 people.

27. Rework Exercise 15, starting with 1000 people.

28. Rework Exercise 14, starting with 100 people.

29. Rework Exercise 21(a), starting with 250 students.

30. Rework Exercise 22, starting with 200 pens.

Solutions to Check Your Understanding 6.6

1. The problem asks for $\Pr(B_2 \mid A)$. Bayes' theorem gives this probability as a quotient with numerator $\Pr(B_2) \Pr(A \mid B_2)$ and the same denominator as in the solution to Example 2. Therefore,

$$\Pr(B_2 \mid A) = \frac{\Pr(B_2) \Pr(A \mid B_2)}{.02} = \frac{(.10)(.02)}{.02} = .10.$$

2. Beginning with a group of 30 people, $\frac{1}{2} \cdot 30 = 15$ people will be from Philadelphia, and $\frac{1}{2} \cdot 30 = 15$ people will be from Pittsburgh. Of the 15 people from Philadelphia, $\frac{2}{5} \cdot 15 = 6$ will favor the Republican candidate. The remainder, $15 - 6 = 9$, will favor the Democratic candidate. (*Note:* We could have obtained the number 9 by performing the multiplication $\frac{3}{5} \cdot 15$, but subtracting is easier than multiplying.) Of the 15 people from Pittsburgh, $\frac{2}{3} \cdot 15 = 10$ will favor the Republican candidate, and the rest, $15 - 10 = 5$, will favor the Democratic candidate. These numbers are displayed in Fig. 5.

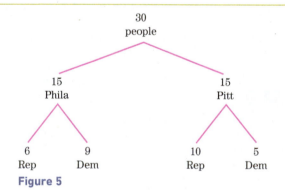

Figure 5

The tree diagram shows that $6 + 10 = 16$ people favor the Republican candidate and that 6 of those people are from Philadelphia. Therefore, the probability that a person who favors the Republican candidate is from Philadelphia is $\frac{6}{16} = \frac{3}{8}$.

6.7   Simulation

Simulation is a method of imitating an experiment by using an artificial device to substitute for the real thing. The technique is used often in industrial and scientific applications. For example, in the 1970s, Deaconess Hospital in St. Louis was planning to add an extension to the hospital, with 144 medical-surgical beds. The planners knew that an increase in the number of beds would require additional operating rooms and recovery room beds. By studying the pattern of patients already being treated in the existing hospital, Homer H. Schmitz and N. K. Kwak constructed a mathematical model to imitate the rate of flow of patients through the planned hospital (with the additional beds) to see what the typical operating suite schedule would look like. (*Source: Operations Research.*) But they did not use patients! They used random numbers and a computer in their simulation model. They built a model in which they could vary the number of operating rooms and adjust the schedule. By repeating the experiment many times, they found the optimal number of new operating rooms and recovery room beds to complement the added beds.

Graphing calculators have a command called **randInt** that can be used to select a number at random from a specified set of numbers. Each number in the set is just as likely to be selected as any other. The command **randInt(m,n)** generates a random integer from m through n, and **randInt(m,n,r)** generates a list of r random numbers from m through n. (The command **randInt** is the fifth item on the MATH/PROB menu.) After the list of random numbers has been assigned to a list variable, the command **SortA(list)** sorts the list in ascending order. (**SortA** is found on the LIST/OPS menu of the TI-84 Plus). Some examples of the use of these commands follow:

1. *Simulate the roll of a single die*:

    randInt(1,6)

2. *Simulate the sum for a roll of a pair of dice*:

    randInt(1,6)+randInt(1,6)

3. *Simulate 10 rolls of a pair of dice*:

    randInt(1,6,10)+randInt(1,6,10)

4. *Simulate the selection of a ball from an urn containing 7 white balls and 3 red balls*: Think of the white balls as numbered from 1 to 7 and the red balls as numbered from 8 to 10.

    randInt(1,10)

5. *Simulate the outcome of a free throw by Michael Jordan, who had an 83% free-throw average*: Consider each shot as a random whole number from 1 to 100, where a number from 1 to 83 represents a successful free throw and a number from 84 to 100 represents a miss.

    randInt(1,100)

On an Excel spreadsheet, the formula =RANDBETWEEN(*m,n*) produces a randomly selected whole number from m to n. The Random Number Generation routine that is part of the Data Analysis tool generates a random sample from a probability distribution. (See Appendix C for details.) *Note:* The RANDBETWEEN function is available only if the Analysis ToolPak has been installed.

★WolframAlpha The Wolfram|Alpha instruction **RandomInteger [{m, n}, r]** generates a set of r random integers from m to n. So does the instruction **Find r random integers from m to n**.

EXAMPLE 1   **Heads and Tails** Simulate seven tosses of a fair coin. Then count the number of heads and tails.

SOLUTION   We will generate and sort seven numbers that are each either 1 or 2. The number 1 will be interpreted as a heads and the number 2 will be interpreted as a tails. See Fig. 1. This simulation of 7 coin tosses yields 3 heads and 4 tails.

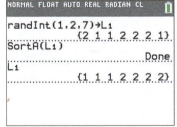

```
NORMAL FLOAT AUTO REAL RADIAN CL
randInt(1,2,7)→L₁
            {2 1 1 2 2 2 1}
SortA(L₁)
                      Done
L₁
            {1 1 1 2 2 2 2}
```

Figure 1

In Fig. 2, each entry in the range A1:G1 was generated with the equation =RANDBETWEEN(1,2). If we interpret **1** as a Heads and **2** as a Tails, then 3 Heads (and therefore 4 Tails) were produced. The COUNTIF function counts the number of ones appearing in the range of cells.

A2		f_x	=COUNTIF(A1:G1,"=1")			
	A	B	C	D	E	F	G
1	2	1	1	2	1	2	2
2	3						

Figure 2 Example 1 solved with a spreadsheet

★WolframAlpha Execute the instruction **RandomInteger [{1, 2}, 7]**, and then count the number of ones and twos.

EXAMPLE 2

Seventy-Two Rolls of a Die Simulate 72 rolls of a fair die, and tabulate the results. Compare your results with the theoretical probabilities.

SOLUTION

Generate the 72 random whole numbers from 1 to 6 and store them in $\mathbf{L_1}$. You may tally the results by setting the STATPLOT as shown in Fig. 3 and setting the WINDOW to $[0, 8]$ *by* $[-10, 30]$ to allow the maximum frequency to show in the bar graph. Such a result is shown in Fig. 4. The GRAPH key will sketch a special type of bar graph indicating the number of tosses resulting in each of the possible outcomes, 1, 2, 3, 4, 5, and 6. Use the TRACE command and the cursor to determine the number of items in each of the bars. In Fig. 4, for example, we can read from the graph that there are $n = 12$ rolls that are greater than or equal to 2 and less than 3. (For a die, this means exactly 2.)

Since the probability of each outcome is $\frac{1}{6}$ and the number of tosses is 72, we would expect that each of the outcomes would occur $\left(\frac{1}{6}\right)(72) = 12$ times. Observations of the graph or the generated list will give us the actual number of occurrences in each simulation of 72 tosses.

In Fig. 5, each cell in column B has the content **=1/6**. The Random Number Generation routine from the Data Analysis tool used the probability distribution in A1:B6 to generate the random sample in D1:L8. Then the Histogram routine from the Analysis ToolPak used the random sample and the numbers in column A to create the frequency distribution table in N1:O7. (See Appendix C for details.)

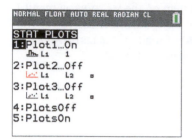

Figure 3

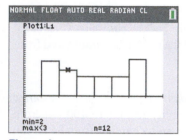

Figure 4

	A	B	C	D	E	F	G	H	I	J	K	L	M	N	O
1	1	0.166667		5	5	6	4	3	1	3	3	4		Bin	Frequency
2	2	0.166667		3	2	3	1	1	3	5	1	5		1	13
3	3	0.166667		1	4	4	6	3	4	5	4	4		2	9
4	4	0.166667		3	6	6	3	3	6	1	3	4		3	21
5	5	0.166667		3	4	2	2	2	3	5	3	2		4	13
6	6	0.166667		3	5	4	5	3	3	1	3	3		5	9
7	Prob. Dist.			6	4	2	2	5	4	3	4	1		6	7
8				1	1	1	6	3	1	2	1	2			

Figure 5 Example 2 solved with a spreadsheet

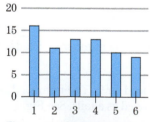

Figure 6 Histogram generated by Wolfram|Alpha

❋ WolframAlpha Execute the instruction **RandomInteger[{1, 6}, 72]**, and then count the number of times each integer 1–6 appears. Or, scroll down the page and use the histogram to obtain the number of times each integer occurs. (Fig. 6 shows a sample histogram.)

» Now Try Exercise 1

EXAMPLE 3

Simulation of Recovery Room Patients having surgery fall into three categories. Sixty percent of them require two hours in the recovery room, 30% require one hour in the recovery room, and the remainder require a half-hour. Simulate the number of hours of recovery room time required by 25 patients.

SOLUTION

One technique is to generate 25 random whole numbers from 1 to 10. Numbers 1 to 6 represent a patient who requires two hours in the recovery room, numbers 7 to 9 represent a patient who requires one hour in the recovery room, and the number 10 represents a patient needing a half-hour in the recovery room. After the 25 random numbers are generated, they can be sorted and then displayed. The right-arrow key can be used to scroll along the list so that we can count those random entries from 1 to 6, those from 7 to 9, and the 10s. Here is one possible outcome:

1, 1, 2, 2, 2, 4, 4, 4, 4, 5, 5, 5, 5, 5, 6,   7, 7, 8, 8, 8, 9,   10, 10, 10, 10, giving 15 patients who require two hours, 6 patients who require one hour, and 4 patients who require a half-hour of recovery time. The total amount of time in the recovery room needed by these patients is $15(2) + 6(1) + 4(\frac{1}{2}) = 38$ hours.

In Fig. 7, the Random Number Generation routine from the Data Analysis tool used the probability distribution in A1:B3 to generate the random sample in D1:H5. Then, the Histogram routine from the Analysis ToolPak used the random sample and the numbers in column A to create the frequency distribution table in J1:K4. The total amount of time in the recovery room needed by these patients is $3(\frac{1}{2}) + 4(1) + 18(2) = 41.5$ hours.

	A	B	C	D	E	F	G	H	I	J	K
1	1/2	.1		1	2	2	2	2		Bin	Frequency
2	1	.3		2	.5	2	1	2		1/2	3
3	2	.6		2	2	2	2	1		1	4
4	Prob. Dist.			2	2	1	.5	2		2	18
5				2	.5	2	2	2			

Figure 7 Example 3 solved with a spreadsheet

WolframAlpha Execute the instruction **RandomInteger[{1, 10}, 25]** to generate 25 random whole numbers from 1 to 10. Then interpret the output as in the previous graphing calculator discussion. **» Now Try Exercise 5**

EXAMPLE 4

Simulation of a Queue Customers steadily arrive at a bank during the hour from 9 A.M. to 10 A.M. so that the line of customers is never empty. There are three tellers, and each customer requires a varying amount of time with a teller. For simplicity, we assume that 40% of the customers need 3 minutes, 50% need 5 minutes, and 10% need 8 minutes. Each customer enters the queue at the end and goes to the first available teller when reaching the front of the queue. Simulate the service process.

(a) Show how many of the first 20 customers each of the tellers is able to service on a random day and in that hour.

(b) If all 20 customers were at the bank when it opened at 9 A.M., what was the average time spent per customer at the bank once it opened?

SOLUTION

(a) or **WolframAlpha** We generate 20 random whole numbers from 1 to 10. We consider numbers 1 to 4 as representing customers requiring 3 minutes, numbers 5 to 9 as customers requiring 5 minutes, and 10 as customers needing 8 minutes. We do not sort them, because we want to preserve the randomness of their arrival. Here is a typical list:

Cust. #:   1, 2, 3, 4, 5, 6, 7, 8, 9, 10, 11, 12, 13, 14, 15, 16, 17, 18, 19, 20
Random #: 10, 7, 2, 3, 1, 6, 4, 3, 1, 10, 6, 5, 2, 3, 7, 7, 10, 10, 5, 5

To determine the schedule, let the tellers be A, B, and C. Then the first customer goes to teller A. Since their random number is 10, they require 8 minutes, occupying teller A until 9:08. Meanwhile, the second customer, with random number 7, goes to teller B, where they need 5 minutes. They occupy teller B until 9:05. Meanwhile customer #3, with random number 2, goes to teller C until 9:03. Since teller C finishes first, customer #4 steps up to C at 9:03. That customer (with random number 3) requires 3 minutes and leaves teller C at 9:06. In case two tellers are free at the same moment, let us use the convention that the tellers are chosen in alphabetical order, with A first. From Table 1 on the next page we see that teller A served 6 of the first 20 customers, teller B served 6, and C served 8 customers. They completed the first 20 transactions at 9:34 A.M.

Table 1

Customer #	Random #	Time Req.	Teller	Start Time	End Time
1	10	8	A	9:00	9:08
2	7	5	B	9:00	9:05
3	2	3	C	9:00	9:03
4	3	3	C	9:03	9:06
5	1	3	B	9:05	9:08
6	6	5	C	9:06	9:11
7	4	3	A	9:08	9:11
8	3	3	B	9:08	9:11
9	1	3	A	9:11	9:14
10	10	8	B	9:11	9:19
11	6	5	C	9:11	9:16
12	5	5	A	9:14	9:19
13	2	3	C	9:16	9:19
14	3	3	A	9:19	9:22
15	7	5	B	9:19	9:24
16	7	5	C	9:19	9:24
17	10	8	A	9:22	9:30
18	10	8	B	9:24	9:32
19	5	5	C	9:24	9:29
20	5	5	C	9:29	9:34

In Fig. 8, the Random Number Generation routine from the Data Analysis tool could have used the probability distribution in A1:B3 to generate the random sample in D2:E21, the same sequence of random numbers presented in the third column of Table 1. Table 1 can be used as before (with the second column removed) to obtain the same answer.

	A	B	C	D	E	F	G
					Time	Start time (minutes	End time (minutes
1	3	0.4		Customer #	required	after 9 am)	after 9 am)
2	5	0.5		1	8	0	8
3	8	0.1		2	5	0	5
4	Prob. Dist			3	3	0	3
5				4	3	3	6
6				5	3	5	8
7				6	5	6	11
8				7	3	8	11
9				8	3	8	11
10				9	3	11	14
11				10	8	11	19
12				11	5	11	16
13				12	5	14	19
14				13	3	16	19
15				14	3	19	22
16				15	5	19	24
17				16	5	19	24
18				17	8	22	30
19				18	8	24	32
20				19	5	24	29
21				20	5	29	34

Figure 8 The random sample in Example 4(a) generated with a spreadsheet

(b) We can find the average amount of time spent per customer after the bank opened by totaling the time spent by all customers (use the number of minutes after 9 A.M. in the "End Time" column) and dividing by 20, the number of customers. Using the formula **=AVERAGE(G2:G21)** in the spreadsheet in Fig. 8, we calculate an average of 17.25 minutes. Thus, on the average, a person who was at the bank at 9 A.M. required 17.25 minutes after the bank opened to be served and to complete his transaction.   **≫ Now Try Exercise 7**

The simulation should be repeated many times to determine the typical outcome. So you might simulate 50 days from 9 A.M. to 10 A.M. to see how long it takes to service the first 20 customers and, on the average, how long a customer spends in the bank if they are one of the first 20 people there when the bank opens. It might be useful to see the effect of using four tellers or see what happens when the probabilities of the service times are defined differently.

In practice, the arrival times of the customers are also random and can be built into the simulation. A time-and-motion study would be used to determine the appropriate probabilistic model of the arrival process.

EXERCISES 6.7

1. **Rolling a Die** Simulate 36 rolls of a fair die. Give the relative frequency and the corresponding theoretical probability of each of the outcomes, and compare them.

2. **Rolling Dice** Simulate 96 rolls of a pair of dice where the sum is observed. Give the relative frequency and the corresponding theoretical probability of each of the outcomes. Make a table showing your results. Repeat the experiment 6 times, and consider the total as if 576 rolls were simulated. *Note:* (6)(96) = 576.

3. **Free-Throws** Simulate 10 free-throws for Kobe Bryant, whose career free throw average is 84%. How many of the shots were successful?

4. **Baseball** A baseball player is a .331 hitter. Simulate 10 at-bats for this player, and tell how many hits he gets. (Being a .331 hitter means that 33.1% of his at-bats result in a hit.)

5. **Test Taking** A student who has not studied for a 10-question multiple-choice test, with 4 choices among the answers (a, b, c, d) for each question, decides to simulate such a test and answer the questions according to a simulation in which each choice of answer has the same probability. Use technology to generate a simulated answer sheet. Assume that the correct answers are a, b, b, c, d, d, a, c, b, a. What is the student's score?

6. **Balls in an Urn** In sampling 4 balls at random from an urn containing 30 balls, *without replacement* after each draw, we consider the balls as numbered 1 to 30. In selecting random whole numbers from 1 to 30, we ignore any number that has already been selected and continue the selection until we obtain a sample of size 4. Assume that there are 20 red balls and 10 green balls in the urn. Draw 10 samples of size 4, and tabulate the number of red balls in each sample. Compare your results with the theoretical probability.

7. **Registration Queue** Students are queued up at the registrar's office when the registration windows open at 8 A.M. There are four open windows; students approach the first open window as they advance to the front of the queue. Assume that 10% of the students require 5 minutes of service time, 30% of the students require 7 minutes of service time, 40% require 10 minutes, and 20% require 15 minutes. Simulate the service of the first 20 students in a random queue. Show the schedule of service at the four windows (A, B, C, D), determine how long it takes to process these students, and give the average time from 8 A.M. to leaving the service window.

8. **Bank Queue** Simulate the bank queue of Example 4, using four tellers. Give the time needed to process the first 20 customers and the average time spent by each customer in the bank after 9 A.M.

9. **Gas Queue** A gas station with four self-serve pumps has determined that 80% of all customers completely fill their gas tanks and the remaining 20% fill their tank with a fixed dollar amount's worth of fuel. Suppose that it takes an average of 5 minutes for a complete fill-up and 3 minutes for a partial fill-up. Suppose also that, from 5 P.M. until 6 P.M., customers arrive steadily so that there is always a line and that the next customer in line proceeds to the next available pump. Simulate this process for 30 customers.

10. **Rolling Three Dice** Simulate 108 rolls of three dice, and show the frequency of each possible sum of the faces: 3, 4, . . . , 18.

CHAPTER 6 Summary

KEY TERMS AND CONCEPTS	EXAMPLES

6.1 Experiments, Outcomes, Sample Spaces, and Events

The **sample space** for an **experiment** is the set of all possible **outcomes**. An **event** is a subset of the sample space.

Experiment: roll a die, and observe the topmost number.

Possible outcomes: 1, 2, 3, 4, 5, 6

Sample space: $\{1, 2, 3, 4, 5, 6\}$

$E = \{2, 4, 6\}$ is the *event* that the outcome is an even number.

The events E and F are **mutually exclusive** if $E \cap F = \varnothing$.

In the experiment above, the events $E = \{2, 4, 6\}$ and $F = \{3\}$ are mutually exclusive.

6.2 Assignment of Probabilities

Properties of probability:

1. $0 \le \Pr(\text{an outcome}) \le 1$.

2. Sum of probabilities of all outcomes $= 1$.

3. $\Pr(E) = $ sum of probabilities of outcomes in E.

Inclusion–Exclusion Principle:

$$\Pr(A \cup B) = \Pr(A) + \Pr(B) - \Pr(A \cap B)$$

Select a card at random from a deck of 52 cards. Let A be the event that the card is red, and B the event that it is a king. $\Pr(A) = \frac{1}{2}$, $\Pr(B) = \frac{1}{13}$, and $\Pr(A \cap B) = \frac{1}{26}$. Then

$$\Pr(A \cup B) = \frac{1}{2} + \frac{1}{13} - \frac{1}{26} = \frac{7}{13}.$$

Odds: Find the ratio (reduced to lowest terms) of the probability that an event will happen to the probability that it will not happen. If the ratio is a/b, then the odds in favor of the event are a to b.

Find the odds of drawing a king from a deck of cards.

$$\frac{\Pr(\text{king})}{\Pr(\text{not king})} = \frac{\frac{1}{13}}{\frac{12}{13}} = \frac{1}{12}.$$

Therefore, the odds of drawing a king are 1 to 12.

6.3 Calculating Probabilities of Events

If S is a sample space with **equally likely outcomes** and E is an event in S, then

$$\Pr(E) = \frac{n(E)}{n(S)}.$$

A number is chosen at random from $S = \{1, 2, 3, 4, 5\}$. What is the probability that the number is greater than 3?

The event E is $\{4, 5\}$, $n(S) = 5$, and $n(E) = 2$.

Therefore, $\Pr(E) = \dfrac{2}{5}$.

Complement rule: $\Pr(E) = 1 - \Pr(E')$

With E and S as above, $E' = \{1, 2, 3\}$ and $\Pr(E') = \frac{3}{5}$.

Therefore, $\Pr(E) = 1 - \dfrac{3}{5} = \dfrac{2}{5}$.

Health statistics: The effectiveness of an intervention can be quantified by the *absolute risk reduction* (ARR), *relative risk reduction* (RRR), and *number needed to treat* (NNT) methods.

If an intervention resulted in 39 people out of 1000 developing a certain condition, but 50 people out of 1000 developed the condition in a control group, then $ARR = \frac{50 - 39}{1000} = 1.1\%$, $RRR = \frac{50 - 39}{50} = 22\%$, and $NNT = \frac{1}{ARR} = 91$.

6.4 Conditional Probability and Independence

Conditional probability:

$$\Pr(E \mid F) = \frac{\Pr(E \cap F)}{\Pr(F)}, \Pr(F) \ne 0$$

Roll a die. What is the probability that the number showing is 2, given that the number is even?

Let E be the event that the number is 2.

Let F be the event that the number is even.

$$\Pr(E \mid F) = \frac{\frac{1}{6}}{\frac{1}{2}} = \frac{1}{3}.$$

KEY TERMS AND CONCEPTS	EXAMPLES
Conditional probability for equally likely outcomes: $$\Pr(E\mid F) = \frac{n(E\cap F)}{n(F)}, n(F) \neq 0$$	For the situation above, $$\Pr(E\mid F) = \frac{n(E\cap F)}{n(F)} = \frac{1}{3}.$$
Product rule: $$\Pr(E\cap F) = \Pr(F)\cdot\Pr(E\mid F), \Pr(F) \neq 0$$	For the situation above, $$\Pr(E\cap F) = \frac{1}{2}\cdot\frac{1}{3} = \frac{1}{6}.$$
Independent events: Definition: $\Pr(E\cap F) = \Pr(E)\cdot\Pr(F)$ $\Pr(E\mid F) = \Pr(E)$ and $\Pr(F\mid E) = \Pr(F)$	For the situation above, $$\Pr(E\cap F) = \frac{1}{6} \text{ and } \Pr(E)\cdot\Pr(F) = \frac{1}{6}\cdot\frac{1}{2} = \frac{1}{12}.$$ Therefore, the two events are not independent.

6.5 Tree Diagrams

Tree diagrams are used to determine probabilities of combined outcomes in a sequence of experiments.

An urn contains four red balls and six white balls. If two balls are drawn in succession without replacement, what is the probability that the second ball is red?

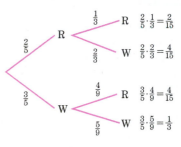

$$\Pr(\text{2nd ball red}) = \frac{2}{15} + \frac{4}{15} = \frac{2}{5}.$$

Medical screening: Tree diagrams are used to obtain the *positive predicted value* of a test [PPV = Pr(patient has condition|test positive)], given the *sensitivity* (true positive rate), the *specificity* (true negative rate), and the percentage of the population having the condition.

For a condition affecting .02% of the population, a test with 98% sensitivity and 99% specificity has PPV $\approx$ .02.

6.6 Bayes' Theorem, Natural Frequencies

Bayes' theorem ($n = 2$): If B_1 and B_2 are mutually exclusive events and $B_1 \cup B_2 = S$, then

$$\Pr(B_i\mid A) = \frac{\Pr(B_i)\cdot\Pr(A\mid B_i)}{\Pr(B_1)\cdot\Pr(A\mid B_1) + \Pr(B_2)\cdot\Pr(A\mid B_2)}.$$

For the situation in 6.5, find the probability that the first ball is red, given that the second ball is red.

Let B_1 be the event that the first ball is red.
Let B_2 be the event that the first ball is white.
Let A be the event that the second ball is red.

$$\Pr(B_1\mid A) = \frac{\frac{2}{5}\cdot\frac{1}{3}}{\frac{2}{5}\cdot\frac{1}{3} + \frac{3}{5}\cdot\frac{4}{9}} = \frac{\frac{2}{15}}{\frac{6}{15}} = \frac{1}{3}.$$

Natural Frequencies: Many types of probability problems that can be solved with Bayes' theorem also can be solved with the natural frequencies method.

6.7 Simulation

The ability to use technology to generate random numbers allows us to **simulate** the outcomes of experiments.

(Calculator) **randInt(*m,n,r*)** generates *r* random integers from *m* to *n*.

(Calculator) **randInt(1,5,3)** might generate {2,4,1}.

(Spreadsheet) The formula **=RANDBETWEEN(*m,n*)** generates a random integer from *m* to *n*.

(Spreadsheet) **=RANDBETWEEN(5,10)** might generate 6.

(Wolfram|Alpha) **RandomInteger[{*m,n*},*r*]** generates *r* random integers from *m* to *n*.

(Wolfram|Alpha) **RandomInteger[{5,10},4]** might generate {9, 9, 6, 8}.

1. What is the sample space of an experiment?

2. Using the language of sets and assuming that A and B are events in a sample space S, write the following events in set notation: (A or B); (A and B); not A.

3. In a sample space, what is the probability of the empty set?

4. What subset in a sample space corresponds to the certain event?

5. Write a formula for the probability of the event $A \cup B$, assuming that you know $\Pr(A)$, $\Pr(B)$, and $\Pr(A \cap B)$.

6. Explain the difference between mutually exclusive events and independent events.

7. State the addition principle.

8. Suppose that the probability of an event is k/n. What are the odds that the event will occur?

9. Suppose that the odds that an event occurs are a to b. What is the probability that the event will occur?

10. State the inclusion–exclusion principle for two events.

11. What is the definition of $\Pr(E \mid F)$?

12. What is Bayes' Theorem?

13. What is a tree diagram?

1. **Coins** A box contains a penny, a nickel, a dime, a quarter, and a half dollar. You select two coins at random from the box.
   (a) Construct a sample space for this situation.
   (b) List the elements of the event E in which the total value of the coins you have selected is an even number of cents.

2. **Candidates for Office** Some of the candidates for president of the computer club at Riverdale High are seniors, and the rest are juniors. Let J be the event in which a junior is elected, and let F be the event in which a female is elected. Describe the following events:
   (a) $J \cap F'$ (b) $(J \cap F)'$ (c) $J \cup F'$

3. Suppose that E and F are events with $\Pr(E) = .4$, $\Pr(F) = .3$, and $\Pr(E \cup F) = .5$. Find $\Pr(E \cap F)$.

4. Suppose that E and F are mutually exclusive events with $\Pr(E) = .5$ and $\Pr(F) = .3$. Find $\Pr(E \cup F)$.

5. **Languages** Of the 120 students in a class, 30 speak Chinese, 50 speak Spanish, 75 speak French, 12 speak Spanish and Chinese, 30 speak Spanish and French, and 15 speak Chinese and French. Seven students speak all three languages. A student is chosen at random. What is the probability that they speak none of these languages?

6. **Physical Fitness** Of 50 fitness buffs surveyed, 15 like bicycling, 20 like jogging, and 5 like bicycling and jogging. What is the probability that a fitness buff selected at random likes only one of the two sports?

7. **Commuting Time** The odds of an American worker living within 20 minutes of work are 13 to 12. What is the probability that a worker selected at random lives within 20 minutes of work? Lives more than 20 minutes from work?

8. **Golf** According to the Professional Golfers Association (PGA), the odds of a professional golfer scoring a hole-in-one in a single round of a PGA tournament are 1 to 3708. What is the probability of a professional golfer scoring a hole-in-one in a single round of a PGA tournament?

9. **Demographics** Twenty-six percent of all Americans are under 18 years old. What are the odds that a person selected at random is under 18? Eighteen years old or over?

10. **Baseball** The probability that the average major league baseball player will get a hit when at bat is .25. What are the odds that the average major league baseball player will get a hit?

11. **Committee Selection** A committee consists of five men and five women. If three people are selected at random from the committee, what is the probability that they all will be men?

12. **Matching Socks** A drawer contains two red socks and two blue socks. If two socks are drawn randomly from the drawer, what is the probability that the two socks have the same color?

13. **Barrel of Apples** Five of the apples in a barrel of 100 apples are rotten. If four apples are selected from the barrel, what is the probability that at least two of the apples are rotten?

14. **Opinion Sampling** Of the nine city council members, four favor school vouchers and five are opposed. If a subcommittee of three council members is selected at random, what is the probability that exactly two of them favor school vouchers?

15. **Exam Questions** Prior to taking an essay examination, students are given 10 questions to prepare. Six of the ten will appear on the exam. One student decides to prepare only eight of the questions. Assume that the questions are equally likely to be chosen by the professor.
    (a) What is the probability that they have prepared every question appearing on the test?
    (b) What is the probability that both questions that they did not prepare appear on the test?

16. **Craps** In the casino game of *craps*, a player rolls two dice and observes their sum. One of the most common bets in craps is the *pass line bet*. The rules of the pass line bet are as follows:
    - If the sum of the dice is 7 or 11, the player wins the pass line bet. If the sum is 2, 3, or 12, the player loses.
    - If the sum is a number other than 2, 3, 7, 11, or 12, a *point* is established and the player continues to roll. If that sum is rolled again before a 7 is rolled, the player wins. If a 7 is rolled before the sum is rolled again, then the player loses.
    (a) What is the probability of a player winning the pass line bet on the first roll? What is the probability of a player losing on the first roll?

(b) Suppose the player rolls a sum of 6 on their first roll. What is the probability that they will win on the next roll? What is the probability that they will lose on the next roll?

17. **Coin Tosses** A coin is to be tossed five times. What is the probability of obtaining at least one head?

18. **Coin Tosses** Two players each toss a coin three times. What is the probability that they get the same number of tails?

19. **Olympic Swimmers** In an Olympic swimming event, two of the seven contestants are American. The contestants are randomly assigned to lanes 1 through 7. What is the probability that the Americans are assigned to the first two lanes?

20. **Final Four** Sixty-eight men's college basketball teams compete in the NCAA championship tournament. They are divided into four regions of seventeen teams each. The winners of each region advance to the Final Four. If a fan selects four teams at random, what are the odds against those teams being in the Final Four? *Note:* Of course, the fan will select only one team per region.

21. **Code Words** A collection of code words consists of all strings of seven characters, where each of the first three characters can be any letter or digit and each of the last four characters must be a digit. For example, 7A32765 is allowed, but 7A3B765 is not.
   (a) What is the probability that a code word chosen at random begins with ABC?
   (b) What is the probability that a code word chosen at random ends with 6578?
   (c) What is the probability that a code word chosen at random consists of three letters followed by four even digits?

22. **Drawing Cards** A card is drawn at random from a deck of cards. Then the card is replaced, and the deck is thoroughly shuffled. This process is repeated two more times.
   (a) What is the probability that all three cards are aces?
   (b) What is the probability that at least one of the cards is an ace?

23. **Dice** What is the probability of having each of the numbers one through six appear in six consecutive rolls of a die?

24. **Dice** Find the odds in favor of getting four different numbers when tossing four dice.

25. **Birthdays** What is the probability that, out of a group of five people, at least two people have the same birthday? *Note:* Assume that there are 365 days in a year.

26. **Birthdays** Four people are chosen at random. What is the probability that at least two of them were born on the same day of the week?

27. Let E and F be events with $\Pr(E) = .4, \Pr(F) = .3$, and $\Pr(E \cup F) = .5$. Find $\Pr(E \mid F)$.

28. Let E and F be events with $\Pr(E \cap F) = \frac{1}{10}$ and $\Pr(E \mid F) = \frac{1}{7}$. Find $\Pr(F)$.

29. **Coin Tosses** When a coin is tossed three times, what is the probability of at least one tail appearing, given that at least one head appeared?

30. **Dice** Suppose that a pair of dice is rolled. Given that the two numbers are different, what is the probability that one die shows a three?

31. **Gender and Unemployment** Consider Table 1, with figures in millions, pertaining to the 2015 American civilian labor force (age 20+). Find the probability that a person selected at random from the American civilian labor force (age 20+)
   (a) is employed.
   (b) is male.
   (c) is female, given that the person is employed.
   (d) is employed, given that the person is female.

Table 1

	Employed	Unemployed	Total
Male	76.78	3.96	80.74
Female	70.70	3.37	74.07
Total	147.48	7.33	154.81

Source: www.bls.gov

32. **Public and Private Colleges** Out of the 50 colleges in a certain state, 25 are private, 15 offer engineering majors, and 5 are private colleges offering engineering majors. Find the probability that a college selected at random from the state
   (a) offers an engineering major.
   (b) offers an engineering major, given that it is public.
   (c) is private, given that it offers an engineering major.
   (d) is public, given that it offers an engineering major.

33. **Premed Majors** Suppose that a certain college contains an equal number of female and male students and that 8% of the female population are premed majors. What is the probability that a randomly selected student is a female premed major?

34. **Balls in an Urn** An urn contains 10 red balls and 20 green balls. If four balls are drawn one at a time without replacement, what is the probability that the sequence of colors will be red, green, green, red?

35. **Dice** A red die and a green die are rolled as a pair. Let E be the event that "the red die shows a 2" and let F be the event that "the sum of the numbers is 8." Are the events E and F independent?

36. **Coin Tosses** Suppose that we toss a coin three times and observe the sequence of heads and tails. Let E be the event that "the first toss lands heads" and F the event that "there are more heads than tails." Are E and F independent?

37. **Balls in an Urn** An urn contains four red balls labeled 1, 2, 3, and 4 and six green balls labeled 5, 6, 7, 8, 9, and 10. A ball is selected at random from the urn. Are the events "the ball is red" and "the number on the ball is even" independent events?

38. **Cards** Two cards are drawn in succession (without replacement) from a deck of cards. Are the events "the first card is a king" and "the second card is red" independent events?

39. **Archery** Two archers shoot at a moving target. One can hit the target with probability 1/4 and the other with probability 1/3. Assuming that their efforts are independent events, what is the probability that
   (a) both will hit the target?
   (b) at least one will hit the target?

40. **Final Exam** Fred will do well on his final exam if the exam is easy or if he studies hard. Suppose that the probability is .4

that the exam is easy, the probability is .75 that Fred will study hard, and that the two events are independent. What is the probability that
(a) the exam is easy and that Fred will study hard?
(b) Fred will do well on his final exam?

41. Let A and B be independent events for which the probability that at least one of them occurs is 1/2 and the probability that B occurs but A does not occur is 1/3. Find $\Pr(A)$.

42. Let A and B be independent events with $\Pr(A) = .3$ and $\Pr(B) = .4$. What is the probability that exactly one of the events A or B occurs?

43. **Prizes** Each box of a certain brand of candy contains either a toy airplane or a toy boat. If one-third of the boxes contain an airplane and two-thirds contain a boat, what is the probability that a person who buys two boxes of candy will receive both an airplane and a boat?

44. **Balls in an Urn** An urn contains three balls numbered 1, 2, and 3. Balls are drawn one at a time without replacement until the sum of the numbers drawn is four or more. Find the probability of stopping after exactly two balls are drawn.

45. **Carnival Game** A carnival huckster has placed a coin under one of three cups and asks you to guess which cup contains the coin. After you select a cup, they remove one of the unselected cups, which they guarantee does not contain the coin. You may now either stay with your original choice or switch to the other remaining cup. What decision will give you the greater probability of winning?

46. **Political Poll** Of a group of people surveyed in a political poll, 60% said that they would vote for candidate R. Of those who said that they would vote for R, 90% actually voted for R, and of those who did not say that they would vote for R, 5% actually voted for R. What percent of the group voted for R?

47. **Left-Handedness** According to a geneticist at Stanford University, the chances of having a left-handed child are 4 in 10 if both parents are left-handed, 2 in 10 if one parent is left-handed, and only 1 in 10 if neither parent is left-handed. Suppose that a left-handed child is chosen at random from a

population in which 25% of the adults are left-handed. What is the probability that the child's parents are both left-handed?

48. **Tax Audits** An auditing procedure for income tax returns has the following characteristics: If the return is incorrect, the probability is 90% that it will be rejected; if the return is correct, the probability is 95% that it will be accepted. Suppose that 80% of all income tax returns are correct. If a return is audited and rejected, what is the probability that the return was actually correct?

49. **Weighing Produce** A supermarket has three employees who package and weigh produce. Employee A records the correct weight 98% of the time. Employees B and C record the correct weight 97% and 95% of the time, respectively. Employees A, B, and C handle 40%, 40%, and 20% of the packaging, respectively. A customer complains about the incorrect weight recorded on a package that they purchased. What is the probability that the package was weighed by employee C?

50. **Dragons** An island contains an equal number of one-headed, two-headed, and three-headed dragons. If a dragon head is picked at random, what is the likelihood of its belonging to a one-headed dragon?

Conceptual Exercises

51. Explain why it makes sense that if E and F are independent events, then so are E and F'.

52. What additional information would allow you to compute $\Pr(E \cap F)$ if you already know $\Pr(E)$ and $\Pr(F)$?

53. Explain why two independent events with nonzero probabilities cannot be mutually exclusive.

54. Explain why two mutually exclusive events with nonzero probabilities cannot be independent.

55. (True or False) When a conditional probability is calculated, the probability of the event that is given is placed in the denominator.

56. Can two events E and F be mutually exclusive if $\Pr(E) = .6$ and $\Pr(F) = .7$?

Two Paradoxes

First Paradox: *Under certain circumstances, you have your best chance of winning a tennis tournament if you play most of your games against the best possible opponent.*

Alice and her two sisters, Betty and Carol, are avid tennis players. Betty is the best of the three sisters, and Carol plays at the same level as Alice. Alice defeats Carol 50% of the time but only defeats Betty 40% of the time.

Alice's mother offers to give her $100 if she can win two consecutive games when playing three alternating games against her two sisters. Since the games will alternate, Alice has two possibilities for the sequence of opponents. One possibility is to play the first game against Betty, followed by a game with Carol, and then another game with Betty. We will refer to this sequence as BCB. The other possible sequence is CBC.

1. Make a guess of the best sequence for Alice to choose—the one having the majority of the games against the weaker opponent or the one having the majority of the games against the stronger opponent.

2. Calculate the probability of Alice getting the $100 reward if she chooses the sequence CBC.

3. Calculate the probability of Alice getting the $100 reward if she chooses the sequence BCB.

4. Which sequence should Alice choose?

5. How would you explain to someone who didn't know probability why the sequence that you chose is best?

Second Paradox: *The probability of a male applicant being admitted to a graduate school can be higher than the probability for a female applicant, even though for each department the probability of a female being admitted is higher.* (This apparent contradiction is known as **Simpson's paradox**.)

To simplify matters, consider a university with two professional graduate programs, medicine and law. Suppose that last year 1000 men and 1000 women applied, and the outcome was as shown in Table 1.

Table 1

	Men			Women		
	Applied	Accepted	Rejected	Applied	Accepted	Rejected
Law	700	560	140	400	340	60
Medicine	300	40	260	600	160	440

6. What is the probability that a male applicant was accepted to a professional program? A female applicant?

7. Which gender does the university appear to be favoring?

8. What is the probability that a male applicant was accepted to law school? A female applicant?

9. What is the probability that a male applicant was accepted to medical school? A female applicant?

10. Which gender do the individual professional schools appear to be favoring?

11. Without using probability, justify the apparent contradiction between the answers for part 7 and part 10.

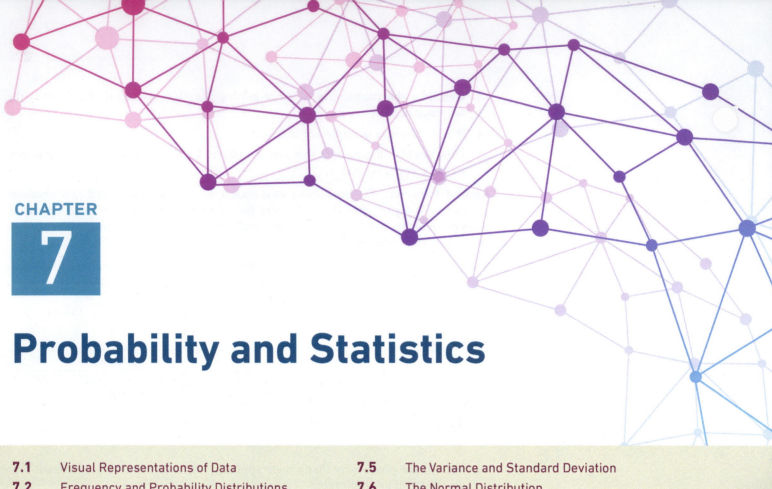

CHAPTER 7

Probability and Statistics

Statistics is the branch of mathematics that deals with data: its collection, description, analysis, and use in prediction. In this chapter, we present some topics in statistics that can be used as a springboard to further study. Since we are presenting a series of topics rather than a comprehensive survey, we will bypass large areas of statistics without saying anything about them. However, the discussion should give you some feeling for the subject.

7.1   Visual Representations of Data

Data can be presented in raw form or organized and displayed in tables or charts. In this section, we use various types of charts to visualize and analyze data.

In the fall of 2015, freshmen at 199 baccalaureate colleges and universities in the United States answered an extensive questionnaire. The detailed results of the questionnaire are given in *The American Freshman*: *National Norms for Fall 2015* (Los Angeles: Higher Education Research Institute, 2016, UCLA). One question asked students to give the highest degree they planned to pursue. See Table 1. Such a table is often referred to as a **frequency table**, since it presents the frequency with which each response occurs.

Table 1 Frequency Table

Highest Degree Planned	Number	Percent
Bachelor's	29,791	21.1
Master's	59,441	42.1
Doctorate	26,826	19.0
Medical	15,813	11.2
Law	5,930	4.2
Other	3,388	2.4
Total	141,189	100.0

Bar and Pie Charts

The numbers from the example in Table 1 are displayed in the **bar chart** of Fig. 1. This pictorial display gives a good feel for the relative number of students planning to earn each degree.

The percentages in the right column of Table 1 represent the students out of this group of 141,189 who plan to pursue each degree. The bar chart for the percentages is shown in Fig. 2. It looks exactly like the bar chart in Fig. 1. The only difference is the labeling of the tick marks along the *y*-axis.

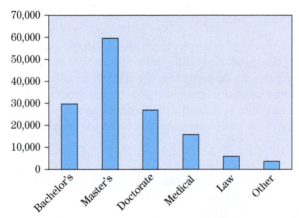

Figure 1 Bar chart for highest degree planned

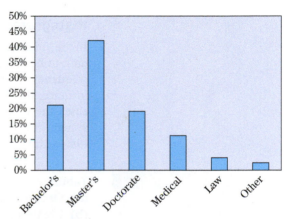

Figure 2 Bar chart for highest degree planned

Another popular type of chart that can be used to display data consisting of several categories is the **pie chart**. It consists of a circle subdivided into sectors (slices of pie), where each sector corresponds to a category. The area of each sector is proportional to the percentage of items in that category. This is accomplished by making the central angle of each sector equal to 360° times the percentage associated with the segment.

EXAMPLE 1   **Freshman Aspirations** Create a pie chart for the "highest degree planned" data. Label each sector with its category and percentage.

SOLUTION   **Step 1**   Use the rightmost column of Table 1 to obtain the central angles for the sectors. See Table 2.

Table 2

Highest Degree Planned	Percent	360° × percent
Bachelor's	21.1	76.0°
Master's	42.1	151.6°
Doctorate	19.0	68.4°
Medical	11.2	40.3°
Law	4.2	15.1°
Other	2.4	8.6°

Step 2   Draw a circle, draw a vertical radius line extending from the center of the circle, and then draw an angle of measure approximately 76° with the radius line as the initial side of the angle. See Fig. 3(a).

Step 3   Draw an angle of approximately 151.6°, using the terminal side of the angle drawn in step 2 as the initial side. See Fig. 3(b).

Step 4   Continue as in step 3 to draw each sector, and then label the sectors with their categories and percentages.

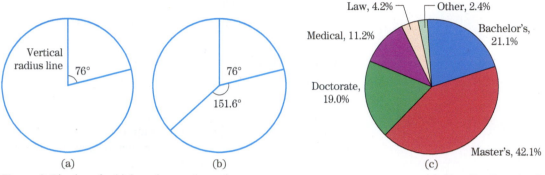

Figure 3 Pie chart for highest degree planned.   » *Now Try Exercise 7*

Histograms

In the "highest degree planned" example, the responses to the question were words. The data to be organized consisted of six different words, where each occurred with a high frequency. In many cases to be analyzed, the data is a collection of numbers. For instance, the data could consist of ages, weights, or test scores of individuals. In such cases, the *x*-axis in a bar chart is labeled with numbers, as in an ordinary *x*-*y* coordinate system, the data is referred to as *numerical data*, and the charts themselves are called **histograms**.

EXAMPLE 2   **Tabulating Quiz Scores**   Figure 4 gives the quiz scores for a class of 25 students.
(a) Organize the data into a frequency table.
(b) Create a histogram for the data.

8 7 6 10 5 10 7 1 8 0 10 5 9 3 8 6 10 4 9 10 7 0 9 5 8
Figure 4

SOLUTION   (a) An easy way to count the number of exams for each score is to write down the numbers from 0 through 10, and considering the quiz papers one at a time, make a slash mark alongside the score for each paper. Such a tabulation produces Table 3(a). In Table 3(b), the slash marks have been totaled.

(b) The histogram (Fig. 5) is drawn on an *x*-*y* coordinate system. Note that each bar is centered over its corresponding score.

Table 3 Tabulation of Quiz Scores

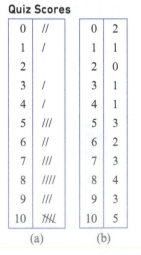

0	//		0	2
1	/		1	1
2			2	0
3	/		3	1
4	/		4	1
5	///		5	3
6	//		6	2
7	///		7	3
8	////		8	4
9	///		9	3
10	𝓗𝓗		10	5
(a)			(b)	

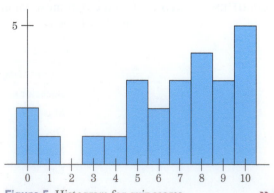

Figure 5 Histogram for quiz scores   » *Now Try Exercise 13*

Median, Quartiles, and Box Plots

When an instructor returns exam papers, they usually give students some indication of how they did overall. Occasionally, the instructor gives the average of the grades and perhaps the standard deviation. (These two topics are discussed extensively in the remainder of this chapter.) However, most instructors state the **median** of the grade distribution. The median grade is the grade that separates the lower half of the grades from the upper half. To find the median of a set of N numbers, first arrange the numbers in increasing or decreasing order. The median is the middle number if N is odd and the average of the two middle numbers if N is even.

EXAMPLE 3

Medians Find the medians of the following two sets of data:

(a) Danny Willett's scores on four rounds of golf in the 2016 U.S. Masters tournament:

$$70 \quad 74 \quad 72 \quad 67$$

(b) The 25 quiz scores discussed in Example 2

SOLUTION

(a) Here $N = 4$, an even number. Arranged in increasing order, the four scores are

$$67, 70, 72, 74.$$

The middle two scores are 70 and 72. The median is their average. Therefore,

$$\text{median} = \frac{70 + 72}{2} = \frac{142}{2} = 71.$$

(b) Here $N = 25$, an odd number. The position of the middle number is $\dfrac{25 + 1}{2} = 13$.

The tabulation of quiz scores in Table 3 can be used instead of an ordering of the scores. The median will be the 13th highest score. Adding up the numbers of scores of 10s, 9s, and 8s gives 12 scores. Therefore, the 13th score must be a 7. That is, the median is 7.   ≪

Graphing calculators can display a picture, called a **box plot**, that analyzes a set of data and shows not only the median but also the **quartiles**. The quartiles are the medians of the sets of data to the left and right of the median. The median of the numbers less than the median is called the first quartile and is denoted Q_1. The median of the numbers greater than the median is called the third quartile and is denoted Q_3. A box plot also is useful in showing pictorially the spread of the data. Figure 6 shows the five pieces of information given by a box plot. This information is referred to as the **five-number summary** of the data. The median is also called the second quartile and is denoted Q_2. Essentially, the three quartiles divide the data into four approximately equal parts, each part consisting of roughly 25% of the numbers. The length of the rectangular part of the box plot, which is $Q_3 - Q_1$, is called the **interquartile range**. The quartiles provide information about the dispersion of the data. The interquartile range is the length of the interval in which approximately the middle 50% of the data lie.

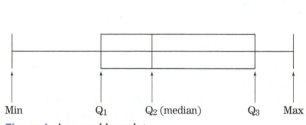

Min Q_1 Q_2 (median) Q_3 Max

Figure 6 A general box plot

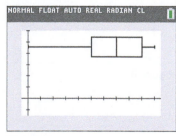

NORMAL FLOAT AUTO REAL RADIAN CL

Figure 7

Figure 7 gives the box plot for the set of 25 quiz scores presented in Example 2. The steps for obtaining the box plot will be shown in the Incorporating Technology discussion at the end of this section.

EXAMPLE 4 **Five-Number Summary and Interquartile Range** Find the five-number summary and the interquartile range for the following set of numbers: 1 3 6 10 15 21 28 36 45 55.

SOLUTION The numbers are given in ascending order, and there are 10 numbers. Immediately, we see that min = 1 and max = 55. The next number to be found is the median. Since 10 is an even number, the median is the average of the middle two numbers.

$$1 \quad 3 \quad 6 \quad 10 \quad \mathbf{15} \quad \mathbf{21} \quad 28 \quad 36 \quad 45 \quad 55$$

That is,

$$Q_2 = \text{median} = \frac{15 + 21}{2} = 18.$$

The numbers to the left of the median are 1 3 6 10 15, and the numbers to the right of the median are 21 28 36 45 55. These lists have medians 6 and 36, respectively. Therefore, $Q_1 = 6$, $Q_3 = 36$, and the interquartile range is $Q_3 - Q_1 = 30$. **≫ Now Try Exercise 19**

INCORPORATING
TECHNOLOGY

Histograms and Box Plots The steps required to create a histogram similar to the one in Fig. 5 are shown in the four figures that follow. The details of carrying out the task are presented in Appendix B. The window settings for the histogram will depend on the data.

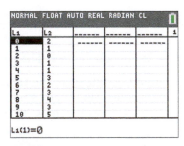

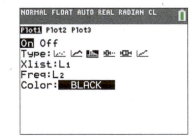

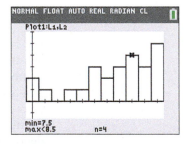

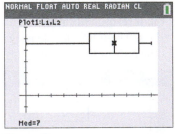

Figure 8

The steps for obtaining a box plot are the same as for the histogram, with the exception that, in the second screen, the fifth icon is selected. Figure 8, which shows the median, appears when TRACE is pressed. Pressing the left and right arrow keys reveals the smallest value (**minX**), the largest value (**maxX**), and the first and third quartiles.

Histograms and Pie Charts A histogram similar to the one in Fig. 5 can be drawn with the following steps using Excel.

1. Place the data from both columns of Table 3(b) into two columns of a spreadsheet, and select the data. (A **Quick Analysis** button will appear at the bottom-right corner of the selection.)
2. Click on the **Quick Analysis** button. (A rectangle of options will appear.)
3. Click on **Charts** in the top row of the rectangle and then click on the **Clustered Column** button. (A coordinate system showing a histogram will appear.)
4. On the **Design** tab, in the **Chart Layouts** group, click on **Quick Layout**, then click on Layout 8 (the second button in the third row).
5. Click on each of the legends labeled "Chart Title" and "Axis Title," and press the **Del** key.

A pie chart similar to the one in Fig. 3(c) can be drawn with the following steps using Excel.

1. Place the data from the left and center columns of Table 1 into two columns of a spreadsheet, and select the data.
2. Click on the **Quick Analysis** button.
3. Click on **CHARTS** in the top row of the rectangle and then click on the **Pie** button.
4. Click on the legend labeled "Chart Title," and press the **Del** key.

✻WolframAlpha A bar chart similar to the one in Fig. 1 and a pie chart similar to the one in Fig. 3(c) can be drawn with the following instructions.

bar chart {29791,59441,26826,15813,5930,3388}

pie chart {29791,59441,26826,15813,5930,3388}

Consider the set of numbers $\{a, b, \ldots, c\}$. The statement **mean** $\{a, b, \ldots, c\}$ gives the average value of the numbers. Replacing the word *mean* with **median**, **minimum**, or **maximum** gives other descriptors of the data. When there is an even number of numbers, the word **mean** can be replaced with **quartiles** or **interquartile** range. (Wolfram|Alpha's definition of quartile differs from ours when there is an odd number of numbers.)

Check Your Understanding 7.1

Solutions can be found following the section exercises.

Suppose that a list consists of 17 numbers in increasing order. Clearly, the first number is the min, and the last number is the max.

1. Which number in the list is the median?

2. Which numbers in the list are used to obtain the first quartile?

3. Which numbers in the list are used to obtain the third quartile?

EXERCISES 7.1

In Exercises 1–4, display the data in a bar chart that shows frequencies on the *y*-axis.

1. **2016 School Enrollments**

Type	Enrollment (in millions)
Elementary	39.0
Secondary	16.1
Postsecondary	20.5

Source: U.S. National Center for Education Statistics, *Digest of Education Statistics.*

2. **2015 U.S. Defense Employees**

Branch	Officers and Enlistees (in thousands)
Army	477.9
Navy	323.7
Marine Corps	183.2
Air Force	307.0

Source: U.S. Dept. of Defense.

3. **Areas of the Great Lakes**

Lake	Area (sq mi)
Superior	31,700
Michigan	22,300
Huron	23,100
Erie	9,910
Ontario	7,550

Source: Encyclopedia Britannica.

4. **Bachelor's Degrees Earned in 2013, by Field**

Field of Study	Number of Degrees
Business	357,823
Social sciences	177,778
Education	104,647
Health services	181,144
Psychology	114,450
Engineering	85,980
Other	815,342

Source: U.S. National Center for Education Statistics, *Digest of Education Statistics.*

5. **School Enrollments** Display the data from the table in Exercise 1 in a bar chart showing percentages on the *y*-axis.

6. **U.S. Defense Employees** Display the data from the table in Exercise 2 in a bar chart showing percentages on the *y*-axis.

7. **Interest on Public Debt** In 2015, the interest on the public debt accounted for about 10% of the federal budget. If the federal budget is displayed in a pie chart, what should be the size of the central angle of the sector corresponding to the interest on the public debt?

8. **U.S. Defense Employees** Display the data from the table in Exercise 2 in a pie chart.

9. **Great Lakes** Display the data from the table in Exercise 3 in a pie chart.

10. **Bachelor's Degrees** Display the data from the table in Exercise 4 in a pie chart.

Freshman Aspirations Exercises 11 and 12 refer to the pie chart in Fig. 3(c).

11. What is the probability that a freshman selected at random in the fall of 2015 planned to obtain a master's or doctorate as their highest degree?

12. What is the probability that a freshman selected at random in the fall of 2015 planned to obtain a medical or law degree?

13. Vice Presidential Tie Breakers The number of tie-breaking votes cast by each of the 21 vice presidents of the United States who served during the twentieth century are shown. Draw a histogram for this data.

0, 0, 4, 10, 0, 2, 3, 3, 4, 1, 7, 8, 0, 4, 2, 0, 0, 1, 7, 0, 4

14. Presidential Ages The ages at inauguration of the 44 presidents from George Washington to Barack Obama are shown. Draw a histogram for the ages.

57, 61, 57, 57, 58, 57, 61, 54, 68, 51, 49, 64, 50, 48, 65, 52, 56, 46, 54, 49, 50, 47, 55, 55, 54, 42, 51, 56, 55, 51, 54, 51, 60, 62, 43, 55, 56, 61, 52, 69, 64, 46, 54, 47

Quiz Scores The bar charts in Exercises 15 and 16 give quiz scores for two different classes. In each exercise, find the median score.

15.

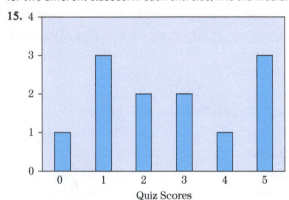
Quiz Scores

16.

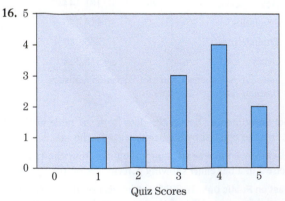
Quiz Scores

In Exercises 17 and 18, draw the box plot corresponding to the given five-number summary.

17. min $= 2$, $Q_1 = 5$, $Q_2 = 7$, $Q_3 = 10$, max $= 15$

18. min $= 10$, $Q_1 = 13$, $Q_2 = 17$, $Q_3 = 20$, max $= 25$

In Exercises 19–22, find the five-number summary and the interquartile range for the given set of numbers, and then draw the box plot.

19. 10, 11, 13, 14, 16, 17, 19, 20, 21, 23, 24

20. 3, 6, 8, 9, 11, 14, 18

21. 20, 25, 31, 38, 42, 47, 51, 54, 56

22. 7, 17, 26, 34, 41, 47, 52, 56, 59, 61

23. Match each of the histograms in column I with the corresponding box plot in column II.

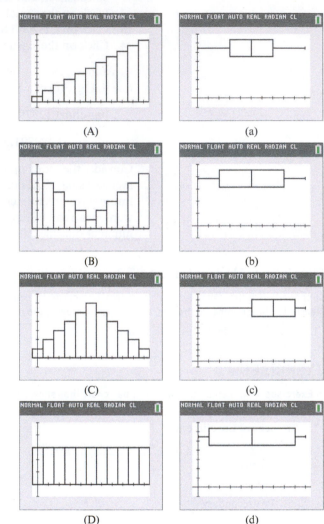

(A) (a)

(B) (b)

(C) (c)

(D) (d)

24. Food Cost The box plot for the price (in cents) of a can of tomato soup is shown.

40   70 90   130   258

(a) Approximately what percentage of the soups is priced below 70 cents?

(b) Approximately what percentage of the soups is priced above 130 cents?

(c) Approximately what percentage of the soups is priced below 90 cents?

(d) Approximately what percentage of the soups is priced from 90 to 130 cents?

(e) What is the median price of a can of soup?

25. Test Scores Consider the following box plot of scores on a standardized test:

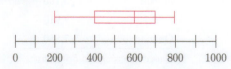
0   200   400   600   800   1000

(a) Give the five-number summary of the data.

(b) Approximately what percentage of the scores is below 400?

(c) Approximately what percentage of the scores is between 400 and 600?

(d) Approximately what percentage of the scores is higher than 600?

(e) Approximately what percentage of the scores is between 200 and 700?

26. **Batting Averages** The data shown gives the top 10 batting averages at the end of the 2015 season for the American League and the National League. Compare the data, using a box plot, and discuss which of the leagues seems to have the better players.

    American League: .338, .320, .313, .310, .307, .305, .303, .302, .299, .297

    National League: .333, .330, .321, .318, .315, .314, .314, .312, .303, .301

27. **Batting Averages** On April 16, 2016, the Toronto Blue Jays played a game against the Boston Red Sox. The batting averages of each team's players on that day were as follows:

    Blue Jays: .188, .313, .297, .304, .119, .045, .091, .214, .200
    Red Sox: .239, .302, .262, .333, .293, .333, .267, .429, .290

    Draw a box plot for each team, and comment on which team seems more likely to win. Explain your answer.

TECHNOLOGY EXERCISES

In Exercises 28–31, use technology to produce either a bar chart or pie chart for the data in the specified exercise.

28. Exercise 1

29. Exercise 2

30. Exercise 3

31. Exercise 4

Solutions to Check Your Understanding 7.1

1. Since there are an odd number of numbers in the list, the middle number (that is, the ninth number) is the median. *Note:* When N, the number of numbers, is odd, the median is the $\frac{N+1}{2}$th number.

2. There are eight numbers in the truncated list consisting of the numbers to the left of the median. Since 8 is even, the median of this truncated list is the average of the middle two numbers. That is, the first quartile is the average of the fourth and fifth numbers. *Note:* When N is even, the median of a list of N numbers is the average of the $\frac{N}{2}$th number and the number following the $\frac{N}{2}$th number.

3. There are eight numbers in the truncated list consisting of the numbers to the right of the median. Since 8 is even, the median of this truncated list is the average of the middle two numbers of the truncated list. So the third quartile is the average of the fourth and fifth numbers of the truncated list; that is, the average of the thirteenth and fourteenth numbers of the original list.

7.2 Frequency and Probability Distributions

Our goal in this section is to describe a given set of data in terms that allow for interpretation and comparison. As we shall see, both graphical and tabular displays of data can be useful for this purpose.

To get an idea of the problems considered in statistics, let us consider a concrete example. Ms. Jones is interested in purchasing a car dealership. Two dealerships are for sale, and each dealer has provided Ms. Jones with data describing past car sales. Dealership A provided one year's worth of data, dealership B, two years' worth. The data is summarized in Table 1. The problem confronting Ms. Jones is that of analyzing the data to determine which car dealership to buy.

Table 1 Weekly Sales of Two Dealerships

Weekly Sales	Number of Occurrences	
	Dealership A	Dealership B
5	2	20
6	2	0
7	13	0
8	20	10
9	10	12
10	4	50
11	1	12

This data is presented in a form often used in statistical surveys. For each possible value of a statistical variable (in this case, the number of cars sold weekly), we have tabulated the number of occurrences. Such a tabulation is called a **frequency distribution.** Although a frequency distribution is a very useful way of displaying and summarizing survey data, it is by no means the most efficient form in which to analyze such data. For example, it is difficult to compare dealership A with dealership B by using only Table 1.

Comparisons are much more easily made if we use proportions rather than actual numbers of occurrences. For example, instead of recording that dealership A had weekly sales of 5 cars during 2 weeks of the year, let us record that the proportion of the observed weeks in which dealership A had weekly sales of 5 was $\frac{2}{52} \approx .04$. Similarly, by dividing each of the entries in the middle column by 52, we obtain the second column of Table 2, describing the sales of dealership A. For simplification, we shall round off the data of this example to two decimal places. We similarly can construct the third column of Table 2 for dealership B.

Table 2 Comparison of Proportions

	Proportion of Occurrences	
Weekly Sales	Dealership A	Dealership B
5	$\frac{2}{52} \approx .04$	$\frac{20}{104} \approx .19$
6	$\frac{2}{52} \approx .04$	0
7	$\frac{13}{52} \approx .25$	0
8	$\frac{20}{52} \approx .38$	$\frac{10}{104} \approx .10$
9	$\frac{10}{52} \approx .19$	$\frac{12}{104} \approx .12$
10	$\frac{4}{52} \approx .08$	$\frac{50}{104} \approx .48$
11	$\frac{1}{52} \approx .02$	$\frac{12}{104} \approx .12$

These tables are called **relative frequency distributions.** In general, consider an experiment with the numerical outcomes $x_1, x_2, \ldots, x_r$. Suppose that the number of occurrences of x_1 is f_1, the number of occurrences of x_2 is f_2, and so forth (Table 3). The frequency distribution lists all of the outcomes of the experiment and the number of times each occurred. (For the sake of simplicity, we usually arrange $x_1, x_2, \ldots, x_r$ in increasing order.)

Table 3

Outcome	Frequency	Relative Frequency
x_1	f_1	$\dfrac{f_1}{n}$
x_2	f_2	$\dfrac{f_2}{n}$
$\vdots$	$\vdots$	$\vdots$
x_r	f_r	$\dfrac{f_r}{n}$
Total n		1

Suppose that the total number of occurrences is n. Then the **relative frequency** of outcome x_1 is f_1/n, the relative frequency of outcome x_2 is f_2/n, and so forth. The relative frequency distribution pairs each outcome with its relative frequency. The sum of the frequencies in a frequency distribution is n. The sum of the relative frequencies in a relative frequency distribution is 1.

The frequency or relative frequency distribution is obtained directly from the performance of an experiment and the collection of data observed at each trial of the experiment.

EXAMPLE 1   **Coin Tossing** An experiment consists of tossing five coins and counting the number of heads. (On each performance of the experiment, we will observe 0, 1, 2, 3, 4, or 5 heads.) Table 4 gives the frequency distribution after 90 repetitions of this experiment. Determine the relative frequency distribution associated with these data.

Table 4 Results of Coin-Tossing Experiment

Number of Heads	Frequency
0	3
1	14
2	23
3	27
4	17
5	6

SOLUTION   Since the number of occurrences is $n = 90$, we divide each frequency by 90 to obtain the relative frequency distribution (see Table 5). Note that the sum of the frequency column is 90 and the sum of the relative frequency column is 1.

Table 5 Relative Frequencies for Coin-Tossing Experiment

Number of Heads	Frequency	Relative Frequency
0	3	$\frac{3}{90} \approx .03$
1	14	$\frac{14}{90} \approx .16$
2	23	$\frac{23}{90} \approx .26$
3	27	$\frac{27}{90} \approx .30$
4	17	$\frac{17}{90} \approx .19$
5	6	$\frac{6}{90} \approx .07$
Total	**90**	$\frac{90}{90} = 1.00$

«

It is often possible to gain useful insight into an experiment by representing its relative frequency distribution graphically as a histogram.

Returning to the car sales example, comparison of the histograms of Fig. 1 reveals significant differences between the two dealerships. On the one hand, dealership A is very consistent. Most weeks, its sales are in the middle range of 7, 8, or 9. On the other hand, dealership B can often achieve very high sales (it had sales of 10 in 48% of the weeks) at the expense of a significant number of weeks of low sales (sales of 5 in 19% of the weeks).

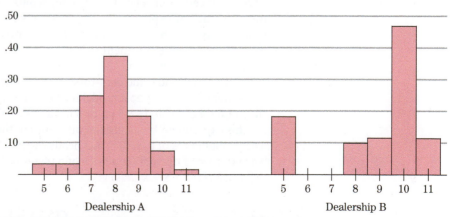

Figure 1 Relative frequency distributions for weekly car sales

EXAMPLE 2	**Histograms** Draw the histogram for the relative frequency distribution in Example 1.
SOLUTION	The histogram is given in Fig. 2.

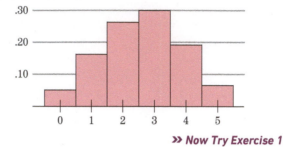

Figure 2 Relative frequency distribution for the number of heads in 5 tosses—experimental results

>> Now Try Exercise 1

It is possible to use histograms to represent the relative frequencies of events as areas, using the following fact.

> In a histogram, the probability of an event E is the sum of the areas of the rectangles corresponding to the outcomes in E.

To illustrate the procedure, consider the histogram for dealership A in Fig. 1. The event $E =$ "sales between 7 and 10, inclusive" consists of the set of outcomes $\{7, 8, 9, 10\}$, so its relative frequency is the sum of the respective relative frequencies of the outcomes 7, 8, 9, and 10. Therefore, the relative frequency of the event E is the area of the blue region of the histogram in Fig. 3.

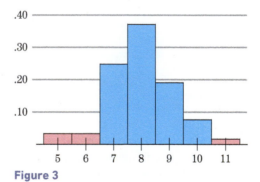

Figure 3

Probability Distributions

An important thing to notice is that, so far, we have made the tables and the histograms for actual, as opposed to theoretical, experiments. That is, the tables and the histograms were produced from collections of sample data that were obtained by actually recording the outcomes of experiments. We shall now look at theoretical experiments and continue to explore the important notion of a **probability distribution.** In many cases, data of an actual experiment is best interpreted when we can construct a theoretical model for the experiment.

Let us reconsider the coin-tossing experiment from Example 1. If the coins are fair coins, then we can set up a model for the experiment by noting once again that the possible outcomes are 0, 1, 2, 3, 4, or 5 heads. We will construct the probability distribution for this experiment by listing the outcomes in the sample space with their probabilities. The probabilities of 0, 1, 2, 3, 4, 5 heads can be obtained with the methods of Chapter 6. The number of distinct sequences of 5 tosses is 2^5, or 32. The number of sequences having k heads (and $5 - k$ tails) is

$$C(5, k) = \binom{5}{k}.$$

Table 6 Probability Distribution for Coin-Tossing Experiment

Number of Heads	Probability
0	$\dfrac{\binom{5}{0}}{2^5} = \dfrac{1}{32}$
1	$\dfrac{\binom{5}{1}}{2^5} = \dfrac{5}{32}$
2	$\dfrac{\binom{5}{2}}{2^5} = \dfrac{10}{32}$
3	$\dfrac{\binom{5}{3}}{2^5} = \dfrac{10}{32}$
4	$\dfrac{\binom{5}{4}}{2^5} = \dfrac{5}{32}$
5	$\dfrac{\binom{5}{5}}{2^5} = \dfrac{1}{32}$

Thus,

$$\Pr(k \text{ heads}) = \frac{\binom{5}{k}}{2^5} \qquad (k = 0, 1, 2, 3, 4, 5).$$

The probability distribution for the experiment is shown in Table 6.

The histogram for a probability distribution is constructed in the same way as the histogram for a relative frequency distribution. (See Fig. 4.)

We note that the histogram in Fig. 4 is based on a theoretical model of coin tossing, whereas the histogram in Fig. 2 was drawn from experimental results available only after the experiment was actually performed and observed.

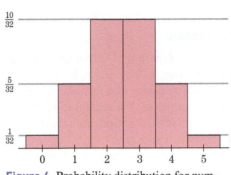

Figure 4 Probability distribution for number of heads in 5 tosses—theoretical results

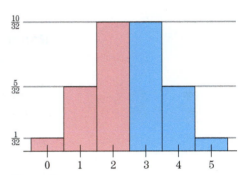

Figure 5

Just as we used the histogram for the relative frequency distribution to picture the relative frequency of an event, we may also use the histogram of a probability distribution to picture the probability of an event. For instance, to find the probability of at least 3 heads on 5 tosses of a fair coin, we need only add the probabilities of the outcomes: 3 heads, 4 heads, 5 heads. That is,

$$\Pr(\text{at least 3 heads}) = \Pr(3 \text{ heads}) + \Pr(4 \text{ heads}) + \Pr(5 \text{ heads}).$$

Since each of these probabilities is equal to the area of a rectangle in the histogram of Fig. 4, the area of the blue region in Fig. 5 equals the probability of the event.

EXAMPLE 3   **Probabilities and Histograms** The histogram of a probability distribution is given in Fig. 6. Calculate the probability of the event "more than 3."

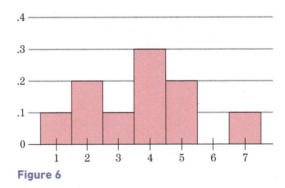

Figure 6

SOLUTION   The event "more than 3" is the set of outcomes {4, 5, 6, 7}. We indicate in blue the portion of the histogram corresponding to these outcomes (Fig. 7 on the next page). The probability of the event "more than 3" is the area of the blue region,

$$.3 + .2 + 0 + .1 = .6.$$

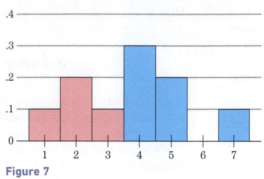

Figure 7

» *Now Try Exercise 11*

Random Variables

Consider a theoretical experiment with numerical outcomes. Denote the outcome of the experiment by the letter X. For example, if the experiment consists of observing the number of heads in five tosses of a fair coin, then X assumes one of the six values 0, 1, 2, 3, 4, 5. Since the values of X are determined by the unpredictable random outcomes of the experiment, X is called a **random variable** or, more specifically, the *random variable associated with the experiment*.

The random variable notation is often convenient and is commonly used in probability and statistics texts. In considering several different experiments, it is sometimes necessary to use letters other than X to stand for random variables. It is customary, however, to use only capital letters, such as X, Y, Z, W, U, V, for random variables.

If k is one of the possible outcomes of the experiment with associated random variable X, then we denote the probability of the outcome k by

$$\Pr(X = k).$$

For example, in the coin-tossing experiment described previously, if X is the number of heads in the five tosses, then

$$\Pr(X = 3)$$

denotes the probability of getting 3 heads. The probability distribution of the random variable X is shown in Table 7.

Rather than speak of the probability distribution associated with the model of an experiment, we can speak of the *probability distribution associated with the corresponding random variable*. Such a probability distribution is a table listing the various values of X (i.e., outcomes of the experiment) and their associated probabilities with $p_1 + p_2 + \cdots + p_r = 1$:

Table 7 Probability Distribution of X

k	$\Pr(X = k)$
0	$\frac{1}{32}$
1	$\frac{5}{32}$
2	$\frac{10}{32}$
3	$\frac{10}{32}$
4	$\frac{5}{32}$
5	$\frac{1}{32}$

k	$\Pr(X = k)$
x_1	p_1
x_2	p_2
$\vdots$	$\vdots$
x_r	p_r

EXAMPLE 4   **Selecting Balls from an Urn** Consider the urn of Example 1 of Section 6.3, in which there are eight white balls and two green balls. A sample of three balls is chosen at random from the urn. Let X denote the number of green balls in the sample. Find the probability distribution of X.

SOLUTION   There are $C(10, 3) = 120$ equally likely outcomes, and X can be 0, 1, or 2. Reasoning as in Section 6.3, we have

$$\Pr(X = 0) = \frac{C(2, 0) \cdot C(8, 3)}{120} = \frac{1 \cdot 56}{120} = \frac{56}{120} = \frac{7}{15}$$

$$\Pr(X = 1) = \frac{C(2, 1) \cdot C(8, 2)}{120} = \frac{2 \cdot 28}{120} = \frac{56}{120} = \frac{7}{15}$$

$$\Pr(X = 2) = \frac{C(2, 2) \cdot C(8, 1)}{120} = \frac{1 \cdot 8}{120} = \frac{8}{120} = \frac{1}{15}.$$

The probability distribution for X is given by the following table:

k	$\Pr(X = k)$
0	$\frac{7}{15}$
1	$\frac{7}{15}$
2	$\frac{1}{15}$

>> *Now Try Exercise 7*

EXAMPLE 5   **Rolling a Pair of Dice**   Let X denote the random variable defined as the sum of the upper faces appearing when two dice are rolled. Determine the probability distribution of X, and draw its histogram.

SOLUTION   The experiment of rolling two dice leads to 36 possibilities, each having probability $\frac{1}{36}$.

$$
\begin{array}{cccccc}
(1, 1) & (1, 2) & (1, 3) & (1, 4) & (1, 5) & (1, 6) \\
(2, 1) & (2, 2) & (2, 3) & (2, 4) & (2, 5) & (2, 6) \\
(3, 1) & (3, 2) & (3, 3) & (3, 4) & (3, 5) & (3, 6) \\
(4, 1) & (4, 2) & (4, 3) & (4, 4) & (4, 5) & (4, 6) \\
(5, 1) & (5, 2) & (5, 3) & (5, 4) & (5, 5) & (5, 6) \\
(6, 1) & (6, 2) & (6, 3) & (6, 4) & (6, 5) & (6, 6)
\end{array}
$$

The sum of the numbers in each pair gives the value of X. For example, the pair $(3, 1)$ corresponds to $X = 4$. Note that the pairs corresponding to a given value of X lie on a diagonal, as shown in the preceding chart, where we have indicated all pairs corresponding to $X = 4$. It is now easy to calculate the number of pairs corresponding to a given value k of X and from it, the probability $\Pr(X = k)$. For example, there are three pairs adding to 4, so

$$\Pr(X = 4) = \tfrac{3}{36} = \tfrac{1}{12}.$$

Performing this calculation for all k from 2 to 12 gives the probability distribution in Table 8 and the histogram in Fig. 8.

Table 8 Probability Distribution of X

k	$\Pr(X = k)$
2	$\frac{1}{36}$
3	$\frac{1}{18}$
4	$\frac{1}{12}$
5	$\frac{1}{9}$
6	$\frac{5}{36}$
7	$\frac{1}{6}$
8	$\frac{5}{36}$
9	$\frac{1}{9}$
10	$\frac{1}{12}$
11	$\frac{1}{18}$
12	$\frac{1}{36}$

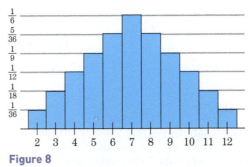

Figure 8

>> *Now Try Exercise 5*

The advantage of the random variable notation is that the variable X can be treated algebraically and we can consider expressions such as X^2. This is just the random variable corresponding to the experiment whose outcomes are the squares of the outcomes of the original experiment. Similarly, we can consider random variables such as $X + 3$ and $(X - 2)^2$. One important example of algebraic manipulation of random variables appears in Section 7.5. There are many others.

EXAMPLE 6   **Probability Distribution of X^2**   Suppose that a random variable X has the probability distribution given in Table 9. Determine the probability distribution of the random variable X^2.

Table 9 Probability Distribution of X

k	$\mathrm{Pr}(X = k)$
-1	.2
0	.3
1	.1
2	.4

SOLUTION   The outcomes of X^2 are the squares of the outcomes of X. The possible outcomes of X^2 are $k = 1$ (which results from the case $X = -1$ or from the case $X = 1$), $k = 0$, and $k = 4$. The probabilities of the outcomes of X^2 are determined by the probabilities of the outcomes of X. Since the outcomes $X = -1$ and $X = 1$ correspond to $X^2 = 1$,

$$\mathrm{Pr}(X^2 = 1) = \mathrm{Pr}(X = -1) + \mathrm{Pr}(X = 1).$$

The probability distribution of X^2 is as follows.

Table 10 Probability Distribution of X^2

k	$\mathrm{Pr}(X = k)$
0	.3
1	.3
4	.4

» Now Try Exercise 13

**INCORPORATING
TECHNOLOGY**

Histograms Appendix B shows how to draw histograms with a graphing calculator.

Histograms The Incorporating Technology discussion at the end of Section 7.1 shows how to use Excel to create histograms.

WolframAlpha A histogram similar to the one in Fig. 4 can be drawn with the instruction

bar chart {1/32, 5/32, 10/32, 10/32, 5/32, 1/32}.

Check Your Understanding 7.2

Solutions can be found following the section exercises.

1. **Carnival Game** In a certain carnival game, a wheel is divided into five equal parts, of which two are red and three are white. The player spins the wheel until the marker lands on red or until three spins have occurred. The number of spins is observed. Determine the probability distribution for this experiment.

2. **Carnival Game** Refer to the carnival game of Problem 1. Suppose that the player pays $1 to play this game and receives 50 cents for each spin. Determine the probability distribution for the experiment of playing the game and observing the player's earnings.

EXERCISES 7.2

1. **Final Grades** Table 11 gives the frequency distribution for the final grades in a course. (Here, A = 4, B = 3, C = 2, D = 1, F = 0.) Determine the relative frequency distribution associated with this data, and draw the associated histogram.

Table 11

Grade	Number of Occurrences
0	2
1	4
2	8
3	6
4	5

2. **Gas Queue** The number of cars waiting to be served at a gas station was counted at the beginning of every minute during the morning rush hour. The frequency distribution is given in Table 12. Determine the relative frequency distribution associated with this data, and draw the histogram.

Table 12

Number of Cars Waiting	Number of Occurrences
0	0
1	9
2	21
3	15
4	12
5	3

3. **Weather Reports** A local news website counted the number of visits to their weather page each minute on a rainy morning from 5 A.M. to 6 A.M. The frequency distribution is given in Table 13. Determine the relative frequency distribution associated with this data.

Table 13

Number of Visits During Minute	Number of Occurrences
20	3
21	3
22	0
23	6
24	18
25	12
26	0
27	9
28	6
29	3

4. **Production Level** A production manager counted the number of items produced each hour during a 40-hour work-week. The frequency distribution is given in Table 14. Determine the relative frequency distribution associated with this data.

Table 14

Number Produced During Hour	Number of Occurrences
50	2
51	0
52	6
53	8
54	12
55	6
56	4
57	0
58	0
59	2

5. **Coin Tosses** A fair coin is tossed three times, and the number of heads is observed. Determine the probability distribution for this experiment, and draw its histogram.

6. **Archery** An archer can hit the bull's-eye of the target with probability 1/3. She shoots until she hits the bull's-eye or until four shots have been taken. The number of shots is observed. Determine the probability distribution for this experiment, and draw its histogram.

7. **Selecting Balls from an Urn** An urn contains three red balls and four white balls. A sample of three balls is selected at random and the number of red balls observed. Determine the probability distribution for this experiment, and draw its histogram.

8. **Rolling a Die** A die is rolled, and the number on the top face is observed. Determine the probability distribution for this experiment, and draw its histogram.

9. **Carnival Game** In a certain carnival game, the player selects two balls at random from an urn containing two red balls and four white balls. The player receives $5 if he draws two red balls and $1 if he draws one red ball. He loses $1 if no red balls are in the sample. Determine the probability distribution for the experiment of playing the game and observing the player's earnings.

10. **Carnival Game** In a certain carnival game, a player pays $1 and then tosses a fair coin until either a heads occurs or he has tossed the coin four times. He receives 50 cents for each toss. Determine the probability distribution for the experiment of playing the game and observing the player's earnings.

11. Figure 9 is the histogram for a probability distribution. What is the probability that the outcome is between 5 and 7, inclusive?

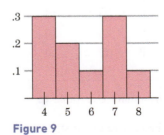

Figure 9

12. Figure 10 on the next page is the histogram for a probability distribution. To what event do the blue rectangles correspond?

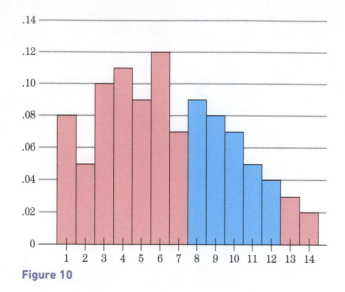

Figure 10

Let the random variables X and Y have the probability distributions listed in Table 15. Determine the probability distributions of the random variables in Exercises 13–20.

Table 15

k	$\Pr(X = k)$	k	$\Pr(Y = k)$
0	.1	5	.3
1	.2	10	.4
2	.4	15	.1
3	.1	20	.1
4	.2	25	.1

13. X^2 **14.** Y^2 **15.** $X - 1$

16. $Y - 15$ **17.** $\frac{1}{5}Y$ **18.** $2X^2$

19. $(X + 1)^2$ **20.** $(\frac{1}{5}Y + 1)^2$

21. Grade Distributions Two classes take the same examination, and the grades are recorded in Table 16. We assigned the integers 0 through 4 to the grades F through A, respectively. Table 16 gives the frequency distribution of grades in each class. Find the relative frequency distribution and the histogram for each class. Describe the difference in the grade distributions for the two classes.

Table 16

Grade		Number of Students	
		9 A.M. Class	10 A.M. Class
F	(0)	10	17
D	(1)	15	20
C	(2)	20	17
B	(3)	10	20
A	(4)	5	26

22. Grade Distributions Use Table 16 to answer the following questions:
(a) What percentage of the students in the 9 A.M. class have grades of C or less?
(b) What percentage of the 10 A.M. class have grades of C or less?

(c) What percentage of the 9 A.M. class have grades of D or F?
(d) What percentage of the students in both classes combined have grades of C or better?

Exercises 23–26 refer to the tables and relative frequency distributions associated with Exercises 1–4.

23. Grade Distribution For the data in Table 11, determine the percentage of students who get C or higher.

24. Gas Queue The data in Table 12 has been tabulated for every minute of the (60-minute) rush hour. What percentage of the time is the waiting line 4 or more cars?

25. Weather Reports Use the data in Table 13.
(a) For what percentage of the 60 minutes from 5 A.M. to 6 A.M. are there either fewer than 22 or more than 27 visits to the weather page?
(b) For what percentage of the hour are there between 23 and 25 visits (inclusive)?
(c) Draw the relative frequency histogram.
(d) What would be your estimate of the average number of visits during a minute of the hour for which the data has been tabulated? Explain.

26. Production Level Use the data in Table 14.
(a) For what percentage of the workweek were there more than 54 items produced?
(b) For what percentage of the workweek were there between 53 and 55 (inclusive) items produced?
(c) Draw the relative frequency histogram.
(d) What is the highest number of items produced in any one hour?
(e) Estimate the average number of items produced per hour in this week. Explain.

27. Table 17 gives the probability distribution of the random variable U.

Table 17

k	$\Pr(U = k)$
0	$\frac{4}{15}$
1	$\frac{2}{15}$
2	$\frac{4}{15}$
3	$\frac{3}{15}$
4	?

(a) Determine the probability that $U = 4$.
(b) Find $\Pr(U \geq 2)$.
(c) Find the probability that U is at most 3.
(d) Find the probability that $U + 2$ is less than 4.
(e) Draw the histogram of the distribution of U.

28. Table 18 gives the probability distribution of the random variable X.

Table 18

k	$\Pr(X = k)$
1	.30
2	.40
3	.20
4	.10

(a) Draw the histogram for X.

(b) Find $\Pr(X = 2 \text{ or } 3)$

(c) Find the probability that X is at least 2.

(d) Find the probability that $X + 2$ is at least 5.

(e) Find the probability distribution of $2X$.

29. Table 19 gives the probability distribution of the random variable Y.

Table 19

k	$\Pr(Y = k)$
1	.20
2	.20
3	.20
4	.40

(a) Draw the histogram for Y.

(b) Find $\Pr(Y = 2 \text{ or } 3)$

(c) Find the probability that Y^2 is at most 9.

(d) Find the probability that Y is at most 10.

(e) Find the probability distribution of $(Y + 2)^2$.

30. Refer back to Tables 18 and 19. Which of the random variables X and Y has the higher average value? Why would you think so?

Solutions to Check Your Understanding 7.2

1. Since the outcomes are the numbers of spins, there are three possible outcomes: one, two, and three spins. The probabilities for each of these outcomes can be computed from a tree diagram (Fig. 11). For instance, the outcome two (spins) occurs if the first spin lands on white and the second spin on red. The probability of this outcome is $\frac{3}{5} \cdot \frac{2}{5} = \frac{6}{25}$.

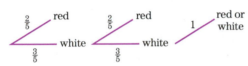

Figure 11

k	$\Pr(X = k)$
1	$\frac{2}{5}$
2	$\frac{6}{25}$
3	$\frac{9}{25}$

2. The same game is being played as in Problem 1, except that now the outcome that we are concentrating on is the player's financial situation at the end of the game. The player's earnings depend on the number of spins as follows: one spin results in $-\$.50$ earnings (i.e., a loss of 50 cents); two spins result in $\$0$ earnings (i.e., breaking even); and three spins result in $\$.50$ earnings (i.e., the player ends up ahead by 50 cents). The probabilities for these three situations are the same as before.

Earnings	Probability
$-\$.50$	$\frac{2}{5}$
0	$\frac{6}{25}$
$\$.50$	$\frac{9}{25}$

Binomial Trials

In this section, we fix our attention on the simplest experiments: those with just two outcomes. These experiments, called **binomial trials** (or *Bernoulli trials*), occur in many applications. Here are some examples of binomial trials:

1. Toss a coin, and observe the outcome, heads or tails.

2. Administer a drug to a sick individual, and classify the reaction as "effective" or "ineffective."

3. Manufacture a lightbulb, and classify it as "nondefective" or "defective."

The outcomes of a binomial trial are usually called *success* and *failure*. Of course, the labels *success* and *failure* need have no connection with the usual meanings of these words. For example, in experiment 2, we might label the outcome "ineffective" as *success* and "effective" as *failure*.

> We will always denote the probability of success by p and probability of failure by q. Since a binomial trial has only two outcomes, we have $p + q = 1$, or
>
> $$q = 1 - p.$$

Consider a particular binomial trial and the following experiment: Repeat the binomial trial n times, and observe the number of successes that occur. Assume that the n successive trials are independent of one another. The fundamental problem of the theory of binomial trials is to calculate the probabilities of the outcomes of this experiment.

Let X be a random variable associated with the experiment. X is the number of successes in the n trials of the experiment. For example, if we toss a coin 20 times and assume that heads is a success, then $X = 3$ means that the experiment resulted in 3 heads and 17 tails. In an experiment of n trials, the number of successes can be any one of the numbers $0, 1, 2, \ldots, n$. These are the possible values of X.

We write $\Pr(X = k)$ to denote the probability that $X = k$, namely, the probability that k of the n trials result in success. As we saw in the coin-tossing experiment in Section 7.2, we can find the probability distribution of X by using the methods of counting and basic probability principles developed earlier in the book.

Binomial Probability Formula   If X is the number of successes in n independent trials, where in each trial the probability of a success is p, then (with $q = 1 - p$)

$$\Pr(X = k) = \binom{n}{k} p^k q^{n-k} \tag{1}$$

for $k = 0, 1, 2, \ldots, n$. Note: $\binom{n}{k}$ is the same as $C(n, k)$; that is,

$$\binom{n}{k} = \frac{n!}{k!(n-k)!}$$

Note that the right side of (1) is one of the terms in the binomial expansion of $(p + q)^n$. (See Section 5.7.) We say that X is a **binomial random variable** with parameters n and p. The derivation of (1) is given at the end of this section.

Let X be the number of heads in five tosses of a fair coin. Then X is a binomial random variable with parameters $p = \frac{1}{2}$ and $n = 5$. The calculation for this particular variable appears in Section 7.2. The probability distribution is

k	$\Pr(X = k)$
0	$\frac{1}{32}$
1	$\frac{5}{32}$
2	$\frac{10}{32}$
3	$\frac{10}{32}$
4	$\frac{5}{32}$
5	$\frac{1}{32}$

By (1),

$$\Pr(X = k) = \binom{5}{k} \left(\frac{1}{2}\right)^k \left(\frac{1}{2}\right)^{5-k}.$$

Substitution of the values of k (0, 1, 2, 3, 4, 5) gives the probabilities in the table.

EXAMPLE 1   **Quality Control**   A plumbing-supplies manufacturer produces faucet washers, which are packaged in boxes of 300. Quality control studies have shown that 2% of the washers are defective. What is the probability that a box of washers contains exactly 9 defective washers?

SOLUTION Deciding whether a single washer is or is not defective is a binomial trial. Since we wish to consider the number of defective washers in a box, let "success" be the outcome "defective." Then

$$p = .02 \qquad q = 1 - .02 = .98 \qquad n = 300.$$

The probability that 9 out of 300 washers are defective equals

$$\Pr(X = 9) = \binom{300}{9}(.02)^9(.98)^{291} \approx .07.$$

>> *Now Try Exercise 17*

EXAMPLE 2 **Veterinary Medicine** The recovery rate for a certain cattle disease is 25%. If 40 cattle are afflicted with the disease, what is the probability that exactly 10 will recover?

SOLUTION In this example, the binomial trial consists of observing a single cow, with recovery as "success." Then

$$p = .25 \qquad q = 1 - .25 = .75 \qquad n = 40.$$

The probability of 10 successes is

$$\Pr(X = 10) = \binom{40}{10}(.25)^{10}(.75)^{30} \approx .14.$$

>> *Now Try Exercise 5*

NOTE Binomial probabilities can be difficult to calculate by hand. Instead, it is recommended that you use technology to compute probabilities like those in Examples 1 and 2. The Incorporating Technology discussion at the end of this section presents ways to do this. «

EXAMPLE 3 **Baseball** Each time that a baseball player is at bat, the probability that he gets a hit is .300. He comes up to bat four times in a game. Assume that his times at bat are independent trials. Find the probability that he gets
(a) exactly two hits.   (b) at least two hits.

SOLUTION Each at-bat is considered an independent binomial trial. A "success" is a hit. So $p = .300$, $q = 1 - p = .700$, and $n = 4$. Therefore, X is the number of hits in four at-bats or the number of successes in four trials.

(a) We need to determine $\Pr(X = 2)$.

$$\Pr(X = 2) = \binom{4}{2}(.300)^2(.700)^{4-2} \quad \text{Formula (1) with } k = 2$$

$$= 6(.09)(.49) = .2646.$$

(b) "At least two hits" means $X \geq 2$.

$$\Pr(X \geq 2) = \Pr(X = 2) + \Pr(X = 3) + \Pr(X = 4)$$

$$= \binom{4}{2}(.300)^2(.700)^2 + \binom{4}{3}(.300)^3(.700)^1 + \binom{4}{4}(.300)^4(.700)^0$$

$$= 6(.09)(.49) + 4(.027)(.700) + 1(.0081)(1)$$

$$= .2646 + .0756 + .0081 = .3483.$$

So the batter can be expected to get at least two hits out of four at-bats in about 35% of the games.

>> *Now Try Exercise 51*

EXAMPLE 4

College Acceptance According to the study *The American Freshman: National Norms 2015* conducted by the Higher Education Research Institute, 76% of college freshmen in 2015 had been accepted by their first-choice college. If five freshmen are selected at random, what is the probability that at least two were accepted by their first-choice college? Assume that each selection is an independent binomial trial.

SOLUTION

Let "success" be "accepted by first-choice college." Then

$$p = .76, q = 1 - p = .24, \text{ and } n = 5.$$

Let X be the number of freshmen (out of the five selected) who were accepted by their first-choice college. Then

$$\Pr(X \geq 2) = 1 - \Pr(X = 0) - \Pr(X = 1) \qquad \text{Complement Rule}$$

$$= 1 - \binom{5}{0}(.76)^0(.24)^5 - \binom{5}{1}(.76)^1(.24)^4 \quad \text{Formula (1) with } k = 0, 1$$

$$\approx 1 - .0008 - .0126 = .9866.$$

Thus, in a group of five randomly selected freshmen, there is a 98.66% chance that at least two of them were accepted by their first-choice college.   **» Now Try Exercise 25**

EXAMPLE 5

Coin Tosses How many times must a person toss a coin so that the probability of obtaining at least one head is greater than 90%?

SOLUTION

Note that the event "at least one toss is a head" is the complement of the event "every toss is a tail." Therefore, let "success" be the event "a tail is tossed." Then

$$p = \frac{1}{2} \quad q = \frac{1}{2}$$

and, if the coin is tossed n times, the probability of obtaining at least one head is

$$1 - \Pr(X = n) = 1 - \binom{n}{n}\left(\frac{1}{2}\right)^n\left(\frac{1}{2}\right)^0$$

$$= 1 - \left(\frac{1}{2}\right)^n.$$

We wish to determine the smallest value of n for which $1 - (\frac{1}{2})^n$ is at least .90. Table 1 shows the probabilities for increasing values of n.

A minimum of $n = 4$ tosses is needed to guarantee a probability greater than 90%.

» Now Try Exercise 57

Table 1 Probability of At Least One Head in n Tosses

n	$1 - (\frac{1}{2})^n$
1	$\frac{1}{2} = .5$
2	$\frac{3}{4} = .75$
3	$\frac{7}{8} = .875$
4	$\frac{15}{16} = .9375$

Summing Binomial Probabilities

Suppose that, in Example 2, we wanted to find the probability that 16 or more cattle recover. Using formula (1) to compute the probabilities that 16, 17, . . . , 40 cattle recover, the desired probability is

$$\Pr(X = 16) + \Pr(X = 17) + \cdots + \Pr(X = 40)$$

$$= \binom{40}{16}(.25)^{16}(.75)^{24} + \binom{40}{17}(.25)^{17}(.75)^{23} + \cdots + \binom{40}{40}(.25)^{40}(.75)^0. \quad (2)$$

This sum is difficult to compute. In Section 7.7, we will discuss the traditional method of approximating this sum by using a table of areas. The table which is obtained with technology, is incomplete, and the approximation obtained is not very accurate. The more current method is to go directly to technology to obtain the sum. For instance, using the table gives an approximation of 2.74%, but the more accurate value obtained with a graphing calculator, Excel, or Wolfram|Alpha is 2.62%. See the Incorporating Technology discussion at the end of this section for details.

Verification of Formula (1)   Each outcome of n independent trials that contains k successes can be thought of as a sequence of k S's and $(n - k)$ F's. Each such sequence is obtained by selecting k of the n positions for the S's. (The remaining positions will be filled with F's.) Therefore, the number of such sequences is $C(n, k)$—that is, $\binom{n}{k}$. The probability of such a sequence is the product of k p's and $(n - k)$ q's—that is, $p^k q^{n-k}$. Hence, $\Pr(X = k) = \binom{n}{k} p^k q^{n-k}$.   «

INCORPORATING

TECHNOLOGY

Calculating Binomial Probabilities   With a TI-84 Plus, **binompdf**(n, p, x) gives the probability of x successes in n trials, where the probability of success is p. That is, **binompdf**(n, p, x) returns the value of

$$\binom{n}{x} p^x q^{n-x}. \tag{3}$$

Figure 1

(To display the **binompdf** wizard, press 2nd [DISTR] **A**; see Fig. 1.) The first command in Fig. 2 uses this command to solve Example 3(a).

The function **binomcdf**(n, p, x) gives the value of *cumulative* probabilities of a binomial distribution (note the "cdf" rather than "pdf" in this function). So **binomcdf**(n, p, x) yields the sum of x *or fewer* successes, or

$$\binom{n}{0} p^0 q^{n-0} + \binom{n}{1} p^1 q^{n-1} + \cdots + \binom{n}{x} p^x q^{n-x}. \tag{4}$$

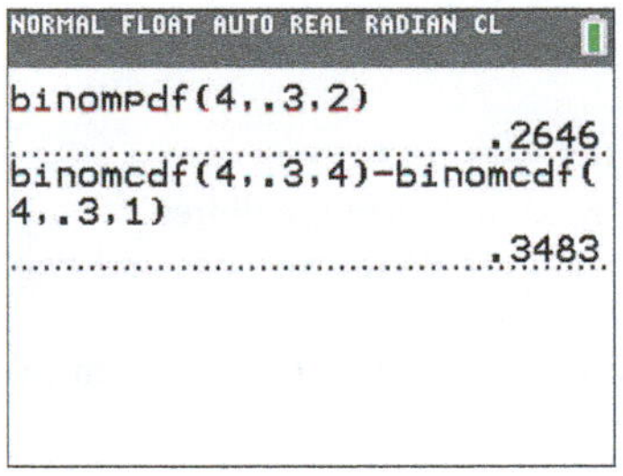

(To display the **binomcdf** wizard, press 2nd [DISTR] **B**.) The second command in Fig. 2 is the solution to Example 3(b). The formula **binomcdf(4, .3, 4) − binomcdf(4, .3, 1)** yields the probability of two, three, or four hits by subtracting the probability of one or fewer hits (no hits or one hit) from the probability of four or fewer hits (zero, one, two, three, or four hits). In general, consecutive probabilities—for instance, $\Pr(X = r)$ through $\Pr(X = s)$—can be calculated with the expression

binomcdf(n, p, s) − **binomcdf**$(n, p, r - 1)$

This expression gives the value of

$$\binom{n}{r} p^r q^{n-r} + \binom{n}{r+1} p^{r+1} q^{n-(r+1)} + \cdots + \binom{n}{s} p^s q^{n-s}. \tag{5}$$

Figure 2

Calculating Binomial Probabilities   In an Excel spreadsheet, the value of $\Pr(X = k)$ for a binomial random variable X is calculated with the function **BINOM.DIST**$(k, n, p, 0)$, where k is the number of successes, n is the number of trials, and p is the probability of success. The value of $\Pr(X \le k)$ is given by the function **BINOM.DIST**$(k, n, p, 1)$. That is, **BINOM.DIST**$(k, n, p, 0)$ and **BINOM.DIST**$(k, n, p, 1)$ return the values of Equations (3) and (4), respectively.

Consecutive probabilities—for instance, $\Pr(X = r)$ through $\Pr(X = s)$—can be calculated with the expression

BINOM.DIST$(s, n, p, 1)$ − **BINOM.DIST**$(r - 1, n, p, 1)$.

This will give the value of Equation (5). For instance, we can use **BINOM.DIST** **(5, 5, .76, 1) − BINOM.DIST(1, 5, .76, 1)** to solve Example 4. This gives the probability of two through five successes in five trials.

✻WolframAlpha Calculating Binomial Probabilities   Consider the binomial random variable X with r trials and probability of success s. The instruction

binomial probabilities n=r, p=s, endpoint=k

gives the values of $\Pr(X < k)$, $\Pr(X = k)$, and $\Pr(X > k)$.

Probabilities:	
$x < 10$	0.43954
$x = 10$	0.144364
$x > 10$	0.416096

Figure 3

The instruction

$$P(a <= X <= b) \text{ for X binomial}(n, p)$$

returns the value of $\Pr(a \le X \le b)$.

For instance, to solve Example 2 we may enter

binomial probabilities n=40, p=.25, endpoint=10,

which returns the table in Fig. 3. The probability that 10 cows recover, or the probability that $X = 10$, is approximately .14.

To calculate the probability of obtaining between two and five heads, inclusive, in seven coin tosses, we may enter

$$P(2 <= X <= 5) \text{ for X binomial}(7, .5)$$

which returns a value of $\frac{7}{8}$.

Check Your Understanding 7.3

Solutions can be found following the section exercises.

1. A number is selected at random from the numbers 0 through 9999. What is the probability that the number is a multiple of 5?

2. If the experiment in Problem 1 is repeated 20 times, with replacement, what is the probability of getting four numbers that are multiples of 5?

3. Consider equation (2) following Example 5. How would you calculate the sum with a graphing calculator, Excel, and Wolfram|Alpha?

EXERCISES 7.3

In Exercises 1–4, calculate $\binom{n}{k} p^k q^{n-k}$ for the given values of n, k, and p.

1. $n = 5, k = 3, p = .3$
2. $n = 6, k = 1, p = .4$
3. $n = 4, k = 3, p = \frac{1}{3}$
4. $n = 3, k = 2, p = \frac{1}{6}$

Coin Tosses A coin is tossed 10 times. In Exercises 5–10, find the probabilities that the number of heads is as stated.

5. Exactly three
6. None
7. Seven or eight
8. Two or three
9. At least one
10. At most seven

Rolling a Die A single die is rolled four times. In Exercises 11–16, find the probabilities that the number of 6s that appear is as stated.

11. Four
12. Exactly two
13. Two or three
14. One or two
15. At most two
16. At least two

Twenty-Somethings Fourteen percent of U.S. residents are in their twenties. Consider a group of eight U.S. residents selected at random. In Exercises 17–22, find the probabilities that the number of people in the group who are in their twenties is as stated.

17. Exactly two
18. None
19. Four or five
20. One or two
21. At least three
22. At most six

Career Training According to the study *The American Freshman: National Norms 2015*, 76.1% of college freshmen said that "to get training for a specific career" was a very important reason for their going to college. Consider a group of seven freshman selected at random. In Exercises 23–26, find the probabilities that the number of people in the group who felt that the reason was very important is as stated.

23. All seven

24. Exactly three of the seven

25. At least six of the seven

26. No more than two of the seven

27. **Children** A family chosen at random has four children.
   (a) What is the probability that there are two boys and two girls?
   (b) What is the probability that there are three children of one gender and one of the other?
   (c) What is the probability that all four children are of the same gender?

28. **New Employees** A manager at a call center notices that 60% of new hires quit within their first year. In a group of 20 new hires, find the probability that
   (a) exactly 15 quit within their first year.
   (b) fewer than 19 quit within their first year.

College Acceptances Exercises 29 and 30 refer to Fig. 4, the histogram for Example 4.

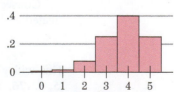

Figure 4 College Acceptances

29. The histogram in Fig. 4 shows that the values of $\Pr(X = 3)$ and $\Pr(X = 5)$ are nearly the same. Calculate the two probabilities to determine which is larger.

30. Use the histogram in Fig. 4 to determine the larger of $\binom{5}{3}(.76)^3(.24)^2$ and $\binom{5}{4}(.76)^4(.24)^1$. Interpret your result in words.

7.4 The Mean

It is important at this point to recognize the difference between a *population* and a *sample*. A **population** is a set of all elements about which information is desired. A **sample** is a subset of a population that is analyzed in an attempt to estimate certain properties of the entire population. For instance, suppose that we are interested in finding out some characteristics of the automobile industry but do not have access to the entire population of dealerships in the United States. We must choose a random sample of these dealerships and concentrate on gathering information from it. Suppose that we are interested in the average sales for all car dealerships in the United States for the month of October 2016. We could choose a random sample of $n = 200$ car dealerships across the country, determine the average October 2016 sales for these dealerships, and then use this value to estimate the average October 2016 sales for all car dealerships in the United States.

The idea of the mean, or average, of a set of values, shows up frequently in real-world situations. Examples include

- measuring a baseball pitcher's effectiveness by calculating his earned run average (ERA);
- determining whether an investment is worthwhile by looking at its average rate of return over a period of time;
- finding the favorability of a casino game by computing a gambler's expected winnings (or losses) over the long run.

The average value of the numbers in a sample is called the **sample mean** and is denoted by the symbol $\bar{x}$. It is calculated by adding together the numbers in the sample and then dividing by the number of numbers in the sample.

> **DEFINITION** The **sample mean** (or *average*) of the sample of n numbers $x_1, x_2, \ldots, x_n$ is
> $$\bar{x} = \frac{x_1 + x_2 + \cdots + x_n}{n}.$$

EXAMPLE 1 **Average Weekly Car Sales** Compute the sample mean of the weekly sales of dealership A of Section 7.2, given here in Table 1.

SOLUTION We may form the sample mean of the weekly sales figures as follows:

$$\frac{1}{52}[(5 + 5) + (6 + 6) + \underbrace{(7 + \cdots + 7)}_{13 \text{ times}} + \underbrace{(8 + \cdots + 8)}_{20 \text{ times}}$$
$$+ \underbrace{(9 + \cdots + 9)}_{10 \text{ times}} + \underbrace{(10 + \cdots + 10)}_{4 \text{ times}} + 11]$$
$$= \frac{5 \cdot 2 + 6 \cdot 2 + 7 \cdot 13 + 8 \cdot 20 + 9 \cdot 10 + 10 \cdot 4 + 11 \cdot 1}{52} \tag{1}$$
$$\approx 7.96.$$

Thus, the sample mean of the weekly sales is approximately 7.96 cars.

>> *Now Try Exercise 7*

Table 1 Weekly Sales at Dealership A

Weekly Sales	Number of Occurrences
5	2
6	2
7	13
8	20
9	10
10	4
11	1

NOTE We considered the data from dealership A to be a sample, since it was only one year's data and we are interested in making comparisons and using them to predict the future sales of the two dealerships. «

Let us reexamine the previous calculation. The sample mean is given by expression (1). We did this calculation by adding together the observed sales for the 52 weeks of the year and then dividing by 52. To make the calculation more efficient, we noticed that we could group all of the weeks in which we observed five sales, all of the weeks in which there were six sales, and so on. Instead of adding 52 numbers, we simply multiplied each observed value of "sales" by its frequency of occurrence and then divided by 52. Another alternative is to multiply each observed value of "sales" by the relative frequency with which it occurred (e.g., $5 \times \frac{2}{52}$), and then add. In other words,

$$\frac{5 \cdot 2 + 6 \cdot 2 + 7 \cdot 13 + 8 \cdot 20 + 9 \cdot 10 + 10 \cdot 4 + 11 \cdot 1}{52}$$

$$= 5 \cdot \frac{2}{52} + 6 \cdot \frac{2}{52} + 7 \cdot \frac{13}{52} + 8 \cdot \frac{20}{52} + 9 \cdot \frac{10}{52} + 10 \cdot \frac{4}{52} + 11 \cdot \frac{1}{52}$$

$$\approx 7.96.$$

If the data for any sample is displayed in a frequency table or a relative frequency table, the sample mean can be calculated in a similar fashion.

Sample Mean   Suppose that an experiment has as outcomes the numbers $x_1, x_2, \ldots, x_r$. Suppose that the frequency of x_1 is f_1, the frequency of x_2 is f_2, and so forth, and that

$$f_1 + f_2 + \cdots + f_r = n.$$

Then

$$\overline{x} = \frac{x_1 f_1 + x_2 f_2 + \cdots + x_r f_r}{n},$$

or

$$\overline{x} = x_1 \left(\frac{f_1}{n} \right) + x_2 \left(\frac{f_2}{n} \right) + \cdots + x_r \left(\frac{f_r}{n} \right).$$

Similar calculations are made when we work with entire populations.

Population Mean   Suppose that population size is N and that $x_1, x_2, \ldots, x_N$ are the population values. The **population mean**, denoted by the Greek letter μ (mu), is given by the formula

$$\mu = \frac{x_1 + x_2 + \cdots + x_N}{N}.$$

If the data has been grouped into a frequency or relative frequency table, a formula analogous to the one for samples is used:

$$\mu = x_1 \left(\frac{f_1}{N} \right) + x_2 \left(\frac{f_2}{N} \right) + \cdots + x_r \left(\frac{f_r}{N} \right)$$

A numerical descriptive measurement made on a sample is called a **statistic**. Such a measurement made on a population is called a **parameter** of the population. For instance, the sample mean $\overline{x}$ is a statistic, and the population mean μ is a parameter. Usually, since we cannot have access to entire populations, we rely on a sample to obtain statistics, and we attempt to use the statistics to estimate the parameters of the population. To help distinguish the parameter from the statistic, all parameters are denoted by Greek letters and all statistics by English letters. We use the common convention of denoting a sample size by lowercase n and a population size by uppercase N.

EXAMPLE 2   **Average Life Expectancy of Deer** An ecologist observes the life expectancy of a certain species of deer held in captivity. On the basis of a population of 1000 deer, they observe the data shown in Table 2. What is the mean life expectancy of this population of deer?

Table 2 Life Expectancy of Deer	
Age at Death (years)	Number Observed
1	0
2	60
3	180
4	250
5	200
6	120
7	50
8	120
9	20

Table 3 Relative Frequencies of Deer Life Expectancies	
Age at Death (years)	Relative Frequency
1	0
2	.06
3	.18
4	.25
5	.20
6	.12
7	.05
8	.12
9	.02

SOLUTION   We convert the given data into a relative frequency distribution by replacing observed frequencies by relative frequencies [= (observed frequency)/1000] (Table 3). The mean of this data is

$$\mu = 1 \cdot 0 + 2 \cdot (.06) + 3 \cdot (.18) + 4 \cdot (.25) + 5 \cdot (.20) + 6 \cdot (.12)$$
$$+ 7 \cdot (.05) + 8 \cdot (.12) + 9 \cdot (.02) = 4.87.$$

So, the mean life expectancy of this population of deer is 4.87 years.

» Now Try Exercise 9

EXAMPLE 3   **Average Weekly Car Sales** Recall the example from the beginning of Section 7.2 where Ms. Jones was interested in buying one of two car dealerships. The sales data from dealership A is in Table 1 of this section, and the sales data from dealership B is in Table 4. Which car dealership should Ms. Jones buy if she wants the one that will, on the average, sell more cars?

Table 4 Relative Frequencies for Dealership B Sales	
Weekly Sales	Relative Frequency
5	$\frac{20}{104}$
6	0
7	0
8	$\frac{10}{104}$
9	$\frac{12}{104}$
10	$\frac{50}{104}$
11	$\frac{12}{104}$

SOLUTION   We have seen in Example 1 that the mean of the sample data for dealership A is $\bar{x}_A \approx 7.96$. On the other hand, associated with the data for dealership B, we find the sample mean:

$$\bar{x}_B = 5 \cdot \left(\tfrac{20}{104}\right) + 6 \cdot 0 + 7 \cdot 0 + 8 \cdot \left(\tfrac{10}{104}\right) + 9 \cdot \left(\tfrac{12}{104}\right) + 10 \cdot \left(\tfrac{50}{104}\right) + 11 \cdot \left(\tfrac{12}{104}\right)$$
$$\approx 8.85.$$

Thus, the average sales of dealership A are 7.96 cars per week, whereas those of dealership B are 8.85 cars per week. If we make the assumption that past sales history predicts future sales, Ms. Jones should buy dealership B.   **» Now Try Exercise 11**

Expected Value

The sample and population means have analogs in the theoretical setting of random variables. Suppose that X is a random variable with the following probability distribution:

x_i	$Pr(X = x_i)$
x_1	p_1
x_2	p_2
$\vdots$	$\vdots$
x_N	p_N

Then the values of X (namely, $x_1, x_2, \ldots, x_N$) are the possible outcomes of an experiment. The expected value of X, denoted $E(X)$, is defined as follows:

DEFINITION The **expected value** of the random variable X is given by

$$E(X) = x_1 p_1 + x_2 p_2 + \cdots + x_N p_N.$$

Since the value of p_i represents the theoretical relative frequency of the outcome x_i, the formula for $E(X)$ is similar to the formula for the mean of a relative frequency distribution. Thus, the expected value of the random variable X is also called the *mean* of the random variable X or of the probability distribution of X and may be denoted either by $E(X)$ or by μ_X. Frequently, $E(X)$ is used interchangeably with the Greek letter μ when the context is clear.

The expected value is often referred to as the *long-term average*. This means that, over the long term of doing an experiment over and over, you would expect this average. It gives a general impression of the behavior of the random variable.

The expected value of a random variable is the center of the probability distribution in the sense that it is the balance point of the histogram. For example, let $X =$ the number of heads in 5 tosses of a fair coin. The probability distribution appears in Table 5 and the histogram in Fig. 1. We can calculate the mean of the random variable X:

$$\mu_X = 0\left(\tfrac{1}{32}\right) + 1\left(\tfrac{5}{32}\right) + 2\left(\tfrac{10}{32}\right) + 3\left(\tfrac{10}{32}\right) + 4\left(\tfrac{5}{32}\right) + 5\left(\tfrac{1}{32}\right) = \tfrac{80}{32} = 2.5.$$

Table 5 $X =$ **Number of Heads in 5 Tosses of a Fair Coin**

k	$Pr(X = k)$	$k \cdot Pr(X = k)$
0	$\frac{1}{32}$	0
1	$\frac{5}{32}$	$\frac{5}{32}$
2	$\frac{10}{32}$	$\frac{20}{32}$
3	$\frac{10}{32}$	$\frac{30}{32}$
4	$\frac{5}{32}$	$\frac{20}{32}$
5	$\frac{1}{32}$	$\frac{5}{32}$
Total 1		$\mu = \frac{80}{32} = 2.5$

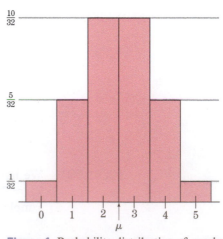

Figure 1 Probability distribution of number of heads

The mean is shown at the bottom of the histogram (Fig. 1). In contrast, we note that the sample mean, $\bar{x}$, for the coin-tossing experiment tabulated in Section 7.2 was 2.66. (See Table 6.) We rarely find that the sample mean $\bar{x}$ is exactly the theoretical value μ_X.

Table 6 Observed Frequency of Heads in 90 Rolls

Number of Heads x_i	Frequency f_i	Relative Frequency $\left(\frac{f_i}{n}\right)$	$x_i \cdot \left(\frac{f_i}{n}\right)$
0	3	$\frac{3}{90}$	0
1	14	$\frac{14}{90}$	.16
2	23	$\frac{23}{90}$	.51
3	27	$\frac{27}{90}$	.90
4	17	$\frac{17}{90}$	.76
5	6	$\frac{6}{90}$	.33
Total $n = 90$		1	$2.66 = \bar{x}$

We use binomial random variables so often that it is helpful to know that we have an easy formula for $E(X)$ in such cases.

> **Expected Value for Binomial Random Variables**   If X is a binomial random variable with parameters n and p, then
>
> $$E(X) = np. \qquad (2)$$

If X is the number of heads in five tosses of a fair coin, then using the formula, we see that $E(X) = 5\left(\frac{1}{2}\right) = 2.5$, which is consistent with our previous calculation. Formula (2) will be verified at the end of this section.

EXAMPLE 4   **Quality Control** Consider the plumbing-supplies manufacturer of Example 1 of Section 7.3. The manufacturer produces faucet washers, which come in boxes of 300. Studies have shown that 2% of the washers are defective. Find the average number of defective washers per box.

SOLUTION   The number of defective washers in each box is a binomial random variable with $n = 300$ and $p = .02$. The average number of defective washers per box is the expected value of X. $E(X) = np = 300 \cdot (.02) = 6$.   **» Now Try Exercise 29**

EXAMPLE 5   **Average Result from Rolling a Pair of Dice** Let the random variable X denote the sum of the faces appearing after rolling two dice. Determine $E(X)$.

SOLUTION   We determined the probability distribution of X in Example 5 of Section 7.2 (Table 7).

Table 7 Probability Distribution of X

k	$\Pr(X = k)$	k	$\Pr(X = k)$
2	$\frac{1}{36}$	8	$\frac{5}{36}$
3	$\frac{1}{18}$	9	$\frac{1}{9}$
4	$\frac{1}{12}$	10	$\frac{1}{12}$
5	$\frac{1}{9}$	11	$\frac{1}{18}$
6	$\frac{5}{36}$	12	$\frac{1}{36}$
7	$\frac{1}{6}$		

Therefore,

$$E(X) = 2 \cdot \tfrac{1}{36} + 3 \cdot \tfrac{1}{18} + 4 \cdot \tfrac{1}{12} + 5 \cdot \tfrac{1}{9} + 6 \cdot \tfrac{5}{36} + 7 \cdot \tfrac{1}{6}$$
$$+ 8 \cdot \tfrac{5}{36} + 9 \cdot \tfrac{1}{9} + 10 \cdot \tfrac{1}{12} + 11 \cdot \tfrac{1}{18} + 12 \cdot \tfrac{1}{36} = 7.$$

Clearly, 7 is the balance point of the histogram shown in Fig. 2.

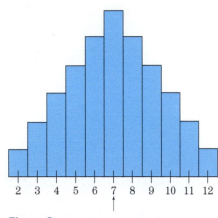

Figure 2   ≫ *Now Try Exercise 23*

The expected value of a random variable may be used to analyze games of chance, as the next two examples show.

EXAMPLE 6   **Chuck-a-Luck** In the carnival game Chuck-a-Luck, the player bets $1 on a number from 1 to 6, and then three dice are rolled. If one or more of the dice show the player's number, then the player receives their bet back plus a dollar for each time their number appears. Find the player's expected winnings or losses.

SOLUTION   Let the random variable X denote the player's winnings after one game. If the player's number does not appear, they lose their $1 bet—that is, their winnings are −$1. If the number appears on at least one die, then their winnings equal the number of dice that contain the number, so $1, $2, or $3. Of the $6^3 = 216$ possible outcomes for the three dice, 1 outcome contains the player's number on all three dice, $3 \cdot 5 = 15$ contain the player's number on exactly two of the dice, $3 \cdot 5 \cdot 5 = 75$ contain the player's number exactly once, and $5^3 = 125$ do not contain the number at all. Table 8 shows the probability distribution for X.

Therefore, $E(X) = -1 \cdot \tfrac{125}{216} + 1 \cdot \tfrac{75}{216} + 2 \cdot \tfrac{15}{216} + 3 \cdot \tfrac{1}{216} = -\tfrac{17}{216} \approx -.0787$. That is, the player will lose nearly 8¢ on each bet.   ≫ *Now Try Exercise 17*

Table 8

k	$\Pr(X = k)$
−1	$\tfrac{125}{216}$
1	$\tfrac{75}{216}$
2	$\tfrac{15}{216}$
3	$\tfrac{1}{216}$

In evaluating a game of chance, we use the expected value of the winnings to determine how fair the game is. The expected value of a completely fair game is zero. Let us compute the expected value of the winnings for two variations of the game roulette. American and European roulette games differ in both the nature of the wheel and the rules for playing.

American roulette wheels have 38 numbers (1 through 36 plus 0 and 00), of which 18 are red, 18 are black, and the 0 and 00 are green [see Fig. 3(a) on the next page]. Many different types of bets are possible. We shall consider the "red" bet. When you bet $1 on red, you win $1 if a red number appears and you lose $1 otherwise.

European roulette wheels have 37 numbers (1 through 36 plus 0) [see Fig. 3(b)]. The rules of European roulette differ from the American rules. One variation is as follows: When you bet $1 on "red," you win $1 if the ball lands on a red number and lose $1 if the ball lands on a black number. However, if the ball lands on the green number (0), then your bet stays on the table (the bet is said to be "imprisoned") and the payoff is determined by the result of the next spin. If a red number appears, you receive your $1 bet back, and if a black number appears, you lose your $1 bet. However, if the green

American Roulette Wheel   **European Roulette Wheel**

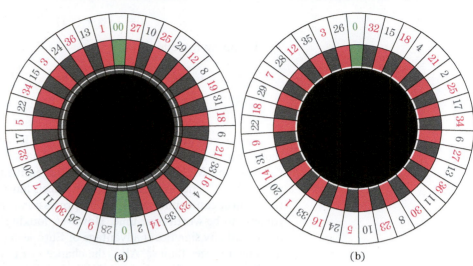

(a)   (b)

Figure 3

number (0) appears, you get back half of your bet, $.50. The tree diagrams for the red bet in American and European roulette are given in Fig. 4.

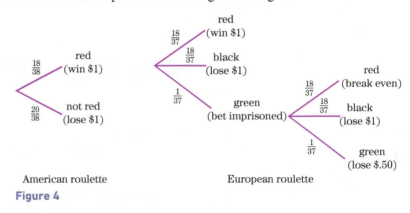

American roulette   European roulette

Figure 4

EXAMPLE 7   **Expected Winnings from Roulette**

(a) Set up the probability distribution tables for the earnings in American roulette and European roulette for the $1 bet on red.

(b) Compute the expected values for the probability distributions in part (a).

SOLUTION   (a) For American roulette, there are only two possibilities: earnings of $1 or of $-$1. These occur with probabilities $\frac{18}{38}$ and $\frac{20}{38}$, respectively.

For European roulette, the possible earnings are $1, $0, $-$.50, or $-$1. There are two ways in which to lose $1, one with probability $\frac{18}{37}$ and the other with probability $\frac{1}{37} \cdot \frac{18}{37} = \frac{18}{1369}$. Therefore,

$$\Pr(\text{lose } \$1) = \tfrac{18}{37} + \tfrac{18}{1369} = \tfrac{684}{1369}.$$

These probability distributions are tabulated in Table 9.

Table 9 Probabilities of Roulette Winnings

American Roulette		European Roulette	
Earnings	Probability	Earnings	Probability
1	$\frac{18}{38}$	1	$\frac{18}{37}$
-1	$\frac{20}{38}$	0	$\frac{18}{1369}$
		$-\frac{1}{2}$	$\frac{1}{1369}$
		-1	$\frac{684}{1369}$

(b) American roulette:

$$\mu = 1 \cdot \tfrac{18}{38} + (-1) \cdot \tfrac{20}{38} = -\tfrac{2}{38} \approx -.0526.$$

European roulette:

$$\mu = 1 \cdot \tfrac{18}{37} + 0 \cdot \tfrac{18}{1369} + \left(-\tfrac{1}{2}\right)\tfrac{1}{1369} + (-1)\tfrac{684}{1369}$$

$$= -\tfrac{1}{74} \approx -.0135. \qquad \text{» Now Try Exercise 15}$$

The secrets of Nicholas "Nick the Greek" Dandolos, one of the famous gamblers of the twentieth century, are revealed by Ted Thackrey Jr. in *Gambling Secrets of Nick the Greek* (Rand-McNally, 1968). The chapter entitled "Roulette" is subtitled "For Europeans Only." Since American roulette wheels have a lower probability of winning than their European counterparts, many professional gamblers do not consider American roulette to be worth their time. However, looking at the probabilities of winning does not reveal any significant advantage of European roulette over American roulette: $\tfrac{18}{37}$ is not much bigger than $\tfrac{18}{38}$. Also, the chance in European roulette to break even is very small. The real difference between the two games is revealed by the expected values. Someone playing American roulette will lose, on the average, about $5\tfrac{1}{4}$ cents per \$1 bet, whereas for European roulette, the average loss is about $1\tfrac{1}{3}$ cents. In both cases, you expect to lose money in the long run, but in American roulette, you lose nearly four times as much.

NOTE Casinos refer to the percent of each bet that they expect to keep as the *house percentage* or the *house advantage*. The house percentage is 5.26% on all but one American roulette bet. «

In summary, there are three types of mean: sample mean ($\overline{x}$), population mean (μ), and expected value of a random variable [$E(X)$, μ, or μ_X].

Verification of Formula (2)   Note that

$$\mu = 0 \cdot \Pr(X = 0) + 1 \cdot \Pr(X = 1) + \cdots + n \cdot \Pr(X = n)$$

$$= 0 \cdot \binom{n}{0} p^0 q^{n-0} + 1 \cdot \binom{n}{1} p^1 q^{n-1} + \cdots + n \cdot \binom{n}{n} p^n q^{n-n}.$$

Moreover, note that

$$k\binom{n}{k} = k \cdot \frac{n(n-1)\cdot\,\cdots\,\cdot(n-k+1)}{k(k-1)\cdot\,\cdots\,\cdot 2 \cdot 1} = n \cdot \frac{(n-1)\cdot\,\cdots\,\cdot(n-k+1)}{(k-1)\cdot\,\cdots\,\cdot 2 \cdot 1}$$

$$= n\binom{n-1}{k-1} \qquad (k = 1, 2, 3, \ldots, n).$$

Therefore,

$$\mu = 1 \cdot \binom{n}{1} p^1 q^{n-1} + 2 \cdot \binom{n}{2} p^2 q^{n-2} + \cdots + n \cdot \binom{n}{n} p^n q^0$$

$$= n\binom{n-1}{0} p^1 q^{n-1} + n\binom{n-1}{1} p^2 q^{n-2} + \cdots + n\binom{n-1}{n-1} p^n q^0$$

$$= np\left[\binom{n-1}{0} p^0 q^{n-1} + \binom{n-1}{1} p^1 q^{n-2} + \cdots + \binom{n-1}{n-1} p^{n-1} q^0 \right]$$

$$= np(p + q)^{n-1} \quad \text{(by the binomial theorem)}$$

$$= np \qquad \text{(since } p + q = 1\text{).}$$

«

INCORPORATING
TECHNOLOGY

Mean In Fig. 5, the sample mean $\overline{x}$ is calculated for the car dealership data from Table 1. The weekly sales were entered into list L_1 and the frequencies into list L_2 (see Fig. 6). To produce the screen in Fig. 5, enter into the 1-Var Stats wizard by pressing STAT ▷ 1 and enter L_1 and L_2 as shown in Fig. 7.

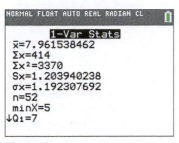

Figure 5

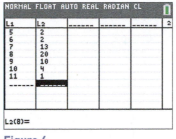

Figure 6

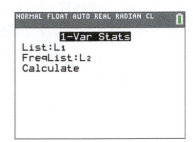

Figure 7

Mean After the numbers in a sample have been placed into a range, the sample mean can be calculated by evaluating the function **AVERAGE**. For example, **AVERAGE(A1:A12)** will calculate the mean of the values in cells **A1** through **A12**.

✳ WolframAlpha An instruction consisting of the word **mean** followed by a sequence of numbers separated by commas produces the average of the numbers.

Check Your Understanding 7.4

Solutions can be found following the section exercises.

1. **Life Insurance** A 74-year-old man pays $1000 for a one-year life insurance policy, which pays $20,000 in the event that he dies during the next year. According to life insurance tables, the probability of a 74-year-old man living one additional year is .95. Write down the probability distribution for the possible financial outcome, and determine its expected value.

2. **Life Insurance** According to life insurance tables, the probability that a 74-year-old man will live an additional five years is .7. How much should a 74-year-old be willing to pay for a policy that pays $20,000 in the event of death any time within the next 5 years?

EXERCISES 7.4

In Exercises 1–4, determine whether the quantity described is a sample mean, a population mean, or an expected value.

1. A class of 30 students is given an exam. The instructor wants to know how the class performed, so they calculate the mean exam grade for the entire class.

2. One hundred randomly selected residents of a city participate in a census. The census bureau wants to understand the financial health of the city, so the mean income for those 100 residents is calculated.

3. A community center sells 500 raffle tickets for $1 each, with a prize of $100 for the winning ticket. The center determines that the average participant will lose $.80 per ticket.

4. A gambler plays a carnival game four times, with winnings of $5, $10, −$3, and −$7 in the four games. They conclude that their mean winnings were $0.

5. Find the expected value for the probability distribution in Table 10.

Table 10

Value	Probability
0	.25
1	.2
2	.1
3	.25
4	.2

6. Find the expected value for the probability distribution in Table 11.

Table 11

Value	Probability
−1	.1
$-\frac{1}{2}$	.4
0	.25
$\frac{1}{2}$	.2
1	.05

7. **Grades** A college student received the following course grades for 10 (three-credit) courses during his freshman year: 4, 4, 4, 4, 3, 3, 2, 2, 2, 1.
   (a) Find his grade point average by adding the grades and dividing by 10.
   (b) Write down the relative frequency table.
   (c) Find the mean of the relative frequency distribution in part (b).

8. **Gymnastic Scores** An Olympic gymnast received the following scores from six judges: 9.8, 9.8, 9.4, 9.2, 9.2, 9.0.
   (a) Find the average score by adding the scores and dividing by 6.
   (b) Write down the relative frequency table.
   (c) Find the mean of the relative frequency distribution in part (b).

9. **Comparing Toothpastes** Table 12 gives the relative frequency of the number of cavities for two groups of children trying different brands of toothpaste. Calculate the sample means to determine which group had fewer cavities.

Table 12

Number of Cavities	Relative Frequency	
	Group A	Group B
0	.3	.2
1	.3	.3
2	.2	.3
3	.1	.1
4	0	.1
5	.1	0

10. **Investments** Table 13 gives the possible returns of two different investments and their probabilities. Calculate the means of the probability distributions to determine which investment has the greater expected return.

Table 13

Investment A		Investment B	
Return	Probability	Return	Probability
$1000	.2	−$2000	.2
$2000	.6	$0	.2
$3000	.2	$4000	.6

In Exercises 11–14, find the expected value for the random variable having the specified histogram.

11.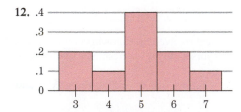

12.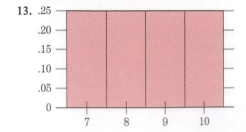

13.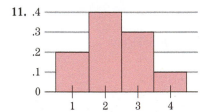

14.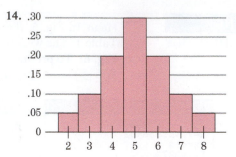

15. **Roulette** In American roulette, a bettor may place a $1 bet on any one of the 38 numbers on the roulette wheel. They win $35 (plus the return of their bet) if the ball lands on their number; otherwise, they lose their bet. Write down the probability distribution for the earnings from this type of bet, and find the expected value.

16. **Roulette** In American roulette, a dollar may be bet on a pair of numbers. The expected earnings for this type of bet is $-\$\frac{1}{19}$. How much money does the bettor receive if the ball lands on one of the two numbers?

17. **Carnival Game** In a carnival game, the player selects balls one at a time, without replacement, from an urn containing two red and four white balls. The game proceeds until a red ball is drawn. The player pays $1 to play the game and receives $.50 for each ball drawn. Write down the probability distribution for the player's earnings, and find its expected value.

18. **Carnival Game** In a carnival game, the player selects two coins from a bag containing two $1 coins and six worthless coins. Write down the probability distribution for the winnings, and determine how much the player would have to pay so that they would break even, on the average, over many repetitions of the game.

19. **Life Insurance** Using life insurance tables, a retired man determines that the probability of living 5 more years is .9. He decides to take out a life insurance policy that will pay $10,000 in the event that he dies during the next 5 years. How much should he be willing to pay for this policy? (Do not take account of interest rates or inflation.)

20. **Life Insurance** Using life insurance tables, a retired couple determines that the probability of living 5 more years is .9 for the man and .95 for the woman. They decide to take out a life insurance policy that will pay $10,000 if either one dies during the next 5 years and $15,000 if both die during that time. How much should they be willing to pay for this policy? (Assume that their life spans are independent events.)

21. **Batting Averages** Five members of a baseball team have batting averages of .300, and the other four members of the team have batting averages of .350. What is the overall batting average among members of the team?

22. **Exam Scores** Three members of a study group each earned a score of 91 on an exam, and the other five members of the group each earned a score of 87. What is the average score earned by the members of this study group?

23. **Dice** A pair of dice is rolled, and the larger of the two numbers showing is recorded. Find the expected value of this experiment.

24. **Dice** A pair of dice is rolled, and the smaller of the two numbers showing is recorded. Find the expected value of this experiment.

25. **Chuck-a-Luck** Redo Example 6, where the player wins $1 when they hit once, $3 when they hit twice, and $5 when they hit on all three dice.

26. **Chuck-a-Luck** Consider Example 6 with six players, where each plays a different number. Determine the house's expected winnings. First, insert the three missing values into the following table:

Outcome	House's Winnings	Probability
3 different numbers appear	0	—
2 different numbers appear	—	$\frac{5}{12}$
Same number on all three dice	2	—

27. **Bachelor's Degrees** According to the National Center for Education Statistics, 20.5% of all bachelor's degrees are awarded to students majoring in business. If 15 students receiving bachelor's degrees are selected at random, what is the expected number of business majors in the group?

28. **Faculty Degrees** According to the American Association of Community Colleges, 71% of the full-time faculty at community colleges have master's degrees. If a group of 30 full-time community college faculty are selected at random, what is the expected number of master's degree holders in the group?

29. **Rolling a Die** A die is rolled 30 times. What is the expected number of times that a 5 or a 6 will appear?

30. **Baseball** Ted is a consistent .275 hitter. How many hits is he expected to have in his next 40 at-bats?

31. **Basketball** A basketball player makes 40% of their three-point shots and 60% of their free throws. If they are fouled while taking a three-point shot, they are given three free throws. Which is greater: the expected number of points from taking a three-point shot or the expected number of points from taking three free throws?

32. **Binomial Trials** What is the probability of success for a binomial random variable with 20 trials whose expected value is 3?

Exercises 33–36 require the use of counting techniques.

33. **Committee Selection** A committee of three people is to be selected at random from a county council consisting of five Democrats and four Republicans. What is the expected number of Democrats on the committee? Republicans?

34. **Selecting Balls from an Urn** Four balls are selected at random from an urn containing six red balls and four green balls. What is the expected number of red balls? Green balls?

35. **Selecting Balls from an Urn** What is the expected number of red balls when three balls are selected at random, with replacement, from an urn containing four red balls and three green balls? Without replacement?

36. **Cards** What is the expected number of hearts when two cards are selected at random, with replacement, from a deck of 52 cards? Without replacement?

37. **Rain Insurance** The promoter of a football game is concerned that it will rain. They have the option of spending $8000 on insurance that will pay $40,000 if it rains. They estimate that the revenue from the game will be $60,000 if it does not rain and $25,000 if it does rain. What must the chance of rain be

if buying the policy has the same expected return as not buying it?

38. **Theft Insurance** Bob wishes to insure a priceless family heirloom against theft. The annual premium for policy A is $150, and it will pay $75,000 if the heirloom is stolen. Policy B will pay $100,000, but the annual premium is $250. Bob estimates the probability that the heirloom will be stolen in any given year and concludes that both policies have the same expected return. What is this estimated probability?

39. **Ticket Sales** Last weekend, a movie theater sold x adult tickets at $10 each and y children's tickets at $8 each. The average (arithmetic mean) revenue per ticket was

(a) $\dfrac{80xy}{x+y}$. (b) $\dfrac{10x+8y}{x+y}$. (c) $\dfrac{10x+8y}{18}$.

(d) $\dfrac{80xy}{18}$. (e) $\dfrac{10x+8y}{xy}$.

40. **Candle Sales** A store sold an average (arithmetic mean) of x candles per day for k days and then sold y candles on the next day. What is the average number of candles sold daily for the $(k+1)$-day period?

(a) $x+\dfrac{y}{k}$ (b) $\dfrac{kx+y}{k+1}$ (c) $\dfrac{k(x+y)}{k+1}$

(d) $\dfrac{x+ky}{k+1}$ (e) $x+\dfrac{y}{k+1}$

41. **Weather** The average daily temperature during April in Washington, D.C., is 56.1°. Suppose that, in a certain year, the average daily temperature for the first 16 days of April is 54°. What temperature must the remaining 14 days average in order for the overall average to reach the historic average of 56.1°?

42. **Grades** José wants to earn a 90% overall average in his mathematics class. His current grades are 83%, 92%, and 89%. What grade does José need to average on the next three assessments to earn a 90% overall average?

43. **Mean** If 5, 6, and x have the same average as 2, 7, and 9, then what does x equal?

44. **Game of Chance** Table 14 gives the probability distribution of a player's earnings from one play of a carnival game. For what value of x is the player's expected winnings equal to zero?

Table 14

Earnings	Probability
−2	.1
−1	.2
0	.6
x	.1

45. **Factory Production** An assembly line produces 12 bicycles per day over the course of five days. If production is increased to 24 bicycles per day, how many additional days will it take for average production to reach 19 bicycles per day?

46. **Weekly Revenue** A small business had an average weekly revenue of $14,000 over the past three weeks. The revenue for the first week was twice the revenue of the third week, and the revenue for the second week was half the revenue of the third week. What was the revenue for the first week?

47. Truck Capacity A truck can carry a maximum of 75,000 pounds of cargo. How many cases of cargo can it carry if half of the cases have an average weight of 20 pounds and the other half have an average weight of 30 pounds?

48. Magazine Sales Half of the magazines at a newsstand sell for an average of $2.00, and the other half sell for an average of $2.50. If the total retail value of the magazines is $135.00, how many magazines are there at the newsstand?

Solutions to Check Your Understanding 7.4

1. There are two possibilities. If the man lives until the end of the year, he loses $1000. If he dies during the year, his estate gains $19,000 (the $20,000 settlement minus the $1000 premium).

Outcome	Probability
−$1000	.95
$19,000	.05

$$\mu = (-1000)(.95) + (19,000)(.05) = 0$$

(Thus, if the insurance company insures a large number of people, it should break even. Its profits will result from the interest that it earns on the money being held.)

2. Let x denote the cost of the policy. The probability distribution is as follows.

Outcome	Probability
−x	.7
$20,000 − x$	.3

$$\mu = (-x)(.7) + (20,000 - x)(.3)$$
$$= -.7x + 6000 - .3x$$
$$= 6000 - x$$

The expected value will be zero if $x = 6000$. Therefore, the man should be willing to pay up to $6000 for his policy.

7.5 The Variance and Standard Deviation

In the previous section, we introduced three related concepts: the mean of a sample ($\overline{x}$); the mean of a population (μ); and the mean, or expected value, of a random variable [$E(X)$]. The mean is probably the single most important number that can be used to describe a sample, a population, or a probability distribution of a random variable. The next most important number is the **variance**.

Roughly speaking, the variance measures the dispersal, or spread, of a distribution about its mean. The more closely concentrated the distribution about its mean is, the smaller the variance will be; the more spread out, the larger the variance. Thus, for example, the probability distribution whose histogram is drawn in Fig. 1(a) has a smaller variance than that in Fig. 1(b).

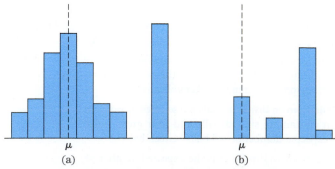

Figure 1 Histograms for two probability distributions

Let us now define the variance of a probability distribution of a random variable. Suppose that X is a random variable with values $x_1, x_2, \ldots, x_N$ and respective probabilities $p_1, p_2, \ldots, p_N$. Suppose that the mean is μ. Then the deviations of the various outcomes from the mean are given by the N differences

$$x_1 - \mu, \quad x_2 - \mu, \ldots, \quad x_N - \mu.$$

Since we want to give weight to the various deviations according to their likelihood of occurrence, it is tempting to multiply each deviation by its probability of occurrence.

However, this will not lead to a very satisfactory measure of deviation from the mean. This is because some of the differences will be positive and others negative. In the process of addition, deviations from the mean (both positive and negative deviations) will combine to yield a zero total deviation. To correct this, we consider instead the squares of the differences,

$$(x_1 - \mu)^2, \quad (x_2 - \mu)^2, \dots, \quad (x_N - \mu)^2,$$

which are all ≥ 0. To obtain a measure of deviation from the mean, we multiply each of these expressions by the probability of the corresponding outcome. The number thus obtained is called the *variance of the probability distribution* (or of the associated random variable) and is denoted $\text{Var}(X)$. That is, the variance of a probability distribution is given by the following formula:

DEFINITION Variance of a Probability Distribution

$$\text{Var}(X) = (x_1 - \mu)^2 p_1 + (x_2 - \mu)^2 p_2 + \cdots + (x_N - \mu)^2 p_N.$$

EXAMPLE 1   **Computing Variance** Compute the variance of the following probability distribution:

Outcome	Probability
0	.1
1	.3
2	.5
3	.1

SOLUTION   The mean is given by

$$\mu = 0 \cdot (.1) + 1 \cdot (.3) + 2 \cdot (.5) + 3 \cdot (.1) = 1.6.$$

In the notation of random variables, the calculations of the variance may be summarized as follows:

k	$\Pr(X = k)$	$k - \mu$	$(k - \mu)^2$	$(k - \mu)^2 \Pr(X = k)$
0	.1	$0 - 1.6 = -1.6$	2.56	.256
1	.3	$1 - 1.6 = -.6$	.36	.108
2	.5	$2 - 1.6 = .4$	.16	.080
3	.1	$3 - 1.6 = 1.4$	1.96	.196

$$\text{Var}(X) = .256 + .108 + .080 + .196 = .64.$$

»» Now Try Exercise 1

There is an alternative formula for the variance that can simplify its calculation. The alternative formula for the variance can be derived from the fact that $\text{Var}(X) = \text{E}[(X - \mu)^2]$.

Alternative Formula for Variance   If X is a random variable, then

$$\text{Var}(X) = \text{E}(X^2) - [\text{E}(X)]^2.$$

EXAMPLE 2   **Using the Alternative Formula for Variance** Use the alternative formula to find the variance of the random variable in Example 1.

SOLUTION By Example 1, we know that $E(X) = 1.6$. The probability distribution of X^2 is as follows:

k	$\Pr(X^2 = k)$
0	.1
1	.3
4	.5
9	.1

$$E(X^2) = 0 \cdot (.1) + 1 \cdot (.3) + 4 \cdot (.5) + 9 \cdot (.1) = 3.2$$

Therefore,

$$\text{Var}(X) = E(X^2) - [E(X)]^2 = 3.2 - 1.6^2 = 3.2 - 2.56 = .64.$$

» Now Try Exercise 21

An advantage of this alternative formula is that, if the expected value is an unwieldy number, we are spared the tedious task of subtracting it from the values of the variable, which sometimes produces even more unwieldy numbers that must be squared and summed.

Actually, a much more commonly used measure of dispersal about the mean is the **standard deviation**, which is just the square root of the variance and is denoted σ_X. (σ is the Greek letter sigma, and σ_X is pronounced "sigma sub X.")

DEFINITION Standard Deviation

$$\sigma_X = \sqrt{\text{Var}(X)}$$

Thus, for example, for the probability distribution of Example 1, we have

$$\sigma_X = \sqrt{.64} = .8.$$

The reason for using the standard deviation as opposed to the variance is that the former is expressed in the same units of measurement as X, whereas the latter is not. The standard deviation is used in statistics to determine the margin of error in polling data, and in finance to measure the volatility of an investment. For example, consider the following two dice games:

Game A: A player rolls a die. If the result is a 1, 2, 3, 4, or 5, the player wins \$6. If it is a 6, the player wins nothing.

Game B: A player rolls a die. If the result is a 1, the player wins \$30.

Let the random variables X and Y represent a player's winnings from one play of games A and B, respectively. Then the probability distributions of X and Y are as follows.

Game A		Game B	
k	$\Pr(X = k)$	k	$\Pr(Y = k)$
0	$\frac{1}{6}$	0	$\frac{5}{6}$
6	$\frac{5}{6}$	30	$\frac{1}{6}$

Note that both games have the same expected winnings, that is, $E(X) = E(Y) = 5$.

The probability distributions for X^2 and Y^2 are as follows.

Game A		Game B	
k	$\Pr(X^2 = k)$	k	$\Pr(Y^2 = k)$
0	$\frac{1}{6}$	0	$\frac{5}{6}$
36	$\frac{5}{6}$	900	$\frac{1}{6}$

So, $E(X^2) = 36 \cdot \frac{5}{6} = 30$ and $E(Y^2) = 900 \cdot \frac{1}{6} = 150$.

It follows that

$$\sigma_X = \sqrt{E(X^2) - [E(X)]^2} = \sqrt{30 - 5^2} = \sqrt{5} \approx 2.24$$
$$\sigma_Y = \sqrt{E(Y^2) - [E(Y)]^2} = \sqrt{150 - 5^2} = \sqrt{125} \approx 11.18.$$

Intuitively, game A will have more consistent results while game B will be much more volatile. The fact that X has a smaller standard deviation than Y quantifies these relative volatilities.

Variance and Standard Deviation of a Binomial Random Variable

In the binomial trials section, we saw that, if X is a binomial random variable with parameters n and p, then $E(X) = np$. The variance and standard deviation of a binomial random variable also have simple formulas.

> **Variance and Standard Deviation of a Binomial Random Variable**   If X is a binomial random variable with parameters n and p, then with $q = 1 - p$,
>
> $$\text{Var}(X) = npq$$
> $$\sigma_X = \sqrt{npq}.$$

EXAMPLE 3   Compute the mean and standard deviation for the binomial random variable X, the number of ones appearing in 180 rolls of a die.

SOLUTION   Here, $n = 180$, $p = \frac{1}{6}$, and $q = 1 - \frac{1}{6} = \frac{5}{6}$. Therefore,

$$\mu = np = 180 \cdot \frac{1}{6} = 30$$

$$\sigma_X = \sqrt{npq} = \sqrt{180 \cdot \frac{1}{6} \cdot \frac{5}{6}} = \sqrt{25} = 5.$$

» Now Try Exercise 9

Population Variance and Standard Deviation

In Section 7.4, we developed formulas for the mean of a population (μ). Similar formulas are available for the variance and standard deviation of a population. In this case, the standard deviation is denoted σ and the variance is denoted σ^2.

> **DEFINITION** **Population Variance and Standard Deviation**   Consider a population of N numbers $x_1, x_2, \dots, x_N$ having mean μ. Then the variance is defined as
>
> $$\sigma^2 = \frac{(x_1 - \mu)^2 + (x_2 - \mu)^2 + \cdots + (x_N - \mu)^2}{N}.$$
>
> If the numbers have been grouped into the values $x_1, x_2, \dots, x_r$ with frequencies $f_1, f_2, \dots, f_r$, then the variance is
>
> $$\sigma^2 = \frac{1}{N}[(x_1 - \mu)^2 f_1 + (x_2 - \mu)^2 f_2 + \cdots + (x_r - \mu)^2 f_r]$$
>
> or
>
> $$\sigma^2 = (x_1 - \mu)^2 \left(\frac{f_1}{N}\right) + (x_2 - \mu)^2 \left(\frac{f_2}{N}\right) + \cdots + (x_r - \mu)^2 \left(\frac{f_r}{N}\right).$$
>
> The **standard deviation**, σ, is defined as the square root of the variance.

EXAMPLE 4	**Computing Variance and Standard Deviation** Compute the variance and the standard deviation for the population of scores on a five-question quiz as tabulated in Table 1.

SOLUTION   We first find μ.

$$\mu = \tfrac{1}{60}\,[0(4) + 1(9) + 2(6) + 3(14) + 4(18) + 5(9)] = \tfrac{180}{60} = 3.$$

We find σ^2 by subtracting 3 from each of the test scores, squaring the differences, weighting each with its frequency, and dividing the resulting sum by $N = 60$. The computation is shown in Table 2. Therefore, $\sigma^2 = \tfrac{132}{60} = 2.2$. The standard deviation, which is found by taking the square root of the variance, is $\sigma \approx 1.48$.

Table 1

Score	Frequency
0	4
1	9
2	6
3	14
4	18
5	9
Total	**60**

Table 2

x_i	f_i	$x_i - \mu$	$(x_i - \mu)^2$	$(x_i - \mu)^2\,(f_i)$
0	4	-3	9	36
1	9	-2	4	36
2	6	-1	1	6
3	14	0	0	0
4	18	1	1	18
5	9	2	4	36
Total	**60**			132

>> Now Try Exercise 7

Sample Variance and Standard Deviation

As we discussed in Section 7.4, we sometimes have only sample values at our disposal. If so, we must use sample statistics to estimate the population parameters. For example, we can use the sample mean $\bar{x}$ as an estimate of the population mean μ, and the sample variance as an estimate of the population variance, σ^2. In fact, if samples of size n were chosen repeatedly from a population and the sample mean were computed for each sample, then the average of these means should be close to the value of the population mean, μ. This is a desirable property for any statistic used to estimate a parameter—the averages of the statistic get arbitrarily close to the actual value of the parameter as the number of samples increases. Such an estimate is said to be **unbiased**. Thus, $\bar{x}$ is an unbiased estimate of μ.

The situation with the variance is a little trickier. In order to have an unbiased estimate of the population variance σ^2, we must define the **sample variance**, s^2, in a slightly peculiar way.

> **DEFINITION Sample Variance and Standard Deviation**   Consider a sample of n numbers $x_1, x_2, \ldots, x_n$ having mean $\bar{x}$. Then the **variance** is
>
> $$s^2 = \frac{(x_1 - \bar{x})^2 + (x_2 - \bar{x})^2 + \cdots + (x_n - \bar{x})^2}{n - 1}.$$
>
> If the numbers have been grouped into the values $x_1, x_2, \ldots, x_r$ with frequencies $f_1, f_2, \ldots, f_r$, then
>
> $$s^2 = \frac{1}{n-1}\,[(x_1 - \bar{x})^2 f_1 + (x_2 - \bar{x})^2 f_2 + \cdots + (x_r - \bar{x})^2 f_r].$$
>
> The **standard deviation**, s, is defined as the square root of the variance.

This is the usual definition and the way in which most statistical calculators do the computation of the sample variance. Note that the divisor is one less than the sample size. With this definition, s^2 is an unbiased estimate of σ^2.

EXAMPLE 5   **Computing Sample Standard Deviation** Compute the sample standard deviations for the frequency distributions of sales in car dealerships A and B. Interpret the results.

SOLUTION   The frequency distribution for dealership A is given by Table 3. In Example 1 of Section 7.4, we found the sample mean of the weekly sales to be $\bar{x}_A = 7.96$. Recall that we are treating the data collected from these dealerships as samples (one year's data from dealership A and two years' data from dealership B). Therefore, the sample variance for dealership A is given by

$$s_A^2 = \tfrac{1}{51}[(5 - 7.96)^2 \cdot 2 + (6 - 7.96)^2 \cdot 2 + (7 - 7.96)^2 \cdot 13$$
$$+ (8 - 7.96)^2 \cdot 20 + (9 - 7.96)^2 \cdot 10 + (10 - 7.96)^2 \cdot 4$$
$$+ (11 - 7.96)^2 \cdot 1] \approx 1.45.$$

Table 3

Weekly Sales	Frequency
5	2
6	2
7	13
8	20
9	10
10	4
11	1
Total	$n = 52$

The sample standard deviation, s_A, for dealership A is given by

$$s_A = \sqrt{s_A^2} \approx \sqrt{1.45} \approx 1.20 \text{ cars.}$$

In a similar way, we find that the sample standard deviation of dealership B is $s_B \approx 2.03$ cars. Since s_A is smaller than s_B, dealership B exhibited greater variation than dealership A during the time that the sales were observed. On average, dealership B had higher weekly sales, but those of dealership A showed greater consistency. The sample statistics might help Ms. Jones decide which dealership to buy. She will have to decide whether consistency is more important than the size of long-term average sales per week.

>> Now Try Exercise 5

Chebychev's Inequality

We can use the value of the sample mean to estimate the population mean. The sample standard deviation helps us to determine the degree of accuracy of our estimate. If the sample standard deviation is small, indicating that the population is not widely dispersed about its mean, the estimated value of μ is likely to be close to the actual value of μ.

What does the standard deviation tell us about the dispersal of the data about the mean of a probability distribution? *Chebychev's inequality* helps us to see that, the larger the standard deviation, the more likely it is that we find extreme values in the data. The probability that an outcome falls more than c units away from the mean is at most σ^2/c^2.

> **Chebychev's Inequality**   Suppose that a probability distribution with numerical outcomes has expected value μ and standard deviation σ. Then the probability that a randomly chosen outcome lies between $\mu - c$ and $\mu + c$, inclusive, is at least $1 - (\sigma^2/c^2)$.

A verification of the Chebychev inequality can be found in most elementary statistics texts.

EXAMPLE 6   **Applying Chebychev's Inequality** Suppose that a probability distribution has mean 5 and standard deviation 1. Use the Chebychev inequality to estimate the probability that an outcome lies from 3 to 7.

SOLUTION   Here, $\mu = 5, \sigma = 1$. Since we wish to estimate the probability of an outcome lying between 3 and 7, we set $\mu - c = 3$ and $\mu + c = 7$. Thus, $c = 2$. Then by the Chebychev inequality, the desired probability is at least

$$1 - \frac{\sigma^2}{c^2} = 1 - \frac{1}{4} = .75.$$

That is, if the experiment is repeated a large number of times, we expect at least 75% of the outcomes to be between 3 and 7. Also, we expect at most 25% of the outcomes to fall below 3 or above 7.

>> Now Try Exercise 13(a)

The Chebychev inequality has many practical applications, one of which is illustrated in the next example.

EXAMPLE 7   **Quality Control**   Apex Drug Supply Company sells bottles containing 100 capsules of penicillin. Due to the bottling procedure, not every bottle contains exactly 100 capsules. Assume that the average number of capsules in a bottle is indeed 100 ($\mu = 100$) and the standard deviation is 2 ($\sigma = 2$). If the company ships 5000 bottles, estimate the number having between 95 and 105 capsules, inclusive.

SOLUTION   By Chebychev's inequality, the proportion of bottles having between $100 - 5$ and $100 + 5$ capsules should be at least

$$1 - \frac{2^2}{5^2} = \frac{21}{25} = .84.$$

That is, we expect at least 84% of the 5000 bottles, or 4200 bottles, to be in the desired range.   **≫ Now Try Exercise 15**

INCORPORATING TECHNOLOGY

Standard Deviation   In Figs. 2, 3, and 4, the sample standard deviation **Sx** and population standard deviation are calculated for the car dealership data from Table 3. Note that the steps are the same as in the Section 7.4 Incorporating Technology discussion.

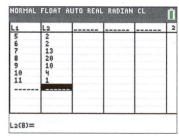

Figure 2

Figure 3

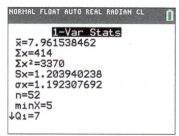

Figure 4

Variance and Standard Deviation   After the numbers in a sample have been placed into a range, the sample variance and standard deviation can be calculated by evaluating the functions VAR.S and STDEV.S of the range. For example, **STDEV.S(A1:A12)** will calculate the sample standard deviation of the values in cells **A1** through **A12**. The functions VAR.P and STDEV.P calculate the population variance and standard deviation.

✳ WolframAlpha   An instruction consisting of **variance** or **standard deviation** followed by a sequence of numbers separated by commas produces the sample variance or sample standard deviation of the numbers. Population variances and population standard deviations are produced when the instructions are preceded with the word **population**.

Check Your Understanding 7.5

Solutions can be found following the section exercises.

1. (a) Compute the variance of the probability distribution in Table 4.
   (b) Using Table 4, find the probability that the outcome is from 22 to 24.

2. Refer to the probability distribution in Table 4. Use the Chebychev inequality to approximate the probability that the outcome is from 22 to 24.

Table 4

Outcome	Probability
21	$\frac{1}{16}$
22	$\frac{1}{8}$
23	$\frac{5}{8}$
24	$\frac{1}{8}$
25	$\frac{1}{16}$

EXERCISES 7.5

1. Compute the variance of the probability distribution in Table 5.

Table 5

Outcome	Probability
70	.5
71	.2
72	.1
73	.2

2. Compute the variance of the probability distribution in Table 6.

Table 6

Outcome	Probability
-1	$\frac{1}{8}$
$-\frac{1}{2}$	$\frac{3}{8}$
0	$\frac{1}{8}$
$\frac{1}{2}$	$\frac{1}{8}$
1	$\frac{2}{8}$

3. Determine by inspection which one of the probability distributions, A or B, in Fig. 5 has the greater variance.

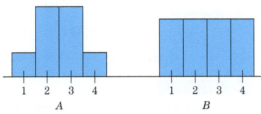

Figure 5

4. Determine by inspection which one of the probability distributions, C or D, in Fig. 6 has the greater variance.

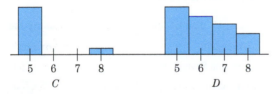

Figure 6

5. **Investment Returns** Table 7 gives the probability distribution for the possible returns from two different investments.
   (a) Compute the mean and the variance for each investment.
   (b) Which investment has the higher expected return (i.e., mean)?
   (c) Which investment is less risky (i.e., has lesser variance)?

Table 7

Investment A		Investment B	
Return ($ millions)	Probability	Return ($ millions)	Probability
-10	.2	0	.3
20	.2	10	.4
25	.6	30	.3

6. **Golf Scores** Two golfers recorded their scores for 20 nine-hole rounds of golf. Golfer A's scores were

   $$39, 39, 40, 40, 40, 40, 40, 40,$$
   $$41, 41, 41, 41, 41, 41, 41, 42, 43, 43, 43, 44.$$

   Golfer B's scores were

   $$40, 40, 40, 41, 41, 41, 41, 42, 42, 42, 42, 42,$$
   $$43, 43, 43, 43, 43, 43, 44, 44.$$

   (a) Compute the sample mean and the variance of each golfer's scores.
   (b) Who is the better golfer? *Note:* The lower the score is, the better.
   (c) Who is the more consistent golfer?

7. **Weekly Sales** Table 8 gives the relative frequency distribution for the weekly sales of two businesses.

Table 8

	Relative Frequency	
Sales	Business A	Business B
100	.1	0
101	.2	.2
102	.3	0
103	0	.2
104	0	.1
105	.2	.2
106	.2	.3

   (a) Compute the population mean and the variance for each business.
   (b) Which business has the better sales record?
   (c) Which business has the more consistent sales record?

8. **Course Grades** Student A received the following course grades during her first year of college:

   $$4, 4, 4, 3, 3, 3, 2, 2, 2, 1$$

   Student B received the following course grades during her first year:

   $$4, 4, 4, 4, 4, 4, 3, 1, 1, 1$$

   (a) Compute the population means and variances.
   (b) Which student had the better grade point average?
   (c) Which student was more consistent?

9. **Coin Tosses** Suppose that a coin is tossed 10 times. Find the mean and standard deviation for the number of heads.

10. **Dice** Suppose that a pair of dice is rolled 720 times. Find the mean and standard deviation for the number of times the sum of seven appears.

11. **Quality Control** A manufacturer produces smart thermostats that are packaged in boxes of 200. The probability of a thermostat being defective is .015. Find the mean and standard deviation for the number of defective thermostats in a box.

12. **Basketball** A basketball player makes each free throw with probability 3/5. Find the mean and standard deviation for the number of successes in 20 tries.

13. **Chebychev Inequality** Suppose that a probability distribution has mean 35 and standard deviation 5. Use the Chebychev inequality to estimate the probability that an outcome will lie from
   (a) 25 to 45.   (b) 20 to 50.   (c) 29 to 41.

14. **Chebychev Inequality** Suppose that a probability distribution has mean 8 and standard deviation .4. Use the Chebychev inequality to estimate the probability that an outcome will lie from
   (a) 6 to 10.   (b) 7.2 to 8.8.   (c) 7.5 to 8.5.

15. **Bulb Lifetimes** For certain types of fluorescent lights, the number of hours a bulb will burn before requiring replacement has a mean of 3000 hours and a standard deviation of 250 hours. Suppose that 5000 of such bulbs are installed in an office building. Use the Chebychev inequality to estimate the number that will require replacement between 2000 and 4000 hours, inclusive, from the time of installation.

16. **Quality Control** An electronics firm determines that the number of defective circuit boards in each batch averages 15 with standard deviation 10. Suppose that 100 batches are produced. Use the Chebychev inequality to estimate the number of batches having from 0 to 30 defective circuit boards.

17. **Chebychev Inequality** Suppose that a probability distribution has mean 75 and standard deviation 6. Use the Chebychev inequality to find the value of c for which the probability that the outcome lies between $75 - c$ and $75 + c$, inclusive, is at least 7/16.

18. **Chebychev Inequality** Suppose that a probability distribution has mean 17 and standard deviation .2. Use the Chebychev inequality to find the value of c for which the probability that the outcome lies between $17 - c$ and $17 + c$, inclusive, is at least 15/16.

19. **Dice** The probability distribution for the sum of numbers obtained from rolling a pair of dice is given in Table 9.
   (a) Compute the mean and the variance of this probability distribution.
   (b) Using the table, give the probability that the number is between 4 and 10, inclusive.
   (c) Use the Chebychev inequality to estimate the probability that the number is between 4 and 10, inclusive.

Table 9

Sum	Probability
2	$\frac{1}{36}$
3	$\frac{2}{36}$
4	$\frac{3}{36}$
5	$\frac{4}{36}$
6	$\frac{5}{36}$
7	$\frac{6}{36}$
8	$\frac{5}{36}$
9	$\frac{4}{36}$
10	$\frac{3}{36}$
11	$\frac{2}{36}$
12	$\frac{1}{36}$

20. **Dice** The probability distribution, rounded to the nearest thousandth, for the number of ones obtained from rolling 12 dice is given in Table 10. This probability distribution has mean 2 and standard deviation 1.291 ($\sigma^2 = \frac{5}{3}$).
   (a) Using the table, give the probability that the number of ones rolled is between 0 and 4, inclusive.
   (b) Use the Chebychev inequality to estimate the probability that the number of ones is between 0 and 4, inclusive.

Table 10

Ones	Probability
0	.112
1	.269
2	.296
3	.197
4	.089
5	.028
6	.007
7	.001
8	.000
9	.000
10	.000
11	.000
12	.000

21. **Variance** Use the alternative formula for variance to calculate the variance of the random variable in Table 11.

Table 11

k	$\Pr(X = k)$
-2	.2
-1	.2
0	.2
1	.2
2	.2

22. If X is a random variable, then the variance of X equals the variance of $X - a$ for any number a. Redo Exercise 1, using this result with $a = 70$.

23. If X is a random variable, then the variance of aX equals a^2 times the variance of X. Verify this result for the random variable in Exercise 2 with $a = 2$.

24. If X is a random variable, then

$$E(X - a) = E(X) - a \quad \text{and} \quad E(aX) = aE(X)$$

for any number a. Give intuitive justifications of these results.

TECHNOLOGY EXERCISES

25. **College Enrollments** Table 12 gives the fall 2015 enrollments of 10 of the largest university campuses in the United States. Determine the population mean and standard deviation for these enrollments. Which schools have enrollments within one standard deviation of the mean?

Table 12 Ten Large Public University Campuses

University	Enrollment
Arizona State University	60,168
University of Central Florida	59,770
Ohio State University	58,663
Florida International University	52,980
Texas A&M University	52,449
University of Minnesota	51,147
University of Texas at Austin	51,145
Michigan State University	50,085
University of Florida	49,042
Indiana University	46,817

26. **Priciest Colleges** Table 13 gives the tuition, fees, room, and board for eight of the priciest colleges in the United States for the academic year 2015–16. Determine the population mean and standard deviation of these costs. Which schools have costs at least one standard deviation greater than the mean?

Table 13 Eight Pricy U.S. Colleges

College	Tuition, Fees, Room, and Board
Harvey Mudd College	$67,255
Columbia University	$66,383
New York University	$65,860
Sarah Lawrence College	$65,630
University of Chicago	$64,965
Bard College	$64,519
University of Southern California	$64,482
Claremont McKenna College	$64,325

27. Ph.D. Degrees Table 14 summarizes the number of Ph.D. degrees awarded at two universities during the past 25 years. For example, three Ph.D. degrees were awarded at University A during 5 of the last 25 years.

Table 14

University A		University B	
Number of Degrees	Number of Years	Number of Degrees	Number of Years
3	5	3	5
4	7	4	10
5	8	5	3
6	2	6	3
7	1	7	0
8	2	8	4

(a) Find the population mean and standard deviation for the number of degrees awarded each year at each university.

(b) Which university produces more Ph.D.'s per year on average?

(c) Which university produces more consistent numbers of Ph.D.'s per year?

28. Games of Chance Table 15 gives the probability distributions for the possible earnings from two games of chance.

Table 15

Game A		Game B	
Earnings	Probability	Earnings	Probability
−5	.23	−5	.32
−1	.32	−1	.10
1	.35	1	.40
5	.07	5	.13
10	.03	10	.05

(a) Find the mean and standard deviation for the earnings in each game.

(b) Which game is more favorable for a gambler in the long run?

(c) Which game produces more consistent results?

Solutions to Check Your Understanding 7.5

1. (a)

k	$\Pr(X = k)$	$k - \mu$	$(k - \mu)^2$	$(k - \mu)^2 \Pr(X = k)$
21	$\frac{1}{16}$	−2	4	$\frac{4}{16}$
22	$\frac{1}{8}$	−1	1	$\frac{1}{8}$
23	$\frac{5}{8}$	0	0	0
24	$\frac{1}{8}$	1	1	$\frac{1}{8}$
25	$\frac{1}{16}$	2	4	$\frac{4}{16}$

$$\mu = 21 \cdot \tfrac{1}{16} + 22 \cdot \tfrac{1}{8} + 23 \cdot \tfrac{5}{8} + 24 \cdot \tfrac{1}{8} + 25 \cdot \tfrac{1}{16}$$
$$= \tfrac{21}{16} + \tfrac{44}{16} + \tfrac{230}{16} + \tfrac{48}{16} + \tfrac{25}{16} = \tfrac{368}{16} = 23$$

$$\mathrm{Var}(X) = \tfrac{4}{16} + \tfrac{1}{8} + 0 + \tfrac{1}{8} + \tfrac{4}{16}$$
$$= \tfrac{2}{8} + \tfrac{1}{8} + 0 + \tfrac{1}{8} + \tfrac{2}{8} = \tfrac{6}{8} = \tfrac{3}{4}$$

(b) $\frac{7}{8}$. The probability that the outcome is from 22 to 24 is

$$\Pr(22) + \Pr(23) + \Pr(24) = \tfrac{1}{8} + \tfrac{5}{8} + \tfrac{1}{8} = \tfrac{7}{8}.$$

2. Probability $\geq \frac{1}{4}$. Here, $\mu = 23$, $\sigma^2 = \frac{3}{4}$, and $c = 1$. By the Chebychev inequality, the probability that the outcome is between $23 - 1$ and $23 + 1$ is at least $1 - (\frac{3}{4}/1^2) = 1 - \frac{3}{4} = \frac{1}{4}$. [From 1(b), we obtained the actual probability of $\frac{7}{8}$, which is much greater than $\frac{1}{4}$. In the next sections, we will study a technique that gives better estimates. However, this technique holds only for a special type of probability distribution.]

7.6	# The Normal Distribution

In this section, we will see that the histogram for a binomial random variable can be approximated by a region under a smooth curve called a *normal curve*.

Toss a coin 20 times, and observe the number of heads. By using formula (1) of Section 7.3 with $n = 20$ and $p = .5$, we can calculate the probability of k heads. The results, rounded to the nearest ten-thousandth, are displayed in Table 1. The data of Table 1 can be displayed in histogram form, as in Fig. 1.

Table 1 **Probability of k Heads (to four decimal places)**

k	Probability of k Heads	k	Probability of k Heads
0	.0000	10	.1762
1	.0000	11	.1602
2	.0002	12	.1201
3	.0011	13	.0739
4	.0046	14	.0370
5	.0148	15	.0148
6	.0370	16	.0046
7	.0739	17	.0011
8	.1201	18	.0002
9	.1602	19	.0000
		20	.0000

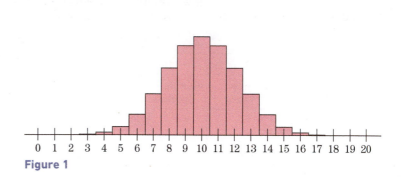

Figure 1

As we have seen in Section 7.2, various probabilities may be interpreted as areas. For example, the probability that at most 9 heads occur is equal to the sum of the areas of all of the rectangles to the left of the central one (Fig. 2). The shape of the histogram in Figs. 1 and 2 suggests that we might be able to approximate such areas by using a smooth bell-shaped curve. The curve shown in Fig. 3 is a good candidate. It is called a **normal curve** and plays an important role in statistics and probability. For instance, the area of the blue rectangles in Fig. 2 is approximately the same as the area under the normal curve to the left of 9.5 shown in Fig. 4. As another example, the probability of

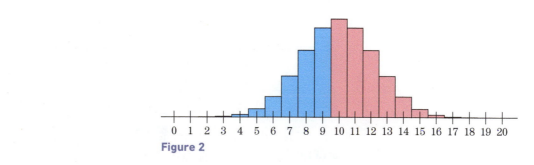

Figure 2

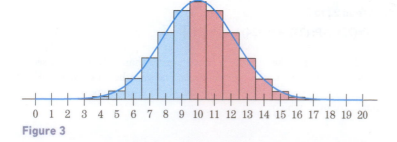

Figure 3

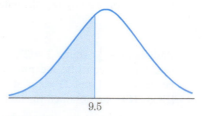

Figure 4

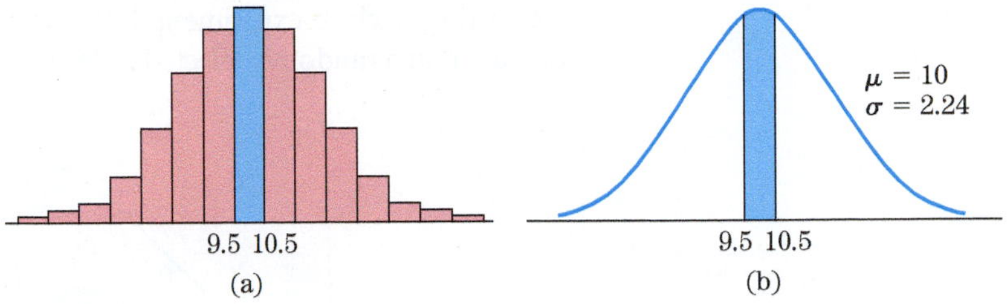

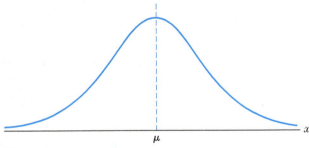

μ = 10
σ = 2.24

9.5 10.5
(a)

9.5 10.5
(b)

Figure 5

obtaining exactly 10 heads in 20 tosses of a coin is the area of the blue rectangle in Fig. 5(a), which is approximated by the area under the normal curve shown in Fig. 5(b).

To be able to use normal curves in our computations, we need to study them more closely. We shall focus on appropriate experiments—namely, experiments with **normally distributed outcomes**. For such experiments, the probabilities of events are computed as areas under normal curves. It is no exaggeration to say that experiments with normally distributed outcomes are among the most significant in probability theory. Such experiments abound in the world around us. Here are a few examples:

1. Choose an individual at random, and observe their IQ.
2. Choose a 1-day-old infant, and observe their weight.
3. Choose an 8-year-old male at random, and observe his height.
4. Choose a leaf at random from a particular tree, and observe its length.
5. A lumber mill is cutting planks that are supposed to be 8 feet long; choose a plank at random, and observe its actual length.

Associated to each of these experiments is the normal curve, as shown in Fig. 6. The curve is symmetric about a vertical line drawn through its highest point. This line of symmetry indicates the mean value of the corresponding experiment. The mean value is denoted, as usual, by the Greek letter μ. For example, if in experiment 4, the average length of the leaves on the tree is 5 inches, then $\mu = 5$ and the corresponding bell-shaped curve is symmetric about the line $x = 5$.

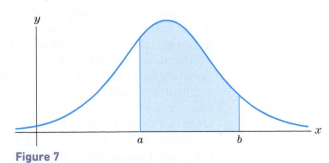

μ

x

Figure 6

The connection between an experiment with normally distributed outcomes and its associated normal curve is as follows: The probability that the experimental outcome is between a and b equals the area under the associated normal curve from $x = a$ to $x = b$. (This is the shaded region in Fig. 7.)

y

a b

x

Figure 7

Referring back to experiment 4, the area of the shaded region in Fig. 8 is equal to the probability that a randomly selected leaf is between 4 and 7 inches long, if the mean is 5 inches.

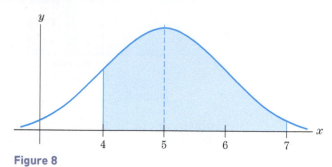

Figure 8

The total area under a normal curve is always 1. This is due to the fact that the probability that the variable X corresponding to the distribution takes on some numerical value on the x-axis is 1.

EXAMPLE 1   **Shading Regions under a Normal Curve**   A certain experiment has normally distributed outcomes with mean $\mu = 1$. Shade the region corresponding to the probabilities of the following outcomes:

(a) The outcome lies between 1 and 3.
(b) The outcome lies between 0 and 2.
(c) The outcome is less than .5.
(d) The outcome is greater than 2.

SOLUTION   The outcomes are plotted along the x-axis. We then shade the appropriate region under the curve.

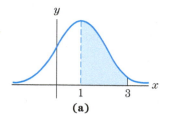 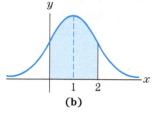 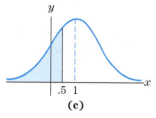 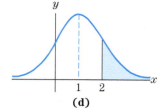

     (a)   **(b)**   **(c)**   **(d)**   **《**

There are many different normal curves with the same mean. For instance, in Fig. 9, we show three normal curves, all with $\mu = 0$. Roughly speaking, the difference between these normal curves is in the width of the center "hump."

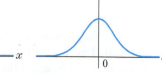

Figure 9

A sharper hump indicates that the outcomes are more likely to be close to the mean. A flatter hump indicates a greater likelihood for the outcomes to be spread out. As we have seen, the spread of the outcomes about the mean is described by the standard deviation, denoted by the Greek letter σ. In the case of a normal curve, the standard deviation has a simple geometric meaning: The normal curve "twists" (or, in calculus terminology, "inflects") at a distance σ on either side of the mean (Fig. 10). More specifically, a normal curve may be thought of as made up of two pieces: a "cap," which looks like an upside-down bowl; and a pair of legs, which curve in the opposite direction from the cap. The places at which the cap and legs are joined are at a distance σ from the mean. Thus, it is clear that the size of σ controls the sharpness of the hump.

Figure 10

A normal curve is completely described by its mean μ and standard deviation σ. In fact, given μ and σ, we may write the equation of the associated normal curve as

$$y = \frac{1}{\sigma\sqrt{2\pi}}\, e^{-\left(\frac{1}{2}\right)\left(\frac{x-\mu}{\sigma}\right)^2},$$

where $\pi \approx 3.1416$ and $e \approx 2.7183$. Fortunately, we will not need this rather complicated formula in what follows. But it is only fair to say that all theoretical work on the normal curve ultimately rests on this equation.

Areas of regions under normal curves can be calculated with technology or with tables. The Incorporating Technology discussion at the end of this section shows how to find areas with graphing calculators, Excel, and Wolfram|Alpha. The remainder of this section explains how to use tables to find areas of regions under normal curves. One might expect that a separate table would be needed for each normal curve, but such is not the case. Only one table is needed: the table corresponding to the **standard normal curve**, which is the one for which $\mu = 0$ and $\sigma = 1$. So let us begin our discussion of areas under normal curves by considering the standard normal curve.

We usually use the letter Z to denote a random variable having the standard normal distribution. Let z be any number, and let $A(z)$ denote the area under the standard normal curve to the left of z (Fig. 11). Table 2 gives $A(z)$ for various values of z, with the values of $A(z)$ rounded to four decimal places. Thus, $A(z) = \Pr(Z \le z)$. We could have said that $A(z) = \Pr(Z < z)$. However, since the region strictly to the left of z and that region with the line segment at z adjoined have the same area, $\Pr(Z < z)$ and $\Pr(Z \le z)$ are the same. We always use the $\le$ symbol. A more extensive table can be found in Appendix A. The efficient use of these tables depends on the following three facts:

1. The standard normal curve is symmetric about the y-axis.
2. The total area under the standard normal curve is 1.
3. The probability that the standard normal variable Z lies to the left of the number z is the area $A(z)$ in Fig. 11.

These facts allow us to use the tables to find the areas of various types of regions.

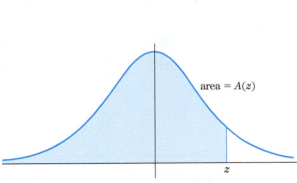

Figure 11

Table 2 Areas under the Normal Curve

z	A(z)	z	A(z)	z	A(z)
−4.00	.0000	−1.25	.1056	1.50	.9332
−3.75	.0001	−1.00	.1587	1.75	.9599
−3.50	.0002	−.75	.2266	2.00	.9772
−3.25	.0006	−.50	.3085	2.25	.9878
−3.00	.0013	−.25	.4013	2.50	.9938
−2.75	.0030	0	.5000	2.75	.9970
−2.50	.0062	.25	.5987	3.00	.9987
−2.25	.0122	.50	.6915	3.25	.9994
−2.00	.0228	.75	.7734	3.50	.9998
−1.75	.0401	1.00	.8413	3.75	.9999
−1.50	.0668	1.25	.8944	4.00	1.0000

EXAMPLE 2 **Determining Areas of Regions under the Standard Normal Curve** Use Table 2 to determine the areas of the regions under the standard normal curve pictured in Fig. 12. Interpret your results.

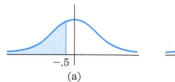
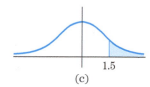

Figure 12

SOLUTION USING TABLES

(a) This region is just the portion of the curve to the left of −.5. So its area is $A(-.5)$. Looking down the middle pair of columns of the table, we find that $A(-.5) = .3085$. This means that

$$\Pr(Z \le -.5) = .3085.$$

In other words, 30.85% of values are less than .5 standard deviations below the mean.

(b) This region results from beginning with the region to the left of 2 and subtracting the region to the left of 1. We obtain an area of

$$A(2) - A(1) = .9772 - .8413 = .1359.$$

Thus,

$$\Pr(1 \le Z \le 2) = .1359.$$

That is, 13.59% of values are between one and two standard deviations above the mean.

(c) This region can be thought of as the entire region under the curve, with the region to the left of 1.5 removed. Therefore, the area is

$$1 - A(1.5) = 1 - .9332 = .0668.$$

So

$$\Pr(Z \ge 1.5) = .0668.$$

That is, 6.68% of values are greater than 1.5 standard deviations above the mean.

SOLUTION

As we will discuss in the Incorporating Technology discussion at the end of this section, the TI-84 Plus command **normalcdf(a, b, μ, σ)** produces the area under a normal curve with mean μ and standard deviation σ from $x = a$ to $x = b$. Since we are working with the standard normal curve, $\mu = 0$ and $\sigma = 1$.

(a) Refer to the TI-84 Plus screens in Figs. 13 and 14. To approximate the area to the left of the upper bound −.5, we choose a large negative number for the lower bound.

(b) Refer to the TI-84 Plus screen in Fig. 15. Choose a lower bound of 1 and an upper bound of 2.

(c) Refer to the TI-84 Plus screen in Fig. 16. To approximate the area to the right of the lower bound 1.5, we choose a large positive number for the upper bound.

Figure 13

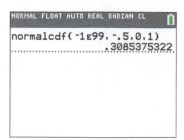

Figure 14

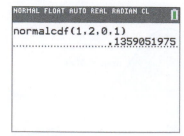

Figure 15

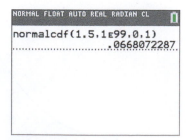

Figure 16

>> *Now Try Exercises 1, 3, and 5*

NOTE The areas in Table 2 have been rounded to four decimal places. Because of this, it is possible for answers to differ slightly depending on whether technology or tables are used. «

EXAMPLE 3 **Finding a Region under the Standard Normal Curve** Find the value of z for which $\Pr(Z \geq z) = .1056$. (See Fig. 17.)

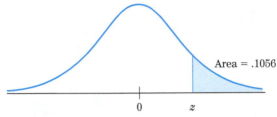

Area = .1056

0   z

Figure 17

SOLUTION Since the area under the standard normal curve is 1 and the curve is symmetric about $z = 0$, the area of the portion to the right of 0 must be .5. We draw a sketch of the standard normal curve, placing z on the axis to the right of 0. (This way, the area to the right of z will be less than .5.) Table 2 gives the values of $A(z)$, which are areas to the left of z. The area to the left of our z is

$$A(z) = 1 - .1056 = .8944.$$

From Table 2, we find that the value of z for which $A(z) = .8944$ is 1.25.

» Now Try Exercise 9

Percentiles

In large-scale testing, scores are frequently reported as percentiles rather than as raw scores. What does it mean to say that a score is "the 90th percentile"? It means that, roughly speaking, the score separates the bottom 90% of the scores from the top 10%.

> **DEFINITION** If a score S is the **pth percentile** of a normal distribution, then $p\%$ of all scores fall below S, and $(100 - p)\%$ of all scores fall above S.

EXAMPLE 4 **Determining a Percentile of the Standard Normal Distribution** What is the 50th percentile of the standard normal distribution?

SOLUTION The standard normal curve is symmetric about $z = 0$, and the total area under the curve is 1. Thus, 50% of the values of the standard normal variable fall below 0, and $(100 - 50)\% = 50\%$ of its values fall above 0. So 0 is the 50th percentile of the standard normal distribution. «

EXAMPLE 5 **Determining a Percentile of the Standard Normal Distribution** What is the 95th percentile of the standard normal distribution? Interpret your result.

SOLUTION USING TABLES We shall call the value that we seek z_{95} to remind us that it is a score and that the probability that an outcome is to the left of it is 95%. (See Fig. 18.)

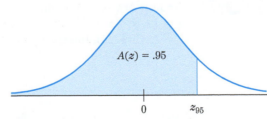

$A(z) = .95$

0   z_{95}

Figure 18

Since Appendix A gives areas to the left of values of z, we should search the column marked $A(z)$ for the area we need—.95. We find that the closest value to .95 is .9505, and $A(1.65) = .9505$. Hence, $z_{95} \approx 1.65$. This means that 95% of the time, the standard normal variable falls below 1.65. Since $\mu = 0$ and $\sigma = 1$, another way of stating the result is that, in the standard normal distribution, 95% of the values are less than 1.65 standard deviations above the mean.

SOLUTION　The TI-84 Plus command **invNorm** (y, μ, σ) produces the value of x for which the area to the left of x under the normal curve with mean μ and standard deviation σ equals y. Refer to the TI-84 Plus screens in Figs. 19 and 20. The desired area is $y = .95$. Since we are working with the standard normal curve, $\mu = 0$ and $\sigma = 1$.

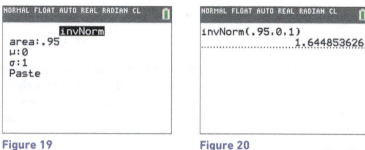

Figure 19　　　　　Figure 20

>> *Now Try Exercise 13*

The problem of finding the area of a region under *any* normal curve can be reduced to finding the area of a region under the standard normal curve. To illustrate the computation procedure, let us consider a numerical example.

EXAMPLE 6　**Finding Areas of Regions under Normal Curves**　Find the area under the normal curve with $\mu = 3, \sigma = 2$ from $x = 1$ to $x = 5$. This represents $\Pr(1 \le X \le 5)$ for a random variable X having the given normal distribution.

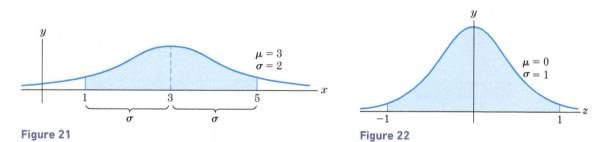

Figure 21　　　　　　　　　　Figure 22

SOLUTION　We have sketched the described region in Fig. 21. It extends from one standard deviation below the mean to one standard deviation above. Draw the corresponding region under the standard normal curve. That is, draw the region from one standard deviation below to one standard deviation above the mean (Fig. 22). It is the case that this new region has the same area as the original one. So the area in Fig. 22 may be computed from Table 2 as $A(1) - A(-1)$. Our desired area is

$$A(1) - A(-1) = .8413 - .1587 = .6826.$$ >> *Now Try Exercise 25*

EXAMPLE 7　**Finding Areas of Regions under a Normal Curve**　Consider the normal curve with $\mu = 12, \sigma = 1.5$. Find the area of the region under the curve between $x = 11.25$ and $x = 15$ (Fig. 23). Interpret your result.

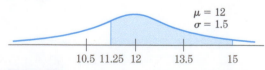

Figure 23

SOLUTION   Expressed as a probability, we want to find $\Pr(11.25 \leq X \leq 15)$ for a random variable X having a normal distribution with $\mu = 12$ and $\sigma = 1.5$. The number 11.25 is .75 below the mean 12. And .75 is $.75/1.5 = .5$ standard deviations. The number 15 is 3 above the mean. And 3 is $3/1.5 = 2$ standard deviations. Therefore, the region has the same area as the region under the standard normal curve from $-.5$ to 2, which is

$$A(2) - A(-.5) = .9772 - .3085 = .6687.$$

In other words, 66.87% of the values are greater than .5 standard deviations below the mean and less than 2 standard deviations above the mean.   **» Now Try Exercise 23**

Suppose that a normal curve has mean μ and standard deviation σ. Then the area under the curve from $x = a$ to $x = b$ is

$$A\left(\frac{b - \mu}{\sigma}\right) - A\left(\frac{a - \mu}{\sigma}\right).$$

The numbers $b - \mu$ and $a - \mu$, respectively, measure the distances of b and a from the mean. The numbers $(b - \mu)/\sigma$ and $(a - \mu)/\sigma$ express these distances as multiples of the standard deviation σ. So the area under the normal curve from $x = a$ to $x = b$ is computed by expressing x in terms of standard deviations from the mean and then treating the curve as if it were the standard normal curve.

We summarize the procedure.

If X is a random variable having a normal distribution with mean μ and standard deviation σ, and Z has the standard normal distribution, then

$$\Pr(a \leq X \leq b) = \Pr\left(\frac{a - \mu}{\sigma} \leq Z \leq \frac{b - \mu}{\sigma}\right) = A\left(\frac{b - \mu}{\sigma}\right) - A\left(\frac{a - \mu}{\sigma}\right)$$

$$\Pr(X \leq x) = \Pr\left(Z \leq \frac{x - \mu}{\sigma}\right) = A\left(\frac{x - \mu}{\sigma}\right)$$

$$\Pr(X \geq x) = \Pr\left(Z \geq \frac{x - \mu}{\sigma}\right) = 1 - A\left(\frac{x - \mu}{\sigma}\right)$$

Graphically, the three equations mean that the shaded areas under the normal curves in Figs. 24(a), (b), and (c) are equal to the shaded areas under the standard normal curves in Figs. 25(a), (b), and (c), respectively.

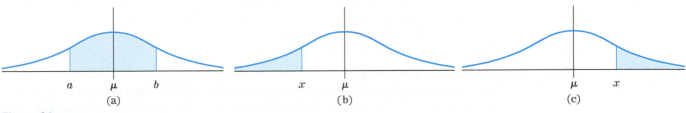

Figure 24

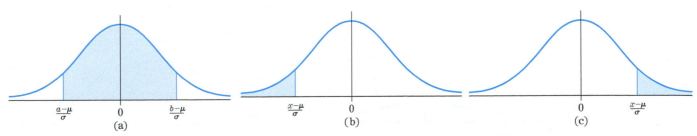

Figure 25

Let us now use our knowledge of areas under normal curves to calculate probabilities arising in some applied problems.

EXAMPLE 8   **Birth Weights of Infants** Suppose that, for a certain population, the birth weights of infants in pounds are normally distributed with $\mu = 7.75$ and $\sigma = 1.25$. Find the probability that an infant's birth weight is more than 9 pounds, 10 ounces. *Note:* 9 pounds, 10 ounces $= 9\frac{5}{8}$ pounds.

SOLUTION   Let $X =$ infant's birth weight. Then X is a random variable having a normal distribution with $\mu = 7.75$ and $\sigma = 1.25$ pounds. $\Pr(X \geq 9\frac{5}{8})$ is given by the area under the appropriate normal curve to the right of $9\frac{5}{8}$—that is, the area shaded in Fig. 26. Since $9\frac{5}{8} = 9.625$, the number $9\frac{5}{8}$ lies $9.625 - 7.75 = 1.875$ units above the mean. In turn, this is $1.875/1.25 = 1.5$ standard deviations. We can find the corresponding z-value in one calculation by finding that

$$z = \frac{x - \mu}{\sigma} = \frac{9.625 - 7.75}{1.25} = 1.5.$$

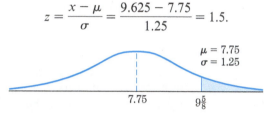

Figure 26

Thus, 9.625 is 1.5 standard deviations above the mean. The area we seek is sketched under the standard normal curve in Fig. 27 and is shown as

$$1 - A(1.5) = 1 - .9332 = .0668.$$

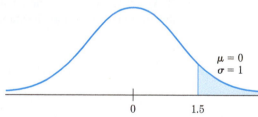

Figure 27

So the probability that an infant weighs more than 9 pounds, 10 ounces is .0668.

>> Now Try Exercise 31

EXAMPLE 9   **Monthly Rents** A property management company finds that the monthly rents (in dollars) for two-bedroom apartments in a town are normally distributed with $\mu = 1200$ and $\sigma = 100$. Find the probability that a randomly selected two-bedroom apartment in this town has a monthly rent less than \$1000.

SOLUTION   Let $X =$ the rent for a two-bedroom apartment. Since rents are normally distributed, the desired probability, $\Pr(X \leq 1000)$, is the area to the left of 1000 in the normal curve drawn in Fig. 28. The number 1000 is 2 standard deviations below the mean; that is, $x = 1000$ corresponds to a z-value of

$$z = \frac{x - \mu}{\sigma} = \frac{1000 - 1200}{100} = -2.$$

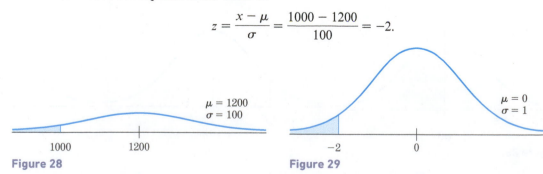

Figure 28   **Figure 29**

Therefore, the area we seek is $A(-2) = .0228$, shown in Fig. 29. The probability that the monthly rent is less than \$1000 is .0228.

>> Now Try Exercise 36(a)

EXAMPLE 10   **Apartment Hunting** A prospective renter wants to be 99% sure that they can afford a two-bedroom apartment in the town from Example 9. How much should they budget for monthly rent?

SOLUTION   Let x be the amount that the tenant should budget for rent. Since they want to be 99% sure that the monthly rent does not exceed x, we must find the 99th percentile of a normal distribution with $\mu = 1200$ and $\sigma = 100$. To help us remember what x really is, we will rename it x_{99}. The corresponding value for the standard normal random variable is z_{99}. Figures 30 and 31 show the appropriate areas—first under the given normal curve and then under the standard normal curve.

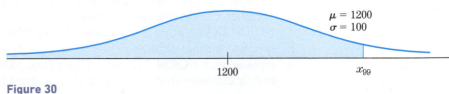

Figure 30

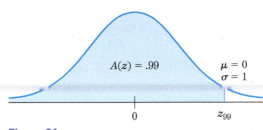

Figure 31

The area, $A(z)$, that we seek under the standard normal curve is .99. Referring to Appendix A, we get closest with $A(z) = .9906$, corresponding to $z_{99} = 2.35$. The value z_{99} is 2.35 standard deviations above the mean of its distribution. We conclude that x_{99} is also 2.35 standard deviations above its mean. Hence,

$$x_{99} = 1200 + (2.35)(100) = \$1435$$

Therefore, we expect that 99% of two-bedroom apartments will have monthly rents no greater than $1435.   **» Now Try Exercise 36(b)**

We summarize the technique for finding percentiles of normal distributions.

> If x_p is the pth percentile of a normal distribution with mean μ and standard deviation σ, then
>
> $$x_p = \mu + z_p \cdot \sigma,$$
>
> where z_p is the pth percentile of the standard normal distribution.

INCORPORATING

TECHNOLOGY

With a TI-84 Plus, the command **normalcdf(a, b, μ, σ)** produces the area under the normal curve with mean μ and standard deviation σ from $x = a$ to $x = b$. Press 2nd DISTR 2 to enter the **normalcdf** wizard (see Fig. 32 on the next page). To find the area of an infinite region, give a large negative value for the lower bound or a large positive value for the upper bound. If y is a number between 0 and 1, then **invNorm(y, μ, σ)** gives the value of x for which the area to the left of x under the specified normal curve is y. *Note:* **inv** is short for "inverse." Press 2nd DISTR 3 to enter the **invNorm** wizard.

Figure 33 shows how the **normalcdf** and **invNorm** commands can be used to solve Examples 2(b), 9, and 10.

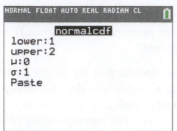

Figure 32

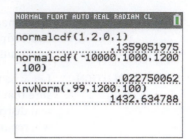

Figure 33

If x is any number, then NORM.DIST(x, μ, σ, TRUE) gives the area to the left of x under the specified normal curve. If y is a number between 0 and 1, then NORM.INV(y, μ, σ) is the value of x for which the area to the left of x under the specified normal curve is y. For the standard normal curve, these commands can be replaced with NORM.S.DIST(x, TRUE) and NORM.S.INV(y). In Fig. 34, the answers to four examples from this section are calculated in an Excel spreadsheet.

	A	B	C	D
1	Example	Answer		Formula Used
2	2(b)	0.135905122		NORM.S.DIST(2,TRUE)-NORM.S.DIST(1,TRUE)
3	5	1.644853627		NORM.S.INV(0.95)
4	7	0.668712329		NORM.DIST(15,12,1.5,TRUE)-NORMDIST(11.25,12,1.5,TRUE)
5	10	1432.634787		NORM.INV(.99,1200,100)

Figure 34

❋WolframAlpha The area under the normal curve with mean μ and standard deviation σ from $x = a$ to $x = b$ is given by

$$P(a < x < b) \text{ for x normal}(\mu, \sigma).$$

The area to the left and the area to the right of $x = a$ under the normal curve with mean μ and standard deviation σ are respectively given by

$$P(x < a) \text{ for x normal}(\mu, \sigma) \quad \text{and} \quad P(x > a) \text{ for x normal}(\mu, \sigma).$$

For instance, the three areas in Example 2 can be computed with the instructions

$$P(x < -.5) \text{ for x normal}(0,1)$$

$$P(1 < x < 2) \text{ for x normal}(0,1)$$

$$P(x > 1.5) \text{ for x normal}(0,1).$$

The area in Example 7 can be calculated with the instruction

$$P(11.25 < x < 15) \text{ for x normal}(12,1.5).$$

If y is a number between 0 and 1, then the value of x for which the area to the left of x under the specified normal curve is y is given by

$$\text{inverse normal probability y, mean} = \mu, \text{std dev} = \sigma.$$

Check Your Understanding 7.6

Solutions can be found following the section exercises.

1. Refer to Fig. 35(a). Find the value of z for which the area of the shaded region is .0802.

2. Refer to the normal curve in Fig. 35(b). Express the following numbers in terms of standard deviations from the mean:

    (a) 90   (b) 82   (c) 94   (d) 104

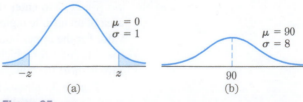

Figure 35

EXERCISES 7.6

In Exercises 1-8, use the table for $A(z)$ (Table 2) to find the areas of the shaded regions under the standard normal curve.

1.

1.25

2.

−.75   1

3.

.25

4.

−1   1

5.

.5 1.5

6.

−1

7.
−.5 .5

8.
−1.25

In Exercises 9-12, find the value of z for which the area of the shaded region under the standard normal curve is as specified.

9. Area is .0401.

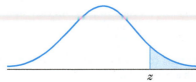
z

10. Area is .0456.

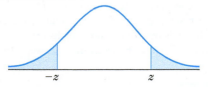
$-z$ z

11. Area is .5468.

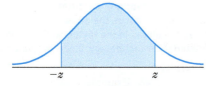
$-z$ z

12. Area is .6915.

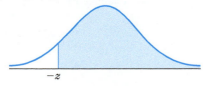
$-z$

13. What is the 80th percentile of the standard normal distribution?

14. What is the 40th percentile of the standard normal distribution?

In Exercises 15-18, determine μ and σ by inspection.

15.
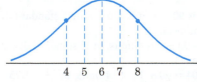
4 5 6 7 8

16.
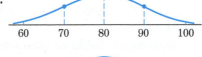
60   70   80   90   100

17.
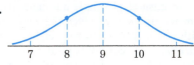
7   8   9   10   11

18.
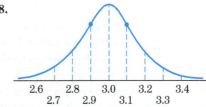
2.6   2.8   3.0   3.2   3.4
  2.7   2.9   3.1   3.3

Exercises 19–22 refer to the normal curve with $\mu = 8, \sigma = \frac{3}{4}$.

19. Convert 4 into standard deviations from the mean.

20. Convert $9\frac{1}{4}$ into standard deviations from the mean.

21. What value is exactly 10 standard deviations above the mean?

22. What value is exactly 2 standard deviations below the mean?

In Exercises 23-26, find the areas of the shaded regions under the given normal curves.

23.
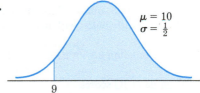
$\mu = 10$
$\sigma = \frac{1}{2}$
9

24.
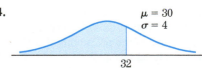
$\mu = 30$
$\sigma = 4$
32

25.
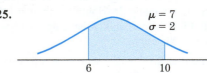
$\mu = 7$
$\sigma = 2$
6   10

26.
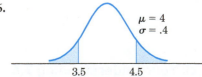
$\mu = 4$
$\sigma = .4$
3.5   4.5

27. What is the probability that an outcome of a normal random variable is within two standard deviations of the mean?

28. What is the probability that an outcome of a normal random variable is within 2.5 standard deviations of the mean?

29. Find the value of σ for a normal random variable X having $\mu = 5$, if $\Pr(X \le 6) = .9772$

30. Find the value of σ for a normal random variable X having $\mu = 10$, if $\Pr(14.5 \le X) = .0013$.

31. Elephant Heights Suppose that the height (at the shoulder) of adult bull African bush elephants is normally distributed with $\mu = 3.3$ meters and $\sigma = .2$ meter. The elephant on display at the Smithsonian Institution has height 4 meters and is the largest elephant on record. What is the probability that an adult bull African bush elephant has height 4 meters or more?

32. Bottling Reliability At a soft-drink bottling plant, the amount of cola put into the bottles is normally distributed with $\mu = 16\frac{3}{4}$ ounces and $\sigma = \frac{1}{4}$. What is the probability that a bottle will contain fewer than 16 ounces?

33. Manufacturing Reliability Bolts produced by a machine are acceptable provided that their length is within the range from 5.95 to 6.05 centimeters. Suppose that the lengths of the bolts produced are normally distributed with $\mu = 6$ centimeters and $\sigma = .04$. What is the probability that a bolt will be of an acceptable length?

34. Heights In a certain population, heights (in inches) are normally distributed with $\mu = 67$ and $\sigma = 3$. Find the probability that a person selected at random has a height between 63 and 71 inches.

35. Manufacturing Reliability In a certain manufacturing process, lengths (in cm) of bolts are normally distributed with $\mu = 5.4$ and $\sigma = .6$. Find the probability that a bolt selected at random has a length greater than 5.832 cm.

36. IQ Scores As measured with the Stanford-Binet Intelligence Scale, IQ scores are normally distributed with mean 100 and standard deviation 16.
  (a) What percent of the population has an IQ score of 140 or more?
  (b) Find the 90th percentile of IQ scores.

37. Production Quotas The number of barrels of oil produced yearly by a specific oil well is normally distributed with $\mu = 7500$ barrels and $\sigma = 1000$. If the well owner has a yearly quota of 9750 barrels, what is the probability that the well's production will meet or exceed this quota?

38. Lightbulb Lifetimes Suppose that the lifetimes of a certain type of light bulb are normally distributed with $\mu = 1200$ hours and $\sigma = 160$. Find the probability that a light bulb will burn out in less than 1000 hours.

39. SAT Scores Assume that SAT verbal scores for a first-year class at a university are normally distributed with mean 520 and standard deviation 75.

  (a) The top 10% of the students are placed into the honors program for English. What is the lowest score for admittance into the honors program?
  (b) What is the range of the middle 90% of the SAT verbal scores at this university?
  (c) Find the 98th percentile of the SAT verbal scores.

40. Mailing Bags A mail-order house uses an average of 300 mailing bags per day. The number of bags needed each day is approximately normally distributed with $\sigma = 50$. How many bags must the company have on hand at the beginning of a day to be 95% certain that all orders can be filled?

41. Tire Lifetimes The lifetime of a certain brand of tires is normally distributed with mean $\mu = 30,000$ miles and standard deviation $\sigma = 5000$ miles. The company has decided to issue a warranty for the tires but does not want to replace more than 2% of the tires that it sells. At what mileage should the warranty expire?

42. Soft-Drink Dispenser Let X be the amount of soda released by a soft-drink dispensing machine into a 6-ounce cup. Assume that X is normally distributed with $\sigma = .25$ ounces and that the average "fill" can be set by the vendor.
  (a) At what quantity should the average "fill" be set so that no more than .5% of the releases overflow the cup?
  (b) Using the average "fill" found in part (a), determine the minimal amount that will be dispensed in 99% of the cases.

43. (True or False) A normal curve with a large value of σ will be flatter than one with a small value of σ.

44. Let X be a random variable with $\mu = 4$ and $\sigma = .5$.
  (a) Use the Chebychev inequality to estimate $\Pr(3 \le X \le 5)$.
  (b) If X were normally distributed, what would be the exact probability that X is between 3 and 5, inclusive?
  (c) Reconcile the difference between the answers to (a) and (b).

TECHNOLOGY EXERCISES

In Exercises 45-48, give a graphing calculator, Excel, or Wolfram|Alpha statement that produces the answer to the specified example. *Note:* Values calculated in the textbook differ slightly from the values calculated with technology since the values in Table A are limited.

45. Example 6   **46.** Example 8

47. Example 9   **48.** Example 10

Solutions to Check Your Understanding 7.6

1. 1.75. Due to the symmetry of normal curves, each piece of the shaded region has area $\frac{1}{2}(.0802) = .0401$. Therefore, $A(-z) = .0401$ and so by Table 2, $-z = -1.75$. Thus $z = 1.75$.

2. (a) 0. Since 90 *is* the mean, it is 0 standard deviations from the mean.

  (b) -1. Since 82 is $90 - 8$, it is 8 units, or 1 standard deviation, below the mean.
  (c) .5. Here, $94 = 90 + 4$ is 4 units, or .5 standard deviation, above the mean.
  (d) 1.75. Here, $104 = 90 + 14$ is 14 units, or $\frac{14}{8} = 1.75$ standard deviations, above the mean.

Normal Approximation to the Binomial Distribution

In the binomial trials section, we saw that complicated and tedious calculations can arise from the binomial probability distribution. For instance, determining the probability of getting at least 20 threes in 100 rolls of a die requires the computation

$$\binom{100}{20}\left(\frac{1}{6}\right)^{20}\left(\frac{5}{6}\right)^{80} + \binom{100}{21}\left(\frac{1}{6}\right)^{21}\left(\frac{5}{6}\right)^{79} + \cdots + \binom{100}{100}\left(\frac{1}{6}\right)^{100}\left(\frac{5}{6}\right)^{0}.$$

The Incorporating Technology discussion at the end of that section showed how to calculate such a sum with a graphing calculator, Excel, and Wolfram|Alpha. In this section, we use normal curves to approximate sums arising from binomial trials.

Consider the histograms of the binomial distributions in Figs. 1 and 2. The number of trials, n, increases from 5 to 40, with p fixed at .3. As n increases, the shape of the histogram more closely conforms to the shape of the region under a normal curve.

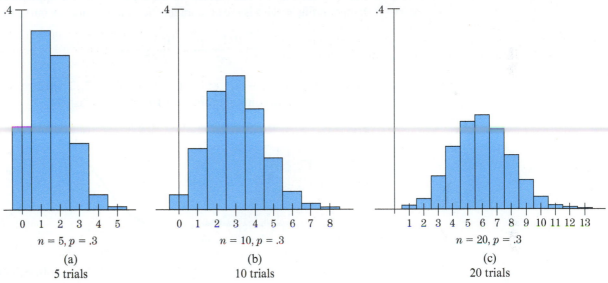

(a)	(b)	(c)
5 trials	10 trials	20 trials

$n = 5, p = .3$ $n = 10, p = .3$ $n = 20, p = .3$

Figure 1 Binomial distribution

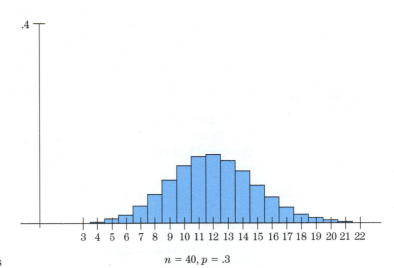

$n = 40, p = .3$

Figure 2 Binomial distribution, 40 trials

We have the following result:

> Suppose that we perform a sequence of n binomial trials with probability of success p and probability of failure q and observe the number of successes. Then the histogram for the resulting probability distribution may be approximated by the normal curve with $\mu = np$ and $\sigma = \sqrt{npq}$.

> **NOTE** ▶   This approximation is very accurate when both $np > 5$ and $nq > 5$. ◀◀

EXAMPLE 1   **Quality Control** Refer to Example 1 of Section 7.3. A plumbing-supplies manufacturer produces faucet washers that are packaged in boxes of 300. Quality-control studies have shown that 2% of the washers are defective. What is the probability that more than 10 of the washers in a single box are defective?

SOLUTION   Let $X =$ the number of defective washers in a box. Then X is a binomial random variable with $n = 300$ and $p = .02$. We will use the approximating normal curve with

$$\mu = np = 300(.02) = 6$$
$$\sigma = \sqrt{npq} = \sqrt{300(.02)(.98)} \approx 2.425.$$

The probability that more than 10 of the washers in a single box are defective is the sum of the areas of the blue rectangles centered at 11, 12, . . . , 300 in the histogram for the random variable X. (See Fig. 3.) The corresponding region under the approximating normal curve is shaded in Fig. 4. This is the area under the standard normal curve to the right of

$$z = \frac{10.5 - \mu}{\sigma} = \frac{10.5 - 6}{2.425} \approx 1.85.$$

The area of the region is $1 - A(1.85) = 1 - .9678 = .0322$. Therefore, approximately 3.22% of the boxes should contain more than 10 defective washers.

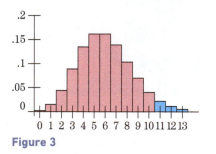

Figure 3

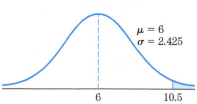

Figure 4

>> *Now Try Exercise 1(c)*

> **NOTE** ▶   In Fig. 4, we shaded the region to the right of 10.5 rather than to the right of 10. This gives a better approximation to the corresponding area under the histogram, since the rectangle corresponding to 11 "successes" has its left endpoint at 10.5. ◀◀

Let us consider an application to medical research.

EXAMPLE 2   **Veterinary Medicine** Consider the cattle disease of Example 2 of Section 7.3, from which 25% of the cattle recover. A veterinarian discovers a serum to combat the disease. In a test of the serum, they observe that 16 of a herd of 40 recover. Suppose that the serum had not been used. What is the likelihood that at least 16 cattle would have recovered?

SOLUTION   Let X be the number of cattle that recover. Then X is a binomial random variable with $n = 40$ independent trials. If the serum is not used, $p = .25$. The approximating normal curve has

$$\mu = np = 40(.25) = 10, \qquad \sigma = \sqrt{npq} = \sqrt{40(.25)(.75)} \approx 2.74.$$

The likelihood that at least 16 cattle would have recovered is $\Pr(X \geq 16)$. This corresponds to the area under the normal curve to the right of 15.5 (Fig. 5). The area to the right of 15.5 under a normal curve with $\mu = 10$ and $\sigma = 2.74$ is the same as the area under the standard normal curve to the right of

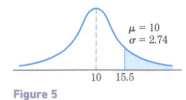

Figure 5

$$z = \frac{15.5 - \mu}{\sigma} = \frac{15.5 - 10}{2.74} \approx 2.01.$$

We find that $1 - A(2.01) \approx 1 - A(2.00) = 1 - .9772 = .0228$. Thus, if the serum were not used, the veterinarian would expect about a 2% chance that 16 or more cattle would recover. Thus, the 16 observed recoveries probably did not occur by chance. The veterinarian can reasonably conclude that the serum is effective against the disease.

>> **Now Try Exercise 3**

EXAMPLE 3   **Heads and Tails** Assume that a fair coin is tossed 100 times. Find the probability of observing exactly 50 heads.

SOLUTION   Let X be the number of heads on $n = 100$ binomial trials with $p = .5$ probability of "success" on each trial. We are to find the area of the rectangle extending from 49.5 to 50.5 on the x-axis in the histogram for X. The actual probability is

$$\binom{100}{50}(.5)^{50}(.5)^{50} = \binom{100}{50}(.5)^{100}.$$

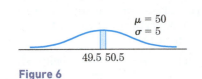

Figure 6

We approximate the probability with an area under the normal curve with $\mu = np = 100(.5) = 50$ and $\sigma = \sqrt{npq} = \sqrt{100(.5)(.5)} = 5$. The area we seek is sketched in Fig. 6. It is the area under the standard normal curve from

$$z = \frac{49.5 - 50}{5} = -.10 \quad \text{to} \quad z = \frac{50.5 - 50}{5} = .10.$$

Using Appendix A, we find that

$$A(.10) - A(-.10) = .5398 - .4602 = .0796.$$

So the likelihood of getting exactly 50 heads in 100 tosses of a fair coin is quite small—about 8%.

>> **Now Try Exercise 1(a)**

EXAMPLE 4   **Rolling a Die** Find the probability that in 100 rolls of a fair die, we observe at least 20 threes.

SOLUTION   Let X be the number of threes observed in $n = 100$ trials, with $p = \frac{1}{6}$ the probability of a three on each trial. Then we need to find the area to the right of 19.5 under the normal curve with

$$\mu = np = 100\left(\tfrac{1}{6}\right) \approx 16.7 \quad \text{and} \quad \sigma = \sqrt{npq} = \sqrt{100\left(\tfrac{1}{6}\right)\left(\tfrac{5}{6}\right)} \approx 3.73.$$

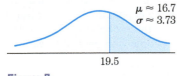

Figure 7

The area we seek is sketched in Fig. 7. This area is

$$1 - A\left(\frac{19.5 - 16.7}{3.73}\right) \approx 1 - A(.75) = 1 - .7734 = .2266.$$

Therefore, the probability of observing at least 20 threes is about .2266.

>> **Now Try Exercise 7**

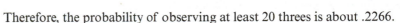

Check Your Understanding 7.7
Solutions can be found following the section exercises.

1. **Drug Testing** A new drug is being tested on laboratory mice. The mice have been given a disease for which the recovery rate is 1/2.

   (a) In the first experiment, the drug is given to five of the mice and all five recover. Find the probability that the success of this experiment was due to luck. That is, find the probability that five out of five mice recover in the event that the drug has no effect on the illness.

   (b) In a second experiment, the drug is given to 25 mice and 18 recover. Find the probability that 18 or more recover in the event that the drug has no effect on the illness.

2. **Drug Testing** What conclusions can be drawn from the results in Problem 1?

EXERCISES 7.7

In Exercises 1–22, use the normal curve to approximate the probability.

1. An experiment consists of 25 binomial trials, each having probability 1/5 of success. Use an approximating normal curve to estimate the probability of
   (a) exactly 5 successes.
   (b) between 3 and 7 successes, inclusive.
   (c) less than 10 successes.

2. An experiment consists of 18 binomial trials, each having probability 2/3 of success. Use an approximating normal curve to estimate the probability of
   (a) exactly 10 successes.
   (b) between 8 and 16 successes, inclusive.
   (c) more than 12 successes.

3. **Drug Testing** Laboratory mice are given an illness for which the usual recovery rate is 1/6 . A new drug is tested on 20 of the mice, and 8 of them recover. What is the probability that 8 or more would have recovered if the 20 mice had not been given the drug?

4. **ESP** A person claims to have ESP (extrasensory perception). A coin is tossed 16 times, and each time, the person is asked to predict in advance whether the coin will land heads or tails. The person predicts correctly 75% of the time (i.e., on 12 tosses). What is the probability of being correct 12 or more times by pure guessing?

5. **Wine Tasting** A wine-taster claims that she can usually distinguish between domestic and imported wines. As a test, she is given 100 wines to test and correctly identifies 63 of them. What is the probability that she accomplished that good of a record by pure guessing? That is, what is the probability of being correct 63 or more times out of 100 by pure guessing?

6. **Roulette** In American roulette, the probability of winning when betting "red" is 9/19. What is the probability of being ahead after betting the same amount 90 times?

7. **Basketball** A basketball player makes each free throw with probability 3/4. What is the probability of making 68 or more shots out of 75 trials?

8. **Bookstore Customers** A bookstore determines that 2/5 of the people who come into the store make a purchase. What is the probability that of the 54 people who come into the store during a certain hour, less than 14 make a purchase?

9. **Baseball** A baseball player gets a hit with probability .310. Find the probability that they get at least 6 hits in 20 times at bat.

10. **Advertising Campaign** An advertising agency, which reached 25% of its target audience with its old campaign, has devised a new advertising campaign. In a sample of 1000 people, it finds that 290 people have been reached by the new advertising campaign. What is the probability that at least 290 people would have been reached by the old campaign? Does the new campaign seem to be more effective?

11. **Equipment Reliability** A washing machine manufacturer finds that 2% of its washing machines break down within the first year. Find the probability that less than 15 out of a lot of 1000 washers break down within 1 year.

12. **Color Blindness** The incidence of color blindness among the men in a certain country is 20%. Find the expected number of color-blind men in a random sample of 70 men. What is the probability of finding exactly that number of color-blind men in a sample of size 70?

13. **Product Reliability** The probabilities of failure for each of three independent components in a device are .01, .02, and .01, respectively. The device fails only if all three components fail. Out of a lot of 1 million devices, how many would be expected to fail? Find the probability that more than three devices in the lot fail.

14. **Smartphones** In a random sample of 250 college students, 175 of them own a smartphone. Estimate the probability that a college student chosen at random owns a smartphone. If actually 75% of all college students own a smartphone, what is the probability that, in a random survey of 250 students, at most 175 of them own an a smartphone?

15. **Marksman** A marksman hits a target with probability .35. Estimate the probability of hitting the target from 30 to 40 times in 100 attempts.

16. **Airline Reservations** An airline accepts 150 reservations for a flight on an airplane that holds 140 passengers. If the probability of a passenger for this flight cancelling is .14, estimate the probability that some passengers will have to be bumped.

17. **Ski Tour** A travel agent is arranging a tour for the local 1000-member ski club. They need a minimum of 29 people to register and think that the probability of a member registering for the tour is .03. Estimate the probability that enough members will register.

18. **Coin Tosses** A fair coin is tossed 100 times. Estimate the probability that more than 65 heads or more than 65 tails appear.

19. **Coin Tosses** In 100 tosses of a fair coin, let X be the number of heads. Estimate $\Pr(49 \le X \le 51)$.

20. **Rolling a Die** Let X be the number of 4s in 120 rolls of a fair die. Estimate $\Pr(17 \le X \le 21)$.

21. **Name Recognition** Say that there is a 20% chance that a person chosen at random from the population has never heard of John Steinbeck. Estimate the probability that, in 150 people, we find exactly 30 people who have not heard of Steinbeck.

22. **Male Heights** About 5% of American males are 6 feet 2 inches tall or taller. Estimate the probability that, of 150 men at a business meeting, no more than five are 6 feet 2 inches tall or taller.

TECHNOLOGY EXERCISES

In Exercises 23–26, find the more precise answer to the specified example that is produced with technology, and give a graphing calculator, Excel, or Wolfram|Alpha statement that produces the more precise answer.

23. Example 1
24. Example 2
25. Example 3
26. Example 4

Solutions to Check Your Understanding 7.7

1. (a) Giving the drug to a single mouse is a binomial trial with "recovery" as "success" and "death" as "failure." If the drug has no effect, then the probability of success is $\frac{1}{2}$. The probability of five successes in five trials is given by formula (1) of Section 7.3, with $n = 5$, $p = \frac{1}{2}$, $q = \frac{1}{2}$, $k = 5$.

$$\Pr(X = 5) = \binom{5}{5}\left(\frac{1}{2}\right)^5\left(\frac{1}{2}\right)^0$$

$$= \left(\frac{1}{2}\right)^5$$

$$= \frac{1}{32} = .03125$$

(b) As in part (a), this experiment is a binomial experiment with $p = \frac{1}{2}$. However, now $n = 25$. The probability that 18 or more mice recover is

$$\Pr(X = 18) + \Pr(X = 19) + \cdots + \Pr(X = 25).$$

This probability is the area of the blue portion of the histogram in Fig. 7(a). The histogram can be approximated by the normal curve with

$$\mu = 25 \cdot \frac{1}{2} = 12.5$$

and

$$\sigma = \sqrt{25 \cdot \frac{1}{2} \cdot \frac{1}{2}} = \sqrt{\frac{25}{4}} = \frac{5}{2} = 2.5$$

[Fig. 7(b)]. Since the blue portion of the histogram begins at the point 17.5, the desired probability is approximately the area of the shaded region under the normal curve. The number 17.5 is

$$\frac{17.5 - 12.5}{2.5} = \frac{5}{2.5} = 2$$

standard deviations to the right of the mean. Hence, the area under the curve is

$$1 - A(2) = 1 - .9772 = .0228.$$

Therefore, the probability that 18 or more mice recover is approximately .0228.

2. Both experiments offer convincing evidence that the drug is helpful in treating the illness. The likelihood of obtaining the results by pure chance is slim. The second experiment might be considered more conclusive than the first, since the result, if due to chance, has a lower probability.

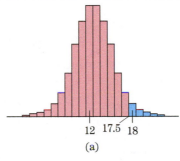

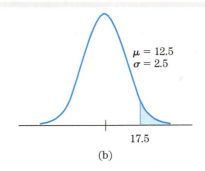

12 17.5 18

(a)

17.5

(b)

$\mu = 12.5$
$\sigma = 2.5$

Figure 7

CHAPTER 7   Summary

KEY TERMS AND CONCEPTS	EXAMPLES
7.1 Visual Representation of Data **Bar charts, pie charts, histograms,** and **box plots** turn raw data into visual forms that often allow us to see patterns in data quickly.	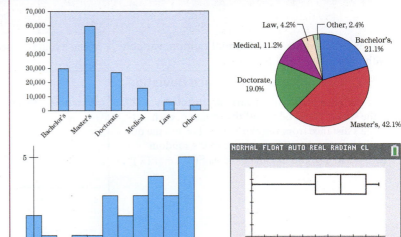

KEY TERMS AND CONCEPTS	EXAMPLES
The **median** is a number separating the higher half of a set of ordered data from the lower half.	The median of {1, 4, 6, 9, 10} is 6. The median of {1, 4, 6, 9} is 5.

The **first quartile**, $Q1$, for an ordered list of data is the median of the list of data items below the median, and the **third quartile**, $Q3$, is the median of the list of data items above the median. The difference of the third and first quartiles is called the **interquartile range**. The sequence of numbers consisting of the lowest number, $Q1$, the median, $Q3$, and the highest number is called the **five-number summary**.

The set {1, 3, 6, 10, 15, 21, 28, 36, 45, 55} has the following five-number summary:

1	6	18	36	55
↑	↑	↑	↑	↑
lowest	Q_1	median	Q_3	highest

7.2 Frequency and Probability Distributions

The **probability distribution** for a random variable can be displayed in a histogram or a table. With a histogram, the probability of an event is the sum of the areas of the rectangles corresponding to the outcomes in the event.

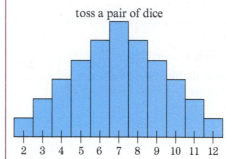

toss a pair of dice

k	$\Pr(X = k)$	k	$\Pr(X = k)$
2	$\frac{1}{36}$	8	$\frac{5}{36}$
3	$\frac{1}{18}$	9	$\frac{1}{9}$
4	$\frac{1}{12}$	10	$\frac{1}{12}$
5	$\frac{1}{9}$	11	$\frac{1}{18}$
6	$\frac{5}{36}$	12	$\frac{1}{36}$
7	$\frac{1}{6}$		

7.3 Binomial Trials

A **binomial trial** is an experiment with two outcomes—usually called *success* and *failure*.

A **binomial experiment** is a sequence of a specified number of independent binomial trials. If p is the probability of success on each trial, then

$\Pr(k \text{ successes in } n \text{ trials}) = \binom{n}{k} p^k (1 - p)^{n-k}.$

A single die is rolled 10 times. Find the probability that 3 twos appear.

Answer: Let *success* be the occurrence of a two when the die is rolled. So $n = 10, k = 3, p = \frac{1}{6}, (1 - p) = \frac{5}{6}.$

$$\Pr(3 \text{ twos in 10 rolls}) = \binom{10}{2}\left(\frac{1}{6}\right)^3\left(\frac{5}{6}\right)^7 \approx 0.58$$

7.4 The Mean

The **mean** (or **average**) of a set of n numbers is the sum of the numbers divided by n.

The **expected value** of a random variable, denoted $E(X)$, is the sum of the products of each outcome with its probability.

The mean of {2, 5, 8, 11} is $\frac{26}{4} = 6.5.$

American Roulette

Earnings	Probability
1	$\frac{18}{38}$
−1	$\frac{20}{38}$

Let X be the "red bet" earnings.

$$E(X) = 1 \cdot \frac{18}{38} + (-1) \cdot \frac{20}{38}$$
$$\approx -.0526$$

The **expected value of a binomial random variable** with parameters n and p is np.

When a coin is tossed 11 times, the expected number of heads is $11 \cdot \frac{1}{2} = 5.5.$

7.5 The Variance and Standard Deviation

The **variance** of a random variable, denoted $\text{Var}(X)$, is the sum of the products of the square of each outcome's distance from the expected value and the outcome's probability. The variance of the random variable X also can be calculated as $E(X^2) - [E(X)]^2$.

Outcome	Probability
0	.1
1	.3
2	.5
3	.1

Let X be the outcome.

$\mu = E(X) = 0 \cdot .1 + 1 \cdot .3 + 2 \cdot .5 + 3 \cdot .1 = 1.6$

$\text{Var}(X) = (0 - 1.6)^2 \cdot .1 + (1 - 1.6)^2 \cdot .3 +$
$\qquad (2 - 1.6)^2 \cdot .5 + (3 - 1.6)^2 \cdot .1 = .64$

$E(X^2) = 0 \cdot .1 + 1^2 \cdot .3 + 2^2 \cdot .5 + 3^2 \cdot .1 = 3.2$

Alternatively,

$$\text{Var}(X) = 3.2 - 1.6^2 = .64$$

KEY TERMS AND CONCEPTS	EXAMPLES
The **variance of a binomial random variable** with parameters n and p is $np(1-p)$.	When a coin is tossed 11 times, the variance for the number of heads is $11 \cdot \frac{1}{2} \cdot \frac{1}{2} = 2.75$.
The **standard deviation** of a random variable is the square root of the variance.	The standard deviation of the random variable X on the previous page is $\sqrt{.64} = .8$.
The **variance of a population** of N numbers, denoted σ^2, is $\frac{1}{N}$ times the sum of the products of the square of each number's distance from the mean. The standard deviation is σ.	
The **variance of a sample** of N numbers, denoted s^2, is $\frac{1}{N-1}$ times the sum of the products of the square of each number's distance from the mean. The standard deviation is s.	
Chebychev's inequality states that the probability that an outcome of an experiment is within c units of the mean is at least $1 - \frac{\sigma^2}{c^2}$, where σ is the standard deviation.	Consider the previous random variable X, and let $c = 1$. Then $\Pr(.6 \le X \le 2.6) \ge 1 - \frac{.8^2}{1^2} = .36$.

7.6 The Normal Distribution

A **normal curve** is identified by its mean (μ) and its standard deviation (σ). Areas of regions under a normal curve can be obtained with a table, a graphing calculator, Excel, or Wolfram|Alpha.	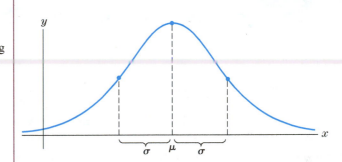
The **standard normal curve** has $\mu = 0$ and $\sigma = 1$. Areas of regions under the standard normal curve can be obtained with technology or with the table in Appendix A.	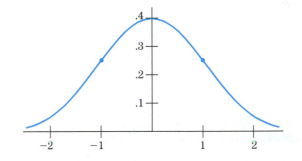
A random variable is said to be **normally distributed** if the probability that an outcome lies between a and b is the area of the region under a normal curve from $x = a$ to $x = b$. A sought-after probability can be obtained directly with technology. Alternatively, it can be calculated with the table in Appendix A after the numbers a and b are converted to standard deviations from the mean.	Suppose that the lifetimes of light bulbs manufactured by the Royal Corporation are normally distributed with an average lifetime of 1300 hours and a standard deviation of 75 hours. Find the probability that a lightbulb will last more than 1400 hours. *Answer:* Let X = lifetime of a lightbulb. 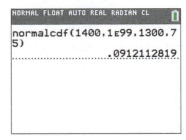 $\Pr(X \ge 1400) \approx .0912$

KEY TERMS AND CONCEPTS	EXAMPLES

7.7 Normal Approximation to the Binomial Distribution

Probabilities associated with a **binomial random variable** having parameters n and p can be approximated with the use of a normal curve having $\mu = np$ and $\sigma = \sqrt{np(1-p)}$. $\Pr(a \le X \le b)$ is approximately the area under the normal curve from $x = a - .5$ to $x = b + .5$.

Estimate the probability of obtaining between 45 and 55 heads when tossing a coin 100 times.

Answer: $\mu = 100 \cdot \frac{1}{2} = 50$, $\sigma = \sqrt{100 \cdot \frac{1}{2} \cdot \frac{1}{2}} = 5$

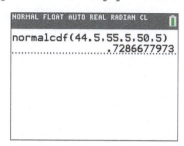

$\Pr(45 \le X \le 55) \approx .7287$

CHAPTER 7 Fundamental Concept Check Exercises

1. What is a bar chart? A pie chart? A histogram? A box plot?

2. What is the median of a list of numbers? The first quartile? The third quartile? The interquartile range? The five-number summary?

3. What is a frequency distribution? A relative frequency distribution? A probability distribution?

4. How is a histogram constructed from a distribution?

5. What is a random variable?

6. What is meant by the probability distribution of a random variable?

7. What are the identifying features of a binomial random variable?

8. What is the formula for the probability of k successes in n independent binomial trials?

9. What is intuitively meant by the expectation (or expected value) of a random variable? Variance? Standard deviation?

10. What does the Chebychev inequality do?

11. What is meant by a normal random variable?

12. What is meant by the pth percentile of a normal random variable?

13. How are binomial probabilities approximated with the normal distribution?

CHAPTER 7 Review Exercises

1. **U.S. Population** Display the data from Table 1 in a bar chart that shows frequencies on the y-axis. Then display the data in a pie chart.

Table 1 U.S. Population by Region (in millions)

Region	Population
Northeast	56.2
Midwest	67.9
South	121.2
West	76.0

Source: U.S. Census Bureau, 2015.

2. Find the five-number summary and the interquartile range for the following set of numbers, and then draw the box plot:

1, 2, 3, 4, 5, 9, 14, 23

3. **Supermarket Queue** The manager of a supermarket counts the number of customers waiting in the express checkout line at random times throughout the week. Her observations are found in the following frequency table:

Number Waiting in Line	Frequency
0	2
1	5
2	9
3	13
4	11
5	7
6	3

Construct the corresponding relative frequency table, and use it to estimate the probability that at most three customers are waiting in line.

4. **Coin Tosses** A fair coin is tossed twice. Let X be the number of heads.
   (a) Determine the probability distribution of X.
   (b) Determine the probability distribution of $2X + 5$.

5. An experiment consists of three binomial trials, each having probability 1/3 of success.
   (a) Determine the probability distribution table for the number of successes.
   (b) Use the table to compute the mean and the variance of the probability distribution.

6. **Archery** An archer has probability .3 of hitting a certain target. What is the probability of hitting the target exactly two times in four attempts?

7. **Guessing on an Exam** A true–false exam consists of ten 10-point questions. The instructor informs the students that six of the answers are *true* and four are *false*. An unprepared student decides to guess the answer to each question with the use of the spinner in Fig. 1, which gives *true* 60% of the time. Determine the student's expected score. Can you think of a better strategy?

Figure 1

8. **Dice** A pair of fair dice is rolled 12 times. *Note:* For each roll, the probability of getting a seven is 1/6.
   (a) What is $\Pr(\text{the result is seven exactly twice})$?
   (b) What is $\Pr(\text{the result is seven at least twice})$?
   (c) What is the expected number of times that the result is seven?

9. Table 2 gives the probability distribution of the random variable X. Compute the mean and the variance of the random variable.

Table 2

k	$\Pr(X = k)$
0	.2
1	.3
5	.1
10	.4

10. **Balls in an Urn** An urn contains four red balls and four white balls. An experiment consists of selecting at random a sample of four balls and recording the number of red balls in the sample. Set up the probability distribution, and compute its mean and variance.

11. The probability distribution of a random variable X is given in the table. Determine the mean, variance, and standard deviation of X.

k	$\Pr(X = k)$
-2	.3
0	.1
1	.4
3	.2

12. **Dice Game** Lucy and Ethel play a game of chance in which a pair of fair dice is rolled once. If the result is 7 or 11, then Lucy pays Ethel $10. Otherwise, Ethel pays Lucy $3. In the long run, which player comes out ahead, and by how much?

13. Suppose that a probability distribution has mean 10 and standard deviation 1/3. Use the Chebychev inequality to estimate the probability that an outcome will lie between 9 and 11.

14. Suppose that a probability distribution has mean 50 and standard deviation 8. Use the Chebychev inequality to estimate the probability that an outcome will lie between 38 and 62.

15. Find the area of the shaded region under the normal curve with $\mu = 5$, $\sigma = 3$ shown in Fig. 2(a).

16. Find the area of the shaded region under the standard normal curve shown in Fig. 2(b).

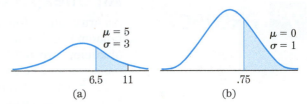

Figure 2

17. **Heights of Adult Males** The height of adult males in the United States is normally distributed with $\mu = 5.75$ feet and $\sigma = .2$ feet. What percent of the adult male population has height of 6 feet or greater?

18. Figure 3(a) is a standard normal curve. Find the value of z for which the area of the shaded region is .7734.

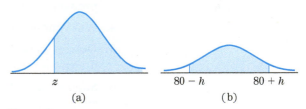

Figure 3

19. Figure 3(b) is a normal curve with $\mu = 80$ and $\sigma = 15$. Find the value of h for which the area of the shaded region is .8664.

20. **IQ Scores** As measured with the Wechsler Adult Intelligence Scale, IQ scores are normally distributed with mean 100 and standard deviation 15.
   (a) What percent of the adult population has an IQ score of 133 or more?
   (b) Find the 95th percentile of IQ scores.

21. **Jury Selection** In a certain city, 2/5 of the registered voters are women. Out of a group of 54 voters allegedly selected at random for jury duty, 13 are women. A local civil liberties group has charged that the selection procedure discriminated against women. Use the normal curve to estimate the probability of 13 or fewer women being selected in a truly random selection process.

22. **Quality Control** In a complicated production process, 1/4 of the items produced have to be readjusted. Use the normal curve to estimate the probability that out of a batch of 75 items, between 8 and 22 (inclusive) of the items require readjustment.

Conceptual Exercises

23. Give an example of a grade distribution for a class of students in which
  (a) scoring in the 3rd quartile is not very good.
  (b) scoring in the 3rd quartile corresponds to a perfect grade.

24. Give an example of a distribution of 10 grades for which
  (a) the mean and median are equal.
  (b) the mean is less than the median.
  (c) the median is less than the mean.

25. What is the difference between a population mean and a sample mean?

26. Explain in your own words the meaning of *sample mean*.

27. If each number in a set of numbers is increased by 5, will the mean increase by 5?

28. If each number in a set of numbers is doubled, will the standard deviation be doubled?

29. Explain the type of probability situations for which the binomial distribution applies.

30. Give an example of a sequence of repeated trials that does not produce a binomial distribution.

An Unexpected Expected Value

An urn contains four red balls and six white balls. Suppose that two balls are drawn at random from the urn, and let X be the number of red balls drawn. The probability of obtaining two red balls depends on whether the balls are drawn with or without replacement. The purpose of this project is to show that the expected number of red balls drawn is not affected by whether or not the first ball is replaced before the second ball is drawn. The idea for this project was taken from the article "An Unexpected Expected Value," by Stephen Schwartzman, which appeared in the February 1993 issue of *The Mathematics Teacher*.

1. Do you think that the probability that both balls are red is higher if the first ball is replaced before the second ball is drawn or if the first ball is not replaced?

2. Suppose that the first ball is replaced before the second ball is drawn. Find the probability that both balls are red.

3. Suppose that the first ball is not replaced before the second ball is drawn. Find the probability that both balls are red.

4. Was your intuitive guess in part 1 correct?

5. Do you think that the expected number of red balls drawn is higher if the balls are drawn with replacement or without replacement?

6. Suppose that the first ball is replaced before the second ball is drawn. Find the expected number of red balls that will be drawn.

7. Suppose that the first ball is not replaced before the second ball is drawn. Find the expected number of red balls that will be drawn.

8. Was your intuitive guess in part 5 correct?

9. Pretend that the 10 balls are ping-pong balls, that they have been finely ground up, and that the red and white specks have been thoroughly mixed. Forty percent of the specks will be red, and 60% will be white. Suppose that you stir the specks and use a tablespoon to scoop out 10% of the specks. That is, suppose that the tablespoon holds a quantity of specks corresponding to one ball.
  (a) What percentage of a red ball is contained in the spoon?
  (b) What percentage of the remaining specks in the urn are red?
  (c) If the spoonful of specks is replaced, does the percentage of red specks in the urn change?
  (d) Use the results from parts (b) and (c) to explain why the expected number of red balls as calculated in parts 6 and 7 is the same with and without replacement.

Suppose that we perform, one after the other, a sequence of experiments that have the same set of outcomes. The probabilities of the various outcomes of a particular experiment of the sequence may depend in some way on the outcomes of preceding experiments. The nature of such a dependency may be very complicated. In the extreme, the outcome of the current experiment may depend on the entire history of the outcomes of preceding experiments. However, there is a simple type of dependency that occurs frequently in applications and that we can analyze with fair ease. Namely, we suppose that the probabilities of the various outcomes of the current experiment depend (at most) on the outcome of the preceding experiment. In this case, the sequence of experiments is called a **Markov process**. In this chapter, we present some of the most elementary ideas concerning Markov processes and their applications.

8.1   The Transition Matrix

Here are some Markov processes that arise in applications.

EXAMPLE 1   **Investment**   A particular utility stock is very stable, and in the short run, the probability that it increases or decreases in price depends only on the result of the preceding day's trading. The price of the stock is observed at 4:00 PM each day and is recorded as "increased," "decreased," or "unchanged." The sequence of observations forms a Markov process.   **«**

EXAMPLE 2   **Medicine**   A doctor tests the effect of a new drug on high blood pressure. Based on the effects of metabolism, a given dose is eliminated from the body in 24 hours. Blood pressure is measured once a day and is recorded as "high," "low," or "normal." The sequence of measurements forms a Markov process.   **«**

EXAMPLE 3   **Sociology** A sociologist postulates that the likelihood that, in certain countries, a woman will enter the labor force depends primarily on whether the woman's mother worked. They design an experiment to test this hypothesis by viewing the sequence of career choices of a woman, her daughters, her granddaughters, her great-granddaughters, and so on as a Markov process.   ≪

Let us now introduce some vocabulary and mathematical machinery with which to study Markov processes. The experiments are performed at regular time intervals and have the same set of outcomes. These outcomes are called *states*, and the outcome of the current experiment is referred to as the *current state* of the process. After each time interval, the process may change its state. The transition from state to state can be described by tree diagrams, as is shown in the next example.

EXAMPLE 4   **Using a Tree Diagram to Represent Transitions** Refer to the utility stock of Example 1. Suppose that, if the stock increases one day, the probability that on the next day it increases is .3, remains unchanged is .2, decreases is .5. On the other hand, if the stock is unchanged one day, the probability that on the next day it increases is .6, remains unchanged is .1, decreases is .3. If the stock decreases one day, the probability that it increases the next day is .3, is unchanged is .4, decreases is .3. Represent the possible transitions between states and their probabilities by tree diagrams.

SOLUTION   The Markov process has three states: "increases," "unchanged," and "decreases." The transitions from the first state ("increases") to the other states are

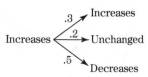

Note that each branch of the tree has been labeled with the probability of the corresponding transition. Similarly, the tree diagrams corresponding to the other two states are

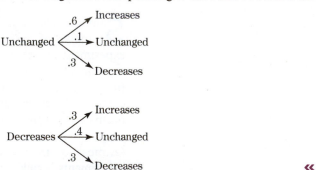

The three tree diagrams of Example 4 may be summarized in a single matrix. We insert the probabilities from a given tree down a column of the matrix, so that each column of the matrix records the information about transitions from one particular state. So the first column of the matrix is

$$\begin{bmatrix} .3 \\ .2 \\ .5 \end{bmatrix},$$

corresponding to transitions from the state "increases." The complete matrix is

		Current state		
		Increases	Unchanged	Decreases
Next state	Increases	.3	.6	.3
	Unchanged	.2	.1	.4
	Decreases	.5	.3	.3

This matrix, which records all data about transitions from one state to another, is called the **transition matrix** of the Markov process.

EXAMPLE 5   **Women in the Labor Force** Census studies from the 1960s reveal that, in the United States, 80% of the daughters of women in the labor force also worked outside of the home and that 30% of the daughters of women not in the labor force worked outside of the home. Assume that this trend remains unchanged from one generation to the next. Determine the corresponding transition matrix.

SOLUTION   There are two states, which we refer to as "work" and "don't work." The first column corresponds to transitions from the first state—that is, from "work." The probability that the daughter of a working woman chooses *not* to work is $1 - .8 = .2$. Therefore, the first column is

$$\begin{bmatrix} .8 \\ .2 \end{bmatrix}.$$

In similar fashion, the second column is

$$\begin{bmatrix} .3 \\ .7 \end{bmatrix}.$$

The transition matrix is, therefore,

$$\text{Next generation} \begin{array}{c} \\ \text{Work} \\ \text{Don't work} \end{array} \overset{\begin{array}{c} \text{Current generation} \\ \overline{\text{Work} \quad \text{Don't work}} \end{array}}{\begin{bmatrix} .8 & .3 \\ .2 & .7 \end{bmatrix}}.$$

>> *Now Try Exercise 25(a)*

Here is the form of a general transition matrix for a Markov process:

$$\text{Next state} \begin{array}{c} \\ \text{State 1} \\ \vdots \\ \text{State } i \\ \vdots \\ \text{State } r \end{array} \overset{\begin{array}{c} \text{Current state} \\ \overline{\text{State 1} \quad \cdots \quad \text{State } j \quad \cdots \quad \text{State } r} \end{array}}{\begin{bmatrix} & & & \\ & & & \\ & \Pr(\text{next } i \mid \text{current } j) & \\ & & & \\ & & & \end{bmatrix}}.$$

Note that this matrix satisfies the following two properties:

1. All entries are greater than or equal to 0.
2. The sum of the entries in each column is 1.

Any square matrix satisfying properties 1 and 2 is called a **stochastic matrix**. (The word "stochastic" derives from the Greek word *stochastices*, which means "a person who predicts the future.")

Let us examine further the Markov process of Example 5. In 1960, about 40% of U.S. women worked outside the home and 60% did not. This distribution is described by the column matrix

$$\begin{bmatrix} .4 \\ .6 \end{bmatrix}_0,$$

which is called a **distribution matrix**. The subscript 0 is added to denote that this matrix describes generation 0. Their daughters constitute generation 1, and their granddaughters generation 2. There is a distribution matrix for each generation. The distribution matrix for generation n is

$$\begin{bmatrix} p_W \\ p_{DW} \end{bmatrix}_n,$$

where p_W is the percentage of women in generation n who work and p_{DW} is the percentage who don't work. Of course, the numbers p_W and p_{DW} are also probabilities. The number p_W is the probability that a woman selected at random from generation n works. Similarly, p_{DW} is the probability that a woman selected at random from generation n does not work. Shortly, we shall give a method for calculating the distribution matrix for generation n.

In general, whenever a Markov process applies to a group with members in r possible states, a distribution matrix of the form

$$\begin{bmatrix} p_1 \\ p_2 \\ \vdots \\ p_r \end{bmatrix}_0$$

gives the initial percentages of members in each of the r states. Similarly, a matrix of the same type (with the subscript n) gives the percentages of members in each of the r states after n time periods. Note that each of the percentages in the distribution matrix is also a probability—the probability that a randomly selected member will be in the corresponding state.

EXAMPLE 6   **Projecting Women in the Labor Force**   In 1960, census figures showed that 40% of American women worked outside of the home. Use the stochastic matrix of Example 5 to determine the percentage of working women in each of the next two generations.

SOLUTION   Let us denote by A the stochastic matrix of Example 5:

		Current generation	
		Work	Don't work
Next generation	Work	.8	.3
	Don't work	.2	.7

The initial distribution matrix is

$$\begin{bmatrix} .4 \\ .6 \end{bmatrix}_0.$$

Our goal is to compute the distribution matrices for generations 1 and 2. With an eye toward performing similar calculations for other countries, let us use letters rather than specific numbers in these distribution matrices. Let

$$\begin{bmatrix} \; \end{bmatrix}_0 = \begin{bmatrix} x_0 \\ y_0 \end{bmatrix}, \qquad \begin{bmatrix} \; \end{bmatrix}_1 = \begin{bmatrix} x_1 \\ y_1 \end{bmatrix}, \qquad \begin{bmatrix} \; \end{bmatrix}_2 = \begin{bmatrix} x_2 \\ y_2 \end{bmatrix}.$$

The state transitions from generation 0 to generation 1 may be displayed in a tree diagram.

Generation 0	*Generation 1*	*Probabilities*
x_0 — Work	.8 — Work	$.8x_0$
	.2 — Don't work	$.2x_0$
y_0 — Don't work	.3 — Work	$.3y_0$
	.7 — Don't work	$.7y_0$

Adding together the probabilities for the two paths leading to "work" in generation 1 gives

$$x_1 = .8x_0 + .3y_0. \tag{1}$$

Similarly, adding together the probabilities for the two paths leading to "don't work" in generation 1 gives

$$y_1 = .2x_0 + .7y_0. \tag{2}$$

Thus, (1) and (2) show that x_1 and y_1 can be computed from this pair of equations.

$$x_1 = .8x_0 + .3y_0$$
$$y_1 = .2x_0 + .7y_0$$

This system of equations is equivalent to the one matrix equation

$$\begin{bmatrix} .8 & .3 \\ .2 & .7 \end{bmatrix} \begin{bmatrix} x_0 \\ y_0 \end{bmatrix} = \begin{bmatrix} x_1 \\ y_1 \end{bmatrix}. \tag{3}$$

Or, to write equation (3) symbolically,

$$A \begin{bmatrix} \\ \end{bmatrix}_0 = \begin{bmatrix} \\ \end{bmatrix}_1. \tag{4}$$

Now it is easy to do the arithmetic to compute the distribution matrix for generation 1 of American women:

$$A \begin{bmatrix} \\ \end{bmatrix}_0 = \begin{bmatrix} .8 & .3 \\ .2 & .7 \end{bmatrix} \begin{bmatrix} .4 \\ .6 \end{bmatrix}_0 = \begin{bmatrix} .5 \\ .5 \end{bmatrix}_1$$

That is, 50% of American women in generation 1 will work, and 50% will not.

To compute the distribution matrix for generation 2, we use the same reasoning as before, when we showed that to get

$$\begin{bmatrix} \\ \end{bmatrix}_1 \quad \text{from} \quad \begin{bmatrix} \\ \end{bmatrix}_0,$$

just multiply by A.

Similarly,

$$A \begin{bmatrix} \\ \end{bmatrix}_1 = \begin{bmatrix} \\ \end{bmatrix}_2.$$

However, by (4), we have a formula for $\begin{bmatrix} \\ \end{bmatrix}_1$, which we can insert into the last equation, getting

$$\begin{bmatrix} \\ \end{bmatrix}_2 = A \begin{bmatrix} \\ \end{bmatrix}_1 = A\left(A \begin{bmatrix} \\ \end{bmatrix}_0\right) = A^2 \begin{bmatrix} \\ \end{bmatrix}_0.$$

In other words,

$$A^2 \begin{bmatrix} \\ \end{bmatrix}_0 = \begin{bmatrix} \\ \end{bmatrix}_2. \tag{5}$$

A simple calculation gives

$$A^2 = \begin{bmatrix} .70 & .45 \\ .30 & .55 \end{bmatrix},$$

so that we can now compute the distribution for generation 2 of American women:

$$\begin{bmatrix} .70 & .45 \\ .30 & .55 \end{bmatrix} \begin{bmatrix} .4 \\ .6 \end{bmatrix}_0 = \begin{bmatrix} .55 \\ .45 \end{bmatrix}_2.$$

That is, after two generations, 55% of American women work and 45% do not.

» Now Try Exercise 19

The fact that successive distribution matrices are obtained by multiplying on the left by the matrix A can be used to show that

$$A^3 \begin{bmatrix} \\ \end{bmatrix}_0 = \begin{bmatrix} \\ \end{bmatrix}_3, \quad A^4 \begin{bmatrix} \\ \end{bmatrix}_0 = \begin{bmatrix} \\ \end{bmatrix}_4, \quad A^5 \begin{bmatrix} \\ \end{bmatrix}_0 = \begin{bmatrix} \\ \end{bmatrix}_5, \quad \dots$$

A condensed notation is

$$A^n \begin{bmatrix} \\ \end{bmatrix}_0 = \begin{bmatrix} \\ \end{bmatrix}_n \qquad (n = 1, 2, 3, \dots). \tag{6}$$

That is, to compute the distribution matrix for generation n, merely compute the product of A^n times the distribution matrix for generation 0. Equation (6) can be used to predict the distribution matrix for any number of generations into the future, starting from any given distribution matrix. Let us now look at an entirely different type of situation that can be described by a Markov process.

EXAMPLE 7   **Taxi Zones** Taxis pick up and deliver passengers in a city that is divided into three zones. Records kept by the drivers show that, of the passengers picked up in zone I, 50% are taken to a destination in zone I, 40% to zone II, and 10% to zone III. Of the passengers picked up in zone II, 40% go to zone I, 30% to zone II, and 30% to zone III. Of the passengers picked up in zone III, 20% go to zone I, 60% to zone II, and 20% to zone III. Suppose that, at the beginning of the day, 60% of the taxis are in zone I, 10% in zone II, and 30% in zone III. What is the distribution of taxis in the various zones after all have had one passenger? Two passengers?

SOLUTION   This situation is an example of a Markov process. The states are the zones. The initial distribution of the taxis gives the zeroth distribution matrix:

$$\begin{bmatrix} .6 \\ .1 \\ .3 \end{bmatrix}_0$$

The stochastic matrix associated with the process is the one giving the probabilities of taxis starting in any one zone and ending up in any other. There is one column for each zone.

		From zone I	II	III
To zone	I	.5	.4	.2
	II	.4	.3	.6
	III	.1	.3	.2

$= A.$

After all taxis have had one passenger, the distribution matrix is

$$A \begin{bmatrix} .6 \\ .1 \\ .3 \end{bmatrix}_0 = \begin{bmatrix} .5 & .4 & .2 \\ .4 & .3 & .6 \\ .1 & .3 & .2 \end{bmatrix} \begin{bmatrix} .6 \\ .1 \\ .3 \end{bmatrix}_0 = \begin{bmatrix} .40 \\ .45 \\ .15 \end{bmatrix}_1.$$

That is, 40% of the taxis are in zone I, 45% in zone II, and 15% in zone III. After all taxis have had two passengers, the distribution matrix is

$$A^2 \begin{bmatrix} .6 \\ .1 \\ .3 \end{bmatrix}_0,$$

which, after some arithmetic, can be shown to be

$$\begin{bmatrix} .410 \\ .385 \\ .205 \end{bmatrix}_2.$$

That is, after two passengers, 41% of the taxis are in zone I, 38.5% are in zone II, and 20.5% are in zone III. **» Now Try Exercise 27**

The crucial formula used in both Examples 6 and 7 is

$$A \begin{bmatrix} \end{bmatrix}_0 = \begin{bmatrix} \end{bmatrix}_1. \tag{7}$$

From this one follows the more general formula

$$A^n \begin{bmatrix} \end{bmatrix}_0 = \begin{bmatrix} \end{bmatrix}_n.$$

We carefully proved (7) in the special case of Example 6. Let us do the same for Example 7. Suppose that the initial distribution of taxis is

$$\begin{bmatrix} x_0 \\ y_0 \\ z_0 \end{bmatrix}.$$

How many taxis end up in zone I after one passenger? The taxis in zone I come from three sources—zones I, II, and III. And the first row of the stochastic matrix A gives the percentages of taxis starting out in each of the zones and ending up in zone I:

[percent of taxis going to zone I] $= [.5] \cdot$ [percent of taxis in zone I]

$+ [.4] \cdot$ [percent of taxis in zone II]

$+ [.2] \cdot$ [percent of taxis in zone III]

$= .5x_0 + .4y_0 + .2z_0.$

Indeed, the first entry, x_1, in the product

$$\begin{bmatrix} .5 & .4 & .2 \\ .4 & .3 & .6 \\ .1 & .3 & .2 \end{bmatrix} \begin{bmatrix} x_0 \\ y_0 \\ z_0 \end{bmatrix} = \begin{bmatrix} x_1 \\ \\ \end{bmatrix}$$

is $.5x_0 + .4y_0 + .2z_0$. Similarly, the proportions of taxis in zones II and III coincide with the other two entries in the matrix product. So equation (7) holds.

Interpretation of the Entries of A^n

Label the columns and rows of the transition matrix A and A^n with the states as in the paragraph following Example 5. Then the entry in the ith row and jth column of A is the probability of the transition from state j to state i after one time period.

> The entry in the ith row and jth column of the matrix A^n is the probability of the transition from state j to state i after n time periods.

Transition Diagrams

The probabilities for a Markov process can be visually displayed in a transition diagram. Figure 1 shows the stochastic matrix for women in the labor force and the corresponding transition diagram, and Fig. 2 shows the same for the taxi zone example. The transition diagrams have an oval for each state, and each arrow is labeled with the probability of the corresponding transition.

$$
\begin{array}{c}
\text{Next generation} \\
\begin{array}{r}
\text{Work} \\
\text{Don't work}
\end{array}
\end{array}
\quad
\begin{array}{c}
\text{Current generation} \\
\begin{array}{cc}
\text{Work} & \text{Don't work}
\end{array} \\
\left[
\begin{array}{cc}
.8 & .3 \\
.2 & .7
\end{array}
\right]
\end{array}
$$

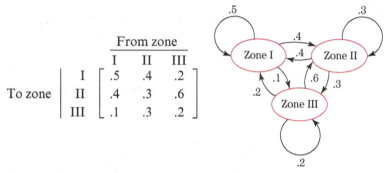

Figure 1 Women in the labor force

$$
\begin{array}{c}
\text{To zone} \\
\begin{array}{r}
\text{I} \\
\text{II} \\
\text{III}
\end{array}
\end{array}
\quad
\begin{array}{c}
\text{From zone} \\
\begin{array}{ccc}
\text{I} & \text{II} & \text{III}
\end{array} \\
\left[
\begin{array}{ccc}
.5 & .4 & .2 \\
.4 & .3 & .6 \\
.1 & .3 & .2
\end{array}
\right]
\end{array}
$$

Figure 2 Taxi zones

EXAMPLE 8   **Forming a Stochastic Matrix** Write a stochastic matrix corresponding to the following transition diagram:

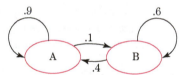

SOLUTION   Since the transition diagram has two ovals labeled A and B, there are two states (referred to as A and B). The corresponding stochastic matrix will have two columns and two rows labeled A and B. The arrow labeled .9 in the transition diagram represents the probability of going from state A to state A. Therefore, the probability in the first row, first column, of the corresponding stochastic matrix is .9. Placing the probability for each arrow into the appropriate location in the matrix results in the following stochastic matrix:

$$
\begin{array}{c}
\begin{array}{r}
\text{A} \\
\text{B}
\end{array}
\end{array}
\begin{array}{c}
\begin{array}{cc}
\text{A} & \text{B}
\end{array} \\
\left[
\begin{array}{cc}
.9 & .4 \\
.1 & .6
\end{array}
\right]
\end{array}.
$$

»» Now Try Exercise 7

INCORPORATING TECHNOLOGY

 Let

$$
A = \begin{bmatrix} .8 & .3 \\ .2 & .7 \end{bmatrix}
$$

be the 2×2 matrix of Examples 5 and 6, and let

$$
B = \begin{bmatrix} .4 \\ .6 \end{bmatrix}
$$

be the initial distribution matrix. Successive distribution matrices are easy to obtain with a graphing calculator. The most recently displayed list, number, or matrix becomes the value of the variable **Ans**. (To obtain **Ans**, press 2nd[ANS].) In Fig. 3, after matrix B is displayed, it becomes the value of the variable **Ans**. Therefore, the value of **[A]*Ans** is the matrix product AB. Each time ENTER is pressed, the instruction **[A]*Ans** is repeated. That is, the preceding matrix is multiplied on the left by the matrix A. Figure 4, which results after ENTER is pressed two more times, shows the additional matrices A^2B and A^3B.

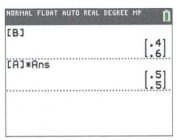

Figure 3

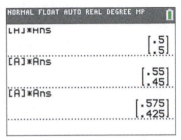

Figure 4

Figure 5, which shows successive distribution matrices, can be created as follows:

1. Give the name **A** to the 2 × 2 matrix on the left.
2. Enter **.4** and **.6** into cells D2 and D3.
3. Select the range E2:E3, type in **=MMULT(A,D2:D3)**, and press Ctrl+Shift+Enter.
4. Select the range E2:E3, and drag its fill handle right to column I.

NOTE ▶ In Step 3, the column matrix $\begin{bmatrix} .4 \\ .6 \end{bmatrix}$ was referred to as the range of cells D2:D3 rather than by a name. This allowed AutoFill to use the pattern specified in E2:E3 as a basis for filling additional cells. «

	A	B	C	D	E	F	G	H	I
1	**Matrix A**		**Generation**	0	1	2	3	4	5
2	.8	.3		.4	.5	.55	.575	.5875	.59375
3	.2	.7		.6	.5	.45	.425	.4125	.40625

Figure 5

✹WolframAlpha The nth distribution matrix from Example 6 can be computed with the instruction **({{.8,.3},{.2,.7}}^n)*{{.4},{.6}}**.

Check Your Understanding 8.1

Solutions can be found following the section exercises.

1. Is $\begin{bmatrix} \frac{2}{5} & 1 \\ \frac{3}{5} & .2 \\ \frac{2}{5} & -.3 \end{bmatrix}$ a stochastic matrix?

2. **Learning Process** An elementary learning process consists of subjects participating in a sequence of events. Experiment shows that, of the subjects not conditioned to make the correct response at the beginning of any event, 40% will

be conditioned to make the correct response at the end of the event. Once a subject is conditioned to make the correct response, they stay conditioned.

(a) Set up the 2 × 2 stochastic matrix with columns and rows labeled N (not conditioned) and C (conditioned) that describes this situation.

(b) Compute A^3.

(c) If, initially, all the subjects are not conditioned, what percent of them will be conditioned after three events?

EXERCISES 8.1

In Exercises 1–6, determine whether or not the matrix is stochastic.

1. $\begin{bmatrix} 1 & .8 \\ 0 & .2 \end{bmatrix}$ **2.** $\begin{bmatrix} \frac{1}{3} & \frac{1}{3} \\ \frac{2}{3} & \frac{2}{3} \end{bmatrix}$

3. $\begin{bmatrix} .4 & .3 & .2 \\ .6 & .7 & .8 \end{bmatrix}$ **4.** $\begin{bmatrix} .4 & .5 & .1 \\ .3 & .4 & 0 \\ .3 & .2 & .9 \end{bmatrix}$

5. $\begin{bmatrix} \frac{1}{6} & \frac{5}{12} & 0 \\ \frac{1}{2} & \frac{1}{4} & \frac{1}{2} \\ \frac{1}{3} & \frac{1}{3} & \frac{1}{2} \end{bmatrix}$ **6.** $\begin{bmatrix} 1 & 0 & 0 \\ 0 & 1 & 0 \\ 0 & 0 & 1 \end{bmatrix}$

In Exercises 7–12, write a stochastic matrix corresponding to the transition diagram.

7.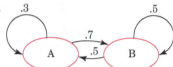

8.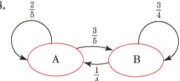

9.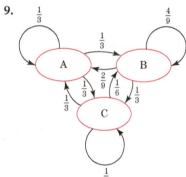

10.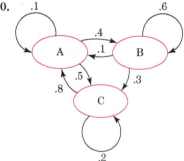

11.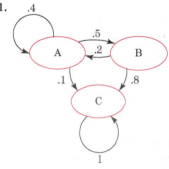

12.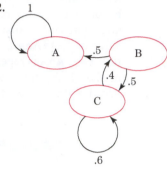

In Exercises 13–18, draw a transition diagram corresponding to the stochastic matrix.

13. $\begin{array}{c} \\ A \\ B \end{array} \begin{array}{cc} A & B \\ \begin{bmatrix} \frac{2}{7} & \frac{1}{8} \\ \frac{5}{7} & \frac{7}{8} \end{bmatrix} \end{array}$ **14.** $\begin{array}{c} \\ A \\ B \end{array} \begin{array}{cc} A & B \\ \begin{bmatrix} 1 & .3 \\ 0 & .7 \end{bmatrix} \end{array}$

15. $\begin{array}{c} \\ A \\ B \\ C \end{array} \begin{array}{ccc} A & B & C \\ \begin{bmatrix} .4 & .1 & .5 \\ 0 & .7 & .1 \\ .6 & .2 & .4 \end{bmatrix} \end{array}$ **16.** $\begin{array}{c} \\ A \\ B \\ C \end{array} \begin{array}{ccc} A & B & C \\ \begin{bmatrix} 0 & \frac{2}{3} & \frac{1}{8} \\ \frac{1}{5} & 0 & \frac{1}{4} \\ \frac{4}{5} & \frac{1}{3} & \frac{5}{8} \end{bmatrix} \end{array}$

17. $\begin{array}{c} \\ A \\ B \\ C \end{array} \begin{array}{ccc} A & B & C \\ \begin{bmatrix} .2 & 0 & .4 \\ 0 & 1 & .5 \\ .8 & 0 & .1 \end{bmatrix} \end{array}$ **18.** $\begin{array}{c} \\ A \\ B \\ C \end{array} \begin{array}{ccc} A & B & C \\ \begin{bmatrix} .2 & .7 & 0 \\ .3 & 0 & 1 \\ .5 & .3 & 0 \end{bmatrix} \end{array}$

19. Women in the Labor Force Referring to Example 5, consider a typical group of French women, of whom 47% currently work outside of the home. Assume that the same percentage of daughters follow in their mothers' footsteps as with the American women—that is, that given by the matrix

$$A = \begin{bmatrix} .8 & .3 \\ .2 & .7 \end{bmatrix}.$$

Use A and A^2 to determine the proportion of French women who work outside of the home in the next two generations. (Round to the nearest whole percent.)

20. Women in the Labor Force Repeat Exercise 19 for the women of China, of whom 44% currently work outside of the home. (Round to the nearest whole percent.)

21. Cell Phone Usage A cell phone provider classifies its customers as Low users (less than 400 minutes per month) or High users (400 or more minutes per month). Studies have shown that 80% of the people who were Low users one month will be Low users the next month, and that 70% of the people who were High users one month will be High users the next month.
  (a) Set up the 2×2 stochastic matrix with columns and rows labeled L and H that displays these transitions.
  (b) Suppose that, during the month of January, 50% of the customers were Low users. What percent of the customers will be Low users in February? In March?

22. Health Plan Option A university faculty health plan offers an optional dental plan. During the open enrollment period each year, 90% of the people who currently have the dental plan reenroll for it and 10% opt out. Of the people who do not have the dental plan, 40% enroll for it and 60% stay out of the plan.

(a) Draw a transition diagram with ovals labeled "Dental plan" and "No dental plan" for this Markov process.
(b) Set up the 2×2 stochastic matrix (with columns and rows labeled D and N) for the Markov process.
(c) Compute the second power of the matrix in part (b).
(d) Suppose that, for the year 2017, 70% of the faculty were in the dental plan and 30% were not in the plan. That is, the initial distribution is given by the column matrix $\begin{bmatrix} .7 \\ .3 \end{bmatrix}$.
Use the matrices in parts (b) and (c) to find the distribution matrices for 2018 and 2019.

23. Population Movement The Southwestern states were popular destinations in 2015. Suppose that Fig. 6 shows the percentages of people who moved in and out of the Southwest during 2015. At the beginning of 2015, about 12% of the U.S. population lived in the Southwest.

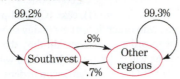

Figure 6

(a) Set up the stochastic matrix that displays the transitions.
(b) Assuming that these transition rates persisted for several years, use the matrix from part (a) to estimate the percent of the U.S. population living in the Southwestern states at the beginning of 2016 and 2017.

24. Voter Patterns For a certain group of states, it was observed that 70% of the Democratic governors were succeeded by Democrats and 30% by Republicans. Also, 40% of the Republican governors were succeeded by Democrats and 60% by Republicans.
(a) Set up the 2×2 stochastic matrix with columns and rows labeled D and R that displays these transitions.
(b) Compute A^2 and A^3.
(c) Suppose that all current governors are Democrats. Assuming that the current trend holds for three elections, what percent of the governors will then be Democrats?

25. T-Maze Each day, mice are put into a T-maze (a maze shaped like a "T"; Fig. 7). In this maze, they have the choice of turning to the left (are rewarded with cheese) or to the right (receive cheese along with mild shock). After the first day, their decision whether to turn left or right is influenced by what happened on the previous day. Of those that go to the left on a certain day, 90% go to the left on the next day and 10% go to the right. Of those that go to the right on a certain day, 70% go to the left on the next day and 30% go to the right.

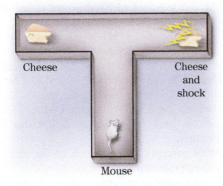

Figure 7

(a) Set up the 2×2 stochastic matrix with columns and rows labeled L and R that describes this situation.
(b) Compute the second power of the matrix in part (a).
(c) Suppose that, on the first day (day 0), 50% go to the left and 50% go to the right. So, the initial distribution is given by the column matrix $\begin{bmatrix} .5 \\ .5 \end{bmatrix}_0$. Using the matrices in parts (a) and (b), find the distribution matrices for the next two days, days 1 and 2.
(d) Make a guess as to the percentage of mice that will go to the left after 50 days. (Do not compute.)

26. Analysis of a Poem In 1913, Markov analyzed a long poem written by a Russian author. He found that vowels were followed by consonants 87.2% of the time (either in the same word or the next word) and that consonants were followed by vowels 66.3% of the time. (*Source: Bulletin de l'Académie Imperiale des Sciences de St Petersburg.* Published by l'Académie.démie.)
(a) Set up the 2×2 stochastic matrix, with columns and rows labeled V and C, that describes this situation.
(b) Find the probability that the second letter following a vowel is also a vowel.

27. Taxi Zones Refer to Example 7 (taxi zones). If, originally, 40% of the taxis start in zone I, 40% in zone II, and 20% in zone III, how will the taxis be distributed after each has taken one passenger?

28. Fitness A group of physical fitness devotees works out in the gym every day. The workouts vary from strenuous to moderate to light. When their exercise routine was recorded, the following observation was made: Of the people who work out strenuously on a particular day, 40% will work out strenuously on the next day and 60% will work out moderately. Of the people who work out moderately on a particular day, 50% will work out strenuously and 50% will work out lightly on the next day. Of the people working out lightly on a particular day, 30% will work out strenuously on the next day, 20% moderately, and 50% lightly.
(a) Set up the 3×3 stochastic matrix with columns and rows labeled S, M, and L that describes these transitions.
(b) Suppose that, on a particular Monday, 80% have a strenuous, 10% a moderate, and 10% a light workout. What percent will have a strenuous workout on Wednesday?

29. Political Views According to the Higher Education Research Institute, 33% of students at baccalaureate-granting colleges who entered college in 2015 characterize their political views as Liberal, 45% as Middle-of-the-road, and 22% as Conservative. Suppose that, each year, these students changed their political views as described by the following matrix. (*Source: The American Freshman: National Norms Fall 2015.*)

		From view:		
		L	M	C
To view:	L	.94	.02	.01
	M	.05	.96	.04
	C	.01	.02	.95

(a) What percentage of the students who held conservative political views as freshmen held middle-of-the-road views as sophomores?
(b) Explain the meaning of the .96 appearing in the center of the matrix.

(c) Draw the transition diagram for this Markov process.

(d) What percentage of the students held middle-of-the-road political views as sophomores? As juniors?

30. **Student Residences** According to the Higher Education Research Institute, 80% of students at baccalaureate-granting colleges who entered college in 2015 lived in College residence halls, 15% lived with Family, and 5% lived in Other types of housing. Suppose that each year, these students changed their residences as described by the following matrix. (*Source: The American Freshman: National Norms Fall 2015.*)

$$
\begin{array}{c}
\text{From residence:}\\
\begin{array}{cccc}
 & \text{C} & \text{F} & \text{O}\\
\text{To residence:} &
\begin{array}{c}\text{C}\\\text{F}\\\text{O}\end{array}
\left[\begin{array}{ccc}
.9 & .1 & .2\\
.05 & .8 & .1\\
.05 & .1 & .7
\end{array}\right]
\end{array}
\end{array}
$$

(a) What percentage of the students who lived with family as freshmen lived in college residence halls as sophomores?

(b) Explain the meaning of the .8 appearing in the center of the matrix.

(c) Draw the transition diagram for this Markov process.

(d) What percentage of the students lived in college residence halls as sophomores? As juniors?

31. **Population Movement** A sociologist studying living patterns in a certain region determines that, each year, the population shifts between urban, suburban, and rural areas as shown in Fig. 8.

(a) Set up a stochastic matrix that displays these transitions.

(b) What percentage of people who live in urban areas in 2017 will live in rural areas in 2019?

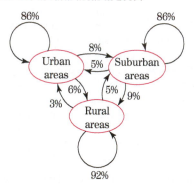

Figure 8

32. **Brand Loyalty** A retailer stocks three brands of breakfast cereal. A survey is taken of 5000 people who purchase cereal weekly from this retailer. Each week, Crispy Flakes loses 12% of its customers to Crunchy Nuggets and 19% to Toasty Cinnamon Twists. Crunchy Nuggets loses 16% of its customers to Crispy Flakes and 10% of its customers to Toasty Cinnamon Twists, and Toasty Cinnamon Twists loses 20% of its customers to Crispy Flakes and 14% to Crunchy Nuggets.

(a) Set up a stochastic matrix displaying these transitions.

(b) Suppose that, this week, 1500 people buy Crispy Flakes, 1500 buy Crunchy Nuggets, and 2000 people buy Toasty Cinnamon Twists. How many people will buy Crispy Flakes next week? How many will buy Toasty Cinnamon Twists in two weeks?

33. **Birth Weights** Birth weights can be classified as Low (less than 6 pounds), Average (between 6 and 8 pounds), and High (greater than 8 pounds). Birth weights of female babies tend to be influenced by the birth weights of their mothers. Suppose that this tendency is given by the following matrix:

$$
\begin{array}{cc}
 & \text{Mother}\\
 & \begin{array}{ccc}\text{H} & \text{A} & \text{L}\end{array}\\
\text{Daughter}\;
\begin{array}{c}\text{H}\\\text{A}\\\text{L}\end{array}
&
\left[\begin{array}{ccc}
.5 & .25 & .2\\
.4 & .5 & .2\\
.1 & .25 & .6
\end{array}\right].
\end{array}
$$

(a) If the initial distribution of birth weights is given by the matrix $\begin{bmatrix}.4\\.4\\.2\end{bmatrix}$, what is the distribution after one generation?

(b) What is the probability that a woman of High birth weight will have a granddaughter of High birth weight? Of Average birth weight?

34. **Dessert Choices** A college cafeteria offers three choices of dessert for lunch: ice cream (I), cake (C), and pie (P). A study of the students who order dessert every day showed that, of those students who order I on a certain day, the next day 60% of these students will order I and the others will be equally likely to order C or P. Of the students who order C on a certain day, the next day 40% of these students will order C and the others will be equally likely to order I or P. Of the students who order P on a certain day, the next day 80% of these students will order P and the others will be equally likely to order I or C.

(a) Set up the 3×3 stochastic matrix that describes this situation.

(b) Suppose that today 20% of the students ordered ice cream, 30% ordered cake, and 50% ordered pie. What percent of the students will order cake tomorrow? The day after tomorrow?

35. **Ehrenfest Urn Model** The Ehrenfest urn model was originally proposed as a model for diffusion of gases, but has since come to be applied in a wide variety of fields. This problem is a simplified version of the model. Consider two urns, Urn A and Urn B, containing a total of four balls. At each time interval, a ball is selected at random and transferred into the other urn. Let the states be the number of balls in Urn A. Therefore, the different states are 0, 1, 2, 3, and 4.

(a) Set up the stochastic matrix for the process.

(b) If, at some point in time, Urn A contains three balls, what is the probability that it will contain three balls after two time intervals?

36. **Occupational Mobility** A study of occupational mobility in a certain community classified occupations as P (professional), WC (white collar), SM (skilled manual), and UM (unskilled manual). Currently, 4% of the adults are P, 20% are WC, 40% are SM, and 36% are UM. Studies predict that, of the children of P parents, 40% become P, 45% become WC, 10% become SM, and 5% become UM. Of the children of WC parents, 20% become P, 60% become WC, 10% become SM, and 10% become UM. Of the children of SM parents, 10% become P, 45% become WC, 30% become SM, and 15% become UM. Of the children of UM parents, 5% become P, 15% become WC, 25% become SM, and 55% become UM.

(a) Set up the 4×4 stochastic matrix that describes this situation.

(b) What percent of the children of the current generation will be professionals?

(c) What percent of the grandchildren of the current generation will be professionals?

In Exercises 37 and 38, find the third and fourth distribution matrices for the given stochastic matrix and initial distribution. (Round entries to two decimal places.)

37. $\begin{bmatrix} .5 & .4 \\ .5 & .6 \end{bmatrix}, \begin{bmatrix} .3 \\ .7 \end{bmatrix}_0$

38. $\begin{bmatrix} .3 & .9 \\ .7 & .1 \end{bmatrix}, \begin{bmatrix} .5 \\ .5 \end{bmatrix}_0$

In Exercises 39–44, compute the first five powers of each matrix. (Round to two decimal places.)

39. $\begin{bmatrix} \frac{1}{3} & \frac{1}{3} \\ \frac{2}{3} & \frac{2}{3} \end{bmatrix}$ **40.** $\begin{bmatrix} 1 & \frac{1}{2} \\ 0 & \frac{1}{2} \end{bmatrix}$ **41.** $\begin{bmatrix} .1 & .3 \\ .9 & .7 \end{bmatrix}$

42. $\begin{bmatrix} 0 & 1 \\ 1 & 0 \end{bmatrix}$ **43.** $\begin{bmatrix} .3 & .3 & .3 \\ .1 & .1 & .1 \\ .6 & .6 & .6 \end{bmatrix}$ **44.** $\begin{bmatrix} .2 & .2 & .2 \\ .3 & .3 & .3 \\ .5 & .5 & .5 \end{bmatrix}$

Stochastic matrices for which some power contains no zero entries are called *regular* matrices. In Exercises 45 and 46, conjecture whether or not the given matrix is regular, by looking at the first few powers of the matrix.

45. $\begin{bmatrix} .5 & 0 \\ .5 & 1 \end{bmatrix}$ **46.** $\begin{bmatrix} .7 & 1 \\ .3 & 0 \end{bmatrix}$

TECHNOLOGY EXERCISES

47. Let A be the stochastic matrix $\begin{bmatrix} .1 & .6 \\ .9 & .4 \end{bmatrix}$, and let the initial distribution be $B = \begin{bmatrix} .5 \\ .5 \end{bmatrix}_0$.

(a) Generate the next four distribution matrices.
(b) Calculate $A^4 B$, and confirm that it is the same as the fourth distribution matrix.

48. Repeat Exercise 47 for the matrices $\begin{bmatrix} .4 & .2 \\ .6 & .8 \end{bmatrix}$ and $\begin{bmatrix} .7 \\ .3 \end{bmatrix}_0$.

49. Consider the matrices of Exercise 47. Beginning with the initial distribution matrix, generate 10 more distributions. Continue to generate 10 more. The matrices will get closer and closer to a certain 2×1 matrix. What is that matrix?

50. Repeat Exercise 49 for the matrices of Exercise 48.

51. Generate 35 successive powers of the matrix A from Exercise 47. (With a graphing calculator, the instruction **[A]*Ans** can be used to compute successive powers of a square matrix.) The matrices will get closer and closer to a certain 2×2 matrix. What is that matrix? How is that matrix related to the 2×1 matrix found in Exercise 49?

52. Repeat Exercise 51 for the square matrix of Exercise 48.

Solutions to Check Your Understanding 8.1

1. No. It fails on all three conditions. The matrix is not square, the entry $-.3$ is not ≥ 0, and the sum of the entries in the first (and second) column is not equal to 1.

2. (a) $\begin{array}{c} \\ \text{N} \\ \text{C} \end{array} \begin{array}{cc} \text{N} & \text{C} \\ \begin{bmatrix} .6 & 0 \\ .4 & 1 \end{bmatrix} \end{array}$. Since there are only two possibilities and 40% of those not conditioned become conditioned, the remaining 60% stay not conditioned. After each event, 100% of the conditioned stay conditioned, and therefore, 0% become not conditioned.

(b) $A^2 = \begin{bmatrix} .6 & 0 \\ .4 & 1 \end{bmatrix}\begin{bmatrix} .6 & 0 \\ .4 & 1 \end{bmatrix} = \begin{bmatrix} .36 & 0 \\ .64 & 1 \end{bmatrix}$,

$A^3 = A^2 \cdot A = \begin{bmatrix} .36 & 0 \\ .64 & 1 \end{bmatrix}\begin{bmatrix} .6 & 0 \\ .4 & 1 \end{bmatrix}$

$= \begin{bmatrix} .216 & 0 \\ .784 & 1 \end{bmatrix}$.

(c) Here, $\begin{bmatrix} \\ \end{bmatrix}_0 = \begin{bmatrix} 1 \\ 0 \end{bmatrix}$. Therefore,

$$\begin{bmatrix} \\ \end{bmatrix}_3 = A^3 \begin{bmatrix} 1 \\ 0 \end{bmatrix} = \begin{bmatrix} .216 & 0 \\ .784 & 1 \end{bmatrix}\begin{bmatrix} 1 \\ 0 \end{bmatrix} = \begin{bmatrix} .216 \\ .784 \end{bmatrix}.$$

So 78.4% will be conditioned after three events.

8.2 Regular Stochastic Matrices

In the preceding section, we studied the percentages of women who work outside of the home in various generations in the United States. We showed that if $\begin{bmatrix} \\ \end{bmatrix}_0$ is the initial distribution matrix, then the distribution matrix $\begin{bmatrix} \\ \end{bmatrix}_n$ for the nth generation is given by

$$\begin{bmatrix} \\ \end{bmatrix}_n = A^n \begin{bmatrix} \\ \end{bmatrix}_0, \tag{1}$$

where A is the stochastic matrix

$$A = \begin{bmatrix} .8 & .3 \\ .2 & .7 \end{bmatrix}.$$

In this section, we are interested in determining long-term trends in Markov processes. To get an idea of what is meant, consider the example of Egyptian women in the labor force.

EXAMPLE 1   **Women in the Labor Force** In Egypt, 23% of the women currently work outside of the home. The effect of maternal influence of mothers on their daughters is given by the matrix

$$\begin{bmatrix} .6 & .2 \\ .4 & .8 \end{bmatrix}.$$

(a) How many women will work outside of the home after 1, 2, 3, ..., 11 generations?
(b) Estimate the long-term trend.
(c) Answer the same questions (a) and (b) for Spain, assuming that 46% of all Spanish women currently work outside of the home and that the effect of maternal influence is the same as for Egypt.

SOLUTION   (a) The percentages of women working outside of the home in generation n (for $n = 1, 2, 3, ...$) can be determined from equation (1):

$$\begin{bmatrix} \\ \end{bmatrix}_n = \begin{bmatrix} .6 & .2 \\ .4 & .8 \end{bmatrix}^n \begin{bmatrix} .23 \\ .77 \end{bmatrix}_0$$

After the mildly tedious job of raising the stochastic matrix to various powers, we obtain the results shown in Table 1.

Table 1

Generation	Percent of Women Working	Generation	Percent of Women Working
0	23	6	33.29
1	29.2	7	33.32
2	31.7	8	33.33
3	32.7	9	33.33
4	33.07	10	33.33
5	33.23	11	33.33

(b) It appears from the accompanying table that the long-term trend is for one-third, or $33\frac{1}{3}\%$, of all Egyptian women to work outside of the home.
(c) The corresponding results for Spain can be computed by replacing the initial distribution matrix $\begin{bmatrix} .23 \\ .77 \end{bmatrix}_0$ with $\begin{bmatrix} .46 \\ .54 \end{bmatrix}_0$, reflecting that, initially, 46% of all Spanish women work outside of the home. The results of the calculations are shown in Table 2. Again, the long-term trend is for one-third of the women to work outside of the home.

Table 2

Generation	Percent of Women Working	Generation	Percent of Women Working
0	46	6	33.39
1	38.4	7	33.35
2	35.4	8	33.34
3	34.14	9	33.34
4	33.66	10	33.33
5	33.46	11	33.33

From Example 1, one might begin to suspect the following: The long-term trend is always for one-third of the women to work outside of the home, independent of the initial distribution.

Verification

To see why this rather surprising fact should hold, it is useful to examine the powers of A:

$$A^2 = \begin{bmatrix} .44 & .28 \\ .56 & .72 \end{bmatrix} \qquad A^3 = \begin{bmatrix} .376 & .312 \\ .624 & .688 \end{bmatrix} \qquad A^4 = \begin{bmatrix} .3504 & .3248 \\ .6496 & .6752 \end{bmatrix}$$

$$A^5 = \begin{bmatrix} .3402 & .3299 \\ .6598 & .6701 \end{bmatrix} \qquad A^6 = \begin{bmatrix} .3361 & .3320 \\ .6639 & .6680 \end{bmatrix} \qquad A^7 = \begin{bmatrix} .3344 & .3328 \\ .6656 & .6672 \end{bmatrix}$$

$$A^8 = \begin{bmatrix} .3338 & .3331 \\ .6662 & .6669 \end{bmatrix} \qquad A^9 = \begin{bmatrix} .3335 & .3332 \\ .6665 & .6668 \end{bmatrix} \qquad A^{10} = \begin{bmatrix} .3334 & .3333 \\ .6666 & .6667 \end{bmatrix}$$

As A is raised to further powers, the matrices approach

$$\begin{bmatrix} \frac{1}{3} & \frac{1}{3} \\ \frac{2}{3} & \frac{2}{3} \end{bmatrix}. \tag{2}$$

Now suppose that, initially, the proportion of women working is x_0. That is, the initial distribution matrix is

$$\begin{bmatrix} x_0 \\ 1 - x_0 \end{bmatrix}.$$

Then, after n generations, the distribution matrix is

$$\begin{bmatrix} \quad \\ \quad \end{bmatrix}_n = A^n \begin{bmatrix} \quad \\ \quad \end{bmatrix}_0 = A^n \begin{bmatrix} x_0 \\ 1 - x_0 \end{bmatrix}.$$

But after many generations, n is large, so A^n is approximately the matrix (2). Thus,

$$\begin{bmatrix} \quad \\ \quad \end{bmatrix}_n \approx \begin{bmatrix} \frac{1}{3} & \frac{1}{3} \\ \frac{2}{3} & \frac{2}{3} \end{bmatrix} \begin{bmatrix} x_0 \\ 1 - x_0 \end{bmatrix} = \begin{bmatrix} \frac{1}{3}x_0 + \frac{1}{3}(1 - x_0) \\ \frac{2}{3}x_0 + \frac{2}{3}(1 - x_0) \end{bmatrix} = \begin{bmatrix} \frac{1}{3} \\ \frac{2}{3} \end{bmatrix}.$$

In other words, after n generations, approximately one-third of the women work outside of the home and two-thirds do not. ◀◀

From the preceding calculations, we see that the stochastic matrix A possesses a number of very special properties. First, as n gets large, A^n approaches the matrix

$$\begin{bmatrix} \frac{1}{3} & \frac{1}{3} \\ \frac{2}{3} & \frac{2}{3} \end{bmatrix}$$

Second, any initial distribution approaches the distribution $\begin{bmatrix} \frac{1}{3} \\ \frac{2}{3} \end{bmatrix}$ after many generations.

The limiting matrix

$$\begin{bmatrix} \frac{1}{3} & \frac{1}{3} \\ \frac{2}{3} & \frac{2}{3} \end{bmatrix}.$$

is called the **stable matrix** of A, and the limiting distribution

$$\begin{bmatrix} \frac{1}{3} \\ \frac{2}{3} \end{bmatrix}$$

is called the **stable distribution** of A. Finally, note that all of the columns of the stable matrix are the same and are equal to the stable distribution.

The matrices that share the aforementioned properties with A are very important. For these matrices, one can predict a long-term trend, and this trend is independent of the initial distribution. An important class of matrices with these properties is the class of *regular stochastic matrices*.

> **DEFINITION** A stochastic matrix is said to be **regular** if some power has all positive entries.

EXAMPLE 2 **Identifying Regular Stochastic Matrices** Which of the following stochastic matrices are regular?

(a) $\begin{bmatrix} .6 & .2 \\ .4 & .8 \end{bmatrix}$ (b) $\begin{bmatrix} 0 & .5 \\ 1 & .5 \end{bmatrix}$ (c) $\begin{bmatrix} 0 & 1 \\ 1 & 0 \end{bmatrix}$

SOLUTION (a) All entries are positive, so the matrix is regular.

(b) Here, a zero occurs in the first power. However,

$$\begin{bmatrix} 0 & .5 \\ 1 & .5 \end{bmatrix}^2 = \begin{bmatrix} .5 & .25 \\ .5 & .75 \end{bmatrix},$$

which has all positive entries. So the original matrix is regular.

(c) Note that

$$\begin{bmatrix} 0 & 1 \\ 1 & 0 \end{bmatrix}^2 = \begin{bmatrix} 1 & 0 \\ 0 & 1 \end{bmatrix} \quad \begin{bmatrix} 0 & 1 \\ 1 & 0 \end{bmatrix}^3 = \begin{bmatrix} 0 & 1 \\ 1 & 0 \end{bmatrix}$$

$$\begin{bmatrix} 0 & 1 \\ 1 & 0 \end{bmatrix}^4 = \begin{bmatrix} 1 & 0 \\ 0 & 1 \end{bmatrix} \quad \begin{bmatrix} 0 & 1 \\ 1 & 0 \end{bmatrix}^5 = \begin{bmatrix} 0 & 1 \\ 1 & 0 \end{bmatrix}.$$

The even powers of the matrix are the 2×2 identity matrix, and the odd powers are the original matrix. Every power has a zero in it, so the matrix is not regular.

» Now Try Exercises 1, 3, and 5

Regular matrices share all of the properties observed in the special case. Moreover, there is a simple technique for computing the stable distribution (see property 4 in the next box) that spares us from having to multiply matrices.

> **Stable Matrix and Distribution** Let A be a regular stochastic matrix.
>
> 1. The powers A^n approach a certain matrix as n gets large. This limiting matrix is called the **stable matrix** of A.
>
> 2. For any initial distribution $\begin{bmatrix} \\ \end{bmatrix}_0$, $A^n \begin{bmatrix} \\ \end{bmatrix}_0$ approaches a certain distribution $\begin{bmatrix} \\ \end{bmatrix}$. This limiting distribution is called the **stable distribution** of A.
>
> 3. All columns of the stable matrix are the same; they equal the stable distribution.
>
> 4. The stable distribution $X = [\ ]$ can be determined by solving the system of linear equations
>
> $$\begin{cases} \text{sum of the entries of } X = 1 \\ AX = X. \end{cases}$$

EXAMPLE 3 **Finding the Stable Distribution of a Regular Stochastic Matrix** Use property 4 to determine the stable distribution of the regular stochastic matrix

$$A = \begin{bmatrix} .6 & .2 \\ .4 & .8 \end{bmatrix}.$$

SOLUTION Let $X = \begin{bmatrix} x \\ y \end{bmatrix}$ be the stable distribution. The condition "sum of the entries of $X = 1$" yields the equation

$$x + y = 1.$$

The condition $AX = X$ gives the equations

$$\begin{bmatrix} .6 & .2 \\ .4 & .8 \end{bmatrix} \begin{bmatrix} x \\ y \end{bmatrix} = \begin{bmatrix} x \\ y \end{bmatrix} \quad \text{or} \quad \begin{cases} .6x + .2y = x \\ .4x + .8y = y. \end{cases}$$

So we have the system

$$\begin{cases} x + y = 1 \\ .6x + .2y = x \\ .4x + .8y = y. \end{cases}$$

Combining terms in the second and third equations and eliminating the decimals by multiplying by 10, we have

$$\begin{cases} x + y = 1 \\ -4x + 2y = 0 \\ 4x - 2y = 0. \end{cases}$$

Note that the second and third equations are the same, except for a factor of -1, so the last equation may be omitted. Now the system reads

$$\begin{cases} x + y = 1 \\ -4x + 2y = 0. \end{cases}$$

The diagonal form is obtained by the Gauss–Jordan elimination method:

$$\begin{bmatrix} 1 & 1 & | & 1 \\ -4 & 2 & | & 0 \end{bmatrix} \xrightarrow{R_2 + 4R_1} \begin{bmatrix} 1 & 1 & | & 1 \\ 0 & 6 & | & 4 \end{bmatrix}$$

$$\xrightarrow{\frac{1}{6}R_2} \begin{bmatrix} 1 & 1 & | & 1 \\ 0 & 1 & | & \frac{2}{3} \end{bmatrix}$$

$$\xrightarrow{R_1 + (-1)R_2} \begin{bmatrix} 1 & 0 & | & \frac{1}{3} \\ 0 & 1 & | & \frac{2}{3} \end{bmatrix}.$$

Thus,

$$x = \tfrac{1}{3}, \qquad y = \tfrac{2}{3}.$$

So the stable distribution is

$$\begin{bmatrix} x \\ y \end{bmatrix} = \begin{bmatrix} \frac{1}{3} \\ \frac{2}{3} \end{bmatrix},$$

as we observed before. Note that, once the stable distribution is determined, the stable matrix is easy to find. Just place the stable distribution in every column:

$$\begin{bmatrix} \frac{1}{3} & \frac{1}{3} \\ \frac{2}{3} & \frac{2}{3} \end{bmatrix}.$$

≫ Now Try Exercise 7

EXAMPLE 4 **Finding the Stable Taxi Distribution** In Section 8.1, we studied the distribution of taxis in three zones of a city. The movement of taxis from zone to zone was described by the regular stochastic matrix

$$A = \begin{bmatrix} .5 & .4 & .2 \\ .4 & .3 & .6 \\ .1 & .3 & .2 \end{bmatrix}.$$

In the long run, what percentage of taxis will be in each of the zones?

SOLUTION Let $X = \begin{bmatrix} x \\ y \\ z \end{bmatrix}$ be the stable distribution of A. Then x is the long-term percentage of taxis in zone I, y the percentage in zone II, and z the percentage in zone III. X is determined by the equations

$$\begin{cases} x + y + z = 1 \\ \qquad AX = X, \end{cases}$$

or, equivalently,

$$\begin{cases} x + y + z = 1 \\ .5x + .4y + .2z = x \\ .4x + .3y + .6z = y \\ .1x + .3y + .2z = z. \end{cases}$$

Rewriting the equations with all terms involving variables on the left, we get

$$\begin{cases} x + y + z = 1 \\ -.5x + .4y + .2z = 0 \\ .4x - .7y + .6z = 0 \\ .1x + .3y - .8z = 0. \end{cases}$$

Applying the Gauss–Jordan elimination method to this system, we get the solution $x = .4$, $y = .4$, $z = .2$. Thus, after many trips, approximately 40% of the taxis are in zone I, 40% in zone II, and 20% in zone III. **>> Now Try Exercise 11**

We have come full circle. We began Chapter 2 by solving systems of linear equations. Matrices were developed as a tool for solving such systems. Then we found matrices to be interesting in their own right. Finally, in the current chapter, we have used systems of linear equations to answer questions about matrices.

INCORPORATING TECHNOLOGY

If A is a square matrix, then graphing calculators can compute A^n. (The exponent n can be as large as 255. However, even higher powers can be obtained with instructions such as **Ans^255**.) For a regular stochastic matrix A, A^{255} should be an excellent approximation to the stable matrix of A. In Figs. 1 and 2, A is the matrix of Example 3 and C is the matrix of Example 4. *Note:* **▶Frac**, which stands for "display as a fraction," is obtained by pressing MATH **1**.

Calculators can be used to obtain a stable distribution by solving the system of linear equations or can be used to confirm a stable distribution by carrying out a matrix multiplication. The system of linear equations from Example 3 is solved in Fig. 3.

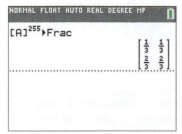

Figure 1

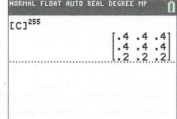

Figure 2

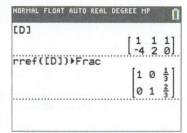

Figure 3

Successive powers of a matrix can be generated on a spreadsheet in much the same way that successive distribution matrices were generated in Fig. 5 of Section 8.1. The stable distribution can be found by using Solver to solve the appropriate system of linear equations.

❋WolframAlpha Consider the matrix from Example 1. A high power of the matrix can be computed with an instruction such as {{.6,.2},{.4,.8}}^1000, and the stable distribution can be computed with the instruction **solve** $x + y = 1, -4x + 2y = 0$.

APPENDIX Verification of Method for Obtaining the Stable Distribution

In this appendix, we verify that the stable distribution X can be obtained by solving the system of equations

$$\begin{cases} \text{sum of the entries of } X = 1 \\ \qquad\qquad\qquad AX = X. \end{cases}$$

Let A be a regular stochastic matrix. Suppose that we take as the initial distribution its stable distribution X. Then the nth distribution matrix, or $A^n X$, approaches the stable distribution (by property 2) so that

$$X \approx A^n X \quad \text{for large } n.$$

Therefore,

$$AX \approx A \cdot A^n X = A^{n+1} X.$$

But $A^{n+1} X$ is the $(n + 1)$st distribution matrix, which is also approximately X. Thus, AX is approximately X. But the approximation can be made closer and closer by making n large. Therefore, AX is arbitrarily close to X, or

$$AX = X. \tag{3}$$

This is a matrix equation in X. There is one other condition on X: X is a distribution matrix. Therefore,

$$\text{sum of the entries of } X = 1. \tag{4}$$

So, by (3) and (4), we find this system of linear equations for the entries of X:

$$\begin{cases} \text{sum of the entries of } X = 1 \\ \qquad\qquad\qquad AX = X \end{cases}$$

Check Your Understanding 8.2

Solutions can be found following the section exercises.

1. Is $\begin{bmatrix} 0 & .2 & .5 \\ .5 & 0 & .5 \\ .5 & .8 & 0 \end{bmatrix}$ a regular stochastic matrix? Explain your answer.

2. Find the stable matrix for the regular stochastic matrix in Problem 1.

3. **Cigarette Smokers** A study of cigarette smokers determined that, of the people who smoked menthol cigarettes on a particular day, 10% smoked menthol the next day and 90% smoked nonmenthol. Of the people who smoked nonmenthol cigarettes on a particular day, the next day 30% smoked menthol and 70% smoked nonmenthol. In the long run, what percent of the people will be smoking nonmenthol cigarettes on a particular day?

EXERCISES 8.2

In Exercises 1–6, determine whether or not the matrix is a regular stochastic matrix.

1. $\begin{bmatrix} \frac{1}{4} & \frac{2}{7} \\ \frac{3}{4} & \frac{5}{7} \end{bmatrix}$

2. $\begin{bmatrix} .4 & .9 \\ .6 & .1 \end{bmatrix}$

3. $\begin{bmatrix} .3 & 1 \\ .7 & 0 \end{bmatrix}$

4. $\begin{bmatrix} 1 & .3 \\ 0 & .7 \end{bmatrix}$

5. $\begin{bmatrix} .3 & 0 & .5 \\ .4 & 1 & .5 \\ .3 & 0 & 0 \end{bmatrix}$

6. $\begin{bmatrix} 0 & .8 & 0 \\ 1 & .1 & .5 \\ 0 & .1 & .5 \end{bmatrix}$

In Exercises 7–12, find the stable distribution for the given regular stochastic matrix.

7. $\begin{bmatrix} .5 & .1 \\ .5 & .9 \end{bmatrix}$

8. $\begin{bmatrix} .4 & 1 \\ .6 & 0 \end{bmatrix}$

9. $\begin{bmatrix} .8 & .3 \\ .2 & .7 \end{bmatrix}$

10. $\begin{bmatrix} .1 & .6 \\ .9 & .4 \end{bmatrix}$

11. $\begin{bmatrix} .1 & .4 & .7 \\ .6 & .4 & .2 \\ .3 & .2 & .1 \end{bmatrix}$

12. $\begin{bmatrix} .3 & .1 & .2 \\ .4 & .8 & .6 \\ .3 & .1 & .2 \end{bmatrix}$

13. **Cell Phone Usage** Refer to Exercise 21 of Section 8.1. In the long run, what percentage of the customers will be High users?

14. **Voter Patterns** Refer to Exercise 24 of Section 8.1. In the long run, what percentage of the governors will be Democrats?

15. **T-Maze** Refer to Exercise 25 of Section 8.1. What percentage of the mice will be going to the left after many days?

16. **Computer Reliability** A certain university has a computer room with 219 terminals. Each day, there is a 3% chance that a given terminal will break and a 70% chance that a given broken terminal will be repaired. In the long run, about how many terminals in the room will be working?

17. **Brand Loyalty** Suppose that 60% of people who own a General Motors car buy a GM car as their next car and 90% of people who own a non-GM car buy a non-GM car as their next car. What will General Motors' market share be in the long run?

18. **Transportation Modes** Commuters can get into town by car or bus. Surveys have shown that, for those taking their car on a particular day, 20% take their car the next day and 80% take a bus. Also, for those taking a bus on a particular day, 50% take their car the next day and 50% take a bus. In the long run, what percentage of the people take a bus on a particular day?

19. **Weather Patterns** The changes in weather from day to day on the planet Xantar form a regular Markov process. Each day is either rainy or sunny. If it rains one day, there is a 90% chance that it will be sunny the following day. If it is sunny one day, there is a 60% chance of rain the next day. In the long run, what is the daily likelihood of rain?

20. **Women in the Labor Force** Refer to the stochastic matrix in Example 6 of Section 8.1. In the long run, what percentage of American women will work outside of the home?

21. **Car Rentals** The Day-by-Day car rental agency rents cars only on a daily basis. Rented cars can be returned at the end of the day to any of the agency's three locations—A, B, or C. Figure 4 shows the percentages of cars returned to each of the locations on the basis of where they were picked up. Assume that all of the agency's cars are rented each day and that, initially, 40% of the cars are at location A, 30% at location B, and 30% at location C.

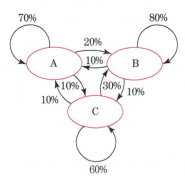

Figure 4

(a) Set up the stochastic matrix that displays these transitions.
(b) Use the matrix from part (a) to estimate the percentage of the cars at location A after one day. After two days.
(c) In the long run, what fraction of the cars will be at each location?

22. **Fitness** Refer to Exercise 28 of Section 8.1. In the long run, what percentage of the people will have a strenuous workout on a particular day?

23. **Genetics** With respect to a certain gene, geneticists classify individuals as dominant, recessive, or hybrid. In an experiment, individuals are crossed with hybrids, then their offspring are crossed with hybrids, and so on. For dominant individuals, 50% of their offspring will be dominant and 50% will be a hybrid. For the recessive individuals, 50% of their offspring will be recessive and 50% hybrid. Hybrid individuals' offspring will be 25% dominant, 25% recessive, and 50% hybrid. In the long run, what percent of the individuals in a generation will be dominant?

24. **Weather Patterns** The day-to-day changes in weather for a certain part of the country form a Markov process. Each day is sunny, cloudy, or rainy. If it is sunny one day, there is a 70% chance that it will be sunny the following day, a 20% chance that it will be cloudy, and a 10% chance of rain. If it is cloudy one day, there is a 30% chance that it will be sunny the following day, a 50% chance that it will be cloudy, and a 20% chance of rain. If it rains one day, there is a 60% chance that it will be sunny the following day, a 20% chance that it will be cloudy, and a 20% chance of rain. In the long run, what is the daily likelihood of rain?

25. Show that $\begin{bmatrix} .6 \\ 0 \\ .4 \end{bmatrix}$ and $\begin{bmatrix} .3 \\ .5 \\ .2 \end{bmatrix}$ are both stable distributions for the stochastic matrix $\begin{bmatrix} .4 & 0 & .9 \\ 0 & 1 & 0 \\ .6 & 0 & .1 \end{bmatrix}$. Explain why this fact does not contradict the main premise of this section with regard to the uniqueness of the stable distribution.

26. As shown in Example 2, $\begin{bmatrix} 0 & 1 \\ 1 & 0 \end{bmatrix}$ is not a regular stochastic matrix. Show that $\begin{bmatrix} .5 \\ .5 \end{bmatrix}$ acts like a stable distribution for this matrix, and explain why this fact does not contradict the main premise of this section.

27. **Birth Weights** Refer to Exercise 33 of Section 8.1. In the long run, what fraction of female babies will have a High birth weight? An Average birth weight?

TECHNOLOGY EXERCISES

28. **Bird Migrations** Figure 5 describes the migration pattern of a species of bird from year to year among three habitats: I, II, and III.

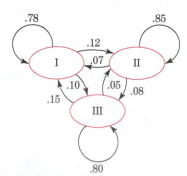

Figure 5

(a) Set up the stochastic matrix that displays these transitions.
(b) If there are 1000 birds in each habitat at the beginning of a year, how many will be in each habitat at the end of the year? At the end of two years?
(c) In the long run, what fraction of the birds will be located at each habitat?

29. Consider the stochastic matrix A, where

$$A = \begin{bmatrix} .85 & .35 \\ .15 & .65 \end{bmatrix}.$$

Approximate the stable matrix of A by raising A to a high power. Then find the exact stable distribution by solving an appropriate system of linear equations. Check your answer by forming the product of A and the stable distribution.

Repeat Exercise 29 for each of the matrices in Exercises 30–32.

30. $\begin{bmatrix} .3 & .1 \\ .7 & .9 \end{bmatrix}$ **31.** $\begin{bmatrix} .1 & .4 & .1 \\ .3 & .2 & .8 \\ .6 & .4 & .1 \end{bmatrix}$ **32.** $\begin{bmatrix} .4 & .2 & .4 \\ .1 & .3 & .1 \\ .5 & .5 & .5 \end{bmatrix}$

Solutions to Check Your Understanding 8.2

1. Yes. It is easily seen to be stochastic. Although it has some zero entries, there are no zero entries in

$$A^2 = \begin{bmatrix} .35 & .40 & .10 \\ .25 & .50 & .25 \\ .40 & .10 & .65 \end{bmatrix}.$$

2. $\begin{cases} x + y + z = 1 \\ \begin{bmatrix} 0 & .2 & .5 \\ .5 & 0 & .5 \\ .5 & .8 & 0 \end{bmatrix}\begin{bmatrix} x \\ y \\ z \end{bmatrix} = \begin{bmatrix} x \\ y \\ z \end{bmatrix} \end{cases}$

or $\begin{cases} x + y + z = 1 \\ \quad .2y + .5z = x \\ .5x \quad + .5z = y \\ .5x + .8y \quad = z \end{cases}$

or $\begin{cases} x + y + z = 1 \\ -x + .2y + .5z = 0 \\ .5x - y + .5z = 0 \\ .5x + .8y - z = 0 \end{cases}$

To simplify the arithmetic, multiply each of the last three equations of the system by 10 to eliminate the decimals. Then, apply the Gauss–Jordan elimination method.

$\begin{cases} x + y + z = 1 \\ -10x + 2y + 5z = 0 \\ 5x - 10y + 5z = 0 \\ 5x + 8y - 10z = 0 \end{cases}$ $\begin{cases} x + y + z = 1 \\ 12y + 15z = 10 \\ -15y = -5 \\ 3y - 15z = -5 \end{cases}$

Next, interchange the second and third equations, and pivot about $-15y$.

$\begin{cases} x + y + z = 1 \\ -15y = -5 \\ 12y + 15z = 10 \\ 3y - 15z = -5 \end{cases}$ $\begin{cases} x + z = \frac{2}{3} \\ y = \frac{1}{3} \\ 15z = 6 \\ -15z = -6 \end{cases}$

Pivoting about $15z$ yields $x = \frac{4}{15}$, $y = \frac{5}{15}$, $z = \frac{6}{15}$, so the stable matrix is

$$\begin{bmatrix} \frac{4}{15} & \frac{4}{15} & \frac{4}{15} \\ \frac{5}{15} & \frac{5}{15} & \frac{5}{15} \\ \frac{6}{15} & \frac{6}{15} & \frac{6}{15} \end{bmatrix}.$$

3. The regular stochastic matrix describing this daily transition is

$$\begin{array}{cc} & \begin{array}{cc} \text{M} & \text{N} \end{array} \\ \begin{array}{c} \text{M} \\ \text{N} \end{array} & \begin{bmatrix} .1 & .3 \\ .9 & .7 \end{bmatrix}. \end{array}$$

The stable distribution is found by solving

$$\begin{cases} x + y = 1 \\ \begin{bmatrix} .1 & .3 \\ .9 & .7 \end{bmatrix}\begin{bmatrix} x \\ y \end{bmatrix} = \begin{bmatrix} x \\ y \end{bmatrix} \end{cases}$$

or $\begin{cases} x + y = 1 \\ .1x + .3y = x \\ .9x + .7y = y \end{cases}$ or $\begin{cases} x + y = 1 \\ -.9x + .3y = 0 \\ .9x - .3y = 0. \end{cases}$

Since the last two equations are essentially the same, we need only solve the system consisting of the first two equations. Multiply the second equation by 10:

$\begin{cases} x + y = 1 \\ -9x + 3y = 0 \end{cases} \rightarrow \begin{cases} x + y = 1 \\ 12y = 9 \end{cases} \rightarrow \begin{cases} x = \frac{1}{4} \\ y = \frac{3}{4}. \end{cases}$

So the stable distribution is $\begin{bmatrix} \frac{1}{4} \\ \frac{3}{4} \end{bmatrix}$. The stable distribution tells us that, in the long run, 25% smoke menthol and 75% smoke nonmenthol cigarettes on any particular day.

8.3 Absorbing Stochastic Matrices

In this section, we study long-term trends for a certain class of matrices that are not regular—the absorbing stochastic matrices. By way of introduction, recall some general facts about stochastic matrices.

Stochastic matrices, such as

$$\begin{bmatrix} .3 & .5 & .1 & 0 \\ .2 & .2 & .8 & 0 \\ .1 & .3 & 0 & 0 \\ .4 & 0 & .1 & 1 \end{bmatrix},$$

describe state-to-state changes in certain processes. Each column of a stochastic matrix describes the transitions (or movements) from one specific state. For example, the first column of the preceding stochastic matrix indicates that, at the end of one time period, the probability is .3 that an object in state 1 stays in state 1, .2 that it goes to state 2, .1 that it goes to state 3, and .4 that it goes to state 4. Similarly, the second column indicates the probabilities for transitions from state 2, the third column from state 3, and the fourth from state 4.

If A is a stochastic matrix, then the columns of A^2 describe the transitions from the various states *after two time periods*. For instance, the third column of A^2 indicates the probabilities of an object starting out in state 3 and ending up in each of the states after two time periods. Similarly, the columns of A^n indicate the transitions from the various states after n time periods.

Consider the stochastic matrix just described. Its fourth column illustrates a curious phenomenon. It indicates that if an object starts out in state 4, after one time period the probabilities of going to states 1, 2, or 3 are 0 and the probability of going to state 4 is 1. In other words, all of the objects in state 4 stay in state 4. A state with this property is called an **absorbing state**. More precisely, an absorbing state is a state that is impossible to leave.

EXAMPLE 1 **Finding Absorbing States of a Stochastic Matrix** Find all absorbing states of the stochastic matrix

$$\begin{bmatrix} 1 & 0 & .3 & 0 \\ 0 & 1 & .1 & 0 \\ 0 & 0 & .5 & 1 \\ 0 & 0 & .1 & 0 \end{bmatrix}.$$

SOLUTION To determine which states are absorbing, one must look at the columns. The first column describes the transitions from state 1. It says that state 1 leads to state 1 100% of the time and to the other states 0% of the time. So state 1 is absorbing. Column 2 describes transitions from state 2. It says that state 2 leads to state 2 100% of the time. So state 2 is absorbing. Clearly, the third column says that state 3 is not absorbing. For example, state 3 leads to state 1 with probability .3. At first glance, column 4 seems to say that state 4 is absorbing. But it is not, because column 4 says that state 4 leads to state 3 100% of the time. **«**

On the basis of Example 1, we can easily determine the absorbing states of any stochastic matrix: First, the corresponding column has a single 1 and the remaining entries 0. Second, the lone 1 must be located on the main diagonal of the matrix. That is, its row and column number must be the same. So, for example, state i is an absorbing state if and only if the ith entry in the ith column is 1 and all of the remaining entries in that column are 0.

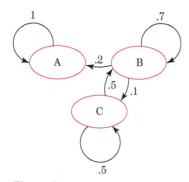

Figure 1

> **DEFINITION** An **absorbing stochastic matrix** is a stochastic matrix in which
>
> 1. there is at least one absorbing state; and
> 2. from any state, it is possible to get to at least one absorbing state, either directly or through one or more intermediate states.

Figure 1 shows the transition diagram for an absorbing stochastic matrix with absorbing state A. Notice that there are no arrows pointing from state A to another state. Also, there is a direct path from state B to state A and an indirect path from state C to state A. The transition diagram in Fig. 2 does not correspond to an absorbing stochastic matrix, since there is no way to get from state C or state D to the absorbing state A.

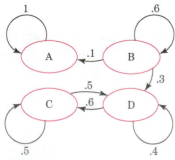

Figure 2

EXAMPLE 2 **Determining Whether a Stochastic Matrix Is Absorbing** Is the matrix

$$\begin{bmatrix} 1 & 0 & .3 & 0 \\ 0 & 1 & .1 & 0 \\ 0 & 0 & .5 & 1 \\ 0 & 0 & .1 & 0 \end{bmatrix}$$

an absorbing stochastic matrix?

SOLUTION The absorbing states are easy to recognize by looking at the matrix. In Example 1, we showed that states 1 and 2 are absorbing. The second part of the definition of an absorbing stochastic matrix is easiest to check by looking at the transition diagram corresponding to the Markov process. Figure 3 shows that there are two direct paths from state 3 to an absorbing state and two indirect paths from state 4 to an absorbing state. Therefore, the matrix is an absorbing stochastic matrix. **» Now Try Exercises 5 and 7**

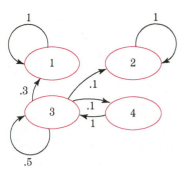

Figure 3

In general, processes described by stochastic matrices can oscillate indefinitely from state to state in such a way that they exhibit no long-term trend. An example was given in Section 2, with the matrix

$$\begin{bmatrix} 0 & 1 \\ 1 & 0 \end{bmatrix}.$$

Absorbing stochastic matrices do not oscillate indefinitely. The main result of this section is that absorbing stochastic matrices exhibit a long-term trend. Further, we can determine this trend by using a simple computational procedure.

When considering an absorbing stochastic matrix, we will always arrange the states so that the absorbing states come first, then the nonabsorbing states. This is called the **standard form** of the absorbing stochastic matrix.

Absorbing   Nonabsorbing

$$\left[\begin{array}{c|c} & \\ & \end{array} \right]$$

When an absorbing stochastic matrix is in standard form, it can be partitioned, or subdivided, into four submatrices.

Absorbing   Nonabsorbing

$$\left[\begin{array}{c|c} I & S \\ \hline 0 & R \end{array} \right]$$

The matrix I is an identity matrix, and 0 denotes a matrix having all entries 0. The matrices S and R are the two pieces corresponding to the nonabsorbing states. For example, in the case of the absorbing stochastic matrix of Example 2, this partition is given by

$$\left[\begin{array}{cc|cc} 1 & 0 & .3 & 0 \\ 0 & 1 & .1 & 0 \\ \hline 0 & 0 & .5 & 1 \\ 0 & 0 & .1 & 0 \end{array} \right].$$

If an absorbing stochastic matrix for a Markov process is not in standard form, you can use its transition diagram to convert it to a standard form.

EXAMPLE 3 **Converting an Absorbing Stochastic Matrix to Standard Form** Convert the following absorbing stochastic matrix to standard form, and identify the matrices S and R:

$$\begin{array}{c} \\ A \\ B \\ C \\ D \end{array} \begin{array}{c} \begin{array}{cccc} A & B & C & D \end{array} \\ \begin{bmatrix} .2 & 0 & .5 & 0 \\ .3 & 1 & 0 & 0 \\ .4 & 0 & .3 & 0 \\ .1 & 0 & .2 & 1 \end{bmatrix} \end{array}$$

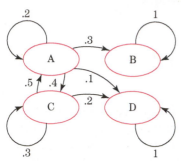

Figure 4

SOLUTION From the transition diagram in Fig. 4, we can confirm that the matrix is an absorbing stochastic matrix with absorbing states B and D. To obtain a stochastic matrix in standard form, label the columns and rows of a 4×4 matrix B, D, A, C, and then read off the probabilities from the transition diagram. We get

$$
\begin{array}{c} \\ B \\ D \\ A \\ C \end{array}
\begin{array}{cccc} B & D & A & C \end{array}
\left[\begin{array}{cc|cc} 1 & 0 & .3 & 0 \\ 0 & 1 & .1 & .2 \\ \hline 0 & 0 & .2 & .5 \\ 0 & 0 & .4 & .3 \end{array}\right].
$$

Therefore, $S = \begin{bmatrix} .3 & 0 \\ .1 & .2 \end{bmatrix}$ and $R = \begin{bmatrix} .2 & .5 \\ .4 & .3 \end{bmatrix}$.

» Now Try Exercise 9

EXAMPLE 4 **Progression of a Disease** The victims of a certain disease are classified into three states: cured, dead from the disease, or sick. Once a person is cured, they are permanently immune. Each year, 70% of those sick are cured, 10% die from the disease, and 20% remain ill.

(a) Determine a stochastic matrix describing the progression of the disease.
(b) Determine the absorbing states.

SOLUTION **(a)** There is one column for each state. Transferring the given data to the matrix gives the result

$$
\begin{array}{c} \\ \text{Cured} \\ \text{Dead} \\ \text{Sick} \end{array}
\begin{array}{ccc} \text{Cured} & \text{Dead} & \text{Sick} \end{array}
\left[\begin{array}{ccc} 1 & 0 & .7 \\ 0 & 1 & .1 \\ 0 & 0 & .2 \end{array}\right].
$$

(b) The absorbing states are "cured" and "dead," since these are the states that always lead back to themselves. The matrix is an absorbing stochastic matrix, since there is at least one absorbing state and the other state, "sick," can lead to an absorbing state. (In fact, in this case it leads to either of the absorbing states.)

» Now Try Exercise 25(a)

EXAMPLE 5 **Long-Term Trend for the Progression of a Disease** Find the long-term trend of the disease described in Example 4.

SOLUTION The stochastic matrix here is

$$
\left[\begin{array}{cc|c} 1 & 0 & .7 \\ 0 & 1 & .1 \\ \hline 0 & 0 & .2 \end{array}\right].
$$

Its first few powers are given by

```
NORMAL FLOAT AUTO REAL DEGREE MP

[A]²⁵⁵▸Frac
          [1 0 7/8]
          [0 1 1/8]
          [0 0  0 ]
```

$$
A^2 = \left[\begin{array}{cc|c} 1 & 0 & .84 \\ 0 & 1 & .12 \\ \hline 0 & 0 & .04 \end{array}\right] \quad A^3 = \left[\begin{array}{cc|c} 1 & 0 & .868 \\ 0 & 1 & .124 \\ \hline 0 & 0 & .008 \end{array}\right]
$$

$$
A^4 \approx \left[\begin{array}{cc|c} 1 & 0 & .874 \\ 0 & 1 & .125 \\ \hline 0 & 0 & .002 \end{array}\right] \quad A^5 \approx \left[\begin{array}{cc|c} 1 & 0 & .875 \\ 0 & 1 & .125 \\ \hline 0 & 0 & .000 \end{array}\right].
$$

It appears that the powers approach the matrix

$$A = \begin{bmatrix} 1 & 0 & \frac{7}{8} \\ 0 & 1 & \frac{1}{8} \\ 0 & 0 & 0 \end{bmatrix}.$$

Of those initially in state 3, "sick," the probability is seven-eighths of eventually being cured, and one-eighth of eventually dying of the disease.   **≪**

Example 5 exhibits three important features of the long-term behavior exhibited by all absorbing stochastic matrices. First, as in the case of regular stochastic matrices, the powers approach a particular matrix. This limiting matrix is called the **stable matrix**. Second, for absorbing stochastic matrices, the long-term trend depends on the initial state. For example, the stable matrix just computed gives different results, depending on the state in which you start. This is reflected by the fact that the three columns are different. (In the case of regular stochastic matrices, the long-term trend does not depend on the initial distribution. All columns of the stable matrix are the same. They all equal the stable distribution.) The third important point to notice is that, no matter what the initial state, all objects eventually go to absorbing states. The absorbing states act like magnets and attract all objects to themselves in the long run.

In Example 5, we computed the stable matrix by raising the given stochastic matrix to various powers. Actually, there is a formal computational procedure for determining the stable matrix. Suppose that we partition an absorbing stochastic matrix into submatrices such that

$$A = \left[\begin{array}{c|c} I & S \\ \hline 0 & R \end{array} \right].$$

> **Stable Matrix**   The stable matrix of A is
>
> $$\left[\begin{array}{c|c} I & S(I - R)^{-1} \\ \hline 0 & 0 \end{array} \right].$$
>
> *Note:* The identity matrix I in $(I - R)^{-1}$ is chosen to be the same size as R in order to make the matrix subtraction permissible.

EXAMPLE 6   **Determining Stable Matrices** Use the preceding formula to determine the stable matrix of

$$\begin{bmatrix} 1 & 0 & .7 \\ 0 & 1 & .1 \\ 0 & 0 & .2 \end{bmatrix}.$$

SOLUTION $S = \begin{bmatrix} .7 \\ .1 \end{bmatrix}$, $R = [.2]$, $I - R = [1] - [.2] = [.8]$, $(I - R)^{-1} = [1/.8]$. Therefore,

$$S(I - R)^{-1} = \begin{bmatrix} .7 \\ .1 \end{bmatrix}[1/.8] = \begin{bmatrix} .7/.8 \\ .1/.8 \end{bmatrix} = \begin{bmatrix} \frac{7}{8} \\ \frac{1}{8} \end{bmatrix}.$$

So the stable matrix is

$$\left[\begin{array}{c|c} I & S(I - R)^{-1} \\ \hline 0 & 0 \end{array} \right] = \begin{bmatrix} 1 & 0 & \frac{7}{8} \\ 0 & 1 & \frac{1}{8} \\ 0 & 0 & 0 \end{bmatrix}.$$

≫ Now Try Exercise 13 (stable part only)

EXAMPLE 7   **Gambler's Ruin** Consider a game of chance with the following characteristics: A person repeatedly bets $1 each play. If he wins, he receives $1. That is, he receives his bet of $1 plus winnings of $1. If he goes broke, he stops playing. Also, if he accumulates $3, he stops playing. On each play, the probability of winning is .4 and of losing .6. What is the probability of eventually accumulating $3 if he starts with $1? With $2?

SOLUTION   There are four states, corresponding to having $0, $3, $1, or $2. The first two are absorbing states. The stochastic matrix is

$$
\begin{array}{c}
\\ \$0 \\ \$3 \\ \$1 \\ \$2
\end{array}
\begin{array}{cccc}
\$0 & \$3 & \$1 & \$2
\end{array}
\left[
\begin{array}{cc|cc}
1 & 0 & .6 & 0 \\
0 & 1 & 0 & .4 \\
\hline
0 & 0 & 0 & .6 \\
0 & 0 & .4 & 0
\end{array}
\right].
$$

The third column is derived in this way: If he has $1, there is a .6 probability of losing $1, which would mean going to state $0. Thus, the first entry in the third column is .6. There is no way to get from $1 to $3 or from $1 to $1 after one play. So the second and third entries are 0. There is a .4 probability of winning $1—that is, of going from $1 to $2. So the last entry is .4. The fourth column is derived similarly.

In this example,

$$
S = \begin{bmatrix} .6 & 0 \\ 0 & .4 \end{bmatrix} \qquad R = \begin{bmatrix} 0 & .6 \\ .4 & 0 \end{bmatrix}.
$$

To compute $S(I - R)^{-1}$, observe that

$$
I - R = \begin{bmatrix} 1 & 0 \\ 0 & 1 \end{bmatrix} - \begin{bmatrix} 0 & .6 \\ .4 & 0 \end{bmatrix} = \begin{bmatrix} 1 & -.6 \\ -.4 & 1 \end{bmatrix}.
$$

To compute $(I - R)^{-1}$, recall that

$$
\begin{bmatrix} a & b \\ c & d \end{bmatrix}^{-1} = \begin{bmatrix} d/D & -b/D \\ -c/D & a/D \end{bmatrix}, \qquad D = ad - bc \neq 0.
$$

So, in this example, $D = 1 \cdot 1 - (-.6)(-.4) = 1 - .24 = .76$ and

$$
(I - R)^{-1} = \begin{bmatrix} 1 & -.6 \\ -.4 & 1 \end{bmatrix}^{-1} \approx \begin{bmatrix} 1.32 & .79 \\ .53 & 1.32 \end{bmatrix}
$$

$$
S(I - R)^{-1} \approx \begin{bmatrix} .6 & 0 \\ 0 & .4 \end{bmatrix}\begin{bmatrix} 1.32 & .79 \\ .53 & 1.32 \end{bmatrix} \approx \begin{bmatrix} .79 & .47 \\ .21 & .53 \end{bmatrix}.
$$

Thus, the stable matrix is

$$
\left[\begin{array}{c|c} I & S(I - R)^{-1} \\ \hline 0 & 0 \end{array} \right] = \left[\begin{array}{cc|cc} 1 & 0 & .79 & .47 \\ 0 & 1 & .21 & .53 \\ \hline 0 & 0 & 0 & 0 \\ 0 & 0 & 0 & 0 \end{array} \right].
$$

We are interested in the probability that the gambler ends up with $3. The percentage is different for each of the two starting amounts $1 and $2. Recall the meanings of the rows and columns:

$$
\begin{array}{c}
\\ \$0 \\ \$3 \\ \$1 \\ \$2
\end{array}
\begin{array}{cccc}
\$0 & \$3 & \$1 & \$2
\end{array}
\left[
\begin{array}{cc|cc}
1 & 0 & .79 & .47 \\
0 & 1 & .21 & .53 \\
\hline
0 & 0 & 0 & 0 \\
0 & 0 & 0 & 0
\end{array}
\right]
$$

Looking at the $1 column, we see that, if the gambler starts with $1, the probability is .21 that he ends up with $3. Looking at the $2 column, the probability is .53 that he ends up with $3.

» Now Try Exercise 29(a)

The Fundamental Matrix

The matrix $(I - R)^{-1}$ used to compute the stable matrix is called the **fundamental matrix** and is denoted by the letter F. Its columns and rows should be labeled with the nonabsorbing states. The fundamental matrix directly provides certain useful probabilities for the process. Specifically, the ijth entry of F is the expected number of times the process will end in nonabsorbing state i if it starts in nonabsorbing state j. The sum of the entries of the jth column of F is the expected number of steps before absorption when the process begins in nonabsorbing state j.

The fundamental matrix for the gambler's ruin example is

$$F = \begin{array}{c} \\ \$1 \\ \$2 \end{array} \begin{array}{c} \$1 \quad\; \$2 \\ \begin{bmatrix} 1.32 & .79 \\ .53 & 1.32 \end{bmatrix} \end{array}.$$

The first column of F indicates that, when the gambler begins with $1, he can expect to play $1.32 + .53 = 1.85$ times before quitting. He should have $1 for an expected number of 1.32 plays and have $2 for an expected number of .53 plays. The second column gives similar information when the gambler begins with $2.

INCORPORATING TECHNOLOGY

A graphing calculator can be used to obtain the stable matrix by raising the original stochastic matrix to a high power or by calculating $S(I - R)^{-1}$. In Figs. 5 and 6, **[A]**, **[B]**, and **[C]** are the matrices A, S, and R, respectively, of Example 7. In Fig. 6, the **MODE FLOAT** option was set to 2.

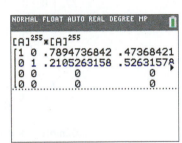

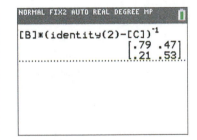

Figure 5   **Figure 6**

Successive powers of an absorbing stochastic matrix can be generated on a spreadsheet in much the same way that successive distribution matrices were generated in Fig. 5 of Section 8.1. The upper-right quadrant of the stable matrix can be obtained with the formula $= \text{MMULT(S,MINVERSE(I−R))}$.

☀ WolframAlpha The stable matrix of Example 7 can be obtained by raising the original stochastic matrix to a high power or by calculating $S*(\textbf{inverse of} (\textbf{identityMatrix(2)} − R))$—that is,

$$\{\{.6,0\},\{0,.4\}\}*\textbf{inverse of } (\textbf{identityMatrix(2)} − \{\{0,.6\},\{.4,0\}\}).$$

Check Your Understanding 8.3

Solutions can be found following the section exercises.

1. When an absorbing stochastic matrix is partitioned, the submatrix in the upper left is an identity matrix, denoted by I. Also, when finding the stable matrix, we subtract the submatrix R from an identity matrix, also denoted by I. Are these two identity matrices the same size?

2. Let A be an absorbing stochastic matrix. Interpret the entries of A^2.

3. Is $\begin{bmatrix} 1 & .4 & 0 \\ 0 & .2 & .1 \\ 0 & .4 & .9 \end{bmatrix}$ an absorbing stochastic matrix?

EXERCISES 8.3

In Exercises 1–4, determine whether the transition diagram corresponds to an absorbing stochastic matrix.

1.

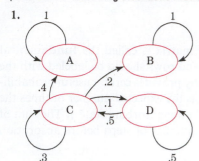

2.

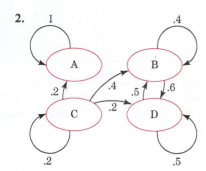

3.

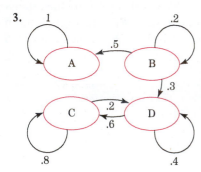

4.

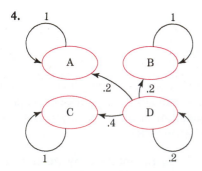

In Exercises 5–8, determine whether the given matrix is an absorbing stochastic matrix.

5. $\begin{bmatrix} 1 & 0 & 0 & 0 \\ 0 & 1 & 0 & 0 \\ 0 & 0 & .8 & .1 \\ 0 & 0 & .2 & .9 \end{bmatrix}$

6. $\begin{bmatrix} 1 & 0 & 0 & .3 \\ 0 & 1 & 0 & .2 \\ 0 & 0 & 1 & .2 \\ 0 & 0 & 0 & .3 \end{bmatrix}$

7. $\begin{bmatrix} 1 & 0 & .4 \\ 0 & .5 & .3 \\ 0 & .5 & .3 \end{bmatrix}$

8. $\begin{bmatrix} 1 & 0 & 0 & 0 \\ 0 & 1 & 0 & 0 \\ 0 & 0 & 0 & 1 \\ 0 & 0 & 1 & 0 \end{bmatrix}$

In Exercises 9–12, convert the absorbing stochastic matrix to standard form.

9. $\begin{array}{c} \\ A \\ B \\ C \end{array} \begin{array}{ccc} A & B & C \\ \begin{bmatrix} .2 & 0 & .5 \\ .3 & 1 & .4 \\ .5 & 0 & .1 \end{bmatrix} \end{array}$

10. $\begin{array}{c} \\ A \\ B \\ C \end{array} \begin{array}{ccc} A & B & C \\ \begin{bmatrix} 1 & .2 & 0 \\ 0 & .3 & 0 \\ 0 & .5 & 1 \end{bmatrix} \end{array}$

11. $\begin{array}{c} \\ A \\ B \\ C \\ D \end{array} \begin{array}{cccc} A & B & C & D \\ \begin{bmatrix} .1 & 1 & .6 & 0 \\ .2 & 0 & .1 & 0 \\ .3 & 0 & .2 & 0 \\ .4 & 0 & .1 & 1 \end{bmatrix} \end{array}$

12. $\begin{array}{c} \\ A \\ B \\ C \\ D \end{array} \begin{array}{cccc} A & B & C & D \\ \begin{bmatrix} 0 & .7 & 0 & 0 \\ 0 & 0 & 0 & .6 \\ 0 & .3 & 1 & .4 \\ 1 & 0 & 0 & 0 \end{bmatrix} \end{array}$

The matrices in Exercises 13–18 are absorbing stochastic matrices in standard form. In each, identify R and S and compute the fundamental matrix and the stable matrix.

13. $\begin{bmatrix} 1 & 0 & .3 \\ 0 & 1 & .2 \\ 0 & 0 & .5 \end{bmatrix}$

14. $\begin{bmatrix} 1 & 0 & \frac{1}{2} \\ 0 & 1 & \frac{1}{6} \\ 0 & 0 & \frac{1}{3} \end{bmatrix}$

15. $\begin{bmatrix} 1 & 0 & \frac{1}{4} & \frac{1}{6} \\ 0 & 1 & \frac{1}{6} & 0 \\ 0 & 0 & \frac{1}{4} & \frac{1}{2} \\ 0 & 0 & \frac{1}{3} & \frac{1}{3} \end{bmatrix}$

16. $\begin{bmatrix} 1 & 0 & .1 & 0 \\ 0 & 1 & .5 & .2 \\ 0 & 0 & .3 & .6 \\ 0 & 0 & .1 & .2 \end{bmatrix}$

17. $\begin{bmatrix} 1 & 0 & 0 & .1 & .2 \\ 0 & 1 & 0 & .3 & 0 \\ 0 & 0 & 1 & 0 & .2 \\ 0 & 0 & 0 & .5 & 0 \\ 0 & 0 & 0 & .1 & .6 \end{bmatrix}$

18. $\begin{bmatrix} 1 & 0 & \frac{1}{4} & 0 & \frac{1}{3} \\ 0 & 1 & \frac{1}{4} & 0 & 0 \\ 0 & 0 & 0 & \frac{1}{3} & \frac{1}{6} \\ 0 & 0 & \frac{1}{2} & \frac{1}{3} & 0 \\ 0 & 0 & 0 & \frac{1}{3} & \frac{1}{2} \end{bmatrix}$

Note: $\begin{bmatrix} 1 & -\frac{1}{3} & -\frac{1}{6} \\ -\frac{1}{2} & \frac{2}{3} & 0 \\ 0 & -\frac{1}{3} & \frac{1}{2} \end{bmatrix}^{-1} = \begin{bmatrix} \frac{3}{2} & 1 & \frac{1}{2} \\ \frac{9}{8} & \frac{9}{4} & \frac{3}{8} \\ \frac{3}{4} & \frac{3}{2} & \frac{9}{4} \end{bmatrix}$.

Gambler's Ruin Exercises 19 and 20 refer to Example 7.

19. Interpret the entry .79 in the fundamental matrix.

20. If the gambler begins with $2, what is the expected number of times that he will play before quitting?

21. Job Mobility The lawyers at a law firm are either associates or partners. At the end of each year, 30% of the associates leave the firm, 20% are promoted to partner, and 50% remain associates. Also, 10% of the partners leave the firm at the end of each year. Assume that a lawyer who leaves the firm does not return.

   (a) Draw the transition diagram for this Markov process. Label the states A, P, and L.
   (b) Set up an absorbing stochastic matrix for the Markov process.
   (c) Find the stable matrix.
   (d) What is the expected number of years that an associate will be in the firm before leaving?
   (e) In the long run, what percent of the lawyers will be associates?

22. Broadband Internet Service Colleges have been rapidly making broadband Internet service available in their residence halls. Of the colleges that offer no broadband Internet service, each year 10% introduce FIOS Internet service, 30% introduce cable Internet service, and 60% continue to offer no broadband Internet service. Once a type of broadband Internet service is established, the type of service is never changed.

(a) Draw the transition diagram for the Markov process.

(b) Set up an absorbing stochastic matrix for the Markov process.

(c) Find the stable matrix.

(d) In the long run, what percent of colleges will provide cable Internet service?

(e) What is the expected number of years required for a college to set up a broadband service if it currently does not provide broadband Internet service?

23. **Mouse in a Maze** A mouse is placed in one of the compartments of the maze shown in Fig. 7. After each minute of looking unsuccessfully for food, the mouse exits through one of the doors at random and moves to an adjacent compartment. The Exit door is one way; that is, the mouse cannot return after it has exited the maze.

Figure 7

(a) Draw the transition diagram for the Markov process.

(b) Set up an absorbing stochastic matrix for the Markov process.

(c) Find the stable matrix.

(d) If the mouse begins in compartment A, what is the expected amount of time that it spends in the maze?

$$Note: \begin{bmatrix} 1 & -\frac{1}{2} & -\frac{1}{2} \\ -\frac{2}{3} & 1 & -\frac{1}{2} \\ -\frac{1}{3} & -\frac{1}{4} & 1 \end{bmatrix}^{-1} = \begin{bmatrix} 4.2 & 3 & 3.6 \\ 4 & 4 & 4 \\ 2.4 & 2 & 3.2 \end{bmatrix}.$$

24. **A Card Game** Heather and Blake play a card game in which they take turns drawing a card from a shuffled deck of 52 cards. Heather wins the game if she draws a heart, and Blake wins the game if he draws a black card. When a player doesn't win on his or her turn, the card is returned to the deck, the deck is reshuffled, and it becomes the other player's turn. The game has four states: Heather Wins, Blake Wins, Heather's Turn, Blake's Turn.

(a) Draw the transition diagram for the Markov process.

(b) Set up an absorbing stochastic matrix for the Markov process.

(c) Find the stable matrix.

(d) What is the probability that Heather wins if she goes first?

(e) What is the expected number of turns if Heather goes first?

25. **Class Standings** Suppose that the following data were obtained from the records of a certain two-year college: Of those who were freshmen (F) during a particular year, 80% became sophomores (S) the next year and 20% dropped out (D). Of those who were sophomores during a particular year, 90% graduated (G) by the next year and 10% dropped out.

(a) Set up the absorbing stochastic matrix with states D, G, F, S that describes this transition.

(b) Find the stable matrix.

(c) Determine the probability that an entering freshman will eventually graduate.

(d) Determine the expected number of years that a student entering as a freshman will attend the college before either dropping out or graduating.

26. **Quality Control** A manufacturer of precise measuring devices carefully tests each device manufactured. Seventy percent of all newly manufactured devices are approved, 20% are sent back to be recalibrated, and 10% are determined to be beyond repair and are destroyed. Eighty percent of the devices that have been recalibrated are approved, 10% are sent back to be recalibrated again, and 10% are destroyed. Therefore, the four states of a devices are New, Approved, Recalibrated, and Destroyed.

(a) Set up the 4×4 absorbing stochastic matrix with columns and rows labeled, A, D, N, and R that describes this transition.

(b) Find the stable matrix.

(c) Determine the probability that a New device will eventually be Approved.

(d) Determine the expected number of times a New device will be tested.

27. **Accounts Payable** A retailer classifies accounts as having one of four possible states: "paid up," "overdue at most 30 days," "overdue less than 60 days but more than 30 days," and "bad." If no payment is made on an overdue account by the end of the month, the status moves to the next state. When a partial payment is made on an overdue account, its improved status depends on the size of the payment. Experience shows that the following matrix describes the changes in status of accounts:

	Paid	≤ 30	< 60	Bad
Paid	1	.4	.1	0
≤ 30	0	.4	.4	0
< 60	0	.2	.4	0
Bad	0	0	.1	1

(a) What is the probability of an account eventually being paid off if it is currently overdue at most 30 days? Less than 60 days?

(b) If an account is currently overdue at most 30 days, what is the expected number of months until it is either paid or bad?

(c) If the company has $2000 in bills in the "≤ 30 day" category and $5000 in bills in the "< 60 day" category, how much can the retailer expect will eventually be paid up and how much will eventually be irretrievable bad debt?

28. **Job Mobility** The managers in a company are classified as top managers, middle managers, and first-line managers. Each year, 10% of top managers retire, 10% leave the company, 60% remain top managers, and 20% are demoted to middle managers. Each year, 5% of middle managers retire, 15% leave the company, 10% are promoted to top managers, 60% remain middle managers, and 10% are demoted to first-line managers. Each year, 5% of first-line managers retire, 25% leave the company, 10% are promoted to middle managers, and 60% remain first-line managers.

(a) What is the probability of a top manager eventually retiring? A middle manager? A first-line manager?

(b) If a person is currently a middle manager, what is the expected number of years that they will be with the company before either leaving or retiring?

Note: $\begin{bmatrix} .4 & -.1 & 0 \\ -.2 & .4 & -.1 \\ 0 & -.1 & 4 \end{bmatrix}^{-1} = \begin{bmatrix} \frac{75}{26} & \frac{10}{13} & \frac{5}{26} \\ \frac{20}{13} & \frac{40}{13} & \frac{10}{13} \\ \frac{5}{13} & \frac{10}{13} & \frac{35}{13} \end{bmatrix}$

29. **Gambler's Ruin** As a variation of the gambler's ruin problem, suppose that on each play, the probability of winning is $\frac{1}{2}$ and the gambler stops playing if he accumulates \$4 or goes broke.

(a) What is the probability of his eventually going broke if he starts with \$1? \$2? \$3?

(b) If the gambler begins with \$2, what is the expected number of times that he will play before quitting?

Note: $\begin{bmatrix} 1 & -\frac{1}{2} & 0 \\ -\frac{1}{2} & 1 & -\frac{1}{2} \\ 0 & -\frac{1}{2} & 1 \end{bmatrix}^{-1} = \begin{bmatrix} \frac{3}{2} & 1 & \frac{1}{2} \\ 1 & 2 & 1 \\ \frac{1}{2} & 1 & \frac{3}{2} \end{bmatrix}$.

30. **Random Walk** The concept of a random walk can be applied to a wide variety of fields such as statistical physics, finance, biology, and psychology. The following problem is a simplified version of a random walk: A particle moves on a line and at any time is located at one of the integers 0, 1, 2, 3, 4, 5. If the particle arrives at 0 or 5, it stays there. Otherwise, after each second, it moves to the right with probability $\frac{1}{2}$ or to the left with the same probability.

(a) Set up the absorbing stochastic matrix, with columns and rows labeled 0, 5, 1, 2, 3, and 4, that describes the transitions.

(b) Find the stable matrix.

(c) What is the probability that a particle beginning at location 3 will eventually be absorbed at location 0?

(d) For a particle beginning at location 4, determine the expected number of times that it will be at location 4 before it is absorbed.

Note: $\begin{bmatrix} 1 & -\frac{1}{2} & 0 & 0 \\ -\frac{1}{2} & 1 & -\frac{1}{2} & 0 \\ 0 & -\frac{1}{2} & 1 & -\frac{1}{2} \\ 0 & 0 & -\frac{1}{2} & 1 \end{bmatrix}^{-1} = \begin{bmatrix} 1.6 & 1.2 & .8 & .4 \\ 1.2 & 2.4 & 1.6 & .8 \\ .8 & 1.6 & 2.4 & 1.2 \\ .4 & .8 & 1.2 & 1.6 \end{bmatrix}$

TECHNOLOGY EXERCISES

31. **Collecting Quotations** A soft drink manufacturer puts one of three different quotations on the inside of each bottle cap. Assuming that each of the quotations is equally likely to appear, what is the probability of receiving all three after purchasing five of the soft drinks? What is the expected number of soft drinks that you have to purchase to have all three quotations?

32. **Tennis** Consider a game of tennis between player A and player B that has reached the state deuce (or 40–40). Continuing from this point, after each rally, the game will be in one of the five states—A wins, B wins, Deuce, Advantage A, or Advantage B. Suppose that the stochastic matrix for the remainder of the game is

	A wins	B wins	D	Ad A	Ad B
A wins	1	0	0	.4	0
B wins	0	1	0	0	.6
Deuce	0	0	0	.6	.4
Ad A	0	0	.4	0	0
Ad B	0	0	.6	0	0

(a) What is the probability that player A will win the game? Player B?

(b) What is the expected number of rallies?

33. Consider the absorbing stochastic matrix A, where

$$A = \begin{bmatrix} 1 & 0 & 0 & 0 & .6 \\ 0 & 1 & 0 & .5 & .1 \\ 0 & 0 & 1 & 0 & .1 \\ 0 & 0 & 0 & .2 & .2 \\ 0 & 0 & 0 & .3 & 0 \end{bmatrix}.$$

Approximate the stable matrix of A by raising A to a high power. Then find the exact stable distribution by calculating $S(I - R)^{-1}$.

34. Repeat Exercise 33 for the matrix

$$\begin{bmatrix} 1 & 0 & .5 & .4 & .2 \\ 0 & 1 & 0 & .3 & .7 \\ 0 & 0 & 0 & .2 & .1 \\ 0 & 0 & .3 & 0 & 0 \\ 0 & 0 & .2 & .1 & 0 \end{bmatrix}.$$

Solutions to Check Your Understanding 8.3

1. Sometimes, they are, but in general, they are not. The size of the first identity matrix equals the number of absorbing states, whereas the size of the second identity matrix is the same as that of the matrix R, which equals the number of nonabsorbing states.

2. Consider the entry in the ith row, jth column of A^2. Suppose that an object begins in state j. Then the entry gives the probability that it ends up in state i after two periods.

3. Yes. (i) Its first state is absorbing. (ii) From each state, it is possible to get to the first state. Note that it is possible to go directly from state 2 to state 1, since $a_{12} = .4 \neq 0$. Since $a_{13} = 0$, it is not possible to go directly from state 3 to state 1. However, this can be accomplished indirectly by going from state 3 to state 2 (possible since $a_{23} = .1 \neq 0$) and then from state 2 to state 1.

CHAPTER 8 Summary

KEY TERMS AND CONCEPTS	EXAMPLES

8.1 The Transition Matrix

A **stochastic matrix** is a square matrix such that:

1. Each entry is between 0 and 1.

2. Sum of the entries in each column is 1.

The rows and columns correspond to states, and the ij^{th} entry is $\Pr(\text{moving to state } i \mid \text{in state } j)$.

The nth **distribution matrix** is a column matrix that gives the percentage of objects in each state after n time periods.

If A is a stochastic matrix and D_n is the nth distribution matrix, then $A \cdot D_n = D_{n+1}$ and $A^n \cdot D_0 = D_n$. The ij^{th} entry of A^n is $\Pr(\text{moving to state } i \text{ after } n \text{ time periods} \mid \text{in state } j)$.

$A = \begin{bmatrix} .8 & .3 \\ .2 & .7 \end{bmatrix}$ is a stochastic matrix.

$\begin{bmatrix} .4 \\ .6 \end{bmatrix}_0 , \begin{bmatrix} .5 \\ .5 \end{bmatrix}_1 , \begin{bmatrix} .55 \\ .45 \end{bmatrix}_2 , \dots$ are distribution matrices.

$\begin{bmatrix} .8 & .3 \\ .2 & .7 \end{bmatrix} \begin{bmatrix} .4 \\ .6 \end{bmatrix}_0 = \begin{bmatrix} .5 \\ .5 \end{bmatrix}_1 , \begin{bmatrix} .8 & .3 \\ .2 & .7 \end{bmatrix} \begin{bmatrix} .5 \\ .5 \end{bmatrix}_1 = \begin{bmatrix} .55 \\ .45 \end{bmatrix}_2 , \dots$

$\begin{bmatrix} .8 & .3 \\ .2 & .7 \end{bmatrix}^2 \begin{bmatrix} .4 \\ .6 \end{bmatrix}_0 = \begin{bmatrix} .55 \\ .45 \end{bmatrix}_2 .$

8.2 Regular Stochastic Matrices

A stochastic matrix is called **regular** if some power of the matrix has only positive entries.

If A is a regular stochastic matrix, then A^n approaches a certain matrix (called the **stable matrix** of A) as n gets large, and the distribution matrices D_n approach a certain column matrix called the **stable distribution** of A. Each column of the stable matrix holds the stable distribution.

The stable distribution can be found by solving

$$\begin{cases} \text{sum of entries of } X = 1 \\ AX = X \end{cases} \text{ for } X.$$

$A = \begin{bmatrix} .8 & .3 \\ .2 & .7 \end{bmatrix}$ is a regular stochastic matrix.

As n gets large, $\begin{bmatrix} .8 & .3 \\ .2 & .7 \end{bmatrix}^n \rightarrow \begin{bmatrix} .6 & .6 \\ .4 & .4 \end{bmatrix}$ and $D_n \rightarrow \begin{bmatrix} .6 \\ .4 \end{bmatrix}$.

The solution of $\begin{cases} x + y = 1 \\ \begin{bmatrix} .8 & .3 \\ .2 & .7 \end{bmatrix} \begin{bmatrix} x \\ y \end{bmatrix} = \begin{bmatrix} x \\ y \end{bmatrix} \end{cases}$ is $\begin{bmatrix} .6 \\ .4 \end{bmatrix}$.

8.3 Absorbing Stochastic Matrices

An **absorbing state** is a state for which the probability of a transition back to itself is 1.

An **absorbing stochastic matrix** has at least one absorbing state, and it is possible to reach some absorbing state from any state.

An absorbing stochastic matrix of the form $\left[\begin{array}{c|c} I & S \\ \hline 0 & R \end{array} \right]$ has the stable matrix $\left[\begin{array}{c|c} I & S(I-R)^{-1} \\ \hline 0 & 0 \end{array} \right]$.

$\left[\begin{array}{cc|c} 1 & 0 & .7 \\ 0 & 1 & .1 \\ \hline 0 & 0 & .2 \end{array} \right]$ is an absorbing stochastic matrix. The states corresponding to the first two columns are absorbing states.

For the preceding matrix,

$S(I-R)^{-1} = \begin{bmatrix} .7 \\ .1 \end{bmatrix} ([1] - [.2])^{-1} = \begin{bmatrix} .7 \\ .1 \end{bmatrix} [\tfrac{1}{.8}] = \begin{bmatrix} \tfrac{7}{8} \\ \tfrac{1}{8} \end{bmatrix}.$

Therefore, the stable matrix is $\left[\begin{array}{cc|c} 1 & 0 & \tfrac{7}{8} \\ 0 & 1 & \tfrac{1}{8} \\ \hline 0 & 0 & 0 \end{array} \right].$

The **fundamental matrix** is the matrix $(I - R)^{-1}$. When its columns and rows are labeled with the nonabsorbing states, its ij^{th} entry is the expected number of times that the process will be in nonabsorbing state i, given that it started in nonabsorbing state j.

The fundamental matrix is $[\tfrac{1}{.8}]$ or $[1.25]$. This means that an object beginning in the nonabsorbing state corresponding to the third column of the original matrix will be in that state an average of 1.25 times.

CHAPTER 8 Fundamental Concept Check Exercises

1. What is a Markov process?

2. What is a transition matrix? A stochastic matrix? A distribution matrix?

3. What is A^n? Give an interpretation of the entries of A^n.

4. How is the nth distribution matrix calculated from the initial distribution matrix?

5. Define *regular* stochastic matrix.

6. Define the stable matrix and the stable distribution of a regular stochastic matrix.

7. Explain how to find the stable distribution of a regular stochastic matrix.

8. What is meant by an absorbing state of a stochastic matrix?

9. What is an absorbing stochastic matrix?

10. Explain how to find the stable matrix of an absorbing stochastic matrix.

11. What is the fundamental matrix of an absorbing stochastic matrix, and how is it used?

CHAPTER 8 Review Exercises

In Exercises 1–6, determine whether or not the given matrix is stochastic. If so, determine if it is regular, absorbing, or neither.

1. $\begin{bmatrix} 1 & .3 & 0 & 0 \\ 0 & .1 & 0 & 0 \\ 0 & .4 & .7 & .4 \\ 0 & .2 & .3 & .6 \end{bmatrix}$

2. $\begin{bmatrix} .1 & .1 & .1 & .1 \\ .2 & .2 & .2 & .2 \\ .3 & .3 & .3 & .3 \\ .4 & .4 & .4 & .4 \end{bmatrix}$

3. $\begin{bmatrix} 0 & .3 \\ 1 & .7 \end{bmatrix}$

4. $\begin{bmatrix} 1 & 0 & 0 \\ 0 & 1 & \frac{1}{3} \\ 0 & 0 & \frac{2}{3} \end{bmatrix}$

5. $\begin{bmatrix} 1 & \frac{1}{2} & 0 \\ 0 & \frac{1}{2} & 0 \\ 0 & \frac{1}{2} & 1 \end{bmatrix}$

6. $\begin{bmatrix} 1 & 0 & 0 & .3 \\ 0 & 1 & 0 & .3 \\ 0 & 0 & .5 & .3 \\ 0 & 0 & .5 & .1 \end{bmatrix}$

7. Find the stable distribution for the regular stochastic matrix
$$\begin{bmatrix} .6 & .5 \\ .4 & .5 \end{bmatrix}.$$

8. Find the stable matrix for the absorbing stochastic matrix
$$\begin{bmatrix} 1 & 0 & 0 & \frac{1}{8} & \frac{1}{4} \\ 0 & 1 & 0 & \frac{1}{8} & 0 \\ 0 & 0 & 1 & 0 & \frac{1}{4} \\ 0 & 0 & 0 & \frac{1}{4} & \frac{1}{2} \\ 0 & 0 & 0 & \frac{1}{2} & 0 \end{bmatrix}.$$

9. **Economic Mobility** In a certain community, currently, 10% of the people are H (high income), 60% are M (medium income), and 30% are L (low income). Studies show that for the children of H parents, 50% also become H, 40% become M, and 10% become L. Of the children of M parents, 40% become H, 30% become M, and 30% become L. Of the children of L parents, 30% become H, 50% become M, and 20% become L.
(a) Set up the 3×3 stochastic matrix that describes this situation.
(b) What percent of the children of the current generation will have high incomes?
(c) In the long run, what percentage of the population will have low incomes?

10. **Quality Control** In a certain factory, some machines are properly adjusted and some need adjusting. Technicians randomly inspect machines and make adjustments. Suppose that, of the machines that are properly adjusted on a particular day, 80% will also be properly adjusted the following day and 20% will need adjusting. Also, of the machines that need adjusting on a particular day, 30% will be properly adjusted the next day and 70% will still need adjusting.
(a) Set up the 2×2 stochastic matrix, with columns labeled P (properly adjusted) and N (need adjusting), that describes this situation.
(b) If, initially, all the machines are properly adjusted, what percent will need adjusting after 2 days?
(c) In the long run, what percent will be properly adjusted each day?

11. Find the stable matrix for the absorbing stochastic matrix
$$\begin{bmatrix} 1 & 0 & \frac{1}{6} & \frac{1}{2} & \frac{2}{5} \\ 0 & 1 & 0 & 0 & \frac{2}{5} \\ 0 & 0 & 0 & 0 & 0 \\ 0 & 0 & \frac{2}{3} & \frac{1}{2} & 0 \\ 0 & 0 & \frac{1}{6} & 0 & \frac{1}{5} \end{bmatrix}.$$

12. **Mouse in a House** Figure 1 gives the layout of a house with four rooms connected by doors. Room I contains a mousetrap, and room II contains cheese. A mouse, after being placed in one of the rooms, will search for cheese; if unsuccessful after one minute, it will exit to another room by selecting one of the doors at random. (For instance, if the mouse is in room III, after one minute, it will go to room II with probability $\frac{1}{3}$ and to room IV with probability $\frac{2}{3}$.) A mouse entering room I will be trapped and therefore no longer move. Also, a mouse entering room II will remain in that room.

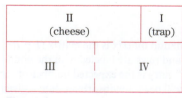

Figure 1

(a) Set up the 4×4 absorbing stochastic matrix that describes this situation.

(b) If a mouse begins in room IV, what is the probability that it will find the cheese after 2 minutes?

(c) If a mouse begins in room IV, what is the probability that it will find the cheese in the long run?

(d) For a mouse beginning in room III, determine the expected number of minutes that will elapse before the mouse either finds the cheese or is trapped.

13. Which of the following is the stable distribution for the

$$\text{regular stochastic matrix } \begin{bmatrix} .4 & .4 & .2 \\ .1 & .1 & .3 \\ .5 & .5 & .5 \end{bmatrix}?$$

(a) $\begin{bmatrix} .6 \\ .4 \\ 1 \end{bmatrix}$ (b) $\begin{bmatrix} .2 \\ .3 \\ .5 \end{bmatrix}$ (c) $\begin{bmatrix} .3 \\ .2 \\ .5 \end{bmatrix}$

14. Listening Preferences A city has two competing news stations. From a survey of regular listeners, it was determined that, of those who listen to station A on a particular day, 90% listen to station A the next day and 10% listen to station B. Of those who listen to station B on a particular day, 20% listen to station A the next day and 80% listen to station B. If today, 50% of the regular listeners listen to each station, what percentage of them would you expect to listen to station A 2 days from now?

15. Traffic Conditions Workday traffic conditions from 9:00 AM to 10:00 AM on the Baltimore Beltway can be characterized as Light, Moderate, and Heavy. The following stochastic matrix describes the day-to-day transitions:

$$\begin{array}{c} \\ L \\ M \\ H \end{array} \begin{array}{ccc} L & M & H \\ \begin{bmatrix} .70 & .20 & .10 \\ .20 & .75 & .30 \\ .10 & .05 & .60 \end{bmatrix} \end{array}$$

(a) Interpret the numbers in the second column of the matrix.

(b) In the long run, what percent of the workdays fall into each category?

(c) Of 20 workdays in a month, how many are expected to have Heavy traffic on the Baltimore Beltway from 9:00 AM to 10:00 AM?

16. Mental Health A mental-health facility rates patients on their ability to live on their own. The state of a person's health is "able to work and considered cured (C)," or "long-term hospitalization or death (L)," or "group home (G)," or "short-term hospital care (S)." The stochastic matrix shown here describes the transitions from month to month. Use the fundamental matrix to determine the expected number of months spent in state G or S before being absorbed into state C or L.

$$\begin{array}{c} \\ C \\ L \\ G \\ S \end{array} \begin{array}{cccc} C & L & G & S \\ \begin{bmatrix} 1 & 0 & .60 & .05 \\ 0 & 1 & .10 & .40 \\ 0 & 0 & .20 & .50 \\ 0 & 0 & .10 & .05 \end{bmatrix} \end{array}$$

17. Reservoir Levels The contents of a reservoir depend on the available rainfall in the region and the demands on the water supply. Suppose that a reservoir holds up to 4 units of water (a unit might be a million gallons), and policy for the use of the water for irrigation and drinking water never allows the contents of the reservoir to drop below 1 unit of water. The (rounded) amount of water in the reservoir from week to week seems to follow the transition matrix

$$\begin{array}{c} \\ 1 \\ 2 \\ 3 \\ 4 \end{array} \begin{array}{cccc} 1 & 2 & 3 & 4 \\ \begin{bmatrix} .20 & .10 & .05 & .05 \\ .30 & .20 & .20 & .30 \\ .40 & .40 & .50 & .40 \\ .10 & .30 & .25 & .25 \end{bmatrix} \end{array}.$$

(a) Determine and interpret the stable distribution for the matrix.

(b) Suppose that the weekly benefits to recreation in the area around the reservoir are estimated to be $4000 when there is 1 unit in the reservoir, $6000 when there are 2 units in the reservoir, $10,000 when there are 3 units, and $3000 when there are 4 units. Determine the average weekly benefits to be realized in the long run.

Conceptual Exercises

18. Explain why the entries in each column of a transition matrix must add up to 1.

19. Suppose that T is a transition matrix and

$$T^4 = \begin{array}{c} \\ \end{array}\begin{array}{cc} A & B \\ \begin{bmatrix} .74 & .18 \\ .26 & .82 \end{bmatrix} \end{array}$$

Interpret the number .26.

20. Explain why A^2 must be a stochastic matrix if A is a stochastic matrix.

21. True or False Every entry in a stochastic matrix is a conditional probability.

22. True or False Every absorbing stochastic matrix is a regular stochastic matrix.

CHAPTER

8 PROJECT

Doubly Stochastic Matrices

A square matrix is said to be **doubly stochastic** if the sum of the entries in each column is 1 and the sum of the entries in each row is 1. Some examples of doubly stochastic matrices are

$$\begin{bmatrix} .4 & .6 \\ .6 & .4 \end{bmatrix}, \quad \begin{bmatrix} .1 & .3 & .6 \\ .6 & .1 & .3 \\ .3 & .6 & .1 \end{bmatrix}, \quad \text{and} \quad \begin{bmatrix} .1 & .2 & .3 & .4 \\ .3 & .4 & .1 & .2 \\ .2 & .3 & .4 & .1 \\ .4 & .1 & .2 & .3 \end{bmatrix}.$$

1. Give another example of a 2 × 2 doubly stochastic matrix.
   (a) Is your matrix symmetric; that is, does it equal its own transpose?
   (b) Prove that every 2 × 2 doubly stochastic matrix is symmetric.
   (c) Show that the product of your matrix and the matrix

   $$\begin{bmatrix} .4 & .6 \\ .6 & .4 \end{bmatrix}$$

   is doubly stochastic.

   (d) Show that $\begin{bmatrix} \frac{1}{2} \\ \frac{1}{2} \end{bmatrix}$ is a stable distribution for your matrix.

2. Give another example of a 3 × 3 doubly stochastic matrix.
   (a) Is your matrix symmetric? Are all 3 × 3 doubly stochastic matrices symmetric?
   (b) Show that the product of your matrix and the matrix

   $$\begin{bmatrix} .1 & .3 & .6 \\ .6 & .1 & .3 \\ .3 & .6 & .1 \end{bmatrix}$$

   is doubly stochastic.

   (c) Show that $\begin{bmatrix} \frac{1}{3} \\ \frac{1}{3} \\ \frac{1}{3} \end{bmatrix}$ is a stable distribution for your matrix.

3. Give another example of a 4 × 4 doubly stochastic matrix.
   (a) Is your matrix symmetric? Are all 4 × 4 doubly stochastic matrices symmetric?
   (b) Show that the product of your matrix and the matrix

   $$\begin{bmatrix} .1 & .2 & .3 & .4 \\ .3 & .4 & .1 & .2 \\ .2 & .3 & .4 & .1 \\ .4 & .1 & .2 & .3 \end{bmatrix}$$

   is doubly stochastic.

   (c) Show that $\begin{bmatrix} \frac{1}{4} \\ \frac{1}{4} \\ \frac{1}{4} \\ \frac{1}{4} \end{bmatrix}$ is a stable distribution for your matrix.

   We will now show that the product of any two $n \times n$ doubly stochastic matrices is doubly stochastic, and that any doubly stochastic $n \times n$ matrix has

   as a stable distribution. Let E_n be the $n \times n$ matrix for which each entry is 1.

4. Show that A is a doubly stochastic $n \times n$ matrix if and only if $AE_n = E_n$ and $E_nA = E_n$.

5. Show that if A and B are doubly stochastic $n \times n$ matrices, then AB is also a doubly stochastic $n \times n$ matrix. *Hint:* Show that $(AB)E_n = E_n$ and $E_n(AB) = E_n$.

6. Show that if A is a doubly stochastic $n \times n$ matrix, then

$$\begin{bmatrix} \frac{1}{n} \\ \frac{1}{n} \\ \vdots \\ \frac{1}{n} \end{bmatrix}$$

is a stable distribution for A . *Hint:* First show that $A \begin{bmatrix} 1 \\ 1 \\ \vdots \\ 1 \end{bmatrix} = \begin{bmatrix} 1 \\ 1 \\ \vdots \\ 1 \end{bmatrix}$.

7. A collection of $n \times n$ matrices is said to be a *convex set* if whenever A and B are in the set and t is a number between 0 and 1, then $tA + (1 - t)B$ is also in the set. Show that the set of all $n \times n$ doubly stochastic matrices is a convex set. *Note:* For a number r and a matrix A, rA is the matrix obtained by multiplying each entry of A by r.

One of the more interesting developments of twentieth-century mathematics has been the theory of games, a branch of mathematics used to analyze competitive phenomena. This theory has been applied extensively in many fields, including business, economics, psychology, and sociology. The 1994 and 2005 Nobel Prizes in economics were awarded to economists for their groundbreaking work in integrating game theory into the study of economic behavior. Game theory has become one of the hottest areas of economics, with applications ranging from how the Federal Reserve sets interest rates to how companies structure incentive pay for employees to how companies bid on lucrative federal contracts. Mathematically, the theory of games blends the theory of matrices with probability theory. Although an extensive discussion is well beyond the scope of this book, we hope to give the flavor of the subject and some indication of the wide range of its applications.

9.1 Games and Strategies

Let us begin our study of game theory by analyzing a typical competitive situation. Suppose that, in a certain town, there are two furniture stores, Reliable Furniture Company and Cut-Rate Furniture Company, which compete for all furniture sales in the town. Each of the stores is planning a Labor Day sale, and each has the option of marking its furniture down by 10% or 20%. The results of their decisions affect the total percentage of the market that each captures. On the basis of an analysis of past consumer tendencies, it is estimated that, if Reliable chooses a 10% discount and so does Cut-Rate, then Reliable will capture 60% of the sales. If Reliable chooses a 10% discount but Cut-Rate chooses 20%, then Reliable will capture only 35% of the sales. On the other hand, if Reliable chooses a 20% discount but Cut-Rate chooses 10%, then Reliable will get 80% of the sales. If Reliable chooses a 20% discount and Cut-Rate also chooses 20%, then Reliable will get 50% of the sales. Each store is able to determine the other store's discount prior to the start of the sale and adjust its own discount accordingly. If you were a consultant to Reliable, what discount would you choose in order to obtain as large a share of the sales as possible?

To analyze the various possibilities, let us summarize the given data in a matrix. For the sake of brevity, denote Reliable by R and Cut-Rate by C. Then the data can be summarized in a matrix showing R's share in each case as follows:

$$
\begin{array}{cc}
 & C \text{ discount} \\
 & \begin{array}{cc} 10\% & 20\% \end{array} \\
R \text{ discount} \quad \begin{array}{c} 10\% \\ 20\% \end{array} & \begin{bmatrix} .6 & .35 \\ .8 & .5 \end{bmatrix}
\end{array}
$$

For example, the number in the second row, first column corresponds to an R discount of 20% and a C discount of 10%. In this case, R will capture 80%, or .8, of the sales.

We may view R's choice of discount as choosing one of the rows of the matrix. Similarly, C's choice of discount amounts to choosing one of the columns of the matrix. Suppose that R and C both act rationally. What will be the result? Let us view things from R's perspective first. In scanning the options, R sees a .8 in the second row. So R's first reaction might be to take a 20% discount and try for 80% of the sales. However, this route is very risky. As soon as C learns that R has chosen a 20% discount, C will set a 20% discount and lower R's share of the sales to 50%. So the result of choosing row 2 will be for R to capture only 50% of the sales. On the other hand, if R chooses row 1, then C will naturally choose a 20% discount to give R a 35% share of the sales. Of the options open to R, the 50% share is clearly the most desirable. So R will choose a 20% discount.

What about C? In setting a discount, C must choose a column of the matrix. Since the entries represent R's share of the sales, C wishes to make a choice resulting in as *small* a number as possible. If C chooses column 1, then R will immediately respond with a 20% discount in order to acquire 80% of the sales, a disaster for C. On the other hand, if C chooses column 2, then R will choose a 20% discount to obtain 50% of the sales. The best option open to C is to choose a 20% discount.

Thus, we see that if both stores act rationally, they will each choose 20% discounts and each will capture 50% of the sales.

The preceding competitive situation is an example of a (mathematical) **game**. In such a game, there are two or more players. In the example, the players are R and C. Each player is allowed to make a move. In the example, the moves are the choices of discount. As the result of a move by each player, there is a payoff to each player. The payoff to each player (store) is the percentage of total sales that they capture. In our example, then, we have solved a problem that can be posed for any game.

> **Fundamental Problem of Game Theory**
>
> How should each player decide their move in order to maximize their gain?

Indeed, in our example, R and C chose moves such that each maximized their own share of sales.

Throughout this chapter, we consider only games with two players, whom we shall denote by R and C. (R and C stand for row and column, respectively.) Suppose that R can make moves $R_1, R_2, \ldots, R_m$ and that C can make moves $C_1, C_2, \ldots, C_n$. Further, suppose that a move R_i by R and C_j by C results in a payoff of a_{ij} to R. Then the game can be represented by the following **payoff matrix**:

$$
\begin{array}{cc}
 & C \text{ moves} \\
 & \begin{array}{cccc} C_1 & C_2 & \cdots & C_n \end{array} \\
R \text{ moves} \quad \begin{array}{c} R_1 \\ R_2 \\ \vdots \\ R_m \end{array} & \begin{bmatrix} a_{11} & a_{12} & \cdots & a_{1n} \\ a_{21} & a_{22} & \cdots & a_{2n} \\ \vdots & \vdots & \cdots & \vdots \\ a_{m1} & a_{m2} & \cdots & a_{mn} \end{bmatrix}
\end{array}.
$$

Note that the payoff matrix is an $m \times n$ matrix with a_{ij} as the entry of the ith row, jth column. In our furniture store example, the payoff matrix was just the matrix we used in our analysis. Note that a move by R corresponds to a choice of a *row* of the payoff matrix, whereas a move by C corresponds to a choice of a *column*.

Suppose that a given game is played repeatedly. The players can adopt various strategies to attempt to maximize their respective gains (or minimize their losses). In what follows, we discuss the problem of determining strategies. The simplest type of strategy is one in which a player, on consecutive plays, consistently chooses the same row (or column). Such strategies are called **pure strategies** and are discussed in this section. Strategies involving varied moves are called **mixed strategies** and are discussed in Section 9.2.

In the furniture example discussed above, the payoffs to C are related in a simple way to the corresponding payoffs to R. The payoff to C is 100% minus the payoff to R. In another common type of game, a payoff to R of a given amount results in a loss to C of the same amount, and vice versa. For such games, the sum of the gains on each play is zero; hence, they are called **zero-sum games**. An illustration of such a game is provided in the next example.

EXAMPLE 1   **Coin-Matching Game** Suppose that R and C play a coin-matching game. Each player can show either heads or tails. If R and C both show heads, then C pays R \$5. If R shows heads and C shows tails, then R pays C \$8. If R shows tails and C shows heads, then C pays R \$3. If R shows tails and so does C, then C pays R \$1.

(a) Determine the payoff matrix of this game.
(b) Suppose that R and C play the game repeatedly.
Determine optimal pure strategies for R and C.

SOLUTION   **(a)** The payoff matrix is

$$
\begin{array}{cc}
 & \begin{array}{cc} C \\ \text{Heads} \quad \text{Tails} \end{array} \\
R \begin{array}{c} \text{Heads} \\ \text{Tails} \end{array} & \begin{bmatrix} 5 & -8 \\ 3 & 1 \end{bmatrix}.
\end{array}
$$

Each entry specifies a payoff from C to R. The entry "-8" denotes a negative payoff to R—that is, a gain of \$8 to C.

(b) R would clearly like to choose heads so as to gain \$5. However, if R consistently chooses heads, C will retaliate by choosing tails, causing R to lose \$8. If R chooses tails, however, then the best C can do is choose tails to give a gain of \$1 to R. So R should clearly choose tails. Now we look at the game from C's point of view. Clearly, C's objective is to minimize the payment to R. If C consistently chooses heads, then R will notice the pattern and choose heads, at a cost to C of \$5. However, if C chooses tails, then the best that R can do is choose tails, thereby costing C \$1. So, clearly, the optimal move for C is to choose tails.   ≪

The reasoning just described is rather cumbersome. There is, however, an easy way to summarize what we have done. Let us first describe R's reasoning. R seeks to choose a row of the matrix that will maximize R's payoff. However, once a row is chosen consistently, R can expect C to counter by choosing the least element of that row. Thus, R should choose a move as follows.

> **Optimal Pure Strategy for R**
>
> 1. For each row of the payoff matrix, determine the least element.
> 2. Choose the row for which this element is as large as possible.

For example, in the game of Example 1, we have circled the least element of each row:

$$
\begin{bmatrix} 5 & \boxed{-8} \\ 3 & \boxed{1} \end{bmatrix}
$$

The largest circled element is 1. So R should choose the second row—that is, tails.

In a similar way, we may describe the optimal strategy for C. C wishes to choose a column of the payoff matrix so as to minimize the payoff to R. However, C can expect R to adjust the choice to the maximum element of the column. Therefore, we can summarize the optimal strategy for C as follows.

Optimal Pure Strategy for C

1. For each column of the payoff matrix, determine the largest element.
2. Choose the column for which this element is as small as possible.

For example, in the game of Example 1, we have circled the largest element in each column:

$$\begin{bmatrix} ⑤ & -8 \\ 3 & ① \end{bmatrix}$$

The smallest circled element is 1, so C should choose the second column, or tails.

EXAMPLE 2   **Optimal Pure Strategy** Determine optimal pure strategies for R and C for the game whose payoff matrix is

$$\begin{bmatrix} -1 & 5 \\ 1 & 4 \\ 0 & -1 \end{bmatrix}.$$

SOLUTION   To determine the strategy for R, we first circle the least element in each row:

$$\begin{bmatrix} ㊀1 & 5 \\ ① & 4 \\ 0 & ㊀1 \end{bmatrix}$$

The largest of these is 1, so R should play the second row. To determine the strategy for C, we circle the largest element in each column.

$$\begin{bmatrix} -1 & ⑤ \\ ① & 4 \\ 0 & -1 \end{bmatrix}$$

The smallest of these is 1, so C should play the first column.   «

All of the games considered so far have an important characteristic in common. There is an entry in the payoff matrix which is *simultaneously* the least element in its row and the largest element in its column. Such an entry is called a **saddle point** for the game. As we have seen in the examples considered previously, if a game possesses a saddle point, then an optimal strategy is for R to choose the row containing the saddle point and for C to choose the column containing the saddle point.

A game need not have a saddle point. For example, the matrix

$$\begin{bmatrix} 2 & -2 \\ 0 & 1 \end{bmatrix}$$

is the payoff matrix of a game with no saddle point. The optimal pure strategy for R is to choose the row with the largest of the circled elements in

$$\begin{bmatrix} 2 & ㊀2 \\ ⓪ & 1 \end{bmatrix}.$$

Thus, R chooses row 2. The optimal pure strategy for C is to choose the column with the least of the circled elements in

$$\begin{bmatrix} ② & -2 \\ 0 & ① \end{bmatrix}.$$

So C chooses column 2. No element is simultaneously the least element in its row and the largest element in its column.

Strictly Determined Games

A game that has a saddle point is called a **strictly determined game**. If v is a saddle point for a strictly determined game, then if each player plays the optimal pure strategy, each repetition of the game will result in a payment of v to player R. The number v is called the **value** of the game. Since v is a payoff to player R, the game favors player R when v is positive and favors player C when v is negative. When $v = 0$, the game is said to be a **fair game**.

EXAMPLE 3   **Value of a Strictly Determined Game** Find the saddle point and the value of the strictly determined game given by the payoff matrix

$$\begin{bmatrix} -1 & -10 & 10 \\ 0 & 7 & 6 \\ 3 & 4 & 11 \\ 2 & 5 & 7 \end{bmatrix}.$$

SOLUTION   The least elements in the various rows are

$$\begin{bmatrix} -1 & \widehat{-10} & 10 \\ ⓪ & 7 & 6 \\ ③ & 4 & 11 \\ ② & 5 & 7 \end{bmatrix}.$$

The largest elements in the columns are

$$\begin{bmatrix} -1 & -10 & 10 \\ 0 & ⑦ & 6 \\ ③ & 4 & ⑪ \\ 2 & 5 & 7 \end{bmatrix}.$$

The element 3 in the third row, first column is the least in its row and the largest in its column and so is a saddle point of the game. The value of the game is therefore 3. Each repetition of the game, assuming optimal strategies, results in a payoff of 3 to R.

≫ Now Try Exercise 5

EXAMPLE 4   **A Children's Game** R and C play a game in which they show 1 or 2 fingers simultaneously. It is agreed that C pays R an amount equal to the total number of fingers shown, less 3 cents. Find the optimal strategy for each player and the value of the game.

SOLUTION   The payoff matrix is given by

$$\begin{matrix} & 1 & 2 \\ 1 & \\ 2 & \end{matrix} \begin{bmatrix} 2-3 & 3-3 \\ 3-3 & 4-3 \end{bmatrix} = \begin{matrix} & 1 & 2 \\ 1 & \\ 2 & \end{matrix} \begin{bmatrix} -1 & 0 \\ 0 & 1 \end{bmatrix}.$$

The saddle point is the element that is simultaneously the least in its row and the largest in its column—so an optimal strategy is for R to show 2 fingers and for C to show 1 finger. The value of the game is 0.

≫ Now Try Exercise 7

A game may have more than one saddle point. Consider the game with payoff matrix

$$\begin{bmatrix} ① & 2 & ① \\ ① & 5 & ① \\ 0 & -7 & -1 \end{bmatrix}.$$

Each circled element is both the least element in its row and the largest element in its column. There are four saddle points representing four optimal strategies. The value of the game, regardless of strategy, is 1. If a game has more than one saddle point, then the value of the game is the same at each of them.

Check Your Understanding 9.1

Solutions can be found following the section exercises.

Which of the given matrices are the payoff matrices of strictly determined games? For those that are, determine the saddle point and optimal pure strategy for each of the players.

1. $\begin{bmatrix} 1 & -1 & -3 \\ 0 & -2 & 3 \end{bmatrix}$ **2.** $\begin{bmatrix} 1 & -1 & 0 \\ 0 & -4 & 5 \end{bmatrix}$ **3.** $\begin{bmatrix} 1 & -2 & 1 \\ -2 & 1 & 1 \\ 1 & 1 & -2 \end{bmatrix}$

EXERCISES 9.1

In Exercises 1–12, determine the optimal pure strategies for the payoff matrix of the game. If the game is strictly determined, give its value.

1. $\begin{bmatrix} 6 & 4 \\ 7 & -5 \end{bmatrix}$ **2.** $\begin{bmatrix} -3 & 5 \\ 2 & 4 \end{bmatrix}$

3. $\begin{bmatrix} 4 & -4 \\ -1 & 2 \end{bmatrix}$ **4.** $\begin{bmatrix} 0 & 2 \\ 4 & -5 \end{bmatrix}$

5. $\begin{bmatrix} 5 & -4 & -1 \\ 4 & -3 & -5 \\ 2 & 1 & 9 \end{bmatrix}$ **6.** $\begin{bmatrix} -4 & 6 & -2 \\ -5 & 9 & -3 \end{bmatrix}$

7. $\begin{bmatrix} -4 & -2 \\ 2 & -3 \\ 1 & 0 \end{bmatrix}$ **8.** $\begin{bmatrix} 8 & -4 \\ 1 & 0 \\ 7 & -1 \end{bmatrix}$

9. $\begin{bmatrix} 2 & -6 & 7 \\ -5 & 3 & 9 \end{bmatrix}$ **10.** $\begin{bmatrix} 2 & -6 & 9 \\ -5 & 2 & 8 \\ 6 & -8 & 0 \end{bmatrix}$

11. $\begin{bmatrix} 1 & 2 \\ 5 & -7 \\ -2 & 0 \\ 4 & 3 \end{bmatrix}$ **12.** $\begin{bmatrix} 5 & -1 & 4 \\ 3 & -2 & 7 \\ 0 & -3 & 8 \\ 2 & 0 & 3 \end{bmatrix}$

13. (True or False) A strictly determined game can have more than one saddle point.

14. (True or False) In a strictly determined game, the value of each saddle point is the same as the value of the game.

15. Pizzerias Rosa's Pizzeria and Carlo's Pizzeria compete for customers. They are each considering either lowering their prices or improving their décor in order to attract more customers. Market research indicates that the results of employing these strategies are as shown in the following matrix.

$$\begin{array}{cc} & \textit{Carlo's} \\ & \begin{array}{cc} \text{Price} & \text{Décor} \end{array} \\ \textit{Rosa's} \begin{array}{c} \text{Price} \\ \text{Décor} \end{array} & \begin{bmatrix} 50 & 30 \\ -25 & 15 \end{bmatrix} \end{array}$$

The entries in the matrix represent the number of pizzas sold each day by Rosa's Pizzeria at the expense of Carlo's Pizzeria. For instance, if Rosa's Pizzeria reduces the price of a pizza and Carlo's Pizzeria improves their décor, Rosa's Pizzeria will sell 30 more pizzas per day and Carlo's Pizzeria will sell 30 fewer pizzas per day. Determine each pizzeria's optimal strategy. Is this game strictly determined?

16. Movie Theaters A small town has two movie theaters, the Roxie and the Cinemart, that compete for customers. They are each considering specializing in either dramas or comedies in order to attract more customers. Market research indicates that the results of specializing in these genres are as shown in the following matrix.

$$\begin{array}{cc} & \textit{Cinemart} \\ & \begin{array}{cc} \text{Comedy} & \text{Drama} \end{array} \\ \textit{Roxie} \begin{array}{c} \text{Comedy} \\ \text{Drama} \end{array} & \begin{bmatrix} 400 & -300 \\ -150 & 250 \end{bmatrix} \end{array}$$

For instance, if they both specialize in comedies, the Roxie will sell 400 more tickets per week and the Cinemart will sell 400 fewer tickets per week. Determine each movie theater's optimal strategy. Is this game strictly determined?

17. Coffee Houses Two competing companies, Rigelbucks and Canisbucks, are deciding in which of three malls (A, B, or C) to move their coffee houses. A market research firm has determined that the number of customers each store gains from or loses to the other per day as a result of the location selected is given by the following payoff matrix.

$$\begin{array}{cc} & \textit{Canisbucks} \\ & \begin{array}{ccc} A & B & C \end{array} \\ \textit{Rigelbucks} \begin{array}{c} A \\ B \\ C \end{array} & \begin{bmatrix} -75 & 600 & -200 \\ 400 & 450 & 300 \\ -100 & 500 & 150 \end{bmatrix} \end{array}$$

For instance, if Rigelbucks locates in mall A and Canisbucks locates in mall B, then Rigelbucks will gain 600 customers per day and Canisbucks will lose 600 customers per day. Determine each company's optimal strategy. Is this game strictly determined? If so, what is its value?

18. **Matching Coins** Rory and Cara each have a coin. If they both show heads, Rory pays \$2 to Cara. If they both show tails, Cara pays \$2 to Rory. If Rory shows heads and Cara shows tails, Cara pays \$3 to Rory. If Rory shows tails and Cara shows heads, then Cara pays \$1 to Rory. Determine each person's optimal strategy. Is this game strictly determined? If so, what is the value of the game?

For each of the games that follow, give the payoff matrix and decide whether the game is strictly determined. For those that are, determine the optimal strategies for R and C.

19. **Matching Coins** Suppose that R and C play a game by matching coins. On each play, C pays R the number of heads shown (0, 1, or 2) minus twice the number of tails shown.

20. **Scissors Paper Stone** In the children's game Scissors Paper Stone, each of two children calls out one of the three words. If they both call out the same word, then the game is a tie. Otherwise, "scissors" beats "paper" (since scissors can cut paper), "paper" beats "stone" (since paper can cover stone), and "stone" beats "scissors" (since stone can break scissors). Suppose that the loser pays a penny to the winner.

21. **Political Action** Two candidates for political office must decide to be for, against, or neutral on a certain referendum. Pollsters have determined that, if candidate R comes out for

the referendum, they will gain 8000 votes if candidate C also comes out for the referendum, will lose 1000 votes if candidate C comes out against, and will gain 1000 votes if candidate C comes out neutral. If candidate R comes out against, they will lose 7000 votes (respectively, gain 4000 votes, lose 2000 votes) if candidate C comes out for (respectively, comes out against, is neutral on) the referendum. If candidate R is neutral, they will gain 3000 votes if C is for or against and will gain 2000 votes if C is neutral.

22. **Program Scheduling** TV stations R and C each have a quiz show and a situation comedy to schedule for their one o'clock and two o'clock time slots. If they both schedule their quiz shows at one o'clock, then station R will take \$3000 in advertising revenue away from station C. If they both schedule their quiz shows at two o'clock, then station C will take \$2000 in advertising revenue from R. If they choose different hours for the quiz show, then R will take \$5000 in advertising from C by scheduling its show at two o'clock, and \$2000 by scheduling it at 1 o'clock.

23. **Card Game** Player R has two cards: a red 5 and a black 10. Player C has three cards: a red 6, a black 7, and a black 8. They each place one of their cards on the table. If the cards are the same color, R receives the difference of the two numbers. If the cards are of different colors, C receives the smaller of the two numbers.

Solutions to Check Your Understanding 9.1

1. Not strictly determined. The least elements of the rows are

$$\begin{bmatrix} 1 & -1 & \boxed{-3} \\ 0 & \boxed{-2} & 3 \end{bmatrix};$$

the largest elements of the columns are

$$\begin{bmatrix} \boxed{1} & \boxed{-1} & -3 \\ 0 & -2 & \boxed{3} \end{bmatrix}.$$

No element is simultaneously the least in its row and the largest in its column.

2. Strictly determined. The least elements of the rows are

$$\begin{bmatrix} 1 & \boxed{-1} & 0 \\ 0 & \boxed{-4} & 5 \end{bmatrix}.$$

The largest elements of the columns are

$$\begin{bmatrix} \boxed{1} & \boxed{-1} & 0 \\ 0 & -4 & \boxed{5} \end{bmatrix}.$$

Thus, -1 is a saddle point. The optimal strategy for R is to choose row 1; the optimal strategy for C is to choose column 2.

3. Not strictly determined. The least elements of the rows are

$$\begin{bmatrix} 1 & \boxed{-2} & 1 \\ \boxed{-2} & 1 & 1 \\ 1 & 1 & \boxed{-2} \end{bmatrix}.$$

The largest element for each column is 1. (Note that there are two choices for each largest element.) But none of the largest column elements is a least row element.

Mixed Strategies

In Section 9.1, we introduced strictly determined games and gave a method for determining optimal strategies for each player. However, not all games are strictly determined. For example, consider the game with payoff matrix

$$\begin{bmatrix} -1 & 5 \\ 2 & -3 \end{bmatrix}.$$

The least entries of the rows are

$$\begin{bmatrix} \boxed{-1} & 5 \\ 2 & \boxed{-3} \end{bmatrix},$$

whereas the largest entries of the columns are

$$\begin{bmatrix} -1 & \boxed{5} \\ \boxed{2} & -3 \end{bmatrix}.$$

Note that no matrix entry is simultaneously the least in its row and the largest in its column. Note also that no simple strategy of the type considered in Section 9.1 is sufficient to both maximize R's winnings and minimize C's losses. To see this, consider the game from R's point of view. Suppose that R repeatedly plays the strategy "first row," thereby attempting to win 5. After a few plays, C will catch on to R's strategy and choose column 1, giving R a loss of 1. Similarly, if R consistently plays the strategy "second row," attempting to win 2, then C can thwart R by choosing the second column, to give R a loss of 3. It is clear that, in order to maximize R's payoff, R should sometimes choose row 1 and sometimes row 2. One might expect that by choosing the rows on a probabilistic basis, R can prevent C from anticipating R's moves and amass enough positive payoffs to counteract the occasional negative ones. Thus, we should investigate strategies of the type

$$A: \begin{cases} \text{Choose row 1 with probability .5} \\ \text{Choose row 2 with probability .5} \end{cases}$$

and

$$B: \begin{cases} \text{Choose row 1 with probability .9} \\ \text{Choose row 2 with probability .1.} \end{cases}$$

Such strategies are called **mixed strategies**. Either of the players can pursue such a strategy.

One way that R can carry out strategy A is to alternate between row 1 and row 2 on successive plays of the game. However, if C is at all clever, C will recognize this pattern and play accordingly. What R should do is toss a coin and play row 1 whenever it lands heads and row 2 whenever it lands tails. Then there is no way that C can anticipate R's choice.

To carry out strategy B, R might use a card with a spinner attached at the center of a circle that is 90% red and 10% white. R would then determine which row to choose by spinning the spinner and choosing row 1 if the spinner lands on the red part of the circle and row 2 if it lands on the white part. (See Fig. 1.)

It will be convenient to write mixed strategies in matrix form. Mixed strategies for R will be row matrices, and mixed strategies for C will be column matrices. Thus, for example, mixed strategy A above (for R) corresponds to the matrix

$$A: [.5 \quad .5],$$

whereas mixed strategy B (for R) corresponds to

$$B: [.9 \quad .1].$$

The mixed strategy in which C chooses column 1 with probability .6 and column 2 with probability .4 corresponds to the column matrix

$$\begin{bmatrix} .6 \\ .4 \end{bmatrix}.$$

In comparing different mixed strategies, we use a number called their **expected value**. This number is just the average amount per game paid to R if the players pursue the given mixed strategies. The next example illustrates the computation of the expected value in a special case.

Figure 1

EXAMPLE 1   **Expected Value of a Mixed Strategy Game** Suppose that a game has payoff matrix

$$\begin{bmatrix} -1 & 5 \\ 2 & -3 \end{bmatrix}.$$

Further, suppose that R pursues the mixed strategy $[.9 \quad .1]$ and that C pursues the mixed strategy $\begin{bmatrix} .6 \\ .4 \end{bmatrix}$. Calculate the expected value of this game.

SOLUTION   Let us view each repetition of the game as an experiment. There are four possible outcomes:

(row 1, column 1)   (row 1, column 2)

(row 2, column 1)   (row 2, column 2)

Let us compute the probability (relative frequency) with which each of the outcomes occurs. For example, consider the outcome (row 1, column 1). According to our assumptions about the strategies of R and C, R chooses row 1 with probability .9 and C chooses column 1 with probability .6. Since R and C make their respective choices independently of one another, the events "row 1" and "column 1" are independent. Therefore, we can compute the probability of the outcome (row 1, column 1) as follows:

$$\Pr(\text{row 1, column 1}) = \Pr(\text{row 1}) \cdot \Pr(\text{column 1}) = (.9)(.6) = .54.$$

That is, the outcome "row 1, column 1" will occur with probability .54. Similarly, we compute the probabilities of the other three outcomes. Table 1 gives the probability and the amount won by R for each outcome. The average amount that R wins per play is then given by

$$(-1)(.54) + (5)(.36) + (2)(.06) + (-3)(.04) = 1.26.$$

Hence, the expected value of the given strategies is 1.26.

Table 1

Outcome	R wins	Probability
Row 1, column 1	−1	$(.9)(.6) = .54$
Row 1, column 2	5	$(.9)(.4) = .36$
Row 2, column 1	2	$(.1)(.6) = .06$
Row 2, column 2	−3	$(.1)(.4) = .04$

» Now Try Exercise 1(a)

We may generalize the preceding computations. To see the pattern, let us first concentrate on 2×2 games. Suppose that a game has payoff matrix

$$\begin{bmatrix} a_{11} & a_{12} \\ a_{21} & a_{22} \end{bmatrix}.$$

Further, suppose that R pursues a strategy $[r_1 \quad r_2]$. That is, R randomly chooses row 1 with probability r_1 and row 2 with probability r_2. Similarly, suppose that C pursues a strategy $\begin{bmatrix} c_1 \\ c_2 \end{bmatrix}$. Then the probabilities of the various outcomes can be tabulated as shown in Table 2. Thus, by following the reasoning used in the preceding special case, we see that the expected value of the strategies is the sum of the products of the payoffs times the corresponding probabilities:

$$a_{11}(r_1 c_1) + a_{12}(r_1 c_2) + a_{21}(r_2 c_1) + a_{22}(r_2 c_2).$$

Table 2

Outcome	Payoff to R	Probability
Row 1, column 1	a_{11}	$r_1 c_1$
Row 1, column 2	a_{12}	$r_1 c_2$
Row 2, column 1	a_{21}	$r_2 c_1$
Row 2, column 2	a_{22}	$r_2 c_2$

On the average, R gains this amount for each play. A somewhat tedious (but easy) calculation shows that

$$[r_1 \quad r_2]\begin{bmatrix} a_{11} & a_{12} \\ a_{21} & a_{22} \end{bmatrix}\begin{bmatrix} c_1 \\ c_2 \end{bmatrix} = [r_1 \quad r_2]\begin{bmatrix} a_{11}c_1 + a_{12}c_2 \\ a_{21}c_1 + a_{22}c_2 \end{bmatrix}$$

$$= [a_{11}(r_1c_1) + a_{12}(r_1c_2) + a_{21}(r_2c_1) + a_{22}(r_2c_2)]. \qquad (1)$$

Formula (1) is a special case of the following general fact:

Expected Value of a Pair of Strategies

Suppose that a game has payoff matrix

$$\begin{bmatrix} a_{11} & a_{12} & \cdots & a_{1n} \\ a_{21} & a_{22} & \cdots & a_{2n} \\ \vdots & \vdots & & \vdots \\ a_{m1} & a_{m2} & \cdots & a_{mn} \end{bmatrix},$$

where each entry is a payoff to R. Suppose that R plays the strategy $[r_1 \quad r_2 \quad \cdots \quad r_m]$ and that C plays the strategy

$$\begin{bmatrix} c_1 \\ c_2 \\ \vdots \\ c_n \end{bmatrix}.$$

Let e be the expected value of the pair of strategies—that is, the average payoff to R. Then

$$[r_1 \quad r_2 \quad \cdots \quad r_m]\begin{bmatrix} a_{11} & a_{12} & \cdots & a_{1n} \\ a_{21} & a_{22} & \cdots & a_{2n} \\ \vdots & \vdots & & \vdots \\ a_{m1} & a_{m2} & \cdots & a_{mn} \end{bmatrix}\begin{bmatrix} c_1 \\ c_2 \\ \vdots \\ c_n \end{bmatrix} = [e].$$

EXAMPLE 2   **Choosing the Advantageous Strategy** Suppose that a game has payoff matrix

$$\begin{bmatrix} 2 & 0 & -1 \\ -1 & 3 & 4 \end{bmatrix}$$

and that R plays the strategy $[.5 \quad .5]$. Which of the following strategies is more advantageous for C?

$$A = \begin{bmatrix} .6 \\ .3 \\ .1 \end{bmatrix} \quad \text{or} \quad B = \begin{bmatrix} .3 \\ .3 \\ .4 \end{bmatrix}$$

SOLUTION   We compare the expected value with C using strategy A to that with C using strategy B. With strategy A, we have

$$[.5 \quad .5]\begin{bmatrix} 2 & 0 & -1 \\ -1 & 3 & 4 \end{bmatrix}\begin{bmatrix} .6 \\ .3 \\ .1 \end{bmatrix} = [.90].$$

Using strategy B, we have

$$[.5 \quad .5]\begin{bmatrix} 2 & 0 & -1 \\ -1 & 3 & 4 \end{bmatrix}\begin{bmatrix} .3 \\ .3 \\ .4 \end{bmatrix} = [1.20].$$

Thus, strategy A will yield an average payment per play of .90 to R, whereas strategy B will yield an average payment per play of 1.20. Clearly, it is to C's advantage to choose strategy A.   «

EXAMPLE 3   **Chemical Company's Dumping Strategy** The Acme Chemical Corporation has two plants, each situated on the banks of the Blue River, 10 miles from one another. A single inspector is assigned to check that the plants do not dump waste into the river. If she discovers plant A dumping waste, Acme is fined \$20,000. If she discovers plant B dumping waste, Acme is fined \$50,000. Suppose that the inspector visits one of the plants each day and that she chooses, on a random basis, to visit plant B 60% of the time. Acme schedules dumping from its two plants on a random basis, one plant per day, with plant B dumping waste on 70% of the days. How much is Acme's average fine per day?

SOLUTION   The competition between Acme and the inspector can be viewed as a non-strictly determined game, whose matrix is

$$\begin{array}{cc} & \text{Inspect } A \quad \text{Inspect } B \\ \begin{array}{c} \text{Plant } A \text{ dumps} \\ \text{Plant } B \text{ dumps} \end{array} & \left[\begin{array}{cc} -20{,}000 & 0 \\ 0 & -50{,}000 \end{array}\right]. \end{array}$$

The strategy of Acme is given by the row matrix $[.3 \quad .7]$. The strategy of the inspector is given by the column matrix $\begin{bmatrix} .4 \\ .6 \end{bmatrix}$. The expected value of the strategies is the matrix product

$$[.3 \quad .7]\begin{bmatrix} -20{,}000 & 0 \\ 0 & -50{,}000 \end{bmatrix}\begin{bmatrix} .4 \\ .6 \end{bmatrix} = [-6000 \quad -35{,}000]\begin{bmatrix} .4 \\ .6 \end{bmatrix}$$

$$= [-23{,}400].$$

In other words, Acme will be fined an average of \$23,400 per day for polluting the river.

» Now Try Exercise 3

In Section 9.3, we will alter payoff matrices by adding a fixed constant to each entry so that all entries become positive numbers. This does not alter the essential character of the game, in that good strategies for the original matrix will also be good strategies for the new matrix. The only difference is that the expected value is increased by the constant added. This procedure enables us to apply the methods of linear programming to the determination of optimal mixed strategies for zero-sum games that are not strictly determined.

EXAMPLE 4   **Payoffs with Constant Increments** In Example 1, we saw that, for the game with payoff matrix

$$\begin{bmatrix} -1 & 5 \\ 2 & -3 \end{bmatrix}$$

and strategies $[.9 \quad .1]$, $\begin{bmatrix} .6 \\ .4 \end{bmatrix}$, the expected value was 1.26. Compute the expected value of those strategies for the matrix obtained by adding 4 to each entry.

SOLUTION   The new matrix is

$$\begin{bmatrix} -1+4 & 5+4 \\ 2+4 & -3+4 \end{bmatrix}, \quad \text{or} \quad \begin{bmatrix} 3 & 9 \\ 6 & 1 \end{bmatrix}.$$

The expected value of the strategies for the new matrix is

$$[.9 \quad .1]\begin{bmatrix} 3 & 9 \\ 6 & 1 \end{bmatrix}\begin{bmatrix} .6 \\ .4 \end{bmatrix} = [3.3 \quad 8.2]\begin{bmatrix} .6 \\ .4 \end{bmatrix} = [5.26].$$

As it should, the expected value has also increased by 4.   «

Suppose that the payoff matrix is

$$\begin{bmatrix} -\frac{1}{2} & \frac{5}{2} \\ 1 & -\frac{3}{2} \end{bmatrix}.$$

What is the expected value of the strategies $[.9 \quad .1]$ and $\begin{bmatrix} .6 \\ .4 \end{bmatrix}$? We see that

$$[.9 \quad .1]\begin{bmatrix} -\frac{1}{2} & \frac{5}{2} \\ 1 & -\frac{3}{2} \end{bmatrix}\begin{bmatrix} .6 \\ .4 \end{bmatrix} = [.63].$$

We note that multiplying each element in the payoff matrix by 2 and adding 4 gives the matrix

$$\begin{bmatrix} 3 & 9 \\ 6 & 1 \end{bmatrix},$$

which with the preceding strategies, gives the expected value 5.26. (See the solution to Example 4.) Multiplying each element of the payoff matrix by 2 and adding 4 to each element produces the same effect on the expected value:

$$5.26 = 2(.63) + 4.$$

Check Your Understanding 9.2

Solutions can be found following the section exercises.

1. Suppose that the payoff matrix of a game is

$$\begin{bmatrix} 4 & -2 \\ -3 & 1 \end{bmatrix}.$$

Suppose that R plays the strategy $[.6 \quad .4]$ Which of the two strategies $\begin{bmatrix} .5 \\ .5 \end{bmatrix}$ or $\begin{bmatrix} .7 \\ .3 \end{bmatrix}$ is better for C?

2. Answer the question in Problem 1 for the game whose payoff matrix is

$$\begin{bmatrix} 9 & 3 \\ 2 & 6 \end{bmatrix}.$$

(This matrix is obtained by adding 5 to each entry of the matrix in Problem 1.)

EXERCISES 9.2

1. Suppose that a game has payoff matrix

$$\begin{bmatrix} 3 & -1 \\ -7 & 5 \end{bmatrix}.$$

Calculate the expected values for the following strategies, and determine which of the following four situations is most advantageous to R:

(a) R plays $[.5 \quad .5]$, C plays $\begin{bmatrix} .5 \\ .5 \end{bmatrix}$.

(b) R plays $[1 \quad 0]$, C plays $\begin{bmatrix} .5 \\ .5 \end{bmatrix}$.

(c) R plays $[.3 \quad .7]$, C plays $\begin{bmatrix} .6 \\ .4 \end{bmatrix}$.

(d) R plays $[.75 \quad .25]$, C plays $\begin{bmatrix} .2 \\ .8 \end{bmatrix}$.

2. Suppose that a game has payoff matrix

$$\begin{bmatrix} 1 & 0 & 2 \\ -1 & 2 & 0 \\ 0 & -1 & -1 \end{bmatrix}.$$

Calculate the expected values for the following strategies, and determine which of the following four situations is most advantageous to C:

(a) R plays $[1 \quad 0 \quad 0]$, C plays $\begin{bmatrix} .5 \\ .4 \\ .1 \end{bmatrix}$.

(b) R plays $[.3 \quad .3 \quad .4]$, C plays $\begin{bmatrix} .4 \\ .4 \\ .2 \end{bmatrix}$.

(c) R plays $[0 \quad .5 \quad .5]$, C plays $\begin{bmatrix} .4 \\ 0 \\ .6 \end{bmatrix}$.

(d) R plays $[.1 \quad .1 \quad .8]$, C plays $\begin{bmatrix} .2 \\ .2 \\ .6 \end{bmatrix}$.

3. **Inspector's Strategy** Refer to Example 3. Suppose that the inspector changes her strategy and visits plant B 80% of the time. How much is Acme's average fine per day?

4. **Inspector's Strategy** Refer to Example 3. Suppose that the inspector visits plant B 30% of the time. How much is Acme's average fine per day?

5. **A Letter Game** Suppose that two players, R and C, write down letters of the alphabet. If both write vowels or both write consonants, then there is no payment to either player. If R writes a vowel and C writes a consonant, then C pays R \$2. If R writes

writes a consonant and C writes a vowel, then R pays C \$1. Suppose that R chooses a consonant 75% of the plays and C chooses a vowel 40% of the plays. What is the average loss (or gain) of R per play?

6. **Flood Insurance** A small business owner must decide whether to carry flood insurance. She may insure her business for \$2 million for \$100,000, \$1 million for \$50,000, or \$.5 million for \$30,000. Her business is worth \$2 million. There is a flood serious enough to destroy her business an average of once every 10 years. In order to save insurance premiums, she decides each year on a probabilistic basis how much insurance to carry. She chooses \$2 million 20% of the time, \$1 million 20% of the time, \$.5 million 20% of the time, and no insurance 40% of the time. What is her average annual loss?

7. Two players, Robert and Carol, play a game with payoff matrix (to Robert)

$$\begin{bmatrix} 3 & -1 \\ -2 & 2 \end{bmatrix}.$$

   (a) Is the game strictly determined? Why?
   (b) Suppose that Robert has strategy $[.3 \quad .7]$. The opponents agree that the game should be fair. Is it possible for this to be a fair game? What would Carol's strategy have to be?

8. Rework Exercise 7 with $[.7 \quad .3]$ as Robert's strategy.

9. Two players, Robert and Carol, play a game with payoff matrix (to Robert)

$$\begin{bmatrix} 5 & -1 \\ -2 & 2 \end{bmatrix}.$$

   (a) Is the game strictly determined? Why?
   (b) Suppose that Robert has strategy $[.7 \quad .3]$. The opponents agree that the game should be fair. Is it possible for this to be a fair game? What would Carol's strategy have to be?

10. Rework Exercise 9 with $[.3 \quad .7]$ as Robert's strategy.

11. Assume that two players, Renée and Carlos, play a game with the following payoff matrix (to Renée):

$$\begin{bmatrix} 1 & 2 & 4 \\ 1 & 0 & 5 \\ 0 & 1 & -1 \end{bmatrix}.$$

   (a) Is the game strictly determined? Determine the strategy for each player.
   (b) What is the value of the game? Is the game fair?

12. The two players of Exercise 11, Renée and Carlos, play the game again, but this time, the payoff matrix (to Renée) is

$$\begin{bmatrix} -3 & -2 & 6 \\ 2 & 0 & 2 \\ 5 & -2 & -4 \end{bmatrix}.$$

   (a) Is the game strictly determined? Determine the strategy for each player.
   (b) Is the game fair? Why?

13. **Football** Suppose that, when the offense calls a running play, they gain an average of 1 yard if the defense anticipates a running play, and gain an average of 3 yards if the defense anticipates a passing play. Suppose that, when the offense calls a passing play, they gain an average of 7 yards if the defense anticipates a running play, and they lose an average of 4 yards if the defense anticipates a passing play.
   (a) What is the payoff matrix for the play?
   (b) Determine the expected number of yards gained if the offense uses the strategy $[.6 \quad .4]$ and the defense uses the strategy $\begin{bmatrix} .7 \\ .3 \end{bmatrix}$.

14. **Baseball** Whenever Randy pitches to Chris, he throws either a fastball or a slider. When Randy pitches a fastball, Chris hits the ball with a probability of .25 when he anticipates a fastball, and hits the ball with a probability of .1 when he anticipates a slider. When Randy pitches a slider, Chris hits the ball with a probability of .15 when he anticipates a fastball, and hits the ball with a probability of .3 when he anticipates a slider.
   (a) What is the payoff matrix for a single pitch where each element of the matrix is a probability?
   (b) Determine the expected probability of a hit if Randy uses the strategy $[.2 \quad .8]$ and Chris uses the strategy $\begin{bmatrix} .5 \\ .5 \end{bmatrix}$.

15. **Two-Finger Morra** Reven and Coddy play a game in which they each simultaneously present a single hand with one or two fingers extended. Reven wins if the total number of fingers extended is even. Otherwise, Coddy wins. The loser pays the winner the number of dollars equal to the total number of fingers extended.
   (a) What is the payoff matrix for the game?
   (b) Determine the expected value of the game if Reven uses the strategy $[\frac{1}{2} \quad \frac{1}{2}]$ and Coddy uses the strategy $\begin{bmatrix} \frac{1}{4} \\ \frac{3}{4} \end{bmatrix}$.

16. **Three-Finger Morra** Reven and Coddy play a game in which they each simultaneously present a single hand with one, two, or three fingers extended. Reven wins if the total number of fingers extended is even. Otherwise, Coddy wins. The loser pays the winner the number of dollars equal to the total number of fingers extended.
   (a) What is the payoff matrix for the game?
   (b) Determine the expected value of the game if Reven uses the strategy $[.4 \quad .5 \quad .1]$ and Coddy uses the strategy $\begin{bmatrix} .3 \\ .2 \\ .5 \end{bmatrix}$.

Solutions to Check Your Understanding 9.2

1. If C plays $\begin{bmatrix} .5 \\ .5 \end{bmatrix}$, the expected value (to R) is

$$[.6 \quad .4]\begin{bmatrix} 4 & -2 \\ -3 & 1 \end{bmatrix}\begin{bmatrix} .5 \\ .5 \end{bmatrix} = [1.2 \quad -.8]\begin{bmatrix} .5 \\ .5 \end{bmatrix} = [.2].$$

If C plays $\begin{bmatrix} .7 \\ .3 \end{bmatrix}$, the expected value (to R) is

$$[.6 \quad .4]\begin{bmatrix} 4 & -2 \\ -3 & 1 \end{bmatrix}\begin{bmatrix} .7 \\ .3 \end{bmatrix} = [1.2 \quad -.8]\begin{bmatrix} .7 \\ .3 \end{bmatrix} = [.6].$$

Thus, in the first case, R gains an average of .2 per play, whereas in the second, R gains .6. Since C wishes to minimize R's winnings, C should clearly play the first strategy.

2. The answer is the same as in Problem 1, since the new expected values will be 5 more than the original expected values and the first strategy will still be better for C.

9.3 Determining Optimal Mixed Strategies

As we have seen, each choice of strategies by R and C results in an expected value, representing the average payoff to R per play. In this section, we shall give a method for choosing the best strategies. Let us begin by clarifying our notion of optimality.

> **DEFINITION** To every choice of a strategy for R, there is a best counterstrategy—that is, a strategy for C that results in the least expected value e. An **optimal mixed strategy for R** is one for which the expected value against C's best counterstrategy is as large as possible.

In a similar way, we can define the optimal strategy for C.

> **DEFINITION** To every choice of a strategy for C, there is a best counterstrategy—that is, a strategy for R that results in the largest expected value e. An **optimal mixed strategy for C** is one for which the expected value against R's best counterstrategy is as small as possible.

It is most surprising that the optimal strategies for R and C in a non–strictly determined game may be determined by using linear programming. To see how this is done, let us consider a particular problem.

EXAMPLE 1   **Games as Linear Programs** Suppose that a game has payoff matrix

$$\begin{bmatrix} 5 & 3 \\ 1 & 4 \end{bmatrix}.$$

Reduce the determination of an optimal strategy for R to a linear programming problem. *Note:* This game is not strictly determined. (Why?)

SOLUTION   Suppose that R plays the strategy $\begin{bmatrix} r_1 & r_2 \end{bmatrix}$. What is C's best counterstrategy? If C plays $\begin{bmatrix} c_1 \\ c_2 \end{bmatrix}$, then the expected value of the game is

$$\begin{bmatrix} r_1 & r_2 \end{bmatrix} \begin{bmatrix} 5 & 3 \\ 1 & 4 \end{bmatrix} \begin{bmatrix} c_1 \\ c_2 \end{bmatrix} = \begin{bmatrix} 5r_1 + r_2 & 3r_1 + 4r_2 \end{bmatrix} \begin{bmatrix} c_1 \\ c_2 \end{bmatrix}$$

$$= [(5r_1 + r_2)c_1 + (3r_1 + 4r_2)c_2].$$

If C pursues the best counterstrategy, then C will try to minimize the expected value of the game. That is, C will try to minimize

$$(5r_1 + r_2)c_1 + (3r_1 + 4r_2)c_2.$$

Since $c_1 \geq 0$, $c_2 \geq 0$, and $c_1 + c_2 = 1$, this expression has as its minimum value the smaller of the terms $5r_1 + r_2$ or $3r_1 + 4r_2$. That is, if $5r_1 + r_2$ is the smaller, then C should choose the strategy $c_1 = 1$, $c_2 = 0$; whereas if $3r_1 + 4r_2$ is the smaller, then C should choose the strategy $c_1 = 0$, $c_2 = 1$. In any case, the expected value of the game if C adopts the best counterstrategy is the smaller of $5r_1 + r_2$ and $3r_1 + 4r_2$. The goal of

R is to maximize this expected value. In other words, the mathematical problem R faces is this:

Maximize the minimum of $5r_1 + r_2$ and $3r_1 + 4r_2$,

where $r_1 \geq 0$, $r_2 \geq 0$, $r_1 + r_2 = 1$.

Let v denote the minimum of $5r_1 + r_2$ and $3r_1 + 4r_2$. Clearly, $v > 0$. Then

$$5r_1 + r_2 \geq v$$
$$3r_1 + 4r_2 \geq v. \tag{1}$$

Maximizing v is the same as minimizing $1/v$. Moreover, the inequalities (1) may be rewritten in the form

$$5\frac{r_1}{v} + \frac{r_2}{v} \geq 1$$

$$3\frac{r_1}{v} + 4\frac{r_2}{v} \geq 1. \tag{2}$$

Moreover, since $r_1 \geq 0$, $r_2 \geq 0$, and $r_1 + r_2 = 1$, we see that

$$\frac{r_1}{v} \geq 0, \qquad \frac{r_2}{v} \geq 0, \qquad \frac{r_1}{v} + \frac{r_2}{v} = \frac{1}{v}. \tag{3}$$

This suggests that we introduce the new variables

$$y_1 = \frac{r_1}{v}, \qquad y_2 = \frac{r_2}{v}.$$

Then (3) and (2) may be rewritten as

$$y_1 + y_2 = \frac{1}{v}$$
$$5y_1 + y_2 \geq 1$$
$$3y_1 + 4y_2 \geq 1$$
$$y_1 \geq 0, \quad y_2 \geq 0.$$

We wish to minimize $1/v$, so we may finally state our original question in terms of a linear programming problem: Minimize $y_1 + y_2$ subject to the constraints

$$\begin{cases} 5y_1 + y_2 \geq 1 \\ 3y_1 + 4y_2 \geq 1 \\ y_1 \geq 0, \quad y_2 \geq 0. \end{cases}$$

» Now Try Exercise 1

In terms of the solution to this linear programming problem, we may calculate R's optimal strategy as follows:

$$r_1 = vy_1 \qquad r_2 = vy_2, \qquad \text{where} \quad v = \frac{1}{y_1 + y_2}.$$

In the preceding derivation, it was essential that the entries of the matrix were positive numbers, for this is how we concluded that $v > 0$. The same reasoning used in Example 1 can be used in general to convert the determination of R's optimal strategy to a linear programming problem, *provided that the payoff matrix has positive entries*. If the payoff matrix does not have positive entries, then just add a sufficiently large positive constant to each of the entries so as to give a matrix with positive entries. The new matrix will have the same optimal strategy as the original one. However, since all of its entries are positive, we may use the previous reasoning to reduce determination of the optimal strategy to a linear programming problem.

Optimal Mixed Strategy for R

Let the payoff matrix of a game be

$$\begin{bmatrix} a_{11} & a_{12} & \cdots & a_{1n} \\ a_{21} & a_{22} & \cdots & a_{2n} \\ \vdots & \vdots & & \vdots \\ a_{m1} & a_{m2} & \cdots & a_{mn} \end{bmatrix},$$

where all entries of the matrix are positive numbers. Let $y_1, y_2, \ldots, y_m$ be chosen so as to minimize

$$y_1 + y_2 + \cdots + y_m,$$

subject to the constraints

$$\begin{cases} y_1 \geq 0, y_2 \geq 0, \ldots, y_m \geq 0 \\ a_{11}y_1 + a_{21}y_2 + \cdots + a_{m1}y_m \geq 1 \\ a_{12}y_1 + a_{22}y_2 + \cdots + a_{m2}y_m \geq 1 \\ \quad\quad\quad\quad \vdots \\ a_{1n}y_1 + a_{2n}y_2 + \cdots + a_{mn}y_m \geq 1. \end{cases}$$

Let

$$v = \frac{1}{y_1 + y_2 + \cdots + y_m}.$$

Then an optimal strategy for R is $\begin{bmatrix} r_1 & r_2 & \cdots & r_m \end{bmatrix}$, where

$$r_1 = vy_1, \quad r_2 = vy_2, \quad \ldots, \quad r_m = vy_m.$$

Furthermore, if C adopts the best counterstrategy, then the expected value is v.

Note that the determination of $y_1, y_2, \ldots, y_m$ is a linear programming problem whose solution can be obtained by using either the graphical method of Chapter 3 (if $m = 2$) or the simplex method of Chapter 4 (any m). The next example illustrates the preceding result.

EXAMPLE 2 **Finding the Optimal Strategy for R** Suppose that a game has payoff matrix

$$\begin{bmatrix} 5 & 3 \\ 1 & 4 \end{bmatrix}.$$

(a) Determine an optimal strategy for R.
(b) Determine the expected payoff to R if C uses the best counterstrategy.

SOLUTION (a) The associated linear programming problem asks us to minimize $y_1 + y_2$, subject to the constraints

$$\begin{cases} y_1 \geq 0, \quad y_2 \geq 0 \\ 5y_1 + y_2 \geq 1 \\ 3y_1 + 4y_2 \geq 1. \end{cases}$$

In Fig. 1, on the next page, we have sketched the feasible set for this problem and evaluated the objective function at each vertex.

The minimum value of $y_1 + y_2$ is $\frac{5}{17}$ and occurs when $y_1 = \frac{3}{17}$ and $y_2 = \frac{2}{17}$. Further,

$$v = \frac{1}{y_1 + y_2} = \frac{1}{\frac{5}{17}} = \frac{17}{5}$$

$$r_1 = vy_1 = \frac{17}{5} \cdot \frac{3}{17} = \frac{3}{5}$$

$$r_2 = vy_2 = \frac{17}{5} \cdot \frac{2}{17} = \frac{2}{5}.$$

Thus, the optimal strategy for R is $\begin{bmatrix} \frac{3}{5} & \frac{2}{5} \end{bmatrix}$.

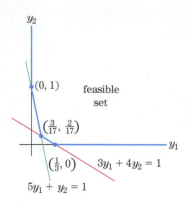

Vertex	$y_1 + y_2$
$(0, 1)$	1
$(\frac{3}{17}, \frac{2}{17})$	$\frac{5}{17}$
$(\frac{1}{3}, 0)$	$\frac{1}{3}$

Figure 1

(b) The expected value against the best counterstrategy is $v = \frac{17}{5}$. «

There is a similar linear programming technique for determining the optimal mixed strategy for C.

Optimal Mixed Strategy for C

Let the payoff matrix of a game be

$$\begin{bmatrix} a_{11} & a_{12} & \cdots & a_{1n} \\ a_{21} & a_{22} & \cdots & a_{2n} \\ \vdots & \vdots & & \vdots \\ a_{m1} & a_{m2} & \cdots & a_{mn} \end{bmatrix},$$

where all entries of the matrix are positive numbers. Let $z_1, z_2, \ldots, z_n$ be chosen so as to maximize

$$z_1 + z_2 + \cdots + z_n,$$

subject to the constraints

$$\begin{cases} z_1 \geq 0, z_2 \geq 0, \ldots, z_n \geq 0 \\ a_{11}z_1 + a_{12}z_2 + \cdots + a_{1n}z_n \leq 1 \\ a_{21}z_1 + a_{22}z_2 + \cdots + a_{2n}z_n \leq 1 \\ \quad\quad\quad \vdots \\ a_{m1}z_1 + a_{m2}z_2 + \cdots + a_{mn}z_n \leq 1. \end{cases}$$

Let $v = 1/(z_1 + z_2 + \cdots + z_n)$. Then an optimal strategy for C is

$$\begin{bmatrix} c_1 \\ c_2 \\ \vdots \\ c_n \end{bmatrix},$$

where $c_1 = vz_1, c_2 = vz_2, \ldots, c_n = vz_n$.

EXAMPLE 3 **Finding the Optimal Strategy for C** Determine the optimal strategy for C for the game with payoff matrix

$$\begin{bmatrix} 5 & 3 \\ 1 & 4 \end{bmatrix}.$$

SOLUTION We must maximize $z_1 + z_2$, subject to the constraints

$$\begin{cases} z_1 \geq 0, \quad z_2 \geq 0 \\ 5z_1 + 3z_2 \leq 1 \\ z_1 + 4z_2 \leq 1. \end{cases}$$

In Fig. 2, we have sketched the feasible set for this problem and evaluated the objective function at each vertex. The maximum value of $z_1 + z_2$ is $\frac{5}{17}$ and occurs when $z_1 = \frac{1}{17}$ and $z_2 = \frac{4}{17}$. Therefore, $v = \frac{17}{5}$, and the optimal strategy for C is $\begin{bmatrix} c_1 \\ c_2 \end{bmatrix}$, where

$$c_1 = vz_1 = \frac{17}{5} \cdot \frac{1}{17} = \frac{1}{5}$$

$$c_2 = vz_2 = \frac{17}{5} \cdot \frac{4}{17} = \frac{4}{5}.$$

»» Now Try Exercise 5

Notice that in both Examples 2 and 3, we obtained $v = \frac{17}{5}$. This was not just a coincidence. This phenomenon always occurs, and the number v is called the **value** of the

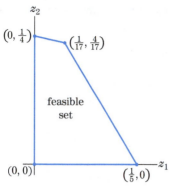

Vertex	$z_1 + z_2$
$(0, 0)$	0
$(0, \frac{1}{4})$	$\frac{1}{4}$
$(\frac{1}{17}, \frac{4}{17})$	$\frac{5}{17}$
$(\frac{1}{5}, 0)$	$\frac{1}{5}$

Figure 2

game. An easy computation shows that, for the matrix of Examples 2 and 3, when R and C each use their optimal strategies, the expected value is $v = \frac{17}{5}$. That is,

$$\begin{bmatrix} \frac{3}{5} & \frac{2}{5} \end{bmatrix} \begin{bmatrix} 5 & 3 \\ 1 & 4 \end{bmatrix} \begin{bmatrix} \frac{1}{5} \\ \frac{4}{5} \end{bmatrix} = \begin{bmatrix} \frac{17}{5} \end{bmatrix}.$$

Let us briefly reconsider the calculations of optimal strategies for R and C. We begin in every case with the matrix for the game, A. Let us assume that A is an $m \times n$ matrix and that each entry of A is a positive number. To find the optimal strategy for C, we find the matrix Z that maximizes the objective function EZ, subject to the constraints $AZ \leq B$ and $Z \geq 0$, where

$$E = \begin{bmatrix} 1 & 1 & 1 & \cdots & 1 \end{bmatrix}$$
$$\underbrace{\phantom{E = \begin{bmatrix} 1 & 1 & 1 & \cdots & 1 \end{bmatrix}}}_{n \text{ entries}}$$

$$Z = \begin{bmatrix} z_1 \\ z_2 \\ \vdots \\ z_n \end{bmatrix} \quad \text{and} \quad B = \left.\begin{bmatrix} 1 \\ 1 \\ \vdots \\ 1 \end{bmatrix}\right\} m \text{ entries.}$$

The **dual** of the linear programming problem associated with finding an optimal strategy for C is to find the matrix Y that minimizes the objective function $B^T Y$, subject to the constraints $A^T Y \geq E^T$ and $Y \geq 0$, where

$$Y = \begin{bmatrix} y_1 \\ y_2 \\ \vdots \\ y_m \end{bmatrix}.$$

But this is exactly the problem of finding an optimal mixed strategy for R.

> **Optimal Strategies**
>
> The problem of finding the optimal strategy for R is a linear programming problem whose dual is the problem of finding an optimal strategy for C, and vice versa.

Our previous work with duality (in Chapter 4) leads us to conclude the following:

1. If there exists an optimal strategy for R, then there exists an optimal strategy for C, and vice versa.

2. The minimum of the objective function $y_1 + y_2 + \cdots + y_m$ and the maximum of the objective function $z_1 + z_2 + \cdots + z_n$ are equal, since the linear programming problems are duals of each other. Hence, the value of the optimal strategy for C is the same as the value of the optimal strategy for R; that is,

$$\frac{1}{z_1 + z_2 + \cdots + z_n} = \frac{1}{y_1 + y_2 + \cdots + y_m}.$$

EXAMPLE 4   **Optimal Strategies When Matrix Entries Are Not Positive** Determine the optimal strategies for R and C for the game with payoff matrix

$$\begin{bmatrix} 1 & -1 \\ -3 & 0 \end{bmatrix}.$$

SOLUTION   We cannot apply our technique directly, since only one of the entries of the given matrix is a positive number. However, if we add 4 to each entry, then the new matrix will be

$$\begin{bmatrix} 5 & 3 \\ 1 & 4 \end{bmatrix},$$

which does have all positive entries. These two payoff matrices have the same optimal strategies. The only difference is that the values in the new matrix are 4 more than that of the given matrix. Now, the optimal strategies and the value of a game with the new matrix were found in Examples 2 and 3 to be

$$\begin{bmatrix} \frac{3}{5} & \frac{2}{5} \end{bmatrix}, \qquad \begin{bmatrix} \frac{1}{5} \\ \frac{4}{5} \end{bmatrix}, \qquad \text{and} \qquad \frac{17}{5}.$$

Therefore, the optimal strategies for the given matrix are

$$\begin{bmatrix} \frac{3}{5} & \frac{2}{5} \end{bmatrix} \qquad \text{and} \qquad \begin{bmatrix} \frac{1}{5} \\ \frac{4}{5} \end{bmatrix},$$

and the value is $\frac{17}{5} - 4 = -\frac{3}{5}$.

>> *Now Try Exercise 7*

EXAMPLE 5   **Finding Optimal Strategies by the Simplex Method** Use the simplex method and the resulting tableau to determine the optimal strategies for the game of Example 4.

SOLUTION   As in Example 4, add 4 to each entry to get a matrix with positive entries, and set up the tableau for finding the optimal strategy for C. (We choose this linear programming problem because it is a maximization problem.) The transformed matrix A is

$$\begin{bmatrix} 5 & 3 \\ 1 & 4 \end{bmatrix}.$$

To find the optimal strategy for C, we need to find the values of z_1 and z_2 that maximize $z_1 + z_2$, subject to the constraints

$$\begin{cases} 5z_1 + 3z_2 \le 1 \\ z_1 + 4z_2 \le 1 \\ z_1 \ge 0, \quad z_2 \ge 0. \end{cases}$$

We set up the tableau by using slack variables t and u and display the initial and final tableaux:

$$\begin{array}{c} \\ t \\ u \\ M \end{array} \begin{array}{c} z_1 \quad z_2 \quad t \quad u \quad M \\ \left[\begin{array}{ccccc|c} 5 & 3 & 1 & 0 & 0 & 1 \\ 1 & 4 & 0 & 1 & 0 & 1 \\ \hline -1 & -1 & 0 & 0 & 1 & 0 \end{array} \right] \end{array} \qquad \begin{array}{c} \\ z_1 \\ z_2 \\ M \end{array} \begin{array}{c} z_1 \quad z_2 \quad t \quad u \quad M \\ \left[\begin{array}{ccccc|c} 1 & 0 & \frac{4}{17} & -\frac{3}{17} & 0 & \frac{1}{17} \\ 0 & 1 & -\frac{1}{17} & \frac{5}{17} & 0 & \frac{4}{17} \\ \hline 0 & 0 & \frac{3}{17} & \frac{2}{17} & 1 & \frac{5}{17} \end{array} \right] \end{array}$$

Thus, the solution is $z_1 = \frac{1}{17}$, $z_2 = \frac{4}{17}$, with $M = z_1 + z_2 = \frac{5}{17}$. Then $v = \frac{17}{5}$, and the optimal strategy for C is

$$\begin{bmatrix} vz_1 \\ vz_2 \end{bmatrix} = \begin{bmatrix} \frac{1}{5} \\ \frac{4}{5} \end{bmatrix}.$$

This agrees with previous solutions, and the value of this game is $\frac{17}{5}$, which is 4 more than the value of the original game. Thus, the value of the game is $\frac{17}{5} - 4 = -\frac{3}{5}$.

But the optimal strategy for R can be read from the final tableau, since y_1 and y_2 are the values of the variables in the dual of the problem that we solved. So $y_1 = t = \frac{3}{17}$, $y_2 = u = \frac{2}{17}$, and $M = \frac{5}{17}$. Then $v = \frac{17}{5}$, and the optimal strategy for R is $\begin{bmatrix} vy_1 & vy_2 \end{bmatrix} = \begin{bmatrix} \frac{3}{5} & \frac{2}{5} \end{bmatrix}$.

>> *Now Try Exercise 17*

We actually have a very useful fact.

Fundamental Theorem of Game Theory

Every two-person zero-sum game has a solution.

Verification of the Fundamental Theorem of Game Theory

If the given two-person game has a saddle point, then the game is strictly determined and optimal strategies for R and C are given by the position of the saddle point.

If the game is not strictly determined, then let us assume that the $m \times n$ payoff matrix A has only positive entries. We let B be an $m \times 1$ column matrix in which each entry is 1, and let E be a $1 \times n$ row matrix of 1's. Then the following apply:

1. There is an optimal feasible solution to the problem

$$\text{Maximize } M = EZ \text{ subject to } AZ \leq B \text{ and } Z \geq 0. \tag{P}$$

2. There is an optimal feasible solution to the problem

$$\text{Minimize } M = B^T Y \text{ subject to } A^T Y \geq E^T \text{ and } Y \geq 0. \tag{D}$$

3. The solutions to (P) and (D) give a solution to the game.

To see that characteristics 1 and 2 hold, we note that there is a feasible solution for the inequalities of the primal problem (P). The $n \times 1$ matrix of zeros, $Z = 0$, satisfies $AZ \leq B$. Also, there is a feasible solution for the inequalities of the dual problem (D). This can be seen by noting that, since every element of the matrix A is positive, we can find an $m \times 1$ matrix $Y \geq 0$ with sufficiently large entries to guarantee that $A^T Y \geq E^T$. Since the inequalities of both (P) and (D) have a feasible solution, the fundamental theorem of duality (Chapter 4) tells us that both (P) and (D) have optimal feasible solutions.

Let Z^* and Y^* be optimal feasible solutions of (P) and (D), respectively. Say that

$$Z^* = \begin{bmatrix} z_1^* \\ z_2^* \\ \vdots \\ z_n^* \end{bmatrix} \quad \text{and} \quad Y^* = \begin{bmatrix} y_1^* \\ y_2^* \\ \vdots \\ y_m^* \end{bmatrix}.$$

The maximum for (P),

$$M = z_1^* + z_2^* + \cdots + z_n^*,$$

equals the minimum for (D),

$$M = y_1^* + y_2^* + \cdots + y_m^*,$$

and

$$AZ^* \leq B \quad \text{and} \quad A^T Y^* \geq E^T.$$

Recall that B is an $m \times 1$ matrix of 1s and E^T is an $n \times 1$ matrix of 1s.

M must be strictly greater than zero, since at least one of the y_i^* must be > 0 in order for $A^T Y \geq E^T$ to hold. Therefore, $1/M$ is defined. We let

$$C = \begin{bmatrix} \dfrac{1}{M} z_1^* \\ \dfrac{1}{M} z_2^* \\ \vdots \\ \dfrac{1}{M} z_n^* \end{bmatrix} = \begin{bmatrix} c_1 \\ c_2 \\ \vdots \\ c_n \end{bmatrix}$$

and

$$R = \begin{bmatrix} \dfrac{1}{M} y_1^* & \dfrac{1}{M} y_2^* & \cdots & \dfrac{1}{M} y_m^* \end{bmatrix} = \begin{bmatrix} r_1 & r_2 & \cdots & r_m \end{bmatrix}.$$

Furthermore, C and R represent optimal strategies for players C and R, respectively. To verify that C and R are legitimate strategies, we note that, since $M > 0$, $Z^* \geq \mathbf{0}$, and $Y^* \geq \mathbf{0}$, every entry in C and R is nonnegative. We need to check only that

$$c_1 + c_2 + \cdots + c_n = 1 \quad \text{and} \quad r_1 + r_2 + \cdots + r_m = 1.$$

This follows directly from the definitions of M, C, and R. ◀◀

Check Your Understanding 9.3

Solutions can be found following the section exercises.

1. Determine the optimal strategy for C for the game with payoff matrix

$$\begin{bmatrix} 2 & 14 \\ 6 & 12 \\ 8 & 6 \end{bmatrix}.$$

2. Determine by inspection the optimal strategies for C for the games whose payoff matrices are given.

(a) $\begin{bmatrix} 0 & 12 \\ 4 & 10 \\ 6 & 4 \end{bmatrix}$ (b) $\begin{bmatrix} 6 & 0 \\ 4 & 3 \\ 8 & -1 \end{bmatrix}$

EXERCISES 9.3

1. Suppose a game has payoff matrix

$$\begin{bmatrix} 1 & 6 \\ 4 & 3 \end{bmatrix}.$$

Reduce the determination of an optimal strategy for R into a linear programming problem. Just set up the problem, showing the constraints and the objective function.

2. Suppose a game has payoff matrix

$$\begin{bmatrix} 10 & 6 \\ 5 & 7 \end{bmatrix}.$$

Reduce the determination of an optimal strategy for R into a linear programming problem. Just set up the problem, showing the constraints and the objective function.

3. Rework Exercise 1 for C instead of R.

4. Rework Exercise 2 for C instead of R.

In Exercises 5–12, determine the value of the game and the optimal strategy for R and for C.

5. $\begin{bmatrix} 2 & 4 \\ 5 & 3 \end{bmatrix}$ 6. $\begin{bmatrix} 2 & 3 \\ 3 & 2 \end{bmatrix}$

7. $\begin{bmatrix} 3 & -6 \\ -5 & 4 \end{bmatrix}$ 8. $\begin{bmatrix} 5 & 2 \\ 7 & 1 \end{bmatrix}$

9. $\begin{bmatrix} 4 & 1 \\ 2 & 4 \end{bmatrix}$ 10. $\begin{bmatrix} 5 & -8 \\ 3 & 6 \end{bmatrix}$

11. $\begin{bmatrix} 5 & -3 \\ -3 & 1 \end{bmatrix}$ 12. $\begin{bmatrix} 4 & -2 \\ -3 & 1 \end{bmatrix}$

In Exercises 13–16, determine the value of the game and the optimal strategy for R.

13. $\begin{bmatrix} 3 & 5 & -1 \\ 4 & -1 & 6 \end{bmatrix}$ 14. $\begin{bmatrix} -2 & 1 & 0 \\ 2 & 0 & 1 \end{bmatrix}$

15. $\begin{bmatrix} 1 & -1 & 6 \\ -1 & 2 & -3 \end{bmatrix}$ 16. $\begin{bmatrix} 8 & 2 & 5 \\ 1 & 8 & 4 \end{bmatrix}$

In Exercises 17–20, determine the value of the game and the optimal strategy for C.

17. $\begin{bmatrix} -3 & 1 \\ 4 & -1 \\ 1 & 0 \end{bmatrix}$ 18. $\begin{bmatrix} 0 & 2 \\ 2 & -1 \\ 1 & 0 \end{bmatrix}$

19. $\begin{bmatrix} 1 & 0 \\ -2 & 4 \\ -1 & -1 \end{bmatrix}$ 20. $\begin{bmatrix} 4 & -2 \\ 1 & 2 \\ 3 & 1 \end{bmatrix}$

21. **Smuggler's Strategy** A rumrunner attempts to smuggle rum into a country having two ports. Each day, the coast guard is able to patrol only one of the ports. If the rumrunner enters via an unpatrolled port, he will be able to sell his rum for a profit of $7000. If he enters the first port and it is patrolled that day, he is certain to be caught and will have his rum (worth $1000) confiscated and be fined $1000. If he enters the second port (which is big and crowded) and it is patrolled that day, he will have time to jettison his cargo and thereby escape a fine.
   (a) What is the payoff matrix for the game? Denote the two ports by I and II, and write the entries of the matrix in thousands of dollars.
   (b) What is the optimal strategy for the rumrunner?
   (c) What is the optimal strategy for the coast guard?
   (d) How profitable is rumrunning? That is, what is the value of the game?

22. **Which Hand?** Ralph puts a coin in one of his hands, and Carl tries to guess which hand holds the coin. If Carl guesses incorrectly, he must pay Ralph $2. If Carl guesses correctly, then Ralph must pay him $3 if the coin was in the left hand and $1 if it was in the right.
   (a) What is the payoff matrix for the game?
   (b) What is the optimal strategy for Ralph?
   (c) What is the optimal strategy for Carl?
   (d) Whom does this game favor?

23. **Football** Suppose that, when the offense calls a running play, they gain an average of 1 yard if the defense anticipates a running play, and gain an average of 3 yards if the defense anticipates a passing play. Suppose that, when the offense calls a

passing play, they gain an average of 7 yards if the defense anticipates a running play, and they lose an average of 4 yards if the defense anticipates a passing play. Determine the optimal strategies for the offense and the defense, and the corresponding expected value.

24. **Baseball** Whenever Randy pitches to Chris, he throws either a fastball or a slider. When Randy pitches a fastball, Chris hits the ball with a probability of .25 when he anticipates a fastball, and hits the ball with a probability of .1 when he anticipates a slider. When Randy pitches a slider, Chris hits the ball with a probability of .15 when he anticipates a fastball, and hits the ball with a probability of .3 when he anticipates a slider. Determine the optimal strategies for Randy and Chris, and the corresponding expected probability that Chris will hit the ball.

25. **Two-Finger Morra** Reven and Coddy play a game in which they each simultaneously present a single hand with one or two fingers extended. Reven wins if the total number of fingers extended is even. Otherwise, Coddy wins. The loser pays the winner the number of dollars equal to the total number of fingers extended.
(a) What is the payoff matrix for the game?
(b) Determine the expected value of the game and the optimal strategies for Reven and Coddy.

26. **Three-Finger Morra** Reven and Coddy play a game in which they each simultaneously present a single hand with one, two, or three fingers extended. Reven wins if the total number of

fingers extended is even. Otherwise, Coddy wins. The loser pays the winner the number of dollars equal to the total number of fingers extended.
(a) What is the payoff matrix for the game?
(b) Determine the expected value of the game and the optimal strategies for Reven and Coddy.

27. **Advertising Strategies** The Carter Company can choose between two advertising strategies (I and II). Its most important competitor, Rosedale Associates, has a choice of three advertising strategies (a, b, c). The estimated payoff to Rosedale Associates away from the Carter Company is given by the payoff matrix

$$\begin{array}{c} & \begin{array}{cc} \text{I} & \text{II} \end{array} \\ \begin{array}{c} a \\ b \\ c \end{array} & \left[\begin{array}{cc} -2 & 1 \\ 2 & -3 \\ 1 & -2 \end{array} \right], \end{array}$$

where the entries represent thousands of dollars per week. Determine the optimal strategies for each company.

TECHNOLOGY EXERCISES

28. **Three-Finger Morra** Use Excel or Wolfram|Alpha to find the expected value and the optimal strategies for the Three-Finger Morra game in Exercise 26.

29. **Two-Finger Morra** Use Excel or Wolfram|Alpha to find the expected value and the optimal strategies for the Two-Finger Morra game in Exercise 25.

Solutions to Check Your Understanding 9.3

1. The associated linear programming problem is as follows: Maximize $z_1 + z_2$ subject to the constraints

$$\begin{cases} z_1 \geq 0, \quad z_2 \geq 0 \\ 2z_1 + 14z_2 \leq 1 \\ 6z_1 + 12z_2 \leq 1 \\ 8z_1 + 6z_2 \leq 1. \end{cases}$$

In Fig. 3 we have sketched the feasible set and evaluated the objective function at each vertex. The maximum value of $z_1 + z_2$ is $\frac{4}{30}$, which is achieved at the vertex $\left(\frac{3}{30}, \frac{1}{30} \right)$. Therefore,

$$v = \frac{1}{z_1 + z_2} = \frac{1}{\frac{4}{30}} = \frac{30}{4}$$

$$c_1 = v \cdot z_1 = \frac{30}{4} \cdot \frac{3}{30} = \frac{3}{4}$$

$$c_2 = v \cdot z_2 = \frac{30}{4} \cdot \frac{1}{30} = \frac{1}{4}.$$

That is, the optimal strategy for C is $\begin{bmatrix} \frac{3}{4} \\ \frac{1}{4} \end{bmatrix}$.

2. (a) $\begin{bmatrix} \frac{3}{4} \\ \frac{1}{4} \end{bmatrix}$. If we add 2 to each entry, we obtain the payoff matrix of Problem 1, so these two games have the same optimal strategies.

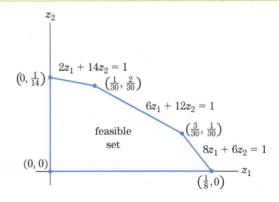

Vertex	$z_1 + z_2$
$(0, 0)$	0
$\left(0, \frac{1}{14}\right)$	$\frac{1}{14}$
$\left(\frac{1}{30}, \frac{2}{30}\right)$	$\frac{3}{30}$
$\left(\frac{3}{30}, \frac{1}{30}\right)$	$\frac{4}{30}$
$\left(\frac{1}{8}, 0\right)$	$\frac{1}{8}$

Figure 3

(b) Always play column 2. This game is strictly determined and has the entry 3 as saddle point. It is a good idea always to check for a saddle point before looking for a mixed strategy.

CHAPTER 9	Summary

KEY TERMS AND CONCEPTS	EXAMPLES

9.1 Games and Strategies

A **zero-sum game** is one in which a payoff to one player is a loss of the same amount to the other.

A **payoff matrix** is a matrix whose rows and columns are labeled with the strategies of the players and whose entries are the payoffs to the row player (equivalently, the losses to the column player).

In the payoff matrix $\begin{bmatrix} 5 & -8 \\ 3 & 1 \end{bmatrix}$, if C selects the 1st column and R selects the 2nd row, then C pays \$3 to R.

A **saddle point** is an entry in a payoff matrix that is simultaneously the least in its row and largest in its column.

In the previous matrix, 1 is a saddle point.

A **strictly determined game** is one with a saddle point. In a strictly determined game, the optimal pure strategy for each player is to choose a row or column containing a saddle point. If a game has more than one saddle point, then the saddle points are equal. The value of the saddle point is said to be the **value** of the game.

The game represented by the previous matrix is strictly determined. C should always select the 2nd column, and R should always select the 2nd row.

9.2 Mixed Strategies

In a **mixed strategy**, each column and row is chosen with a specified probability. The probabilities for R are given in a row matrix, and the probabilities for C are given in a column matrix.

Possible mixed strategies for the game with payoff matrix $\begin{bmatrix} -1 & 5 \\ 2 & -3 \end{bmatrix}$ are $[.9 \ .1]$ and $\begin{bmatrix} .6 \\ .4 \end{bmatrix}$.

The **expected value** is the average payoff to R and is calculated with

$$[R's \ probs] \begin{bmatrix} payoff \\ \hline matrix \end{bmatrix} \begin{bmatrix} C's \\ p \\ r \\ o \\ b \\ s \end{bmatrix}.$$

The expected value of the previous mixed strategy is 1.26, since

$$[.9 \ .1] \begin{bmatrix} -1 & 5 \\ 2 & -3 \end{bmatrix} \begin{bmatrix} .6 \\ .4 \end{bmatrix} = [1.26].$$

9.3 Determining Optimal Mixed Strategies

Optimal mixed strategies for each player are found by solving a pair of linear programming problems. See pages 419 and 420 for details.

CHAPTER 9	Fundamental Concept Check Exercises

1. What do the individual entries of a payoff matrix represent?

2. What is the difference between a pure strategy and a mixed strategy?

3. What is a zero-sum game?

4. Describe the optimal pure strategies for R and for C.

5. When is an entry of a payoff matrix a saddle point?

6. What is a strictly determined game, and what is its value?

7. What is the expected value of a pair of mixed strategies, and how is it computed?

8. What is meant by the optimal mixed strategies of R and C, and how are they computed?

CHAPTER 9 Review Exercises

In Exercises 1–4, state whether or not the games having the given payoff matrices are strictly determined. For those that are, give the optimal pure strategies and the values of the strategies.

1. $\begin{bmatrix} 5 & -1 & 1 \\ -3 & 5 & 1 \\ 4 & 3 & 2 \end{bmatrix}$ **2.** $\begin{bmatrix} 1 & 2 & 3 \\ 3 & 2 & 1 \end{bmatrix}$

3. $\begin{bmatrix} 0 & 1 \\ 1 & 0 \\ 2 & -1 \end{bmatrix}$ **4.** $\begin{bmatrix} 2 & 1 & 2 \\ -1 & 0 & 3 \\ 4 & 1 & -4 \end{bmatrix}$

In Exercises 5–8, determine the expected value of each pair of mixed strategies for the given payoff matrix.

5. $\begin{bmatrix} \frac{3}{4} & \frac{1}{4} \end{bmatrix}$, $\begin{bmatrix} \frac{1}{3} \\ \frac{2}{3} \end{bmatrix}$; $\begin{bmatrix} 0 & 24 \\ 12 & -36 \end{bmatrix}$

6. $\begin{bmatrix} \frac{1}{2} & \frac{1}{2} \end{bmatrix}$, $\begin{bmatrix} \frac{1}{3} \\ \frac{1}{3} \\ \frac{1}{3} \end{bmatrix}$; $\begin{bmatrix} -6 & 6 & 0 \\ 0 & -12 & 24 \end{bmatrix}$

7. $\begin{bmatrix} .2 & .3 & .5 \end{bmatrix}$, $\begin{bmatrix} .4 \\ .6 \end{bmatrix}$; $\begin{bmatrix} 1 & 0 \\ -3 & 1 \\ 0 & 5 \end{bmatrix}$

8. $\begin{bmatrix} .1 & .1 & .8 \end{bmatrix}$, $\begin{bmatrix} .4 \\ .3 \\ .3 \end{bmatrix}$; $\begin{bmatrix} 0 & 1 & 3 \\ -1 & 0 & 2 \\ -3 & -2 & 0 \end{bmatrix}$

Determine the optimal strategies for R and for C for the games with the payoff matrices of Exercises 9 and 10.

9. $\begin{bmatrix} -3 & 4 \\ 2 & -2 \end{bmatrix}$ **10.** $\begin{bmatrix} 3 & -6 \\ -4 & 4 \end{bmatrix}$

11. Determine the optimal strategy for R for the game with payoff matrix

$$\begin{bmatrix} 5 & -2 & 0 \\ 1 & 4 & 1 \end{bmatrix}.$$

12. Determine the optimal strategy for C for the game with payoff matrix

$$\begin{bmatrix} 1 & 3 \\ 3 & 1 \\ 4 & 2 \end{bmatrix}.$$

13. A Card Matching Game Ruth and Carol play the following game. Each has two cards, a two and a six. Each puts one of her cards on the table. If both put down the same rank, Ruth pays Carol $3. Otherwise, Carol pays Ruth as many dollars as the rank of Carol's card.
(a) Find the optimal strategies for Ruth and Carol.
(b) Whom does this game favor?

14. Investment Strategy An investor is considering purchasing one of three stocks. Stock A is regarded as conservative, stock B as speculative, and stock C as highly risky. If the economic growth during the coming year is strong, then stock A should increase in value by $3000, stock B by $6000, and stock C by

$15,000. If the economic growth during the next year is average, then stock A should increase in value by $2000, stock B by $2000, and stock C by $1000. If the economic growth is weak, then stock A should increase in value by $1000 and stocks B and C decrease in value by $3000 and $10,000, respectively.
(a) Set up the 3×3 payoff matrix showing the investing gains for the possible stock purchases and levels of economic growth.
(b) What is the investor's optimal strategy?

Conceptual Exercises

15. Show that, in a two-person game with payoff matrix

$$\begin{bmatrix} a_{11} & a_{12} \\ a_{21} & a_{22} \end{bmatrix},$$

the only games that are *not* strictly determined are those for which either

$$a_{11} > a_{12}, \quad a_{11} > a_{21}, \quad a_{21} < a_{22}, \quad a_{12} < a_{22}$$

or

$$a_{11} < a_{12}, \quad a_{11} < a_{21}, \quad a_{21} > a_{22}, \quad a_{12} > a_{22}.$$

16. In Check Your Understanding Problem 1 in Section 9.3, we determined the optimal strategy for C in a game with payoff matrix

$$\begin{bmatrix} 2 & 14 \\ 6 & 12 \\ 8 & 6 \end{bmatrix}$$

to be

$$\begin{bmatrix} \frac{3}{4} \\ \frac{1}{4} \end{bmatrix}.$$

Consider the effect of changing the entry in the first row, first column of the payoff matrix to $2 + h$, where h is any positive integer. For what values of h does the optimal strategy for C remain the same?

17. Assume that the payoff matrix for a two-person game is given by

$$\begin{bmatrix} a_{11} & a_{12} \\ a_{21} & a_{22} \end{bmatrix}.$$

Then assume that R uses a strategy $[r \quad 1 - r]$. If C chooses column 1, then R can expect a return of $a_{11}r + a_{21}(1 - r)$. If C chooses column 2, then R can expect a return of $a_{12}r + a_{22}(1 - r)$.
(a) Explain why the intersection of the lines

$$y = a_{11}r + a_{21}(1 - r)$$

and

$$y = a_{12}r + a_{22}(1 - r)$$

gives the value of r that is best for R.
(b) How would you find the best strategy for C, using a comparable argument?

18. Using the technique of Exercise 17, find a general formula for the best strategy for R in terms of the entries in the payoff matrix.

Simulating the Outcomes of Mixed-Strategy Games

A mixed strategy requires the use of a device that will randomly select a row (or column) of the payoff matrix subject to a specified probability. For instance, if the strategy for *R* is [.6   .4], the device should select the first row 60% of the time and select the second row 40% of the time. The TI-84 Plus calculator is well suited to handle this task by use of the *randInt* function along with the *relational operators*. *Note:* For this project, we will assume that STATWIZARDS is set to OFF in the MODE menu.

Each time **randInt(1,10)** in MATH PROB menu is called, an integer between 1 and 10 is generated. See Fig. 1. A statement containing a relational operator (found in 2nd [TEST]), often an inequality statement, returns the value 1 when true and the value 0 when false. See Fig. 2. The instructions in Fig. 3 are used to select a row from a matrix with two rows.

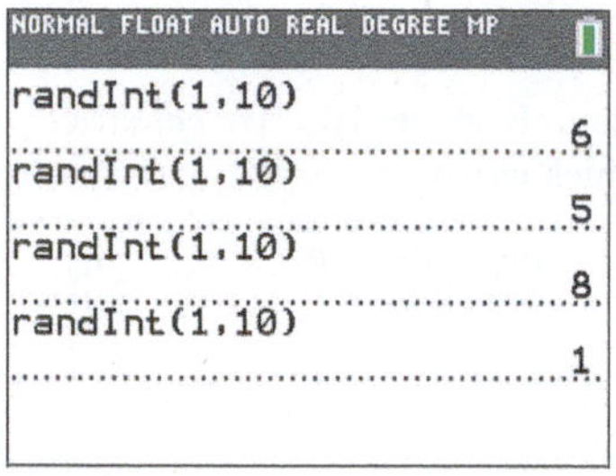

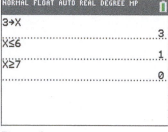

Figure 1

Figure 2

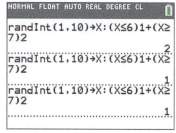

Figure 3

1. Explain why 60% of the time, the instructions in Fig. 3 return a 1 and 40% of the time, they return a 2. Type the instructions into your calculator, press the ENTER key 20 times, and count the number of 1s and 2s. Are there approximately twelve 1s and eight 2s?

2. Explain why the instructions in Fig. 3 should not be replaced with this instruction:

   (randInt(1,10)≤6)1+(randInt(1,10)≥7)2

   Type this line into your calculator, and press the ENTER key several times to convince yourself that it does not always produce 1s and 2s.

   In part 1, you pressed the ENTER key repeatedly and manually kept track of the outcomes. The TI-84 Plus can carry out the instructions many times, store the outcomes in a list, and analyze the list. The first instruction in Fig. 4 generates 100 random numbers between 1 and 10 and places them in the list L_1. The second instruction of Fig. 4 converts each number in the list L_1 to a row number by using the strategy [.6   .4] and places the 100 row choices into the list L_3. Figure 6 uses the window settings shown in Fig. 5 to display the histogram for L_3. We see that the second row was selected 44 times out of 100 times. This is close to the expected 40%.

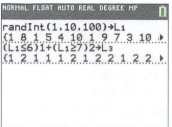

Figure 4

Figure 5

Figure 6

3. Simulate 100 column selections, using the strategy $\begin{bmatrix} .2 \\ .8 \end{bmatrix}$, and create a histogram showing the number of times each column was selected. How many times was the second column selected?

   In Section 9.3, the optimal mixed strategies for the payoff matrix $\begin{bmatrix} 5 & 3 \\ 1 & 4 \end{bmatrix}$ were found to be [.6 .4] for R and $\begin{bmatrix} .2 \\ .8 \end{bmatrix}$ for C, and the value of the game was found to be $\frac{17}{5}$. In Figs. 7 and 8, 100 games are simulated and the payoffs for the games are stored in the list L_3. Fig. 10 shows the histogram for L_3 obtained by using the window settings in Fig. 9. For instance, we see that a payoff of 3 occurred in 59 of the 100 games. Fig. 11 shows that the average of the 100 payoffs, 3.22, is close to the value of the game, 3.4.

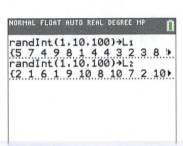

Figure 7

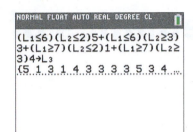

Figure 8

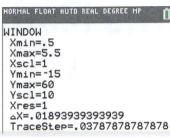

Figure 9

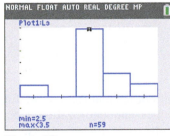

Figure 10

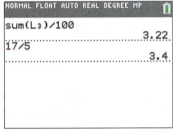

Figure 11

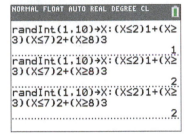

Figure 12

4. Explain in detail what is being calculated in Fig. 8.

5. Create a list of the payoffs for playing 100 games with the payoff matrix $\begin{bmatrix} 5 & 3 \\ 1 & 4 \end{bmatrix}$, using the strategies [.5 .5] for R and $\begin{bmatrix} .3 \\ .7 \end{bmatrix}$ for C, and compute the average of the payoffs.

6. Suppose that a payoff matrix has three rows and that R's strategy is [.2 .5 .3]. Explain why Fig. 12 simulates the selection of a row by R. Type the instructions into your calculator, press the [ENTER] key 40 times, and count the number of 1s, 2s, and 3s. Are there approximately eight 1s, twenty 2s, and twelve 3s?

7. Consider the 3×2 matrix in the first Check Your Understanding problem of Section 9.3. Suppose that the strategies for R and C are [.2 .5 .3] and $\begin{bmatrix} .75 \\ .25 \end{bmatrix}$, respectively. Simulate the payoffs for 100 games, display the histogram for the payoffs, and calculate the average value of the payoffs.

This chapter presents several topics in the mathematics of finance, including compound and simple interest, annuities, and amortization. Computations are carried out in the traditional way, with formulas, and with technology.

10.1 Interest

Compound and Simple Interest

When you deposit money into a savings account, the bank pays you a fee for the use of your money. This fee is called **interest** and is determined by the amount deposited, the duration of the deposit, and the interest rate. The amount deposited is called the **principal** or **present value**, and the amount to which the principal grows (after the addition of interest) is called the **future value** or **balance**.

The entries in a hypothetical bank statement are shown in Table 1. Note the following facts about this statement:

1. The principal is $100.00. The future value after 1 year is $104.06.
2. Interest is being paid four times per year (or, in financial language, *quarterly*).
3. Each quarter, the amount of the interest is 1% of the previous balance. That is, $1.00 is 1% of $100.00, $1.01 is 1% of $101.00, and so on. Since $4 \times 1\%$ is 4%, we say that the money is earning 4% *annual interest compounded quarterly*.

Table 1

Date	Deposits	Withdrawals	Interest	Balance
1/1/16	$100.00			$100.00
4/1/16			$1.00	101.00
7/1/16			1.01	102.01
10/1/16			1.02	103.03
1/1/17			1.03	104.06

As in the statement shown in Table 1, interest rates are usually stated as *annual interest rates*, with the interest to be **compounded** (i.e., computed) a certain number of times per year. Some common frequencies for compounding are listed in Table 2.

Table 2

Number of Interest Periods Per Year	Length of Each Interest Period	Interest Compounded
1	One year	Annually
2	Six months	Semiannually
4	Three months	Quarterly
12	One month	Monthly
52	One week	Weekly
365	One day	Daily

Of special importance is the *interest rate per period*, denoted i, which is calculated by dividing the annual interest rate by the number of interest periods per year. For example, in our statement in Table 1, the annual interest rate is 4%, the interest is compounded quarterly, and the interest rate per period is $4\%/4 = \frac{.04}{4} = .01$.

> **DEFINITION** If interest is compounded m times per year and the annual interest rate is r, then the **interest rate per period** is
>
> $$i = \frac{r}{m}.$$

EXAMPLE 1 **Determining Interest Rate Per Period** Determine the interest rate per period for each of the following annual interest rates.
(a) 3% interest compounded semiannually
(b) 2.4% interest compounded monthly

SOLUTION **(a)** The annual interest rate is 3%, and the number of interest periods is 2. Therefore,

$$i = \frac{3\%}{2} = \frac{.03}{2} = .015.$$

(b) The annual interest rate is 2.4%, and the number of interest periods is 12. Therefore,

$$i = \frac{2.4\%}{12} = \frac{.024}{12} = .002.$$

>> *Now Try Exercise 1*

Consider a savings account in which the interest rate per period is i. Then the interest earned during a period is i times the previous balance. That is, at the end of an interest period, the new balance, B_{new}, is computed by adding this interest to the previous balance, B_{previous}. Therefore,

$$B_{\text{new}} = 1 \cdot B_{\text{previous}} + i \cdot B_{\text{previous}}$$
$$B_{\text{new}} = (1 + i) \cdot B_{\text{previous}}. \tag{1}$$

Formula (1) says that the balances for successive interest periods are computed by multiplying the previous balance by $(1 + i)$.

EXAMPLE 2   **Computing Interest and Balances** Compute the balance for the first two interest periods for a deposit of $1000 at 2% interest compounded semiannually.

SOLUTION   Here, the initial balance is $1000 and $i = 1\% = .01$. Let B_1 be the balance at the end of the first interest period and B_2 be the balance at the end of the second interest period. By formula (1),

$$B_1 = (1 + .01)1000 = 1.01 \cdot 1000 = 1010.$$

Similarly, applying formula (1) again, we get

$$B_2 = 1.01 \cdot B_1 = 1.01 \cdot 1010 = 1020.10.$$

Therefore, the balance is $1010 after the first interest period and $1020.10 after the second interest period.   **»** *Now Try Exercises 37(a), (b)*

A simple formula for the balance after any number of interest periods can be derived from formula (1) as follows:

Principal (present value)	P
Balance after 1 interest period	$(1 + i)P$
Balance after 2 interest periods	$(1 + i) \cdot (1 + i)P$ or $(1 + i)^2 P$
Balance after 3 interest periods	$(1 + i) \cdot (1 + i)^2 P$ or $(1 + i)^3 P$
Balance after 4 interest periods	$(1 + i)^4 P$
$\vdots$	$\vdots$
Balance after n interest periods	$(1 + i)^n P.$

Future Value Formula for Compound Interest The future value F after n interest periods is

$$F = (1 + i)^n P, \tag{2}$$

where i is the interest rate per period in decimal form, and P is the principal (or present value).

EXAMPLE 3   **Computing Future Values** Apply formula (2) to the savings account statement discussed at the beginning of this section, and calculate the future value after **(a)** 1 year and **(b)** 5 years.

SOLUTION   **(a)** $F = (1 + i)^n P$ Future value formula for compound interest

$\qquad\quad = (1.01)^4 \cdot 100$ $n = 1 \cdot 4 = 4, i = \frac{.04}{4} = .01, P = 100$

$\qquad\quad = \$104.06$ Calculate. Round to nearest cent.

(b) $F = (1 + i)^n P$ Future value formula for compound interest

$\qquad\quad = (1.01)^{20} \cdot 100$ $n = 5 \cdot 4 = 20, i = \frac{.04}{4} = .01, P = 100$

$\qquad\quad = \$122.02$ Calculate. Round to nearest cent.   **»** *Now Try Exercise 13*

Table 3 shows the effects of interest rates (compounded quarterly) on the future value of $100.

Table 3

	Future Value	
Interest Rate	5 Years	10 Years
1%	$105.12	$110.50
2%	$110.49	$122.08
3%	$116.12	$134.83
4%	$122.02	$148.89
5%	$128.20	$164.36
6%	$134.69	$181.40
7%	$141.48	$200.16
8%	$148.59	$220.80
9%	$156.05	$243.52
10%	$163.86	$268.51

Principal = $100.00

EXAMPLE 4 **Computing a Present Value** How much money must be deposited now in order to have $1000 after 5 years if interest is paid at a 4% annual interest rate compounded quarterly?

SOLUTION As in Example 3(b), we have $i = .01$ and $n = 20$. However, now we are given F and are asked to solve for P.

$$F = (1 + i)^n P \quad \text{Future value formula for compound interest}$$

$$1000 = (1.01)^{20} P \quad F = 1000, i = \tfrac{.04}{4} = .01, n = 5 \cdot 4 = 20$$

$$P = \frac{1000}{(1.01)^{20}} \quad \text{Divide both sides by } (1.01)^{20}. \text{ Rewrite.}$$

$$P = 819.54 \quad \text{Calculate. Round to two decimal places.}$$

We say that $819.54 is the present value of $1000, 5 years from now, at 4% interest compounded quarterly. The concept of "time value of money" says that, at an interest rate of 4% compounded quarterly, $1000 in 5 years is equivalent to $819.54 now.

» Now Try Exercise 21

Compound interest problems involve the four variables P, i, n, and F. Given the values of any three of the variables, we can find the value of the fourth. As we have seen, the formula used to find the value of F is

$$F = (1 + i)^n P.$$

Solving this formula for P gives the present value formula for compound interest.

Present Value Formula for Compound Interest The present value P of F dollars to be received n interest periods in the future is

$$P = \frac{F}{(1 + i)^n},$$

where i is the interest rate per period in decimal form.

| EXAMPLE 5 | **Computing a Present Value** Determine the present value of a $10,000 payment to be received on January 1, 2027, if it is now May 1, 2018, and money can be invested at 3% interest compounded monthly. |

SOLUTION Here, $n = 104$ (the number of months between the two given dates).

$$P = \frac{F}{(1 + i)^n} \qquad \text{Present value formula for compound interest}$$

$$= \frac{10,000}{(1.0025)^{104}} \qquad F = 10,000, i = \tfrac{.03}{12} = .0025, n = 104$$

$$= 7713.02 \qquad \text{Calculate. Round to two decimal places.}$$

Therefore, $7713.02 invested on May 1, 2018, will grow to $10,000 by January 1, 2027.

» Now Try Exercise 19

The interest that we have been discussing so far is the most prevalent type of interest and is known as **compound interest**. There is another type of interest, called **simple interest**, which is used in some financial circumstances.

Interest rates for simple interest are given as an annual interest rate r. Interest is earned *only* on the principal P, and the interest is rP for each year. Therefore, at the end of the year, the new balance, B_{new} is computed by adding this interest to the previous balance, $B_{previous}$. Therefore,

$$B_{new} = B_{previous} + rP$$

This formula says that the balances for successive years are computed by adding rP to the previous balance. Therefore, the interest earned in n years is nrP. So the future value F after n years is the original amount plus the interest earned. That is,

$$F = P + nrP = 1 \cdot P + nrP = (1 + nr)P.$$

Future Value Formula for Simple Interest   The future value F after n years is

$$F = (1 + nr)P,$$

where r is the interest rate per year and P is the principal (or present value).

| EXAMPLE 6 | **Computing a Balance with Simple Interest** Calculate the future value after 4 years if $1000 is invested at 2% simple interest. |

SOLUTION

$$F = (1 + nr)P \qquad \text{Future value formula for simple interest}$$

$$= [1 + 4(.02)]1000 \qquad n = 4, r = .02, P = 1000$$

$$= (1.08)1000 \qquad \text{Multiply and add.}$$

$$= 1080 \qquad \text{Calculate.}$$

Therefore, the future value is $1080.00. **» Now Try Exercise 41**

In Example 6, had the money been invested at 2% compound interest with annual compounding, then the future value would have been $1082.43. Money invested at simple interest is earning interest only on the principal amount. However, with compound interest, after the first interest period, the interest is also earning interest.

Effective Rate of Interest

The annual rate of interest is also known as the **nominal rate** or the **stated rate**. Its true worth depends on the number of compounding periods. The nominal rate does not help you decide, for instance, whether a savings account paying 3.65% interest compounded quarterly is better than a savings account paying 3.6% interest compounded monthly. The *effective rate of interest* provides a standardized way of comparing investments.

> **DEFINITION** The **effective rate of interest** is the simple interest rate that yields the same amount of interest after one year as the compounded annual (nominal) rate of interest. The effective rate is also known as the *annual percentage yield* (APY) or the *true interest rate*.

Suppose that the annual rate of interest r is compounded m times per year. Then, with compound interest, P dollars will grow to $P(1 + i)^m$ in one year, where $i = r/m$. With simple interest r_{eff}, the balance after one year will be $P(1 + r_{eff})$. Equating the two balances,

$$P(1 + r_{eff}) = P(1 + i)^m$$
$$1 + r_{eff} = (1 + i)^m \qquad \text{Divide both sides by } P.$$
$$r_{eff} = (1 + i)^m - 1. \quad \text{Subtract 1 from both sides.} \qquad (3)$$

> **Effective Rate of Interest Formula**   If interest is compounded m times per year, then
> $$r_{eff} = (1 + i)^m - 1,$$
> where i is the interest rate per period in decimal form.

EXAMPLE 7   **Compare Two Interest Rates** Calculate and compare the effective rate of interest for savings accounts paying
(a) a nominal rate of 3.65% compounded quarterly.
(b) a nominal rate of 3.6% compounded monthly.

SOLUTION   **(a)**

$$r_{eff} = (1 + i)^m - 1 \qquad \text{Effective rate of interest formula}$$
$$= (1.009125)^4 - 1 \quad m = 4, i = \tfrac{r}{m} = \tfrac{.0365}{4} = .009125$$
$$\approx .0370 \qquad \text{Calculate. Round to four decimal places.}$$

(b)

$$r_{eff} = (1 + i)^m - 1 \qquad \text{Effective rate of interest formula}$$
$$= (1.003)^{12} - 1 \quad m = 12, i = \tfrac{r}{m} = \tfrac{.036}{12} = .003$$
$$\approx .0366 \qquad \text{Calculate. Round to four decimal places.}$$

Therefore, the first savings account is better than the second.   **» Now Try Exercise 55**

Let us summarize the key formulas developed so far.

> **Compound Interest**
>
> Future value: $F = (1 + i)^n P$
>
> Present value: $P = \dfrac{F}{(1 + i)^n}$
>
> Effective rate: $r_{eff} = (1 + i)^m - 1$
>
> Here, i is the interest rate per period, n is the total number of interest periods, and m is the number of times interest is compounded per year.

> **Simple Interest**
>
> Future value: $F = (1 + nr)P,$
>
> Here, r is the annual interest rate and n is the number of years.

Display Balances Graphing calculators can easily display successive balances in a savings account. Consider the situation from Table 1, in which $100 is deposited at 4% interest compounded quarterly. In Fig. 1, successive balances are displayed in the home screen. In Fig. 2, successive balances are graphed, and in Fig. 3, they are displayed in a table.

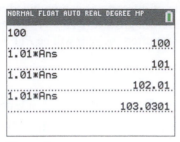

Figure 1

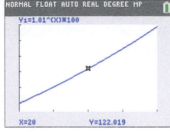

Figure 2

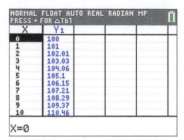

Figure 3

Display Successive Balances on the Home Screen Successive balances can be determined with the relation $B_{\text{new}} = (1 + i)B_{\text{previous}}$. In Fig. 1, after the principal (**100**) is entered, the last displayed value (**100**) is assigned to **Ans**. The instruction **1.01*Ans** generates the next balance (**101**) and assigns it to **Ans**. Each subsequent press of ENTER generates another balance.

Display Balances in a Graph In order to obtain the graph in Fig. 2, **Y₁** was set to **1.01ˣ*100** in the **Y=** editor—that is, $(1 + i)^X \cdot P$. Here, X, instead of n, is used to represent the number of interest periods. Then the window was set to [0, 40] *by* [90, 150], with an **Xscl**= 10 and a **Yscl**= 10. Then, we pressed GRAPH to display the function and pressed TRACE to obtain the trace cursor. *Note:* After **Y₁** has been defined, values of **Y₁** can be displayed on the home screen by entering expressions such as **Y₁(20)**.

Display Balances in a Table In order to obtain the table in Fig. 3, **Y₁** was set to **1.01ˣ*100** in the **Y=** editor. Then, 2nd [TBLSET] was pressed to bring up the TABLE SETUP screen. **TblStart** was set to 0, and **ΔTbl** was set to 1. Finally, 2nd [TABLE] was pressed to bring up the table.

Financial Functions The Finance menu contains the functions Eff and Nom that are used to calculate effective and nominal interest rates. See Appendix B for details.

TVM Solver The details for using this financial calculation tool are presented in Appendix B. In this chapter, TVM Solver screens will be displayed in order to confirm answers to many of the examples. The eight lines of the TVM Solver screen hold the following information.

N: Number of interest periods.

I%:   Annual interest rate given as a percent, such as 6 or 4.5. In this chapter, its value is $100r$.

PV: Present value or principal.

PMT: Periodic payment. It has value 0 in Section 10.1, but will have nonzero values in the rest of the chapter.

FV: Future value.

P/Y: Number of payments per year. For our purposes, it is the same as C/Y.

C/Y: Number of times interest is compounded per year. That is, 1, 2, 4, 12, 52, or 365.

PMT: When payments are paid—the *end* or *beginning* of each interest period. For our purposes, it will always be set to END.

A small black square appears to the left of the computed value.

In Figs. 4, 5, and 6, TVM Solver shows the solutions to Examples 3(b), 4, and 5. *Note:* The values for PV are negative, since they represent money paid to the bank.

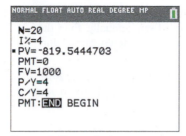

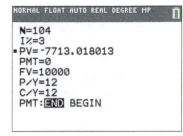

NORMAL FLOAT AUTO REAL DEGREE MP
N=20
I%=4
PV=-100
PMT=0
▪FV=122.019004
P/Y=4
C/Y=4
PMT:**END** BEGIN

Figure 4 Example 3(b)

NORMAL FLOAT AUTO REAL DEGREE MP
N=20
I%=4
▪PV=-819.5444703
PMT=0
FV=1000
P/Y=4
C/Y=4
PMT:**END** BEGIN

Figure 5 Example 4

NORMAL FLOAT AUTO REAL DEGREE MP
N=104
I%=3
▪PV=-7713.018013
PMT=0
FV=10000
P/Y=12
C/Y=12
PMT:**END** BEGIN

Figure 6 Example 5

Excel Spreadsheet The section "Using Excel's Financial Functions" of Appendix C shows how the functions FV, PV, and EFFECT are used to calculate future values, present values, and effective rates of interest.

✳WolframAlpha The formulas presented in this section can be evaluated directly with Wolfram | Alpha. In addition, the answers to examples from this section can be obtained with the following instructions:

> Example 3(b): **compound interest PV=\$100,i=1%,n=20,annual**
> Example 4: **compound interest FV=\$1000,i=1%,n=20,annual**
> Example 6: **simple interest PV=\$1000,i=2%,n=4**
> Example 7(a): **EffectiveInterestRate with nominal=3.65%,n=4**

Suppose \$100 deposited into a savings account grows to \$200 after 36 interest periods. Then the compound interest rate per period is given by **interest rate PV=\$100,FV=\$200, n=36**, and the simple interest rate per period is given by **simple interest rate PV=\$100, FV=\$200,n=36**.

Also, a table similar to the one in Fig. 3 can be generated with an instruction such as **Table[100*(1.01)^x,{x,0,10,1}]**.

Check Your Understanding 10.1

Solutions can be found following the section exercises.

1. Calculate the present value of \$1000 to be received 10 years in the future at 6% interest compounded annually.

2. Calculate the future value after 2 years of \$1 at 26% interest compounded weekly.

3. Calculate the future amount of \$2000 after 6 months if invested at 6% simple interest.

EXERCISES 10.1

In Exercises 1–6, give the values of i and n under the given conditions.

1. 3% interest compounded monthly for 2 years

2. 2% interest compounded quarterly for 5 years

3. 2.2% interest compounded semiannually for 20 years

4. 6% interest compounded annually for 3 years

5. 4.5% interest compounded monthly from January 1, 2013, to July 1, 2016

6. 2.4% interest compounded quarterly from January 1, 2013, to October 1, 2016

In Exercises 7–12, give the values of i, n, P, and F.

7. \$500 invested at 2.8% interest compounded annually grows to \$558.40 in 4 years.

8. \$800 invested on January 1, 2011, at 1.8% interest compounded monthly, grows to \$957.64 by January 1, 2021.

9. \$7174.85 is deposited on January 1, 2011. The balance on July 1, 2020 is \$9000, and the interest is 2.4% compounded semiannually.

10. The amount of money that must be deposited now at 2.6% interest compounded weekly in order to have \$7500 in 1 year is \$7307.56.

11. \$3000 deposited at 6% interest compounded monthly will grow to \$18,067.73 in 30 years.

12. In 1626, Peter Minuit, the first director-general of New Netherlands province, purchased Manhattan Island for trinkets and cloth valued at about \$24. Had this money been invested at 8% interest compounded quarterly, it would have amounted to \$677,454,102,888,106 by 2017.

In Exercises 13–38, solve each problem.

13. **Future Value** Calculate the future value of \$1000 after 2 years if deposited at 2.1% interest compounded monthly.

14. **Future Value** Calculate the future value of \$1000 after 1 year if deposited at 2.19% interest compounded daily.

15. **Future Value** Six thousand dollars is deposited in a savings account at 2.7% interest compounded monthly. Find the balance after 3 years and the amount of interest earned during that time.

16. **Future Value** Two thousand dollars is deposited in a savings account at 4% interest compounded semiannually. Find the balance after 7 years and the amount of interest earned during that time.

17. **Future Value** If you had invested $10,000 on January 1, 2010, at 4% interest compounded quarterly, how much would you have had on January 1, 2016?

18. **Future Value** Ms. Garcia has just invested $100,000 at 2.5% interest compounded annually. How much money will she have in 20 years?

19. **Present Value** Calculate the present value of $100,000 payable in 25 years at 2.4% interest compounded monthly.

20. **Present Value** Calculate the present value of $10,000 payable in 5 years at 2.4% interest compounded semiannually.

21. **Savings Account** Mr. Smith wishes to purchase a $10,000 sailboat upon his retirement in 3 years. He has just won the state lottery and would like to set aside enough cash in a savings account paying 3.4% interest compounded quarterly to buy the boat upon retirement. How much should he deposit?

22. **Investment** In order to have $10,000 on his 25th birthday, how much would a person who just turned 21 have to invest if the money will earn 1.5% interest compounded monthly?

23. **Comparing Payouts** Is it more profitable to receive $1400 now or $1700 in 9 years? Assume that money can earn 2.5% interest compounded annually.

24. **Comparing Payouts** Is it more profitable to receive $7000 now or $10,000 in 9 years? Assume that money can earn 4% interest compounded quarterly.

25. **Interest Rates** In 1999, the NASDAQ Composite Index grew at a rate of 62.2% compounded weekly. Is this rate better or worse than 85% compounded annually?

26. **Interest Rates** Would you rather earn 3% interest compounded annually or 2.92% interest compounded daily?

27. **Interest** If $1000 is deposited into a savings account earning 1.6% interest compounded quarterly, how much interest is earned during the first quarter year? Second quarter year? Third quarter year?

28. **Interest** If $2000 is deposited into a savings account earning 1.2% interest compounded quarterly, how much interest is earned during the first quarter year? Second quarter year? Third quarter year?

29. **Savings Account** If $2000 is deposited into a savings account at 3% interest compounded quarterly, how much interest is earned during the first 2 years? During the third year?

30. **Savings Account** If $5000 is deposited into a savings account at 1.8% interest compounded monthly, how much interest is earned during the first year? During the second year?

31. **Interest** If $1000 is deposited for 5 years in a savings account earning 2.6% interest compounded quarterly, how much interest is earned during the fifth year?

32. **Interest** If $1000 is deposited for 6 years in a savings account earning 3.6% interest compounded monthly, how much interest is earned during the sixth year?

33. **Savings Account** How much money must you deposit at 4% interest compounded quarterly in order to earn $406.04 interest in 1 year?

34. **Savings Account** How much money must you deposit at 2.1% interest compounded monthly in order to earn $347.58 interest in 3 years?

35. **Interest Rate** Clara would like to have $2100 in 7 years to give her granddaughter as a 21st birthday present. She has $1500 to invest in a 7-year certificate of deposit (CD). What rate of interest compounded annually must the CD earn?

36. **Interest Rate** Juan would like to have $2300 in 3 years in order to buy a motorcycle. He has $2000 to invest in a 3-year certificate of deposit (CD). What rate of interest compounded annually must the CD earn?

37. **Savings Account** Consider the following savings account statement:

	1/1/17	2/1/17	3/1/17
Deposit	$10,000.00		
Withdrawal			
Interest		$20.00	$20.04
Balance	$10,000.00	$10,020.00	$10,040.04

(a) What interest rate is this bank paying?
(b) Give the interest and balance on 4/1/17.
(c) Give the interest and balance on 1/1/19.

38. **Savings Account** Consider the following savings account statement:

	1/1/17	4/1/17	7/1/17
Deposit	$10,000.00		
Withdrawal			
Interest		$40.00	$40.16
Balance	$10,000.00	$10,040.00	$10,080.16

(a) What interest rate is this bank paying?
(b) Give the interest and balance on 10/1/17.
(c) Give the interest and balance on 1/1/19.

Exercises 39–52 concern simple interest.

39. **Simple Interest** Determine $r, n, P,$ and F for each of the following situations:
   (a) $500 invested at 1.5% simple interest grows to $503.75 in 6 months.
   (b) In order to have $525 after 2 years at 2.5% simple interest, $500 must be invested.

40. **Simple Interest** Determine $r, n, P,$ and F for each of the following situations:
   (a) At 3% simple interest, $1000 deposited on January 1, 2017, was worth $1015 on June 1, 2017.
   (b) At 4% simple interest, in order to have $3600 in 5 years, $3000 must be deposited now.

41. **Future Value** Calculate the future value after 3 years if $1000 is deposited at 1.2% simple interest.

42. **Future Value** Calculate the future value after 18 months if $2000 is deposited at 2% simple interest.

43. **Present Value** Find the present value of $3000 in 10 years at 2% simple interest.

44. **Present Value** Find the present value of $2000 in 8 years at 3.5% simple interest.

45. Interest Rate Determine the (simple) interest rate at which $980 grows to $1000 in 6 months.

46. Interest Rate At what (simple) interest rate will $1000 grow to $1200 in 5 years?

47. Time Period How many years are required for $500 to grow to $800 at 1.5% simple interest?

48. Time Period How many years are required for $1000 to grow to $1240 at 2.4% simple interest?

49. Time Interval Determine the amount of time required for money to double at 2% simple interest.

50. Interest Rate Derive the formula for the (simple) interest rate r at which P dollars grow to F dollars in n years. That is, express r in terms of P, F, and n.

51. Present Value Derive the formula for the present value P of F dollars in n years at simple interest rate r.

52. Time Interval Derive the formula for the number of years n required for P dollars to grow to F dollars at simple interest rate r.

53. Future Value Compute the future value after 1 year for $100 invested at 4% interest compounded quarterly. What simple interest rate will yield the same amount in 1 year?

54. Future Value Compute the future value after 1 year for $100 invested at 2.7% interest compounded monthly. What simple interest rate will yield the same amount in 1 year?

Effective Rate of Interest In Exercises 55–58, calculate the effective rate of interest corresponding to the given nominal rate.

55. 4% interest compounded semiannually

56. 8.45% interest compounded weekly

57. 4.4% interest compounded monthly

58. 7.95% interest compounded quarterly

59. Savings Account On January 1, 2014, a deposit was made into a savings account paying interest compounded quarterly. The balance on January 1, 2017 was $10,000, and the balance on April 1, 2017 was $10,100. How large was the deposit?

60. Savings Account During the 1990s, a deposit was made into a savings account paying 4% interest compounded quarterly. On January 1, 2017, the balance was $2020. What was the balance on October 1, 2016?

61. Future Value Suppose that a principal of $100 is deposited for 5 years in a savings account at 6% interest compounded semiannually. Which of items (a)–(d) can be used to fill in the blank in the following statement? (*Note:* Before computing, use your intuition to guess the correct answer.)
If the _____ is doubled, then the future value will double.
(a) principal
(b) interest rate
(c) number of interest periods per year
(d) number of years

62. Future Value Suppose that a principal of $100 is deposited for 5 years in a savings account at 6% simple interest. Which of items (a)–(c) can be used to fill in the blank in the following statement? (*Note:* Before computing, use your intuition to guess the correct answer.)
If the _____ is doubled, then the future value will double.

(a) principal
(b) interest rate
(c) number of years

63. Doubling Time If a $1000 investment at compound interest doubles every 6 years, how long will it take the investment to grow to $8000?

64. Doubling Time (True or False) An investment growing at the rate of 12% compounded monthly will double (that is, increase by 100%) in about 70 months. Therefore, the investment should increase by 50% in about 35 months.

65. Future Value If the value of an investment grows at the rate of 4% compounded annually for 10 years, then it grows about _____ over the 10-year period.

(a) 25%   (b) 40%   (c) 44%   (d) 48%

66. Future Value If your stock portfolio gained 20% in 2014 and 30% in 2015, then it gained _____ over the two-year period.

(a) 50%   (b) 56%   (c) 60%   (d) 100%

67. Comparing Investments The same amount of money was invested in each of two different investments on January 1, 2015. Investment A increased by 2.5% in 2015, 3% in 2016, and 8.4% in 2017. Investment B increased by the same amount, r%, in each of the 3 years and was worth the same as investment A at the end of the 3-year period. Determine the value of r.

68. Future Value If you increase the compound interest rate for an investment by 20%, will the future value increase by 20%?

TECHNOLOGY EXERCISES

In Exercises 69–74, give the settings or statements to determine the solution with TVM Solver, Excel, or Wolfram|Alpha.

69. Exercise 13   **70.** Exercise 14

71. Exercise 19   **72.** Exercise 20

73. Exercise 35   **74.** Exercise 36

75. Savings Account One thousand dollars is deposited into a savings account at 2.7% interest compounded annually. How many years are required for the balance to reach $1946.53? After how many years will the balance exceed $2500?

76. Savings Account Ten thousand dollars is deposited into a savings account at 1.8% interest compounded monthly. How many months are required for the balance to reach $10,665.74? After how many months will the balance exceed $11,000?

77. Investment Tom invests $500,000 at 1.9% interest compounded annually. When will Tom be a millionaire?

78. Savings Account How many years are required for $100 to double if deposited at 2.2% interest compounded quarterly?

79. Comparing Investments Consider the following two interest options for an investment of $1000: (A) 4% simple interest, (B) 3% interest compounded annually. After how many years will option B outperform option A?

80. Comparing Investments Consider the following two interest options for an investment of $1000: (A) 4% simple interest, (B) 4% interest compounded daily. After how many years will option B outperform option A by at least $100?

1. Here, we are given the value in the future F and are asked to find the present value P. Interest compounded annually has just one interest period per year.

$$P = \frac{F}{(1 + i)^n} \quad \text{Present value formula for compound interest}$$

$$= \frac{1000}{(1.06)^{10}} \quad F = 1000, i = \frac{.06}{1} = .06, n = 10 \cdot 1 = 10$$

$$= \$558.39 \quad \text{Calculate. Round to nearest cent.}$$

2. Here, we are given the present value, $P = 1$, and are asked to find the value F at a future time. Interest compounded weekly has 52 interest periods each year, so $n = 2 \times 52 = 104$ and

$i = 26\%/52 = (1/2)\% = .005$. Using the future value formula, we have

$$F = (1 + i)^n P = (1.005)^{104} \cdot 1 = \$1.68.$$

3. In simple interest problems, time should be expressed in terms of years. Therefore, 6 months is 1/2 of a year.

$$F = (1 + nr)P \quad \text{Future value formula for simple interest}$$

$$= [1 + \tfrac{1}{2}(.06)]2000 \quad n = \tfrac{1}{2}, r = .06, P = 2000$$

$$= (1.03)2000 \quad \text{Multiply and add.}$$

$$= \$2060 \quad \text{Calculate.}$$

10.2   Annuities

An **annuity** is a sequence of equal payments made at regular intervals of time. Here are two illustrations.

1. As the proud parent of a newborn daughter, you decide to save for her college education by depositing $200 at the end of each month into a savings account paying 2.7% interest compounded monthly. Eighteen years from now, after you make the last of 216 payments, the account will contain $55,547.79.

2. Having just won the state lottery, you decide not to work for the next 5 years. You want to deposit enough money into the bank so that you can withdraw $5000 at the end of each month for 60 months. If the bank pays 2.1% interest compounded monthly, you must deposit $284,551.01.

The periodic payments in the foregoing financial transactions are called **rent**. The amount of a rent payment is denoted by the letter R. Thus, in the preceding examples, we have $R = \$200$ and $R = \$5000$, respectively.

In illustration 1, you make equal payments to a bank in order to generate a large sum of money in the future. This sum, namely $55,547.79, is called the **future value of the annuity**. Since the amount of money in the savings account increases each time a payment is made, this type of annuity is an **increasing annuity.**

In illustration 2, the bank will make equal payments to you in order to pay back (with interest) the sum of money that you currently deposit. The value of the current deposit, namely, $284,551.01, is called the **present value of the annuity**. Since the amount of money in the savings account decreases each time a payment is made, this type of annuity is a **decreasing annuity**.

Increasing Annuities

> **DEFINITION**   The **future value of an increasing annuity** of n equal payments is the value of the annuity after the nth payment.

Consider an annuity consisting of n payments of R dollars, each deposited into an account paying compound interest at the rate of i per interest period, and suppose that each payment is made at the end of an interest period. Let us derive a formula for the future value of the annuity—that is, a formula for the balance of the account immediately after the last payment.

Each payment accumulates interest for a different number of interest periods, so let us calculate the balance in the account as the sum of n future values, one corresponding to each payment (Table 1). Denote the future value of the annuity by F.

Table 1 Increasing Annuity

Payment	Amount	Number of Interest Periods on Deposit	Future Value of Payment
1	R	$n-1$	$(1+i)^{n-1}R$
2	R	$n-2$	$(1+i)^{n-2}R$
$\vdots$	$\vdots$	$\vdots$	$\vdots$
$n-2$	R	2	$(1+i)^2 R$
$n-1$	R	1	$(1+i)R$
n	R	0	R

Then F is the sum of the numbers in the right-hand column:

$$F = R + (1+i)R + (1+i)^2 R + (1+i)^3 R + \cdots + (1+i)^{n-1}R$$

A compact expression for F can be obtained by multiplying both sides of this equation by $(1+i)$ and then subtracting the original equation from the new equation:

$$(1+i)F = \quad (1+i)R + (1+i)^2 R + (1+i)^3 R + \cdots + (1+i)^{n-1}R + (1+i)^n R$$
$$F = R + (1+i)R + (1+i)^2 R + (1+i)^3 R + \cdots + (1+i)^{n-1}R$$
$$\overline{iF = (1+i)^n R - R}$$

The last equation can be written $iF = [(1+i)^n - 1] \cdot R$. Dividing both sides by i yields

$$F = \frac{(1+i)^n - 1}{i} \cdot R.$$

Solving for R yields

$$R = \frac{i}{(1+i)^n - 1} \cdot F.$$

> **Formulas for an Increasing Annuity** The future value F and the rent R of an increasing annuity of n payments with interest compounded at the rate i per interest period are related by the formulas
>
> $$F = \frac{(1+i)^n - 1}{i} \cdot R \quad \text{and} \quad R = \frac{i}{(1+i)^n - 1} \cdot F.$$

EXAMPLE 1   **Calculating the Future Value of an Increasing Annuity** Calculate the future value of an increasing annuity of $100 per month for 5 years at 3% interest compounded monthly.

SOLUTION

$$F = \frac{(1+i)^n - 1}{i} \cdot R \qquad \text{Formula for future value}$$

$$= \frac{1.0025^{60} - 1}{.0025} \cdot 100 \qquad i = \frac{.03}{12} = .0025,\ n = 5 \cdot 12 = 60,\ R = 100$$

$$= 6464.67 \qquad \text{Calculate. Round to two decimal places.}$$

The future value is $6464.67.

>> **Now Try Exercise 5**

We can derive a formula that illustrates exactly how the future value of an increasing annuity changes from period to period. Each new balance can be computed from the previous balance with the formula

$$B_{\text{new}} = (1 + i)B_{\text{previous}} + R.$$

That is, the new balance equals the growth of the previous balance due to interest, plus the amount paid. So, for instance, if B_1 is the balance after the first payment is made, B_2 is the balance after the second payment is made, and so on, then

$$B_1 = R$$
$$B_2 = (1 + i)B_1 + R$$
$$B_3 = (1 + i)B_2 + R$$
$$\vdots$$

Successive balances are computed by multiplying the previous balance by $(1 + i)$ and adding R.

> **Successive Balances**   If R dollars is deposited into an increasing annuity at the end of each interest period with interest rate i per period, then
>
> $$B_{\text{new}} = (1 + i)B_{\text{previous}} + R \qquad (1)$$
>
> can be used to calculate each new balance from the previous balance.

EXAMPLE 2   **Calculating the Future Value of an Increasing Annuity** Consider the annuity of Example 1. Determine the future value after 61 months.

SOLUTION   From Example 1, $i = .0025$, the balance after 60 months is $6464.67, and $R = 100$. Therefore, by (1)

$$[\text{balance after 61 months}] = (1 + .0025)[\text{balance after 60 months}] + 100$$
$$= 1.0025(6464.67) + 100$$
$$= 6580.83.$$

Therefore, the annuity will have accumulated to $6580.83 after 61 months.   **«**

EXAMPLE 3   **Determining the Rent for an Increasing Annuity** Ms. Adams would like to buy a $300,000 airplane when she retires in 8 years. How much should she deposit at the end of each half-year into an account paying 4% interest compounded semiannually so that she will have enough money to purchase the airplane?

SOLUTION

$$R = \frac{i}{(1 + i)^n - 1} \cdot F \qquad \text{Formula for rent}$$

$$= \frac{.02}{1.02^{16} - 1} \cdot 300,000 \qquad i = \frac{.04}{2} = .02, n = 8 \cdot 2 = 16, F = 300,000$$

$$= 16,095.04 \qquad \text{Calculate. Round to two decimal places.}$$

She should deposit $16,095.04 at the end of each half-year period.   **«**

Decreasing Annuities

> **DEFINITION** The **present value of a decreasing annuity** is the amount of money necessary to finance the sequence of annuity payments. That is, the present value of the annuity is the amount you would need to deposit in order to provide the desired sequence of annuity payments and leave a balance of zero at the end of the term.

To find a formula for the present value P of an annuity, consider an annuity consisting of n payments of R dollars made at the end of each interest period, with interest compounded at a rate i per interest period.

Situation 1: Suppose that the P dollars were just left in the account and that the annuity payments were not withdrawn. At the end of the n interest periods, there would be $(1 + i)^n P$ dollars in the account.

Situation 2: Suppose that the payments are withdrawn but are immediately redeposited into another account having the same rate of interest. At the end of the n interest periods, there would be $\dfrac{(1+i)^n-1}{i} \cdot R$ dollars in the new account.

In both of these situations, P dollars is deposited and it, together with all of the interest generated, is earning income at the same interest rate for the same amount of time. Therefore, the final amounts of money in the accounts should be the same. That is,

$$(1 + i)^n P = \frac{(1 + i)^n - 1}{i} \cdot R$$

$$P = (1 + i)^{-n} \cdot \frac{(1 + i)^n - 1}{i} \cdot R \quad \text{\color{blue} Multiply both sides by } (1 + i)^{-n}.$$

$$= \frac{1 - (1 + i)^{-n}}{i} \cdot R \quad \text{\color{blue} Multiply numerator of fraction by } (1 + i)^{-n}$$

Solving for R yields

$$R = \frac{i}{1 - (1 + i)^{-n}} \cdot P.$$

Formulas for a Decreasing Annuity The present value P and the rent R of a decreasing annuity of n payments with interest compounded at the rate i per interest period are related by the formulas

$$P = \frac{1 - (1 + i)^{-n}}{i} \cdot R \quad \text{and} \quad R = \frac{i}{1 - (1 + i)^{-n}} \cdot P.$$

EXAMPLE 4 **Determining the Present Value for a Decreasing Annuity** How much money must you deposit now at 6% interest compounded quarterly in order to be able to withdraw $3000 at the end of each quarter year for 2 years?

SOLUTION We are asked to calculate the present value of the sequence of payments.

$$P = \frac{1 - (1 + i)^{-n}}{i} \cdot R \quad \text{\color{blue} Formula for present value}$$

$$= \frac{1 - 1.015^{-8}}{.015} \cdot 3000 \quad \text{\color{blue} } i = \frac{.06}{4} = .015, n = 2 \cdot 4 = 8, R = 3000$$

$$= 22{,}457.78 \quad \text{\color{blue} Calculate. Round to two decimal places.}$$

The present value is $22,457.78.

» Now Try Exercise 25

EXAMPLE 5 **Determining the Rent for a Decreasing Annuity** If you deposit $10,000 into a fund paying 6% interest compounded monthly, how much can you withdraw at the end of each month for 1 year?

SOLUTION   We are asked to calculate the rent for the sequence of payments.

$$R = \frac{i}{1 - (1 + i)^{-n}} \cdot P \qquad \text{Formula for rent}$$

$$= \frac{.005}{1 - 1.005^{-12}} \cdot 10{,}000 \qquad i = \frac{.06}{12} = .005, n = 1 \cdot 12 = 12, P = 10{,}000$$

$$= 860.66 \qquad \text{Calculate. Round to two decimal places.}$$

The rent is $860.66.   **» Now Try Exercise 11**

NOTE▸ In this section, we have considered only annuities with payments made at the end of each interest period. Such annuities are called *ordinary annuities*. Annuities that have payments at the beginning of the interest period are called *annuities due*. Annuities whose payment period is different from the interest period are called *general annuities*. «

Two key formulas in this section are as follows:

1. Increasing Annuity: $F = \dfrac{(1 + i)^n - 1}{i} \cdot R$

2. Decreasing Annuity: $P = \dfrac{1 - (1 + i)^{-n}}{i} \cdot R$

The following tips help us recall these formulas:

- Both formulas have i in the denominator and R on the right side.
- Since an increasing annuity builds up a future value, formula 1 involves F.
- Since a decreasing annuity depletes a present value, formula 2 involves P.
- Associate *increasing* with *positive* and *decreasing* with *negative*. Formula 1 has a positive exponent, and formula 2 has a negative exponent.

NOTE▸ The following two formulas, obtained by solving formulas 1 and 2 for R, are frequently used in annuity computations.

3. Increasing Annuity: $R = \dfrac{i}{(1 + i)^n - 1} \cdot F$

4. Decreasing Annuity: $R = \dfrac{i}{1 - (1 + i)^{-n}} \cdot P$ «

EXAMPLE 6   **Determining the Time for a Future Value to Exceed a Specified Amount** Consider the situation in which $100 is deposited monthly at 6% interest compounded monthly. Use technology to determine when the balance will exceed $10,000.

SOLUTION   There are three ways to answer this question with a graphing calculator—with a graph, with a table, and with TVM Solver.

1. (Graph) Set $\mathtt{Y_2}$=$\mathtt{10000}$, and find its intersection with the graph of

$$\mathtt{Y_1} = \frac{1.005^X - 1}{.005} * 100.$$

See Fig. 1. (The window was set to [0, 120] *by* [−4000, 17000] with an X-scale of 10 and a Y-scale of 1000.)

2. (Table) Scroll down the table for $\mathtt{Y_1}$ until the balance exceeds 10,000. See Fig. 2.
3. (TVM Solver) See Fig. 3. *Note:* The value for PMT is negative, since it represents money paid to the bank.

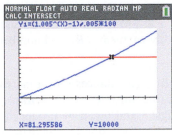

Figure 1

Figure 2

Figure 3

Entering $=\mathbf{NPER(0.005, -100, 0, 10000)}$ into a cell produces 81.29558565. *Note:* The payment is negative, since it represents money paid to the bank.

WolframAlpha The instruction **annuityperiods, i = .5%, pmt = $100, FV = $10000** produces "number of periods is 81.3."

In each case, we see that the balance will exceed $10,000 after 82 months—that is, after 6 years and 10 months. ≪

INCORPORATING
TECHNOLOGY

Display Successive Balances Graphing calculators can easily display successive balances in an annuity. (We limit our discussion here to increasing annuities. The corresponding analysis for decreasing annuities is similar.) Consider the situation in which $100 is deposited at the end of each month at 6% interest compounded monthly. In Fig. 4, successive balances are displayed in the home screen. In Fig. 5, successive balances are graphed, and in Fig. 6, they are displayed in a table.

These figures were obtained with the same processes as Figs. 1–3 in Section 10.1. Successive balances were determined with the relation $B_{\text{new}} = (1 + i)B_{\text{previous}} + R$. The function to be graphed was specified as

$$Y_1 = \frac{1.005^X - 1}{.005} * 100.$$

The window was set to [0, 120] *by* [−4000, 17000] with an X-scale of 10 and a Y-scale of 1000. In the TABLE SETUP menu, both **TblStart** and **ΔTbl** were set to 1.

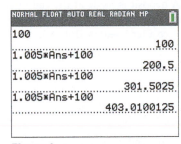

Figure 4

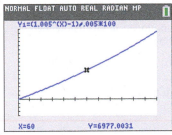

Figure 5

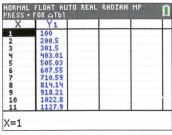

Figure 6

TVM Solver In Figs. 7, 8, and 9, TVM Solver is used to solve Examples 1, 3, and 4. PMT, the monthly payment, is the value we have been calling *rent*. The value of PV for an increasing annuity is 0 since initially there is no money in the bank. The value for PV for a decreasing annuity is negative, since it represents money paid to the bank.

Figure 7 Example 1

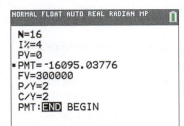

Figure 8 Example 3

Figure 9 Example 4

Excel Spreadsheet The "Using Excel's Financial Functions" section in Appendix C shows how the functions FV, PV, PMT, and NPER are used to calculate future values (F), present values (P), monthly payments (R), and number of payments (n) for annuities.

✳ WolframAlpha A graph and the first 10 balances for an increasing annuity can be generated with an instruction of the form $B(n) = (1 + i)B(n-1) + R, B(0) = 0$. For instance, the graph and table in Figs. 5 and 6 can be produced with the instruction $B(n) = 1.005B(n-1) + 100, B(0) = 0$. The corresponding instruction for a decreasing annuity is $B(n) = (1 + i)B(n-1) - R, B(0) = P$.

The formulas presented in this section can be evaluated directly. In addition, the answers to the examples from this section can be obtained with the instructions that follow. (*Note:* The answers will sometimes be rounded.)

Example 1: **annuity FV,i = .25%,n = 60,pmt = $100**
Example 2: **annuity FV,i = .25%,n = 61,pmt = $100**
Example 3: **annuity pmt,FV = $300000,i = 2%,n = 16**
Example 4: **annuity PV,i = 1.5%,n = 8,pmt = $3000**
Example 5: **annuity pmt,PV = $10000,i = .5%,n = 12**

Check Your Understanding 10.2

Solutions can be found following the section exercises.

Decide whether or not each of the given annuities is an ordinary annuity—that is, the type of annuity considered in this section. If so, identify n, i, and R, and calculate the present value or the future value, whichever is appropriate.

1. You make a deposit at 9% interest compounded monthly into a fund that pays you $1 at the end of each month for 5 years.

2. At the end of each week for 2 years, you deposit $10 into a savings account earning 6% interest compounded monthly.

3. At the end of each month for 2 years, you deposit $10 into a savings account earning 6% interest compounded monthly.

EXERCISES 10.2

Exercises 1 and 2 describe increasing annuities. Specify i, n, R, and F.

1. If, at the end of each month, $100 is deposited into a savings account paying 2.1% interest compounded monthly, the balance after 10 years will be $13,340.09.

2. Mr. Smith is saving to buy a $200,000 yacht in January 2020. Since January 2010, he has been depositing $8231.34 at the end of each half year into a fund paying 4% interest compounded semiannually.

Exercises 3 and 4 describe decreasing annuities. Specify i, n, R, and P.

3. In order to receive $2000 at the end of each quarter year beginning in 2015 until the end of 2019, Ms. Williams deposited $36,642.08 into an investment paying 3.4% interest compounded quarterly.

4. A retiree deposits $185,288.07 into an investment paying 2.7% interest compounded monthly and withdraws $1000 at the end of each month for 20 years.

In Exercises 5 and 6, calculate the future value of the increasing annuity.

5. At the end of each half year, for 5 years, $1500 is deposited into an investment paying 2.6% interest compounded semiannually.

6. At the end of each quarter year, for 6 years, $1200 is deposited into an investment paying 3.4% interest compounded quarterly.

In Exercises 7 and 8, calculate the rent of the increasing annuity.

7. A deposit is made at the end of each month into a savings account paying 1.8% interest compounded monthly. The balance after 1 year is $1681.83.

8. Money is deposited at the end of each week into an investment paying 2.6% interest compounded weekly. The balance after 3 years is $16,382.52.

In Exercises 9 and 10, calculate the present value of the decreasing annuity.

9. At the end of each month, for two years, $3000 will be withdrawn from a savings account paying 1.5% interest compounded monthly.

10. At the end of each month, for one year, $1000 will be withdrawn from a savings account paying 1.2% interest compounded monthly.

In Exercises 11 and 12, calculate the rent of the decreasing annuity.

11. A withdrawal is made at the end of each quarter year for 3 years from a savings account paying 1.6% interest compounded quarterly. The account initially contained $47,336.25.

12. A withdrawal is made at the end of each half year for 5 years from a savings account paying 1.8% interest compounded semiannually. The account initially contained $57,781.39.

13. **Savings Account** Ethan deposits $500 into a savings account at the end of every month for 4 years at 3% interest compounded monthly.
    (a) Find the balance at the end of 4 years.
    (b) How much money did Ethan deposit into the account?
    (c) How much interest did Ethan earn during the 4 years?
    (d) Prepare a table showing the balance and interest for the first 3 months.

14. **Savings Account** Emma deposits $2000 into a savings account at the end of every quarter year for 5 years at 3% interest compounded quarterly.
    (a) Find the balance at the end of 5 years.
    (b) How much money did Emma deposit into the account?
    (c) How much interest did Emma earn during the 5 years?
    (d) Prepare a table showing the balance and interest for the first 3 quarters.

15. **Savings Account** A person deposits $10,000 into a savings account at 2.2% interest compounded quarterly and then withdraws $1000 at the end of each quarter year. Prepare a table showing the balance and interest for the first 3 quarters.

16. **Savings Account** A person deposits $5000 into a savings account at 2.4% interest compounded monthly and then withdraws $300 at the end of each month. Prepare a table showing the balance and interest for the first 4 months.

In Exercises 17–20, determine the amount of interest earned by the specified annuity.

17. The increasing annuity in Exercise 1

18. The increasing annuity in Exercise 2

19. The decreasing annuity in Exercise 3

20. The decreasing annuity in Exercise 4

21. **Comparing Payouts** Is it more profitable to receive $1000 at the end of each month for 10 years or to receive a lump sum of $140,000 at the end of 10 years? Assume that money can earn 3% interest compounded monthly.

22. **Comparing Payouts** Is it more profitable to receive a lump sum of $9,000 at the end of 3 years or to receive $750 at the end of each quarter-year for 3 years? Assume that money can earn 2.2% interest compounded quarterly.

23. **Comparing Bonus Plans** When Bridget takes a new job, she is offered the choice of a $2300 bonus now or an extra $200 at the end of each month for the next year. Assume money can earn an interest rate of 3.3% compounded monthly.
    (a) What is the future value of payments of $200 at the end of each month for 12 months?
    (b) Which option should Bridget choose?

24. **Comparing Lottery Payouts** A lottery winner is given two options:

    Option I: Receive a $20 million lump-sum payment.

    Option II: Receive 25 equal annual payments totaling $60 million, with the first payment occurring immediately.

    If money can earn 6% interest compounded annually during that long period, which option is better? *Hint:* With each option, calculate the amount of money earned at the end of 24 years if all of the funds are to be deposited into a savings account as soon as they are received.

25. **College Allowance** During Jack's first year at college, his father had been sending him $200 per month for incidental expenses. For his sophomore year, his father decided instead to make a deposit into a savings account on August 1 and have his son withdraw $200 on the first of each month from September 1 to May 1. If the bank pays 1.8% interest compounded monthly, how much should Jack's father deposit?

26. **Magazine Subscription** Suppose that a magazine subscription costs $45 per year and that you receive a magazine at the end of each month. At an interest rate of 2.1% compounded monthly, how much are you actually paying for each issue?

27. **Savings Account** Suppose that $1000 was deposited on January 1, 1989, into a savings account paying 8% interest compounded quarterly, and an additional $100 was deposited into the account at the end of each quarter year. How much would have been in the account on January 1, 2000?

28. **Savings Account** Suppose that you opened a savings account on January 1, 2014 and made a deposit of $100. In 2016, you began depositing $10 into the account at the end of each month. If the bank pays 2.7% interest compounded monthly, how much money will be in the account on January 1, 2020?

29. **Savings Account** Ms. Jones deposited $100 at the end of each month for 10 years into a savings account paying 2.1% interest compounded monthly. However, she deposited an additional $1000 at the end of the seventh year. How much money was in the account at the end of the 10th year?

30. **Savings Account** Redo Exercise 29 for the situation in which Ms. Jones withdrew $1000 at the end of the seventh year instead of depositing it.

31. **Savings Account** How much money must you deposit into a savings account at the end of each year at 2% interest compounded annually in order to earn $3400 interest during a 10-year period? (*Hint:* The future value of the annuity will be $10R + $3400.)

32. **Savings Account** How much money must you deposit into a savings account at the end of each quarter at 4% interest compounded quarterly in order to earn $403.80 interest during a 5-year period? (*Hint:* The future value of the annuity will be $20R + $403.80.)

33. **Annuity** Suppose that you deposit $600 every 6 months for 5 years into an annuity at 4% interest compounded semiannually. Which of items (a)–(d) can be used to fill in the blank in the statement that follows? (*Note:* Before computing, use your intuition to guess the correct answer.)

    If the _____ doubles, then the amount accumulated will double.
    (a) rent
    (b) interest rate
    (c) number of interest periods per year
    (d) number of years

34. **Total Rent** Suppose that you deposit $100,000 into an annuity at 4% interest compounded semiannually and withdraw an equal amount at the end of each interest period so that the account is depleted after 10 years. Which of items (a)–(d) can be used to fill in the blank in the statement that follows? (*Note:* Before computing, use your intuition to guess the correct answer.)

If the _____ doubles, then the total amount withdrawn each year will double.
(a) amount deposited
(b) interest rate
(c) number of interest periods per year
(d) number of years

35. **Municipal Bond** A municipal bond pays 4% interest compounded semiannually on its face value of $5000. The interest is paid at the end of every half-year period. Fifteen years from now, the face value of $5000 will be returned. How much should you pay for the bond?

36. **Present Value** What is the present value of a loan that pays $800 at the end of every quarter for 4 years and pays an additional $10,000 at the end of the 4 years? Assume an interest rate of 3% compounded quarterly.

37. **Business Loan** A business loan for $200,000 carries an interest rate of 9% compounded monthly. Suppose that the business pays only interest for the first 5 years and then repays the loan amount plus interest in equal monthly installments for the next 5 years.
(a) How much money will be paid each month during the first 5 years?
(b) Calculate the monthly payments during the second 5-year period in order to pay off the $200,000 still owed.

38. **Lottery Payoff** A lottery winner is to receive $1000 a month for the next 5 years. How much is this sequence of payments worth today if interest rates are 1.8% compounded monthly? How is the difference between this amount and the $60,000 paid out beneficial to the agency running the lottery?

A *sinking fund* is an increasing annuity set up by a corporation or government to repay a large debt at some future date. Exercises 39–44 concern such annuities.

39. **Sinking Fund** A city has a debt of $1,000,000 due in 15 years. How much money must it deposit at the end of each half year into a sinking fund at 4% interest compounded semiannually in order to pay off the debt?

40. **Sinking Fund** A corporation wishes to deposit money into a sinking fund at the end of each half year in order to repay $50 million in bonds in 10 years. It can expect to receive a 6% (compounded semiannually) return on its deposits to the sinking fund. How much should the deposits be?

41. **Sinking Fund** The Federal National Mortgage Association (Fannie Mae) puts $30 million at the end of each month into a sinking fund paying 4.8% interest compounded monthly. The sinking fund is to be used to repay bonds that mature 15 years from the creation of the fund. How large is the face amount of the bonds, assuming that the sinking fund will exactly pay it off?

42. **Saving for an Upgrade** A corporation sets up a sinking fund to replace some aging machinery. It deposits $100,000 into the fund at the end of each month for 10 years. The annuity earns 12% interest compounded monthly. The equipment originally cost $13 million. However, the cost of the equipment is rising 6% each year. Will the annuity be adequate to replace the equipment? If not, how much additional money is needed?

43. **Saving for a New Headquarters** A corporation borrows $5 million to erect a new headquarters. The financing is arranged by the use of bonds, to be repaid in 15 years. How much should the corporation deposit into a sinking fund at the end of each quarter if the fund earns 2.6% interest compounded quarterly?

44. **Saving for a New Warehouse** A corporation sets up a sinking fund to replace an aging warehouse. The cost of the warehouse today would be $8 million. However, the corporation plans to replace the warehouse in 5 years. It estimates that the cost of the warehouse will increase by 5% annually. The sinking fund will earn 3.6% interest compounded monthly. What should be the monthly payment to the fund?

A *perpetuity* is similar to a decreasing annuity, except that the payments continue forever. Exercises 45 and 46 concern such annuities.

45. **Scholarship** A grateful alumnus decides to donate a permanent scholarship of $12,000 per year. How much money should be deposited in the bank at 5% interest compounded annually in order to be able to supply the $12,000 for the scholarship at the end of each year so that the amount of money in the account remains constant?

46. **Deposit** Show that establishing a perpetuity paying R dollars at the end of each interest period requires a deposit of R/i dollars, where i is the interest rate per interest period.

A *deferred annuity* is a type of decreasing annuity whose term is to start at some future date. Exercises 47–50 concern such annuities.

47. **College Fund** On her 10th birthday, Emma inherits $20,000, which is to be used for her college education. The money is deposited into a trust fund at 3% interest compounded annually that will pay her R dollars on her 18th, 19th, 20th, and 21st birthdays.
(a) How much money will be in the trust fund on Emma's 17th birthday?
(b) Find the value of R. *Note:* R is the rent on a decreasing annuity with a duration of 4 years.

48. **College Fund** Refer to Exercise 47. Find the size of the inheritance that would result in $10,000 per year during the college years (ages 18–21, inclusive).

49. **Trust Fund** On December 1, 2014, a philanthropist set up a permanent trust fund to buy Christmas presents for needy children. The fund will provide $90,000 each year beginning on December 1, 2024. How much must have been set aside in 2014 if the money earns 3% interest compounded annually?

50. **Rent** Show that the rent paid by a deferred annuity of n payments that are deferred by m interest periods is given by the formula

$$R = \frac{i(1+i)^{n+m}}{(1+i)^n - 1} \cdot P.$$

TECHNOLOGY EXERCISES

In Exercises 51–58, give the settings or statements to determine the solution with TVM Solver, Excel, or Wolfram|Alpha.

51. Exercise 5   52. Exercise 6

53. Exercise 7   54. Exercise 8

55. Exercise 9   56. Exercise 10

57. Exercise 11   58. Exercise 12

59. **Time Interval** A person deposits $1000 at the end of each year into an annuity earning 5% interest compounded annually.

How many years are required for the balance to reach $30,539? After how many years will the balance exceed $50,000?

60. **Time Interval** A person deposits $800 at the end of each quarter-year into an annuity earning 2.2% interest compounded quarterly. How many quarters are required for the balance to reach $17,754? After how many quarters will the balance exceed $20,000?

61. **Bicycle Fund** Jane has taken a part-time job to save for a $503 racing bike. If she puts $15 each week into a savings account paying 5.2% interest compounded weekly, when will she be able to buy the bike?

62. **Home Repairs Fund** Bob needs $3064 to have some repairs done on his house. He decides to deposit $150 at the end

of each month into an annuity earning 2.7% interest compounded monthly. When will he be able to do the repairs?

63. **Comparing Expense Ratios** Some stock funds charge an annual fee, called the *expense ratio,* that usually ranges from about .3% to about 1.5% of the amount of money in the account at the end of the year. For instance, if the expense ratio is .5% and you have $10,000 in the account at the end of the year, then you would be charged $.005 \cdot 10000 = 50 in fees. Suppose that you put $5000 at the beginning of each year into an IRA and that the funds are invested in a stock fund earning 6% per year, before expenses. How much more will you have at the end of 10 years if the stock fund has an expense ratio of .3%, as opposed to 1.5%?

Solutions to Check Your Understanding 10.2

1. An ordinary decreasing annuity with $n = 60$, $i = (3/4)\% = .0075$, and $R = 1$. You will make a deposit now, in the present, and then withdraw money each month. The amount of this deposit is the present value of the annuity.

$$P = \frac{1 - (1 + i)^{-n}}{i} \cdot R$$

$$= \frac{1 - 1.0075^{-60}}{.0075} \cdot 1$$

$$= \$48.17$$

Note: For this transaction, the future value of the annuity needn't be computed. At the end of 5 years, the fund will have a balance of 0.

2. Not an ordinary annuity, since the payment period (1 week) is different from the interest period (1 month).

3. An ordinary increasing annuity with $n = 24$, $i = .5\% = .005$, and $R = \$10$. There is no money in the account now, in the present. However, in 2 years, in the future, money will have accumulated. So for this annuity, only the future value is of concern.

$$F = \frac{(1 + i)^n - 1}{i} \cdot R$$

$$= \frac{1.005^{24} - 1}{.005} \cdot 10$$

$$= \$254.32$$

10.3 Amortization of Loans

In this section, we analyze the mathematics of paying off loans. The loans that we shall consider will be repaid in a sequence of equal payments at regular time intervals, with the payment intervals coinciding with the interest periods. The process of paying off such a loan is called **amortization**. In order to obtain a feeling for the amortization process, let us consider a particular case, the amortization of a $563 loan to buy a high-definition TV. Suppose that this loan charges interest at a 12% rate with interest compounded monthly on the unpaid balance and that the monthly payments are $116 for 5 months. The repayment process is summarized in Table 1.

Table 1 Payments on a Loan

Payment Number	Amount	Interest	Applied to Principal	Unpaid Balance
1	$116	$5.63	$110.37	$452.63
2	116	4.53	111.47	341.16
3	116	3.41	112.59	228.57
4	116	2.29	113.71	114.85
5	116	1.15	114.85	0.00

Note the following facts about the financial transactions:

1. Payments are made at the end of each month. The payments have been carefully calculated to pay off the debt, with interest, in the specified time interval.

2. Since $i = 1\%$, the interest to be paid each month is 1% of the unpaid balance at the end of the previous month. That is, 5.63 is 1% of 563, 4.53 is 1% of 452.63, and so on.

3. Although we write just one check each month for $116, we regard part of the check as being applied to payment of that month's interest. The remainder (namely, $116 -$ [interest]) is regarded as being applied to repayment of the principal amount.

4. The unpaid balance at the end of each month is the previous unpaid balance minus the portion of the payment applied to the principal. A loan can be paid off early by paying the current unpaid balance.

The four factors that describe the amortization process just described are as follows:

the principal	$563
the interest rate	12% compounded monthly
the term	5 months
the monthly payment	$116

The important fact to recognize is that the sequence of payments in the preceding amortization constitutes a decreasing annuity for the bank. That is, when thinking of the loan as a decreasing annuity, the roles of the bank and the person are reversed. The person receives a sum of money and returns it in a sequence of equal payments (including interest) to the bank. Therefore, the mathematical tools developed in Section 10.2 suffice to analyze the amortization. In particular, we can determine the monthly payment or the principal once the other three factors are specified.

EXAMPLE 1 **Calculating Values Associated with a Loan** Suppose that a loan has an interest rate of 12% compounded monthly and a term of 5 months.
(a) Given that the principal is $563, calculate the monthly payment.
(b) Given that the monthly payment is $116, calculate the principal.
(c) Given that the monthly payment is $116, calculate the unpaid balance after 3 months.

SOLUTION The sequence of payments constitutes a decreasing annuity with the monthly payments as rent and the principal as the present value.

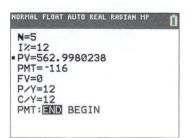

NORMAL FLOAT AUTO REAL RADIAN MP

```
N=5
I%=12
PV=563
■PMT=-116.0004072
FV=0
P/Y=12
C/Y=12
PMT:END BEGIN
```

Figure 1 The value for PMT is negative since it denotes money paid to the bank.

(a) $R = \dfrac{i}{1 - (1 + i)^{-n}} \cdot P$ Formula for rent

$\quad = \dfrac{.01}{1 - 1.01^{-5}} \cdot 563$ $i = \dfrac{.12}{12} = .01, n = 5, P = 563$

$\quad = 116.00$ Calculate. Round to two decimal places.

The rent is $116.00. (The TVM Solver screen in Fig. 1 confirms this result.)

NORMAL FLOAT AUTO REAL RADIAN MP

```
N=5
I%=12
■PV=562.9980238
PMT=-116
FV=0
P/Y=12
C/Y=12
PMT:END BEGIN
```

Figure 2 The value for PMT is negative since it denotes money paid to the bank.

(b) $P = \dfrac{1 - (1 + i)^{-n}}{i} \cdot R$ Formula for present value (that is, the principal)

$\quad = \dfrac{1 - 1.01^{-5}}{.01} \cdot 116$ $i = \dfrac{.12}{12} = .01, n = 5, R = 116$

$\quad = 563.00$ Calculate. Round to two decimal places.

The principal is $563.00. (The TVM Solver screen in Fig. 2 confirms this result.)

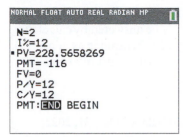

Figure 3 The value for PMT is negative since it denotes money paid to the bank.

(c) The unpaid balance is most easily calculated by regarding it as the amount necessary to retire the debt. Therefore, it must be sufficient to generate the sequence of two remaining payments. That is, the unpaid balance is the present value of a decreasing annuity of two payments of $116.

$$P = \frac{1 - (1 + i)^{-n}}{i} \cdot R \quad \text{Formula for present value (that is, the unpaid balance)}$$

$$= \frac{1 - 1.01^{-2}}{.01} \cdot 116 \quad i = \frac{.12}{12} = .01, n = 2, R = 116$$

$$= 228.57 \quad \text{Calculate. Round to two decimal places.}$$

The unpaid balance after 3 months is $228.57. (The TVM Solver screen in Fig. 3 confirms this result.)
»> Now Try Exercises 1 and 5

A **mortgage** is a long-term loan used to purchase real estate. The real estate is used as collateral to guarantee that the loan will be repaid.

EXAMPLE 2   **Calculating Values Associated with a Mortgage**   On December 31, 1996, a house was purchased with the buyer taking out a 30-year, $112,475 mortgage at 9% interest compounded monthly. The mortgage payments are made at the end of each month. (*Note:* Computations will be made under the assumption that the loan was not refinanced.)
(a) Calculate the amount of the monthly payment.
(b) Calculate the unpaid balance of the loan on December 31, 2022, just after the 312th payment.
(c) How much interest will be paid during the month of January 2023?
(d) How much of the principal will be paid off during the year 2022?
(e) How much interest will be paid during the year 2022?
(f) What is the total amount of interest paid during the 30 years?

SOLUTION   (a) $R = \dfrac{i}{1 - (1 + i)^{-n}} \cdot P \quad \text{Formula for rent (that is, the monthly payment)}$

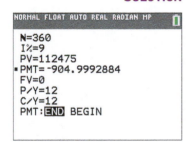

Figure 4 The value for PMT is negative since it denotes money paid to the bank.

$$= \frac{.0075}{1 - 1.0075^{-360}} \cdot 112,475 \quad i = \frac{.09}{12} = .0075, n = 360, P = 112,475$$

$$= 905.00 \quad \text{Calculate. Round to two decimal places.}$$

The monthly payment is $905.00. (The TVM Solver screen in Fig. 4 confirms this result.)

(b) The remaining payments constitute a decreasing annuity of 48 payments.

$$[\text{unpaid balance}] = \frac{1 - (1 + i)^{-n}}{i} \cdot R \quad \text{Formula for present value (that is, the unpaid balance)}$$

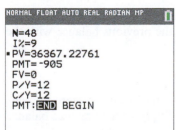

Figure 5 The value for PMT is negative since it denotes money paid to the bank.

$$= \frac{1 - 1.0075^{-48}}{.0075} \cdot 905 \quad i = \frac{.09}{12} = .0075, n = 48, R = 905$$

$$= 36,367.23 \quad \text{Calculate. Round to two decimal places.}$$

The unpaid balance after the 312th payment is $36,367.23. (The TVM Solver screen in Fig. 5 confirms this result.)

(c) The interest paid during 1 month is i, the interest rate per month, times the unpaid balance at the end of the preceding month. Therefore,

$$[\text{interest for January 2023}] = \tfrac{3}{4}\% \text{ of } \$36,367.23$$

$$= .0075 \cdot 36,367.23$$

$$= \$272.75.$$

(d) Since the portions of the monthly payments applied to repay the principal have the effect of reducing the unpaid balance, this question may be answered by calculating how much the unpaid balance will be reduced during 2022. Reasoning as in part (b), we determine that the unpaid balance on December 31, 2021 (just after the 300th payment), is equal to $43,596.90. Therefore,

[amount of principal repaid in 2022]

$$= \text{[unpaid balance Dec. 31, 2021]} - \text{[unpaid balance Dec. 31, 2022]}$$

$$= \$43,596.90 - \$36,367.23$$

$$= \$7229.67.$$

(e) During the year 2022, the total amount paid is $12 \times 905 = \$10,860$. But by part (d), $7229.67 is applied to repayment of principal, the remainder being applied to interest.

[interest in 2022] = [total amount paid] − [principal repaid in 2022]

$$= \$10,860 - \$7229.67$$

$$= \$3630.33$$

(f) The total amount of money paid during the 30 years is

$$360 \cdot 905 = \$325,800.$$

Since the principal is $112,475, the rest of the total amount paid is interest. That is,

[total amount of interest paid] $= 325,800 - 112,475$

$$= \$213,325. \qquad \textbf{\textit{» Now Try Exercise 17}}$$

In the early years of a mortgage, most of each monthly payment is applied to interest. For the mortgage described in Example 2, the interest portion will exceed the principal portion until the 23rd year. Figure 6 shows how the monthly payment is apportioned between interest and repayment of principal. Figure 7 shows the decrease in the balance for the duration of the mortgage.

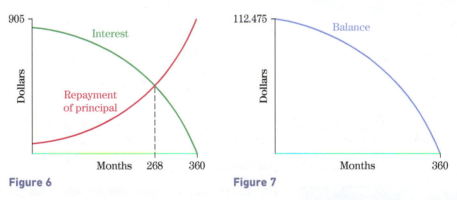

Figure 6   **Figure 7**

Each new (unpaid) balance can be computed from the previous balance with the formula

$$B_{\text{new}} = (1 + i)B_{\text{previous}} - R.$$

That is, the new balance equals the growth of the previous balance due to interest, minus the amount paid. Successive balances are computed by multiplying the previous balance by $(1 + i)$ and subtracting R.

> **Successive Balances** If P dollars is borrowed at interest rate i per period and R dollars is paid back at the end of each interest period, then the formula
>
> $$B_{\text{new}} = (1 + i)B_{\text{previous}} - R \qquad (1)$$
>
> can be used to calculate each new balance, B_{new}, from the previous balance, B_{previous}.

The Excel spreadsheet in Table 2 calculates the payment per period and shows the successive balances for the annuity of Example 2. The entry in cell B4 was set to `=-PMT(B1,B2,B3)`. The form of the payment function is `PMT(rate,nper,pv)`. The value of the payment function is a negative number, since the money is paid to the bank. B10 was set to B3, and B11 was set to `=(1+$B$1)*B10-$B$4`. The remainder of the B column was created by selecting the cell B11 and dragging its fill handle. C11 was set to `=$B$1*B10`, and D11 was set to `=$B$4-C11`. The remainder of the C and D columns was created by dragging fill handles.

Table 2 Amortization Table

	A	B	C	D
1	Interest rate	.75%		
2	Number of periods	360		
3	Principal	$112,475.00		
4	Payment per period	$905.00		
5				
6		Amortization Table		
7				
8		Unpaid		Applied to
9	Payment number	balance	Interest	principal
10	0	$112,475.00		
11	1	$112,413.56	$843.56	$61.44
12	2	$112,351.67	$843.10	$61.90
13	3	$112,289.30	$842.64	$62.36
14	4	$112,226.47	$842.17	$62.83
15	5	$112,163.17	$841.70	$63.30
16	6	$112,099.40	$841.22	$63.78
17	7	$112,035.14	$840.75	$64.25
18	8	$111,970.41	$840.26	$64.74
19	9	$111,905.19	$839.78	$65.22
20	10	$111,839.48	$839.29	$65.71
21	11	$111,773.27	$838.80	$66.20
22	12	$111,706.57	$838.30	$66.70

	A	B	C	D
359	349	$9,521.20	$77.61	$827.38
360	350	$8,687.61	$71.41	$833.59
361	351	$7,847.77	$65.16	$839.84
362	352	$7,001.63	$58.86	$846.14
363	353	$6,149.14	$52.51	$852.49
364	354	$5,290.26	$46.12	$858.88
365	355	$4,424.94	$39.68	$865.32
366	356	$3,553.13	$33.19	$871.81
367	357	$2,674.78	$26.65	$878.35
368	358	$1,789.84	$20.06	$884.94
369	359	$898.26	$13.42	$891.58
370	360	$0.00	$6.74	$898.26
371	Payment number	Unpaid	Interest	Applied to
372		balance		principal

EXAMPLE 3 **Calculating the Unpaid Balance on a Mortgage** Refer to Example 2. Compute the unpaid balance of the loan on January 31, 2023, just after the 313th payment.

SOLUTION By formula (1),

$$B_{\text{Jan}} = (1 + i)B_{\text{Dec}} - R$$
$$= (1.0075) \cdot 36,367.23 - 905$$
$$= \$35,734.98. \qquad \text{«}$$

Sometimes, amortized loans stipulate a **balloon payment** at the end of the term. For instance, you might pay $200 at the end of each quarter year for 3 years and an additional $1000 at the end of the third year. The $1000 is a balloon payment. Balloon payments are more common in commercial real estate than in residential real estate.

EXAMPLE 4 **Analyzing the Apportionment of Mortgage Payments** Consider the situation from Example 2 in which $112,475 is borrowed at 9% interest compounded monthly and repaid with 360 monthly payments of $905. Use technology to determine when the repayment of the principal portion of the payment will surpass the interest portion.

SOLUTION The question can be answered with a graph. Define Y_1, Y_2, and Y_3 as follows:

$$Y_1 = \frac{1 - 1.0075^{X-360}}{.0075} * 905 \qquad \text{Unpaid balance}$$

$$Y_2 = .0075 * Y_1(X - 1) \qquad \text{Interest on the previous balance}$$

$$Y_3 = 905 - Y_2 \qquad \text{Repayment of principal}$$

Then, set the window as in Fig. 8, graph Y_2 and Y_3, and find their point of intersection as shown in Fig. 9.

Figure 8

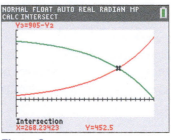

Figure 9

✳ WolframAlpha The instruction

solve 905(1-1.0075^(x-361)) = 905-905(1-1.0075^(x-361))

yields 268.234.

In each case, we see that principal repayment will exceed interest after 269 months—that is, after 22 years and 5 months. **>> Now Try Exercise 51**

EXAMPLE 5 **Determining How Much You Can Afford to Borrow** How much money can you borrow at 8% interest compounded quarterly if you agree to pay $200 at the end of each quarter year for 3 years and, in addition, a balloon payment of $1000 at the end of the third year?

SOLUTION Here, you are borrowing in the present and repaying in the future. The amount of the loan will be the present value of *all* future payments. The future payments consist of an annuity and a lump-sum payment. Let us calculate the present values of each of these separately.

NORMAL FLOAT AUTO REAL RADIAN MP

```
N=12
I%=8
■PV=2115.068244
PMT=-200
FV=0
P/Y=4
C/Y=4
PMT:END BEGIN
```

Figure 10

$$[\text{present value of annuity}] = \frac{1 - (1 + i)^{-n}}{i} \cdot R \qquad \text{Formula for present value}$$

$$= \frac{1 - 1.02^{-12}}{.02} \cdot 200 \qquad i = \frac{.08}{4} = .02, n = 3 \cdot 4 = 12, R = 200$$

$$= 2115.07 \qquad \text{Calculate. Round to two decimal places.}$$

The present value of the annuity is $2115.07. (The TVM Solver screen in Fig. 10 confirms this result.)

NORMAL FLOAT AUTO REAL RADIAN MP

```
N=12
I%=8
■PV=-788.4931756
PMT=0
FV=1000
P/Y=4
C/Y=4
PMT:END BEGIN
```

Figure 11

$$[\text{present value of balloon payment}] = \frac{F}{(1 + i)^n} \qquad \text{Formula for present value}$$

$$= \frac{1000}{1.02^{12}} \qquad F = 1000, i = \frac{.08}{4} = .02, n = 3 \cdot 4 = 12$$

$$= 788.49 \qquad \text{Calculate. Round to two decimal places.}$$

The present value of the balloon payment is $788.49. (The TVM Solver screen in Fig. 11 confirms this result.)

Therefore, the amount that you can borrow is

$$\$2115.07 + \$788.49 = \$2903.56. \qquad \text{«}$$

INCORPORATING TECHNOLOGY

Display Balances Graphing calculators can easily display successive balances for a loan. Consider the situation from Example 2 in which $112,475 is borrowed at 9% interest compounded monthly and repaid with 360 monthly payments of $905. In Fig. 12, successive balances are displayed in the home screen. In Fig. 13, successive balances are graphed, and in Fig. 14, they are displayed as part of an amortization table. (The Y_2 column gives the interest portion of the monthly payment.)

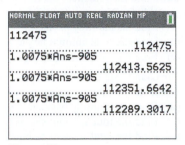

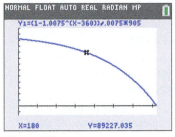

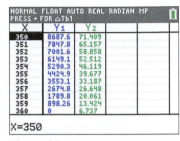

| **Figure 12** | **Figure 13** | **Figure 14** |

Figures 12–14 were obtained with the same processes as Figures 1–3 in Section 10.1. Successive balances were determined with the relation

$$B_{\text{new}} = (1 + i)B_{\text{previous}} - R.$$

The unpaid balance function to be graphed was specified as

$$Y_1 = \frac{1 - 1.0075^{X-360}}{.0075} * 905.$$

The window was set to [0, 360] *by* [−15000, 130000], with an X-scale of 50 and a Y-scale of 10000. In the TABLE SETUP menu, **TblStart** was set to **350**. The second function was defined as $Y_2 = .0075 * Y_1(X-1)$—that is, the interest on the previous balance. The up-arrow key can be used to generate earlier balances and interest payments.

Excel Spreadsheet The section "Using Excel's Financial Functions" of Appendix C shows how the functions IPMT and PPMT are used to calculate the interest and principal reduction portions of a specific payment—that is, the values in columns C and D of Table 2. The annuity functions FV, PV, PMT, and NPER also apply to mortgages.

WolframAlpha Since a mortgage is a decreasing annuity with the borrower serving as the bank, questions about mortgages can be answered with the annuity instructions presented in Section 10.2. For instance, the following instructions give the solutions to Example 1:

(a) annuity pmt,PV = $563,i = 1%,n = 5

(b) annuity PV,pmt = $116,i = 1%,n = 5

(c) annuity PV,pmt = $116,i = 1%,n = 2

The instruction

mortgage $112475,30 years,9%

generates an amortization table for the mortgage of Example 2. The table gives the monthly payment and displays values at the end of each year. The instruction also produces graphs similar to the ones in Figs. 6 and 7.

Check Your Understanding 10.3   Solutions can be found following the section exercises.

1. The word "amortization" comes from the French "à mort," meaning "at the point of death." Justify the word.

2. **Mortgage** Determine the y-coordinate of the intersection point in Fig. 6 on page 452.

3. Explain why only present values and not future values arise in amortization problems.

EXERCISES 10.3

1. **Loan Payment** A car loan of $10,000 is to be repaid with quarterly payments for 5 years at 6.4% interest compounded quarterly. Calculate the quarterly payment.

2. **Loan Payment** A loan of $5000 is to be repaid with quarterly payments for 2 years at 5.6% interest compounded quarterly. Calculate the quarterly payment.

3. **Loan Payment** A loan of $4000 is to be repaid with semiannual payments for 3 years at 5.2% interest compounded semiannually. Calculate the semiannual payment.

4. **Loan Payment** A loan of $3000 is to be repaid with semiannual payments for 4 years at 4.8% interest compounded semiannually. Calculate the semiannual payment.

5. **Loan Amount** The weekly payment on a 2-year loan at 7.8% compounded weekly is $23.59. Calculate the amount of the loan.

6. **Loan Amount** The quarterly payment on a 5-year loan at 6.8% compounded quarterly is $235.82. Calculate the amount of the loan.

7. **Mortgage Payment** Find the monthly payment on a $100,000, 25-year mortgage at 5.4% interest compounded monthly.

8. **Mortgage Payment** Find the monthly payment on a $250,000, 30-year mortgage at 4.8% interest compounded monthly.

9. **Mortgage Amount** Find the amount of a 30-year mortgage at 4.5% interest compounded monthly where the monthly payment is $724.56.

10. **Mortgage Amount** Find the amount of a 25-year mortgage at 4.2% interest compounded monthly where the monthly payment is $1121.00.

11. **Mortgage Balance** A 30-year mortgage at 4.2% interest compounded monthly with a monthly payment of $1019.35 has an unpaid balance of $10,000 after 350 months. Find the unpaid balance after 351 months.

12. **Mortgage Balance** A 25-year mortgage at 4.5% interest compounded monthly with a monthly payment of $258.96 has an unpaid balance of $5,000 after 280 months. Find the unpaid balance after 281 months.

13. **Loan Interest** A loan with a weekly payment of $100 has an unpaid balance of $2000 after 5 weeks and an unpaid balance of $1903 after 6 weeks. If interest is compounded weekly, find the interest rate.

14. **Loan Interest** A loan with a quarterly payment of $1500 has an unpaid balance of $10,000 after 30 quarters and an unpaid balance of $8670 after 31 quarters. If interest is compounded quarterly, find the interest rate.

15. **Amortization Schedule** Write out a complete amortization schedule (as in Table 1 at the beginning of this section) for the amortization of a $830 loan with monthly payments of $210.10 at 6% interest compounded monthly for 4 months.

16. **Amortization Schedule** Write out a complete schedule (as in Table 1) for the amortization of a $945 loan with payments of $173.29 every 6 months at 5.6% interest compounded semiannually for 3 years.

17. **Mortgage** Consider a $204,700, 30-year mortgage at interest rate 4.8%, compounded monthly, with a $1073.99 monthly payment.

(a) How much interest is paid the first month?

(b) How much of the first month's payment is applied to paying off the principal?

(c) What is the unpaid balance after 1 month?

18. **Mortgage** James buys a house for $370,000. He puts $70,000 down and then finances the rest at 6.3% interest compounded monthly for 25 years.

(a) Find his monthly payment.

(b) Find the total amount he pays for the house.

(c) Find the total amount of interest he pays.

19. **Car Loan** Susie takes out a car loan for $9480 for a term of 3 years at 6% interest compounded monthly.

(a) Find her monthly payment.

(b) Find the total amount she pays for the car.

(c) Find the total amount of interest she pays.

(d) Find the amount she still owes after 1 year.

(e) Find the amount she still owes after 2 years.

(f) Find the total interest she pays in year 2.

(g) Prepare the amortization schedule for the first 4 months.

20. **Loan** Consider a $21,281.27 loan for 7 years at 8% interest compounded quarterly with a payment of $1000 per quarter-year.

(a) Compute the unpaid balance after 5 years.

(b) How much interest is paid during the fifth year?

(c) How much principal is repaid in the first payment?

(d) What is the total amount of interest paid on the loan?

(e) Prepare the amortization schedule for the first 4 months.

21. **Comparing Financing Options** In a recent year, Toyota was offering the choice of a .9% loan for 60 months or $500 cash back on the purchase of an $18,000 Toyota Corolla.

(a) If you take the .9% loan offer, how much will your monthly payment be?

(b) If you take the $500 cash-back offer and can borrow money from your local bank at 6% interest compounded monthly for 5 years, how much will your monthly payment be?

(c) Which of the two offers is more favorable for you?

22. **Comparing Financing Options** In a recent year, Ford was offering the choice of a 1.9% loan for 36 months or $1000 cash back on the purchase of a $28,000 Taurus.

(a) If you take the 1.9% loan offer, how much will your monthly payment be?

(b) If you take the $1000 cash-back offer and can borrow money from your local bank at 6% interest compounded monthly for 3 years, how much will your monthly payment be?

(c) Which of the two offers is more favorable for you?

23. **Comparing Financing Options** A bank makes the following two loan offers to its credit card customers:

Option I: Pay 0% interest for 5 months and then 9% interest compounded monthly for the remainder of the duration of the loan.

Option II: Pay 6% interest compounded monthly for the duration of the loan.

With either option, there will be a minimum payment of $100 due each month. Suppose that you would like to borrow $5500 and pay it back in a year and a half. For which option do you pay the least amount to the bank?

24. **Buy Now or Later?** According to an article in the *New York Times* on August 9, 2008, economists were predicting that the average interest rate for a 30-year mortgage would increase from 6.7% to 7.1% during the next year and home prices would decline by 9% during that same period. In August 2008, the Johnson family was considering purchasing a $400,000 home. They intended to make a 20% down payment and finance the rest of the cost. What would their monthly payment have been if they bought the house in August 2008? A comparable house in August 2009?

25. **Balloon Payment** A loan is to be amortized over an 8-year term at 6.4% interest compounded semiannually, with payments of $905.33 every 6 months and a balloon payment of $5,000 at the end of the term. Calculate the amount of the loan.

26. **Balloon Payment** A loan of $127,000.50 is to be amortized over a 5-year term at 5.4% interest compounded monthly, with monthly payments and a $10,000 balloon payment at the end of the term. Calculate the monthly payment.

27. **Car Loan** A car is purchased for $6287.10, with $2000 down and a loan to be repaid at $100 a month for 3 years, followed by a balloon payment. If the interest rate is 6% compounded monthly, how large will the balloon payment be?

28. **Cash Flow for a Rental Property** You are considering the purchase of a condominium to use as a rental property. You estimate that you can rent the condominium for $1500 per month and that taxes, insurance, and maintenance costs will run about $300 per month. If interest rates are 4.8% compounded monthly, how large of a 25-year mortgage can you assume and still have the rental income cover all of the monthly expenses?

29. **Repayment of Principal** If you take out a 30-year mortgage at 6.8% interest compounded monthly, what percentage of the principal will be paid off after 15 years?

30. **Repayment of Principal** If you take out a 20-year mortgage at 6.4% interest compounded monthly, what percentage of the principal will be paid off after 10 years?

31. **Terminating a Mortgage** In 2006, Emma purchased a house and took out a 25-year, $50,000 mortgage at 6% interest compounded monthly. In 2016, she sold the house for $150,000. How much money did she have left after she paid the bank the unpaid balance on the mortgage?

32. **Refinancing a Mortgage** A real estate speculator purchases a tract of land for $1 million and assumes a 25-year mortgage at 4.2% interest compounded monthly.
    (a) What is his monthly payment?
    (b) Suppose that at the end of 5 years the mortgage is changed to a 10-year term for the remaining balance. What is the new monthly payment?
    (c) Suppose that after 5 more years, the mortgage is required to be repaid in full. How much will then be due?

33. **Loan Payment** The total interest paid on a 3-year loan at 6% interest compounded monthly is $1,085.16. Determine the monthly payment for the loan. (*Hint:* Use the fact that the loan amount equals $36 \cdot R - $ [total interest].)

34. **Loan Payment** The total interest paid on a 5-year loan at 8% interest compounded quarterly is $2,833.80. Determine the quarterly payment for the loan. (*Hint:* Use the fact that the loan amount equals $20 \cdot R - $ [total interest].)

35. **Total of Loan Payment** Suppose that you borrow $10,000 at 4% interest compounded semiannually and pay off the loan in 7 years. Which of items (a)–(d) can be used to fill in the blank in the statement that follows? (*Note:* Before computing, use your intuition to guess the correct answer.)

    If the _____ is doubled, then the total amount paid will double.
    (a) amount of the loan
    (b) interest rate
    (c) number of interest periods per year
    (d) term of the loan

36. **Loan Amount** Suppose that you borrow money at 4% interest compounded semiannually and pay off the loan in payments of $1000 per interest period for 5 years. Which of items (a)–(d) can be used to fill in the blank in the statement that follows? (*Note:* Before computing, use your intuition to guess the correct answer.)

    If the _____ is doubled, then the amount that can be borrowed will double.
    (a) amount paid per interest period
    (b) interest rate
    (c) number of interest periods per year
    (d) term of the loan

37. Let $B_n = $ balance of a loan after n payments, $I_n = $ the interest portion of the nth payment, and $Q_n = $ the portion of the nth payment applied to the principal. Equation (1) states that $B_n = (1 + i)B_{n-1} - R$.
    (a) Use the fact that $I_n + Q_n = R$ and $I_n = iB_{n-1}$ to show that

    $$B_{n-1} = \frac{R - Q_n}{i}.$$

    (b) Use equation (1) and the results that

    $$B_{n-1} = \frac{R - Q_n}{i} \quad \text{and} \quad B_n = \frac{R - Q_{n+1}}{i}$$

    from part (a) to derive the formula $Q_{n+1} = (1 + i)Q_n$.
    (c) State in your own words the formula obtained in part (b).
    (d) Suppose that, for the 10th monthly payment on a loan at 1% interest per month, $100 was applied to the principal. How much of the 11th and 12th payments will be applied to the principal?

38. Let Q_n and I_n be as defined in Exercise 37.
    (a) Use the result in part (b) of Exercise 37 and the facts that $Q_{n+1} = R - I_{n+1}$ and $Q_n = R - I_n$ to show that $I_{n+1} = (1 + i)I_n - iR$.
    (b) State in your own words the formula obtained in part (a).
    (c) Suppose that, for a loan with a monthly payment of $400 and an interest rate of 1% per month, $100 of the 9th monthly payment goes toward paying off interest. How much of the 10th and 11th payments will be used to pay off interest?

TECHNOLOGY EXERCISES

In Exercises 39–46, give settings or statements to determine the solution with TVM Solver, Excel, or Wolfram|Alpha.

39. Exercise 1   40. Exercise 4

41. Exercise 5   42. Exercise 6

43. Exercise 7

44. Exercise 8

45. Exercise 9

46. Exercise 10

47. **Financing a Computer** Bill buys a top-of-the-line computer for $4193.97 and pays off the loan (at 4.8% interest compounded monthly) by paying $100 at the end of each month. After how many months will the loan be paid off?

48. **Financing Medical Equipment** Alice borrows $20,000 to buy some medical equipment and pays off the loan (at 7% interest compounded annually) by paying $4195.92 at the end of each year. After how many years will the loan be paid off?

49. **Debt Reduction** A loan of $10,000 at 9% interest compounded monthly is repaid in 80 months with monthly payments of $166.68. After how many months will the loan be one-quarter paid off? One-half? Three-quarters?

50. **Debt Reduction** A loan of $4000 at 6% interest compounded monthly is repaid in 8 years with monthly payments of $52.57. After how many months will the loan be one-quarter paid off? One-half? Three-quarters?

51. **Debt Reduction** A 25-year mortgage of $124,188.57 at 8.5% interest compounded monthly has monthly payments of $1000. After how many months will at least 75% of the monthly payment go toward debt reduction?

52. **Debt Reduction** A 30-year mortgage of $118,135.40 at 8.4% interest compounded monthly has monthly payments of $900. After how many months will the amount applied to the reduction of the debt be more than twice the amount applied to interest?

Solutions to Check Your Understanding 10.3

1. A portion of each payment is applied to reducing the debt, and by the end of the term the debt is totally annihilated.

2. 452.5. Since the *amount of interest* equals the *repayment of principal* at the point of intersection and the two numbers add up to 905, they must each have the value 905/2; that is, 452.5

3. The debt is formed when the creditor gives you a lump sum of money now, in the present. The lump sum of money is gradually repaid by you, with interest, thereby generating the decreasing annuity. At the end of the term, in the future, the loan is totally paid off, so there is no more debt. That is, the future value is always zero!

NOTE You are actually functioning like a savings bank, since you are paying the interest. Think of the creditor as depositing the lump sum with you and then making regular withdrawals until the balance is 0. **«**

10.4 Personal Financial Decisions

You will be faced with several financial decisions shortly after graduating from college. Financial advisers suggest that you open an individual retirement account as soon as you start earning an income. You will probably buy a car and some major appliances with consumer loans. You will consider buying a condominium or a house and will need a mortgage. This section will help you answer the following questions:

- How important is it to open an individual retirement account as soon as possible? What kind of IRA should you open?

- What is the difference between the finance charges on a consumer loan that employs the add-on method and one that employs the compounding method of Section 10.3?

- How do discount points work, and when are they a good choice?

- How do you choose between two mortgages if they have different interest rates and up-front costs?

- What are interest-only and adjustable-rate mortgages and when are they a good choice?

Individual Retirement Accounts

Money earned in an ordinary savings account is subject to federal, state, and local income taxes. However, a special type of savings account, called an **individual retirement account** (IRA), provides a shelter from these taxes. IRAs are highly touted by financial planners and are available to any individual whose income does not exceed a certain limit. The maximum annual contribution for persons under 50 in 2016 was about $5500. (However, the amount contributed must not exceed the person's earned income.) Typically, money in an IRA cannot be withdrawn without penalty until the person reaches $59\frac{1}{2}$ years of age. However, funds may be withdrawn earlier under certain circumstances.

The two main types of IRAs are known as a **traditional IRA** and a **Roth IRA**. Contributions to a traditional IRA are tax deductible, but all withdrawals are taxed. Interest earned is not taxed until it is withdrawn. In contrast, contributions to a Roth IRA are not tax deductible, but withdrawals are not taxed. Therefore, interest earned is never taxed.

EXAMPLE 1   **Calculating Values Associated with a Traditional IRA** Suppose that, for the year 2016, you deposit $5000 of earned income into a traditional IRA, you earn an annual interest rate of 6% compounded annually, and you are in a 30% marginal tax bracket for the duration of the account. (That is, for each additional dollar you earn, you pay an additional $.30 in taxes. Likewise, for each dollar you can deduct, you save $.30 in taxes.) We recognize that interest rates and tax brackets are subject to change over a long period of time, but some assumptions must be made in order to evaluate the investment.

(a) How much income tax on earnings will you save for the year 2016?

(b) Assuming that no additional deposits are made, how much money will be in the account after 48 years?

(c) Suppose that you withdraw all of the money in the IRA after 48 years (assuming that you are older than $59\frac{1}{2}$). How much will you have left after you pay the taxes on the money?

SOLUTION   (a) [income tax saved] = [tax bracket] · [amount]

$$= \quad .30 \quad \cdot \quad 5000$$

$$= 1500$$

Therefore, $1500 in income taxes will be saved. You will have the full $5000 of earned income to deposit into your traditional IRA.

(b) [balance after 48 years] = $P(1 + i)^n$

$$= 5000(1.06)^{48}$$

$$= 81,969.36$$

Therefore, after 48 years, the balance in the IRA will be $81,969.36.

(c) When you withdraw the money, 30% of the money will be paid in taxes. Hence, 70% will be left for you.

$$[\text{amount after taxes}] = .70(81,969.36)$$

$$= \$57,378.55$$ ≪

EXAMPLE 2   **Comparing a Savings Account with a Traditional IRA** Rework Example 1 with the money going into an ordinary savings account.

SOLUTION   (a) None. With regard to a savings account, there are no savings on income tax, since you pay the taxes as money is earned on the account. Hence,

$$[\text{earnings after income tax}] = [1 - \text{tax bracket}] \cdot [\text{amount}]$$

$$= \quad .70 \quad \cdot \quad 5000$$

$$= 3500.$$

Therefore, $3500 will be deposited into the savings account.

(b) The interest earned each year will be taxed during that year. Therefore, each year, 30% of the interest will be paid in taxes, and hence, only 70% will be kept. Since 70% of 6% is 4.2%, you will effectively earn just 4.2% interest per year. Hence, the future value of your investment is as follows:

$$[\text{balance after 48 years}] = P(1 + i)^n = 3500(1.042)^{48} = \$25,218.46$$

(c) $25,218.46, since the interest was taxed in the year that it was earned.   ≪

Comparing the results from Examples 1 and 2, we see why financial planners encourage people to set up IRAs. The balance upon withdrawal is $32,160.09 greater with the IRA. The amount of money in the traditional IRA is more than 2.25 times the amount of money in an ordinary savings account.

EXAMPLE 3	**Calculating the Future Value of a Roth IRA** Consider the following variation of Example 1. Suppose that you earned $5000, paid your taxes, and then deposited the remainder into a Roth IRA. How much money would you have if you closed out the account after 48 years?

SOLUTION

$$[\text{earnings after income tax}] = [1 - \text{tax bracket}] \cdot [\text{amount}]$$
$$= \quad .70 \quad \cdot \quad 5000$$
$$= \$3500$$

Therefore, $3500 will be deposited into the Roth IRA.

$$[\text{balance after 48 years}] = P(1 + i)^n = 3500(1.06)^{48} = \$57{,}378.55$$

Therefore, after 48 years, the balance will be $57,378.55, the same amount found in part (c) of Example 1. Since you do *not* pay taxes on the interest earned, you will have all of this money available to you upon withdrawal. «

The result of Example 3 suggests the following general conjecture, which is easily verified:

> **Equivalence of Traditional and Roth IRAs**   With the assumption that your tax bracket does not change, the net earnings upon withdrawal from contributing P dollars to a traditional IRA account is the same amount as would result from paying taxes on the P dollars but then contributing the remaining money to a Roth IRA.

Verification of the Equivalence Conjecture   Suppose that P dollars is deposited into a traditional IRA at interest rate r compounded annually for n years. Also, suppose that your marginal tax bracket is k. (In Example 1, $k = .30$.) In general, if taxes are paid on A dollars, then the amount remaining after taxes is

$$A - k \cdot A = A \cdot (1 - k).$$

With a traditional IRA, the future value (or balance) after n years is $[P(1 + r)^n]$ and the amount left after taxes are paid is $[P(1 + r)^n] \cdot (1 - k)$.

With a Roth IRA, taxes are paid initially on the P dollars, so the amount deposited is $[P \cdot (1 - k)]$. The future value (or balance) after n years is obtained by multiplying the amount deposited by $(1 + r)^n$. That is, the balance is $[P \cdot (1 - k)] \cdot (1 + r)^n$.

Each of the preceding amounts is the product of the same three numbers; only the order is different. Therefore, the two investments have the same future value. «

Financial planners encourage clients not only to establish IRAs, but also to begin making contributions at an early age. The following example illustrates the advantage of beginning as soon as possible:

EXAMPLE 4	**The Value of Starting an IRA Early** Earl and Larry each begin full-time jobs in January 2017 and plan to retire in January 2065 after working for 48 years. Assume that any money that they deposit into IRAs earns 6% interest compounded annually. **(a)** Suppose that Earl opens a traditional IRA account immediately and deposits $5000 into his account at the end of each year for 12 years. After that, he makes no further deposits and just lets the money earn interest. How much money will Earl have in his account when he retires in January 2065?

(b) Suppose that Larry waits 12 years before opening his traditional IRA and then deposits $5000 into the account at the end of each year until he retires. How much money will Larry have in his account when he retires in January 2065?

(c) Who paid more money into his IRA?

(d) Who had more money in his account upon retirement?

SOLUTION   **(a)** Earl's contributions consist of an increasing annuity with $R = 5000$, $i = .06$, and $n = 12$. The future value (or balance) in the account on January 1, 2029 will be

$$F = \frac{(1 + i)^n - 1}{i} \cdot R = \frac{1.06^{12} - 1}{.06} \cdot 5000 = \$84{,}349.71.$$

This money then earns interest compounded annually for 36 years. Using the formula $F = P(1 + i)^n$, it grows to

$$84{,}349.71 \cdot (1.06)^{36} = \$687{,}218.34.$$

(b) Larry's contributions consist of an increasing annuity with $R = 5000$, $i = .06$, and $n = 36$. The balance in the account on January 1, 2065 will be

$$F = \frac{(1 + i)^n - 1}{i} \cdot R = \frac{1.06^{36} - 1}{.06} \cdot 5000 = \$595{,}604.33.$$

(c) Larry pays in $36 \cdot 5000 = \$180{,}000$ versus the $12 \cdot 5000 = \$60{,}000$ paid in by Earl. That is, Larry pays in three times as much money as Earl does.

(d) Earl has $\$91{,}614.01$ more than Larry. Figure 1 shows the growths of the accounts.

≫ Now Try Exercise 9

Figure 1

Consumer Loans and the Truth-in-Lending Act

Consumer loans are paid off with a sequence of monthly payments in much the same way as mortgages. However, the most widely used method for determining finance charges on consumer loans is the **add-on method**, which results in an understated interest rate. When an add-on method interest rate is stated as r per year, the term of the loan is t years, and the amount of the loan is P, then the total interest to be paid is $P \cdot r \cdot t$. The monthly payment is then determined by dividing the sum of the principal and the interest by the total number of months. For instance, suppose that you take out a two-year loan of $1000 at a stated annual interest rate of 6%.

$$[\text{monthly payment}] = \frac{1000 + 1000 \cdot .06 \cdot 2}{12 \cdot 2}$$

$$= \frac{1000(1 + .06 \cdot 2)}{24}$$

$$= \$46.67$$

The general formula for the monthly payment R is

$$R = \frac{P(1 + rt)}{12t}.$$

This formula can be solved for r to obtain

$$r = \frac{\frac{12Rt - P}{t}}{P} = \frac{12Rt - P}{Pt}.$$

Since $12Rt - P$ is the amount of interest paid for the duration of the loan, $\frac{12Rt - P}{t}$ is the amount of interest paid per year and $\frac{12Rt - P}{Pt}$ is the annual percent of the principal that consists of the interest, namely, r.

Figure 2

If this loan were amortized as in Section 10.3, the interest rate would be $i = .06/12 = .005$. The monthly payment would be

$$[\text{monthly payment}] = \frac{i}{1 - (1 + i)^{-n}} \cdot P \qquad \begin{array}{l}\text{Formula for rent (that is, the monthly} \\ \text{payment)}\end{array}$$

$$= \frac{.005}{1 - 1.005^{-24}} \cdot 1000 \qquad i = \frac{.06}{12} = .005, n = 2 \cdot 12 = 24, P = 1000$$

$$= 44.32 \qquad \text{Calculate. Round to two decimal places.}$$

The monthly payment is $44.32. (The TVM Solver screen in Fig. 2 confirms this result.)

This means that the add-on method results in the borrower being charged an additional $46.67 - 44.32 = 2.35$ dollars in interest per month. The problem with the add-on method is that each month, the borrower is being charged interest on the entire principal even though, after the first month, part of the principal has already been repaid.

In 1968, Congress passed the federal Truth-in-Lending Act as a part of the Consumer Protection Act. The law's stated goal is

> to assure a meaningful disclosure of credit terms so that the consumer will be able to compare more readily the various credit terms available . . . .

The Truth-in-Lending Act requires a lender to disclose what is known as the **annual percentage rate** (or **APR**) for any advertised loan. For the preceding loan, the monthly payment of $46.67 corresponds to an interest rate of 11.134% when the payment is calculated as in Section 10.3. Therefore, the APR is 11.134%.

EXAMPLE 5   **Calculating the Monthly Payment for a Loan with the Add-on Method**   A one-year loan for $5000 is advertised at 8% by the add-on method. What is the monthly payment?

SOLUTION   The interest on the loan is $.08 \cdot (5000) = \$400$. The total amount (principal plus interest) of $5400 is to be paid in 12 monthly payments of $5400/12 = \$450$.

>> *Now Try Exercise 11*

EXAMPLE 6   **Determining the Interest Rate on a Consumer Loan**   Suppose that a consumer loan of $100,000 for 10 years has an APR of 8% compounded monthly and a monthly payment of $1213.28. What interest rate would be stated with the add-on method?

SOLUTION   The total amount paid by the borrower is $120 \cdot 1213.28 = \$145,593.60$. Therefore, the interest paid on the loan is $145,593.60 - 100,000 = \$45,593.60$. The interest paid per year is $45,593.60/10 = \$4559.36$, which is about 4.56% of $100,000. Therefore, with the add-on method, the interest rate would be given as 4.56%.   >> *Now Try Exercise 15*

The computation of the APR is easily accomplished with a spreadsheet or graphing calculator. For instance, with Excel, the APR of 11.134% discussed previously is calculated with the RATE function as 12*RATE(24,−46.67,1000,0). This same value is easily obtained on a TI-84 Plus calculator with TVM Solver. To determine the APR on an arbitrary graphing calculator, graphically solve the equation $1000 = \frac{1 - (1 + X)^{-24}}{X} \cdot 46.67$ for X, and then multiply by 12.

Mortgages with Discount Points

Home mortgages have a stated annual interest rate, called the **contract rate**. However, some loans also carry **points** or **discount points** to reduce the contract rate. Each point requires that you pay up-front additional interest equal to 1% of the stated loan amount. For instance, for a $200,000 mortgage with 3 discount points you must pay $6000 immediately. This has the effect of reducing the loan to $194,000, yet requiring the same monthly payment as a $200,000 mortgage having no points.

EXAMPLE 7 **Discount Points and Monthly Payments for Mortgages** Compute the monthly payment for each of the following 30-year mortgages:

Mortgage A: $200,000 at 6% interest compounded monthly, one point

Mortgage B: $198,000 at 6.094% interest compounded monthly, no points

SOLUTION Mortgage A:

Figure 3

Figure 4

$$R = \frac{i}{1 - (1 + i)^{-n}} \cdot P \qquad \text{Formula for rent (that is, the monthly payment)}$$

$$= \frac{.005}{1 - 1.005^{-360}} \cdot 200,000 \qquad i = \frac{.06}{12} = .005, n = 30 \cdot 12 = 360, P = 200,000$$

$$= 1199.10 \qquad \text{Calculate. Round to two decimal places.}$$

The monthly payment is $1199.10. (The TVM Solver screen in Fig. 3 confirms this result.)

Mortgage B:

$$R = \frac{i}{1 - (1 + i)^{-n}} \cdot P \qquad \text{Formula for rent (that is, the monthly payment)}$$

$$= \frac{\frac{.06094}{12}}{1 - \left(1 + \frac{.06094}{12}\right)^{-360}} \cdot 198,000 \qquad i = \frac{.06094}{12}, n = 30 \cdot 12 = 360, P = 198,000$$

$$= 1199.10 \qquad \text{Calculate. Round to two decimal places.}$$

The monthly payment is $1199.10. (The TVM Solver screen in Fig. 4 confirms this result.) «

The two monthly payments in Example 7 are the same. With mortgage A, the borrower is essentially borrowing $198,000 (since $2000 is paid back to the lender immediately) and then making the same monthly payments as on a $198,000 loan at 6.094%. The Truth-in-Lending Act requires the lender of mortgage A to specify that the loan has an APR of 6.094%. Purchasing discount points has the effect of changing the APR, which can be calculated by following the steps shown next.

> **APR** The following steps calculate the APR for a mortgage loan having a term of n months:
>
> 1. Calculate the monthly payment (call it R) on the stated loan amount.
> 2. Subtract the up-front costs (such as points) from the stated loan amount. Denote the result by P.
> 3. Find the interest rate that produces the monthly payment R for a loan of P dollars to be repaid in n months.

The interest rate discussed in step 3 can be calculated in Excel with the RATE function as

$$12 * \text{RATE}(n, -R, P, 0).$$

To find the interest rate in step 3 with a TI-84 Plus calculator, use TVM Solver. On a graphing calculator or with Wolfram|Alpha, find the intersection of the graphs of $Y_1 = P$ and $Y_2 = \frac{1 - (1 + X)^{-n}}{X} \cdot R$, and multiply the X-coordinate by 12.

EXAMPLE 8 **Calculating the APR for a Mortgage with Discount Points** Use the three steps just given to calculate the APR for mortgage A of Example 7.

TECHNOLOGY SOLUTION 1. From Example 7, the monthly payment is $1199.10.
2. $P = 200,000 - 2000 = \$198,000$, since the up-front costs consist of one discount point.

3. 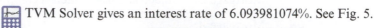 The interest rate can be calculated in Excel with the RATE function. The value of

$$12*\text{RATE}(360, -1199.10, 198000, 0)$$

is .06093981.

 TVM Solver gives an interest rate of 6.093981074%. See Fig. 5.

```
NORMAL FLOAT AUTO REAL RADIAN MP
 N=360
•I%=6.093981074
 PV=198000
 PMT=-1199.1
 FV=0
 P/Y=12
 C/Y=12
 PMT:END BEGIN
```

Figure 5

Therefore, the APR for mortgage A is 6.094%.   **》 Now Try Exercise 27**

The APR allows you to compare loans that are kept for their full terms. However, the typical mortgage is refinanced or terminated after approximately five years—that is, 60 months. Mortgage analysts often rely on the **effective mortgage rate** that takes into account the length of time the loan will be held and the unpaid balance at that time. For the $200,000 mortgage A discussed in Example 7, the monthly payment is $1199.10, and we show in Example 9 that the unpaid balance after 60 months is $186,108.55. With a 5-year lifetime, the effective mortgage rate is the interest rate for which a decreasing annuity with a beginning balance of $198,000 ($200,000 loan − $2000 cost of the point) and monthly payment of $1199.10 will decline to $186,108.55 after 60 months. A graphing calculator or a spreadsheet can be used to determine that the effective rate of interest for mortgage A is 6.24%. If terminated after 5 years, mortgage A is equivalent to a mortgage of $200,000 at 6.24% interest compounded monthly and having no discount points. That is, with the assumption that mortgage A will be held for only 5 years, the interest rate is effectively 6.24%.

EXAMPLE 9   **Calculating Values Associated with Mortgages** Consider mortgage A of Example 7. The mortgage had an up-front payment of $2000, an interest rate of 6% compounded monthly, and a monthly payment of $1199.10. Let mortgage C be a 360-month loan of $200,000 at 6.24% compounded monthly, with no points.

(a) Find the future value of $2000 after 60 months at 6.24% compounded monthly.

(b) Find the future value of an increasing annuity consisting of 60 monthly payments of $1199.10 with an interest rate of 6.24%.

(c) Find the unpaid balance on mortgage A after 60 months.

(d) Find the monthly payment for mortgage C.

(e) Find the future value of an increasing annuity consisting of 60 monthly payments of the amount found in part (d), with an interest rate of 6.24%.

(f) Find the unpaid balance on mortgage C after 60 months.

(g) Compare the sum of the three amounts found in parts (a), (b), and (c) with the sum of the two amounts found in parts (e) and (f).

SOLUTION   **(a)** $F = (1 + i)^n P$ Formula for future value

$\qquad = (1.0052)^{60} \cdot 2000$ $n = 60, i = \dfrac{.0624}{12} = .0052, P = 2000$

$\qquad = 2730.10$ Calculate. Round to two decimal places.

The future value of $2000 after 60 months is $2730.10. (The TVM Solver screen in Fig. 6 confirms this result.)

```
NORMAL FLOAT AUTO REAL RADIAN MP
 N=60
 I%=6.24
 PV=-2000
 PMT=0
•FV=2730.101483
 P/Y=12
 C/Y=12
 PMT:END BEGIN
```

Figure 6

Figure 7

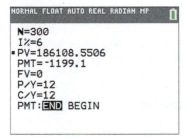

Figure 8

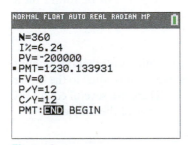

Figure 9

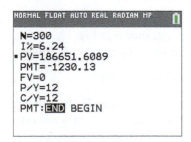

Figure 10

(b) $F = \dfrac{(1+i)^n - 1}{i} \cdot R$ Formula for future value of an increasing annuity

$= \dfrac{1.0052^{60} - 1}{.0052} \cdot 1199.10$ $i = \dfrac{.0624}{12} = .0052,\ n = 60,\ R = 1199.10$

$= 84{,}179.30$ Calculate. Round to two decimal places.

The future value after 60 months is \$84,179.30. (The TVM Solver screen in Fig. 7 confirms this result.)

(c) After 5 years, there are $360 - 60 = 300$ months remaining. The unpaid balance is calculated as the present value of a sequence of $n = 300$ payments. Therefore,

$[\text{unpaid balance}] = \dfrac{1 - (1+i)^{-n}}{i} \cdot R$ Formula for present value

$= \dfrac{1 - 1.005^{-300}}{.005} \cdot 1199.10$ $i = \dfrac{.06}{12} = .005,\ R = 1199.10$

$= 186{,}108.55$ Calculate. Round to two decimal places.

The unpaid balance after 60 months is \$186,108.55. (The TVM solver screen in Fig. 8 confirms this result.)

(d) $R = \dfrac{i}{1 - (1+i)^{-n}} \cdot P$ Formula for rent (that is, monthly payment)

$= \dfrac{.0052}{1 - 1.0052^{-360}} \cdot 200{,}000$ $i = \dfrac{.0624}{12} = .0052,\ n = 360,\ P = 200{,}000$

$= 1230.13$ Calculate. Round to two decimal places.

The monthly payment is \$1230.13. (The TVM Solver screen in Fig. 9 confirms this result.)

(e) $F = \dfrac{(1+i)^n - 1}{i} \cdot R$ Formula for future value of an increasing annuity

$= \dfrac{1.0052^{60} - 1}{.0052} \cdot 1230.13$ $i = \dfrac{.0624}{12} = .0052,\ n = 60,\ R = 1230.13$

$= 86{,}357.67$ Calculate. Round to two decimal places.

The future value after 60 months is \$86,357.67. (The TVM Solver screen in Fig. 10 confirms this result.)

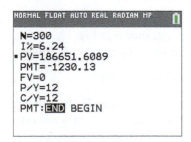

Figure 11

(f) Mortgage C has $360 - 60 = 300$ months to go.

$[\text{unpaid balance}] = \dfrac{1 - (1+i)^{-n}}{i} \cdot R$ Formula for present value

$= \dfrac{1 - 1.0052^{-300}}{.0052} \cdot 1230.13$ $i = \dfrac{.0624}{12} = .0052,\ n = 300,\ R = 1230.13$

$= 186{,}651.61$ Calculate. Round to two decimal places.

The unpaid balance after 60 months is \$186,651.61. (The TMV Solver screen in Fig. 11 confirms this result.)

(g) $[\text{sum of amounts found for mortgage } A] = 2730.10 + 84{,}179.30 + 186{,}108.55$
$= \$273{,}017.95$

$[\text{sum of amounts found for mortgage } C] = 86{,}357.67 + 186{,}651.61$
$= \$273{,}009.28$

The two sums are very close. This demonstrates that the two loans are essentially equivalent after 60 months. (The difference of $8.67 is due to round-off error.) ≪

> **Effective Mortgage Rate** The following steps calculate the effective mortgage rate for a mortgage loan expected to be held for m months:
>
> 1. Calculate the monthly payment (call it R) on the stated loan amount.
> 2. Subtract the up-front costs (such as points) from the stated loan amount. Denote the result by P.
> 3. Determine the unpaid balance on the stated loan amount after m months. Denote the result by B.
> 4. Find the interest rate that causes a decreasing annuity with a beginning balance of P dollars and monthly payment R to decline to B dollars after m months.

The interest rate discussed in step 4 can be calculated in Excel with the RATE function as

$$12*\text{RATE}(m, -R, P, -B).$$

To calculate the interest rate with TVM Solver, set $N = m$, $PV = P$, $PMT = -R$, $FV = -B$, $P/Y = 12$ and solve for I%. The interest rate also can be found on a calculator by graphically solving the equation

$$R - iB = (R - iP) \cdot (1 + i)^m$$

for i and then multiplying the value of i by 12. The preceding equation says that the amount of the monthly payment applied to principal in the $(m + 1)$st payment is equal to the future value of the amount applied to principal in the first payment. There are several equations that involve m, R, P, and i. We chose this equation because it is one of the simplest.

EXAMPLE 10 **Calculating Effective Mortgage Rates** Use the four steps just given to calculate the effective mortgage rate for mortgage A of Example 7, assuming that the mortgage will be held for 5 years.

SOLUTION
1. The monthly payment $R = \$1199.10$ was calculated in Example 7.
2. $P = 200{,}000 - 2000 = 198{,}000$, since the up-front fee consists only of the one discount point.
3. The unpaid balance after 60 months, $B = \$186{,}108.55$, was computed in part (c) of Example 9.
4. The interest rate can be calculated in Excel with the RATE function. The value of

$$12*\text{RATE}(60, -1199.10, 198000, -186108.55)$$

is .0624069.

TVM Solver gives an interest rate of 6.240691672%, as shown in Fig. 12. With a graphing calculator or Wolfram|Alpha, the interest rate is found by graphically solving

$$1199.10 - X(186108.55) = (1199.10 - X(198000))(1 + X)^{60}$$

for X and then multiplying the answer by 12. The value obtained is .06240696. Therefore, the effective mortgage rate for mortgage A is $\approx 6.24\%$, which is the rate of mortgage C in Example 9. **≫ Now Try Exercise 31**

```
NORMAL FLOAT AUTO REAL DEGREE MP

 N=60
•I%=6.240691672
 PV=198000
 PMT=-1199.1
 FV=-186108.55
 P/Y=12
 C/Y=12
 PMT:END BEGIN
```

Figure 12

NOTE Mortgage analysts have a rule of thumb for calculating the effective mortgage rate for a mortgage with points. For mortgages that will be held for 4 to 6 years, each discount point adds about ¼% to the stated interest rate. For instance, for the mortgage in Example 10, the rule of thumb predicts that the effective mortgage rate should be ¼% higher than the stated rate for a mortgage having one discount point. This is very close to the actual increase, .24%. Table 1 extends the rule of thumb to other lifetimes. ≪

Table 1 Rule of Thumb for the Significance of Discount Points

Lifetime	Difference between Effective Mortgage Rate and Stated Interest Rate, Per Discount Point
1 year	1 percentage point
2 years	$\frac{1}{2}$ of a percentage point
3 years	$\frac{1}{3}$ of a percentage point
4 to 6 years	$\frac{1}{4}$ of a percentage point
7 to 9 years	$\frac{1}{6}$ of a percentage point
10 to 12 years	$\frac{1}{7}$ of a percentage point
More than 12 years	$\frac{1}{8}$ of a percentage point

Example 11 shows another way to take the expected lifetime of a loan into account when deciding between two mortgages. If you intend to either sell the house or refinance the loan within a few years, then you should avoid a loan carrying excess points, since the large amount of money paid up front will not be recovered. The longer you keep the mortgage, the less expensive points become.

EXAMPLE 11 **Using Expected Lifetime to Compare Mortgages** Suppose that a lender gives you a choice between the following two 30-year mortgages of $100,000:

Mortgage A: 6.1% interest compounded monthly, two points, monthly payment of $605.99

Mortgage B: 6.6% interest compounded monthly, no points, monthly payment of $638.66

The best choice of mortgage depends on the length of time that you expect to hold the mortgage. Mortgage B is the better choice if you plan to terminate or refinance the mortgage fairly early. Assuming that you can invest money at 3.6% interest compounded monthly, determine the length of time that you must retain the mortgage in order for mortgage A to be the better choice.

SOLUTION The difference in the monthly payments is $638.66 - 605.99 = \$32.67$. Although mortgage A saves $32.67 per month, it carries two points and therefore requires the payment of an additional $2000 up front. With mortgage B, the $2000 could be used to generate an income stream of $32.67 per month for a certain number of months, n, by investing it in an annuity at 3.6% compounded monthly.

The number of months, n, can be calculated in Excel with the NPER function. The value of NPER(.003, 32.67, −2000, 0) is approximately 67.74. (*Note:* .003 = .036/12.)

TVM Solver gives a duration of 67.7407041 months, as shown in Fig. 13. The number of months, n, also can be found graphically on a calculator by solving

$$P = \frac{1 - (1 + i)^{-n}}{i} \cdot R$$

for n. This equation can be shown to be equivalent to

$$\left(\frac{1}{1 + i}\right)^n = 1 - \frac{iP}{R}.$$

(See Exercise 40 for details.) With the settings $i = .003$, $P = 2000$, and $R = 32.67$, the equation has solution $n = 67.740704$.

Therefore, if you plan to hold the loan for 68 months (about $5\frac{2}{3}$ years) or more, you should choose mortgage A. **» Now Try Exercise 53**

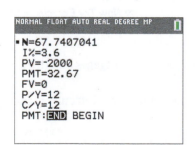

```
NORMAL FLOAT AUTO REAL DEGREE MP
▪N=67.7407041
 I%=3.6
 PV=-2000
 PMT=32.67
 FV=0
 P/Y=12
 C/Y=12
 PMT:END BEGIN
```

Figure 13

Interest-Only and Adjustable-Rate Mortgages

With an **interest-only mortgage**, the monthly payment for a certain number of years (usually, five or ten years) consists only of interest payments. That is, there is no amount applied to paying off the principal. At the end of the interest-only period, the monthly payment is determined by the number of months remaining. We will assume that the interest rate for the second period is the same as the interest rate for the first period. (With some interest-only mortgages, the interest rate is reset to conform to prevailing interest rates at that time.)

EXAMPLE 12   **Calculating Monthly Payments for an Interest-Only Mortgage** Consider a 30-year mortgage of $400,000 at 6.6% interest compounded monthly, where the loan is interest-only for 10 years.
(a) What is the monthly payment during the first 10 years?
(b) What is the monthly payment during the last 20 years?
(c) Compare the monthly payment to the monthly payment of $2554.64 for a conventional 30-year mortgage at 6.6% interest compounded monthly.

SOLUTION   Here, $P = 400,000$ and $i = \frac{.066}{12} = .0055$.

(a) During the first 10 years, the monthly payment is

$$i \cdot P = .0055(400,000)$$
$$= \$2200.$$

(b) At the beginning of the 11th year, the principal is still $P = 400,000$, and there are $20 \cdot 12 = 240$ months left to pay off the loan.

Figure 14

$$[\text{monthly payment}] = \frac{i}{1-(1+i)^{-n}} \cdot P$$
$$= \frac{.0055}{1 - 1.0055^{-240}} \cdot 400,000$$
$$= 3005.89$$

Formula for rent (that is, monthly payment)

$i = \dfrac{.066}{12} = .0055, n = 240,$
$P = 400,000$

Calculate. Round to two decimal places.

Therefore, beginning in the 11th year, the monthly payment is $3005.89. (The TVM Solver screen in Fig. 14 confirms this result.)

(c) Since $2554.64 - 2200 = 354.64$, the borrower's payment is $354.64 a month less than a conventional loan for the first 10 years. However, since $3005.89 - 2554.64 = 451.25$, the borrower pays an additional $451.25 per month for the last 20 years.

                     » Now Try Exercise 41

Some disadvantages of interest-only mortgages are as follows:

- Interest-only mortgages are riskier for lenders than fixed-rate mortgages and therefore tend to have slightly higher interest rates.
- When the higher monthly payment kicks in, the borrower might have difficulty making the payments.

Some advantages of interest-only mortgages are as follows:

- Borrowers with a modest current income who are reasonably certain that their income will increase in the future can afford to buy a more expensive house.
- In the early years of the mortgage, the savings can be contributed to an IRA that otherwise might not have been affordable.

Another type of mortgage in which monthly payments change is the **adjustable-rate mortgage** (ARM). With an ARM, the interest rate changes periodically as determined by a measure of current interest rates, called an **index**. The most common index in the United States, called the 1-year CMT (constant-maturity Treasury), is based on the yield of one-year Treasury bills. (During the past 10 years, the CMT index has ranged from about .10% to about 5.22%.) The interest rate on the ARM is reset by adding a fixed percent called the **margin** to the index percent. For instance, if the current value of the CMT index is 3% and the margin is 2.7%, then the adjusted interest rate for the mortgage would be 5.7%. The margin usually remains fixed for the duration of the loan.

Of the many types of ARMs, we will consider the so-called 5/1 ARM. With this type of mortgage, the interest rate is fixed for the first 5 years, and then is readjusted each year depending on the value of an index.

EXAMPLE 13   **Calculating Values Associated with an ARM**   Consider a $200,000 5/1 ARM that has a 2.7% margin, is based on the CMT index, and has a 30-year maturity. Suppose that the interest rate is initially 5.7% and the value of the CMT index is 4.5% 5 years later when the rate adjusts. Assume that all interest rates use monthly compounding.
(a) Calculate the monthly payment for the first 5 years.
(b) Calculate the unpaid balance at the end of the first 5 years.
(c) Calculate the monthly payment for the 6th year.

SOLUTION   (a) The monthly payment for the first 5 years is calculated as follows:

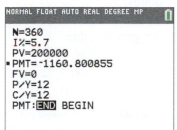

Figure 15

$$R = \frac{i}{1 - (1 + i)^{-n}} \cdot P \qquad \text{Formula for rent (that is, monthly payment)}$$

$$= \frac{.00475}{1 - 1.00475^{-360}} \cdot 200{,}000 \qquad i = \frac{.057}{12} = .00475, \, n = 360, \, P = 200{,}000$$

$$= 1160.80 \qquad \text{Calculate. Round to two decimal places.}$$

The monthly payment for the first 5 years is $1160.80. (The TVM Solver screen in Fig. 15 confirms this result.)

(b) After 5 years, there are $360 - 60 = 300$ months remaining. The unpaid balance is calculated as the present value of a sequence of 300 payments of $1160.80 each.

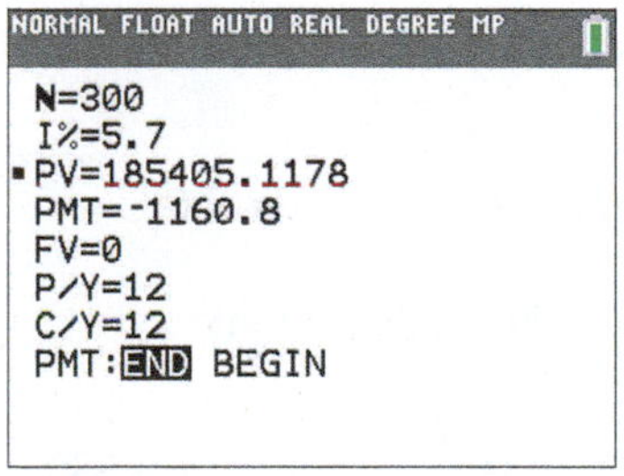

$$[\text{unpaid balance}] = \frac{1 - (1 + i)^{-n}}{i} \cdot R \qquad \text{Formula for present value}$$

$$= \frac{1 - 1.00475^{-300}}{.00475} \cdot 1160.80 \qquad i = \frac{.057}{12} = .00475, \, n = 300,$$

$$\phantom{= \frac{1 - 1.00475^{-300}}{.00475} \cdot 1160.80 \qquad} R = 1160.80$$

$$= 185{,}405.12 \qquad \text{Calculate. Round to two decimal places.}$$

The balance after 5 years is $185,405.12. (The TVM Solver screen in Fig. 16 confirms this result.)

Figure 16

(c) Here, $P = \$185{,}405.12$ and $n = 300$. The annual rate of interest for the 6th year will be $4.5\% + 2.7\% = 7.2\%$. The new monthly payment is calculated as before.

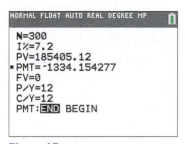

Figure 17

$$R = \frac{i}{1 - (1 + i)^{-n}} \cdot P \qquad \text{Formula for rent (that is, monthly payment)}$$

$$= \frac{.006}{1 - 1.006^{-300}} \cdot 185{,}405.12 \qquad i = \frac{.072}{12} = .006, \, n = 300, \, P = 185{,}405.12$$

$$= 1334.15 \qquad \text{Calculate. Round to two decimal places.}$$

Therefore, the monthly payment for the 6th year is $1334.15. (The TVM Solver screen in Fig. 17 confirms this results.)   **≫ Now Try Exercise 43**

Caps   There are several types of caps that can be specified for ARMs. **Periodic adjustment caps** limit the amount that the interest rate can adjust from one adjustment period to the next. **Lifetime caps** limit the interest-rate increase over the lifetime of the loan. **Payment caps** limit the amount that the monthly payment may increase from one adjustment to the next. (Periodic adjustment caps and payment caps usually do not apply to the first adjustment.) We will focus on payment caps.

Although payment caps limit the increase to the monthly payment, they can cause a slowdown in the repayment of the principal. The payment cap can even cause the unpaid balance to rise, a phenomenon called **negative amortization**.

EXAMPLE 14   **Calculating Values Associated with an ARM** Consider the mortgage discussed in Example 13, and assume that it carries a payment cap of 8%. (That is, the amount of the monthly payment in the seventh year can be at most 8% higher than the monthly payment for the sixth year.) Also, assume that the CMT index is 6.9% at the beginning of the seventh year.

(a) Calculate the unpaid balance of the loan at the beginning of the seventh year—that is, at the beginning of the 73rd month.

(b) Calculate the monthly payment for the seventh year, without using the payment cap.

(c) Calculate the monthly payment for the seventh year, using the payment cap.

(d) How much interest is due for the 73rd month?

(e) Determine the unpaid balance at the end of the 73rd month.

SOLUTION   (a) $\text{[unpaid balance]} = \dfrac{1 - (1 + i)^{-n}}{i} \cdot R$ Formula for present value

$$= \dfrac{1 - 1.006^{-288}}{.006} \cdot 1334.15 \qquad i = \dfrac{.072}{12} = .006, \ n = 360 - 72 = 288,$$
$$R = 1334.15$$
$$= 182{,}654.27 \qquad \text{Calculate. Round to two decimal places.}$$

```
NORMAL FLOAT AUTO REAL DEGREE MP

N=288
I%=7.2
■PV=182654.2693
PMT=-1334.15
FV=0
P/Y=12
C/Y=12
PMT:END BEGIN
```

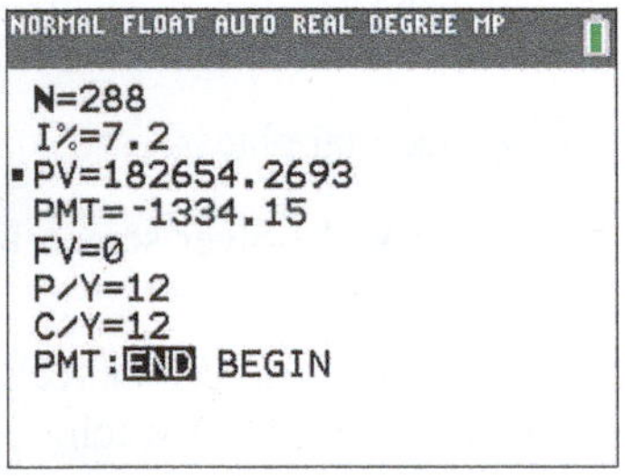

Figure 18

The balance after 5 years is $182,654.27. (The TVM Solver screen in Fig. 18 confirms this result.)

```
NORMAL FLOAT AUTO REAL DEGREE MP

N=288
I%=9.6
PV=182654.27
■PMT=-1624.998658
FV=0
P/Y=12
C/Y=12
PMT:END BEGIN
```

Figure 19

(b) The annual rate of interest for the seventh year will be 6.9% + 2.7% = 9.6%.

$$\text{[new monthly payment]} = \dfrac{i}{1 - (1 + i)^{-n}} \cdot P \qquad \text{Formula for rent}$$

$$= \dfrac{.008}{1 - 1.008^{-288}} \cdot 182{,}654.27 \qquad i = \dfrac{.096}{12} = .008, \ n = 288,$$
$$P = 182{,}654.27$$
$$= 1625.00 \qquad \begin{array}{l}\text{Calculate. Round to two}\\ \text{decimal places.}\end{array}$$

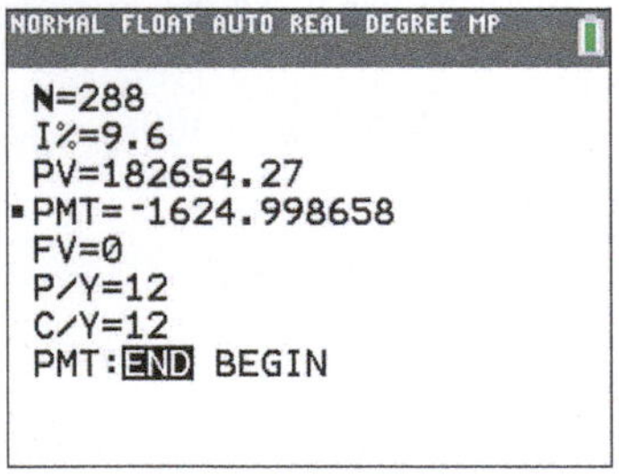

The new monthly payment for the seventh year (without using the payment cap) is $1625.00. (The TVM Solver screen in Fig. 19 confirms this result.)

(c) Without the payment cap, the percentage increase in the monthly payment from the sixth year to the seventh year would be

$$\dfrac{1625.00 - 1334.15}{1334.15} \approx .2180 = 21.8\%.$$

Since 21.8% is greater than the payment cap of 8%, the payment cap kicks in to mandate that the monthly payment for the seventh year be

$$1.08 \cdot 1334.15 = \$1440.88.$$

(d) The interest due for the 73rd month is

$$.008 \cdot 182{,}654.27 = \$1461.23.$$

(e) The interest due is greater than the monthly payment by

$$1461.23 - 1440.88 = \$20.35.$$

Therefore, negative amortization occurs. The $20.35 is added to the previous unpaid balance. The unpaid balance at the end of the 73rd month is

$$182,654.27 + 20.35 = \$182,674.62.$$

The increased unpaid balance will be taken into account when the next adjustment of the monthly payment is calculated.   **>> Now Try Exercise 45**

Here are a few words of caution about ARMs:

- Usually, monthly payments go up when interest rates rise and go down when interest rates fall—but not always.
- Some lenders offer ARMs with an artificially low interest rate, called a *discounted interest rate* or *teaser rate*. Such loans can produce payment shock when the interest rate adjusts.
- If you want to pay off your ARM early to avoid higher monthly payments, you might have to pay a penalty.

Check Your Understanding 10.4

Solutions can be found following the section exercises.

1. Rework Example 11 under the assumption that the $2000 will not earn any interest.

2. Consider the two loans discussed in Example 7. Calculate the unpaid balance of each loan after 1 month.

3. Which is the better deal for a 30-year mortgage: (a) 6% interest plus three discount points (6.286% APR), or (b) 6.3% interest plus one discount point (6.396% APR)?

EXERCISES 10.4

In Exercises 1 and 2, fill in the blank with the word "free" or "deferred."

1. Interest earned in a traditional IRA is tax _____.

2. Interest earned in a Roth IRA is tax _____.

3. **Traditional IRA** Carlos is 60 years old, is in the 45% marginal tax bracket, and has $300,000 in his traditional IRA. How much money will he have after taxes if he withdraws all of the money from the account?

4. **Roth IRA** Rework Exercise 3 for a Roth IRA.

5. **Traditional IRA** If you are 18 years old, deposit $5000 each year into a traditional IRA for 52 years, at 6% interest compounded annually, and retire at age 70, how much money will be in the account upon retirement?

6. **Comparing IRAs** Rework Examples 1 and 3 for the case where the marginal tax bracket is 30% when the money is contributed but is only 25% when the balance is withdrawn. Which type of IRA is most advantageous?

7. **Roth IRA** Rework Exercise 5 for a Roth IRA.

8. **Comparing IRAs** Rework Examples 1 and 3 for the case where the marginal tax bracket is 30% when the money is contributed but rises to 35% when the balance is withdrawn. Which type of IRA is most advantageous?

9. **Value of Starting an IRA Early** Redo Example 4 where Earl and Larry are each in a 40% marginal tax bracket, have Roth IRAs, and contribute the remainder of $5000 after taxes are deducted.

10. **Advantage of Prompt IRA Contributions** The contribution into an IRA for a particular year can be made any time from January 1 of that year to April 15 of the following year. Suppose that Enid and Lucy both set up traditional IRA accounts on January 1 of 2017 and each contributes $5000 into her account for 10 years at 6% interest compounded annually. Assume that Enid makes her contributions as soon as possible and Lucy makes her contributions 1 year later. Calculate the balances in the two accounts at the time Lucy makes her final contribution.

In Exercises 11–14, use the add-on method to determine the monthly payment.

11. $4000 loan at 10% stated interest rate for 1 year

12. $6000 loan at 7% stated interest rate for 1 year

13. $3000 loan at 9% stated interest rate for 3 years

14. $10,000 loan at 8% stated interest rate for 2 years

In Exercises 15–18, give the add-on interest rate.

15. A $2000 loan for 1 year at 5% APR with a monthly payment of $171.21

16. A $5000 loan for 1 year at 6% APR with a monthly payment of $430.33

17. A $20,000 loan for 3 years at 6% APR with a monthly payment of $608.44

18. A $10,000 loan for 2 years at 7% APR with a monthly payment of $447.73

19. **Discount Method** Another method used by lenders to determine the monthly payment for a loan is the *discount method*. In this case, the stated interest rate is called the *discount rate* and the formula for determining the total amount paid is

$$[\text{total payment}] = \frac{[\text{loan amount}]}{1 - rt},$$

where r is the discount rate and t is the term of the loan in years. The monthly payment is then [total payment]$/12t$.

(a) Calculate the monthly payment for a 2-year loan of $880 with a discount rate of 6%.

(b) If the add-on method were used instead, would the monthly payment be greater than or less than the amount found in part (a)?

20. **True or False** For any mortgage with discount points, the APR will be greater than the stated interest rate.

21. **True or False** For any mortgage that will be terminated before its full term, the effective mortgage rate will be greater than the stated rate.

22. **True or False** If a mortgage is expected to be held for its entire term, the effective mortgage rate will be the same as the APR.

23. **True or False** The longer a loan with discount points is expected to be held, the greater its effective mortgage rate will be.

24. **True or False** Increasing the discount points for a mortgage increases its APR and its effective mortgage rate.

25. **True or False** Changing the amount of a loan while keeping the stated interest rate, the term, and the up-front costs the same will have no effect on the APR.

26. **True or False** Refer to Example 10. If the stated interest rate were 9% instead of 6%, then the effective mortgage rate would be 9.24%. That is, the interest rate would increase by the same amount as before, .24%.

APR In multiple-choice Exercises 27–30, assume that the loan amount is $250,000 and confirm your answer by calculating monthly payments. (*Note:* The two monthly payments will not be identical, due to round-off error.)

27. The APR for a 25-year mortgage at 9% interest compounded monthly and with two discount points is
   (a) 8.75%.   (b) 9.05%.   (c) 12%.   (d) 9.25%.

28. The APR for a 30-year mortgage at 5.9% interest compounded monthly and having one discount point is
   (a) 9%.   (b) 6%.   (c) 5.7%.

29. The APR for a 20-year mortgage at 5.5% interest compounded monthly and with four discount points is
   (a) 6%.   (b) 5%.   (c) 6.5%.

30. The APR for a 25-year mortgage at 9% interest compounded monthly and having four discount points is
   (a) 8.5%.   (b) 9.5%.   (c) 11%.

Effective Mortgage Rate In multiple-choice Exercises 31–34, assume that the loan amount is $100,000 and confirm your answer by calculating sums as in Example 9.

31. The effective mortgage rate for a 30-year mortgage at 5.71% interest compounded monthly with one discount point that is expected to be kept for 4 years is
   (a) 6%.   (b) 5.54%.   (c) 6.56%.   (d) 7%.

   (*Hint:* For a 5.71%, 30-year mortgage, $R = \$581.03$ and the unpaid balance after 4 years is $94,341.50.)

32. The effective mortgage rate for a 15-year mortgage at 5.481% interest compounded monthly, with three discount points, that is expected to be kept for 10 years is
   (a) 5%.   (b) 6%.   (c) 6.5%.   (d) 7%.

   (*Hint:* For a 5.481%, 15-year mortgage, $R = \$816.08$ and the unpaid balance after 10 years is $42,743.72.)

33. The effective mortgage rate for a 30-year mortgage at 6% interest compounded monthly, with three discount points, that is expected to be kept for 7 years is
   (a) 5.5%.   (b) 6.2%.   (c) 6.56%.   (d) 9%.

34. The effective mortgage rate for a 30-year mortgage at 9% interest compounded monthly, with two discount points, that is expected to be kept for 7 years is
   (a) 9%.   (b) 8.5%.   (c) 9.4%.   (d) 10%.

In Exercises 35 and 36, give the two Excel formulas that can be used to obtain the APR and the effective mortgage rate for the mortgage. When calculating the effective mortgage rate, assume that the mortgage will be held for 10 years.

35. A 20-year mortgage of $80,000 carrying a stated interest rate of 6% and three points

36. A 15-year mortgage of $180,000 carrying a stated interest rate of 9% and two points

In Exercises 37 and 38, give the two equations that can be solved for i with a graphing calculator to obtain the monthly interest rates corresponding to the APR and the effective mortgage rate. Assume that the loans will be held for 5 years.

37. A 15-year mortgage of $120,000 carrying a stated interest rate of 9% and one point

38. A 20-year mortgage of $150,000 carrying a stated interest rate of 6% and three points

39. **Loan Quandary** You decide to buy a $1250 entertainment system for your apartment. The salesman asks you to pay 20% down and is willing to finance the remaining $1000 with a 2-year loan having an APR of 5% and a monthly payment of $43.87. (The total of your payments on the loan will be $1052.88.) Suppose that you actually have $1000 in your savings account (after the down payment) and can invest it at 4% compounded monthly. You must decide whether or not to take the loan. The salesman tells you that you will come out ahead if you take the loan, since $1000 invested at 4% interest compounded monthly grows to $1083.14, which is greater than the total payments on the loan. This doesn't make sense to you since you would be borrowing money at 5% and only earning 4% on the money you invest. What is wrong with the salesman's reasoning? (*Note:* Ignore taxes.)

40. Perform algebraic manipulations on the equation $P = \frac{1 - (1 + i)^{-n}}{i} \cdot R$ to obtain $\left(\frac{1}{1+i}\right)^n = 1 - \frac{iP}{R}$. *Hint:* Multiply the equation by $\frac{i}{R}$, replace $(1 + i)^{-n}$ by $\left(\frac{1}{1+i}\right)^n$, and then solve for $\left(\frac{1}{1+i}\right)^n$.

41. **Interest-Only Mortgage** Consider a 15-year mortgage of $200,000 at 6.9% interest compounded monthly, where the loan is interest-only for 5 years. What is the monthly payment during the first 5 years? Last 10 years?

42. **Interest-Only Mortgage** Consider a 15-year mortgage of $300,000 at 7.2% interest compounded monthly, where the loan is interest-only for 5 years. What is the monthly payment during the first 5 years? Last 10 years?

43. **Adjustable-Rate Mortgage** Consider a 25-year $250,000 5/1 ARM having a 2.5% margin and based on the CMT index. Suppose that the interest rate is initially 6% and the value of the CMT index is 4.4% five years later. Assume that all interest rates use monthly compounding.

(a) Calculate the monthly payment for the first 5 years.
(b) Calculate the unpaid balance at the end of the first 5 years.
(c) Calculate the monthly payment for the sixth year.

44. **Adjustable-Rate Mortgage** Consider a 15-year $300,000 5/1 ARM having a 3% margin and based on the CMT index. Suppose that the interest rate is initially 6.3% and the value of the CMT index is 3% 5 years later. Assume that all interest rates use monthly compounding.
(a) Calculate the monthly payment for the first 5 years.
(b) Calculate the unpaid balance at the end of the first 5 years.
(c) Calculate the monthly payment for the sixth year.

45. **Adjustable-Rate Mortgage** Consider the mortgage discussed in Exercise 43, and assume that it carries a payment cap of 7%. Also, assume that the CMT index is 7.7% at the beginning of the seventh year.
(a) Calculate the unpaid balance of the loan at the beginning of the seventh year.
(b) Calculate the monthly payment for the seventh year, without using the payment cap.
(c) Calculate the monthly payment for the seventh year, using the payment cap.
(d) How much interest is due for the 73rd month?
(e) Determine the unpaid balance at the end of the 73rd month.

46. **Adjustable-Rate Mortgage** Consider Example 14. What is the unpaid balance at the end of the 74th month?

TECHNOLOGY EXERCISES

47. Find the APR on a $10,000 loan with an add-on method interest rate of 6% compounded monthly for 3 years.

48. Find the APR on a $15,000 loan with an add-on method interest rate of 8% compounded monthly for 2 years.

49. Find the APR on a 25-year mortgage of $300,000 carrying a stated interest rate of 6.5% compounded monthly and having three points.

50. Find the APR on a 30-year mortgage of $250,000 carrying a stated interest rate of 6% compounded monthly and having two points.

51. Find the effective mortgage rate for a 25-year $140,000 mortgage at 6.5% interest compounded monthly, with three discount points. Assume that the mortgage will be held for 7 years.

52. Find the effective mortgage rate for a 30-year $250,000 mortgage at 6% interest compounded monthly, with two discount points, that is expected to be kept for 6 years.

53. **Comparing Mortgages** Suppose that a lender gives you a choice between the following two 25-year mortgages of $175,000:

Mortgage A: 6.3% interest compounded monthly, with three points, monthly payment of $1159.84

Mortgage B: 6.5% interest compounded monthly, with two points, monthly payment of $1181.61

Assuming that you can invest money at 2.4% interest compounded monthly, determine the length of time that you must retain the mortgage in order for mortgage A to be the better choice.

54. **Comparing Mortgages** Suppose that a lender gives you a choice between the following two 15-year mortgages of $250,000:

Mortgage A: 6.5% interest compounded monthly, with one point, monthly payment of $2177.77

Mortgage B: 6.1% interest compounded monthly, with two points, monthly payment of $2123.17

Assuming that you can invest money at 3.6% interest compounded monthly, determine the length of time that you must retain the mortgage in order for mortgage B to be the better choice.

55. **Comparing Mortgages** A lender gives you a choice between the following two 30-year mortgages of $200,000:

Mortgage A: 6.65% interest compounded monthly, with one point, monthly payment of $1283.93

Mortgage B: 6.8% interest compounded monthly, with no points, monthly payment of $1303.85

Assuming that you can invest money at 4.8% interest compounded monthly, determine the length of time that you must retain the mortgage in order for mortgage A to be the better choice.

56. **Comparing Mortgages** A lender gives you a choice between the following two 30-year mortgages of $235,000:

Mortgage A: 6.9% interest compounded monthly, with one point, monthly payment of $1547.71

Mortgage B: 6.5% interest compounded monthly, with three points, monthly payment of $1,485.36

Assuming that you can invest money at 3.3% interest compounded monthly, determine the length of time that you must retain the mortgage in order for mortgage B to be the better choice.

Solutions to Check Your Understanding 10.4

1. In this case, we merely have to solve $n \cdot (32.67) = 2000$.

$$n = \frac{2000}{32.67} = 61.218$$

Therefore, take the additional points only if you expect to hold onto the loan for more than 61 months (that is, a little more than 5 years).

NOTE This computation is much simpler than the one in Example 11 and can be used to get a rough estimate of the proper number of months. «

2. Successive balances can be calculated with the formula $B_{new} = (1 + i)B_{previous} - R$. Mortgage A was a $200,000 loan at 6% interest with a monthly payment of $1199.10.

[balance after 1 month]
$= (1.005) \cdot 200,000 - 1199.10 = \$199,800.90$

Mortgage B was a $198,000 loan at 6.094% interest with a monthly payment of $1199.10. The monthly interest $i = .06094/12 = .0050783333$.

[balance after 1 month]
$= (1.0050783333) \cdot 198,000 - 1199.10 = \$197,806.41$

The balance for mortgage *A* is $1994.49 higher. The balance will remain higher until the final payment is made.

3. It depends on how long you intend to keep the mortgage. If you expect to hold the mortgage for 30 years, then mortgage *A* is clearly superior because it has a lower APR.

On the other hand, if you expect to terminate the mortgage after two years, then Table 1 on page 467 says that the effective rates are approximately 7.5% $[3 \cdot (\frac{1}{2}\%) + 6\%]$ and 6.8% $[1 \cdot (\frac{1}{2}\%) + 6.3\%]$, respectively, obviously favoring mortgage *B*.

10.5 A Unifying Equation

In this chapter we discuss a number of topics from the mathematics of finance: compound interest, simple interest, mortgages, and annuities. As we shall see, all such financial transactions can be described by a single type of equation, called a *difference equation*. Furthermore, the same type of difference equation can be used to model many other phenomena, such as the spread of information, radioactive decay, and population growth, to mention just a few.

NOTE If you have not read the first three sections of this chapter, or would like a refresher of the financial terms discussed there, read the appendix at the end of this section. «

In the first three sections of this chapter, we encountered the following four formulas describing the change in the balance *B* at the end of each interest period. (Here, *r* is the stated annual interest rate, *i* is the interest rate per interest period, *P* is the principal, and *R* is the payment made at the end of each interest period.)

Compound interest: $\quad B_{\text{new}} = (1 + i)B_{\text{previous}}$

Simple interest: $\quad B_{\text{new}} = B_{\text{previous}} + rP$

Increasing annuity: $\quad B_{\text{new}} = (1 + i)B_{\text{previous}} + R$

Decreasing annuity: $\quad B_{\text{new}} = (1 + i)B_{\text{previous}} - R$

Each formula has the form

$$B_{\text{new}} = a \cdot B_{\text{previous}} + b, \tag{1}$$

where *a* and *b* are numbers. For compound interest, $a = (1 + i)$, $b = 0$; for simple interest, $a = 1$, $b = rP$; for an increasing annuity, $a = (1 + i)$, $b = R$; and for a decreasing annuity, $a = (1 + i)$, $b = -R$. *Note:* A loan is actually a decreasing annuity with the borrower paying the interest.

If we let y_n represent the balance after *n* interest periods, then formula (1) can be written as

$$y_n = a \cdot y_{n-1} + b. \tag{2}$$

Formula (2) can be used to generate successive balances once the beginning balance, y_0, is given. The combination of formula (2) and the value of y_0 is called a **difference equation**. The number y_0 is called the **initial value** of the difference equation. (Specifically, the difference equation is called a **linear difference equation with an initial value**.) Table 1 summarizes the financial difference equations.

Table 1 Summary of Financial Difference Equations

Compound interest	$y_n = (1 + i) \cdot y_{n-1}$	$y_0 = $ initial amount deposited
Simple interest	$y_n = y_{n-1} + ry_0$	$y_0 = $ initial amount deposited
Increasing annuity	$y_n = (1 + i) \cdot y_{n-1} + R$	$y_0 = $ initial amount deposited
Decreasing annuity (or loan)	$y_n = (1 + i) \cdot y_{n-1} - R$	$y_0 = $ initial amount deposited (or borrowed)

EXAMPLE 1   **Generate Values of a Difference Equation** Determine the first five values generated by the difference equation $y_n = .2y_{n-1} + 4.8$, $y_0 = 1$.

SOLUTION   Beginning with the value of y_0, formula (2) can be used to successively obtain y_1, y_2, y_3, and so on. The first five values are as follows:

$$y_0 = 1$$
$$y_1 = .2y_0 + 4.8 = .2(1) + 4.8 = 5$$
$$y_2 = .2y_1 + 4.8 = .2(5) + 4.8 = 5.8$$
$$y_3 = .2y_2 + 4.8 = .2(5.8) + 4.8 = 5.96$$
$$y_4 = .2y_3 + 4.8 = .2(5.96) + 4.8 = 5.992$$

≫ Now Try Exercise 15(a)

Not only can we generate as many values as we like for a difference equation, there are formulas that directly give any specific value. Such a formula is called a **solution** of the difference equation. The two formulas below are derived at the end of the section.

Solution of a Difference Equation   The solution of the difference equation $y_n = a \cdot y_{n-1} + b$, y_0 is

$$y_n = \frac{b}{1-a} + \left(y_0 - \frac{b}{1-a}\right)a^n \qquad \text{when } a \neq 1$$

$$y_n = y_0 + bn \qquad \text{when } a = 1$$

EXAMPLE 2   **Solve a Difference Equation** Determine the solution of the difference equation in Example 1 and use it to calculate y_3.

SOLUTION   The difference equation is $y_n = .2y_{n-1} + 4.8$, $y_0 = 1$. Here $a = .2$ and $b = 4.8$. Therefore,

$$\frac{b}{1-a} = \frac{4.8}{1-.2} = \frac{4.8}{.8} = 6$$

and

$$y_n = \frac{b}{1-a} + \left(y_0 - \frac{b}{1-a}\right)a^n \qquad \text{General solution}$$
$$= 6 + (1 - 6)(.2)^n \qquad \frac{b}{1-a} = 6, y_0 = 1, a = .2$$
$$= 6 + (-5)(.2)^n \qquad \text{Subtract.}$$
$$= 6 - 5(.2)^n \qquad \text{Rewrite.}$$

Therefore, the solution of the difference equation is $y_n = 6 - 5(.2)^n$.

When $n = 3$,

$$y_3 = 6 - 5(.2)^3 = 6 - 5(.008) = 5.96$$

Notice that the value of y_3 agrees with the value found in Example 1.

≫ Now Try Exercise 15(b)

EXAMPLE 3   **Financial Difference Equations** Determine the difference equation for each of the following financial transactions where y_n is the balance after n interest periods.
(a) $1000 is deposited into a savings account earning 4% interest compounded quarterly.
(b) $1000 is deposited into a savings account earning 3% annual simple interest.

(c) At the end of each week, $100 is deposited into a savings account earning 2.6% interest compounded weekly.

(d) A $304,956.29 mortgage has an interest rate of 4.8% compounded monthly and a monthly payment of $1600.

SOLUTION

(a) $y_n = (1 + i)y_{n-1}, y_0$ given   Difference equation for compound interest

$y_n = (1 + .01)y_{n-1}, y_0 = 1000$ $i = \frac{.04}{4} = .01, y_0 = 1000$

$y_n = (1.01)y_{n-1}, y_0 = 1000$ Add.

(b) $y_n = y_{n-1} + ry_0, y_0$ given   Difference equation for simple interest

$y_n = y_{n-1} + .03 \cdot 1000, y_0 = 1000$ $r = .03, y_0 = 1000$

$y_n = y_{n-1} + 30, y_0 = 1000$ Multiply.

(c) $y_n = (1 + i)y_{n-1} + R, y_0$ given   Difference equation for increasing annuity

$y_n = (1 + .0005)y_{n-1} + 100, y_0 = 0$ $i = \frac{.026}{52} = .0005, R = 100$

$y_n = 1.0005y_{n-1} + 100, y_0 = 0$ Add.

(d) $y_n = (1 + i)y_{n-1} - R, y_0$ given   Difference equation for decreasing annuity

$y_n = (1 + .004)y_{n-1} - 1600,$ $i = \frac{.048}{12} = .004, R = 1600, y_0 = 304{,}956.29$

$\qquad y_0 = 304{,}956.29$

$y_n = 1.004y_{n-1} - 1600,$ Add.

$\qquad y_0 = 304{,}956.29$

» Now Try Exercises 1, 3, 5, 9

The financial formulas presented in the first part of this chapter can be derived from difference equations.

EXAMPLE 4   **Future Value of an Increasing Annuity** Use a difference equation to derive the formula for the future value of an increasing annuity in which R dollars is deposited into a savings account at the end of each interest period and the interest rate per period is i.

SOLUTION   The difference equation for an increasing annuity is $y_n = (1 + i)y_{n-1} + R, y_0 = 0$. Now find the solution to this equation.

$$\frac{b}{1 - a} = \frac{R}{1 - (1 + i)} = \frac{R}{-i} = -\frac{R}{i}$$

Therefore,

$$y_n = \frac{b}{1 - a} + \left(y_0 - \frac{b}{1 - a}\right)a^n \qquad \text{General solution}$$

$$= -\frac{R}{i} + \left[0 - \left(-\frac{R}{i}\right)\right](1 + i)^n \qquad \text{Replace } \frac{b}{1 - a} \text{ with } -\frac{R}{i}, \text{ and } a \text{ with } (1 + i).$$

$$= -\frac{R}{i} + \frac{R}{i}(1 + i)^n \qquad 0 - \left(-\frac{R}{i}\right) = \frac{R}{i}$$

$$= \frac{R}{i}(1 + i)^n - \frac{R}{i} \cdot 1 \qquad \text{Reorder terms. Replace } \frac{R}{i} \text{ with } \frac{R}{i} \cdot 1.$$

$$= \frac{R}{i}[(1 + i)^n - 1] \qquad \text{Factor out } \frac{R}{i}.$$

$$= \frac{(1 + i)^n - 1}{i} \cdot R \qquad \text{Rewrite.}$$

» Now Try Exercise 27

EXAMPLE 5

Balance in a Mortgage The amortization table in Section 10.3 shows that for a \$112,475 mortgage at 9% interest compounded monthly with a monthly payment of \$905, the balance after 12 months is \$111,706.57. Use a difference equation to derive a formula for the balances of the mortgage and use the formula to calculate the balance after 12 months.

SOLUTION

Let y_n be the balance after n months.

$y_n = (1 + i)y_{n-1} - R,\ y_0 =$ amount borrowed $\qquad$ Difference equation for a loan

$\quad = (1.0075)y_{n-1} - 905,\ y_0 = 112,475 \qquad i = \frac{.09}{12} = .0075,\ R = 905,\ y_0 = 112,475$

Therefore,

$$\frac{b}{1-a} = \frac{-905}{1-(1.0075)} = \frac{-905}{-.0075} = \frac{905}{.0075} \qquad \text{Calculate } \frac{b}{1-a} \text{ with } a = 1.0075,\ b = -905.$$

$$y_n = \frac{b}{1-a} + \left(y_0 - \frac{b}{1-a}\right)a^n \qquad \text{General solution}$$

$$= \frac{905}{.0075} + \left(112,475 - \frac{905}{.0075}\right)(1.0075)^n \qquad \text{Substitute values for } \frac{b}{1-a},\ y_0,\ \text{and } a.$$

The balance after 12 months is

$$y_{12} = \frac{905}{.0075} + \left(112,475 - \frac{905}{.0075}\right)(1.0075)^{12} \qquad \text{Substitute 12 for } n.$$

$$= \$111,706.57 \qquad \text{Calculate. Round to two decimal places.}$$

≫ Now Try Exercise 33

NOTE The solution to Example 5 also can be obtained by solving the general difference equation to obtain $y_n = \dfrac{R}{i} + \left(y_0 - \dfrac{R}{i}\right)(1 + i)^n$, and then substituting the given values. **≪**

EXAMPLE 6

Loan Payment A car loan of \$10,000 is to be repaid with quarterly payments for 5 years at 6.4% interest compounded quarterly. Calculate the quarterly payment R.

SOLUTION

Let y_n be the balance after n quarters. The balance owed on the loan after n quarterly payments satisfies the difference equation

$$y_n = (1.016)y_{n-1} - R,\ y_0 = 10,000 \qquad 1 + i = 1 + \tfrac{.064}{4} = 1 + .016 = 1.016$$

Since

$$\frac{b}{1-a} = \frac{-R}{1-1.016} = \frac{-R}{-.016} = \frac{R}{.016}, \qquad \text{Calculate } \frac{b}{1-a} \text{ with } a = 1.016,\ b = -R.$$

the solution of the difference equation is

$$y_n = \frac{b}{1-a} + \left(y_0 - \frac{b}{1-a}\right)a^n \qquad \text{General solution}$$

$$y_n = \frac{R}{.016} + \left(10,000 - \frac{R}{.016}\right)(1.016)^n \qquad \frac{b}{1-a} = \frac{R}{.016},\ y_0 = 10,000,\ a = 1.016$$

After 20 quarters, the loan will be paid off (that is, $y_{20} = 0$) and the equation becomes

$$\frac{R}{.016} + \left(10,000 - \frac{R}{.016}\right)(1.016)^{20} = 0 \qquad n = 20$$

Now solve this equation for R.

$$\frac{R}{.016} + 10{,}000(1.016)^{20} - \left(\frac{R}{.016}\right)(1.016)^{20} = 0 \quad \text{Multiply values in parentheses by } (1.016)^{20}.$$

$$\frac{R}{.016}[1 - (1.016)^{20}] + 10{,}000(1.016)^{20} = 0 \quad \text{Subtract like terms and factor out } \frac{R}{.016}.$$

$$\frac{R}{.016}[1 - (1.016)^{20}] = -10{,}000(1.016)^{20} \quad \text{Subtract } 10{,}000(1.016)^{20}.$$

$$\frac{R}{.016} = \frac{-10{,}000(1.016)^{20}}{1 - (1.016)^{20}} \quad \text{Divide by } [1 - (1.016)^{20}].$$

$$R = .016\left(\frac{-10{,}000(1.016)^{20}}{1 - (1.016)^{20}}\right) \quad \text{Multiply by } .016.$$

$$R = 588.22 \quad \text{Calculate. Round to two decimal places.}$$

Therefore, the quarterly payment is \$588.22. **» Now Try Exercise 37**

Nonfinancial Applications

The difference equation developed in this section can be used to model topics in many fields—such as demographics, economics, biology, psychology, sociology, and physics.

EXAMPLE 7 **Population Growth** Suppose the population of a certain country is currently 6 million. The growth of this population attributable to an excess of births over deaths is 2% per year. Further, the country is experiencing immigration at the rate of 40,000 people per year. Let y_n denote the population (in millions) of the country after n years.
(a) Determine a difference equation for y_n.
(b) Solve the difference equation in part (a) and use it to calculate the population of the country after 35 years.

SOLUTION (a) $y_0 = 6$. The growth of the population in year n due to an excess of births over deaths is $.02y_{n-1}$. There are $.04$ (million) immigrants each year. Therefore,

$$
\begin{array}{ccccccc}
y_n & = & y_{n-1} & + & .02y_{n-1} & + & .04 \\
\begin{bmatrix} \text{new} \\ \text{population} \end{bmatrix} & = & \begin{bmatrix} \text{previous} \\ \text{population} \end{bmatrix} & + & \begin{bmatrix} \text{natural} \\ \text{increase} \end{bmatrix} & + & \begin{bmatrix} \text{number of} \\ \text{immigrants} \end{bmatrix}
\end{array}
$$

Thus, the difference equation for the population is

$$y_n = 1.02y_{n-1} + .04, \quad y_0 = 6.$$

(b) $$\frac{b}{1-a} = \frac{.04}{1-(1.02)} = \frac{.04}{-.02} = -2 \qquad b = .04, \ a = 1.02$$

Therefore, the solution of the difference equation is

$$y_n = \frac{b}{1-a} + \left(y_0 - \frac{b}{1-a}\right)a^n \quad \text{General solution}$$
$$= -2 + [6 - (-2)](1.02)^n \qquad \frac{b}{1-a} = -2, \ y_0 = 6, \ a = 1.02$$
$$= -2 + 8(1.02)^n. \qquad \text{Subtract.}$$

The population after 35 years will be

$$y_{35} = -2 + 8(1.02)^{35} \quad \text{Substitute 35 for } n.$$
$$= 14. \quad \text{Calculate. Round to a whole number.}$$

That is, after 35 years the population will be about 14 million.

» Now Try Exercise 41

EXAMPLE 8   **Supply and Demand** This year's level of production and price for most agricultural products affects the level of production and price for the next year. Suppose that the current crop of soybeans in a certain country is 80 million bushels. Let q_n denote the quantity of soybeans (in millions of bushels) grown n years from now, and let p_n denote the market price (in dollars per bushel) in n years. Suppose that experience has shown that q_n and p_n are related by the following equations:

$$p_n = 20 - .1q_n \qquad q_n = 5p_{n-1} - 10.$$

(a) Determine a difference equation for q_n.

(b) Solve the difference equation in part (a) and use it to calculate the number of bushels of soybeans produced after 4 years.

SOLUTION   **(a)**

$$
\begin{aligned}
q_0 &= 80 && \text{Initial value}\\
q_n &= 5p_{n-1} - 10 && \text{Given equation}\\
&= 5(20 - .1q_{n-1}) - 10 && \text{Substitute } (20 - .1q_{n-1}) \text{ for } p_{n-1}.\\
&= 100 - .5q_{n-1} - 10 && \text{Multiply.}\\
&= -.5q_{n-1} + 90 && \text{Combine numbers.}
\end{aligned}
$$

Therefore, the difference equation is $q_n = -.5q_{n-1} + 90$, $q_0 = 80$.

(b) Since $a = -.5$ and $b = 90$,

$$\frac{b}{1-a} = \frac{90}{1-(-.5)} = \frac{90}{1.5} = 60.$$

Therefore, the solution of the difference equation is

$$
\begin{aligned}
q_n &= \frac{b}{1-a} + \left(q_0 - \frac{b}{1-a}\right)a^n && \text{General solution}\\
&= 60 + (80 - 60)(-.5)^n && \frac{b}{1-a} = 60, q_0 = 80, a = -.5\\
&= 60 + 20(-.5)^n. && \text{Subtract.}\\
q_4 &= 60 + 20(-.5)^4 && \text{Quantity produced after 4 years}\\
&= 61.25 && \text{Calculate.}
\end{aligned}
$$

That is, after 4 years, 61.25 million bushels of soybeans will be produced.   «

INCORPORATING TECHNOLOGY

Display Successive Values Graphing calculators can easily generate successive values of a difference equation. Consider the difference equation $y_n = .2y_{n-1} + 4.8$, $y_0 = 1$, of Examples 1 and 2. In Fig. 1, successive values are displayed in the home screen, and in Fig. 2, they are displayed in a table.

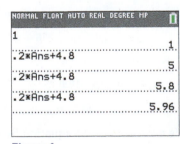

Figure 1

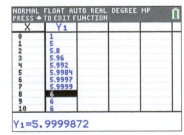

Figure 2

Display Successive Values on the Home Screen In Fig. 1, after the initial value (1) is entered, the last displayed value (1) is assigned to **Ans**. The instruction **.2*Ans+4.8** generates the next value (5) and assigns it to **Ans**. Each subsequent press of ENTER generates another value.

Display Successive Values in a Table To obtain the table in Fig. 2, set $\mathtt{Y_1=6-5*.2^x}$ in the $\mathtt{Y=}$ editor. Then press [2nd] [TBLSET] to bring up the TABLE SETUP screen. Set **TblStart** to 0, and set $\mathbf{\Delta Tbl}$ to 1. Finally, press [2nd] [TABLE] to bring up the table. The down-arrow key can be used to generate further values.

The table in Fig. 3 is easily created in an Excel spreadsheet. The following steps create the two columns of the table.

1. To enter the numbers 0 through 6 in the first column, enter 0 into cell A2, enter 1 into cell A3, select the two cells, and drag the fill handle down to A8.
2. Type 1 into cell B2, and press [ENTER].
3. Type =.2*B2+4.8 into cell B3, and press [ENTER].
4. Click on cell B3, and drag its fill handle down to B8.

	A	B
1	n	y_n
2	0	1
3	1	5
4	2	5.8
5	3	5.96
6	4	5.992
7	5	5.9984
8	6	5.99968

Figure 3

n	$y(n)$
0	1
1	5
2	5.8
3	5.96
4	5.992
5	5.9984
6	5.99968

Figure 4

✳ WolframAlpha The instruction $y(n) = .2y(n-1) + 4.8$, $y(0) = 1$ generates a table of values for the difference equation. See Fig. 4.

Derivation of the Solution to the Difference Equation for $a \neq 1$ From the difference equation $y_n = ay_{n-1} + b$, we get

$$y_1 = ay_0 + b$$
$$y_2 = ay_1 + b = a(ay_0 + b) + b = a^2y_0 + ab + b$$
$$y_3 = ay_2 + b = a(a^2y_0 + ab + b) + b = a^3y_0 + a^2b + ab + b$$
$$y_4 = ay_3 + b = a(a^3y_0 + a^2b + ab + b) + b = a^4y_0 + a^3b + a^2b + ab + b$$

The pattern that clearly develops is

$$y_n = a^ny_0 + a^{n-1}b + a^{n-2}b + \cdots + a^2b + ab + b. \tag{3}$$

Multiply both sides of (3) by a, and then subtract the new equation from (3). Notice that many terms drop out.

$$y_n = \quad a^ny_0 + \qquad a^{n-1}b + a^{n-2}b + \cdots + a^2b + ab + b$$
$$\underline{ay_n = a^{n+1}y_0 + a^nb + a^{n-1}b + a^{n-2}b + \cdots + a^2b + ab}$$
$$y_n - ay_n = a^ny_0 - a^{n+1}y_0 - a^nb + b$$

The last equation can be written as

$$(1-a)y_n = (1-a)a^ny_0 - ba^n + b.$$

Now, divide both sides of the equation by $(1-a)$. *Note:* Since $a \neq 1$, $1 - a \neq 0$.

$$y_n = y_0a^n - \frac{b}{1-a} \cdot a^n + \frac{b}{1-a}$$

$$= \frac{b}{1-a} + \left(y_0 - \frac{b}{1-a}\right)a^n$$

Derivation of the Solution to the Difference Equation for $a = 1$ The preceding reasoning also gives the solution in the case $a = 1$—namely, formula (3) holds for any value of a. In particular, for $a = 1$, formula (3) reads

$$\begin{aligned}
y_n &= a^n y_0 + a^{n-1}b + a^{n-2}b + \cdots + a^2b + ab + b \\
&= 1^n y_0 + 1^{n-1}b + 1^{n-2}b + \cdots + 1^2b + 1b + b \\
&= y_0 + b + b + \cdots + b + b + b \\
&= y_0 + bn
\end{aligned}$$

«

APPENDIX Financial Terms

Saving Accounts

Money in a savings account grows with either compound interest or simple interest. The money initially deposited is called the **principal** (denoted P) and the money in the account at a specified time in the future is called the **future value** (denoted F) of the money.

With **compound interest**, the year is divided into interest periods. The interest rate per period is

$$i = \frac{r}{m} = \frac{\text{annual rate of interest}}{\text{number of times compounded during year}}.$$

The rate r is expressed as a percentage or a decimal (usually between 0 and 1). The most common values of m are 1 (annual compounding), 2 (semiannual compounding), 4 (quarterly compounding), and 12 (monthly compounding). The interest earned during each interest period is $i \cdot B_{\text{previous}}$ where B_{previous} is the balance at the end of the previous interest period. Therefore, the new balance at the end of each interest period is given by the formula

$$B_{\text{new}} = B_{\text{previous}} + i \cdot B_{\text{previous}} = (1 + i)B_{\text{previous}}.$$

With **simple interest**, the interest earned each year is rP. Therefore

$$B_{\text{new}} = B_{\text{previous}} + rP.$$

Annuities

An **increasing annuity** is a bank account into which a sequence of equal deposits are made at the end of regular time periods. The successive balances in the account increase due to the compound interest earned and the deposits. We will assume that a deposit of R dollars is made at the end of each interest period and i is the interest per period. Therefore, the new balance at the end of each interest period is given by the formula

$$B_{\text{new}} = B_{\text{previous}} + i \cdot B_{\text{previous}} + R = (1 + i)B_{\text{previous}} + R.$$

| balance at beginning of interest period | interest earned for the interest period | amount deposited |

A **decreasing annuity** is a bank account into which an amount of m*time periods. and then a sequence of equal withdrawals are made at the end of ref each interest We will assume that the withdrawal of R dollars is made at the interest earned period, and that the amount withdrawn is greater than the co*decrease. The new during the interest period. The successive balances in the ⌐ balance at the end of each interest period is given by the f*us − R.

$$B_{\text{new}} = B_{\text{previous}} + i \cdot B_{\text{previous}} - R = (1$$

| balance at beginning of interest period | interest earned f the interest p⌐ |

Derivation of the Solution to the Difference Equation for $a = 1$ The preceding reasoning also gives the solution in the case $a = 1$—namely, formula (3) holds for any value of a. In particular, for $a = 1$, formula (3) reads

$$\begin{aligned}
y_n &= a^n y_0 + a^{n-1}b + a^{n-2}b + \cdots + a^2 b + ab + b \\
&= 1^n y_0 + 1^{n-1}b + 1^{n-2}b + \cdots + 1^2 b + 1b + b \\
&= y_0 + b + b + \cdots + b + b + b \\
&= y_0 + bn
\end{aligned}$$

$\ll$

APPENDIX Financial Terms

Saving Accounts

Money in a savings account grows with either compound interest or simple interest. The money initially deposited is called the **principal** (denoted P) and the money in the account at a specified time in the future is called the **future value** (denoted F) of the money.

With **compound interest**, the year is divided into interest periods. The interest rate per period is

$$i = \frac{r}{m} = \frac{\text{annual rate of interest}}{\text{number of times compounded during year}}.$$

The rate r is expressed as a percentage or a decimal (usually between 0 and 1). The most common values of m are 1 (annual compounding), 2 (semiannual compounding), 4 (quarterly compounding), and 12 (monthly compounding). The interest earned during each interest period is $i \cdot B_{\text{previous}}$ where B_{previous} is the balance at the end of the previous interest period. Therefore, the new balance at the end of each interest period is given by the formula

$$B_{\text{new}} = B_{\text{previous}} + i \cdot B_{\text{previous}} = (1 + i)B_{\text{previous}}.$$

With **simple interest**, the interest earned each year is rP. Therefore

$$B_{\text{new}} = B_{\text{previous}} + rP.$$

Annuities

An **increasing annuity** is a bank account into which a sequence of equal deposits are made at the end of regular time periods. The successive balances in the account increase due to the compound interest earned and the deposits. We will assume that a deposit of R dollars is made at the end of each interest period and i is the interest per period. Therefore, the new balance at the end of each interest period is given by the formula

$$B_{\text{new}} = B_{\text{previous}} + i \cdot B_{\text{previous}} + R = (1 + i)B_{\text{previous}} + R.$$

| balance at beginning of interest period | interest earned for the interest period | amount deposited |

A **decreasing annuity** is a bank account into which an amount of money is deposited and then a sequence of equal withdrawals are made at the end of regular time periods. We will assume that the withdrawal of R dollars is made at the end of each interest period, and that the amount withdrawn is greater than the compound interest earned during the interest period. The successive balances in the account decrease. The new balance at the end of each interest period is given by the formula

$$B_{\text{new}} = B_{\text{previous}} + i \cdot B_{\text{previous}} - R = (1 + i)B_{\text{previous}} - R.$$

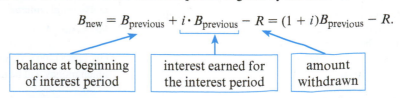

| balance at beginning of interest period | interest earned for the interest period | amount withdrawn |

A **mortgage** is a long-term loan that is used to purchase real estate and is paid back (with interest) by a sequence of equal payments made at the end of each interest period. We can think of it as a decreasing annuity where the borrower functions as the bank.

Check Your Understanding 10.5

Solutions can be found following the section exercises.

Mortgage The difference equation for the monthly balance of a mortgage is

$$y_n = (1.004)y_{n-1} - 1015,$$
$$y_0 = 177,138.81$$

and the solution of the difference equation is

$$y_n = 253,750 - 76,611.19(1.004)^n.$$

1. What are the amount, the annual rate of interest, and the monthly payment of the mortgage?

2. Is the duration of the mortgage 15 years, 25 years, or 30 years?

EXERCISES 10.5

In Exercises 1–10, give a difference equation for y_n, the balance after n interest periods.

1. **Compound Interest** $1000 is deposited into a savings account paying 4% interest compounded quarterly.

2. **Compound Interest** $1250 is deposited into a savings account paying 3% interest compounded monthly.

3. **Simple Interest** $2500 is deposited into a savings account paying 2.5% simple interest.

4. **Simple Interest** $500 is deposited into a savings account paying 2.75% simple interest.

5. **Mortgage** A $204,700 mortgage with interest rate 4.8% compounded monthly is repaid with monthly payments of $1073.99.

6. **Mortgage** A $143,000 mortgage with interest rate 4.5% compounded monthly with monthly payments of $724.56.

7. **Decreasing Annuity** $25,000 is deposited into a savings account paying 3.9% interest compounded weekly and $100 is withdrawn from the account at the end of each week.

8. **Loan** A car loan of $9480 with interest rate 6% compounded monthly is repaid in 3 years with monthly payments of $288.40.

9. **Increasing Annuity** $2500 is deposited into a savings account paying 2.8% interest compounded semiannually and $1000 is added to the account at the end of each half year.

10. **Increasing Annuity** $1000 is deposited into a savings account paying 5.2% interest compounded weekly and $25 is added to the account at the end of each week.

In Exercises 11 and 12, answer the questions.

11. **Increasing Annuity** The difference equation for the quarterly balance in an increasing annuity with interest compounded quarterly and money added at the end of each quarter year is

$$y_n = 1.008y_{n-1} + 1196, \quad y_0 = 37,780$$

and the solution of the difference equation is

$$y_n = -149,500 + 187,280(1.008)^n.$$

(a) What are the amount initially deposited into the annuity, the annual rate of interest, and the quarterly amount deposited into the annuity?

(b) Will the value of the annuity reach $100,000 after 7 years, 8 years, or 9 years?

12. **Decreasing Annuity** The difference equation for the weekly balance in a decreasing annuity with interest compounded weekly and money withdrawn at the end of each week is

$$y_n = 1.0005y_{n-1} - 200, \quad y_0 = 24,792$$

and the solution of the difference equation is

$$y_n = 400,000 - 375,208(1.0005)^n.$$

(a) What are the amount initially deposited into the annuity, the annual rate of interest, and the weekly amount withdrawn from the annuity?

(b) Will the annuity be depleted after 128 weeks or 138 weeks?

In Exercises 13–18, **(a)** determine the first five values generated by the difference equation, and **(b)** find the solution of the difference equation.

13. $y_n = y_{n-1} + 5, y_0 = 1$

14. $y_n = y_{n-1} - 2, y_0 = 50$

15. $y_n = .4y_{n-1} + 3, y_0 = 7$

16. $y_n = 3y_{n-1} - 12, y_0 = 10$

17. $y_n = -5y_{n-1}, y_0 = 2$

18. $y_n = -.7y_{n-1} + 3.4, y_0 = 3$

In Exercises 19–38, use difference equations to answer the question.

19. **Compound Interest** Calculate the future value of $1000 after 2 years if deposited at 2.1% interest compounded monthly.

20. **Compound Interest** Calculate the future value of $3000 after 4 years if deposited at 2.1% interest compounded monthly.

21. **Compound Interest** Calculate the present value of $40,100 payable in 3 years at 4.4% interest compounded quarterly.

22. **Compound Interest** Calculate the present value of $101,850 payable in 6 years at 4% interest compounded quarterly.

23. **Simple Interest** Calculate the future value after 3 years if $1000 is deposited at 4.5% simple interest.

24. **Simple Interest** Calculate the future value after 5 years if $4000 is deposited at 3% simple interest.

25. **Simple Interest** Calculate the present value of $2000 in 10 years at 2.5% simple interest.

26. **Simple Interest** Calculate the present value of $1000 in 7 years at 4% simple interest.

27. **Increasing Annuity** Seventeen thousand dollars is deposited into a savings account at 5% interest compounded semiannually and $1500 is deposited at the end of each half year. How much money will be in the account after 5 years?

28. Increasing Annuity Suppose P dollars is deposited into a savings account at 3.2% interest compounded quarterly and then $500 is added to the account at the end of each quarter year. For what value of P will the account contain $41,000 after 6 years?

29. Increasing Annuity Suppose you deposit P dollars into a savings account at 4.5% interest compounded quarterly and then add $1500 to the account at the end of each quarter. For what value of P will the account contain $73,000 after 4 years?

30. Increasing Annuity Thirty-one thousand dollars is deposited into a savings account paying 3.9% interest compounded semiannually and $1000 is deposited at the end of each half year. How much money is in the account after 5 years?

31. Increasing Annuity Suppose you deposit $10,500 into a savings account paying 3.75% interest compounded annually. How much money should you deposit into the account at the end of each year in order to have $20,000 in the account at the end of 7 years?

32. Increasing Annuity Suppose you deposit $12,700 into a savings account paying 3.9% interest compounded weekly. How much money should you deposit into the account at the end of each week in order to have $25,000 in the account at the end of 6 years?

33. Decreasing Annuity Forty-three thousand dollars is deposited into a savings account paying 3.2% interest compounded quarterly and $700 is withdrawn at the end of each quarter year. How much money is in the account after 5 years?

34. Mortgage Calculate the monthly payment for a 30-year mortgage for $300,062 at 4.5% interest compounded monthly.

35. Mortgage Calculate the monthly payment for a 25-year mortgage for $300,080 at 5.1% interest compounded monthly.

36. Decreasing Annuity Thirty-one thousand dollars is deposited into a savings account paying 3% interest compounded quarterly and $1000 is withdrawn at the end of each quarter. How much money is in the account after 4 years?

37. Loan Suppose you borrow $2710 at 5% interest compounded quarterly. How much money should you pay back each quarter in order to pay back the loan in 3 years?

38. Loan Suppose you borrow $49,000 at 6.8% interest compounded quarterly. How much money should you pay back each quarter in order to pay back the loan in 15 years?

39. Present Value of a Decreasing Annuity Use a difference equation to derive the formula for the present value of a decreasing annuity in which R dollars is withdrawn from a savings account at the end of each interest period and where the interest rate per period is i.

40. Parachuting A parachutist opens her parachute after reaching a speed of 100 feet per second. Suppose that y_n, her speed n seconds after opening the parachute, satisfies the difference equation $y_n = .1y_{n-1} + 14.4$, $y_0 = 100$.
(a) Solve the difference equation
(b) Her speed will get closer and closer to what speed?

41. Population Dynamics A small city with current population 50,000 is experiencing a departure of 600 people each year. Assume that each year the increase in population due to natural causes is 1% of the population at the start of that year.

(a) Find the difference equation for y_n, the population after n years.
(b) Solve the difference equation from part (a).
(c) Use the solution from part (b) to calculate the population after 10 years.

42. Elevation and Atmospheric Pressure The atmospheric pressure at sea level is 14.7 pounds per square inch. Suppose that, at any elevation, an increase of 1 mile results in a decrease of 20% of the atmospheric pressure at that elevation.
(a) Find the difference equation for y_n, the atmospheric pressure at elevation n miles.
(b) Solve the difference equation from part (a).
(c) Use the solution from part (b) to calculate the atmospheric pressure at 12 miles above sea level.

43. Spread of Information A sociological study was made to examine the process by which doctors decide to adopt a new drug. Certain doctors who had little interaction with other physicians were called *isolated*. Out of 100 isolated doctors, the number who adopted the new drug at the end of each month was 8% of those who had not yet adopted the drug at the beginning of the month. (*Source:* James S. Coleman, Elihu Katz, and Herbert Menzel, "The Diffusion of an Innovation Among Physicians.")
(a) Find a difference equation for y_n, the number of isolated physicians adopting the drug after n months.
(b) Solve the difference equation from part (a).
(c) Use the solution from part (b) to calculate the number of isolated doctors who had adopted the drug after 11 months.

44. Solute Concentration A cell is put into a fluid containing an 8 milligram/liter concentration of a solute. (This concentration stays constant throughout.) Initially, the concentration of the solute in the cell is 3 milligrams/liter. The solute passes through the cell membrane at such a rate that each minute the increase in concentration in the cell is 40% of the difference between the outside concentration and the inside concentration.
(a) Find the difference equation for y_n, the concentration of the solute in the cell after n minutes.
(b) Solve the difference equation from part (a).
(c) Use the solution from part (b) to calculate the concentration of the solute in the cell after 7 minutes.

45. Learning Curve Psychologists have found that, in certain learning situations in which there is a maximum amount that can be learned, the additional amount learned each minute is 30% of the amount yet to be learned at the beginning of that minute. Let 12 units of information be the maximum amount that can be learned.
(a) Find the difference equation for y_n, the amount learned after n minutes.
(b) Solve the difference equation from part (a).
(c) Use the solution from part (b) to calculate the amount learned after 6 minutes.

46. Genetics Consider two genes A and a in a population, where A is a dominant gene and a is a recessive gene controlling the same genetic trait. (That is, A and a belong to the same locus.) Suppose that, initially, 80% of the genes are A and 20% are a. Suppose that, in each generation, .003% of gene A mutate to gene a.

(a) Find the difference equation for y_n, the percentage of gene a after n generations. [*Note:* The percentage of gene $A = 100 - $ (the percentage of gene a).]

(b) Solve the difference equation from part (a).

(c) Use the solution from part (b) to calculate the percentage of gene a after 1800 generations.

47. Thermodynamics When a cold object is placed in a warm room, each minute its increase in temperature is 20% of the difference between the room temperature and the temperature of the object at the beginning of the minute. Suppose that the room temperature is 70°F and the initial temperature of the object is 40°F.

(a) Find the difference equation for y_n, the temperature of the object after n minutes.

(b) Solve the difference equation from part (a).

(c) Use the solution from part (b) to calculate the temperature of the object after 5 minutes.

48. Electricity Usage Suppose that the annual amount of electricity used in the United States will increase at a rate of 1.8% each year and that, this year, 3.2 trillion kilowatt-hours are being used.

(a) Find the difference equation for y_n, the number of trillion kilowatt-hours to be used during the year that is n years from now.

(b) Solve the difference equation from part (a).

(c) Use the solution from part (b) to calculate the number of trillion kilowatt-hours that will be used during the year that is 7 years from now.

49. Probability Suppose a coin has probability p of landing on heads. What is the probability of obtaining heads an even number of times in n tosses of the coin? *Note:* Let y_n be the probability of obtaining heads an even number of times in n tosses. Then $y_0 = 1$ since, if you make no tosses, you get zero heads, and zero is an even number. The tree diagram in Fig. 4 displays the probabilities associated with n tosses for $n \geq 1$.

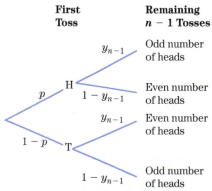

Figure 4 Tree diagram for probability problem with H = heads, T = tails

(a) Use the tree diagram to obtain a difference equation for y_n.

(b) Solve the difference equation in part (a).

50. Drug Absorption The popular sleep-aid drug Zaleplon has a half-life of one hour. That is, the amount of the drug present in the body is halved each hour. Suppose 10 mg of Zaleplon is taken at bedtime.

(a) Find the difference equation for y_n, the amount of the drug present in the body after n hours.

(b) Solve the difference equation from part (a).

(c) Use the solution from part (b) to calculate the amount of the drug present in the body after 8 hours.

TECHNOLOGY EXERCISES

51. Compound Interest From 1984 to 2014, the value of Berkshire Hathaway stock grew at a rate of about 18% compounded annually. Give the difference equation for an investment of $1000 earning 18% compounded annually, and determine the number of years required for the investment to more than triple.

52. Compound Interest A loan of $5614 at 6% interest compounded monthly is paid off with monthly payments of $818.12. Give the difference equation for the monthly balances and determine the number of months until the loan is repaid.

53. Thermodynamics A steel rod of temperature 600°F is immersed in a large vat of water at temperature 75°F. At the end of each minute, its decrease in temperature is 30% of the difference between the water temperature and the temperature of the rod at the beginning of the minute.

(a) Give the difference equation for y_n, the temperature of the rod after n minutes.

(b) What is the temperature after 2 minutes?

(c) When will the temperature drop below 80°F?

54. Population Dynamics The birth rate in a certain city is 3.5% per year, and the death rate is 2% per year. Also, there is a net movement of population out of the city at a steady rate of 300 people per year. The population in 2016 was 5 million.

(a) Give the difference equation for y_n, the population after n years.

(b) Estimate the 2021 population.

(c) When will the population exceed 6 million?

(d) When will the population have doubled to 10 million?

55. Spread of Information Consider the sociological study of Exercise 43. Out of a group of *nonisolated* doctors, let y_n be the percent who adopted the new drug after n months. Then y_n satisfies the difference equation

$$y_n = .0025 y_{n-1}(500 - y_{n-1}), \quad y_0 = 3.$$

That is, initially 3% had adopted the drug.

(a) What percent had adopted the drug after 1 year?

(b) When had over half of the doctors adopted the drug?

(c) When had over 99% adopted the drug?

56. Caffeine Absorption After caffeine is absorbed into the body, 13% is eliminated from the body each hour. Assume that a person drinks an 8-oz cup of brewed coffee containing 130 mg of caffeine, and the caffeine is absorbed immediately into the body.

(a) Find the difference equation for y_n, the amount of caffeine in the body after n hours.

(b) After how many hours will 65 mg (one-half the original amount) remain in the body?

(c) How much caffeine will be in the body 24 hours after the person drank the coffee?

57. Caffeine Absorption Refer to Exercise 56. Suppose that the person drinks a cup of coffee at 7 a.m. and then drinks a cup of coffee at the end of each hour until 7 a.m. the next day.

(a) Find the difference equation for y_n, the amount of caffeine in the body after n hours.

(b) How much caffeine will be in the body at the end of the 24 hours?

Solutions to Check Your Understanding 10.5

1. The general form of the difference equation for the balances in a mortgage is

$$y_n = (1 + i)y_{n-1} - R,$$

$$y_0 = \text{amount borrowed}$$

Therefore, the amount of the mortgage is \$177,138.81. The monthly payment is the value of R, namely \$1015. Using the fact that $1 + i = 1.004 = 1 + .004$, we see that the monthly rate of interest (in decimal form) is .004. Therefore, the annual rate of interest is $12 \cdot .004 = .048$ or 4.8%.

2. We are given that the duration of the mortgage is either 15 years (that is, 180 months), 25 years (that is, 300 months), or 30 years (that is, 360 months). To determine which duration is correct, we should set $n = 180, 300,$ and 360 in the solution of the difference equation and see which one gives a balance of 0. Rounded to two decimal places, $y_{180} = 96,583.31$, $y_{300} = 0.00$, and $y_{360} = -68,675.08$. Therefore, the mortgage will be paid off after 25 years.

CHAPTER 10 | Summary

KEY TERMS AND CONCEPTS	EXAMPLES
10.1 Interest	
Interest is money paid for the use of money.	
Compound interest is calculated each period on the original principal and all interest accumulated during past periods. When P dollars, the **principal**, is invested at annual interest rate r compounded m times per year, then the **balance** after n interest periods is $P(1 + i)^n$, where $i = r/m$ is the **interest per period**.	Find the balance after 10 years when \$1000 is deposited at interest rate 4% compounded quarterly. *Answer:* $P = 1000$, $i = \frac{.04}{4} = .01$, and $n = 4 \cdot 10 = 40$. Therefore, the balance is $1000(1.01)^{40} = \$1488.86$.
Present value is the value today of an amount of money in the future. The present value P of F dollars n interest periods in the future is $$P = \frac{F}{(1 + i)^n}.$$	Find the present value of \$1000 to be received five years from now at interest rate 3% compounded monthly. *Answer:* $F = 1000$, $i = \frac{.03}{12} = .0025$, and $n = 12 \cdot 5 = 60$. Therefore, the present value is $\frac{1000}{1.0025^{60}} = \$860.87$.
The **effective rate** is the simple interest rate that produces the same amount of interest as the compound interest rate. $r_{\text{eff}} = (1 + i)^m - 1$, where interest is compounded m times per year.	Find the effective rate of 3% interest compounded quarterly. *Answer:* $i = \frac{.03}{4} = .0075$, and $m = 4$. Therefore, $r_{\text{eff}} = 1.0075^4 - 1 = .030339 \approx 3.034\%$.
Simple interest is calculated on the original principal only. When P dollars is invested at a simple interest rate r for n years, the balance is $(1 + nr)P$.	Find the balance after 10 years when \$1000 is deposited at simple interest rate 4%. *Answer:* $P = 1000$, $n = 10$, and $r = .04$. Therefore, the balance is $(1 + 10 \cdot .04)1000 = (1.4)1000 = \$1400$.
10.2 Annuities	
An **increasing (decreasing) annuity** is a sequence of equal deposits (withdrawals) made at the ends of regular time periods.	
Successive **balances in an increasing annuity** are calculated with the formula $$B_{\text{new}} = (1 + i)B_{\text{previous}} + R,$$ where i is the interest rate per time period and R is the amount of the regular deposit. The **future value of an increasing annuity** is calculated with the formula $$F = \frac{(1 + i)^n - 1}{i} \cdot R.$$	If \$100 is deposited into an account at the end of each month paying an interest rate of 3% compounded monthly ($i = .0025$), then successive monthly balances can be calculated with the formula $$B_{\text{new}} = (1.0025)B_{\text{previous}} + 100.$$ The balance after 5 years (60 months) is $$\frac{1.0025^{60} - 1}{.0025} \cdot 100 = \$6464.67.$$

KEY TERMS AND CONCEPTS	EXAMPLES
Successive **balances in a decreasing annuity** are calculated with the formula $$B_{new} = (1 + i)B_{previous} - R,$$ where i is interest per time period and R is the amount of the regular withdrawal. The **present value of a decreasing annuity** is calculated with the formula $$P = \frac{1 - (1 + i)^{-n}}{i} \cdot R.$$	If $100 is withdrawn from an account at the end of each month paying an interest rate of 3% compounded monthly, then successive monthly balances can be calculated with the formula $$B_{new} = (1.0025)B_{previous} - 100.$$ Find the present value of a decreasing annuity from which $100 will be withdrawn at the end of each month for 5 years (60 months). *Answer:* $R = 100$, $n = 60$, and $i = .0025$. Therefore, the present value is $$\frac{1 - 1.0025^{-60}}{.0025} \cdot 100 = \$5565.24.$$

10.3 Amortization of Loans

Amortization is a method of retiring a loan (such as a mortgage). It consists of a steady stream of equal payments at regular periods that pays down the loan with interest. The formula $$P = \frac{1 - (1 + i)^{-n}}{i} \cdot R$$ relates the principal P to the regular payment R.	Find the amount of a 30-year mortgage at 4.6% compounded monthly with a monthly payment of $1025. Here, $R = 1025$, $i = \frac{.046}{12} = .0038333333$, and $n = 12 \cdot 30 = 360$. Therefore, the amount of the mortgage is $$\frac{1 - 1.0038333333^{-360}}{.003833333} \cdot 1025 \approx \$200,000.$$

10.4 Personal Financial Decisions

An **individual retirement account** (IRA) is an increasing annuity in which the annual interest earned is either tax free (Roth IRA) or tax-deferred (traditional IRA). Contributions are tax deductible only with a traditional IRA.	If $5000 is deposited annually into a traditional IRA at 6% interest compounded annually, the balance after 36 years will be $595,604.33.
When the **add-on method** is used to calculate finance charges on a consumer loan, the interest paid each month is a fixed percentage of the principal.	With the add-on method, the monthly payment on a two-year loan of $1000 at a stated annual interest rate of 6% is $46.67. (The payment would be $44.32 if calculated as in Section 10.3.)
A **discount point** accompanying a mortgage loan requires the borrower to pay additional interest up front equal to 1% of the amount borrowed.	A $200,000 mortgage with three discount points requires an up-front payment of $6000.
The **APR** of a mortgage is determined by calculating the interest rate corresponding to the loan amount after deducting certain up-front costs.	The APR for a mortgage of $200,000 with one point at 6% interest compounded monthly is 6.094%.
The **effective mortgage rate** is the interest rate corresponding to an annuity that decreases the loan amount (after the deduction of certain up-front costs) to the balance after the expected lifetime of the loan.	With the assumption that the aforementioned mortgage will be held for 5 years, the effective mortgage rate is 6.24%.
The **interest-only mortgage** and the **adjustable-rate mortgage** (ARM) are alternatives to the standard fixed-rate mortgage. The interest rate of an ARM is reset periodically, depending on a floating interest rate determined by an index.	A 30-year mortgage of $400,000 at 6.6% interest compounded monthly that is interest-only for 10 years has a monthly payment of $2200 for the first 10 years and $3005.89 thereafter.

10.5 A Unifying Equation

The **difference equation** $y_n = ay_{n-1} + b$ with y_0 given and $a \neq 1$ has the **solution** $$y_n = \frac{b}{1 - a} + \left(y_0 - \frac{b}{1 - a}\right)a^n.$$ The difference equation $y_n = y_{n-1} + b$ with y_0 given has the **solution** $$y_n = y_0 + bn.$$	Solve $y_n = .2y_{n-1} + 4.8$, $y_0 = 1$. Since $\frac{b}{1 - a} = \frac{4.8}{1 - .2} = \frac{4.8}{.8} = 6,$ $$y_n = 6 + (1 - 6)(.2)^n = 6 - 5(.2)^n$$ The solution of $y_n = y_{n-1} + 2$, $y_0 = 3$ is $$y_n = 3 + 2n.$$

KEY TERMS AND CONCEPTS	EXAMPLES
The **interest formulas** $$y_n = y_0 + (iy_0)n \quad \text{simple interest}$$ $$y_n = y_0(1 + i)^n \quad \text{compound interest}$$ each give the balance after n interest periods when y_0 dollars is deposited at an interest rate i per interest period.	If $100 is deposited at 2% interest compounded annually, the balance after 10 years will be $$y_{10} = 100 + 2 \cdot 10 = \$120.00 \quad \text{simple interest}$$ $$y_{10} = 100(1.02)^{10} = \$121.90 \quad \text{compound interest}$$
Successive balances in an **increasing annuity** can be calculated with the difference equation $$y_n = (1 + i)y_{n-1} + R, y_0 = 0$$ where i is the interest rate per period and R is the periodic deposit.	If $100 is deposited into an account at the end of each month paying an interest rate of 3% compounded monthly, then successive monthly balances can be calculated with the difference equation $$y_n = (1.0025)y_{n-1} + 100, y_0 = 0.$$
Successive balances in a **decreasing annuity** (or a loan) can be calculated with the difference equation $$y_n = (1 + i)y_{n-1} - R, y_0 = \text{initial balance}$$ where i is the interest rate per period and R is the periodic payment.	A 30-year mortgage of $200,000 at 4.6% compounded monthly has a monthly payment of $1025. Successive balances are calculated with the difference equation $$y_n = (1.003833)y_{n-1} - 1025, y_0 = 200,000.$$
Phenomena in physics, biology, sociology, and economics can be described by difference equations.	Radioactive decay, growth in a bacteria culture, spread of information, and supply and demand can be modeled with difference equations.

CHAPTER 10 Fundamental Concept Check Exercises

1. What is meant by *principal*?

2. What is the difference between compound interest and simple interest?

3. What is meant by the *balance* in a savings account? *Future value*?

4. How is the interest rate per period determined from the annual interest rate?

5. Explain how compound interest works.

6. Explain how simple interest works.

7. What is meant by the *present value* of a sum of money to be received in the future?

8. Explain the difference between the *nominal* and *effective* rates for compound interest.

9. What is an annuity?

10. What is meant by the *future value* of an annuity? *Present value*? *Rent*?

11. Describe the two types of annuities discussed in this chapter. In each case, identify the present and future values.

12. Give the formula for computing a new balance from a previous balance for each type of annuity.

13. Give a formula relating F and R in an increasing annuity.

14. Give a formula relating P and R in a decreasing annuity.

15. What are the components of an amortization table of a loan?

16. Give the formula for computing a new balance from a previous balance for a loan.

17. What is a balloon payment?

18. Explain how *traditional* and *Roth IRAs* work.

19. How are finance charges on a consumer loan calculated with the *add-on method*?

20. What are *discount points*?

21. What is the difference between the effective mortgage rate of a mortgage and the APR?

22. What is an interest-only mortgage?

23. What is an adjustable-rate mortgage?

24. Explain how a sequence of numbers is generated by a difference equation of the form $y_n = ay_{n-1} + b$, y_0 given.

25. What is meant by an *initial value for a difference equation*?

26. Give the solution of the difference equation $y_n = ay_{n-1} + b$, y_0 given, with $a \neq 1$. With $a = 1$.

CHAPTER 10 Review Exercises

1. **Future Valve** If $100 earns 3% interest compounded annually, find the future value after 10 years.

2. **Saving for Retirement** Mr. West wishes to purchase a condominium for $240,000 cash upon his retirement 10 years from now. How much should he deposit at the end of each month

into an annuity paying 2.7% interest compounded monthly in order to accumulate the required savings?

3. **Mortgage Considerations** The income of a typical family in a certain city is currently $39,216 per year. Family finance experts recommend that mortgage payments not exceed 25%

of a family's income. Assuming a current mortgage interest rate of 4.2% compounded monthly for a 30-year mortgage with monthly payments, how large a mortgage can the typical family in that city afford?

4. **Future Value** Calculate the future value of $50 after a year if it is deposited at 2.19% compounded daily.

5. **Comparing Payouts** Which is a better investment: 3% compounded annually or 2.92% compounded daily?

6. **Bond Fund** Ms. Smith deposits $200 at the end of each month into a bond fund yielding 3% interest compounded monthly. How much are her holdings worth after 5 years?

7. **Nonstandard Mortgage** A real estate investor takes out a $200,000 mortgage subject to the following terms: For the first 5 years, the payments will be the same as the monthly payments on a 15-year mortgage at 4.5% interest compounded monthly. The unpaid balance will then be payable in full.
   (a) What are the monthly payments for the first 5 years?
   (b) What balance will be owed after 5 years?

8. **College Expenses** College expenses at a private college currently average $35,000 per year. It is estimated that these expenses are increasing at the rate of ½% per month. What is the estimated cost of a year of college 10 years from now?

9. **Present Value** What is the present value of $50,000 in 10 years at 3% interest compounded monthly?

10. **Investment Value** An investment will pay $10,000 in 2 years and then $5000 1 year later. If the current market interest rate is 2.7% compounded monthly, what should a rational person be willing to pay for the investment?

11. **Car Loan** A woman purchases a car for $12,000. She pays $3,000 as a down payment and finances the remaining amount at 6% interest compounded monthly for 4 years. What is her monthly car payment?

12. **Nonstandard Loan** A businessman buys a $100,000 piece of manufacturing equipment on the following terms: Interest will be charged at a rate of 4% compounded semiannually, but no payments will be made until 2 years after purchase. Starting at that time, equal semiannual payments will be made for 5 years. Determine the semiannual payment.

13. **Retirement Account** A retired person has set aside a fund of $105,003.50 for his retirement. This fund is in a bank account paying 2.4% interest compounded monthly. How much can he draw out of the account at the end of each month so that there is a balance of $30,000 at the end of 15 years? (*Hint:* First compute the present value of the $30,000.)

14. **Balloon Payment** A business loan of $500,000 is to be paid off in monthly payments for 10 years with a $100,000 balloon payment at the end of the 10th year. The interest rate on the loan is 6% compounded monthly. Calculate the monthly payment.

15. **Savings Plan** Ms. Jones saved $100 per month for 30 years at 6% interest compounded monthly. How much were her accumulated savings worth?

16. **Loan** An apartment building is currently generating an income of $2000 per month. Its owners are considering a 10-year loan at 4.5% interest compounded monthly in order to pay for repairs. How large a loan can the income of the apartment house support?

17. **Comparing Investments** Investment *A* generates $1000 at the end of each year for 10 years. Investment *B* generates $5000 at the end of the fifth year and $5000 at the end of the tenth year. Assume a market rate of interest of 2.5% compounded annually. Which is the better investment?

18. **Bond Value** A 5-year bond has a face value of $1000 and is currently selling for $800. The bond pays $5 interest at the end of each month and, in addition, will repay the $1000 face value at the end of the fifth year. The market rate of interest is currently 4.8% compounded monthly. Is the bond a bargain? Why or why not?

19. **Effective Rate** Calculate the effective rate for 2.2% interest compounded semiannually.

20. **Effective Rate** Calculate the effective rate for 2.7% interest compounded monthly.

21. **Annuities with Extra Deposit** A person makes an initial deposit of $10,000 into a savings account and then deposits $1000 at the end of each quarter year for 15 years. If the interest rate is 2.2% compounded quarterly, how much money will be in the account after 15 years?

22. **Car Loan** A $10,000 car loan at 6% interest compounded monthly is to be repaid with 36 equal monthly payments. Write out an amortization schedule for the first 6 months of the loan.

23. **Savings Plan** A person pays $200 at the end of each month for 10 years into a fund paying .15% interest per month compounded monthly. At the end of the 10th year, the payments cease, but the balance continues to earn interest. What is the value of the fund at the end of the 20th year?

24. **Savings Fund** A savings fund currently contains $300,000. It is decided to pay out this amount with 1.8% interest compounded monthly over a 5-year period. What are the monthly payments?

25. **Mortgage Payment** What is the monthly payment on a $150,000, 30-year mortgage at 4.8% interest compounded monthly?

26. **Traditional IRA** Elisa, age 60, is currently in the 30% tax bracket and has $30,000 in a traditional IRA that earns 6% interest compounded annually. She anticipates being in the 35% tax bracket 5 years from now.
   (a) How much money will Elisa have after paying taxes if she withdraws her money now?
   (b) How much will Elisa have after paying taxes if she waits 5 years and then withdraws the money from the account?

27. **Roth IRA** Rework Exercise 26 for a Roth IRA.

28. **Consumer Loan** A consumer loan of $5000 for 2 years has an APR of 9% compounded monthly and a monthly payment of $228.42. What interest rate would be stated with the add-on method?

29. **Comparing Loans** Spike is considering purchasing a new computer and has decided to take one of two consumer loans:

   Consumer Loan *A*: 1 year, $3000, 10% APR, monthly payment of $263.75
   Consumer Loan *B*: 1 year, $3000, 6% add-on interest

   Which loan should Spike take and why?

30. **Mortgage with Points** Consider a 15-year mortgage of $90,000 at 6% interest compounded monthly with two discount points

and a monthly payment of $759.47. The APR for the mortgage is obtained by solving $P = \frac{1 - (1 + i)^{-n}}{i} \cdot R$ for i and then multiplying by 12. What are the values of P, n, and R?

31. **Mortgage with Points** Consider a 20-year mortgage of $100,000 at 6% interest compounded monthly with three discount points and a monthly payment of $716.43. Assume that the loan is expected to be held for 6 years, at the end of which time the unpaid balance will be $81,298.44. The effective mortgage rate is obtained in Excel by evaluating 12*RATE(m, $-R$, P, $-B$). What are the values of m, R, P, and B?

32. **Comparing Mortgages** Suppose that a lender gives you a choice between the following two 25-year mortgages of $200,000:

   Mortgage A: 6.5% interest compounded monthly, two points, monthly payment of $1350.41
   Mortgage B: 7% interest compounded monthly, one point, monthly payment of $1413.56

   Assume that you can invest money at 3.5% compounded monthly. The length of time that you must retain the mortgage in order for mortgage A to be the better choice is obtained in Excel by evaluating NPER(i, R, $-P$). What are the values of i, R, and P?

33. **Mortgage with Points** A 20-year mortgage of $250,000 at 5.75% interest compounded monthly, with two discount points, has a monthly payment of $1755.21. Show that the APR for this mortgage is 6%.

34. **Mortgage with Points** A 30-year mortgage of $100,000 at 5.5% interest compounded monthly, with three discount points, has a monthly payment of $567.79. Assume that the loan is expected to be terminated after 8 years, at which time the unpaid balance will be $86,837.98. Show that the effective mortgage rate is 6%.

35. **Interest-Only Mortgage** Consider a 25-year mortgage of $380,000 at 6.9% interest compounded monthly, where the loan is interest-only for 10 years. What is the monthly payment during the first 10 years? Last 15 years?

36. **Adjustable-Rate Mortgage** Consider a 25-year $220,000 5/1 ARM with a 2.8% margin and which is based on the CMT index. Suppose the value of the CMT index is 3.5% when the loan is initiated and is 4.55% 5 years later. Assume that all interest rates use monthly compounding.
   (a) Calculate the monthly payment for the first 5 years.
   (b) Calculate the unpaid balance at the end of the first 5 years.
   (c) Calculate the monthly payment for the sixth year.

37. Consider the difference equation $y_n = -3y_{n-1} + 8$, $y_0 = 1$.
   (a) Generate y_1, y_2, y_3 from the difference equation.
   (b) Solve the difference equation.
   (c) Use the solution in part (b) to obtain y_4.

38. Consider the difference equation $y_n = y_{n-1} - \frac{3}{2}$, $y_0 = 10$.
   (a) Generate y_1, y_2, y_3 from the difference equation.
   (b) Solve the difference equation.
   (c) Use the solution in part (b) to obtain y_6.

In Exercises 39–43, use difference equations to answer the question.

39. **Account Balance** How much money would you have to deposit into a savings account initially at 3.05% interest compounded quarterly in order to have $2474 after 7 years?

40. **Account Balance** How much money would you have in the bank after 2 years if you deposited $1000 at 5.2% interest compounded weekly?

41. **Annuity** How much money must be deposited at the end of each week into an annuity at 2.6% interest compounded weekly in order to have $36,000 after 21 years?

42. **Mortgage Payments** Find the monthly payment on a $33,100 20-year mortgage at 6% interest compounded monthly.

43. **State Legislature** Suppose that 100 people were just elected to a certain state legislature and that after each term, 8% of those still remaining from this original group will either retire or not be reelected. Let y_n be the number of legislators from the original group of 100 who are still serving after n terms. Find a difference equation for y_n.

Conceptual Exercises

44. If you decrease the interest rate for an investment by 10%, will the future value decrease by 10%?

45. If interest is compounded semiannually, will the effective rate be higher or lower than the nominal rate?

46. If you increase the number of months in which to pay off a loan, will you increase or decrease the monthly payment? The total amount of interest paid? Give an example to justify your answers.

47. Consider a decreasing annuity. If the amount withdrawn each month increases by 5%, will the duration decrease by 5%? Give an example to justify your answer.

48. Give an intuitive explanation for why successive payments for a mortgage contribute steadily more toward repayment of the principal.

CHAPTER

10 PROJECT

Two Items of Interest

Item 1: Successive Interest Computations

1. Suppose that a $100,000 investment grows 3% during the first year and 4% during the second year. By what percent will it have grown after the 2-year period? (*Note:* The answer is not 7%.)

2. Rework Exercise 1 for the case in which the investment earns 4% during the first year and 3% during the second year. Is the answer to Exercise 2 greater than, less than, or equal to the answer to Exercise 1?

3. Consider an annuity in which $100,000 is invested and $10,000 is withdrawn at the end of each year. Suppose that the interest rate is 3% during the first year and 4% during the second year. What is the balance at the end of the second year?

4. Rework Exercise 3, where the interest rate is 4% during the first year and 3% during the second year. Is the balance at the end of the second year greater than, less than, or equal to the answer to Exercise 3? Explain why your answer to this question makes sense.

5. Show that, if an investment of P dollars declines by 4% during a year, the balance at the end of the year is $P \cdot (1 - .04)$—that is, $P \cdot (.96)$.

6. Show that, if an investment of P dollars declines by r% during a year, the balance at the end of the year is $P \cdot (1 - r/100)$. We say that the investment earned $-r$%.

7. Show that, if an investment of P dollars earns r% one year and s% the following year, then the balance after the 2-year period is $P(1 + r/100)(1 + s/100)$. Conclude that the order of the two numbers r and s does not affect the balance after 2 years. (*Note:* The numbers r and s can be positive or negative.)

8. Which of the following two statements is true?
   (a) Suppose that an investment of P dollars earns 4% one year and loses 4% the next year. Then the value of the investment at the end of the 2-year period will be P dollars.
   (b) Suppose that investment A earns 4% one year and then loses 3% the next year, and investment B loses 3% one year and then gains 4% the next year. Then the balances of the two investments will be the same after the 2-year period.

Item 2: Rule of 72

9. Show that, if $1000 is invested at 8% interest compounded annually, then it will double in about 9 years. (*Note:* $9 = 72/8$.)

10. Show that, if $1000 is invested for 6 years, then it will approximately double in that time if it appreciates at 12% per year. (*Note:* $12 = 72/6$.)

> **Rule of 72**   If money is invested at r% interest compounded annually, then it will double in about $72/r$ years. Alternatively, if money is invested for n years, then it will double during that time if it appreciates by about $(72/n)$% per year.

Table 1

m	Number Associated with Increasing m-Fold
2	72
3	114
4	144
5	167
6	186

11. Conclude from the Rule of 72 that, for an interest rate of r%, $(1 + r/100)^{72/r} \approx 2$. Also, conclude that for a number of years, n, $(1 + .72/n)^n \approx 2$.

We needn't restrict ourselves to doubling. For instance, the *Rule of 114* says that, if money is invested at r% interest, then it will triple in about $114/r$ years. Table 1 gives the numbers associated with several different multiples. Let us denote the number associated with m by N_m. That is, $N_2 = 72$ and $N_3 = 114$.

12. Notice that $N_6 = N_2 + N_3$. Explain why this makes sense.

13. Explain why $N_{n \cdot p} = N_n + N_p$ for any positive numbers n and p.

14. Find N_{10}, and then use that number to estimate the amount of time required for money to increase tenfold if invested at 8% interest compounded annually. Check your answer by raising (1.08) to that power.

15. Use the values of N_2 and N_3 to find $N_{1.5}$, the number associated with money increasing by one-half. Estimate the amount of time required for $1000 to grow to $1500 when invested at 7% interest compounded annually.

In this chapter, we introduce logic: the foundation for mathematics, coherent argument, and computing. The idea of logical argument was introduced in writing by Aristotle around 350 BCE. For centuries, his principles have formed the basis for systematic thought, communication, debate, law, mathematics, and science. More recently, the concept of a computing machine and the essence of programming have become applications of the ancient ideas that we present here. Also, new developments in mathematical logic have helped to promote significant advances in artificial intelligence.

11.1   Introduction to Logic

The building blocks of logic are statements, connectives, and the rules for calculating the truth or falsity of compound statements. We begin with statements.

DEFINITION   A **statement** is a declarative sentence that is either true or false.

EXAMPLE 1   **Statements**   Determine whether each of the following sentences is a statement.
(a) George Washington was the first president of the United States.
(b) The Miami Heat won the NBA Championship in 2015.
(c) The number of atoms in the universe is 10^{75}.
(d) If x is a positive number, and $x^2 = 9$, then $x = 3$.
(e) He is a really nice guy.
(f) Do your homework!
(g) $x + 3 = 2x$
(h) Is this course fun?

SOLUTION   To be a statement, each sentence must meet two conditions:
(i) it is a declarative sentence;   **(ii)** it is TRUE or FALSE.

	Declarative Sentence?	TRUE or FALSE	Conclusion
(a)	Yes	TRUE	**Statement**
(b)	Yes	FALSE	**Statement**
(c)	Yes	It is TRUE or FALSE, although we are not sure of which.	**Statement**
(d)	Yes	TRUE	**Statement**
(e)	Yes	Neither, since the person "he" is not specified.	Not a statement
(f)	No	—	Not a statement
(g)	Yes	Neither; the value of x is not specified.	Not a statement
(h)	No	—	Not a statement

» Now Try Exercise 1

The **truth value** of a statement is either *true* or *false*. To simplify things, we will often use symbols to represent statements, and declare the statement to be true or false. When dealing with real-world examples, however, determining whether a statement is true or false may not be so straightforward. The assignment of a truth value to a statement may be obvious, as in the statement "George Washington was the first president of the United States." The truth value of the statement "The number of atoms in the universe is 10^{75}," on the other hand, is much more problematic.

We combine statements naturally when we speak, using **connectives**, words like "and" and "or." We state implications with the words "if" and "then." Frequently, we negate statements with the word "not." A **compound statement** is formed by combining statements by use of the words "and," "or," "not," or "if, then." A **simple statement** is a statement that is not a compound statement. A compound statement is analyzed as a combination of simple statements.

EXAMPLE 2   **Simple Statements** Give the simple statements in each of the following compound statements:
(a) The number 6 is even and the number 5 is odd.
(b) Tom Jones does a term paper or takes the final exam.
(c) If the United Kingdom is in the European Union, then the British eat Spanish oranges and Italian melons.

SOLUTION   **(a)** The simple statements are "The number 6 is even" and "The number 5 is odd."
(b) The simple statements are "Tom Jones does a term paper" and "Tom Jones takes the final exam."
(c) The simple statements are "The United Kingdom is in the European Union," "The British eat Spanish oranges," and "The British eat Italian melons."

It is sometimes helpful to add English words to clarify the implied meaning in simple statements. For example, the phrase "The British eat Spanish oranges and Italian melons" can be rewritten as "The British eat Spanish oranges and the British eat Italian melons." The second formulation makes it much easier to identify the simple statements.

» Now Try Exercise 16

To be able to develop the rules of logic and logical argument, we need to deal with any logical statement, rather than specific examples. We use the letters p, q, r, and so on to denote simple statements. These letters represent variables for which a statement may be substituted. For example, we can let p denote a statement in general. Then, if we wish, we can specify that, for the moment, p will denote the statement "The number 5 is odd." In this particular case, p has the truth value *true*. We might decide to let p denote the statement "There are 13 states in the United States today." Then the truth value of the statement p is *false*.

The use of these logical variables is similar to the use of variables x, y, z, and so on to denote unspecified numbers in algebra. In algebra, we manipulate the symbols (such as x, y, $+$, and the implied multiplication and exponentiation) in expressions according to specified rules to establish identities such as

$$(x + y)^2 = x^2 + 2xy + y^2,$$

which are true no matter which numbers are substituted for x and y. In logic, we will manipulate the symbols in compound statements (p, q, r, "and," "or," "not," and "if, then") to establish rules that hold no matter which statements are substituted for p, q, r, and so on.

It is useful to be able to write a compound statement in terms of its component parts and to use symbols to represent the connectives. Three such symbols correspond to the words "and," "or," and "not":

Conjunction	The symbol $\land$ represents the English word "and."
Disjunction	The symbol $\lor$ represents the English word "or."
Negation	The symbol $\sim$ represents the English word "not," in that the symbolic statement $\sim p$ can be read as "It is not the case that p."

EXAMPLE 3   **Connectives** Let p denote the statement "Ina likes popcorn" and q denote the statement "Fred likes peanuts." Write each of the following compound statements symbolically.
(a) Ina likes popcorn and Fred likes peanuts.
(b) Ina likes popcorn or Fred likes peanuts.
(c) Ina does not like popcorn.

SOLUTION
(a) $p \land q$
(b) $p \lor q$
(c) $\sim p$

>> *Now Try Exercise 18*

In the previous example, note that we could write the sentence "Ina likes popcorn and Fred likes peanuts" as "Fred likes peanuts and Ina likes popcorn" without changing its meaning. Therefore we could denote the statement symbolically as $q \land p$. Similarly, the symbolic statement $p \lor q$ can be expressed as $q \lor p$.

EXAMPLE 4   **Symbolic Form of Compound Statements** Let p denote "Fred likes Cindy," and let q denote "Cindy likes Fred." Use the connectives $\land$, $\lor$, and $\sim$ to denote the following compound sentences:
(a) Fred and Cindy like each other.
(b) Fred likes Cindy, but Cindy does not like Fred.
(c) Fred and Cindy dislike each other.
(d) Fred likes Cindy or Cindy likes Fred.

SOLUTION
(a) "Fred and Cindy like each other" should be rewritten as "Fred likes Cindy and Cindy likes Fred." This is expressed symbolically as $p \land q$.
(b) "Fred likes Cindy, but Cindy does not like Fred" can be written $p \land \sim q$. Note that the word "but" here means "and."
(c) "Fred and Cindy dislike each other" can be rewritten as "Fred dislikes Cindy and Cindy dislikes Fred." This is written symbolically as $\sim p \land \sim q$.
(d) "Fred likes Cindy or Cindy likes Fred" is simply $p \lor q$.   >> *Now Try Exercise 19*

EXAMPLE 5   **Translating from Symbols to English** Let p denote the statement "The interest rate is 10%," and let q denote the statement "The Dow Jones average is over 10,000." Write the English statements corresponding to each of the following:

(a) $p \vee q$ **(b)** $p \wedge q$ **(c)** $p \wedge \sim q$ **(d)** $\sim p \vee \sim q$

SOLUTION   **(a)** "The interest rate is 10% or the Dow Jones average is over 10,000."
(b) "The interest rate is 10% and the Dow Jones average is over 10,000."
(c) "The interest rate is 10% and the Dow Jones average is less than or equal to 10,000."
(d) "The interest rate is not 10% or the Dow Jones average is less than or equal to 10,000."   **» Now Try Exercise 25**

We represent implications, using the words "if" and "then," by one additional symbol:

> **Implication**   The symbol $\rightarrow$ represents the English connective "If . . . , then," in that the symbolic statement $p \rightarrow q$ can be read as "If p, then q."

Using the statements from Example 5, the English sentence "If the interest rate is 10%, then the Dow Jones average is over 10,000" can be written as $p \rightarrow q$.

EXAMPLE 6   **Translating from English to Symbols** Let w denote the statement "The train stops in Washington," and let n denote the statement "The train stops in New York." Write the following statements in symbolic form:

(a) The train stops in New York and Washington.
(b) The train stops in Washington but not in New York.
(c) The train does not stop in Washington.
(d) The train stops in New York or Washington.
(e) The train stops in New York or Washington but not in both.
(f) If the train stops in New York, then it does not stop in Washington.

SOLUTION   **(a)** $n \wedge w$
(b) $w \wedge \sim n$
(c) $\sim w$
(d) $n \vee w$
(e) $(n \vee w) \wedge \sim(n \wedge w)$. We can rewrite this sentence as "The train stops in New York or Washington, but the train does not stop in New York and Washington." The parentheses make the statement clear.
(f) $n \rightarrow \sim w$ **» Now Try Exercise 27**

How are truth values assigned to compound statements? The assignment depends on the truth values of the simple statements and the connectives in the compound form. The rules are discussed in the next section.

Check Your Understanding 11.1

Solutions can be found following the section exercises.

1. Determine which of the following sentences are statements.
   (a) The earnings of IBM went up from 2014 to 2015.
   (b) The national debt of the United States is $10 trillion.
   (c) What an exam that was!
   (d) Abraham Lincoln was the 16th president of the United States.
   (e) Lexington is the capital of Kentucky or Albany is the capital of the United States.
   (f) When was the Civil War?

2. Let p denote the statement "Sally is the class president," and let q denote "Sally is an accounting major." Translate the symbolic statements into proper English.
   (a) $p \wedge q$ **(b)** $\sim p$
   (c) $p \vee q$ **(d)** $\sim p \vee q$
   (e) $p \wedge \sim q$ **(f)** $\sim p \vee \sim q$

EXERCISES 11.1

In Exercises 1–15, determine which sentences are statements.

1. The number 3 is odd.

2. The 1939 World's Fair was held in Miami.

3. The price of coffee depends on the rainfall in Brazil.

4. The Nile River flows through Asia.

5. What a way to go!

6. If snow falls on the Rockies, people are skiing in Aspen.

7. Why is the sky blue?

8. Moisture in the atmosphere determines the type of cloud formation.

9. No United States aircraft carrier is assigned to the Indian Ocean.

10. The number of stars in the universe is 10^{60}.

11. $x + 1 = 6$.

12. She is a good student.

13. Let us pray.

14. The Louvre and the Metropolitan Museum of Art contain paintings by Leonardo da Vinci.

15. If a United States coin is fair, then the chance of getting a head is $\frac{1}{3}$.

In Exercises 16 and 17, give the simple statements in each of the compound statements.

16. Lagos is the largest city in Nigeria or Alpha Centauri is the nearest star to Earth.

17. $2 + 2 = 5$ and $4 < 7$.

In Exercises 18 and 19, give the simple statements in each of the compound statements, and write the compound statement symbolically.

18. China is in Asia and Chicago is in North America.

19. The Phelps Library is in New York or in Dallas.

20. If a U.S. citizen travels overseas, then they need a valid passport.

21. The Smithsonian Museum of Natural History has displays of rocks and bugs.

In Exercises 22 and 23, write the compound statements in symbolic form, using p and q for each of the simple statements.

22. The number 7 is odd and the number 14 is even.

23. No Amtrak trains go to Chicago or Cincinnati.

24. Let p denote the statement "Paris is called the City of Lights," and let q denote the statement "The Eiffel Tower is located in Paris." Write the following statements as English sentences:
   (a) $\sim p$ (b) $\sim p \vee q$ (c) $p \rightarrow q$
   (d) $p \vee q$ (e) $\sim p \wedge \sim q$ (f) $\sim(p \vee q)$

25. Let p denote the statement "Ozone is opaque to ultraviolet light," and let q denote the statement "Life on Earth requires ozone." Write the following statements as English sentences:
   (a) $p \wedge q$ (b) $\sim p \vee q$ (c) $\sim p \vee \sim q$ (d) $\sim q \rightarrow p$

26. Let p denote the statement "Papyrus is the earliest form of paper," and let q denote "The papyrus reed is found in Africa." Put the following statements into symbolic form:
   (a) Papyrus is not the earliest form of paper.
   (b) The papyrus reed is not found in Africa or papyrus is the earliest form of paper.
   (c) If papyrus is the earliest form of paper, then the papyrus reed is not found in Africa.

27. Let a denote the statement "Florida borders Alabama," and let f denote the statement "Florida borders Mississippi." Put the following into symbolic form:
   (a) Florida borders Alabama or Mississippi.
   (b) Florida borders Alabama but not Mississippi.
   (c) Florida borders Mississippi but not Alabama.
   (d) Florida borders neither Alabama nor Mississippi.

Solutions to Check Your Understanding 11.1

1. Statements appear in (a), (b), (d), and (e). Both (c) and (f) are not statements.

2. (a) "Sally is the class president and Sally is an accounting major."
   (b) "Sally is not the class president."
   (c) "Sally is the class president or Sally is an accounting major."
   (d) "Sally is not the class president or Sally is an accounting major."
   (e) "Sally is the class president and Sally is not an accounting major."
   (f) "Sally is not the class president or Sally is not an accounting major."

11.2 Truth Tables

This section discusses how the truth values of the statements $p \wedge q$, $p \vee q$, and $\sim p$ depend on the truth values of p and q.

> **DEFINITION** A **statement form** is an expression formed from simple statements and connectives according to the following rules:
>
> 1. A simple statement is a statement form.
> 2. If p is a statement form, then $\sim p$ is a statement form.
> 3. If p and q are statement forms, then so are $p \wedge q$, $p \vee q$, and $p \rightarrow q$.
>
> In other words, a statement form is a symbolic representation of a simple or compound statement.

EXAMPLE 1   **Defining Statement Forms** Show that each item is a statement form according to the definition. Assume that p, q, and r are simple statements.

(a) $(p \land \sim q) \rightarrow r$

(b) $\sim[p \rightarrow (q \lor \sim r)]$

SOLUTION

(a) Since p, q, and r are simple statements, they are statement forms. Since q is a statement form, so is $\sim q$. Then $p \land \sim q$ is a statement form. Since r is a statement form, $(p \land \sim q) \rightarrow r$ is a statement form.

(b) Since r is a statement form, so is $\sim r$. Since both q and $\sim r$ are statement forms, $q \lor \sim r$ is a statement form. However, p is also a statement form; thus, $p \rightarrow (q \lor \sim r)$ is a statement form. The negation of a statement form is a statement form, so $\sim[p \rightarrow (q \lor \sim r)]$ is also.   **>> Now Try Exercise 1**

The main mechanism for determining the truth values of statement forms is a **truth table**.

Consider any simple statement p. Then p has one of the two truth values T (TRUE) or F (FALSE). We can list the possible values of p in a simple table:

p
T
F

Clearly, $\sim p$ (the negation of p) is a form derived from p and has two possible values also. When p has the truth value T, $\sim p$ has the truth value F, and vice versa.

We represent the truth value of $\sim p$ in a truth table.

Truth Table for the Negation

p	$\sim p$
T	F
F	T

The statement form $p \land q$ (p and q) is made up of two statements denoted by p and q and the connective $\land$. The statement p can have one of the truth values T or F. Similarly, q can have one of the truth values T or F. Hence, by the multiplication principle, there are four possible pairs of truth values for p and q.

The **conjunction** of p and q, $p \land q$ (p and q), is true if and only if both p is true *and q is true*. The truth table for the conjunction is

Truth Table for the Conjunction

p	q	$p \land q$
T	T	T
T	F	F
F	T	F
F	F	F

The **disjunction** of p and q, $p \lor q$ (p or q), is true if either p is true *or* q is true or both p and q are true.

Truth Table for the Disjunction

p	q	$p \lor q$
T	T	T
T	F	T
F	T	T
F	F	F

In English, the word "or" is ambiguous; we have to distinguish between the exclusive and inclusive "or." For example, in the sentence "Ira will go to either Princeton or Stanford," the word "or" is assumed to be exclusive, since Ira will choose one of the schools but not both. On the other hand, in the sentence "Diana is smart or Diana is rich," the "or" is inclusive, since either Diana is smart, Diana is rich, or Diana is both smart and rich. The mathematical statement form $p \lor q$ uses the inclusive "or."

What we have described so far is a system of calculating the truth value of a statement form on the basis of the truth of its component statement forms. The beauty of truth tables is that they can be used to determine the truth values of more elaborate statement forms by reapplying the basic rules.

EXAMPLE 2	**Constructing a Truth Table** Construct a truth table for the statement form

$$\sim(p \lor q).$$

SOLUTION   We write all of the components of the statement form so that they can be evaluated for the four possible pairs of values for p and q. Note that we enter the truth values for p and q in the same order as in the previous tables. It is a good idea to use this order all the time:

p	q
T	T
T	F
F	T
F	F

The truth table for the form $\sim(p \lor q)$ contains a column for each calculation.

p	q	$p \lor q$	$\sim(p \lor q)$
T	T	T	F
T	F	T	F
F	T	T	F
F	F	F	T

The third column gives the truth values of the disjunction of the first two columns, and the fourth column gives the negation of the third column. The statement form $\sim(p \lor q)$ is TRUE in one case—when both p and q are FALSE. Otherwise, $\sim(p \lor q)$ is FALSE.

» Now Try Exercise 5

EXAMPLE 3	**Constructing a Truth Table** Construct a truth table for the statement form

$$\sim(p \land \sim q) \lor p.$$

FIRST SOLUTION

p	q	$\sim q$	$p \land \sim q$	$\sim(p \land \sim q)$	$\sim(p \land \sim q) \lor p$
T	T	F	F	T	T
T	F	T	T	F	T
F	T	F	F	T	T
F	F	T	F	T	T
(1)	(2)	(3)	(4)	(5)	(6)

We fill in each column by using the rules already established. Column 3 is the negation of the truth values in column 2. Column 4 gives the conjunction of columns 1 and 3. Column 5 is the negation of the statement form whose values appear in column 4. Finally, column 6 is the disjunction of columns 5 and 1.

SECOND SOLUTION

There is a more efficient way to prepare the truth table of Example 3. Use the same order for entering possible truth values of p and q. Put the statement form at the top of the table. Fill in columns under each operation as you need them, working from the inside out. We label the columns in the order in which we fill them in.

p	q	$\sim(p \wedge \sim q) \vee p$
T	T	T F F T
T	F	F T T T
F	T	T F F T
F	F	T F T T
(1)	(2)	(5) (4) (3) (6)

We entered the values of p and q in columns 1 and 2 and then entered the values for $\sim q$ in column 3. We used the conjunction of columns 1 and 3 to fill in column 4, applied the negation to column 4 to get column 5, and applied the disjunction of columns 5 and 1 to get the values in column 6.

We see that this statement form has truth value TRUE no matter what the truth values of the statements p and q are. Such a statement is called a tautology.

» Now Try Exercise 7

DEFINITION A **tautology** is a statement form that always has truth value TRUE, regardless of the truth values of the individual statement variables it contains.

DEFINITION A **contradiction** is a statement form that always has truth value FALSE, regardless of the truth values of the individual statement variables it contains.

EXAMPLE 4 **Constructing Truth Tables** Construct the truth table for the statement form

$$(\sim p \vee \sim q) \wedge (p \wedge \sim q).$$

SOLUTION

Again, we will use the short form of the table, putting the statement form at the top of the table and labeling each column in the order in which it was completed.

p	q	$(\sim p \vee \sim q) \wedge (p \wedge \sim q)$
T	T	F F F F T F F
T	F	F T T T T T T
F	T	T T F F F F F
F	F	T T T F F F T
(1)	(2)	(3) (5) (4) (7) (6)

There are some unlabeled columns. These were just recopied for convenience and clarity. It is not necessary to include them. Column 7, the conjunction of columns 5 and 6, tells us that the statement form is TRUE only when p is TRUE and q is FALSE.

» Now Try Exercise 9

NOTE We will often use the short forms of truth tables in this chapter. In this form, the truth values for the compound statement appear in the highest-numbered column. If using the short version, it is important that the final truth values are clearly marked, for example, by numbering the columns or circling or shading the final column. **«**

EXAMPLE 5   **Compressing Truth Tables** Find the truth table for the statement form

$$(p \lor q) \land {\sim}(p \land q).$$

SOLUTION

p	q	$(p \lor q) \land {\sim}(p \land q)$			
T	T	T	F F	T
T	F	T	T T	F
F	T	T	T T	F
F	F	F	F T	F
(1)	(2)	(3)	(6)(5)	(4)

Column 6, the conjunction of columns 3 and 5, gives the truth values of the statement form.   «

The statement form in Example 5 can be read as "p or q but not both p and q." This is the **exclusive or**. It a useful connective and is denoted by $\oplus$ or $\veebar$. The statement form $p \oplus q$ is true exactly when p is true or q is true but not when both p and q are true. The truth table of $p \oplus q$ is

Truth Table for the Exclusive Or

p	q	$p \oplus q$
T	T	F
T	F	T
F	T	T
F	F	F

If there are three simple statements in a statement form and we denote them p, q, and r, then each could take on the truth values TRUE OR FALSE. Hence, by the multiplication principle, there are $2 \times 2 \times 2 = 8$ different assignments to the three variables together. There are eight lines in a truth table for such a statement form. Again, in the truth table, we will list all of the Ts and then all of the Fs for the first variable. Then we alternate TT followed by FF in the second column. Finally, for the third variable, we alternate T with F. This ensures that we have listed all possibilities and facilitates comparison of the final results.

The formal evaluation of the truth table of a statement form can help in deciding the truth value when particular sentences are substituted for the statement variables.

EXAMPLE 6   **Logically True or False** Let p denote the statement "London is the capital of England," and let q denote the statement "Venice is the capital of Italy." Determine the truth value of each of the following statements:

(a) $p \land q$ (b) $p \lor q$ (c) ${\sim}p \land q$
(d) ${\sim}p \lor {\sim}q$ (e) ${\sim}(p \land q)$ (f) $p \oplus q$

SOLUTION   First, we note that p has the truth value TRUE and q has the truth value FALSE (Rome is the capital of Italy).

(a) Since q is FALSE, $p \land q$ is FALSE.
(b) Since p is TRUE, $p \lor q$ is TRUE.
(c) Both ${\sim}p$ and q are FALSE, so ${\sim}p \land q$ is FALSE.
(d) Since ${\sim}q$ is TRUE, ${\sim}p \lor {\sim}q$ is TRUE.
(e) From (a), we see that $p \land q$ is FALSE. Therefore, ${\sim}(p \land q)$ is TRUE.
(f) Since p is TRUE and q is FALSE, $p \oplus q$ is TRUE.

» Now Try Exercise 35

EXAMPLE 7	**Constructing Truth Tables for Three Simple Statements** Construct a truth table for $(p \vee q) \wedge [(p \vee r) \wedge \sim r]$.

SOLUTION   Note the order in which the Ts and Fs are listed in the first three columns. The columns are numbered in the order in which they were filled.

p	q	r	\multicolumn{5}{c}{$(p \vee q) \wedge [(p \vee r) \wedge \sim r]$}
T	T	T	T	F	T	F	F
T	T	F	T	T	T	T	T
T	F	T	T	F	T	F	F
T	F	F	T	T	T	T	T
F	T	T	T	F	T	F	F
F	T	F	T	F	F	F	T
F	F	T	F	F	T	F	F
F	F	F	F	F	F	F	T
(1)	(2)	(3)	(4)	(8)	(5)	(7)	(6)

The statement form has truth value TRUE if and only if p is TRUE and r is FALSE (regardless of the truth value of q).   **>> Now Try Exercise 25**

Logical Equivalence

Different statement forms may have the same truth tables. For example, consider the truth tables for the statement forms $\sim(p \wedge q)$ and $\sim p \vee \sim q$:

p	q	$(p \wedge q)$	$\sim p$	$\sim q$	$\sim(p \wedge q)$	$\sim p \vee \sim q$
T	T	T	F	F	F	F
T	F	F	F	T	T	T
F	T	F	T	F	T	T
F	F	F	T	T	T	T
(1)	(2)	(3)	(4)	(5)	(6)	(7)

Comparing columns (6) and (7), we see that the two statement forms have the same final column in their truth tables.

> **DEFINITION** **Logical Equivalence** Two statement forms that have the same truth tables are called **logically equivalent.**

The fact that $\sim(p \wedge q)$ and $\sim p \vee \sim q$ are equivalent is important and is known as one of *De Morgan's laws*. We will explore logical equivalence further in Section 11.4. Most notably, two logically equivalent statement forms can be used to represent the same English sentence. Therefore, logical equivalence is an important tool in understanding the relationship between English and symbolic statements.

Logic and Computer Languages

In many computer languages, the logical connectives are incorporated into programs that depend on logical decision making. The symbols are replaced by their original English words in some computer languages; for example,

AND	$\wedge$	
OR	$\vee$	
NOT	$\sim$	
XOR	$\oplus$	(XOR stands for "exclusive or").

If the statements p and q are assigned truth values T and F, then the truth tables for the connectives AND, OR, XOR, and NOT are as shown in Fig. 1.

p	q	p AND q	p OR q	p XOR q
T	T	T	T	F
T	F	F	T	T
F	T	F	T	T
F	F	F	F	F

p	NOT p
T	F
F	T

Figure 1

INCORPORATING TECHNOLOGY

Graphing calculators can evaluate logical expressions involving the following four logical operators: AND, OR, XOR, NOT. See Fig. 2, which was obtained on a TI-84 Plus by pressing ⎡2nd⎤ [TEST] ▶. TRUE and FALSE are represented by the numbers 1 and 0, respectively. See Figs. 3 and 4.

Figure 2

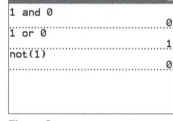

Figure 3

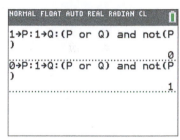

Figure 4

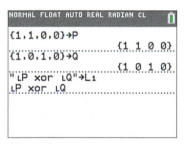

Figure 5

An entire truth table can be displayed by creating lists of zeros and ones for the variables and attaching a formula containing logical operators to a list name (see Fig. 6). (The lists **P** and **Q** can be created from the home screen.) A formula is attached to **L1** by entering a logical expression as in Fig. 5. The list names **LP** and **LQ** are displayed from the LIST menu. The column headers in Fig. 6 were specified by pressing ⎡STAT⎤ 5 to display the word SetUpEditor and then executing **SetUpEditor P,Q,L1**.

An Excel spreadsheet can display an entire truth table for logical expressions involving AND, OR, NOT, and XOR. If x, y, ..., z are addresses of cells containing the words TRUE or FALSE, then the truth values of the four logical functions are given in Table 1.

Figure 6

Table 1

Function	Truth Value	Interpretation
AND(x,y,...,z)	(value of x) $\wedge$ (value of y) $\wedge \ldots \wedge$ (value of z)	Returns TRUE if $x, y, \ldots, z$ are all TRUE; returns FALSE otherwise
OR(x,y,...,z)	(value of x) $\vee$ (value of y) $\vee \ldots \vee$ (value of z)	Returns FALSE if $x, y, \ldots, z$ are all FALSE; returns TRUE otherwise
NOT(x)	$\sim$(value of x)	Returns TRUE if x is FALSE; returns FALSE if x is TRUE
XOR(x,y,...,z)	(value of x) $\oplus$ (value of y) $\oplus \ldots \oplus$ (value of z)	Returns TRUE if an odd number of $x, y, \ldots, z$ are TRUE; returns FALSE otherwise

Figure 7 shows the truth table for the logical expression in Example 5. After the formula was entered into cell C1, the cell was selected and its fill handle was dragged down to C4.

C1			▾ ●	f_x	=AND(OR(A1,B1),NOT(AND(A1,B1)))		
◢	A	B	C	D	E	F	G
1	TRUE	TRUE	FALSE				
2	TRUE	FALSE	TRUE				
3	FALSE	TRUE	TRUE				
4	FALSE	FALSE	FALSE				

Figure 7

✳ WolframAlpha The truth table in Fig. 8 can be generated with the instruction

p Xor ((p Or q) And (Not p)).

p	q	$p \veebar ((p \vee q) \wedge \neg p)$
T	T	T
T	F	T
F	T	T
F	F	F

Figure 8

The function BooleanMinimize can often find a statement form that is simpler but logically equivalent to a given statement form. (In computer languages, expressions involving logical connectives like AND, OR, or XOR are called **Boolean**.) For example, the value of

BooleanMinimize((p Or q) And (p Or Not q))

is **p**.

Check Your Understanding 11.2

Solutions can be found following the section exercises.

1. Construct the truth table for $(p \vee \sim r) \wedge q$.

2. Construct the truth table for $p \oplus \sim q$.

3. Let p denote "May follows April," and let q denote "June follows May." Determine the truth values of the following:
   (a) $p \wedge \sim q$ (b) $\sim(p \vee \sim q)$
   (c) $p \oplus q$ (d) $\sim[(p \wedge q) \oplus \sim q]$

EXERCISES 11.2

In Exercises 1–4, show that the expressions are statement forms according to the definition. Assume that p, q, and r are simple statements.

1. $\sim(p \wedge \sim r) \vee q$ 2. $(p \rightarrow \sim q) \wedge r$

3. $(\sim p \vee r) \rightarrow (q \wedge r)$ 4. $\sim[(p \wedge r) \rightarrow q]$

In Exercises 5–28, construct truth tables for the given statement forms.

5. $p \wedge \sim q$ 6. $\sim(p \vee \sim q) \wedge (p \wedge \sim q)$

7. $(p \vee \sim q) \wedge q$ 8. $(p \wedge q) \vee (p \wedge \sim q)$

9. $\sim[(p \vee q) \wedge (p \wedge q)]$ 10. $(p \vee \sim q) \oplus \sim p$

11. $p \oplus (\sim p \vee q)$ 12. $(p \wedge q) \wedge r$

13. $(p \wedge \sim r) \oplus q$ 14. $\sim(p \wedge r) \vee q$

15. $\sim[(p \wedge r) \vee q]$ 16. $p \wedge \sim p$

17. $p \vee \sim p$ 18. $(p \vee q) \oplus \sim r$

19. $p \oplus (q \oplus r)$ 20. $p \vee (q \wedge r)$

21. $(p \vee q) \wedge (p \vee r)$ 22. $(p \vee q) \wedge (p \vee \sim r)$

23. $(p \vee q) \wedge \sim(p \vee q)$ 24. $(\sim p \vee q) \wedge r$

25. $\sim(p \vee q) \wedge r$ 26. $\sim[(p \vee q) \wedge r]$

In Exercises 27–30, determine whether the statement forms are logically equivalent.

27. $\sim p \wedge \sim q$ and $\sim(p \wedge q)$

28. $p \wedge p$ and p

29. $\sim(p \oplus q)$ and $(p \wedge q) \vee \sim(p \wedge q)$

30. $(p \wedge q) \vee r$ and $p \wedge (q \vee r)$

31. How many possible truth tables can you construct for statement forms involving two variables p and q? (*Hint:* Consider the number of ways to complete the last column of a truth table for each of the four possible pairs of values for p and q.)

32. How many possible truth tables can you construct for statement forms involving the three variables p, q, and r?

33. The Sheffer Stroke (NAND) Define $p \mid q$ by the following truth table:

p	q	$p \mid q$
T	T	F
T	F	T
F	T	T
F	F	T

This connective is called the *Sheffer stroke* (or NAND). Construct truth tables for the following:

(a) $p \mid p$ (b) $(p \mid p) \mid (q \mid q)$

(c) $(p \mid q) \mid (p \mid q)$ (d) $p \mid [(p \mid q) \mid q]$

34. The Symbol ⊖ (XNOR) Define the connective ⊖ by the following truth table:

p	q	$p \ominus q$
T	T	T
T	F	F
F	T	F
F	F	T

Construct the truth tables of the following:

(a) $p \ominus \sim q$ (b) $(p \ominus q) \ominus r$

(c) $p \ominus (q \ominus r)$ (d) $\sim(p \ominus q) \wedge (p \oplus q)$

35. Let p denote "John Lennon was a member of the Beatles," and let q denote the statement "The Beatles came from Spain." Determine the truth values of the following:

(a) $p \vee \sim q$ (b) $\sim p \wedge q$

(c) $p \oplus q$ (d) $\sim p \oplus q$

(e) $\sim(p \oplus q)$ (f) $(p \vee q) \oplus \sim q$

36. Let m denote the statement "The Magna Carta was signed in 1995," and let p denote "The Parthenon is located in Italy." Determine the truth value of each of the following:

(a) $m \wedge p$ (b) $m \vee \sim p$

(c) $\sim m \wedge \sim p$ (d) $\sim m \oplus \sim p$

37. Suppose $p \vee q$ and $p \vee \sim q$ are both TRUE. What is the truth value of p?

38. Suppose $\sim p \oplus q$ is TRUE and $p \wedge q$ is FALSE. What is the truth value of p?

39. Suppose $p \wedge r$ is FALSE and $q \wedge r$ is TRUE. What is the truth value of p?

40. Suppose $p \oplus \sim r$ is TRUE and $p \wedge q$ is TRUE. What is the truth value of r?

TECHNOLOGY EXERCISES

In Exercises 41 and 42, what value will be displayed when the Enter key is pressed?

41.

```
NORMAL FLOAT AUTO REAL RADIAN CL

1→P:0→Q:(P xor Q) and not(
Q)
```

42.

```
NORMAL FLOAT AUTO REAL RADIAN CL

1→P:0→Q:(P and Q) or not(P
)
```

In Exercises 43–46, use technology to determine the value of the logical expression when **(a)** p is TRUE, q is TRUE, and **(b)** p is TRUE, q is FALSE.

43. $(p \vee q) \wedge \sim p$ **44.** $(p \wedge q) \vee \sim q$

45. $\sim q \wedge (\sim p \wedge q)$ **46.** $\sim p \wedge (p \vee \sim q)$

In Exercises 47–50, display a truth table for the given expression.

47. $p \wedge \sim q$ **48.** $\sim(p \vee q)$

49. $p \wedge (\sim q \vee r)$ **50.** $(p \wedge \sim r) \vee (\sim q \wedge r)$

51. Use the Wolfram|Alpha function BooleanMinimize to find a simpler Boolean expression equivalent to the Boolean expression in Exercise 43.

52. Use the Wolfram|Alpha function BooleanMinimize to find a simpler Boolean expression equivalent to the Boolean expression in Exercise 44.

Solutions to Check Your Understanding 11.2

1.

p	q	r	$(p \vee \sim r) \wedge q$
T	T	T	T F T
T	T	F	T T T
T	F	T	T F F
T	F	F	T T F
F	T	T	F F F
F	T	F	T T T
F	F	T	F F F
F	F	F	T T F
(1)	(2)	(3)	(5) (4) (6)

2.

p	q	$p \oplus \sim q$
T	T	T F
T	F	F T
F	T	F F
F	F	T T
(1)	(2)	(4) (3)

3. Both p and q are TRUE. Thus, (a) is FALSE. Item (b) is the negation of the statement $(p \vee \sim q)$, which is TRUE, since p is TRUE. Hence, the statement in (b) is FALSE. Since both p and q are TRUE, $p \oplus q$ is FALSE. To analyze (d), we consider that $(p \wedge q)$ is TRUE and $\sim q$ is FALSE, so $[(p \wedge q) \oplus \sim q]$ is TRUE. Its negation is FALSE.

11.3   Implication

We are quite familiar with implications. For example, "If Jay is caught smoking in the restroom, then he is suspended from school" is an implication. Note that, although Jay may avoid being caught smoking in the restroom, he still might be suspended for some other infraction. Symbolically, we can denote the statement "Jay is caught smoking in the restroom" by p, and can denote "Jay is suspended from school" by q. The implication is represented by the **conditional connective** $\rightarrow$. Recall the definition of this connective from Section 12.1:

> **Implication**   The symbol $\rightarrow$ represents the English words "If . . . , then," in that the symbolic statement $p \rightarrow q$ can be read as "If p, then q."

The truth table for the statement form using the conditional is as follows:

Truth Table for the Conditional	p	q	$p \rightarrow q$
	T	T	T
	T	F	F
	F	T	T
	F	F	T

There are several ways to read $p \rightarrow q$ in English:

1. p implies q.
2. If p, then q.
3. p only if q.
4. q, if p.
5. p is sufficient for q.
6. q is a necessary condition for p.

We call p the **hypothesis** and q the **conclusion**. The conditional statement form $p \rightarrow q$ has truth value FALSE only when the hypothesis has truth value TRUE and the conclusion has truth value FALSE.

CAUTION   The last two rows of the truth table may seem paradoxical. They state that, whenever the hypothesis has truth value FALSE, the whole conditional has truth value TRUE. For instance, the statement "If elephants can fly, then the Washington Monument is a million feet tall" has truth value TRUE. «

In the statement "If Jay is caught smoking in the restroom, then he is suspended from school," there is a causal relationship between the hypothesis and the conclusion. That is, Jay's suspension resulted from his smoking in the restroom. However, in general, there need not be any relationship between the two. For instance, the statement "If Chicago is in Illinois, then Dallas is in Texas" has truth value TRUE. However, Chicago being in Illinois did not cause Dallas to be in Texas.

EXAMPLE 1   **Hypothesis and Conclusion**   In each of the following statements, determine the hypothesis and the conclusion:
(a) Bill goes to the party only if Greta goes to the party.
(b) Sue goes to the party if Craig goes to the party.
(c) For 6 to be even, it is sufficient that its square, 36, be even.

SOLUTION   (a) The statement is in the form of *p* only if *q*. Thus, the hypothesis is "Bill goes to the party" and the conclusion is "Greta goes to the party." We could rewrite the statement as "If Bill goes to the party, then Greta goes to the party." The statement "If Bill goes to the party, then Greta goes to the party" seems to mean that Greta is following Bill around, whereas the statement "Bill goes to the party only if Greta goes to the party" makes the romance seem quite the opposite. However, this last statement should be interpreted as follows: If "Bill goes to the party" is true, then that means that Greta must also have gone—for that was the ONLY reason he would go. The unemotional interpretation of the logical form shows the equivalence of the two statements.

(b) The statement is of the form *q*, if *p*. It has hypothesis "Craig goes to the party" and conclusion "Sue goes to the party."

(c) This is of the form *p* is sufficient for *q*, with the hypothesis "The square of the integer 6 is even" and the conclusion "The integer 6 is even."

>> *Now Try Exercise 39*

EXAMPLE 2   **Truth Value of an Implication**   Determine whether each of the following statements is true or false:
(a) If Paris is in France, then the Louvre is in Paris.
(b) If the Louvre is in Paris, then $2 + 3 = 7$.
(c) If $2 + 3 = 7$, then the Louvre is in Paris.
(d) If $2 + 3 = 7$, then Paris is in Spain.
(e) Paris is in Spain only if $2 + 3 = 7$.

SOLUTION   The statement (a) is TRUE because both the hypothesis and the conclusion are TRUE. The statement (b) is FALSE because the hypothesis is TRUE but the conclusion is FALSE. The last three statements are TRUE because the hypothesis in each is FALSE. Note that the last statement has hypothesis "Paris is in Spain" and conclusion "$2 + 3 = 7$."

>> *Now Try Exercise 37*

EXAMPLE 3   **Using a Truth Table for an Implication**   Construct the truth table of the statement "If the president dies or becomes incapacitated, then the vice president becomes president."

SOLUTION   Let *p* be the statement "The president dies." Let *q* be the statement "The president becomes incapacitated." Let *r* be the statement "The vice president becomes president." The statement form is $(p \lor q) \to r$. A truth table for the statement form is as follows:

p	*q*	*r*	$p \lor q$	$(p \lor q) \to r$
T	T	T	T	T
T	T	F	T	F
T	F	T	T	T
T	F	F	T	F
F	T	T	T	T
F	T	F	T	F
F	F	T	F	T
F	F	F	F	T

The statement is FALSE only in the cases in which the president dies or becomes incapacitated but the vice president does not become president.   >> *Now Try Exercise 5*

It is important to note that $p \to q$ is not logically equivalent to $q \to p$. We should not confuse the hypothesis and the conclusion in an implication. An example will demonstrate the difference between $p \to q$ and $q \to p$. Consider the implication "If the truck carries ice cream, then it is refrigerated." If the implication is in the form $p \to q$, then $q \to p$ is the implication "If the truck is refrigerated, then it carries ice cream." These implications need not have the same truth values. The first implication is probably true; the second need not be. The truth tables of $p \rightarrow q$ and $q \rightarrow p$ show the differences.

p	q	$p \rightarrow q$	$q \rightarrow p$
T	T	T	T
T	F	F	T
F	T	T	F
F	F	T	T

The implication $q \rightarrow p$ is called the **converse** of the implication $p \rightarrow q$. Thus, given the statement "If the Los Angeles Dodgers won the pennant, then some games of the World Series are in California," its converse is "If some games of the World Series are in California, then the Los Angeles Dodgers won the pennant." While the original statement is true, the converse is not. There are situations in which the hypothesis of the converse is true, but the conclusion is false. Some games of the World Series may be in California because the San Francisco Giants, the Oakland Athletics, the San Diego Padres, or the Los Angeles Angels of Anaheim, rather than the Los Angeles Dodgers, won the pennant.

EXAMPLE 4

The Conjunction of a Statement and Its Converse Construct the truth table for the conjunction of $p \rightarrow q$ and its converse. That is, find the truth table of the statement form $(p \rightarrow q) \wedge (q \rightarrow p)$.

SOLUTION

p	q	$p \rightarrow q$	$\wedge$	$q \rightarrow p$
T	T	T	T	T
T	F	F	F	T
F	T	T	F	F
F	F	T	T	T
(1)	(2)	(3)	(5)	(4)

We note that the statement form has truth value TRUE whenever p and q have the same truth value: either both TRUE or both FALSE. **» Now Try Exercise 41**

The statement form $(p \rightarrow q) \wedge (q \rightarrow p)$ is referred to as the **biconditional**, which we write as $p \leftrightarrow q$. The statement $p \leftrightarrow q$ is read as "p if and only if q" or "p is necessary and sufficient for q." The truth table for the biconditional is as follows:

Truth Table for the Biconditional

p	q	$p \leftrightarrow q$
T	T	T
T	F	F
F	T	F
F	F	T

The biconditional is a statement form simply because it can be expressed as $(p \rightarrow q) \wedge (q \rightarrow p)$. In some sense, which we will discuss further in the next section, the biconditional expresses a kind of equivalence of p and q. That is, the biconditional statement form is TRUE whenever p and q are both TRUE or both FALSE. The biconditional is FALSE whenever p and q have different truth values.

Implication and Common Language

The mathematical uses of the conditional and the biconditional are very precise. However, everyday speech is rarely as precise. For example, consider the statement "Peter gets dessert only if he eats his broccoli." Technically, this means that if Peter gets dessert, then he eats his broccoli. However, Peter and his parents probably interpret the statement to mean that Peter gets dessert if and only if he eats his broccoli. This imprecision in language is sometimes confusing; we make every effort to avoid confusion in mathematics by adhering to strict rules for using the conditional and biconditional.

Order of Precedence

Using connectives, we have seen how to string together several simple statements into compound and fairly complex statement forms. Such forms should not be confusing: What is the meaning of $\sim p \vee q$? Do we mean $\sim(p \vee q)$, or do we mean $(\sim p) \vee q$?

| EXAMPLE 5 | **Comparing Truth Tables** Compare the truth tables of $(\sim p) \vee q$ and $\sim(p \vee q)$. |

SOLUTION

p	q	$(\sim p) \vee q$	$\sim(p \vee q)$
T	T	F T	F T
T	F	F F	F T
F	T	T T	F T
F	F	T T	T F
(1)	(2)	(3) (4)	(6) (5)

Note that columns 4 and 6 are different. Also, notice that column 4 shows that $p \rightarrow q$ has the same truth table as $(\sim p) \vee q$.   **» Now Try Exercise 13**

To clarify matters and to avoid the use of too many parentheses, we define an *order of precedence* for the connectives. This dictates which of the connectives should be applied first.

> **DEFINITION** The **order of precedence** for logical connectives is
>
> $$\sim, \quad \wedge, \quad \vee, \quad \rightarrow, \quad \leftrightarrow.$$

One applies $\sim$ first, then $\wedge$, and so on. If there is any doubt, insert parentheses to clarify the statement form. Using the order of precedence, we see that $\sim p \vee q$ is $(\sim p) \vee q$.

| EXAMPLE 6 | **Order of Precedence** Insert parentheses in the statement to show the proper order for the application of the connectives. |

$$p \wedge q \vee r \rightarrow \sim s \wedge r$$

SOLUTION   Reading from the left, apply the $\sim$ first, so $\sim s \wedge r$ becomes $(\sim s) \wedge r$. Then scan for the connective $\wedge$, since that is next in the precedence list. So $p \wedge q \vee r$ becomes $(p \wedge q) \vee r$. Apply the $\rightarrow$ last. The statement can be written as

$$[(p \wedge q) \vee r] \rightarrow [(\sim s) \wedge r]. \qquad \text{**» Now Try Exercise 15**}$$

Parentheses have highest priority in the precedence. Clearly,

$$p \wedge (q \vee (r \rightarrow \sim s)) \wedge r$$

has the same symbols as the statement of Example 6, but the parentheses have defined a different statement form.

That the statement form $p \vee q \vee r$ may be written either as $(p \vee q) \vee r$ or as $p \vee (q \vee r)$ is shown in Section 11.4. The statement form $p \wedge q \wedge r$ can be written as $(p \wedge q) \wedge r$ or as $p \wedge (q \wedge r)$, which is also shown in Section 11.4.

Implications and Computer Languages

The logical connective → is used in writing computer programs, although in some programming languages, it is expressed with the words IF, THEN. Thus, we might see an instruction of the form "IF $a = 3$ THEN LET $s = 0$." Strictly speaking, this is not a statement form, because "LET $s = 0$" is not a statement. However, that command means "the computer will set $s = 0$." Thus, if the value of a is 3, the computer sets s to 0. If $a \neq 3$, the program does not change the value of s. The instruction

"IF . . . THEN . . . ELSE"

allows for a branch in the program. For example,

"IF $a = 3$ THEN LET $s = 0$ ELSE LET $s = 1$"

assigns value 0 to s if $a = 3$ and assigns value 1 to s if $a \neq 3$.

EXAMPLE 7 **Implication and Computing** For the given input values of A and B, use the program to determine the value of C. The asterisk (∗) denotes multiplication.

IF $(A * B) + 6 \geq 10$

THEN LET $C = A * B$

ELSE LET $C = 10$

(a) A = −2, B = −7 (b) A = −2, B = 3 (c) A = 2, B = 2

SOLUTION (a) We determine that $(A * B) + 6 = 14 + 6 = 20$. Hence, we set $C = A * B = 14$.
(b) In this case, $(A * B) + 6 = 0$. So we set $C = 10$.
(c) $(A * B) + 6 = 10$. Hence, $C = A * B = 4$.   **» Now Try Exercise 47**

INCORPORATING
TECHNOLOGY

A truth table for the implicational connective can be displayed as in Fig. 5 of the previous section by specifying **L1="not(LP) or LQ"**.

A truth table for the implicational connective can be displayed as in Fig. 7 of the previous section by using the setting **=OR(NOT(A1),B1)** for C1.

✳WolframAlpha A truth table for the implicational connective can be displayed with the instruction **p Implies q** or the instruction **if p then q**.

Check Your Understanding 11.3

Solutions can be found following the section exercises.

1. Let p denote the statement "A square is a rectangle" and let q denote the statement "A rectangle has four sides." Write the statements in symbolic form. Name the hypothesis and the conclusion in each statement.
   (a) A square is a rectangle if a rectangle has four sides.
   (b) A rectangle has four sides if a square is a rectangle.
   (c) A rectangle has four sides only if a square is a rectangle.
   (d) A square is a rectangle is sufficient to show that a rectangle has four sides.

   (e) For a rectangle to have four sides, it is necessary for a square to be a rectangle.

2. Let p denote the statement "There are 48 states in the United States," let q denote the statement "The American flag is red, white, and blue," and let r denote the statement "Maine is on the East Coast." Determine the truth value of each of the following statement forms:
   (a) $p \wedge q \rightarrow r$ (b) $p \vee q \rightarrow r$
   (c) $\sim p \wedge q \leftrightarrow r$ (d) $p \wedge \sim q \rightarrow r$

EXERCISES 11.3

Construct a truth table for each of the statement forms in Exercises 1–12.

1. $\sim p \rightarrow \sim q$
2. $p \vee (q \rightarrow \sim r)$
3. $(p \oplus q) \rightarrow q$
4. $\sim q \rightarrow \sim p$
5. $(\sim p \wedge q) \rightarrow r$
6. $\sim (p \rightarrow q)$
7. $(p \rightarrow q) \leftrightarrow (\sim p \vee q)$
8. $p \oplus (q \rightarrow r)$
9. $(p \rightarrow q) \rightarrow r$
10. $p \rightarrow (q \rightarrow r)$
11. $(p \vee q) \leftrightarrow (p \wedge q)$
12. $[(\sim p \wedge q) \oplus q] \rightarrow p$

For each of the Exercises 13–16, insert parentheses in accordance with the precedence of logical operations.

13. $\sim p \wedge \sim q \rightarrow \sim p \wedge q$ **14.** $p \wedge q \rightarrow p \vee \sim q$

15. $\sim p \wedge \sim q \vee r \rightarrow \sim q \wedge r$ **16.** $p \vee \sim q \wedge r \rightarrow p \vee r$

Let p denote the TRUE statement "Abraham Lincoln was the 16th president of the United States," and let q denote the FALSE statement "The battle of Gettysburg took place at the O.K. Corral." Determine the truth value of each of the statement forms in Exercises 17–26.

17. $\sim p \rightarrow q$ **18.** $p \rightarrow q$

19. $q \rightarrow p$ **20.** $p \rightarrow \sim q$

21. $(p \oplus q) \rightarrow p$ **22.** $p \leftrightarrow q$

23. $(p \wedge \sim q) \rightarrow (\sim p \oplus q)$ **24.** $(p \vee q) \leftrightarrow q$

25. $p \rightarrow [p \wedge (p \oplus q)]$ **26.** $[p \wedge (\sim q \rightarrow \sim p)] \oplus q$

In Exercises 27–34, write the statement forms in symbols, using the conditional ($\rightarrow$) or the biconditional ($\leftrightarrow$) connective. Name the hypothesis and the conclusion in each conditional form. Let p be the statement "Sally studied" and q the statement "Sally passes."

27. Sally studied if and only if Sally passes.

28. If Sally studied, then Sally passes.

29. Sally passes only if Sally studied.

30. If Sally passes, then Sally studied.

31. Sally's studying is necessary for Sally to pass.

32. That Sally studied is sufficient for Sally to pass.

33. If Sally did not study, then Sally does not pass.

34. Sally passes implies that Sally studied.

In Exercises 35–38, let p denote the TRUE statement "The die is fair" and let q denote the TRUE statement "The probability of a 2 is $\frac{1}{6}$." Write each of the statement forms in symbols, name the hypothesis and the conclusion in each, and determine whether the statement is TRUE or FALSE.

35. If the die is not fair, the probability of a 2 is not $\frac{1}{6}$.

36. If the die is fair, the probability of a 2 is $\frac{1}{6}$.

37. The die is not fair if the probability of a 2 is not $\frac{1}{6}$.

38. The probability of a 2 is $\frac{1}{6}$ only if the die is fair.

39. Give the hypothesis and the conclusion in each statement.
   (a) I will run a marathon only if Amy watches my dogs.
   (b) In order for a student to pass the course, it is sufficient that they earn a B.
   (c) The football team wears green uniforms if it is my favorite.
   (d) Isaac has wrinkled fingers implies that he was in the bathtub.

40. Give the hypothesis and the conclusion in each statement.
   (a) If I drive too fast, I receive a traffic ticket.
   (b) I drive too fast only if I receive a traffic ticket.
   (c) Having a valid license is necessary for being a delivery driver.
   (d) A cow has clear eyes and a wet nose if it is healthy.

41. State the converse of each of the following statements:
   (a) City Sanitation collects the garbage if the mayor calls.
   (b) The price of beans goes down only if there is no drought.
   (c) If goldfish swim in Lake Erie, then Lake Erie is fresh water.
   (d) If tap water is not salted, then it boils slowly.

42. State the converse of each of the following statements:
   (a) If Jane runs 20 miles, then Jane is tired.
   (b) Cindy loves Fred only if Fred loves Cindy.
   (c) Jon cashes a check if the bank is open.
   (d) Errors are clear only if the documentation is complete.
   (e) Sally's eating the vegetables is a necessary condition for Sally's getting dessert.
   (f) Sally's eating the vegetables is sufficient for Sally's getting dessert.

43. Determine the output value of A in the program for the given input values of X and Y.

$$\text{LET } Z = X + Y$$
$$\text{IF } (Z \neq 0) \text{ AND } (X > 0)$$
$$\text{THEN LET } A = 6$$
$$\text{ELSE LET } A = 4$$

   (a) X = 0, Y = 0   **(b)** X = 8, Y = −8
   (c) X = −3, Y = 3   **(d)** X = −3, Y = 8
   (e) X = 8, Y = −3   **(f)** X = 3, Y = −8

44. For the given input values of A and B, use the program to find the output values of X.

$$\text{LET } C = A + B$$
$$\text{IF } ((A > 0) \text{ OR } (B > 0)) \text{ AND } (C > 0)$$
$$\text{THEN LET } X = 100$$
$$\text{ELSE LET } X = -100$$

   (a) A = 2, B = 2   **(b)** A = 2, B = −2
   (c) A = 2, B = −5   **(d)** A = −2, B = −5
   (e) A = −5, B = 3   **(f)** A = −5, B = 8

45. For the input values of A and B, use the program to determine the value of Y.

$$\text{LET } C = A * B$$
$$\text{IF } ((C \geq 10) \text{ AND } (A < 0)) \text{ OR } (B < 0)$$
$$\text{THEN LET } Y = 7$$
$$\text{ELSE LET } Y = 0$$

   (a) A = −2, B = −6   **(b)** A = −1, B = −6
   (c) A = −2, B = 6   **(d)** A = 6, B = −1
   (e) A = 4, B = 3   **(f)** A = 3, B = 1

46. For the given input values of X and Y, use the program to determine the output value of Z.

$$\text{IF } Y \neq 0$$
$$\text{THEN LET } Z = X/Y$$
$$\text{ELSE LET } Z = -1{,}000{,}000$$

   (a) X = 6, Y = 2   **(b)** X = 10, Y = 5
   (c) X = 5, Y = 10   **(d)** X = 0, Y = 10
   (e) X = 10, Y = 0   **(f)** X = −1,000,000, Y = 1

47. For the given input values of A and B, use the program to determine the value of C.

$$\text{IF } ((A < 0) \text{ AND } (B < 0)) \text{ OR } (B \geq 6)$$
$$\text{THEN LET } C = (A * B) + 4$$
$$\text{ELSE LET } C = 0$$

   (a) A = −1, B = −2   **(b)** A = −2, B = 8
   (c) A = −2, B = 3   **(d)** A = 3, B = −2
   (e) A = 3, B = 8   **(f)** A = 3, B = −3

48. For the given input values for A and B, use the program to determine the output value of C.

(a) A = 7, B = 7	**(b)** A = 7, B = 3
(c) A = 7, B = 12	**(d)** A = 4, B = 7
(e) A = 4, B = 3	**(f)** A = 4, B = 12

$$\text{IF } ((A \geq 5) \text{ OR } (B \geq 5)) \text{ AND } (B \leq 10)$$
$$\text{THEN LET } C = A - B$$
$$\text{ELSE LET } C = -30$$

Solutions to Check Your Understanding 11.3

1. (a) This is of the form $q \rightarrow p$ with hypothesis q and conclusion p. Statement (b) is of the form $p \rightarrow q$, where p is the hypothesis and q is the conclusion. Statement (c) is of the form $q \rightarrow p$, since it states that "If a rectangle has four sides, then a square is a rectangle." Note that "only if" signals the clause that is the conclusion. In (c), q is the hypothesis and p is the conclusion. Statement (d) is of the form $p \rightarrow q$ with hypothesis p and conclusion q. Statement (e) is the converse of (d) and is of the form $q \rightarrow p$ with hypothesis q and conclusion p.

2. First, we note that the statements p, q, and r have truth values FALSE, TRUE, and TRUE, respectively. Use the order of precedence to decide which connectives are applied first.

(a) TRUE. Since $p \wedge q$ is FALSE, the implication $(p \wedge q) \rightarrow r$ is TRUE.

(b) TRUE. Here, $p \vee q$ is TRUE and r is TRUE, so $(p \vee q) \rightarrow r$ is TRUE.

(c) TRUE. Since p is FALSE and q is TRUE, $(\sim p) \wedge q$ is TRUE. Since r is also TRUE, $(\sim p \wedge q) \leftrightarrow r$ is TRUE.

(d) TRUE. Since p is FALSE and $\sim q$ is also FALSE, $p \wedge (\sim q)$ is FALSE. Thus, $(p \wedge \sim q) \rightarrow r$ is TRUE.

11.4 Logical Implication and Equivalence

In Section 11.2, we introduced the concept of *logically equivalent* statement forms, that is, statement forms that look different but have identical truth tables. In logic, any two logically equivalent statements are interchangeable. Suppose, for example, that p and q are simple statements and we construct the truth tables of both $p \rightarrow q$ and $\sim p \vee q$.

p	q	$p \rightarrow q$	$\sim p \vee q$
T	T	T	F T
T	F	F	F F
F	T	T	T T
F	F	T	T T
(1)	(2)	(3)	(4) (5)

Comparing columns 3 and 5 shows that, for any given pair of truth values assigned to p and q, the statement forms $p \rightarrow q$ and $\sim p \vee q$ have the same truth values. For instance, the English sentences "If Jay is caught smoking in the restroom, then he is suspended from school," and "Jay is not caught smoking in the restroom or he is suspended from school," have the same meaning.

When attempting to simplify complex English sentences or analyze logical arguments, it is often useful to express a statement form in a different, but equivalent, way. In this section, we will present a list of logical equivalence rules that we can apply to accomplish this.

Earlier, we defined a *tautology* to be a statement form that has truth value TRUE regardless of the truth values of its component statements. We denote a tautology with the letter t. We defined a *contradiction* to be a statement form that has truth value FALSE regardless of the truth values of its component statements. We use the letter c to denote a contradiction.

EXAMPLE 1   **Tautology and Contradiction** Show that $p \vee \sim p$ is a tautology and that $p \wedge \sim p$ is a contradiction.

SOLUTION   Construct the truth tables for the statement forms.

p	$\sim p$	$p \vee \sim p$	$p \wedge \sim p$
T	F	T	F
F	T	T	F

No matter what the value of p, $p \vee \sim p$ is TRUE and $p \wedge \sim p$ is FALSE.

>> **Now Try Exercise 1**

EXAMPLE 2   **Logical Equivalence** Show that $\sim p \rightarrow \sim q$ is logically equivalent to $q \rightarrow p$.

SOLUTION   Construct a truth table to verify the equivalence.

p	q	$(\sim p \rightarrow \sim q)$	$(q \rightarrow p)$
T	T	F T F	T
T	F	F T T	T
F	T	T F F	F
F	F	T T T	T
(1)	(2)	(3) (5) (4)	(6)

Since columns 5 and 6 are the same, $\sim p \rightarrow \sim q$ and $q \rightarrow p$ have the same truth tables and therefore are logically equivalent.

>> **Now Try Exercise 3**

We saw in Section 11.3 that $q \rightarrow p$ is called the converse of $p \rightarrow q$; the statement $\sim p \rightarrow \sim q$ is called the **inverse** of $p \rightarrow q$. Therefore the inverse and converse of an implication are equivalent to each other.

For convenience, we denote compound statement forms by capital letters. We use P, Q, R, and so on, to denote statement forms such as $p \rightarrow q$, $p \wedge (q \rightarrow \sim r)$, $r \vee [p \wedge \sim (r \vee q)]$, and so on. As we saw in the last section, $P \leftrightarrow Q$ is TRUE whenever P and Q have the same truth values (either both TRUE or both FALSE). And $P \leftrightarrow Q$ is FALSE if the truth values of P and Q differ. Thus, P is logically equivalent to Q if and only if $P \leftrightarrow Q$ is a tautology. We write $P \Leftrightarrow Q$ when P and Q are logically equivalent. In other words,

> P and Q are logically equivalent, and we write $P \Leftrightarrow Q$, whenever $P \leftrightarrow Q$ is a tautology.

CAUTION   While the logical equivalence symbol $\Leftrightarrow$ and the biconditional symbol $\leftrightarrow$ look similar, please note that they have two different meanings. **«**

Using the truth values of $\sim p \rightarrow \sim q$ and $q \rightarrow p$ from Example 2, we verify from column 5 of the table below that $(\sim p \rightarrow \sim q) \leftrightarrow (q \rightarrow p)$ is a tautology. We write $(\sim p \rightarrow \sim q) \Leftrightarrow (q \rightarrow p)$,

p	q	$(\sim p \rightarrow \sim q)$	$\leftrightarrow$	$(q \rightarrow p)$
T	T	T	T	T
T	F	T	T	T
F	T	F	T	F
F	F	T	T	T
(1)	(2)	(3)	(5)	(4)

We summarize some important logical equivalences in Table 1 on the next page.

Table 1 Logical Equivalences

Suppose P and Q are statement forms, t represents any statement form that is a tautology, and c represents any statement form that is a contradiction. Then:

1.	$\sim\sim P \Leftrightarrow P$	Double negation
2a.	$(P \vee Q) \Leftrightarrow (Q \vee P)$	Commutative laws
b.	$(P \wedge Q) \Leftrightarrow (Q \wedge P)$	
c.	$(P \leftrightarrow Q) \Leftrightarrow (Q \leftrightarrow P)$	
3a.	$[(P \vee Q) \vee R] \Leftrightarrow [P \vee (Q \vee R)]$	Associative laws
b.	$[(P \wedge Q) \wedge R] \Leftrightarrow [P \wedge (Q \wedge R)]$	
4a.	$[P \vee (Q \wedge R)] \Leftrightarrow [(P \vee Q) \wedge (P \vee R)]$	Distributive laws
b.	$[P \wedge (Q \vee R)] \Leftrightarrow [(P \wedge Q) \vee (P \wedge R)]$	
5a.	$(P \vee P) \Leftrightarrow P$	Idempotent laws
b.	$(P \wedge P) \Leftrightarrow P$	
6a.	$(P \vee c) \Leftrightarrow P$	Identity laws
b.	$(P \vee t) \Leftrightarrow t$	
c.	$(P \wedge c) \Leftrightarrow c$	
d.	$(P \wedge t) \Leftrightarrow P$	
7a.	$(P \vee \sim P) \Leftrightarrow t$	Negation laws
b.	$(P \wedge \sim P) \Leftrightarrow c$	
8a.	$\sim(P \vee Q) \Leftrightarrow (\sim P \wedge \sim Q)$	De Morgan's laws
b.	$\sim(P \wedge Q) \Leftrightarrow (\sim P \vee \sim Q)$	
9.	$(P \rightarrow Q) \Leftrightarrow (\sim Q \rightarrow \sim P)$	Contrapositive
10a.	$(P \rightarrow Q) \Leftrightarrow (\sim P \vee Q)$	Implication
b.	$(P \rightarrow Q) \Leftrightarrow \sim(P \wedge \sim Q)$	
11.	$(P \leftrightarrow Q) \Leftrightarrow [(P \rightarrow Q) \wedge (Q \rightarrow P)]$	Equivalence
12.	$(P \rightarrow Q) \Leftrightarrow [(P \wedge \sim Q) \rightarrow c]$	Reductio ad absurdum

The logical equivalences are important for understanding common English, as well as for more formal analysis of mathematical statements and applications to computing. For example, the *contrapositive rule* is used frequently in speech. It states the equivalence of $p \rightarrow q$ with $\sim q \rightarrow \sim p$.

DEFINITION The **contrapositive** of the statement form $p \rightarrow q$ is the statement form $\sim q \rightarrow \sim p$.

EXAMPLE 3

Stating the Contrapositive of a Statement State the contrapositive of "If Jay does not play Lotto, then Jay will not win the Lotto jackpot."

SOLUTION Let p denote the statement "Jay does not play Lotto" and q denote the statement "Jay will not win the Lotto jackpot." We use the *double negation* rule from Table 1 to see that the negation of p is the statement "Jay plays Lotto," while the negation of q is the statement "Jay will win the Lotto jackpot." The contrapositive is the statement "If Jay will win the Lotto jackpot, then Jay plays Lotto." It is logically equivalent to the original statement form. (See rule 9.)
                                                              ≫ Now Try Exercise 25

To prove the *contrapositive* rule, use a truth table to show that $(p \rightarrow q) \Leftrightarrow (\sim q \rightarrow \sim p)$.

p	q	$(p \rightarrow q)$	$\leftrightarrow$	$(\sim q \rightarrow \sim p)$
T	T	T	T	F T F
T	F	F	T	T F F
F	T	T	T	F T T
F	F	T	T	T T T
(1)	(2)	(3)	(7)	(4) (6) (5)

A comparison of columns 3 and 6 shows that $p \rightarrow q$ and $\sim q \rightarrow \sim p$ have the same truth tables. Column 7 shows that the biconditional is a tautology. Hence,

$$p \rightarrow q$$

is logically equivalent to its contrapositive

$$\sim q \rightarrow \sim p.$$

EXAMPLE 4   **Stating the Contrapositive of a Statement** A 1985 newsmagazine article quoted U.S. Attorney General Edwin Meese as having said, "If a person is innocent of a crime, then he is not a suspect." State the contrapositive. Do you think that the statement is TRUE or FALSE? (*Source:* articles.chicagotribune.com.)

SOLUTION   The contrapositive of the statement is "If a person is a suspect, then he is guilty of a crime." That seems FALSE, so the original statement is also FALSE.   «

De Morgan's laws are of particular importance because they allow us to negate statements having the disjunctive or conjunctive connectives.

> **De Morgan's laws**
>
> **8a.** $\sim(P \lor Q) \Leftrightarrow (\sim P \land \sim Q)$
>
> **b.** $\sim(P \land Q) \Leftrightarrow (\sim P \lor \sim Q)$

It is important to recognize that when the negation is brought inside the parentheses, it changes the connective from $\lor$ to $\land$, or vice versa.

EXAMPLE 5   **Using De Morgan's Laws** Negate the statement "The earth's orbit is round and a year has 365 days."

SOLUTION   We are asked to negate a statement of the form $p \land q$. By rule 8b, the negation can be written as "The earth's orbit is not round, or a year does not have 365 days." This is TRUE whenever the original statement is FALSE.   **» Now Try Exercise 17**

EXAMPLE 6   **Using De Morgan's Laws** Rewrite the statement "It is false that Jack and Jill went up the hill."

SOLUTION   We use De Morgan's laws to negate a statement of the form $(p \land q)$ to get "Jack did not go up the hill or Jill did not go up the hill."   **» Now Try Exercise 21**

EXAMPLE 7   **Using De Morgan's Laws** Use De Morgan's laws to negate the statement form $(p \land q) \land r$.

SOLUTION $\sim[(p \land q) \land r] \Leftrightarrow [\sim(p \land q) \lor (\sim r)] \Leftrightarrow [(\sim p \lor \sim q) \lor \sim r]$. Note that each time the negation is applied to the conjunction or the disjunction, the connective changes.

The connecting braces above the solution are labeled **Rule 8b** and **Rule 8b**.

» Now Try Exercise 19

The rules in Table 1 allow us to simplify statement forms in a variety of ways. In place of any of the simple statements p, q, and r, we may substitute any compound statement. Provided that we substitute consistently throughout, the equivalence holds. The rules are similar to the rules of algebra in that the need and the purpose for simplification depend on the situation. Frequently, however, we can simplify statement forms containing $\to$ by eliminating the implication. As the implication rule

$$(p \to q) \Leftrightarrow (\sim p \lor q)$$

shows, we can reduce a statement involving $\to$ to one involving other connectives. In fact, as the next examples demonstrate, we may express statements involving any of the connectives by using only $\sim$ and $\lor$ or only $\sim$ and $\land$.

EXAMPLE 8   **Eliminating the $\to$ Symbol** Eliminate the connective $\to$ in the statement form, and simplify as much as possible.

$$(\sim p \land \sim q) \to \sim q.$$

SOLUTION   Let P represent the statement form $(\sim p \land \sim q)$ and Q represent $\sim q$. Then,

$$(P \to Q) \Leftrightarrow (\sim P \lor Q)$$

by the implication rule. We begin by simplifying $\sim P$, or $\sim(\sim p \land \sim q)$:

$$\sim(\sim p \land \sim q) \Leftrightarrow (\sim\sim p \lor \sim\sim q) \quad \text{De Morgan's law}$$
$$\Leftrightarrow (p \lor q) \quad\quad\quad\quad \text{Double negation}$$

Therefore,

$$[(\sim p \land \sim q) \to \sim q] \Leftrightarrow [(p \lor q) \lor \sim q].$$

We now simplify $[(p \lor q) \lor \sim q]$. Since $\lor$ is the only connective that appears in this statement form, it makes sense to use the associative law:

$$[(p \lor q) \lor \sim q] \Leftrightarrow [p \lor (q \lor \sim q)] \quad \text{Associative law}$$

Note the statement form $(q \lor \sim q)$; by the negation law, it is a tautology. So

$$[p \lor (q \lor \sim q)] \Leftrightarrow p \lor t \quad \text{Negation law}$$
$$\Leftrightarrow t \quad\quad\quad \text{Identity law}$$

Thus $[(\sim p \land \sim q) \to \sim q] \Leftrightarrow t$, and the original statement is a tautology.

» Now Try Exercise 13

EXAMPLE 9   **Negating an Implication** The following statement appears in the iTunes store Terms of Service. Negate it using only the connectives $\land$ and $\sim$, and write the negation in English.

"If the song is not successfully matched, your copy of the song will be uploaded to Apple." (From Terms and Conditions-I tunes Match. Copyright © 2016, Apple Inc..el. Used by permission of Apple Inc.)

SOLUTION Let p denote the statement "The song is not successfully matched," and q denote the statement "Your copy of the song will be uploaded to Apple." Then we wish to negate $p \rightarrow q$. From the implication rule

$$(p \rightarrow q) \Leftrightarrow \ \sim(p \wedge \sim q).$$

It follows that

$$\sim(p \rightarrow q) \Leftrightarrow \ \sim\sim(p \wedge \sim q) \Leftrightarrow p \wedge \sim q.$$

Therefore the negation can be written

"The song is not successfully matched and your copy of the song will not be uploaded to Apple."   **» Now Try Exercise 23**

EXAMPLE 10 **Using the Implication Rule** Use the implication rule to show that

$$[(p \wedge \sim r) \rightarrow (q \rightarrow r)] \Leftrightarrow [\sim(p \wedge \sim r) \vee (q \rightarrow r)].$$

SOLUTION First, observe that the statement forms

$$[(p \wedge \sim r) \rightarrow (q \rightarrow r)] \text{ and } [\sim(p \wedge \sim r) \vee (q \rightarrow r)]$$

each contain $(p \wedge \sim r)$ and $(q \rightarrow r)$. Letting P represent $(p \wedge \sim r)$ and Q represent $(q \rightarrow r)$, we write $(p \wedge \sim r) \rightarrow (q \rightarrow r)$ as $P \rightarrow Q$. By the implication rule,

$$(P \rightarrow Q) \Leftrightarrow (\sim P \vee Q),$$

and $(\sim P \vee Q)$ can be rewritten as $\sim(p \wedge \sim r) \vee (q \rightarrow r)$.   **» Now Try Exercise 4**

We have seen in this section how statement forms that look quite different can be logically equivalent. Statement forms sometimes can be simplified by using the logical equivalences of Table 1. Logical argument is based not only on logical equivalences but also on those statements that logically imply others.

> **DEFINITION** Given statement forms P and Q, we say that P **logically implies** Q (written $P \Rightarrow Q$) whenever $P \rightarrow Q$ is a tautology. That is, P logically implies Q if, whenever P is TRUE, Q is also TRUE.

CAUTION While the logical implication symbol $\Rightarrow$ and the conditional symbol $\rightarrow$ look similar, please note that they have two different meanings. **«**

For example, let P denote "My laptop is defective, and if my laptop is defective then the manufacturer will replace it," and let Q denote "The manufacturer will replace my laptop." If P is TRUE, then it must be TRUE that both the laptop is defective and the manufacturer will replace a defective laptop. Therefore, the manufacturer will replace the laptop, so Q must be TRUE as well. To determine whether a statement form P logically implies a statement form Q, we need only to consider those rows of the truth table for which P is TRUE or those for which Q is FALSE.

Table 2 summarizes some useful logical implications.

Table 2 Logical Implications

1.	$p \Rightarrow (p \vee q)$	Addition
2.	$(p \wedge q) \Rightarrow p$	Simplification
3.	$[p \wedge (p \rightarrow q)] \Rightarrow q$	Modus ponens
4.	$[(p \rightarrow q) \wedge \sim q] \Rightarrow \sim p$	Modus tollens
5.	$[(p \vee q) \wedge \sim q] \Rightarrow p$	Disjunctive syllogism
6.	$[(p \rightarrow q) \wedge (q \rightarrow r)] \Rightarrow (p \rightarrow r)$	Hypothetical syllogism
7a.	$[(p \rightarrow q) \wedge (r \rightarrow s)] \Rightarrow [(p \vee r) \rightarrow (q \vee s)]$	Constructive dilemmas
7b.	$[(p \rightarrow q) \wedge (r \rightarrow s)] \Rightarrow [(p \wedge r) \rightarrow (q \wedge s)]$	

EXAMPLE 11 **Logical Implication: Disjunctive Syllogism** Verify disjunctive syllogism: $[(p \lor q) \land \sim q] \Rightarrow p$.

FIRST SOLUTION We construct the truth table for $[(p \lor q) \land \sim q] \to p$ and verify that it is a tautology.

p	q	$p \lor q$	$\sim q$	$(p \lor q) \land \sim q$	$[(p \lor q) \land \sim q] \to p$
T	T	T	F	F	T
T	F	T	T	T	T
F	T	T	F	F	T
F	F	F	T	F	T

SECOND SOLUTION Assume that $(p \lor q) \land \sim q$ is TRUE; to verify the logical implication, we will show that p must also be TRUE. In order for the conjunction $(p \lor q) \land \sim q$ to be TRUE, both $(p \lor q)$ and $\sim q$ must be TRUE. Since $\sim q$ is TRUE, q is FALSE. Then, since the disjunction $(p \lor q)$ is TRUE and q is FALSE, p must be TRUE.

In this example, it was simple to construct the entire truth table, but in general, we can save a lot of work by considering only the relevant cases. **» Now Try Exercise 33**

EXAMPLE 12 **Logical Implication: Hypothetical Syllogism** In *The Art of War*, Sun Tzu writes "If I am quick, then I survive, and if I am not quick, then I am lost."[1] Show that this logically implies "If I am not lost, then I survive."

SOLUTION Let q denote "I am quick." s denote "I survive," and l denote "I am lost." Then

"If I am quick, then I survive, and if I am not quick, then I am lost."

can be expressed symbolically as $(q \to s) \land (\sim q \to l)$. We wish to show that this logically implies $(\sim l \to s)$. This doesn't match any of the logical implications in Table 2; however, we can use the contrapositive equivalence (Table 1, Rule 9) to express $\sim q \to l$ as $\sim l \to q$. Therefore,

$$[(q \to s) \land (\sim q \to l)] \Leftrightarrow (q \to s) \land (\sim l \to q)$$
$$\Leftrightarrow (\sim l \to q) \land (q \to s) \quad \text{Commutative law}$$

which logically implies $(\sim l \to s)$ by the hypothetical syllogism rule.

» Now Try Exercise 7

Check Your Understanding 11.4

Solutions can be found following the section exercises.

1. Show that $(p \to q) \Leftrightarrow (\sim p \lor q)$.

2. Rewrite the statement

$$(p \to \sim q) \to (r \land p),$$

eliminating the connective $\to$ and simplifying if possible.

3. Assume that the statement "Jim goes to the ballgame only if Ted gets tickets" is TRUE.
  (a) Write the contrapositive, and give its truth value.
  (b) Write the converse, and give its truth value.
  (c) Write the negation, and give its truth value.

EXERCISES 11.4

1. Show that $[(p \to q) \land q] \to p$ is not a tautology.

2. Show that the distributive laws hold:
  (a) $[p \lor (q \land r)] \Leftrightarrow [(p \lor q) \land (p \lor r)]$
  (b) $[p \land (q \lor r)] \Leftrightarrow [(p \land q) \lor (p \land r)]$.

3. Show that $p \to q$ is logically equivalent to $\sim (p \land \sim q)$.

4. Without using truth tables, show that

$$\{[(p \lor \sim q) \land r] \to p\} \Leftrightarrow [(p \lor q) \lor \sim r].$$

5. Show that

$$(p \to q) \Leftrightarrow [(p \land \sim q) \to c],$$

using the fact that c denotes a contradiction; that is, c is always FALSE.

6. (a) Show that

$$p \Rightarrow [q \to (p \land q)].$$

  (b) True or false?

$$p \Leftrightarrow [q \to (p \land q)]$$

7. True or false?

$$(p \lor q) \Rightarrow [q \to (p \land q)]$$

---

[1] Sunzi. 2001. *The Art of War: A New Translation.* Boston: Shambhala.

8. True or false?

$$[p \lor (p \land q)] \Leftrightarrow p$$

9. Write a statement equivalent to

$$\sim(p \lor q) \land \sim q,$$

using only $\sim$ and $\rightarrow$.

10. Write a statement equivalent to

$$\sim(p \land \sim q) \rightarrow (r \land p),$$

using only $\sim$ and $\land$.

11. Write an equivalent form of $p \oplus q$, using only $\sim$ and $\lor$.

12. Write each statement, using only the connectives $\sim$ and $\lor$.
    (a) $(p \land \sim q) \rightarrow p$
    (b) $(p \rightarrow r) \land (q \lor r)$
    (c) $p \rightarrow [r \land (p \lor q)]$
    (d) $(p \land q) \rightarrow (\sim q \lor r)$

13. Write an equivalent form of the statement without using implication ($\rightarrow$).

$$(p \lor q) \rightarrow (q \land \sim r)$$

14. Write an equivalent form of the statement without using implication ($\rightarrow$).

$$(q \rightarrow \sim r) \rightarrow (p \land q)$$

15. Write an equivalent form of the statement without using implication ($\rightarrow$).

$$\sim(p \land \sim q) \rightarrow (p \lor \sim r)$$

16. Write an equivalent form of the statement without using implication ($\rightarrow$).

$$\sim(\sim(p \rightarrow q) \rightarrow r)$$

17. Negate the following statements:
    (a) Arizona borders California and Arizona borders Nevada.
    (b) There are tickets available or the agency can get tickets.
    (c) The killer's hat was white or gray.

18. Negate the following statements:
    (a) Montreal is a province in Canada and Ottawa is a province in Canada.
    (b) The salesman goes to the customer or the customer calls the salesman.
    (c) The hospital does not admit psychiatric patients or orthopedic patients.

19. Use De Morgan's laws to negate the statement $p \lor \sim q \lor r$, without using parentheses.

20. Use De Morgan's laws to negate the statement $(p \lor \sim q) \lor (\sim r)$, without using parentheses.

21. Negate the following statements:
    (a) Jeremy takes 12 credits, or Jeremy takes 15 credits this semester.
    (b) Sandra received a gift from Sally and a gift from Sacha.

22. Negate the following statements:
    (a) The plane from California to Maine was on time, and every seat on it was taken.
    (b) Kenneth was not on time nor was he dressed properly.

23. Negate the following statements:
    (a) If I have a ticket to the theater, then I spent a lot of money.

(b) Basketball is played on an indoor court only if the players wear sneakers.
(c) The stock market is going up implies that interest rates are going down.
(d) For humans to stay healthy, it is sufficient that humans have enough water.

24. Negate the following statements:
    (a) Isaac Newton and Henry Ford invented the calculus.
    (b) James Galway or Paul Simon plays the piano.

25. For each of the following statements, give the contrapositive and determine its truth value.
    (a) If the sum $2 + x$ is an odd number, then x is odd.
    (b) If a computer keyboard has the QWERTY layout, then "S" is not next to "K."

26. For each of the following statements, give the contrapositive and determine its truth value.
    (a) If a book does not have a list of synonyms, then it is not a thesaurus.
    (b) The water is salty if it comes from the Atlantic Ocean.

27. Give the contrapositive, the inverse, and the converse of each statement, and then give the truth value of each.
    (a) If a dog is small, then it is a Chihuahua.
    (b) If a traffic light is green, then you do not stop.
    (c) If we live in France, then we are French citizens.

28. For each statement, give the contrapositive, the inverse, the converse, and the negation. Determine the truth value in each case.
    (a) If a rectangle does not have four equal sides, then it is not a square.
    (b) If a coin is fair, then the probability of a head is $\frac{1}{2}$.
    (c) A nation has a red and white flag only if it is Poland.

29. **Presidential Quotation** During a campaign speech in Chicago on February 5, 2008, Barack Obama said the following:

    "Change will not come if we wait for some other person or some other time." (Quote by Barack Obama.)

    (a) Let p represent "Change will come," q represent "We wait for some other person," and r represent "We wait for some other time." Write the quotation symbolically.
    (b) Negate the quotation using only the connectives $\lor$ and $\land$, and write the negation symbolically and in English.

30. **Terms of Service** Suppose that a video upload site's terms of service contains the following statement:

    "If we remove your content due to a copyright violation, and you believe this removal was in error, you have 30 days to file an appeal."

    (a) Let p represent "We remove your content due to a copyright violation," q represent "You believe this removal was in error," and r represent "You have 30 days to file an appeal." Write the above statement symbolically.
    (b) Negate the statement using only the connectives $\sim$ and $\land$, and write the negation symbolically and in English.

31. **Tax Instruction** The following statements can be found in the IRS instructions for Form 1040. Give the inverse, converse, contrapositive, and negation of each. Simplify and use De Morgan's laws when possible. (*Source:* www.irs.gov.)
    (a) If you do not wish to claim the premium tax credit for 2015, you do not need the information in Part II.

(b) You can't take the standard deduction if your spouse itemizes deductions.

(c) If you check a box, your tax won't change or your refund won't change.

32. Code of Conduct The following statements can be found in the American Medical Association's code of conduct. Give the inverse, converse, contrapositive, and negation of each. Simplify and use De Morgan's laws when possible. (*Source:* www.ama-assn.org.)

(a) If the patient cannot assent, physicians should explain the plan of care and tell her what to expect.

(b) If physicians disagree with laws, they should seek to change them.

(c) If a patient refuses care from a resident or a fellow, the attending physician should be notified.

33. Verify modus ponens (logical implication 3 in Table 2).

34. Verify modus tollens (logical implication 4 in Table 2).

Solutions to Check Your Understanding 11.4

1. We show that $(p \rightarrow q) \leftrightarrow (\sim p \vee q)$ is a tautology.

p	q	$(p \rightarrow q)$	$\leftrightarrow$	$(\sim p$	$\vee$	$q)$
T	T	T	T	F	T	
T	F	F	T	F	F	
F	T	T	T	T	T	
F	F	T	T	T	T	
(1)	(2)	(3)	(6)	(4)	(5)	

2. $(p \rightarrow \sim q) \rightarrow (r \wedge p)$

$\Leftrightarrow (\sim p \vee \sim q) \rightarrow (r \wedge p)$ Implication

$\Leftrightarrow [\sim(\sim p \vee \sim q)] \vee (r \wedge p)$ Implication

$\Leftrightarrow (p \wedge q) \vee (r \wedge p)$ De Morgan's law; double negation

$\Leftrightarrow (p \wedge q) \vee (p \wedge r)$ Commutative law

$\Leftrightarrow p \wedge (q \vee r)$ Distributive law

3. (a) Contrapositive: "If Ted does not get tickets, then Jim does not go to the ball game." TRUE.

(b) Converse: "If Ted gets tickets, then Jim goes to the ball game." Truth value is unknown.

(c) We negate a statement of the form $p \rightarrow q$. We use the implication law $(p \rightarrow q) \Leftrightarrow (\sim p \vee q)$, form the negative, and apply De Morgan's law to get $[\sim(p \rightarrow q)] \Leftrightarrow (p \wedge \sim q)$. Negation: "Jim goes to the ballgame, and Ted does not get tickets." The negation of the original true statement is FALSE.

11.5   Valid Argument

In this section, we study methods of valid argument. An argument consists of statements called hypotheses, which we assume to be true, and a statement called a conclusion, which we wish to show follows logically from the hypotheses. We rely heavily on Table 2, the table of logical implications, presented in Section 11.4. Let us begin with an example.

EXAMPLE 1   **Valid Argument** Suppose the statements

"If Marvin studies mathematics, then he is smart" and

"Marvin is not smart" are both true. Show that the conclusion

"Marvin does not study mathematics" must also be true.

SOLUTION   Let p denote "Marvin studies mathematics" and q denote "Marvin is smart." Then, $p \rightarrow q$ and $\sim q$ are both true, so

$$(p \rightarrow q) \wedge \sim q$$

must also be true. Note that

$$[(p \rightarrow q) \wedge \sim q] \Rightarrow \sim p$$

is the rule of modus tollens from Table 2 of the previous section. Thus, $\sim p$, or "Marvin does not study mathematics," is true.   **>> Now Try Exercise 1**

The technique used in the example can be extended to more complex arguments.

DEFINITION An **argument** is a set of statements

$$H_1, H_2, \ldots, H_n$$

each of which is assumed to be true and a statement C that is claimed to have been deduced from them. The statements $H_1, H_2, \ldots, H_n$ are called **hypotheses**, and the statement C is called the **conclusion**. We say that the argument is **valid** if and only if

$$H_1 \wedge H_2 \wedge \cdots \wedge H_n \Rightarrow C;$$

that is, if assuming that all the hypotheses are true forces the conclusion to be true.

An argument is also called a **proof**. The statement $H_1 \wedge H_2 \wedge \cdots \wedge H_n \to C$ is a tautology, provided that it is never false; that is, if each hypothesis is true, the conclusion is also true.

In Example 1, the argument is valid because, for the hypotheses

$$H_1: (p \to q) \quad \text{and} \quad H_2: {\sim}q \quad \text{and the conclusion} \quad C: {\sim}p,$$

we have the logical implication $H_1 \wedge H_2 \Rightarrow C$.

There are several important points to make here. First, although the conclusion may be a true statement, the argument presented may or may not be valid. Also, if one or more of the premises is false, it is possible for a valid argument to result in a conclusion that is false. The logical implications stated earlier in this chapter can be restated in the form of **rules of inference**, as given in Table 1.

Table 1 Rules of Inference

	From:	Conclude:	
1.	P	$P \vee Q$	Addition
2.	$P \wedge Q$	P	Subtraction
3.	$P \wedge (P \to Q)$	Q	Modus ponens
4.	$(P \to Q) \wedge {\sim}Q$	${\sim}P$	Modus tollens
5.	$(P \vee Q) \wedge {\sim}P$	Q	Disjunctive syllogism
6.	$(P \to Q) \wedge (Q \to R)$	$P \to R$	Hypothetical syllogism
7a.	$(P \to Q) \wedge (R \to S)$	$(P \vee R) \to (Q \vee S)$	Constructive dilemmas
b.	$(P \to Q) \wedge (R \to S)$	$(P \wedge R) \to (Q \wedge S)$	

EXAMPLE 2

Multiple-Step Argument Suppose that the following statements are true:

Oil prices stay above $50 a barrel or I will sell my Exxon stock.

If oil prices stay above $50 a barrel, then hybrid car sales are not decreasing.

Hybrid car sales are decreasing.

Prove: I will sell my Exxon stock.

SOLUTION

Let o, s, and h denote the statements

o: "Oil prices stay above $50 a barrel."

s: "I will sell my Exxon stock."

h: "Hybrid car sales are decreasing."

Write the argument step by step. Numbers in parentheses refer to previous steps used in the argument.

1. $o \lor s$ Hypothesis
2. $o \rightarrow \sim h$ Hypothesis
3. h Hypothesis
4. $\sim o$ Modus tollens (2, 3)
5. s Disjunctive syllogism (1, 4)

Step 4 results from the implication $[(o \rightarrow \sim h) \land \sim(\sim h)] \Rightarrow \sim o$, since h is equivalent to $\sim[(\sim h)]$. Step 5 results from the implication $[(o \lor s) \land \sim o] \Rightarrow s$. We have made convenient substitutions in the rules in Table 1.   **» Now Try Exercise 3**

EXAMPLE 3   **Verify Validity** Show that the following argument is valid:

I study mathematics or economics.

If I have to take English, then I do not study economics.

I do not study mathematics.

Therefore, I do not have to take English.

SOLUTION   Let p, q, and r denote the following statements:

$$p: \text{``I study mathematics.''}$$
$$q: \text{``I study economics.''}$$
$$r: \text{``I have to take English.''}$$

The hypotheses are

$$p \lor q, \quad r \rightarrow \sim q, \quad \text{and} \quad \sim p.$$

Write the argument step by step. Numbers in parentheses refer to the previous steps used in the argument.

1. $p \lor q$ Hypothesis
2. $r \rightarrow \sim q$ Hypothesis
3. $\sim p$ Hypothesis
4. q, since $[(p \lor q) \land \sim p] \Rightarrow q$ Disjunctive syllogism (1, 3)
5. $\sim r$, since $[(r \rightarrow \sim q) \land q] \Rightarrow \sim r$ Modus tollens (2, 4)

Thus, we have shown that r can be proven false by a valid argument from the given hypotheses.   **» Now Try Exercises 7**

EXAMPLE 4   **Verify Validity** Verify that the following argument is valid:

If you did not file income taxes before the deadline, then you registered for an extension or owe a penalty.

If you registered for an extension, then you received an additional six months to file.

You did not owe a penalty or the IRS will attempt to contact you.

Therefore, if you did not file income taxes before the deadline, then you received an additional six months to file or the IRS will attempt to contact you.

SOLUTION   Let p, q, r, s, and u represent the following statements:

$$p: \text{``You did not file income taxes before the deadline.''}$$
$$q: \text{``You registered for an extension.''}$$
$$r: \text{``You owe a penalty.''}$$
$$s: \text{``You received an additional six months to file.''}$$
$$u: \text{``The IRS will attempt to contact you.''}$$

The argument proceeds as follows:

1. $p \rightarrow (q \lor r)$ Hypothesis
2. $q \rightarrow s$ Hypothesis

3. $\sim r \lor u$ Hypothesis
4. $r \rightarrow u$, since $(r \rightarrow u) \Leftrightarrow (\sim r \lor u)$ Implication (3)
5. $(q \lor r) \rightarrow (s \lor u)$ Constructive dilemma (2, 4)
6. $p \rightarrow (s \lor u)$ Hypothetical syllogism (1, 5)

From Step 6, we see the conclusion "If you did not file income taxes before the deadline, then you received an additional six months to file or the IRS will attempt to contact you," is valid.

In some cases, we can replace one or more hypotheses with logically equivalent ones in order to apply the rules of inference, as we did in Step 4 above.

>> *Now Try Exercise 9*

Indirect Proof

In trying to deduce that

$$H_1 \land H_2 \land \cdots \land H_n \Rightarrow C,$$

sometimes it is easier to prove that the contrapositive of

$$H_1 \land H_2 \land \cdots \land H_n \rightarrow C$$

is a tautology. Thus, we would need to show that

$$\sim C \rightarrow \sim (H_1 \land H_2 \land \cdots \land H_n)$$

is a tautology. If we assume that $\sim C$ is TRUE, then, of course, the premise is that C is FALSE. The only case that we need to consider to prove the tautology is the case in which $\sim C$ is TRUE. In that case, we must show that $\sim (H_1 \land H_2 \land \cdots \land H_n)$ is TRUE also. This requires that the statement

$$(\sim H_1) \lor (\sim H_2) \lor \cdots \lor (\sim H_n)$$

be TRUE. (Use De Morgan's law.) The disjunction is TRUE if and only if at least one of its components is TRUE. So we must show that $(\sim H_i)$ is TRUE for some subscript i. Hence, we are required to show that H_i is FALSE for at least one i. This is called an *indirect proof*.

We summarize the idea.

DEFINITION Indirect Proof   To prove

$$(H_1 \land H_2 \land \cdots \land H_n) \Rightarrow C,$$

assume that the conclusion C is FALSE and then prove that at least one of the hypotheses H_i must be FALSE.

EXAMPLE 5   **Using an Indirect Proof**   Use an indirect proof to show that the following argument is valid:

If I am happy, then I do not eat too much.

I eat too much, or I spend money.

I do not spend money.

Therefore, I am not happy.

SOLUTION   Let p, q, and r represent the following statements:

p: "I am happy,"

q: "I eat too much," and

r: "I spend money."

We wish to show that $[(p \rightarrow \sim q) \land (q \lor r) \land \sim r] \Rightarrow \sim p$. Assume, by way of indirect proof, that the conclusion is FALSE.

1. p Negation of conclusion
2. $p \rightarrow \sim q$ Hypothesis
3. $\sim q$ Modus ponens (1, 2)
4. $q \vee r$ Hypothesis
5. r Disjunctive syllogism (3, 4)

Therefore, the given hypothesis $\sim r$ is FALSE. We express the proof in words. Let us assume the negative of the conclusion; that is, we assume "I am happy." Then "I do not eat too much" (modus ponens). We were to assume "I eat too much, or I spend money"; hence (disjunctive syllogism), it is true that "I spend money." This means that the hypothesis "I do not spend money" cannot be true. Assumption of the negation of the conclusion leads us to claim that one of the hypotheses is false. Thus, we have a valid indirect proof.

>> *Now Try Exercise 21*

Invalid Argument

In order to prove that an argument is invalid, we must show that the hypotheses do not logically imply the conclusion; that is, we show that the conclusion can be false while all the hypotheses are true.

EXAMPLE 6   **Invalid Argument**   Show that the following argument is invalid:

If Aziz is at the beach, then he is on vacation.

Aziz is on vacation.

Therefore, Aziz is at the beach.

SOLUTION   Let p and q represent the following statements:

p: "Aziz is at the beach"

q: "Aziz is on vacation"

Then, the argument is valid if and only if

$$[(p \rightarrow q) \wedge q] \rightarrow p$$

is a tautology. From Column 7 of the truth table below, we see this is not the case, so the argument is invalid.

p	q	$[(p$	$\rightarrow q)$	$\wedge$	$q]$	$\rightarrow$	p
T	T		T	T	T	T	T
T	F		F	F	F	T	T
F	T		T	T	T	F	F
F	F		T	T	F	F	T	F
(1)	(2)		(3)	(5)	(4)	(7)	(6)

Indeed, if Aziz in on vacation but not at the beach, the hypotheses would be true and the conclusion false.

>> *Now Try Exercise 16*

The invalid argument in the previous example illustrates one common **fallacy,** or invalid argument, known as the **fallacy of the converse**.

Table 2 Common Fallacies

	From:	Attempt to Conclude:	
1.	$(P \rightarrow Q) \wedge \sim P$	$\sim Q$	Fallacy of the inverse
2.	$(P \rightarrow Q) \wedge Q$	P	Fallacy of the converse

EXAMPLE 7 **Testing an Argument** Test the validity of the following argument:

The concentration of lead in Gotham's water supply is not below 15 parts per billion or the state will provide emergency funds to Gotham High School. If the state will not provide emergency funds to Gotham High School, then the school cannot stay open. The concentration of lead in Gotham's water supply is below 15 parts per billion. Therefore, the high school can stay open.

FIRST SOLUTION

We can attempt to reduce the argument to one of the known valid or invalid forms. Let c, s, and o represent the following statements:

c: "The concentration of lead in Gotham's water supply is below 15 parts per billion."

s: "The state will provide emergency funds to Gotham High School."

o: "Gotham High School can stay open."

The argument proceeds as follows.

1. $\sim c \vee s$	Hypothesis
2. $\sim s \rightarrow \sim o$	Hypothesis
3. c	Hypothesis
4. $c \rightarrow s$, since $(c \rightarrow s) \Leftrightarrow (\sim c \vee s)$	Implication (1)
5. s	Modus ponens (3, 4)

Note that, in order for the argument to be valid, we must show that the conclusion o is TRUE whenever $\sim s \rightarrow \sim o$ (Step 2) and s (Step 5) are TRUE. This is the fallacy of the inverse, and therefore the argument is invalid.

SECOND SOLUTION

The argument is valid if and only if assuming the hypotheses to be TRUE forces the conclusion to be TRUE. If c is TRUE, then $\sim c$ is FALSE; therefore, if we assume $\sim c \vee s$ is TRUE, s must be TRUE. Since we assume $\sim s$ is FALSE, the hypothesis $\sim s \rightarrow \sim o$ is TRUE regardless of the truth value of o. Since the conclusion o need not be TRUE, the argument is invalid.

» Now Try Exercise 14

Check Your Understanding 11.5

Solutions can be found following the section exercises.

1. Show that the argument is valid.

   If goldenrod is yellow, then violets are blue.

   Pine trees are not green or goldenrod is yellow.

   Pine trees are green.

   Therefore, violets are blue.

2. Show by indirect proof that the argument is valid.

   If I go to the beach, then I cannot study.

   I study or I work as a waiter.

   I go to the beach.

   Therefore, I work as a waiter.

EXERCISES 11.5

In Exercises 1–10, show that the argument is valid.

1. The auto manufacturer follows environmental regulations or they face legal action. The manufacturer does not follow environmental regulations. Therefore, the manufacturer faces legal action.

2. If the class votes for an oral final, the teacher is glad. If the exam is not scheduled for a Monday, the teacher is sad. The exam is scheduled for a Friday. Therefore, the class doesn't vote for an oral final.

3. If my allowance comes this week and I pay the rent, then my bank account will be in the black. If I do not pay the rent, I will be evicted. I am not evicted and my allowance comes. Therefore, my bank account is in the black.

4. Jane is in sixth grade implies that Jane understands fractions and Jane is in sixth grade implies that Jane is in a remedial math class. Jane is in sixth grade. Therefore, Jane understands fractions or she is in a remedial math class.

5. If the price of oil increases, the OPEC countries are in agreement. If there is no U.N. debate, the price of oil increases. The OPEC countries are in disagreement. Therefore, there is a U.N. debate.

6. If Jill wins, then Jack loses. If Peter wins, then Paul loses. Jill wins or Peter wins. Therefore, Jack loses or Paul loses.

7. If the germ is present, then the rash and the fever are present. The fever is present. The rash is not present. Therefore, the germ is not present.

8. If Sophia does not pay her car insurance, then she must stop driving her car or pay a fine. Sophia does not stop driving her car. She does not pay a fine. Therefore, Sophia pays her car insurance.

9. If the material is cotton or rayon, it can be made into a dress. The material cannot be made into a dress. Therefore, it is not rayon.

10. If there is money in my account and I have a check, then I will pay the rent. If I do not have a check, then I am evicted. Therefore, if I am not evicted and if I do not pay the rent, then there is no money in my account.

In Exercises 11–20, test the validity of the arguments.

11. If the salaries go up, then more people apply. More people apply or the salaries go up. Therefore, the salaries go up.

12. If Rita studies, she gets good grades. Rita gets bad grades. Therefore, Rita does not study.

13. The balloon is yellow or the ribbon is pink. If the balloon is filled with helium, then the balloon is a green one. The balloon is filled with helium. Therefore, the ribbon is pink.

14. If the job offer is for at least $30,000 or has 5 weeks vacation, I will accept the position. If the offer is for less than $30,000, then I will not accept the job and I will owe rent money. I will not accept the job. Therefore, I will owe rent.

15. If the papa bear sits, the mama bear stands. If the mama bear stands, the baby bear crawls on the floor. The baby bear is standing. Therefore, the papa bear is not sitting.

16. If it is snowing, I wear my boots. It is not snowing. Therefore, I am not wearing boots.

17. If wheat prices are steady, exports will increase or the GNP will be steady. Wheat prices are steady and the GNP is steady. Therefore, exports will increase.

18. If we eat out, Mom or Dad will treat. I pay for dinner. Therefore, we do not eat out.

19. If the Boston Red Sox win the American League championship, then they play in the World Series. The Red Sox play in the World Series or they fire their coach. Therefore, if the Red Sox fire their coach, then they won the American League championship.

20. If I pass history, then I do not go to summer school. If I go to summer school, I will take a course in French. Therefore, if I go to summer school, then I do not pass history or I take a course in French.

In Exercises 21–24, use indirect proof to show that the argument is valid.

21. Sam goes to the store only if he needs milk. Sam does not need milk. Therefore, Sam does not go to the store.

22. If it rains hard, there will be no picnic. If Dave brings the Frisbee, the kids will be happy. The kids are not happy and there is a picnic. Therefore, it does not rain hard and Dave did not bring the Frisbee.

23. If the newspaper and television both report a crime, then it is a serious crime. If a person was killed, then the newspaper reports the crime. A person is killed. Television reports the crime. Therefore, the crime is serious.

24. If Linda feels ill, she takes aspirin. If she runs a fever, she does not take a bath. If Linda does not feel ill, she takes a bath. Linda runs a fever. Therefore, she takes aspirin.

Show that each of the arguments in Exercises 25 and 26 is valid, first by using a direct proof and then by using an indirect proof.

25. If Jimmy does not find the keys, he will do his homework. Jimmy does not do his homework. Show that Jimmy found his keys.

26. If spring vacation includes Easter, Simone goes to France. If George goes to France, Simone does not go. Spring vacation includes Easter. Therefore, George does not go to France.

Show that each of the arguments in Exercises 27 and 28 is valid by using a direct proof or by using an indirect proof.

27. If Marissa does not go to the movies, then she is not idle. If Marissa is not in a knitting class, she is idle. Marissa is not in a knitting class. Therefore, Marissa goes to the movies.

28. Sydney collects a penalty payment if Norman leaves a mess in the apartment. Rachel cleans the apartment or Norman leaves a mess in the apartment. Rachel does not clean the apartment. Prove that Sydney collects a penalty payment.

Solutions to Check Your Understanding 11.5

1. Let

g = "Goldenrod is yellow."
v = "Violets are blue."
p = "Pine trees are green."

The following steps show that the argument is valid:

1. $g \rightarrow v$ — Hypothesis
2. $\sim p \vee g$ — Hypothesis
3. p — Hypothesis
4. g — Disjunctive syllogism (2, 3)
5. v — Modus ponens (1, 4)

2. Let

p = "I go to the beach."
q = "I study."
r = "I work as a waiter."

Begin by assuming the negation of the conclusion.

1. $\sim r$ — Negation of conclusion
2. $p \rightarrow \sim q$ — Hypothesis
3. $q \vee r$ — Hypothesis
4. q — Disjunctive syllogism (1, 3)
5. $\sim p$ — Modus tollens (2, 4)

Since one of the hypotheses has been shown to be false, the argument is valid.

11.6 Predicate Calculus

In a previous discussion, we noted that a sentence of the form

"The number x is even"

is not a statement. Although the sentence is a declarative one, we cannot determine whether the statement is TRUE or FALSE because we do not know to what x refers. For example, if $x = 3$, then the statement is FALSE. If $x = 6$, then the statement is TRUE.

> **DEFINITION** An **open sentence** $p(x)$ is a declarative sentence that becomes a statement when x is given a particular value chosen from a universe of values. An open sentence is also known as a **predicate**.

Consider the open sentence $p(x) =$ "If x is even, $x - 6 > 8$." Let us consider as possible values for x all integers greater than zero, so the universe $U = \{1, 2, 3, 4, \ldots\}$. Then the open sentence $p(x)$ represents many statements, one for each value of x chosen from U. Let us state $p(x)$ and record its truth value for several possible values of x. The open sentence $p(x)$ becomes the statement $p(1)$ when we substitute the specific value 1 for the indeterminate letter x. Recall that an implication $p \rightarrow q$ is false if and only if p is true and q is false.

$p(1) =$ "If 1 is even, $-5 > 8$" is TRUE (since the hypothesis is FALSE).

$p(2) =$ "If 2 is even, $-4 > 8$" is FALSE.

$p(3) =$ "If 3 is even, $-3 > 8$" is TRUE.

$p(4) =$ "If 4 is even, $-2 > 8$" is FALSE.

$p(20) =$ "If 20 is even, $14 > 8$" is TRUE.

$p(21) =$ "If 21 is even, $15 > 8$" is TRUE.

In fact, $p(x)$ is TRUE for any odd integer x. And $p(x)$ is TRUE for all values of x such that x is even and $x > 14$. The only values of x for which $p(x)$ is FALSE are x is even and between 2 and 14, inclusive. That is, $p(2)$, $p(4)$, $p(6)$, $p(8)$, $p(10)$, $p(12)$, and $p(14)$ are FALSE. All other values of x from U make $p(x)$ a TRUE statement.

EXAMPLE 1 **Analyzing Predicates** Let

$$p(x) = \text{"If } x > 4, \text{ then } x + 10 > 14\text{"}$$

be an open sentence. Let $x \in U$ (that is, x is an element of U), where $U = \{1, 2, 3, 4, \ldots\}$. Find the truth value of each statement formed when these values are substituted for x in $p(x)$.

SOLUTION $p(1)$ is TRUE because if $x = 1$, the hypothesis is FALSE.
$p(2)$ is TRUE because if $x = 2$, the hypothesis is FALSE.
$p(3)$ is TRUE because if $x = 3$, the hypothesis is FALSE.
$p(4)$ is TRUE because if $x = 4$, the hypothesis is FALSE.
$p(5)$ is TRUE because if $x = 5$, then the hypothesis is TRUE and $x + 10 > 14$.

In fact, $p(x)$ is TRUE for all values of $x \in U$. We say that "for all $x \in U$, if $x > 4$, then $x + 10 > 14$." **>> Now Try Exercise 1**

The statement

For all $x \in U, p(x)$

is symbolized by

$$\forall x \in U \, p(x).$$

The statement $\forall x \in U\, p(x)$ is TRUE if and only if $p(x)$ is TRUE for every $x \in U$.

We call the symbol $\forall$ the **universal quantifier** and read it as "for all," "for every," or "for each." The universal set must be known in order to decide whether $\forall x \in U\, p(x)$ is TRUE or FALSE. We emphasize that, whereas $p(x)$ is an open sentence with no assignable truth value, $\forall x \in U\, p(x)$ is a legitimate logical statement. At times, the notation will be abbreviated to $\forall x\, p(x)$ when the universe U is clear, or to $\forall x\, [p(x)]$ for further clarity.

From the example, we see that the statement "For all $x \in U$, if $x > 4$, then $x + 10 > 14$" is a TRUE statement for $U = \{1, 2, 3, \ldots\}$.

EXAMPLE 2   **Truth of a Predicate in a Fixed Universe** Let $U = \{1, 2, 3, 4, 5, 6\}$. Determine the truth value of the statement

$$\forall x\, [(x - 4)(x - 8) > 0].$$

SOLUTION   Let $p(x)$ denote the open statement "$(x - 4)(x - 8) > 0$." Consider the truth values of $p(1)$, $p(2)$, $p(3)$, $p(4)$, $p(5)$, and $p(6)$. Note that $p(1)$ is TRUE because $(1 - 4)(1 - 8) = (-3)(-7) = 21$ and $21 > 0$. We find that $p(2)$ and $p(3)$ are also TRUE. However, $p(4)$ is FALSE because $(4 - 4)(4 - 8) = 0$. We need not check any other values from U. Already, we know that

$$[\forall x \in U\, p(x)] \text{ is FALSE.}$$

Note that there are values of x in U for which $p(x)$ is TRUE.   **» Now Try Exercise 11(b)**

The statement

There exists an x in U such that $p(x)$

is symbolized by

$$\exists x \in U\, p(x).$$

The statement $\exists x \in U\, p(x)$ is TRUE if and only if there is at least one element $x \in U$ such that $p(x)$ is TRUE.

The symbol $\exists$ is called the **existential quantifier**. We read the existential quantifier as "there exists x such that $p(x)$," "for some x, $p(x)$," or "there is some x for which $p(x)$." We may write $\exists x\, p(x)$ when the universal set is clear. When the universal set does not appear explicitly in the statement, then U must be made clear in order for $\exists x\, p(x)$ to be a statement and to have a truth value.

EXAMPLE 3   **Universal and Existential Quantifiers** Determine the truth value of the following statements, where $U = \{1, 2, 3, 4, 5, 6, 7, 8\}$:
(a) $\forall x\, (x + 3 < 15)$ **(b)** $\exists x\, (x > 5)$
(c) $\exists x\, (x^2 = 0)$ **(d)** $\forall x\, (0 < x < 10)$
(e) $\forall x\, [(x + 2 > 5) \lor (x \leq 3)]$ **(f)** $\exists x\, [(x - 1)(x + 2) = 0]$

SOLUTION   The statement in (a) is TRUE because, for every $x \in U$, $x + 3$ is less than 15. The statement in (b) is TRUE because, for example, $7 \in U$ and $7 > 5$. Statement (c) is FALSE because there is no $x \in U$ that satisfies the equation $x^2 = 0$. Statement (d) is TRUE. Statement (e) is TRUE because, for each $x \in U$, $x + 2 > 5$ or $x \leq 3$. Statement (f) is TRUE because $1 \in U$ and $(1 - 1)(1 + 2) = 0$.   **» Now Try Exercise 9**

Sometimes, it is convenient to let U be a very large set—the real numbers, for example—and to restrict U within the statement itself. Consider the statement

For every positive integer x, x is even or $x + 1$ is even.

One way to write this is to let U be the set of positive integers and to write

$$\forall x \in U \ [(x \text{ is even}) \lor (x + 1 \text{ is even})].$$

An equivalent statement that more explicitly shows the universal set would be

$$\forall x \in U \ [(x \text{ is a positive integer}) \rightarrow ((x \text{ is even}) \lor (x + 1 \text{ is even}))].$$

The universal set can then be the set of all integers. The universe is restricted by using an implication with the hypothesis that x is a positive integer in the open sentence $p(x)$. We can allow U to be any set containing the positive integers and restrict U by using such an implication.

A statement with the existential quantifier can be rephrased to include the universe in the statement. For example, let U be the set of positive integers, and consider the statement

$$\exists x \in U \ [(x - 3)(x + 6) = 0].$$

We can rewrite the statement as

$$\exists x \in U \ [(x \text{ is a positive integer}) \land ((x - 3)(x + 6) = 0)].$$

We could let U be the set of all integers or the reals, if we wish. Any set containing the positive integers would be an acceptable choice for U. The existential operator and the connective $\land$ can be used to restrict the universe to the positive integers.

We frequently have reason to negate a quantified statement. Of course, the negation changes the truth value from TRUE to FALSE or from FALSE to TRUE. The negations should be properly worded.

EXAMPLE 4

Negation of a Universal Statement Negate the statement

"For every positive integer x, x is even or $x + 1$ is even."

SOLUTION

We could say, "It is not the case that for every positive integer x, x is even or $x + 1$ is even." That is not entirely satisfactory, however. What we mean is that there exists a positive integer for which it is not the case that x is even or $x + 1$ is even. Assuming that U is the set of positive integers, this is written symbolically as

$$\sim [\forall x \ ((x \text{ is even}) \lor (x + 1 \text{ is even}))] \Leftrightarrow \exists x \ [\sim((x \text{ is even}) \lor (x + 1 \text{ is even}))]$$
$$\Leftrightarrow \exists x \ [(x \text{ is odd}) \land (x + 1 \text{ is odd})].$$

The last equivalence follows from De Morgan's laws, which allow us to rewrite a statement of the form $\sim(p \lor q)$ as $(\sim p \land \sim q)$. The negation of the original statement then is "There exists a positive integer x such that x is odd and $x + 1$ is odd." **«**

EXAMPLE 5

Negation of an Existential Statement Negate the statement "There exists a positive integer x such that $x^2 - x - 2 = 0$."

SOLUTION

One way to negate the statement is to say, "It is not the case that there exists a positive integer x such that $x^2 - x - 2 = 0$." A better way is to say, "For all positive integers x, $x^2 - x - 2 \neq 0$." Note that

$$\sim [\exists x \ (x^2 - x - 2 = 0)] \Leftrightarrow [\forall x \ (x^2 - x - 2 \neq 0)].$$

» *Now Try Exercise 11*

The negation of a quantified statement can be rewritten by bringing the negation inside the open statement and replacing $\exists$ by $\forall$, or vice versa.

$$\sim[\exists x\, p(x)] \Longleftrightarrow [\forall x \sim p(x)]$$

$$\sim[\forall x\, p(x)] \Longleftrightarrow [\exists x \sim p(x)]$$

These rules are also called *De Morgan's laws*.

EXAMPLE 6

Negation of Statements with Quantifiers Write the negation of each of the following statements:

(a) All university students like football.

(b) There is a mathematics textbook that is both short and clear.

(c) For every positive integer x, if x is even, then $x + 1$ is odd.

SOLUTION

(a) Here, the universe is "university students" and $p(x)$ is the statement "Student x likes football." We negate $\forall x\, p(x)$ to get $\exists x\, [\sim p(x)]$, which in words is "There exists a university student who does not like football."

(b) The universe is the set of all mathematics textbooks. The statement has the form $\exists x\, [p(x) \wedge q(x)]$, where $p(x)$ denotes "The mathematics textbook is short" and $q(x)$ denotes "The mathematics textbook is clear." The negation of the statement is of the form $\sim[\exists x\, (p(x) \wedge q(x))] \Longleftrightarrow [\forall x \sim (p(x) \wedge q(x))] \Longleftrightarrow [\forall x\, (\sim p(x) \vee \sim q(x))]$. This translates into the English sentence "Every mathematics textbook is not short or not clear."

(c) The universe is the set of positive integers. The statement is of the form

$$\forall x\, (p(x) \rightarrow q(x)),$$

where $p(x)$ is the statement "x is even" and $q(x)$ is "$x + 1$ is odd." The negation is

$$\sim[\forall x\, (p(x) \rightarrow q(x))] \Longleftrightarrow [\exists x \sim (p(x) \rightarrow q(x))] \Longleftrightarrow [\exists x \sim (\sim p(x) \vee q(x))]$$
$$\Longleftrightarrow [\exists x\, (p(x) \wedge \sim q(x))].$$

In English, this is "There exists a positive integer x such that x is even and $x + 1$ is also even." $\blacktriangleleft\!\blacktriangleleft$

We sometimes have occasion to use sentences that have two quantified variables. For example, the sentence "For every child, there exists an adult who cares for him or her" has two variables: the child and the adult. We can write the statement symbolically, using two variables x and y. We let the universe for the x variable be the set of all children and the universe for the y variable be the set of all adults. Then we let $p(x, y) =$ "y cares for x." The statement can be written symbolically as

$$\forall x\, \exists y\, p(x, y).$$

Binding each of the variables with the universal or existential quantifier and a universal set gives a legitimate logical statement.

EXAMPLE 7

Predicates with Two Variables Let the universe for each of the variables be the set of all Americans. Let

$$p(x, y) = x \text{ is taller than } y.$$

$$q(x, y) = x \text{ is heavier than } y.$$

Write the English sentences from the symbolic statements.

(a) $\forall x\, \forall y\, [p(x, y) \rightarrow q(x, y)]$

(b) $\forall x\, \exists y\, [p(x, y) \wedge q(x, y)]$

(c) $\forall y\, \exists x\, [p(x, y) \vee q(x, y)]$

(d) $\exists x\, \exists y\, [p(x, y) \wedge q(x, y)]$

(e) $\exists x\, \exists y\, [p(x, y) \wedge q(y, x)]$

SOLUTION   (a) For all Americans x and y, if x is taller than y, then x is heavier than y.

(b) For every American x, there is an American y such that x is taller and heavier than y.

(c) For every American y, there is an American x such that x is taller than y or x is heavier than y.

(d) There are Americans x and y such that x is taller than y and x is heavier than y.

(e) There are Americans x and y such that x is taller than y and y is heavier than x.   **≪**

The rules for the negation of statements with more than one variable are just repeated applications of De Morgan's laws mentioned for single-variable statements. For instance,

$$\sim [\forall x \, \forall y \, p(x, y)] \Leftrightarrow \exists x \, [\sim \forall y \, p(x, y)] \Leftrightarrow \exists x \, \exists y \, [\sim p(x, y)]$$

and

$$\sim [\exists x \, \exists y \, p(x, y)] \Leftrightarrow \forall x \, [\sim \exists y \, p(x, y)] \Leftrightarrow \forall x \, \forall y \, [\sim p(x, y)].$$

We summarize these rules and several others in Table 1.

Table 1

1. $\sim [\forall x \, \forall y \, p(x, y)] \Leftrightarrow \exists x \, \exists y \, [\sim p(x, y)]$ ⎫
2. $\sim [\forall x \, \exists y \, p(x, y)] \Leftrightarrow \exists x \, \forall y \, [\sim p(x, y)]$ ⎪
3. $\sim [\exists x \, \forall y \, p(x, y)] \Leftrightarrow \forall x \, \exists y \, [\sim p(x, y)]$ ⎬ De Morgan's laws
4. $\sim [\exists x \, \exists y \, p(x, y)] \Leftrightarrow \forall x \, \forall y \, [\sim p(x, y)]$ ⎭
5. $\forall x \, \forall y \, p(x, y) \Leftrightarrow \forall y \, \forall x \, p(x, y)$
6. $\exists x \, \exists y \, p(x, y) \Leftrightarrow \exists y \, \exists x \, p(x, y)$
7. $\exists x \, \forall y \, p(x, y) \Rightarrow \forall y \, \exists x \, p(x, y)$

Notice that rule 7 is a logical implication and not a logical equivalence. Any other exchanges of $\exists$ and $\forall$ should be handled very carefully. They are unlikely to give equivalent statements even in specific cases.

EXAMPLE 8   **English Translation and Truth Value for Predicates with Two Variables** Let the universe for both variables be the nonnegative integers $0, 1, 2, 3, 4, \ldots$. Write the statements in words, and decide on the truth value of each.

(a) $\forall x \, \forall y \, [2x = y]$ (b) $\forall x \, \exists y \, [2x = y]$

(c) $\exists x \, \forall y \, [2x = y]$ (d) $\exists x \, \forall y \, [2y = x]$

(e) $\exists y \, \forall x \, [2x = y]$ (f) $\exists x \, \exists y \, [2x = y]$

(g) $\forall y \, \exists x \, [2x = y]$

SOLUTION   (a) For every pair of nonnegative integers x and y, $2x = y$. This is FALSE because, for example, it is FALSE in the case $x = 3$, $y = 10$.

(b) For every nonnegative integer x, there is a nonnegative integer y such that $2x = y$. This is TRUE because, once having chosen any nonnegative number x, we can let y be the double of x.

(c) There exists a nonnegative integer x such that, for every nonnegative integer y, $2x = y$. This statement is FALSE because, no matter what x we choose, $2x = y$ is TRUE only for y equal to twice x and for no other nonnegative number.

(d) There is a nonnegative number x such that, for all nonnegative numbers y, $2y = x$. This is FALSE for similar reasons as in (c).

(e) There exists a nonnegative integer y such that, for all nonnegative integers x, $2x = y$. The statement is FALSE. For it to be TRUE, we would need to find a specific value of y that can be fixed and for which, no matter what nonnegative integer x we choose, $2x = y$. This is the same as (d).

(f) There exist nonnegative integers x and y such that $2x = y$. This is TRUE. One choice that makes the statement $2x = y$ TRUE is $x = 4$ and $y = 8$.

(g) For every nonnegative integer y, there exists a nonnegative integer x such that $2x = y$. This is FALSE because if $y = 5$, there exists no nonnegative integer x such that $2x = 5$. » *Now Try Exercise 15*

A statement of the form $\forall x\, p(x)$ is FALSE if $\exists x\, [\sim p(x)]$ is TRUE. If we want to show that the statement $\forall x\, p(x)$ is FALSE, we need only find a value for x for which $p(x)$ is FALSE. Such an example is called a **counterexample**. We used such a counterexample in parts (a) and (g) of Example 8.

Analogy between Sets and Statements

Actually, the symbols and the truth values in the tables are suggestive of set theory definitions introduced in Chapter 5. Recall that we use U to denote the universe and the letters S and T to denote sets in that universe. In the descriptions of sets that follow, we use the colon (:) as a shorthand for "such that." Then the union of the sets S and T is defined as

$$S \cup T = \{x \in U : x \in S \text{ or } x \in T\}.$$

If we allow $p(x)$ to denote "$x \in S$" and $q(x)$ to denote "$x \in T$," then it is TRUE that $x \in S \cup T$ if and only if the statement $p(x) \vee q(x)$ is TRUE. Thus,

$$\forall x\, [x \in S \cup T \leftrightarrow p(x) \vee q(x)].$$

Similarly, the intersection of the two sets is defined as

$$S \cap T = \{x \in U : x \in S \text{ and } x \in T\}.$$

It is TRUE that $x \in S \cap T$ if and only if the statement $p(x) \wedge q(x)$ is TRUE, where $p(x)$ and $q(x)$ are as shown previously. Thus,

$$\forall x\, [x \in S \cap T \leftrightarrow p(x) \wedge q(x)].$$

Using the symbol $\notin$ for "is not an element of," we have

$$S' = \{x \in U : x \notin S\}.$$

Thus, $x \in S'$ if and only if $\sim p(x)$ is TRUE. Hence,

$$\forall x\, [x \in S' \leftrightarrow \sim p(x)].$$

The *exclusive or* corresponds to the **symmetric difference** of S and T defined by

$$S \oplus T = \{x \in U : x \in S \cup T \text{ but } x \notin S \cap T\}.$$

That is, it is TRUE that $x \in S \oplus T$ if and only if the statement $p(x) \oplus q(x)$ is TRUE, or

$$\forall x\, [x \in S \oplus T \leftrightarrow p(x) \oplus q(x)].$$

As with the connectives $\vee$ and $\wedge$, the connectives $\rightarrow$ and $\leftrightarrow$ have counterparts in set theory. We will assume that the universe is U and that S and T are subsets of U. Let $p(x)$ be the statement "$x \in S$," and let $q(x)$ be the statement "$x \in T$." Then we have seen that S is a subset of T if and only if every element of S is also an element of T. We say that

$$S \subseteq T \text{ if and only if } (x \in S) \rightarrow (x \in T) \text{ is TRUE for all } x,$$

or equivalently,

$$S \subseteq T \text{ if and only if } \forall x\, [p(x) \rightarrow q(x)] \text{ is TRUE}$$

Two sets are equal if they have the same elements. We claim that

$$S = T \text{ if and only if } (x \in S) \leftrightarrow (x \in T) \text{ is TRUE for all } x;$$

that is,

$$S = T \text{ if and only if } \forall x\, [\,p(x) \leftrightarrow q(x)] \text{ is TRUE.}$$

EXAMPLE 9 **Sets and Statements** Let $U = \{1, 2, 3, 4, 5, 6\}$, let $S = \{x \in U : x \le 3\}$, and let $T = \{x \in U : x \text{ divides } 6\}$. Show $S \subseteq T$.

SOLUTION We must show that $\forall x\, [(x \in S) \rightarrow (x \in T)]$ is TRUE. An implication is FALSE only in the case in which the hypothesis is TRUE and the conclusion is FALSE. Let us suppose, then, that the hypothesis is TRUE and show that the conclusion must also be TRUE in that case. Assume that $x \in S$. By the definition of the set S, this means that x is 1, 2, or 3. Then $x \in T$, since in each of the cases, x divides 6 and is therefore an element of T. Hence, $(x \in S) \rightarrow (x \in T)$ cannot be FALSE and must be TRUE for all $x \in U$.

» Now Try Exercise 19

We saw De Morgan's laws in another form in Chapter 5. In that case, we assumed that S and T were sets and showed that

$$(S \cup T)' = S' \cap T' \quad \text{and} \quad (S \cap T)' = S' \cup T'.$$

We know that $x \in S \cup T$ if and only if $(x \in S) \vee (x \in T)$ is TRUE. So $x \notin S \cup T$ if and only if the negation of the statement is TRUE. Thus,

$$x \in (S \cup T)' \text{ if and only if } \sim[(x \in S) \vee (x \in T)] \text{ is TRUE.}$$

By De Morgan's law in Table 1 of Section 11.4, we know that

$$x \in (S \cup T)' \text{ if and only if } \sim(x \in S) \wedge \sim(x \in T) \text{ is TRUE.}$$

Thus,

$$x \in (S \cup T)' \text{ if and only if } (x \notin S) \wedge (x \notin T) \text{ is TRUE.}$$

This means that $x \in S' \cap T'$. The De Morgan laws for unions and intersections of sets are closely related to the De Morgan laws for the connectives $\vee$ and $\wedge$.

Colloquial Usage

Again, we point out that colloquial usage is frequently not precise. For example, the statement "All students in finite mathematics courses do not fail" is strictly interpreted to mean that $\forall x \in U\,[x \text{ does not fail}]$ with $U = $ the set of students in finite mathematics courses. This means, of course, that no one fails. On the other hand, some might loosely interpret this to mean that not every student in finite mathematics courses fails. This is the statement $\exists x \in U\,[x \text{ does not fail}]$. We require precision of language in mathematics and do *not* accept the second interpretation as correct.

Check Your Understanding 11.6

Solutions can be found following the section exercises.

1. Write the following statement symbolically, and write its negation in English: "There exists a flower that can grow in sand and is not subject to mold."

2. For the universe $U = \{0, 1, 2, 3, 4, 5, 6, 7, 8\}$, determine the truth values of the following statements:
   (a) $\forall x\, (x > 2)$
   (b) $\exists x\, (x > 2)$
   (c) $\forall x\, (x^2 < 100)$
   (d) $\exists x\, [(x - 1 = 4) \wedge (3x + 5 = 20)]$

3. Consider the universe for both variables x and y to be the set of nonnegative integers $\{0, 1, 2, 3, 4, \dots\}$. Write the statements in English, and determine the truth value of each.
   (a) $\forall x\, \forall y\, [x < y]$
   (b) $\forall x\, \exists y\, [x < y]$
   (c) $\forall y\, \exists x\, [x < y]$
   (d) $\exists x\, \forall y\, [x < y]$
   (e) $\exists x\, \exists y\, [x < y]$
   (f) $\forall x\, \forall y\, [(x < y) \vee (y < x)]$

EXERCISES 11.6

1. Let $U = \{1, 2, 3, 4, 5, 6\}$. Determine the truth value of

$$p(x) = [(x \text{ is even}) \text{ or } (x \text{ is divisible by } 3)]$$

for the given values of x.
(a) $x = 1$ **(b)** $x = 4$ **(c)** $x = 3$
(d) $x = 6$ **(e)** $x = 5$

2. Determine the truth value of $p(x)$ for the values of x chosen from the universe of all letters of the alphabet, where

$$p(x) = [(x \text{ is a vowel}) \text{ and } (x \text{ is in the word ABLE})].$$

(a) $x = A$ **(b)** $x = D$
(c) $x = B$ **(d)** $x = I$

3. An alert California teacher chided "Dear Abby" (*Baltimore Sun*, March 1, 1989) for her statement "Confidential to Eunice: All men do not cheat on their wives." Let $p(x)$ be the statement "x cheats on his wife" in the universe of all men. Write Abby's statement symbolically. Rewrite the statement, using the existential quantifier. Do you think that the statement is TRUE or FALSE? What do you think Abby really meant to say?

4. Recently, as the Amtrak train pulled into the Baltimore station, the conductor announced, "All doors do not open." Since passengers were permitted to get off at the station, what do you think the conductor meant to say?

5. Let the universe be all university professors. Let $p(x)$ be the open statement "x likes poetry." Write the following statements symbolically:
(a) All university professors like poetry.
(b) Some university professors do not like poetry.
(c) Some university professors like poetry.
(d) Not all university professors like poetry.
(e) All university professors do not like poetry.
(f) No university professor likes poetry.
(g) Are any of the statements in (a) through (f) equivalent? Explain.

6. Consider the universe of all orange juice. Write the following symbolically, using $p(x) = $ "x comes from Florida."
(a) All orange juice does not come from Florida.
(b) Some orange juice comes from Florida.
(c) Not all orange juice comes from Florida.
(d) Some orange juice does not come from Florida.
(e) No orange juice comes from Florida.
(f) Are any of the statements in (a) through (e) equivalent? Explain.

7. Let $U = \{1, 2, 3, 4, 5, 6, 7, 8, 9\}$. Let

$$p(x) = [(x \text{ is prime}) \rightarrow (x^2 + 1 \text{ is even})].$$

Find the truth value of
(a) $\exists x\, p(x)$ **(b)** $\forall x\, p(x)$

8. Let $U = \{0, 1, 2, 3, 4\}$. Let

$$p(x) = [x^2 > 9].$$

Find the truth value of
(a) $\exists x\, p(x)$ **(b)** $\forall x\, p(x)$

9. Let the universe consist of all nonnegative integers. Let $p(x)$ be the statement "x is even." Let $q(x)$ be the statement "x is odd." Determine the truth value of
(a) $\forall x\, [p(x) \vee q(x)]$ **(b)** $[\forall x\, p(x)] \vee [\forall x\, q(x)]$

(c) $\exists x\, [p(x) \vee q(x)]$ **(d)** $\exists x\, [p(x) \wedge q(x)]$
(e) $\forall x\, [p(x) \wedge q(x)]$ **(f)** $[\exists x\, p(x)] \wedge [\exists x\, q(x)]$
(g) $\forall x\, [p(x) \rightarrow q(x)]$ **(h)** $[\forall x\, p(x)] \rightarrow [\forall x\, q(x)]$

10. Let the universe consist of all real numbers. Let

$$p(x) = (x \text{ is positive})$$

and

$$q(x) = (x \text{ is a perfect square}).$$

Determine the truth value of each of the following:
(a) $\forall x\, p(x)$ **(b)** $\forall x\, q(x)$
(c) $\exists x\, p(x)$ **(d)** $\exists x\, q(x)$
(e) $\exists x\, [p(x) \rightarrow q(x)]$ **(f)** $\exists x\, [q(x) \rightarrow p(x)]$
(g) $\forall x\, [p(x) \rightarrow q(x)]$ **(h)** $\forall x\, [q(x) \rightarrow p(x)]$
(i) $[\forall x\, p(x)] \rightarrow [\exists x\, q(x)]$ **(j)** $[\forall x\, p(x)] \rightarrow [\forall x\, q(x)]$

11. Negate each statement by changing existential quantifiers to universal quantifiers, or vice versa.
(a) Every dog has its day.
(b) Some men fight wars.
(c) All women are married.
(d) For every pot, there is a cover.
(e) Not all children have pets.
(f) Not every month has 30 days.

12. Negate each statement by changing existential quantifiers to universal quantifiers, or vice versa.
(a) Every stitch saves time.
(b) All books have hard covers.
(c) Some children are afraid of snakes.
(d) Not all computers have a hard disk.
(e) Some chairs do not have arms.

13. Consider the universe of nonnegative integers $= \{0, 1, 2, 3, 4, \ldots\}$. Write the English sentence for each symbolic statement. Determine the truth value of each statement. If the statement is FALSE, give a counterexample and write its negation out in words.
(a) $\forall x\, \forall y\, [x + y > 12]$ **(b)** $\forall x\, \exists y\, [x + y > 12]$
(c) $\exists x\, \forall y\, [x + y > 12]$ **(d)** $\exists x\, \exists y\, [x + y > 12]$

14. Consider the universe of all subsets of the set $A = \{a, b, c\}$. Let the variables x and y denote subsets of A. Find the truth value of each of the statements, and explain your answer. If the statement is FALSE, give a counterexample.
(a) $\forall x\, \forall y\, [x \subseteq y]$
(b) $\forall x\, \forall y\, [(x \subseteq y) \vee (y \subseteq x)]$
(c) $\exists x\, \forall y\, [x \subseteq y]$
(d) $\forall x\, \exists y\, [x \subseteq y]$

15. Let the universe for both variables x and y be the set $\{1, 2, 3, 4, 5, 6\}$. Let $p(x, y) = $ "x divides y." Give the truth values of each of the statements; explain your answer, and give a counterexample in the case that the statement is FALSE.
(a) $\forall x\, \forall y\, p(x, y)$ **(b)** $\forall x\, \exists y\, p(x, y)$
(c) $\exists x\, \forall y\, p(x, y)$ **(d)** $\exists y\, \forall x\, p(x, y)$
(e) $\forall y\, \exists x\, p(x, y)$ **(f)** $\forall x\, p(x, x)$

16. If $p(x)$ denotes "$x \in S$" and $q(x)$ denotes "$x \in T$," describe the following, using logical statement forms with $p(x)$ and $q(x)$:
(a) $x \in S' \cup T$ **(b)** $x \in S \oplus T'$
(c) $x \in S' \cap T'$ **(d)** $x \in (S \cup T)'$

17. Let the universal set be

$$U = \{0, 1, 2, 3, 4, 5, 6, 7, 8, 9, 10, 11, 12, 13\},$$

let $S = \{x \in U : x \geq 8\}$, and let $T = \{x \in U : x \leq 10\}$.
  (a) What implication must be TRUE if and only if S is a subset of T?
  (b) Is S a subset of T? Explain.

18. Is S a subset of T? Let

$$U = \{1, 2, 3, 4, 5, 6, 7, 8, 9, 10, 11, 12\}$$
$$S = \{x \in U : x \text{ divides } 12 \text{ evenly}\}$$
$$T = \{x \in U : x \text{ is a multiple of } 2\}.$$

19. Prove that S is a subset of T, where

$$U = \{1, 2, 3, 4, 5, 6, 7, 8, 9\}$$
$$S = \{x \in U : x \text{ is a multiple of } 2\}$$
$$T = \{x \in U : x \text{ divides } 24 \text{ evenly}\}.$$

20. Prove that S is a subset of T, where

$$U = \{a, b, c, d, e, f, g, h\}$$
$$S = \{x \in U : x \text{ is a letter in } bad\}$$
$$T = \{x \in U : x \text{ is a letter in } badge\}.$$

Solutions to Check Your Understanding 11.6

1. Let the universe be the collection of all flowers. Let $p(x)$ be the open sentence "x can grow in sand," and let $q(x)$ be "x is subject to mold." The statement is then

$$\exists x \, [p(x) \wedge \sim q(x)].$$

The negation is

$$\forall x \, [\sim p(x) \vee q(x)].$$

This can be stated in English as

"Every flower cannot grow in sand
or is subject to mold."

2. (a) FALSE. As a counterexample, consider $x = 2$.
   (b) TRUE. Consider $x = 3$.
   (c) TRUE. In fact, for every $x \in U$, $x^2 \leq 64$.
   (d) TRUE. Consider $x = 5$.

3. (a) For all nonnegative integers x and y, $x < y$. FALSE.

   (b) For every nonnegative integer x, there exists a nonnegative integer y such that $x < y$. TRUE. For any nonnegative integer x, we can let $y = x + 1$.
   (c) For every nonnegative integer y, there exists a nonnegative integer x such that $x < y$. FALSE. A counterexample is found by letting $y = 0$. Then there is no nonnegative integer x such that $x < y$.
   (d) There exists a nonnegative integer x such that, for all nonnegative integers y, $x < y$. FALSE. No matter what x is selected, letting $y = 0$ gives a counterexample.
   (e) There exist nonnegative integers x and y such that $x < y$. TRUE. Just let $x = 3$ and $y = 79$.
   (f) For every pair of nonnegative integers x and y, $x < y$ or $y < x$. FALSE. Consider $x = y$—say, $x = 4$ and $y = 4$.

11.7 Logic Circuits

Logic circuits are ubiquitous in the devices that we use every day. We find them in computers, telephones, digital clocks, and television sets. These are electrical circuits that are designed to control current flow depending on the input activity. Current flow passes through gates to an output line. The gates operate according to strict rules that we will discuss. There are several simple types of circuits from which more complex circuits can be developed.

The simplest form of circuit has one input line with current input that is either ON (1) or OFF (0). This gate performs the function of negating the input. The output is the inverse of the input, so if the input is ON (1), the output is OFF (0), and vice versa. In the language of logic, we call this a **NOT gate**. In the language of logic circuits, the gate is called an **inverter**.

There are other basic gates, each with two inputs. The first is the **AND gate**, which behaves exactly as the logical connective AND ($\wedge$). Current flows through the gate from two inputs if and only if both input currents are ON. Thus, the output is ON (1) if and only if both inputs are ON.

The **OR gate** behaves exactly as the logical connective OR ($\vee$). Current flows through the OR gate if and only if at least one of the input currents is ON. Thus, the output of the OR gate is ON if and only if at least one of the inputs is ON.

Engineers and designers find it helpful to draw diagrams of the circuits to reflect the connective with an image of a gate-type device. These are pictured in Fig. 1.

Figure 1

The output from the various inputs is reflected by Tables 1 through 3.

Table 1 The NOT Gate	
Input	Output
p	$\sim p$
0	1
1	0

Table 2 The AND Gate		
Input		Output
p	q	$p \wedge q$
0	0	0
0	1	0
1	0	0
1	1	1

Table 3 The OR Gate		
Input		Output
p	q	$p \vee q$
0	0	0
0	1	1
1	0	1
1	1	1

We can construct more complex circuits by combining these basic building blocks.

EXAMPLE 1   **Drawing a Circuit from a Logical Statement** Draw a logic circuit for three inputs, p, q, and r and output $(\sim p) \wedge (q \vee r)$. What is the output when the inputs are $p = 1$, $q = 1$, and $r = 0$?

SOLUTION   We begin at the end with the AND gate and prepare the diagram for input from the two components, $\sim p$ and $q \vee r$. See Fig. 2.

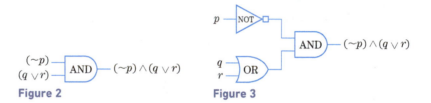

Figure 2   **Figure 3**

Then draw the circuits for each of the components, and complete the diagram as in Fig. 3.

To see the output for the inputs $p = 1$, $q = 1$ and $r = 0$, we note that the inverter changes the input at the first gate to the output of 0. The OR gate converts the inputs $q = 1$ and $r = 0$ to output value 1. The final AND gate has inputs 0 and 1, and therefore, the output from the AND gate is 0.   **» Now Try Exercise 5**

The logic that we have studied to this point helps to simplify complicated circuits. This is especially useful in the efficient design of complicated circuitry for the many devices dependent on electrical current. We repeatedly use the equivalences in the Table of Logical Equivalences (Table 1 of Section 11.4).

EXAMPLE 2   **Finding the Logical Statement from a Circuit** Write the logical statement represented by the circuit drawn in Fig. 4.

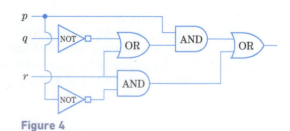

Figure 4

SOLUTION $[p \wedge (\sim q \vee r)] \vee (\sim p \wedge r)$ **» Now Try Exercise 1**

EXAMPLE 3	**Simplifying a Circuit** Simplify the circuit in Example 2.

SOLUTION The expression $[p \wedge (\sim q \vee r)] \vee (\sim p \wedge r)$ can be simplified by using Table 1 of Section 11.4.

$$[p \wedge (\sim q \vee r)] \vee (\sim p \wedge r) \Leftrightarrow (p \wedge \sim q) \vee (p \wedge r) \vee (\sim p \wedge r)$$
$$\Leftrightarrow (p \wedge \sim q) \vee [r \wedge (p \vee \sim p)]$$
$$\Leftrightarrow (p \wedge \sim q) \vee (r \wedge t)$$
$$\Leftrightarrow (p \wedge \sim q) \vee r$$

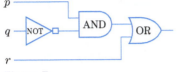

Figure 5

We draw the circuit in its simplified form in Fig. 5.

» Now Try Exercise 9

We say that two logic circuits are **equivalent** if they perform the same function. That is, for every possible input, the output produced in the first circuit is the same as the output produced in the second circuit. We can show this equivalency by using truth tables or by relying on Table 1, Section 11.4.

Circuits with NAND, NOR, and XOR

We can design circuits that behave according to the logical rules of the connectives NAND (Not AND), NOR (Not OR), and XOR (eXclusive OR). The symbols for these gates are given in Tables 4–6. The input and output values are also displayed. Figures 6–8 illustrate the elementary circuits for NAND, NOR, and XOR.

Table 4 The NAND Gate

Input		Output
p	q	p NAND q
0	0	1
0	1	1
1	0	1
1	1	0

Figure 6

Table 5 The NOR Gate

Input		Output
p	q	p NOR q
0	0	1
0	1	0
1	0	0
1	1	0

Figure 7

Table 6 The XOR Gate

Input		Output
p	q	p XOR q
0	0	0
0	1	1
1	0	1
1	1	0

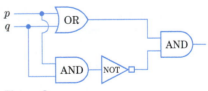

Figure 8

Check Your Understanding 11.7

Solutions can be found following the section exercises.

1. **(a)** Simplify the circuit shown in Fig. 9 by using the Table of Logical Equivalences (Table 1, Section 11.4).
   (b) Check your result with a truth table.

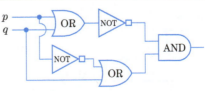

Figure 9

EXERCISES 11.7

1. Write the logic statement represented by Fig. 10.

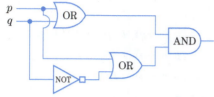

Figure 10

2. Write the logic statement represented by Fig. 11.

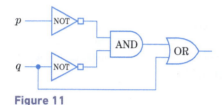

Figure 11

3. Write the logic statement represented by Fig. 12.

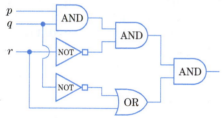

Figure 12

4. Write the logic statement represented by Fig. 13.

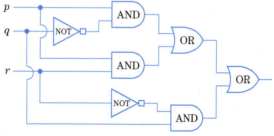

Figure 13

Draw the logic circuit that represents each of the logical statements in Exercises 5–8. Determine the output when the variables have the values given.

5. $p \wedge (\sim q \vee r)$; for $p = 1, q = 0, r = 1$

6. $(p \wedge q) \vee (p \vee r)$; for $p = 1, q = 0, r = 1$

7. $(\sim p \vee \sim q) \wedge (\sim p \wedge r)$; for $p = 1, q = 0, r = 1$

8. $\sim [(p \wedge \sim q) \vee (p \wedge r)]$; $p = 0, q = 0, r = 1$

9. Simplify the logic circuit in Fig. 14 as much as possible.

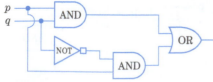

Figure 14

10. Simplify the logic circuit in Fig. 15 as much as possible.

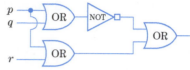

Figure 15

11. Simplify the logic circuit in Fig. 16 as much as possible.

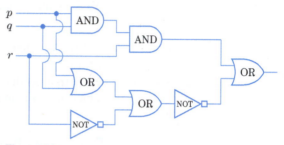

Figure 16

12. Simplify the logic circuit in Fig. 17 as much as possible.

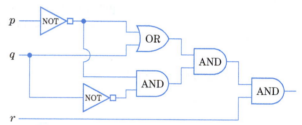

Figure 17

13. Design a logic circuit with two inputs (use p and q) that always has output 1.

14. Design a logic circuit with two inputs (use p and q) that always has output 0.

15. Design a logic circuit that acts as an OR gate for inputs p and q, using three NOR gates.

16. Design a logic circuit that acts as an AND gate for inputs p and q, using three NAND gates.

17. Design a logic circuit that acts as an XOR gate for inputs p and q, using three gates other than XOR or OR.

18. Design a logic circuit for $(p$ NAND $q)$ NAND $(p$ NAND $q)$, and show the outputs for all values of p and q on a truth table.

19. **The Game of Match** In the game of "match," each of two players, A and B, has control of an input switch. At the signal, each player allows the current to flow (ON) or stops the current (OFF). If both players choose the same input (both ON or both OFF), player A wins. If their inputs differ, player B wins. Design a circuit so that the output is ON only when player A wins.

20. **Switch Design for a Lecture Hall** In designing a large four-walled lecture hall, the architect has placed a door on each wall, and next to each door there is a light switch. Design a logic circuit so that the light can be turned on or off from any switch. (*Hint:* The light should go on if an even number of switches are on, and the light should go off if an odd number of switches are on. Why?)

TECHNOLOGY EXERCISES

21. Use the Wolfram|Alpha function BooleanMinimize to carry out the simplification in Example 3.

22. Use the Wolfram|Alpha function BooleanMinimize to find a simpler Boolean expression equivalent to the Boolean expression given by the circuit in Exercise 10.

Solutions to Check Your Understanding 11.7

1. (a) The circuit can be translated into logical symbols:

$$\sim(p \vee q) \wedge (\sim p \vee q)$$

Using the Table of Logical Equivalences (Table 1, Section 11.4), we have

$$\sim(p \vee q) \wedge (\sim p \vee q) \Leftrightarrow (\sim p \wedge \sim q) \wedge (\sim p \vee q)$$
$$\Leftrightarrow ((\sim p \wedge \sim q) \wedge \sim p) \vee ((\sim p \wedge \sim q) \wedge q)$$
$$\Leftrightarrow (\sim p \wedge \sim q) \vee (\sim p \wedge (\sim q \wedge q))$$
$$\Leftrightarrow (\sim p \wedge \sim q) \vee (\sim p \wedge c).$$
$$\Leftrightarrow (\sim p \wedge \sim q) \vee c$$
$$\Leftrightarrow \sim p \wedge \sim q.$$

The circuit equivalent is displayed in Fig. 18.

Figure 18

(b) To check the calculation, we use truth tables.

p	q	$p \vee q$	$\sim(p \vee q)$	$\wedge$	$\sim p$	$\vee$	q	$\sim p$	$\wedge$	$\sim q$
T	T	T	F	F	F	T	T	F	F	F
T	F	T	F	F	F	F	F	F	F	T
F	T	T	F	F	T	T	T	T	F	F
F	F	F	T	T	T	T	F	T	T	T
(1)	(2)	(3)	(4)	(7)	(5)	(6)		(8)		

Compare columns (7) and (8) to see that the statements are equivalent and that the circuits are equivalent.

CHAPTER 11 Summary

KEY TERMS AND CONCEPTS	EXAMPLES
11.1 Introduction to Logic	
A **statement** is a declarative sentence that is either true or false.	Basketball is a sport. Ben Franklin was President of the USA.
The **truth value** of a statement is either TRUE or FALSE.	The first statement has truth value TRUE, and the second has truth value FALSE.
A **compound statement** is formed by combining statements with connectives like "and," "or," and "not." A **simple statement** is a statement that is not a compound statement.	The number 2 is even and the number 5 is odd. The number 2 is even or the number 5 is odd. The number 2 is not even.
We represent simple statements by a single letter, such as p, q, or r. We represent "and" by $\wedge$, "or" by $\vee$, and "not" by $\sim$.	Let p denote "The number 2 is even" and q denote "The number 5 is odd." Then the statement "The number 2 is even and the number 5 is odd" is written $p \wedge q$. The statement "The number 2 is not even" is written $\sim p$.

KEY TERMS AND CONCEPTS	EXAMPLES

11.2 Truth Tables

A **statement form** is an expression formed from simple statements and connectives according to the following rules:

1. A simple statement is a statement form.
2. If p and q are statement forms, then so are $\sim p, p \wedge q$, and $p \vee q$.

A **truth table** is used to determine the truth value of a compound statement. Each simple statement can be either TRUE or FALSE. Each row represents a different possible combination of truth values for the simple statements.

p	q	$\sim p$	$p \wedge q$	$p \vee q$
T	T	F	T	T
T	F	F	F	T
F	T	T	F	T
F	F	T	F	F

A **tautology** is a statement form that has truth value TRUE regardless of the truth values of its component statement forms. The final column of the truth table will consist of only T's.

A **contradiction** is a statement form that has truth value FALSE regardless of the truth values of its component statement forms. The final column of the truth table will consist only of F's.

11.3 Implication

The expression $p \rightarrow q$ is read, "if p, then q." The **conditional** statement form $p \rightarrow q$ has truth value FALSE only when p is true and q is false. We call p the **hypothesis** and q the **conclusion**.

p	q	$p \rightarrow q$
T	T	T
T	F	F
F	T	T
F	F	T

The implication "If $2 + 3 = 7$, then Paris is in France" has the hypothesis "$2 + 3 = 7$" and the conclusion "Paris is in France." Since the hypothesis "$2 + 3 = 7$" is FALSE, the compound statement is TRUE by the truth table regardless of the truth value of the conclusion.

The **converse** of the implication $p \rightarrow q$ is given by interchanging the hypothesis and the conclusion to get $q \rightarrow p$.

"If Paris is in France, then $2 + 3 = 7$" is the converse of the implication above. This statement has truth value FALSE, since the conclusion "$2 + 3 = 7$" is FALSE while the hypothesis "Paris is in France" is TRUE.

The **biconditional** statement form is written $p \leftrightarrow q$ and is TRUE when the truth values of p and q are the same. Read $p \leftrightarrow q$ as "p if and only if q".

p	q	$p \leftrightarrow q$
T	T	T
T	F	F
F	T	F
F	F	T

The **order of precedence** for logical connectives is $\sim, \wedge, \vee, \rightarrow, \leftrightarrow$.

Insert parentheses to show the proper order of the application of the connectives in $p \wedge q \vee r \rightarrow \sim s \wedge r$.

$$[(p \wedge q) \vee r] \rightarrow [(\sim s) \wedge r]$$

KEY TERMS AND CONCEPTS	EXAMPLES

11.4 Logical Implication and Equivalence

Two statement forms that have the same truth table are called **logically equivalent**. If we use capital letters, like P, Q, and R, to represent statement forms, then we write $P \Leftrightarrow Q$ when P and Q are logically equivalent.

Show that $p \to q$ is logically equivalent to $\sim p \vee q$.

p	q	$p \to q$	$\sim p \vee q$
T	T	T	F T
T	F	F	F F
F	T	T	T T
F	F	T	T T
(1)	(2)	(3)	(4) (5)

Since columns 3 and 5 are the same, $p \to q$ is logically equivalent to $\sim p \vee q$.

The **contrapositive** of the statement form $p \to q$ is $\sim q \to \sim p$ and is logically equivalent to $p \to q$.

The contrapositive of "If $2 + 3 = 7$, then Paris is in France" is "If Paris is not in France, then $2 + 3 \neq 7$."

De Morgan's laws are useful logical equivalences for negating conjunctions and disjunctions.

1. $\sim(p \vee q) \Leftrightarrow (\sim p \wedge \sim q)$
2. $\sim(p \wedge q) \Leftrightarrow (\sim p \vee \sim q)$

See Table 1 for a list of logical equivalences.

Negate the statement "The earth's orbit is round and a year has 365 days."

This statement is of the form $p \wedge q$, so De Morgan's law gives the negation as "The earth's orbit is not round or a year does not have 365 days."

Given statement forms P and Q, we say that P **logically implies** Q (written $P \Rightarrow Q$) whenever $P \to Q$ is a tautology.

See Table 2 for a list of logical implications.

Rewrite the statement form $(p \to q) \wedge (q \vee r)$, using only the connectives $\sim$ and $\vee$.

$$(p \to q) \wedge (q \vee r)$$
$\Leftrightarrow (\sim p \vee q) \wedge (q \vee r)$	Implication
$\Leftrightarrow (q \vee \sim p) \wedge (q \vee r)$	Commutative law
$\Leftrightarrow q \vee (\sim p \wedge r)$	Distributive law
$\Leftrightarrow q \vee (\sim p \wedge \sim\sim r)$	Double negation
$\Leftrightarrow q \vee \sim(p \vee r)$	De Morgan's law

11.5 Valid Argument

An **argument**, or **proof**, is a set of statements $H_1, H_2, \ldots, H_n$ (called hypotheses), each of which is assumed to be true, and a statement C (called the conclusion) that is claimed to be deduced from them.

The argument is **valid** if and only if

$$H_1 \wedge H_2 \wedge \ldots \wedge H_n \Rightarrow C.$$

See Table 1 for the Rules of Inference (a restatement of the list of logical implications in Section 11.4).

See Example 3 on pages 520.

Indirect Proof

To prove $H_1 \wedge H_2 \wedge \ldots \wedge H_n \Rightarrow C$, we assume that the conclusion C is FALSE and then prove that at least one of the hypotheses H_i must be FALSE.

See Example 5 on page 521.

KEY TERMS AND CONCEPTS	EXAMPLES
11.6 Predicate Calculus	

An **open sentence** $p(x)$ is a declarative sentence that becomes a statement when x is given a particular value chosen from a universe of values. An open sentence is also called a **predicate**.

$p(x) =$ "If $x > 4$, then $x + 10 > 14$" is an open sentence, where $x \in U = \{1, 2, 3, 4, \ldots\}$. Note that $p(1)$ is true since the hypothesis is FALSE.

We represent "For all $x \in U$, $p(x)$" by the expression $\forall x \in U \, p(x)$. Note that this is TRUE if and only if $p(x)$ is TRUE for every $x \in U$. The symbol $\forall$ is called the **universal qualifier** and is read "for all" or "for every."

Note that since $p(x) =$ "If $x > 4$, then $x + 10 > 14$" is TRUE for every $x \in U = \{1, 2, 3, 4, \ldots\}$, we can write $\forall x \in U \, p(x)$.

We represent "There exists an x in U such that $p(x)$" by $\exists x \in U \, p(x)$. Note that this is TRUE if and only if there is at least one element $x \in U$ such that $p(x)$ is TRUE. The symbol $\exists$ is called the **existential qualifier** and is read "there exists" or "for some."

Let $U = \{1, 2, 3, 4, 5, 6, 7, 8\}$.
  (a) $\exists x (x > 5)$ is TRUE, since $6 \in U$ and $6 > 5$.
  (b) $\forall x (x + 3 < 15)$ is TRUE because, for every $x \in U$, $x + 3$ is less than 15.

Versions of De Morgan's laws are used to negate quantified statements.

1. $\sim [\exists x \, p(x)] \Leftrightarrow [\forall x \sim p(x)]$
2. $\sim [\forall x \, p(x)] \Leftrightarrow [\exists x \sim p(x)]$

See Table 1 for De Morgan's laws applied to predicates with two variables.

Negate the statement "There exists a positive integer x such that $x^2 - x - 2 = 0$."

Answer: "It is not the case that there exists a positive integer x such that $x^2 - x - 2 = 0$." Or "For all positive integers x, $x^2 - x - 2 \neq 0$." Note that $\sim [\exists x (x^2 - x - 2 = 0)] \Leftrightarrow [\forall x (x^2 - x - 2 \neq 0)]$.

11.7 Logic Circuits	

Logic circuits provide a way of visually examining a compound statement and determining the truth value of the statement, given specific truth values of the component simple statements.

The **NOT gate**, or **inverter**, negates the input truth value and corresponds to the logical connective $\sim$.

The **AND gate** corresponds to the logical connective $\wedge$.

The **OR gate** corresponds to the logical connective $\vee$.

CHAPTER 11   Fundamental Concept Check Exercises

1. What is a logical statement?

2. Write down the truth tables of the simple logical connectives $\wedge$, $\vee$, $\sim$, $\rightarrow$, $\leftrightarrow$.

3. When is $p \rightarrow q$ a FALSE statement?

4. What do we mean by "logical equivalence"? Explain how you might use a truth table to determine logical equivalence.

5. State De Morgan's laws. When should you use them?

6. Given the implication $p \rightarrow q$, what is the hypothesis? What is the conclusion?

7. Given the implication $p \rightarrow q$, write down the contrapositive and the converse. If the implication is TRUE, what can we say about the truth of the contrapositive and that of the converse?

8. Give an example (in words) of an implication that is TRUE, and write its contrapositive and converse, making sure that you know the difference.

9. Write the negation of $p \rightarrow q$ without using the arrow symbol.

10. Associate each of the logical connectives AND, OR, NOT, IF . . . THEN, and XOR with a logical symbol.

11. What is a tautology? Describe how you would prove that a statement is a tautology.

12. State De Morgan's laws for quantified statements.

CHAPTER 11   Review Exercises

1. Determine which of the following are statements:
   (a) The universe is 1 billion years old.
   (b) What a beautiful morning!
   (c) Mathematics is an important part of our culture.
   (d) He is a gentleman and a scholar.
   (e) All poets are men.

2. Write each of the following statements in "if . . . , then . . . " format:
   (a) That the lines are perpendicular implies that their slopes are negative reciprocals of each other.
   (b) Goldfish can live in a fishbowl only if the water is aerated.
   (c) Jane uses her umbrella if it rains.
   (d) Only if Morris eats all of his food does Sally give him a treat.

3. Write the contrapositive and the converse of each of the statements.
   (a) The Yankees are playing in Yankee Stadium if they are in New York City.
   (b) If the Richter scale indicates that the earthquake is a 7, then the quake is considered major.
   (c) If a coat is made of fur, it is warm.
   (d) If Jane is in Russia, she is in Moscow.

4. Negate the statements.
   (a) If two triangles are similar, their sides are equal.
   (b) There exists a real number x such that $x^2 = 5$.
   (c) For every positive integer n, if n is even, then n^2 is even also.
   (d) There exists a real number x such that $x^2 + 4 = 0$.

5. Determine which of the statements are tautologies.
   (a) $p \vee \sim p$
   (b) $(p \rightarrow q) \leftrightarrow (\sim p \vee q)$
   (c) $(p \wedge \sim q) \leftrightarrow \sim(\sim p \wedge q)$
   (d) $[p \rightarrow (q \rightarrow r)] \leftrightarrow [(p \rightarrow q) \rightarrow r]$

6. Construct a truth table for each of the following statements:
   (a) $p \rightarrow (\sim q \vee r)$
   (b) $p \wedge [q \leftrightarrow (r \wedge p)]$

7. True or false?
   (a) $[p \wedge (\sim p \vee q)] \Rightarrow q$
   (b) $[(p \rightarrow q) \wedge q] \Rightarrow p$

8. True or false?
   (a) $(\sim q \rightarrow \sim p) \Leftrightarrow (p \rightarrow q)$
   (b) $[(p \rightarrow q) \wedge (r \rightarrow p)] \Leftrightarrow q$

9. True or false?
   (a) $[(p \rightarrow q) \wedge \sim p] \Rightarrow \sim q$
   (b) $[(p \rightarrow q) \wedge \sim q] \Rightarrow \sim p$

10. For the given input for A and B, determine the output for Z.

    LET $C = 3 * A + B$

    IF $((C > 0)$ AND $(B > 3))$

    THEN LET $Z = C$

    ELSE LET $Z = 100$

    (a) A = 4, B = 5
    (b) A = 10, B = 2
    (c) A = −4, B = 5
    (d) A = −10, B = −2

11. For the given inputs, determine the output for Z.

    IF $((X > 0)$ AND $(Y > 0))$ OR $(C \geq 10)$

    THEN LET $Z = X * Y$

    ELSE LET $Z = X + Y$

    (a) X = 5, Y = 10, C = 10
    (b) X = 5, Y = −5, C = 10
    (c) X = −10, Y = −5, C = 2
    (d) X = 2, Y = 5, C = 4

12. Assume that the following statement is TRUE:

    If the deduction is allowed,
    then the tax law has been revised.

    Give the truth value of each of the following statements if it can be determined directly from the truth value of the original:
    (a) If the tax law has been revised, the deduction is allowed.
    (b) If the tax law has not been revised, the deduction is not allowed.
    (c) The deduction is allowed only if the tax law has been revised.
    (d) The deduction is allowed if the tax law has been revised.
    (e) The deduction is allowed or the tax law has been revised.

13. Assume that the following statement is TRUE:

    If the voter is over the age of 21 and has a driver's license, the voter is eligible for free driver education.

    Determine the truth value of each of the following directly from the original, if possible:
    (a) If the voter is eligible for free driver education, the voter is over 21 and has a driver's license.
    (b) If the voter is not eligible for free driver education, the voter is not over 21 and does not have a driver's license.
    (c) If the voter is not eligible for free driver education, the voter either is not over 21 or does not hold a driver's license.

14. Assume that the statement "All mathematicians like rap music" is TRUE. Determine the truth value of the following from this assumption, if possible:
    (a) Some mathematicians like rap music.
    (b) Some mathematicians do not like rap music.
    (c) There exists no mathematician who does not like rap music.

15. Assume that the statement "Some apples are not rotten" is TRUE. Determine the truth value of each of the following, using only that fact, if possible:
    (a) Some apples are rotten.
    (b) All apples are not rotten.
    (c) Not all apples are rotten.

16. Show that the argument is valid: If taxes go up, then I sell the house and move to India. I do not move to India. Therefore, taxes do not go up.

17. Show that the argument is valid: I study mathematics and I study business. If I study business, then I cannot write poetry or I cannot study mathematics. Therefore, I cannot write poetry.

18. Show that the argument is valid: If I shop for a dress, I wear high heels. If I have a sore foot, then I do not wear high heels. I shop for a dress. Therefore, I do not have a sore foot.

19. Show that the argument is valid: Asters or dahlias grow in the garden. If it is spring, then asters do not grow in the garden. It is spring. Therefore, dahlias grow in the garden.

20. Use indirect proof to show that the argument is valid: If the professor gives a test, then Nancy studies hard. If Nancy has a date, she takes a shower. If the professor does not give a test, then Nancy does not take a shower. Nancy has a date. Therefore, Nancy studies hard.

21. Draw the logic circuit corresponding to the expression $(\sim p \vee q) \wedge (q \wedge r)$, and determine the output when the input values are $p = 0$, $q = 0$, $r = 1$.

22. Simplify the logic circuit in Fig. 19 by using the Table of Logical Equivalences (Table 1, Section 11.4) to derive its equivalent.

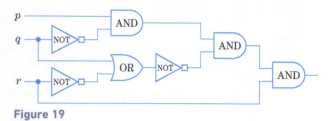

Figure 19

Conceptual Exercises

23. Designing a Voting Machine Design a voting machine for a committee of three people. A motion is presented to the committee for a vote. Each person can vote "yes" by pushing a button (ON) and may vote "no" by not pushing the button. The majority rules. The circuit that you design should light up when the motion passes. Design a similar machine for a committee of five people. An abstention is considered a "no" vote.

24. Design a logic circuit with three inputs for which the output (current) is always OFF (0) regardless of the input.

25. Construct a statement equivalent to p XOR q, using only NOR and AND.

A Logic Puzzle

Denise, Miriam, Sally, Nelson, and Bob are students at the same university. Each is in a different mathematics course, and each is in a different year at the university. From the clues given, determine what mathematics course each is enrolled in and the year (freshman, sophomore, junior, senior, graduate student) each is in.

A. Sally is a sophomore.
B. Miriam (who is neither a junior nor a freshman) is taking a statistics course.
C. Neither the person who is taking calculus (who is not Bob) nor the one taking finite math is the person who is a freshman.
D. The student who is a senior is enrolled in algebra.
E. Neither Nelson (who is not a junior) nor Bob is taking precalculus.

Solve the puzzle by filling in the given chart with O to signify "yes" and X to signify "no." For example, clue A indicates that Sally is a sophomore, so put an O at the intersection of the "Sally" column and the "Sophomore" row. You can conclude that the other four students are not sophomores and that Sally is not a freshman, junior, senior, or graduate student. So you can put eight X's in the chart to represent these conclusions. Solve the problem by working with both the clues and the chart.

	Bob	Denise	Miriam	Nelson	Sally	Freshman	Sophomore	Junior	Senior	Graduate
Algebra										
Finite Math										
Statistics										
Precalculus										
Calculus										
Freshman										
Sophomore										
Junior										
Senior										
Graduate										

Appendices

APPENDIX A ## Areas Under the Standard Normal Curve

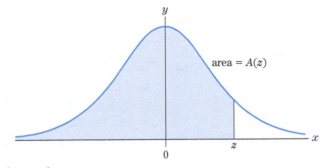

Table 1 Areas under the standard normal curve

z	$A(z)$	z	$A(z)$	z	$A(z)$	z	$A(z)$	z	$A(z)$
−3.50	0.0002	−2.05	0.0202	−0.60	0.2743	0.85	0.8023	2.20	0.9861
−3.45	0.0003	−2.00	0.0228	−0.55	0.2912	0.90	0.8159	2.25	0.9878
−3.40	0.0003	−1.95	0.0256	−0.50	0.3085	0.95	0.8289	2.30	0.9893
−3.35	0.0004	−1.90	0.0287	−0.45	0.3264	1.00	0.8413	2.3263	0.9900
−3.30	0.0005	−1.85	0.0322	−0.40	0.3446	1.05	0.8531	2.35	0.9906
−3.25	0.0006	−1.80	0.0359	−0.35	0.3632	1.10	0.8643	2.40	0.9918
−3.20	0.0007	−1.75	0.0401	−0.30	0.3821	1.15	0.8749	2.45	0.9929
−3.15	0.0008	−1.70	0.0446	−0.25	0.4013	1.20	0.8849	2.50	0.9938
−3.10	0.0010	−1.65	0.0495	−0.20	0.4207	1.25	0.8944	2.55	0.9946
−3.05	0.0011	−1.60	0.0548	−0.15	0.4404	1.2816	0.9000	2.60	0.9953
−3.00	0.0013	−1.55	0.0606	−0.10	0.4602	1.30	0.9032	2.65	0.9960
−2.95	0.0016	−1.50	0.0668	−0.05	0.4801	1.35	0.9115	2.70	0.9965
−2.90	0.0019	−1.45	0.0735	0.00	0.5000	1.40	0.9192	2.75	0.9970
−2.85	0.0022	−1.40	0.0808	0.05	0.5199	1.45	0.9265	2.80	0.9974
−2.80	0.0026	−1.35	0.0885	0.10	0.5398	1.50	0.9332	2.85	0.9978
−2.75	0.0030	−1.30	0.0968	0.15	0.5596	1.55	0.9394	2.90	0.9981
−2.70	0.0035	−1.25	0.1056	0.20	0.5793	1.60	0.9452	2.95	0.9984
−2.65	0.0040	−1.20	0.1151	0.25	0.5987	1.6449	0.9500	3.00	0.9987
−2.60	0.0047	−1.15	0.1251	0.30	0.6179	1.65	0.9505	3.05	0.9989
−2.55	0.0054	−1.10	0.1357	0.35	0.6368	1.70	0.9554	3.10	0.9990
−2.50	0.0062	−1.05	0.1469	0.40	0.6554	1.75	0.9599	3.15	0.9992
−2.45	0.0071	−1.00	0.1587	0.45	0.6736	1.80	0.9641	3.20	0.9993
−2.40	0.0082	−0.95	0.1711	0.50	0.6915	1.85	0.9678	3.25	0.9994
−2.35	0.0094	−0.90	0.1841	0.55	0.7088	1.90	0.9713	3.30	0.9995
−2.30	0.0107	−0.85	0.1977	0.60	0.7257	1.95	0.9744	3.35	0.9996
−2.25	0.0122	−0.80	0.2119	0.65	0.7422	2.00	0.9772	3.40	0.9997
−2.20	0.0139	−0.75	0.2266	0.70	0.7580	2.05	0.9798	3.45	0.9997
−2.15	0.0158	−0.70	0.2420	0.75	0.7734	2.10	0.9821	3.50	0.9998
−2.10	0.0179	−0.65	0.2578	0.80	0.7881	2.15	0.9842		

APPENDIX B Using the TI-84 Plus Graphing Calculator

Functions

Functions are graphed in a rectangular window like the one shown in Fig. 1. The numbers on the *x*-axis range from **Xmin** to **Xmax**, and the numbers on the *y*-axis range from **Ymin** to **Ymax**. The distances between tick marks are **Xscl** and **Yscl** on the *x*- and *y*-axes, respectively. To specify these quantities, press WINDOW and type in the values of the six variables. Figure 2 gives the settings associated with the window in Fig. 4. The axis ranges corresponding to this setting are often denoted by [−4, 4] *by* [−5, 8]. (*Notes:* To enter a negative number, use the (−) key on the bottom row of the calculator. The values of ΔX and **TraceStep** will be set automatically by the calculator.)

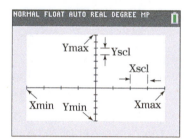

Figure 1 Typical window

```
NORMAL FLOAT AUTO REAL DEGREE MP
WINDOW
  Xmin=-4
  Xmax=4
  Xscl=1
  Ymin=-5
  Ymax=8
  Yscl=1
  Xres=1
 ΔX=.03030303030303
 TraceStep=.06060606060606
```

Figure 2 Settings for Fig. 4

```
NORMAL FLOAT AUTO REAL RADIAN MP
 Plot1  Plot2  Plot3
■\Y1☐X²-2
■\Y2=2X
■\Y3=√1+X
■\Y4=3/4
■\Y5=2^X
■\Y6=5Y3
■\Y7=
```

Figure 3 Y=editor

To specify functions, press the Y= key and type expressions next to the function names Y_1, Y_2, ... (*Note:* To erase an expression, use the arrow keys to move the cursor anywhere on the expression, and press CLEAR.) In the screen of Fig. 3, called the "**Y=** editor," several functions

have been specified. The expressions were produced with the following keystrokes.

Y_1: X,T,θ,n x^2 − 2

Y_2: 2 X,T,θ,n (Notice that a multiplication sign is not needed.)

Y_3: 2nd [√] 1 + X,T,θ,n ▶

Y_4: MATH ◀ 1 3 ▼ 4 ▶

Y_5: 2 ^ X,T,θ,n ▶ (^ is the symbol for exponentiation.)

Y_6: 5 ALPHA [F4] 3 (This is the standard method for entering a function.)

Notice that in Fig. 3, the equal sign in Y_1 is highlighted, whereas the other equal signs are not highlighted. This highlighting can be toggled by moving the cursor to an equal sign and pressing ENTER. Functions with highlighted equal signs are said to be *selected*. Pressing the GRAPH key instructs the calculator to graph all selected functions. (*Note:* The words **Plot1 Plot2 Plot3** on the first row are used for statistical plots. The two symbols preceding each function are used to specify one of fifteen possible colors and eight possible styles for the graph of the function.)

Press GRAPH to obtain Fig. 4. Then Press TRACE and press ▶, the right-arrow, 29 times to obtain Fig. 5. Each time a right- or left-arrow key is pressed, the trace cursor moves along the curve and the coordinates of the trace cursor are displayed.

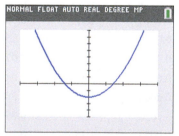

Figure 4 Graph of a function

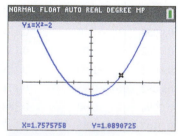

Figure 5 Trace point

To approximate the function value for a specific value of *x*, move the trace cursor as close as possible to the value of *x* and read the *y*-coordinate of the point. For a more precise function value, press 2nd [CALC] 1, type in a value for **X** (such as 2.5), and press ENTER. See Fig. 6. Additional precise function values can be obtained by just typing in the values for **X** without first pressing 2nd [CALC] **1**.

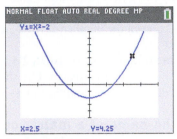

Figure 6 Point locator

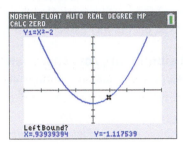

Figure 7 Point to left of a zero

A point at which the graph crosses the *x*-axis is called an *x-intercept*. The coordinates of an *x*-intercept can be approximated by tracing. The *x*-coordinate of an *x*-intercept of the function **Y₁** is called a *zero* of **Y₁** or a *root* of the equation **Y₁ = 0**. Figure 4 shows that **Y₁** has a zero between 1 and 2, and the value of the zero is about 1.5. For a precise value of the zero, press 2nd [CALC] **2** and answer the questions. See Fig. 7. Reply to **"Left Bound?"** by moving the trace cursor to a point whose *x*-coordinate is less than the root and pressing ENTER. Reply to **"Right Bound?"** by moving the trace cursor to a point whose *x*-coordinate is greater than the root and pressing ENTER. See Fig. 8. Reply to **"Guess?"** by moving the trace cursor near the *x*-intercept and pressing ENTER. Figure 9 shows the resulting display. An alternative way to find a zero of a function is to reply to the questions by entering appropriate numbers. For instance, after pressing 2nd [CALC] **2**, respond to **"Left Bound?"** by typing in the number 1 and pressing ENTER, respond to **"Right Bound?"** by typing in the number 2 and pressing ENTER, and respond to **"Guess?"** by typing in the number 1.5 and pressing ENTER. The final screen will be identical to Fig. 9.

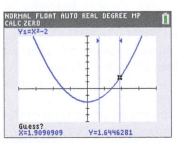

Figure 8 Point to right of a zero

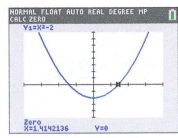

Figure 9 Zero of a function

So far, all operations were carried out while looking at the graph of a function. These same operations also can be carried out in the home screen, which is invoked by pressing 2nd [QUIT]. For instance, the function **Y₁** can be evaluated at 4 by entering **Y₁(4)** and pressing ENTER.

Return to the **Y=** editor by pressing Y= and then select **Y₂** so that now both **Y₁** and **Y₂** are selected. Press GRAPH to obtain the graphs of the two functions. Press TRACE, and then press the right-arrow key several times. Now press the down-arrow key several times. Each time the down-arrow key is pressed, the trace cursor moves from one curve to the other. The identity of the function containing the cursor is given in the top part of the screen. Move the cursor as close as possible to a point of intersection of the two curves to approximate the coordinates of the intersection point. For more precise values, press 2nd [CALC] **5** and answer the questions. Reply to **"First curve?"** by moving the trace cursor to one of the curves and pressing ENTER. Reply to **"Second curve?"** by moving the trace cursor to the other curve and pressing ENTER. Reply to **"Guess?"** by either moving the trace cursor near the point of intersection or typing in a number close to the *x*-coordinate of the point and pressing ENTER. Figure 10 shows the resulting display.

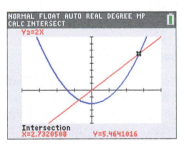

Figure 10 Intersection point

Tables

When you press 2nd [TABLE], a table of function values is displayed with a column for each selected function. (If more than five columns are displayed, you must press the right-arrow key to see the later columns.) Prior to invoking a table, you should press 2nd [TBLSET] to specify certain properties of the table. See Fig. 11. The values for **X** will begin with the setting of **TblStart** and increase by the setting of **ΔTbl**. For our purposes, the settings for **Indpnt** and **Depend** should always be **Auto**. After you press 2nd [TABLE] to display the table (see Fig. 12), you can use the up- and down-arrow keys to generate further values.

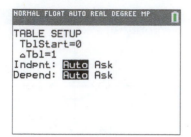
Figure 11 Table setup screen

Figure 12 Table screen

Matrices

The TI-84 Plus can store up to 10 matrices, referred to as [A], [B], [C], ... Pressing 2nd [MATRIX] produces the screen in Fig. 13 with the three menus **NAMES**, **MATH**, and **EDIT**. The **NAMES** menu is used to display the name of a matrix on the home screen, the **MATH** menu is used to perform certain operations on a single matrix, and the **EDIT** menu is used to define a new matrix or alter an existing matrix.

Figure 13 Matrix menu

To create a new matrix Press 2nd [MATRIX] ◄ to call up the **MATRIX EDIT** menu, and then press a number corresponding to one of the unused matrix names to obtain a matrix-entry screen. Type in the number of rows, press ENTER, type in the number of columns, and press ENTER to specify the size of the matrix. Then type in the first entry of the matrix, press ENTER, type in the next entry of the matrix, press ENTER, and so on until all entries have been entered. See Fig. 14.

Figure 14 Matrix editor

To alter an existing matrix Press 2nd [MATRIX] ◄ to call up the matrix **EDIT** menu, and then press the number corresponding to the name of the matrix to be altered. Move the cursor to any entry that you want to change, type in the new number, and press ENTER. You can change as many entries as you like, even the number of rows and columns.

To delete a matrix Press 2nd [MEM] **2 5** to obtain a list of all matrices that have been created. Use the down-arrow key to select the desired matrix, and then press ENTER DEL **2**.

To display the name of a matrix on the home screen Press 2nd [MATRIX] to obtain a list of all matrix names. Use the down-arrow key to select the desired matrix, and then press ENTER. Alternatively, from the matrix **NAMES** menu, type the number associated with the matrix.

To carry out elementary row operations The three elementary row operations are carried out with commands of the following forms from the **MATRIX/MATH** menu.

> **rowSwap** (*matrix, rowA, rowB*)
>   Interchange *rowA* and *rowB* of *matrix*
> ***row** (*value, matrix, row*)
>   Multiply *row* of *matrix* by *value*
> ***row+** (*value, matrix, rowA, rowB*)
>   Add *value * rowA* to *rowB* of *matrix*

When one of these commands is carried out, the resulting matrix is displayed but the stored matrix is not changed. Therefore, when a sequence of commands is executed to carry out the Gauss–Jordan elimination method, each command should be followed with STO ► *matrix* to change the stored matrix. (*Note:* To display the name of a matrix, press 2nd [MATRIX], cursor down to the matrix, and press ENTER.) For instance, if in Example 5 of Section 2.1 the original matrix is named [A], then the first three row operations are carried out with

> ***row(1/3,[A],1)** STO ► **[A]**
> ***row+(-4,[A],1,2)** STO ► **[A]**
> ***row+(2,[A],1,3)** STO ► **[A]**.

Lists

A list can be thought of as a sequence, or an ordered set, of up to 999 numbers. Although lists can have custom names, we will use the built-in names **L₁, L₂, . . . , L₆** that are found above numeric keys. To display **L₁** on the home screen, press 2nd [L₁]. The elements of list **L₁** can be referred to as $L_1(1), L_1(2), L_1(3), \ldots$. A set of numbers can be placed into a list by enclosing them with set braces and storing them in the list. However, often, the stat list editor provides the best way to place numbers into lists and to view lists.

To invoke the stat list editor, press STAT **1**. Initially, the stat list editor shows columns labeled **L₁, L₂, L₃, L₄** and **L₅**. See Fig. 15.

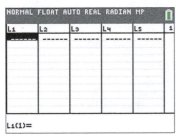

Figure 15 Stat list editor

To place a number into a list in the stat list editor
Move the cursor to the dashed line, type in the number, and press either the down-arrow key or the ENTER key.

To remove a number from a list in the stat list editor   Move the cursor to the number and press DEL.

To delete all numbers in a list from the stat list editor   Move the cursor to the list name at the top of the screen, and press CLEAR ENTER.

To create a scatter plot of a set of points whose x-coordinates are in L_1 and whose y-coordinates are in L_2   Press 2nd [STAT PLOT] **1**, create the screen in Fig. 16, and press GRAPH. (*Note:* Check that the window settings are adequate for displaying the points.)

Figure 16 Stat plot editor

Histograms

The following steps display a frequency histogram with rectangles centered above the integers $a, a + 1, \ldots, b$:

- Press STAT **1** to invoke the list editing screen shown in Fig. 15.
- Enter each number from a to b having nonzero frequency in the L_1 column and its associated frequency to its right in the L_2 column. Or, ignore the L_2 column and place each number (repeated according to its frequency) in the L_1 column.
- Press 2nd [STAT PLOT] **1** to invoke a screen similar to Fig. 16.
- Select "On" in the second line, the histogram icon in the Type line, and L_1 in the Xlist line. If you ignored the L_2 column in the second step, place 1 in the Freq line; otherwise, place L_2 in the Freq line.
- Press WINDOW to invoke the window screen in Fig. 2.

- Set $X\min = a - .5, X\max = b + .5, X\text{scl} = 1, Y\min \approx -.3*$(greatest frequency), $Y\max \approx 1.2*$(greatest frequency), $Y\text{scl} \approx .1*$(greatest frequency).
  Note: These Y settings allow ample space for the display of values while tracing.
- Press GRAPH to view the histogram.
- To view the height of a rectangle, press TRACE and use the arrow keys to move the cursor to the top center of the rectangle. The values of *min*, *max*, and *n* will be the x-coordinate of the left side of the rectangle, the x-coordinate of the right side of the rectangle, and the height of the rectangle, respectively.

A relative frequency histogram can be displayed by placing the relative frequencies in the L_2 column of the list editor. When this is done, the *greatest relative frequency* should be used in place of the *greatest frequency* when setting the values of Ymin, Ymax, and Yscl.

TVM Solver

TVM stands for "Time Value of Money." TVM Solver is a powerful calculating tool than can compute values discussed in Chapter 10, "The Mathematics of Finance." Compound interest problems, annuities, and mortgages involve five numbers: duration, interest rate, periodic payment (0 for compound interest problems), present value, and future value. Once four of the five numbers are known, TVM Solver can be used to calculate the fifth number. The TVM Solver screen, shown in Fig. 17, is invoked by pressing APPS ENTER ENTER.

Figure 17 TVM Solver

The eight lines of TVM Solver hold the following information.

N: The duration in terms of interest periods.

I%: The annual interest rate given as a percent; such as 6 or 4.5. In Chapter 10, its value would be $100r$.

PV: The present value. It is the principal of a compound interest problem and the loan amount of a mortgage. It has the value 0 for an increasing annuity.

PMT: The periodic payment. It is the rent of an annuity or mortgage and has value 0 in a compound interest problem.

FV: The future value. It is the balance in a compound interest problem and an annuity. It has the value 0 for a mortgage.

P/Y: The number of payments per year. For our purposes, it is the same as C/Y.

C/Y: The number of times interest is compounded per year. That is, 1, 2, 4, 12, 52, or 365.

PMT: When payments are paid—the *end* or *beginning* of each interest period. For our purposes, it will always be set to END.

Note: Amounts paid to the bank, such as the principal in a compound interest problem or the rent of a mortgage, are given as negative numbers.

To use TVM Solver:

- Enter all information except for one of the top five values, the value to be calculated.
- Move the cursor to the line containing the value to be calculated.
- Press [ALPHA] [SOLVE]. (*Note:* The green word SOLVE is located above the ENTER key.) A small square will appear to the left of the line, and the calculated value will appear in the line.

Figure 18 shows that the interest rate required for a deposit of $100 to double to $200 in 10 years with interest compounded quarterly is almost 7%. (The value entered for PV is negative, since it represents money paid to the bank.)

Figure 18 Calculate interest

Finance Functions

In addition to containing TVM Solver, the Finance menu contains several functions. The two functions most valuable to us, Eff and Nom, calculate effective and nominal interest rates. (To display one of these functions on the home screen, press [APPS] [ENTER], scroll to the function, and press [ENTER].) If interest rate r compounded m times per year is a nominal rate, then $r_{eff} = \mathtt{Eff(100r,m)/100}$. For instance, the effective rate found in Example 7(a) of Section 10.1 can be calculated as $\mathtt{Eff(3.65,4)/100}$.

Similarly, the value of $\mathtt{Nom(100r_{eff},m)/100}$ is the corresponding nominal rate. For instance, the value of $\mathtt{Nom(3.7,4)/100}$ is approximately 3.65%. *Note:* The values for effective and nominal rates calculated with Eff and Nom will differ slightly from those calculated in Section 10.1 due to rounding.

APPENDIX C Spreadsheet Fundamentals

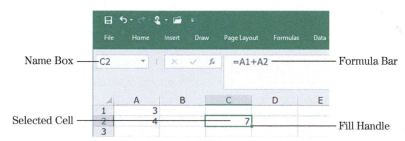

Figure 1 Upper-left corner of a Microsoft Excel spreadsheet

Cell

A *cell* is a small rectangle located at the intersection of a column and a row. Its *address*, or *reference* (also known as its *relative address* or *relative reference*), consists of its column letter followed by its row number. When you click on a cell, its border thickens and its address is displayed in the Name box. Such a cell is said to be *selected* or *active*. After you type text, a number, or a formula into an active cell and press the Enter key, the text, number, or the value of the formula is displayed in the cell. The exact information that you typed (known as the *content* of the cell) is displayed in the Formula bar whenever you reselect the cell.

Range

A *range of cells*, or *range*, is a rectangular array of more than one cell. The *address*, or *reference*, of a range consists of the address of the cell in the upper-left corner of the range, a colon (:), and then the address of the cell in the lower-right corner. To *select* a range, hold down the left mouse button while you drag the cursor from the upper-left cell to the lower-right cell. You can fill a range by entering information into each cell individually. However, often, you will need to fill an entire range with a single formula. If so, you should press Ctrl+Shift+Enter (instead of just Enter) after typing in the formula. Such a formula, called a *range formula* or an *array formula*, is automatically displayed in the Formula bar surrounded by braces.

Moving a Cell or Range

You can move a selected cell or range by dragging its border or by pressing Ctrl+X, clicking on the new location, and pressing Ctrl+V. If the content of another cell or range uses the moved cell's relative address, then the address automatically changes to reflect the new position of the moved cell or range. (You can prevent an address from changing by preceding its column letter and row number with dollar signs. Such an address is called an *absolute address*.)

Naming a Cell or Range

As an alternative to identifying cells or ranges with addresses, you can give a *name* to a cell or range. To specify a name, select the cell or range, click on the Name box, type in the name, and press the Enter key. (*Note:* The name will not be registered if you forget to press the Enter key.) Names can contain up to 255 characters (no spaces), must begin with a letter or underscore character, and cannot look like addresses. Deleting the contents of a cell or range does not delete its name. To delete a name, click Name Manager in the Defined Names group on the Formulas tab, click on the name that you want to delete in the Name Manager dialog box, click the Delete button, and then click OK to confirm the deletion.

Entering a Formula

Formulas are what make a spreadsheet a spreadsheet. To specify that the entry typed into a selected cell or range is a formula, you first type an equal sign (=). The equal sign is followed by a expression containing numbers, addresses, names, functions, and operators such as +, −, *, /, and ^ (exponentiation).

Copying a Formula

Select the cell or range whose content is the original formula. There are two methods to copy the formula. (*Note:* With both methods, all relative addresses will be altered appropriately.)

First method: Press Ctrl+C, select the location(s) to hold the copy or copies, and press Ctrl+V.

Second method: Hover the mouse pointer over the cell's fill handle until the pointer becomes a crosshair (+). Drag the fill handle vertically or horizontally along the cells that are to hold the copy or copies.

Installing Add-Ins

Two Excel tools that we use in this book are called *Solver* and *Data Analysis*. Solver is used to find solutions to systems of linear equations and to linear programming problems. Data Analysis is used to analyze data. To see whether these tools have been installed, click on the Data tab and see whether there is an Analyze (or Analysis) group containing the two items *Solver* and *Data Analysis*. If there is no such group or if the two items are not in the Analyze (or Analysis) group, click the File tab, click Options, click Add-Ins, select Excel Add-ins in the Manage box, click Go, select the check boxes next to Analysis ToolPak and Solver Add-in in the Add-ins window, and click OK.

Entering an Address into a Formula by Pointing

Instead of typing the address of a cell into a formula manually, just click on the cell. A dashed border will surround the cell, the cell's name or address will appear in the formula, and the word *Point* will appear in the status bar below the spreadsheet. To enter the address of a range, click on the upper-left cell of the range and drag the mouse cursor to the lower-right cell of the range. A dashed border will surround the range. When you release the mouse, the range's name or address will appear in the formula.

Converting Formulas to Values

Sometimes, the formulas that were used to fill a cell or range are no longer needed. To replace their contents with the formulas' values, select the cell or range, and press Ctrl+C. Click Paste in the Clipboard group on the Home tab, and then click the first icon under Paste Values.

Summing a Column or Row of Numbers

Click on the top number in the column or row, drag the mouse pointer to one or more cells beyond the column or row of numbers, and click on *AutoSum*. ("∑ AutoSum" is found in the Editing group on the Home tab.) The contents of the last cell will be a formula containing the SUM function, and the value of the cell will be the sum of the numbers in the column or row.

Using Solver to Solve a System of Linear Equations

The following steps use Solver to find the solution to the system of linear equations from Example 5 of Section 2.1.

- Give the cells A1, A2, and A3 the names x, y, and z, respectively. (There is no need to place any values into these cells.)
- In the cells B1, B2, and B3, place the formulas consisting of the left sides of the three equations. For instance, type the formula $=3*x-6*y+9*z$ into cell B1.
- On the Data tab, in the Analyze (or Analysis) group, click on Solver. A window entitled Solver Parameters will appear.
- Clear the contents, if any, of the Set Objective box. Ignore any entries appearing in the To box.

- Type **x,y,z** into the By Changing Variable Cells box.
- Click the Add button. An Add Constraint dialog box will appear.
  a) Enter **$B$1,=,** and **0** into the three boxes, and then press the Add button.
  b) Enter **$B$2,=,** and **−4** into the three boxes, and then press the Add button.
  c) Enter **$B$3,=,** and **7** into the three boxes, and then press the OK button.

A Solver Parameters window will appear. See Fig. 2. *Note:* The entries 0, −4, and 7 are the numbers on the right sides of the equations. Uncheck the Make Unconstrained Variables Non-Negative box if it is checked.

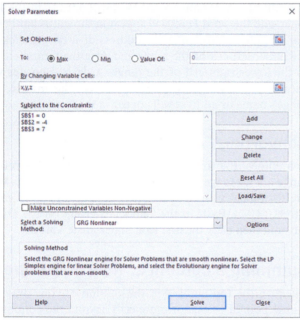

Figure 2 Data Analysis is in the Analyze (or Analysis) group on the Data tab.

- Click the Solve button. A window entitled Solver Results will appear.
- Click the OK button. The solution to the system of linear equations will appear in the x, y, and z cells. *Note:* The values may differ slightly from the true values due to rounding. For instance, the value for y may appear as something like −6.7E-07 (that is, 6.7×10^{-7}) instead of 0.

Simulating a Random Sample from a Probability Distribution

The steps that follow use Excel's Data Analysis tool to create a random sample from a discrete probability distribution. The sample numbers will be displayed in a range of m rows and n columns.

- Place the probability table in the first two columns of the spreadsheet. That is, column A should contain the possible outcomes (which must be numeric), and column B should contain the associated probabilities.

- Click on Data Analysis. Data Analysis is in the Analyze (or Analysis) group on the Data tab.
- Double-click on Random Number Generation in the Data Analysis dropdown list.
- Enter the number of columns (n) of random numbers into the "Number of Variables" box.
- Enter the number of random numbers (m) that you would like to generate in each column into the Number of Random Numbers box.
- Select "Discrete" as the Distribution.
- Place the address of the range containing the probability distribution in the "Value and Probability Input Range" box. This can be accomplished by placing the cursor in the box, clearing the box if necessary, dragging the mouse pointer from cell A1 to the lower-right corner of the probability table, and releasing the mouse button.
- Click on the Output Range circle, and enter the address of an $m \times n$ array of cells.
- Click on the OK button to generate the desired random sample.

Creating a Frequency Distribution Table and Histogram for the Preceding Random Sample Generated

The following steps use Excel's Data Analysis tool to create a frequency distribution and, optionally, a histogram for a range of cells containing the random numbers previously generated. Data Analysis is in the Analyze (or Analysis) group on the Data tab.

- Click on Data Analysis.
- Double-click on Histogram in the Data Analysis list.
- Place into the Input Range box the address of the range of cells containing the random sample.
- Place into the Bin Range box the address of the range of possible outcomes—that is, the first column of the probability table.
- Click the Output Range circle, and key in the address of any cell to the right of the cells already filled.
- If you would like also to display a histogram, click the small rectangle to the left of Chart Output.
- Click the OK button to see the frequency distribution table and possibly the histogram.

Using Goal Seek to Solve an Equation Having One Unknown

Suppose that the contents of cell B1 contains a formula whose value depends on the value of A1, and you know the value that you would like B1 to have. Then the *Goal Seek* tool can be used to determine the value of A1 that will produce the sought-after value in B1. For example, you can use Goal Seek to find the interest rate (compounded annually) for which $100 will grow to $146 after 11 years. Enter

=100*(1+A1)^11 into cell B1, and click on the Goal Seek tool. [Locate Goal Seek by clicking on What-If Analysis in the Forecast (or Data Tools) group on the Data tab.] A Goal Seek window will appear. Fill in the window as shown in Fig. 3, and then click the OK button. The solution, $\approx .035$ (that is, 3.50%), will appear in cell A1 of the spreadsheet. *Note:* Using the Goal Seek tool is sometimes referred to as *backsolving*.

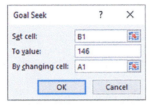

Figure 3

Using the Insert Function Dialog Box

To invoke the Insert Function dialog box, click on the Insert Function button located to the left of the Formula bar. The dialog box allows you to display the more than 300 available functions and descriptions of their arguments and what they do. When you select a function and click on OK (or double-click on the function), a Function Arguments dialog box appears to help you specify the arguments of the function. You can fill in each argument box by typing or pointing. To assist you when using the pointing method, you can click on the icon at the right side of an argument box to temporarily collapse the dialog box and thereby

make more of the spreadsheet visible. After you have filled the boxes for all of the arguments, press OK (or the Enter key) to fill a single cell with the value of the function, or press Ctrl+Shift+Enter to fill a range of cells.

Using Excel's Financial Functions

The five most important financial functions for our purposes are shown in Table 1.

Table 1

Function	Used to calculate	Function and required arguments
FV	future value	FV(rate, nper, pmt, pv)
PV	present value	PV(rate, nper, pmt, fv)
PMT	periodic payment	PMT(rate, nper, pv, fv)
NPER	number of interest periods	NPER(rate, pmt, pv, fv)
EFFECT	effective rate of interest	EFFECT(nominal_rate, npery)

The values of these functions and their parameters are signed according to their flows. Money received by you is given a positive value, and money paid by you is given a negative value. For instance, the values of savings account deposits and loan payments are negative, and the values of savings account withdrawals and loan amounts are positive. Table 2 shows how to use these functions to solve some examples from the text. Table 3 contains some other useful financial functions.

Table 2

Section	Example	Function and required arguments
10.1	3(a)	FV(1%, 4, 0, −100)
10.1	5	PV(0.25%, 104, 0, 10000)
10.1	7(a)	EFFECT(3.65%, 4)
10.2	1	FV(0.25%, 60, −100, 0)
10.2	3	PMT(2%, 16, 0, 300000)
10.2	4	PV(1.5%, 8, 3000, 0)
10.2	5	PMT(0.5%, 12, −10000, 0)
10.2	6	NPER(0.5%, −100, 0, 10000)
10.3	2(a)	PMT(0.75%, 360, 112475, 0)
10.3	5	PV(2%, 12, 200, 1000)

Table 3

Function	Used to calculate	Function and required arguments
RATE	interest rate per period	RATE(nper, pmt, pv, fv)
IPMT	interest portion of loan payment number *per*	IPMT(rate, per, nper, pv, fv)
PPMT	principal reduction portion of loan payment number *per*	PPMT(rate, per, nper, pv, fv)
NOMINAL	nominal rate of interest	NOMINAL(effective_rate, nper)

APPENDIX D   Wolfram|Alpha

Wolfram|Alpha is an answer engine developed by Wolfram Research, one of the world's most respected software companies. It is invoked on a computer by entering WolframAlpha.com in a Web browser. Also, it is available as a tablet or smartphone App. The Wolfram|Alpha screen displays a box similar to the one in Fig. 1. You compute with it on a computer by entering an instruction into the box, and then pressing the ENTER key or clicking on the equal sign. With a tablet or smartphone, you compute by touching the GO key. Wolfram|Alpha is free on computers. The tablet or smartphone App sells for just a few dollars. *Note:* The device must be connected to the Internet in order for Wolfram|Alpha to function.

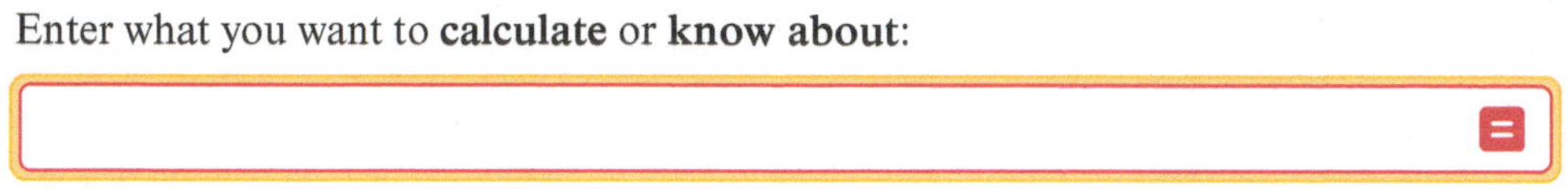

Enter what you want to **calculate** or **know about**:

Figure 1

Selected Answers

CHAPTER 1

Exercises 1.1, page 5

1., 3., 5.
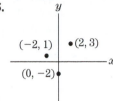

7.
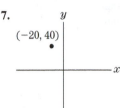

9. $(-2, 2)$ **11.** Yes **13.** No **15.** $m = 5, b = 8$ **17.** $m = 0, b = 3$
19. $y = -2x + 3$ **21.** $x = \frac{5}{3}$ **23.** $(2, 0), (0, 8)$ **25.** $(7, 0)$, none

27.
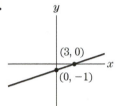

29.
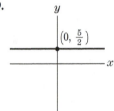

31.
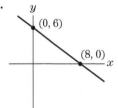

33.
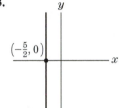

35. a, b, c, e **37. (a)** L_3 **(b)** L_1 **(c)** L_2 **39. (a)** Water at a temperature of 72°F was placed in the kettle.
(b) 162°F **(c)** $4\frac{2}{3}$ minutes or 4 minutes, 40 seconds

41. (a)
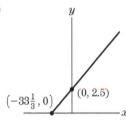
(b) In 1960, 2.5 trillion cigarettes were sold.
(c) 1980
(d) 7.3 trillion

43. (a)
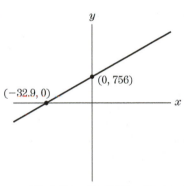
(b) Insurance cost $756 in 1999.
(c) $940
(d) The year 2023

45. (a) In 2000, 4.1% of entering college freshmen intended to major in biology. **(b)** 6.9% **(c)** 2007 **47. (a)** $20,027 **(b)** 2021
49. $y = -\frac{1}{2}x + 8$ **51.** $y = -\frac{5}{4}x + 5$ **53.** $y = 0$ **55.** $x = a$ **57.** $2x - y = -3$ **59.** $2x + 3y = -15$ **63.** $y = x - 9$ **65.** $y = x + 7$
67. $y = x + 2$ **69.** $y = x + 9$ **71.** 12

73. (a)
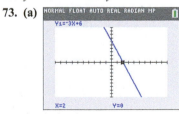
(b) $(2, 0), (0, 6)$
(c) 0

75. (a)
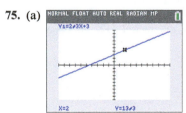
(b) $(-4.5, 0), (0, 3)$
(c) 13/3

77.
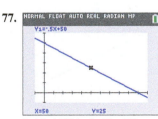
$[-10, 110]$ *by* $[-10, 60]$

Exercises 1.2, page 15

1. $\frac{2}{3}$ **3.** 5 **5.** $-\frac{4}{5}$

7.
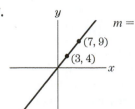

9.
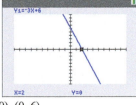

11. Undefined **13.**
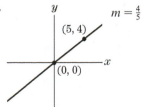

15.
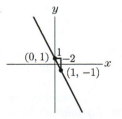

17. $y = -2x + 7$ **19.** $y = -2x + 4$ **21.** $y = \frac{1}{4}x + \frac{3}{2}$ **23.** $y = -x$ **25.** $y = -2x + 1$ **27.** $y = \frac{1}{2}x + 2$ **29.** $y = 3$
31. $(0, 3)$ **33.** $x = 0$ **35.** $y = 4x + 2000$ **37. (a)** $y = -0.00184x + 212$ **(b)** $\approx 158.6°\text{F}$

39. (a) $y = 90x + 5000$ **(d)** y
(b) 5000
(c) 90

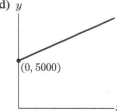
$(0, 5000)$

41. (a) $30{,}000$
(b) 60 coats
(c) $(0, 0)$; If no coats are sold, there is no revenue.
(d) 100; Each additional coat yields an additional $100 in revenue.

43. (a) y

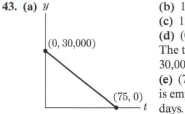
$(0, 30{,}000)$
$(75, 0)$
t

(b) 17,600 gallons
(c) 12,000 gallons
(d) $(0, 30{,}000)$; The tank holds 30,000 gallons.
(e) $(75, 0)$; The tank is empty after 75 days.

45. (a) $y = .1x + 220$ **(b)** 420 **(c)** 3200 **47.** $y = -\frac{1}{2}x$ **49.** $y = -\frac{1}{3}x$ **51.** $y = \frac{1}{2}x - 4$ **53.** $y = -\frac{2}{5}x + 5$ **55.** $y = -x + 2$
57. $y = -1$ **59.** $5; 1; -1$ **61.** $-\frac{5}{4}; -\frac{3}{2}; -\frac{3}{4}$ **63. (a)** (C)   **(b)** (B)   **(c)** (D)   **(d)** (A)   **65.** $y = x + 1$ **67.** $y = 5$ **69.** $y = -\frac{2}{3}x$
71. $F = \frac{9}{5}C + 32$ **73.** $y = \frac{4577}{12}x + 3735; 6786.33$ **75.** $y = -\frac{1}{2}x + 25; 21$ mpg   **77.** $417{,}586$ **79.** $y = \frac{1}{3}x + 3.5; 4.2$ million
81. p

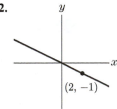
$p = 0.4q + 1$

85. 10.2 **89.** $86°\text{F}$ **91.** $20{,}000$ **93.** $30{,}000$ **95.** 104 **99.** 1 unit down   **101.** 1.4 units up

Exercises 1.3, page 23

1. $(2, 3)$ **3.** $(2, 1)$ **5.** $(12, 3)$ **7.** Yes   **9.** $x = \frac{10}{3}, y = \frac{1}{3}$ **11.** $x = -\frac{7}{9}, y = -\frac{22}{9}$ **13.** $A = (3, 4), B = (6, 2)$
15. $A = (0, 0), B = (2, 4), C = (5, \frac{11}{2}), D = (5, 0)$ **17. (a)** 2.00 **(b)** $.05$ or less   **19.** 29,500 units; 3.00 **21. (a)** 7.5 billion bushels;
11 billion bushels   **(b)** 9.5 billion bushels; $5.50 per bushel   **23.** $-40°$ **25.** $y = 30x + 1200, y = 35x + 500$; 140 shirts; 5400
27. $(18, 63)$; Both plans cost 63 cents for 18-minute calls.   **29.** 6 sq. units   **31.** 325 **33.** $(3.73, 2.23)$ **35.** $(2.68, 1.92)$

Exercises 1.4, page 30
1. 4   **3.** 6.70   **5.** $m = -1.4, b = 8.5$ **7.** $y = 4.5x - 3$ **9.** $y = -2x + 11.5$ **11. (a)** $y = -.5x + 6.5$ **(b)** $y = -.5x + 6.5$ **(c)** The
sum-of-squares error for the line in (b) is $E = 0$.   **13. (a)** $y = \frac{8}{5}x + \frac{22}{5}$ **(b)** $E = 33.64$ **15. (a)** $y = .9257x - .452$ **(b)** 43.06 mpg
(c) 51.26 mpg   **17. (a)** $y = .338x + 21.6$ **(b)** about 393   **19. (a)** $y = 0.451x + 20.6$ **(b)** 31.0   **(c)** 2020
21. (a) $y = 0.141x + 74.84$ **(b)** 79.1   **(c)** 81.9   **(d)** 87.5   **23. (a)** $y = 0.028x + 0.845$ **(b)** 1.21 **(c)** 2021

Chapter 1: Review Exercises, page 36
1. $x = 0$ **2.**
y

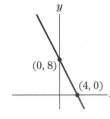
$(2, -1)$
x

3. $(2, -\frac{4}{5})$ **4.** $\frac{3}{4}$ **5.** $y = -\frac{1}{2}x + 5$ **6.** $(3, 5)$ **7.** $y = \frac{1}{5}x + 13$ **8.** 10
9. $(5, 0)$ **10.** $(7, 10)$ **11.** $(0, 7)$ **12.** The rate is $35 per hour plus a flat fee of $20.
13. No   **14.** $y = \frac{2}{3}x - 2$ **15.** 6   **16.** $x = -.3, y = .4$ **17.** $y = -\frac{2}{5}x + \frac{7}{5}$ **18.** No

19. $m = -2$;
y-intercept: $(0, 8)$;
x-intercept: $(4, 0)$;

y
$(0, 8)$
$(4, 0)$
x

21. (a) L_3 **(b)** L_1 **(c)** L_2
22. 300 units; $2
23. (a) In 2004, approximately 28% of the University of Alabama freshmen were from out of state.
(b) 46%   **(c)** 2019

24. (a) $y = 10x - 6000$ **(c)** y
(b) x-intercept: $(600, 0)$,
y-intercept: $(0, -6000)$

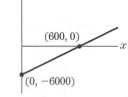
$(600, 0)$
x
$(0, -6000)$

25. (a) A: $y = .1x + 50$, B: $y = .2x + 40$ **(b)** B   **(c)** A   **(d)** 100 miles   **26. (a)** $y = .1875x + 3.20$ **(b)** 2010
27. $y = 13{,}890x + 591{,}300; 674{,}640$ **28.** $5000   **29.** 17.7%   **30.** 102,426   **31. (a)** $y = -.65x + 76.1$ **(b)** 72.85%   **(c)** 2026
32. (a) $y = 1.53x - 36.8$ **(b)** 82.2   **(c)** 79.6   **33. (a)** $y = .152x - 3.063$ **(b)** 21.3   **(c)** 165 grams   **34.** Counterclockwise
35. Up   **36.** Lines having undefined slope are vertical. Lines having zero slope are horizontal.   **37.** When the y-intercept is $(0, 0)$.
38. Yes   **39.** No; no   **40. (a)** Infinitely many   **(b)** Answers will vary

CHAPTER 2

Exercises 2.1, page 47

1. $\xrightarrow{2R_1}$ $\begin{cases} x - 6y = 4 \\ 5x + 4y = 1 \end{cases}$ **3.** $\xrightarrow{R_2 + 5R_1}$ $\begin{cases} x + 2y = 3 \\ 14y = 16 \end{cases}$ **5.** $\xrightarrow{R_3 + (-4)R_1}$ $\begin{cases} x - 2y + z = 0 \\ y - 2z = 4 \\ 9y - z = 5 \end{cases}$ **7.** $\xrightarrow{R_1 + \frac{1}{2}R_2}$ $\begin{bmatrix} 1 & 0 & | & 5 \\ 0 & 1 & | & 4 \end{bmatrix}$

9. $\begin{bmatrix} -3 & 4 & | & -2 \\ 1 & -7 & | & 8 \end{bmatrix}$ **11.** $\begin{bmatrix} 1 & 13 & -2 & | & 0 \\ 2 & 0 & -1 & | & 3 \\ 0 & 1 & 0 & | & 5 \end{bmatrix}$ **13.** $\begin{cases} -2y = 3 \\ x + 7y = -4 \end{cases}$ **15.** $\begin{cases} 3x + 2y = -3 \\ y - 6z = 4 \\ -5x - y + 7z = 0 \end{cases}$

17. Multiply the second row of the matrix by $\frac{1}{3}$. **19.** Change the first row of the matrix by adding to it 3 times the second row. **21.** Interchange rows 2 and 3.

23. $\begin{bmatrix} 1 & 2 & | & 0 \\ 0 & 10 & | & 5 \end{bmatrix}$ **25.** $\begin{bmatrix} 1 & 2 & | & 3 \\ 3 & -2 & | & 0 \end{bmatrix}$ **27.** $\begin{bmatrix} 1 & 3 & | & -5 \\ 0 & 1 & | & 7 \end{bmatrix}$ **29.** $R_2 + 2R_1$ **31.** $R_1 + (-2)R_2$ **33.** $R_1 \leftrightarrow R_2$ or $R_1 \leftrightarrow R_3$

35. $R_1 + (-3)R_3$ or $R_2 + (-2)R_3$ **37.** $\begin{bmatrix} 1 & 1 & -1 & | & 6 \\ 0 & 10 & 2 & | & 18 \\ 0 & -6 & 5 & | & -13 \end{bmatrix}$ **39.** $\begin{cases} x + y = 7 \\ x - y = 1 \end{cases}$; $x = 4, y = 3$ **41.** $\begin{cases} 3x - 4y = -27 \\ x + 2y = 11 \end{cases}$; $x = -1, y = 6$

43. $\begin{cases} 2x + y + 3z = 31 \\ x + y - 2z = 3 \\ 4x - 2y + 5z = 17 \end{cases}$; $x = 3, y = 10, z = 5$ **45.** $\begin{cases} 3x + 7y + 2z = 5 \\ 7x - 6y - 3z = 4 \\ 10x + 9y - 7z = 3 \end{cases}$; $x = 1, y = 0, z = 1$ **47.** $x = -1, y = 1$

49. $x = -\frac{8}{7}, y = -\frac{9}{7}, z = -\frac{3}{7}$ **51.** $x = -1, y = 1$ **53.** $x = 1, y = 2, z = -1$ **55.** $x = -2.5, y = 15$ **57.** $x = 1, y = -6, z = 2$
59. $x = -1, y = -2, z = 5$ **61.** 30 **63.** d **65.** 150 short sleeve, 200 long sleeve **67.** 190 adults, 85 children
69. $x = 3.7, y = 3.9, z = 1.9$ **71.** 3 ounces of Brazilian, 6 ounces of Columbian, 7 ounces of Peruvian **73.** \$25,000 in the bond fund, \$50,000 in the health sciences fund, \$25,000 in the real estate fund **75.** $23\frac{1}{3}$ pounds of first type, 85 pounds of second type, $201\frac{2}{3}$ pounds of third type

77. $\begin{bmatrix} 1 & 0 & | & -5 \\ 0 & 1 & | & 4 \end{bmatrix}$ **79.** $\begin{bmatrix} 1 & 0 & 0 & | & \frac{175}{54} \\ 0 & 1 & 0 & | & \frac{16}{9} \\ 0 & 0 & 1 & | & \frac{26}{27} \end{bmatrix}$ **81.** $x = -2.5, y = 15$ **83.** $x = 1, y = -6, z = 2$

Exercises 2.2, page 56

1. $\begin{bmatrix} 1 & -2 & 3 \\ 0 & 13 & -8 \end{bmatrix}$ **3.** $\begin{bmatrix} 9 & -1 & 0 & -7 \\ -\frac{1}{2} & \frac{1}{2} & 1 & 3 \\ 5 & -1 & 0 & -3 \end{bmatrix}$ **5.** $\begin{bmatrix} 1 & \frac{3}{2} \\ 0 & -9 \\ 0 & \frac{7}{2} \end{bmatrix}$ **7.** $\begin{bmatrix} 4 & 3 & 0 \\ 1 & 1 & 0 \\ \frac{1}{6} & \frac{1}{2} & 1 \end{bmatrix}$ **9.** $\begin{cases} x + y + 4z = 6 \\ 2x + y + z = 10 \end{cases}$; $z =$ any value, $y = 2 - 7z, x = 4 + 3z$

11. $\begin{cases} -5x + 15y - 10z = 5 \\ x - 3y + 2z = 0 \end{cases}$; no solution **13.** $\begin{cases} 2x - y + 5z = 12 \\ -x - 4y + 2z = 3 \\ 8x + 5y + 11z = 30 \end{cases}$; $z =$ any value, $y = z - 2, x = 5 - 2z$

15. $\begin{cases} x + 2y + 3z - w = 4 \\ 2x + 3y + w = -3 \\ 4x + 7y + 6z - w = 5 \end{cases}$; $z =$ any value, $w =$ any value, $y = 11 - 6z + 3w, x = 9z - 5w - 18$

17. $y =$ any value, $x = 3 + 2y$ **19.** No solution **21.** $x = 1, y = 2$ **23.** No solution **25.** No solution **27.** $z =$ any value, $x = -6 - z, y = 5$ **29.** No solution **31.** No solution **33.** $z =$ any value, $w =$ any value, $x = 2z + w, y = 5 - 3w$
35. No solution **37.** Possible answers: $z = 0, x = -13, y = 9; z = 1, x = -8, y = 6; z = 2, x = -3, y = 3$
39. Possible answers: $y = 0, x = 23, z = 5; y = 1, x = 16, z = 5; y = 2, x = 9, z = 5$
41. Food 3: $z =$ any value between 0 and 100, food 2: $y = 100 - z$, food 1: $x = 300 - z$
45. 50 ottomans, 30 sofas, 40 chairs; 5 ottomans, 55 sofas, 35 chairs; 95 ottomans, 5 sofas, 45 chairs
47. 6 floral squares, the other 90 any mix of solid green and solid blue **49.** No solution if $k \neq -12$; infinitely many if $k = -12$
51. None **53.** One; $x = 7, y = 3$ **55.** None **57.** There has been a pivot about the bottom right element.

Exercises 2.3, page 68

1. 2×3 **3.** 1×3, row matrix **5.** 2×2, square matrix **7.** $-4; 0$ **9.** $i = 1, j = 3$

11. $\begin{bmatrix} 9 & 3 \\ 7 & -1 \end{bmatrix}$ **13.** $\begin{bmatrix} 2 & 4 & 2.5 \\ -5.5 & 1 & 1.2 \end{bmatrix}$ **15.** $\begin{bmatrix} 1 & 3 \\ 1 & 2 \\ 4 & -2 \end{bmatrix}$ **17.** $\begin{bmatrix} -7 \\ \frac{1}{6} \end{bmatrix}$ **19.** [11] **21.** [10] **23.** $\begin{bmatrix} 4 & 0 & -\frac{2}{3} \\ -6 & \frac{1}{2} & \frac{1}{3} \end{bmatrix}$

25. $\begin{bmatrix} 3 & 19 \\ 23 & 3 \end{bmatrix}$ **27.** Yes; 3×5 **29.** No **31.** Yes; 3×1 **33.** $\begin{bmatrix} 6 & 17 \\ 6 & 10 \end{bmatrix}$ **35.** $\begin{bmatrix} 21 \\ -4 \\ 8 \end{bmatrix}$

37. $\begin{bmatrix} 5 & 6 \\ 7 & 8 \end{bmatrix}$ **39.** $\begin{bmatrix} .48 & .39 \\ .52 & .61 \end{bmatrix}$ **41.** $\begin{bmatrix} 25 & 17 & 2 \\ 3 & -1 & 2 \\ 1 & 1 & 4 \end{bmatrix}$ **43.** $\begin{bmatrix} \frac{1}{3} & \frac{2}{3} \\ \frac{1}{3} & \frac{2}{3} \end{bmatrix}$ **45.** $[30 \quad 41]$ **47.** $\begin{bmatrix} 10 & 0 \\ 0 & 15 \end{bmatrix}$ **49.** $\begin{bmatrix} 0 & 0 \\ 0 & 0 \end{bmatrix}$ **51.** $\begin{bmatrix} 23 & 24 \\ 25 & 26 \end{bmatrix}$

53. $\begin{cases} 2x + 3y = 6 \\ 4x + 5y = 7 \end{cases}$ **55.** $\begin{cases} x + 2y + 3z = 10 \\ 4x + 5y + 6z = 11 \\ 7x + 8y + 9z = 12 \end{cases}$ **57.** $\begin{bmatrix} 3 & 2 \\ 7 & -1 \end{bmatrix} \begin{bmatrix} x \\ y \end{bmatrix} = \begin{bmatrix} -1 \\ 2 \end{bmatrix}$ **59.** $\begin{bmatrix} 1 & -2 & 3 \\ 0 & 1 & 1 \\ 0 & 0 & 1 \end{bmatrix} \begin{bmatrix} x \\ y \\ z \end{bmatrix} = \begin{bmatrix} 5 \\ 6 \\ 2 \end{bmatrix}$

65. (a) $\begin{bmatrix} 340 \\ 265 \end{bmatrix}$ **(b)** Mike's clothes cost \$340; Don's clothes cost \$265. **(c)** $\begin{bmatrix} 25 \\ 18.75 \\ 62.50 \end{bmatrix}$ **(d)** The costs of the three items of clothing after

a 25% increase **67. (a)** $[2282.50 \quad 2322.50 \quad 3550.50]$, total retail value for the white chocolate-covered, milk chocolate-covered, and dark

chocolate-covered items **(b)** $\begin{bmatrix} 3138.00 \\ 3337.50 \\ 6772.50 \end{bmatrix}$, total revenue from peanuts, raisins, and espresso beans **(c)** $\begin{bmatrix} 94.50 \\ 351.50 \\ 256.50 \end{bmatrix}$, 10% reduction in the number of

pounds sold **69. (a)** I: 2.75, II: 2, III: 1.3 **(b)** A: 74, B: 112, C: 128, D: 64, F: 22 **71.** 10,100 voting Democratic, 7900 voting
Republican **73.** Carpenters: \$2000, bricklayers: \$2100, plumbers: \$1200 **75. (a)** $[162 \quad 150 \quad 143]$, number of units of each nutrient
consumed at breakfast **(b)** $[186 \quad 200 \quad 239]$, number of units of each nutrient consumed at lunch **(c)** $[288 \quad 300 \quad 344]$, number of
units of each nutrient consumed at dinner **(d)** $[5 \quad 8]$, total number of ounces of each food that Mikey eats during a day

(e) $[636 \quad 650 \quad 726]$, number of units of each nutrient consumed per day **77. (a)** $\begin{bmatrix} 720 \\ 646 \end{bmatrix}$ **(b)** \$720

79. (a) $T = \begin{bmatrix} \text{Boston cream pie} & \text{Carrot cake} \\ 30 & 45 \\ 30 & 50 \\ 15 & 10 \end{bmatrix} \begin{matrix} \text{Preparation} \\ \text{Baking} \\ \text{Finishing} \end{matrix}$ **(b)** $S = \begin{bmatrix} 20 \\ 8 \end{bmatrix} \begin{matrix} \text{Boston cream pie} \\ \text{Carrot cake} \end{matrix}$; $TS = \begin{bmatrix} 960 \\ 1000 \\ 380 \end{bmatrix} \begin{matrix} \text{Preparation} \\ \text{Baking} \\ \text{Finishing} \end{matrix}$

(c) Total baking time: 1000 minutes, or $16\frac{2}{3}$ hours; total finishing time: 380 minutes, or $6\frac{1}{3}$ hours

81. (a) $T = \begin{bmatrix} \text{Cutting} & \text{Sewing} & \text{Finishing} \\ 2 & 3 & 2 \\ 1.5 & 2 & 1 \end{bmatrix} \begin{matrix} \text{Huge One} \\ \text{Regular Joe} \end{matrix}$ **(b)** $S = \begin{bmatrix} 32 \\ 24 \end{bmatrix} \begin{matrix} \text{Huge One} \\ \text{Regular Joe} \end{matrix}$

(c) $A = \begin{bmatrix} \text{Huge One} & \text{Regular Joe} \\ 27 & 56 \end{bmatrix}$; $AT = \begin{bmatrix} \text{Cutting} & \text{Sewing} & \text{Finishing} \\ 138 & 193 & 110 \end{bmatrix}$; $AS = \begin{bmatrix} 2208 \end{bmatrix}$ **(d)** 193 hours **(e)** \$2208

85. $\begin{bmatrix} 3 & -2 & 1 \\ -5 & 6 & 7 \end{bmatrix}$ **87.** 4×4 **89.** $[9257 \quad 57,718 \quad 89,389]$ $\begin{bmatrix} 13.9 \\ 14.9 \\ 14.2 \end{bmatrix}$ **91.** $\begin{bmatrix} 6.4 & -2 & -2.7 \\ 20.5 & 22.5 & -2.4 \\ -14 & 17.6 & 16 \end{bmatrix}$

93. $\begin{bmatrix} -171.3 & 40.8 & -31.8 \\ 454.6 & -22.5 & 22.7 \\ -2.6 & 122.3 & 53.56 \end{bmatrix}$ **95.** $\begin{bmatrix} 1.2 & 21 & -9 \\ 57 & 1.5 & 4.8 \\ -27 & 33 & 6 \end{bmatrix}$

Exercises 2.4, page 78

1. $x = 2, y = 0$ **3.** $\begin{bmatrix} 1 & -2 \\ -3 & 7 \end{bmatrix}$ **5.** $\begin{bmatrix} 1 & -1 \\ -\frac{5}{2} & 3 \end{bmatrix}$ **7.** $\begin{bmatrix} 1.6 & -.4 \\ -.6 & 1.4 \end{bmatrix}$

9. $[\frac{1}{3}]$ **11.** $x = 4, y = -\frac{1}{2}$ **13.** $x = 32, y = -6$ **15. (a)** $\begin{bmatrix} .8 & .3 \\ .2 & .7 \end{bmatrix} \begin{bmatrix} x \\ y \end{bmatrix} = \begin{bmatrix} m \\ s \end{bmatrix}$ **(b)** $\begin{bmatrix} x \\ y \end{bmatrix} = \begin{bmatrix} 1.4 & -.6 \\ -.4 & 1.6 \end{bmatrix} \begin{bmatrix} m \\ s \end{bmatrix}$ **(c)** 110,000 married;

40,000 single **(d)** 130,000 married; 20,000 single **17. (a)** $\begin{bmatrix} .7 & .1 \\ .3 & .9 \end{bmatrix} \begin{bmatrix} x \\ y \end{bmatrix} = \begin{bmatrix} u \\ v \end{bmatrix}$ **(b)** $\begin{bmatrix} x \\ y \end{bmatrix} = \begin{bmatrix} \frac{3}{2} & -\frac{1}{6} \\ -\frac{1}{2} & \frac{7}{6} \end{bmatrix} \begin{bmatrix} u \\ v \end{bmatrix}$ **(c)** 8500; 4500

19. $x = 9, y = -2, z = -2$ **21.** $x = 21, y = 25, z = 26$ **23.** $x = 1, y = 5, z = -4, w = 9$ **25.** $x = 4, y = -19, z = 2, w = -4$

29. (a) $\begin{bmatrix} 1 & 2 \\ .9 & 0 \end{bmatrix} \begin{bmatrix} x \\ y \end{bmatrix} = \begin{bmatrix} a \\ b \end{bmatrix}$ **(b)** After 1 year: 1,170,000 in group I and 405,000 in group II. After 2 years: 1,980,000 in group I and

1,053,000 in group II. **(c)** 700,000 in group I and 55,000 in group II. **33.** One possible answer is $\begin{bmatrix} 1 & 1 \\ 1 & 1 \end{bmatrix} \begin{bmatrix} x \\ y \end{bmatrix} = \begin{bmatrix} 2 \\ 3 \end{bmatrix}$.

35. $\begin{bmatrix} -\frac{10}{73} & \frac{75}{292} \\ \frac{25}{73} & -\frac{5}{292} \end{bmatrix}$ **37.** $\begin{bmatrix} \frac{1020}{8887} & \frac{2910}{8887} & -\frac{500}{8887} \\ \frac{3050}{8887} & \frac{860}{8887} & \frac{1990}{8887} \\ \frac{125}{8887} & \frac{618}{8887} & \frac{810}{8887} \end{bmatrix}$ **39.** $x = -\frac{4}{5}, y = \frac{28}{5}, z = 5$ **41.** $x = 0, y = 2, z = 0, w = 2$

Exercises 2.5, page 82

1. $\begin{bmatrix} -2 & 3 \\ 5 & -7 \end{bmatrix}$ **3.** $\begin{bmatrix} \frac{7}{2} & \frac{3}{2} \\ -2 & -1 \end{bmatrix}$ **5.** No inverse **7.** $\begin{bmatrix} -1 & 2 & -4 \\ 1 & -1 & 3 \\ 0 & 0 & 1 \end{bmatrix}$ **9.** No inverse **11.** $\begin{bmatrix} -5 & 6 & 0 & 0 \\ 1 & -1 & 0 & 0 \\ 0 & 0 & -\frac{1}{46} & \frac{1}{46} \\ 0 & 0 & \frac{25}{46} & -\frac{1}{23} \end{bmatrix}$

13. $x = 2, y = -3, z = 2$ **15.** $x = 2, y = 1, z = 3$ **17.** $x = 4, y = -4, z = 3, w = -1$ **19.** $\begin{bmatrix} -3 & 5 \\ 10 & -16 \end{bmatrix}$

21. $x = 42, y = 21, z = 37$ **23.** $x = 82, y = 17, z = 1$

Exercises 2.6, page 88

1. 20 cents **3.** Energy sector **5.** \$6 million **7.** Manufacturing **9.** $AX = \begin{bmatrix} 11.00 \\ 11.50 \\ 9.50 \end{bmatrix}$ **11.** $X = \begin{bmatrix} 12.89 \\ 14.06 \\ 13.08 \end{bmatrix}$

13. Coal: \$8.84 billion; steel: \$3.725 billion; electricity: \$9.895 billion **15.** Computers: \$354 million; semiconductors: \$172 million **17.** \$1.55 billion worth of coal, \$0.86 billion worth of steel, and \$4.55 billion worth of electricity

19. (a) $\begin{array}{c} \\ T \\ E \end{array}\begin{array}{cc} T & E \end{array}\begin{bmatrix} .25 & .30 \\ .20 & .15 \end{bmatrix}$ (b) $\begin{bmatrix} 1.47 & .52 \\ .35 & 1.30 \end{bmatrix}$ (c) Transportation: \$8.91 billion; energy: \$5.65 billion (d) Transportation: \$3.92 billion; energy: \$2.63 billion **21.** Plastics: \$955,000; industrial equipment: \$590,000

23. (a) $\begin{array}{c} \\ W \\ S \\ C \end{array}\begin{array}{ccc} W & S & C \end{array}\begin{bmatrix} .30 & 0 & .10 \\ .20 & .30 & .20 \\ .10 & .20 & .05 \end{bmatrix}$ (b) $\begin{bmatrix} 1.47 & .05 & .16 \\ .49 & 1.54 & .38 \\ .26 & .33 & 1.15 \end{bmatrix}$ (c) Wood: \$1.99; steel: \$7.41; coal: \$3.88 (d) Wood: \$0.98; steel: \$3.39; coal: \$1.87 **25.** Manufacturing: \$398 million; transportation: \$313 million; agriculture: \$452 million

27. Merchant: \$85,000; baker: \$68,000; farmer: \$103,000 **31.** $\begin{bmatrix} 11.91 \\ 15.83 \\ 9.57 \\ 7.26 \end{bmatrix}$

Chapter 2: Review Exercises, page 93

1. $\begin{bmatrix} 1 & -2 & \frac{1}{3} \\ 0 & 8 & \frac{16}{3} \end{bmatrix}$ **2.** $\begin{bmatrix} 1 & 0 & 1 \\ 2 & 1 & 0 \\ -12 & 0 & 7 \end{bmatrix}$ **3.** $x = 4, y = 5$ **4.** $x = 50, y = 2, z = -12$ **5.** $x = -1, y = \frac{2}{3}, z = \frac{1}{3}$ **6.** No solution

7. $z =$ any value, $x = 1 - 3z, y = 4z, w = 5$ **8.** $x = 7, y = 3$ **9.** $\begin{bmatrix} 5 \\ 3 \\ 7 \end{bmatrix}$ **10.** $\begin{bmatrix} 6 & 17 \\ 12 & 26 \end{bmatrix}$ **11.** $\begin{bmatrix} 6 & -\frac{9}{2} \\ \frac{1}{2} & 0 \end{bmatrix}$ **12.** $\begin{bmatrix} .6 & -10 \\ 6.6 & 2 \\ 4 & 11 \end{bmatrix}$

13. 5 **14.** 4 **15.** $x = -2, y = 3$ **16.** (a) $x = 13, y = 23, z = 19$ (b) $x = -4, y = 13, z = 14$ **17.** $\begin{bmatrix} -1 & 3 \\ \frac{1}{2} & -1 \end{bmatrix}$

18. $\begin{bmatrix} 5 & -1 & -1 \\ -3 & 1 & 0 \\ -1 & 0 & 1 \end{bmatrix}$ **19.** Corn: 500 acres; wheat: 0 acres; soybeans: 500 acres **20.** (a) $\begin{bmatrix} 5455 \\ 5275 \end{bmatrix}$; total month's costs for each store

(b) $\begin{bmatrix} 6600 \\ 6360 \end{bmatrix}$; total month's revenue for each store (c) $\begin{bmatrix} 35 \\ 15 \\ 40 \end{bmatrix}$; profit for each piece of equipment (d) $\begin{bmatrix} 1145 \\ 1085 \end{bmatrix}$; total month's profit for each store

21. (a) [10,100 8230 4670]; total amount invested in bonds, stocks, and the conservative fixed income fund, respectively (b) [522.40 1807.30]; total returns on the investments for one year and five years, respectively (c) [10,000 16,000 20,000]; the result of doubling the amounts invested (d) The total amount invested in stocks is \$8230. (e) The total return after one year is \$522.40.

22. (a) $AB = \begin{bmatrix} 328 \\ 336 \\ 323 \\ 326 \end{bmatrix}$; Sara earned \$328, Quinn earned \$336, Tamia earned \$323, and Zack earned \$326. (b) Most: Quinn; least: Tamia

(c) Quinn and Zack both earned \$329. (d) 30 hours **23.** 4 apples, 9 bananas, 5 oranges **24.** (a) A: 9400, 8980; B: 7300, 7510 (b) A: 10,857, 12,082; B: 6571, 5959 **25.** Industry I: 20; industry II: 20 **26.** 4 **27.** (a) True (b) False (c) True **28.** (a) True (b) False

CHAPTER 3

Exercises 3.1, page 103

1. False   **3.** True   **5.** $x \geq 4$ **7.** $x \geq 3$ **9.** $y \leq -2x + 5$ **11.** $y \geq 15x - 18$ **13.** $x \geq -\dfrac{3}{4}$ **15.** Yes

17. No   **19.** Yes   **21.** Yes   **23.** 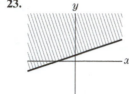   **25.**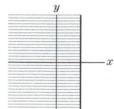

27. $y \leq -x + 3$ **29.** $y \geq \dfrac{1}{2}x - 2$ **31.** 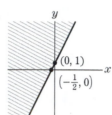   **33.** 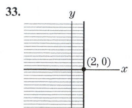   **35.**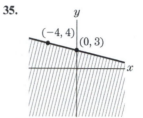

37. 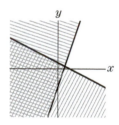   **39.** 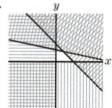   **41.** 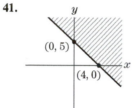   **43.**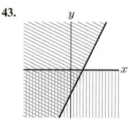

45. 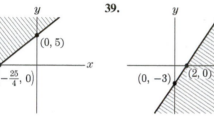   **47.** 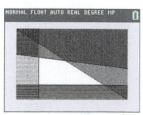   **49.** Yes   **51.** No   **53.** Below   **55.** Above   **57.** $\begin{cases} y \geq 2x - 1 \\ y \leq 2x \end{cases}$ **59.** d

61. (a) $(3.6, 3.7)$ **(b)** Below   **63.**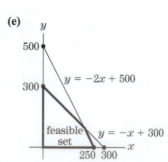

$[-1, 4] \ by \ [-1, 4]$

Exercises 3.2, page 109

1. Yes   **3.** No   **5.** 16

7. (a)

	A	B	Available
Candy bars	2	1	500
Suckers	2	2	600
Profit	40 ¢	30 ¢	

(b) A: $2x + y \leq 500$, **B:** $2x + 2y \leq 600$
(c) $x \geq 0, y \geq 0$ **(d)** $40x + 30y$

(e)

9. (a)

	A	B	Truck Capacity
Volume	4 cubic feet	3 cubic feet	300 cubic feet
Weight	100 pounds	200 pounds	10,000 pounds
Earnings	$13	$9	

(e)

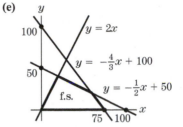

(b) $4x + 3y \le 300$; $100x + 200y \le 10,000$
(c) $y \le 2x$, $x \ge 0$, $y \ge 0$ **(d)** $13x + 9y$

11. (a)

	Essay Questions	Short-Answer Questions	Available
Time to answer	10 minutes	2 minutes	90 minutes
Quantity	10	50	
Required	3	10	
Worth	20 points	5 points	

(e)

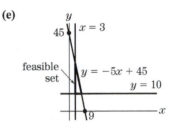

(b) $10x + 2y \le 90$ **(c)** $x \ge 3$, $x \le 10$, $y \ge 10$, $y \le 50$ **(d)** $20x + 5y$

13. (a)

	Alfalfa	Corn	Requirements
Protein	.13 pound	.065 pound	4550 pounds
TDN	.48 pound	.96 pound	26,880 pounds
Vitamin A	2.16 IUs	0 IUs	43,200 IUs
Cost/lb	$.08	$.13	

(c)

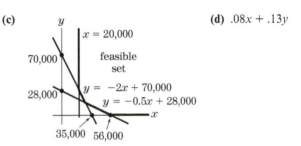

(d) $.08x + .13y$

(b) $.13x + .065y \ge 4550$;
$.48x + .96y \ge 26,880$;
$2.16x \ge 43,200$; $y \ge 0$

Exercises 3.3, page 117

1. (20, 0) **3.** (6, 0) **5.** (0, 5) **7.** (3, 3) **9.** (0, 7) **11.** (2, 1) **13.** 9 cups **15.** 200 packages of assortment A and 100 packages of assortment B; $110 **17.** Ship 75 crates of cargo A and no crates of cargo B. **19.** Answer 3 essay questions and 30 short-answer questions. **21.** The minimum cost of $4060 is achieved by buying 28,000 pounds of alfalfa and 14,000 pounds of corn. **23.** Make 16 chairs and no sofas. **25.** The minimum value is 30 and occurs at (2, 6). **27.** The maximum value is 49 and occurs at (2, 9).
29. The maximum value is 6600 and occurs at (12, 36). **31.** The minimum value is 40 and occurs at (4, 3). **33.** Produce 9 hockey games and 8 soccer games each day. **35.** The farmer should plant 83 1/3 acres of oats and 16 2/3 acres of corn to make a profit of $6933.33. **37.** Produce 49 regular bags and 14 deluxe bags. **39.** Supply 6 tubes of food A and 6 tubes of food B. **41.** Make 400 cans of Fruit Delight and 500 cans of Heavenly Punch. **43. (b)** The farmer should plant 78 acres of oats and 22 acres of corn to make a profit of $7040. Yes, it provides more profit. **45.** 35 Cupid assortments and 10 Patriotic assortments **47.** 200 Pamper Me baskets and 75 Best Friends baskets **49.** The feasible set contains no points. **51.** The minimum value is 30 and occurs at (2, 6).

Exercises 3.4, page 129

1. (a) $y = -\dfrac{3}{2}x + \dfrac{c}{14}$ **(b)** Up **(c)** B **3.** C **5.** D **7.** D **9.** C **11.** $\frac{1}{4} \le k \le 3$ **13.** Feed 1 can of brand A and 3 cans of brand B.
15. Mr. Jones should invest $2000 in low-risk stocks, $3000 in medium-risk stocks, and $4000 in high-risk stocks. **17.** Let (a, b) correspond to a cars shipped from Baltimore to Philadelphia, b cars shipped from Baltimore to Trenton, $4 - a$ cars shipped from NY to Philadelphia, and $7 - b$ cars shipped from NY to Trenton. Then, the minimum cost of $990 is achieved at (0, 5), (4, 1), or anywhere on the line segment connecting these two points. **19.** Produce 90,000 gallons of gasoline, 5000 gallons of jet fuel, and 5000 gallons of diesel fuel. **21.** Buy 9 high-capacity trucks and 21 low-capacity trucks. **23.** Ship 150 pounds of coffee from San José to Salt Lake City, 350 pounds from San José to Reno, and 250 pounds from Seattle to Salt Lake City. **25.** Create 22 of Kit I, 10 of Kit II, and 12 of Kit III. **29.** Possible answer: $5x + y$ **31.** Possible answer: $2x + y$ **33.** Possible answer: $x + 5y$ **35.** Possible answer: $2x + 3y$
37. Rice: [9.33, 42]; soybeans: [7, 31.5]

Chapter 3: Review Exercises, page 133

1. Yes **2.**

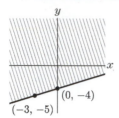

3. $y \geq 2.4\,x - 5.8$ **4.** Use 10 type A planes and 3 type B planes.

5. Use 2 ounces of wheat germ and 1 ounce of enriched oat flour. **6.** Produce 9 hardtops and 16 sports cars. **7.** Make 500 boxes of mixture A and 200 boxes of mixture B. **8.** Publish 60 elementary books, 8 intermediate books, and 4 advanced books. **9.** Package 80 computers at Rochester and 45 computers at Queens. **10.** Transport no refrigerators from warehouse A to outlet I, 200 refrigerators from warehouse A to outlet II, 200 refrigerators from warehouse B to outlet I, and 100 refrigerators from warehouse B to outlet II. **11.** Invest $2000 in the CD, $0 in mutual funds, and $8000 in stocks. **12.** Yes; No **13.** Yes; No

FEASIBLE SETS FOR CHAPTER 3

Exercises 3.3

25.

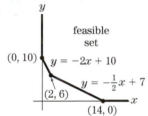

27.

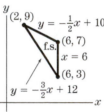

29.

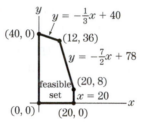

31.

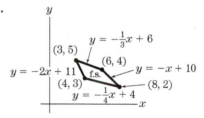

33.

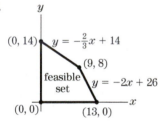

35.

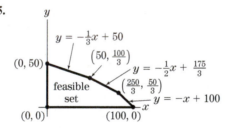

37.

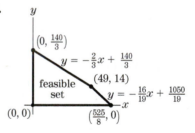

39.

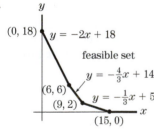

41.

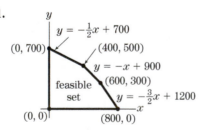

43.

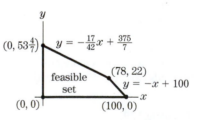

45.

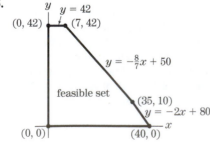

47.

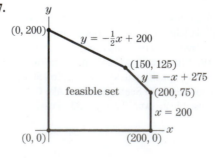

Exercises 3.4

13.

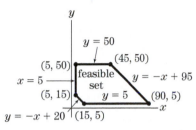

15.

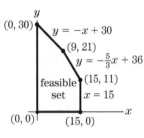

17.

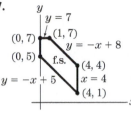

19.

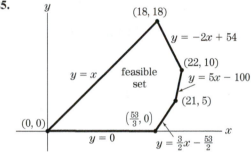

21.

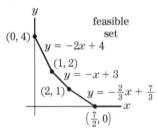

23.

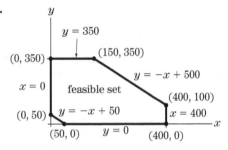

25.

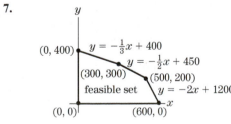

Chapter 3: Review Exercises

4.

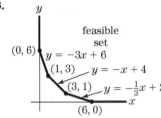

5.

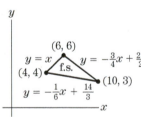

6.

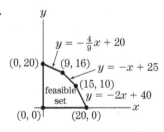

7.

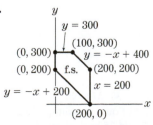

8.

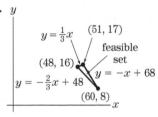

9.

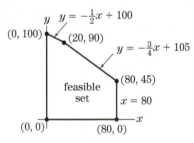

10.

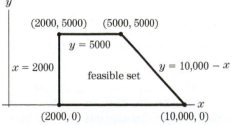

11.

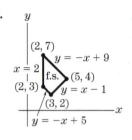

CHAPTER 4

Exercises 4.1, page 142

1. $\begin{cases} 20x + 30y + u & = 3500 \\ 50x + 10y \quad + v & = 5000 \\ -8x - 13y \qquad + M = \quad 0 \end{cases}$ Maximize M given that $x \geq 0, y \geq 0, u \geq 0, v \geq 0$.

3. $\begin{cases} x + \quad y + z + u & = 100 \\ 3x \qquad + z \quad + v & = 200 \\ 5x + 10y \qquad + w & = 100 \\ -x - \quad 2y + 3z \qquad + M = \quad 0 \end{cases}$ Maximize M given that $x \geq 0, y \geq 0, z \geq 0, u \geq 0, v \geq 0, w \geq 0$.

5. $\begin{cases} 4x + 6y - \quad 7z + u & = 16 \\ 3x + 2y \qquad + v & = 11 \\ 9y + \quad 3z \qquad + w & = 21 \\ -3x - 5y - 12z \qquad + M = \ 0 \end{cases}$ Maximize M given that $x \geq 0, y \geq 0, z \geq 0, u \geq 0, v \geq 0, w \geq 0$.

7. (a)
$$\begin{array}{ccccc|c} x & y & u & v & M & \\ 20 & 30 & 1 & 0 & 0 & 3500 \\ 50 & 10 & 0 & 1 & 0 & 5000 \\ -8 & -13 & 0 & 0 & 1 & 0 \end{array}$$

(b) $x = 0, y = 0, u = 3500, v = 5000, M = 0$

9. (a)
$$\begin{array}{ccccccc|c} x & y & z & u & v & w & M & \\ 1 & 1 & 1 & 1 & 0 & 0 & 0 & 100 \\ 3 & 0 & 1 & 0 & 1 & 0 & 0 & 200 \\ 5 & 10 & 0 & 0 & 0 & 1 & 0 & 100 \\ -1 & -2 & 3 & 0 & 0 & 0 & 1 & 0 \end{array}$$

(b) $x = 0, y = 0, z = 0, u = 100, v = 200, w = 100, M = 0$

11. (a)
$$\begin{array}{ccccccc|c} x & y & z & u & v & w & M & \\ 4 & 6 & -7 & 1 & 0 & 0 & 0 & 16 \\ 3 & 2 & 0 & 0 & 1 & 0 & 0 & 11 \\ 0 & 9 & 3 & 0 & 0 & 1 & 0 & 21 \\ -3 & -5 & -12 & 0 & 0 & 0 & 1 & 0 \end{array}$$

13. $x = 15, y = 0, u = 10, v = 0, M = 20$

(b) $x = 0, y = 0, z = 0, u = 16, v = 11, w = 21, M = 0$

15. $x = 14, y = 0, z = 17, u = 0, v = 0, M = 56$ **17.** $x = 42, y = 54, z = 61, u = 0, v = 0, w = 0, M = 604$
19. $x = 10, y = 0, z = 15, u = 23, v = 0, w = 0, M = -11$

21. (a)
$$\begin{array}{ccccc|c} x & y & u & v & M & \\ 1 & \frac{3}{2} & \frac{1}{2} & 0 & 0 & 6 \\ 0 & -\frac{1}{2} & -\frac{1}{2} & 1 & 0 & 4 \\ 0 & -5 & 5 & 0 & 1 & 60 \end{array}$$
$x = 6, y = 0,$
$u = 0, v = 4,$
$M = 60$

(b)
$$\begin{array}{ccccc|c} x & y & u & v & M & \\ \frac{2}{3} & 1 & \frac{1}{3} & 0 & 0 & 4 \\ \frac{1}{3} & 0 & -\frac{1}{3} & 1 & 0 & 6 \\ \frac{10}{3} & 0 & \frac{20}{3} & 0 & 1 & 80 \end{array}$$
$x = 0, y = 4,$
$u = 0, v = 6,$
$M = 80$

(c)
$$\begin{array}{ccccc|c} x & y & u & v & M & \\ 0 & 1 & 1 & -2 & 0 & -8 \\ 1 & 1 & 0 & 1 & 0 & 10 \\ 0 & -10 & 0 & 10 & 1 & 100 \end{array}$$
$x = 10, y = 0,$
$u = -8, v = 0,$
$M = 100$

(d)
$$\begin{array}{ccccc|c} x & y & u & v & M & \\ -1 & 0 & 1 & -3 & 0 & -18 \\ 1 & 1 & 0 & 1 & 0 & 10 \\ 10 & 0 & 0 & 20 & 1 & 200 \end{array}$$
$x = 0, y = 10,$
$u = -18, v = 0,$
$M = 200$

(e) d

23. (a) Group I: x, y, Group II: u, v, M **(b)** (i) feasible; I: y, u; II: x, v, M (ii) feasible; I: x, u; II: y, v, M (iii) not feasible; I: y, v; II: x, u, M (iv) not feasible; I; x, v; II: u, y, M **(c)** Operation (i)

Exercises 4.2, page 151

1. 2 (second row, first column) **3.** $\frac{1}{4}$ (second row, second column) **5.** 1 (third row, second column) **7. (a)** 3

(b)
$$\begin{array}{c}\begin{array}{ccccc|c} & x & y & u & v & M & \\ u & \frac{16}{3} & 0 & 1 & -\frac{2}{3} & 0 & 6 \\ y & \frac{1}{3} & 1 & 0 & \frac{1}{3} & 0 & 2 \\ M & 0 & 0 & 0 & 4 & 1 & 24 \end{array}\end{array}$$
(c) $x = 0, y = 2, u = 6, v = 0; M = 24$

9. (a) 10 **(b)**
$$\begin{array}{c}\begin{array}{ccccc|c} & x & y & u & v & M & \\ u & -13 & 0 & 1 & -\frac{6}{5} & 0 & 6 \\ y & \frac{3}{2} & 1 & 0 & \frac{1}{10} & 0 & \frac{1}{2} \\ M & 7 & 0 & 0 & \frac{1}{5} & 1 & 1 \end{array}\end{array}$$
(c) $x = 0, y = \frac{1}{2}, u = 6, v = 0; M = 1$

11. $x = 0, y = 5; M = 15$ **13.** $x = 12, y = 20; M = 88$ **15.** $x = 0, y = \frac{19}{3}, z = 5; M = 44$ **17.** $x = 0, y = 30; M = 90$
19. $x = 50, y = 100; M = 1300$ **21.** 24 basketballs, 224 footballs **23.** 98 chairs, 4 sofas, 21 tables **25.** 11 hours bicycling, 4 hours swimming, 15 hours jogging **27.** 65 type A restaurants and 5 type C restaurants **29.** 100 bags of mix B and 25 bags of mix C

31. No type A widgets, 90 type B widgets, 15 type C widgets; $2250 **33.** $x = 0, y = 0, z = 600$; $M = 180{,}000$
35. $x = 0, y = 5$; $M = 20$ **37.** $x = 0, y = 4, z = 0$; $M = 8$. Or $x = 0, y = 2, z = 2$; $M = 8$.

Exercises 4.3, page 159

1. Maximize $3x + 4y$ subject to the constraints

$$\begin{cases} -5x - 3y \le -6 \\ 2x - 3y \le 7 \\ x \ge 0, \quad y \ge 0. \end{cases}$$

3. Maximize $-x - y - z$ subject to the constraints

$$\begin{cases} -2x - 3y - z \le -7 \\ -5x - 6y - 7z \le -8 \\ x \ge 0, \quad y \ge 0, \quad z \ge 0. \end{cases}$$

5. -3 (second row, second column) **7.** $-\frac{1}{2}$ (third row, second column) **9.** $x = \frac{3}{5}, y = \frac{22}{5}$; 156 **11.** $x = \frac{5}{2}, y = \frac{1}{2}$; 8
13. $x = 3, y = 5$; 59 **15.** $x = 5, y = 4$; 38 **17.** 1 serving of food A, 3 servings of food B **19.** Stock 100 of brand A, 50 of brand B, and 450 of brand C. **21.** Supply all computers from Chicago at a cost of $8800. **23.** $x = 1, y = 1$; $M = -1$

Exercises 4.4, page 167

1. $x = 4, y = 22$, profit = $1220 **3.** $x = 25, y = 25$; cost = $250 **5.** $-400 \le h \le \frac{400}{3}$

7. $\begin{bmatrix} 9 & 1 & 1 \\ 4 & 8 & -3 \end{bmatrix}$ **9.** $\begin{bmatrix} 7 \\ 6 \\ 5 \\ 1 \end{bmatrix}$ **11.** Yes

13. Minimize $\begin{bmatrix} 7 & 5 & 4 \end{bmatrix} \begin{bmatrix} x \\ y \\ z \end{bmatrix}$ subject to the constraints $\begin{bmatrix} 3 & 8 & 9 \\ 1 & 2 & 5 \\ 4 & 1 & 7 \end{bmatrix} \begin{bmatrix} x \\ y \\ z \end{bmatrix} \ge \begin{bmatrix} 75 \\ 80 \\ 67 \end{bmatrix}$ and $\begin{bmatrix} x \\ y \\ z \end{bmatrix} \ge \begin{bmatrix} 0 \\ 0 \\ 0 \end{bmatrix}$.

15. Maximize $\begin{bmatrix} 30 & 50 \end{bmatrix} \begin{bmatrix} x \\ y \end{bmatrix}$ subject to the constraints $\begin{bmatrix} 3 & 6 \\ 7 & 5 \\ 4 & 3 \end{bmatrix} \begin{bmatrix} x \\ y \end{bmatrix} \le \begin{bmatrix} 90 \\ 138 \\ 120 \end{bmatrix}$ and $\begin{bmatrix} x \\ y \end{bmatrix} \ge \begin{bmatrix} 0 \\ 0 \end{bmatrix}$.

17. Minimize $2x + 3y$ subject to the constraints $\begin{cases} 7x + 4y \ge 33 \\ 5x + 8y \ge 44 \\ x + 3y \ge 55 \\ x \ge 0, y \ge 0 \end{cases}$ **19.** -2; [25, 55]

Exercises 4.5, page 177

1. Minimize $80u + 76v$ subject to the constraints $\begin{cases} 5u + 3v \ge 4 \\ u + 2v \ge 2 \\ u \ge 0, v \ge 0. \end{cases}$

3. Maximize $u + 2v + w$ subject to the constraints $\begin{cases} u - v + 2w \le 10 \\ 2u + v + 3w \le 12 \\ u \ge 0, v \ge 0, w \ge 0. \end{cases}$

5. Maximize $-7u + 10v$ subject to the constraints $\begin{cases} -2u + 8v \le 3 \\ 4u + v \le 5 \\ 6u + 9v \le 1 \\ u \ge 0, v \ge 0. \end{cases}$ **7.** $x = 12, y = 20, M = 88$; $u = \frac{2}{7}, v = \frac{6}{7}, M = 88$

9. $x = 0, y = 2, M = 24$; $u = 0, v = 12, w = 0, M = 24$

11. Maximize $3u + 5v$ subject to the constraints $\begin{cases} u + 2v \le 3 \\ u \le 1 \\ u \ge 0, v \ge 0 \end{cases}$ $x = \frac{5}{2}, y = \frac{1}{2}$, minimum = 8; $u = 1, v = 1$, maximum = 8

13. Minimize $6u + 9v + 12w$ subject to the constraints $\begin{cases} u + 3v \ge 10 \\ -2u + w \ge 12 \\ v + 3w \ge 10 \\ u \ge 0, v \ge 0, w \ge 0 \end{cases}$

$x = 3, y = 12, z = 0$, maximum = 174; $u = 0, v = \frac{10}{3}, w = 12$, minimum = 174
15. Suppose we can hire workers out at a profit of u dollars per hour, sell the steel at a profit of v dollars per unit, and sell the wood at a profit of w dollars per unit. To find the minimum profit at which that should be done, minimize $90u + 138v + 120w$ subject to the

constraints $\begin{cases} 3u + 7v + 4w \ge 30 \\ 6u + 5v + 3w \ge 50 \\ u \ge 0, v \ge 0, w \ge 0. \end{cases}$ **17.** Suppose we can buy anthracite at u dollars per ton, ordinary coal at v dollars per ton, and bituminous coal at w dollars per ton. To find the maximum cost at which this should be done, maximize $80u + 60v + 75w$ subject to the

constraints $\begin{cases} 4u + 4v + 7w \le 150 \\ 10u + 5v + 5w \le 200 \\ u \ge 0, v \ge 0, w \ge 0. \end{cases}$ **19.** \$36.30   **21.** $x = 2, y = 1,$ maximum $= 74$

Chapter 4: Review Exercises, page 182

1. $x = 2, y = 3,$ max. $= 18$ **2.** $x = 0, y = 7,$ max. $= 35$ **3.** $x = 4, y = 5,$ max. $= 23$ **4.** $x = 2, y = 4,$ max. $= 34$
5. $x = 5, y = 1,$ min. $= 6$ **6.** $x = 0, y = 6,$ min. $= 12$ **7.** $x = 4, y = 1,$ min. $= 110$ **8.** $x = 4, y = 3,$ min. $= 41$
9. $x = 1, y = 6, z = 8,$ max. $= 884$ **10.** $x = 60, y = 8, z = 20, w = 0,$ max. $= 312$ **11.** Minimize $14u + 9v + 24w$ subject to the

constraints $\begin{cases} u + v + 3w \ge 2 \\ 2u + v + 2w \ge 3 \\ u \ge 0, v \ge 0, w \ge 0 \end{cases}$ **12.** Maximize $8u + 5v + 7w$ subject to the constraints $\begin{cases} u + v + 2w \le 20 \\ 4u + v + w \le 30 \\ u \ge 0, v \ge 0, w \ge 0 \end{cases}$

13. Primal: $x = 4, y = 5,$ max. $= 23$; dual: $u = 1, v = 1, w = 0,$ min. $= 23$ **14.** Primal: $x = 4, y = 1,$ min. $= 110$;

dual: $u = \frac{10}{3}, v = \frac{50}{3}, w = 0,$ max. $= 110$ **15.** $A = \begin{bmatrix} 1 & 2 \\ 1 & 1 \\ 3 & 2 \end{bmatrix}, B = \begin{bmatrix} 14 \\ 9 \\ 24 \end{bmatrix}, C = [2 \quad 3], X = \begin{bmatrix} x \\ y \end{bmatrix}$ Primal: Maximize CX subject to

$AX \le B, X \ge \mathbf{0}.$ Dual: $U = \begin{bmatrix} u \\ v \\ w \end{bmatrix}$ Minimize $B^T U$ subject to $A^T U \ge C^T, U \ge \mathbf{0}.$

16. $A = \begin{bmatrix} 1 & 4 \\ 1 & 1 \\ 2 & 1 \end{bmatrix}, B = \begin{bmatrix} 8 \\ 5 \\ 7 \end{bmatrix}, C = [20 \quad 30], X = \begin{bmatrix} x \\ y \end{bmatrix}$ Primal: Minimize CX subject to $AX \ge B, X \ge \mathbf{0}.$

Dual: $U = \begin{bmatrix} u \\ v \\ w \end{bmatrix}$ Maximize $B^T U$ subject to $A^T U \le C^T, U \ge \mathbf{0}.$ **17.** 48 type A lenses, 84 type B lenses, and no type C lenses; \$1416

18. 65 pounds of oranges, 8 pounds of cherries, and no blueberries; 17,030 calories   **19. (a)** 30 attack sticks, 40 defense sticks
(b) \$22   **20.** \$210

CHAPTER 5

Exercises 5.1, page 189

1. (a) $\{5, 6, 7\}$ **(b)** $\{1, 2, 3, 4, 5, 7\}$ **(c)** $\{1, 3\}$ **(d)** $\{5, 7\}$ **3. (a)** $\{a, b, c, e, i, o, u\}$ **(b)** $\{a\}$ **(c)** $\varnothing$ **(d)** $\{b, c\}$
5. $\varnothing, \{1\}, \{2\}, \{1, 2\}$ **7. (a)** {all freshman college students who like basketball}   **(b)** {all college students who do not like
basketball}   **(c)** {all college students who are neither freshmen nor like basketball}   **(d)** {all college students who are either freshmen
or like basketball}   **9. (a)** $S = \{1999, 2003, 2006, 2010, 2013\}$ **(b)** $T = \{1996, 1997, 1998, 1999, 2003, 2009, 2013\}$
(c) $S \cap T = \{1999, 2003, 2013\}$ **(d)** $S \cup T = \{1996, 1997, 1998, 1999, 2003, 2006, 2009, 2010, 2013\}$ **(e)** $S' \cap T = \{1996, 1997, 1998, 2009\}$
(f) $S \cap T' = \{2006, 2010\}$ **11.** Between 1996 and 2015, there were only two years in which the Standard and Poor's index increased by
2% or more during the first five days and did not increase by 16% or more for the entire year.   **13. (a)** $\{e, f\}$ **(b)** $\{a, b, c, d, e, f\}$
(c) $\varnothing$ **(d)** $\{a, b\}$ **(e)** $\varnothing$ **(f)** $\{a, b, d, e, f\}$ **(g)** $\{a, b, c\}$ **(h)** $\{a, b\}$ **(i)** $\{d\}$ **15.** S **17.** U **19.** $\varnothing$ **21.** $L \cup T$ **23.** $L \cap P$
25. $P \cap L \cap T$ **27.** S' **29.** $S \cup A \cup D$ **31.** $(A \cap S)' \cap D$ **33.** {students at Mount College who are younger than 35}
35. {people who are both students and teachers at Mount College}   **37.** {people at Mount College who are students or are at most 35}
39. {people at Mount College who are at least 35}   **41.** V' **43.** $V \cap (C \cup S)'$ **45.** $(V \cup C)'$ **47. (a)** $\{B, C, D, E\}$
(b) $\{C, D, E, F\}$ **(c)** $\{A, D, E, F\}$ **(d)** $\{A, C, D, E, F\}$ **(e)** $\{A, F\}$ **(f)** $\{D, E\}$ **49.** 8 ways: no toppings; peppers; onions;
mushrooms; peppers and onions; peppers and mushrooms; onions and mushrooms; all three toppings
51. Possible answer: $\{2\}$ **53.** $S \subseteq T$ **55.** True   **57.** True   **59.** False   **61.** True

Exercises 5.2, page 195

1. 6   **3.** 0   **5.** 8   **7.** $S \subseteq T$ **9.** 11 million   **11.** 10   **13.** 452

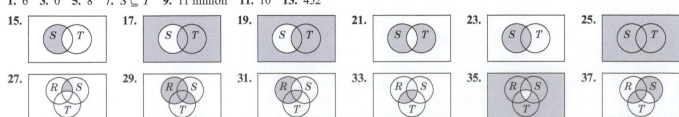

39. $S' \cup T'$ **41.** $S \cap T'$ **43.** U **45.** S' **47.** $R \cap T$ **49.** $R' \cap S \cap T$ **51.** $T \cup (R \cap S')$ **53.** $(R \cap S \cap T) \cup (R' \cap S' \cap T')$
55. Everyone who is not a citizen or is both over the age of 18 and employed   **57.** Everyone over the age of 18 who is unemployed
59. Noncitizens who are 5 years of age or older

Exercises 5.3, page 200

1. 11 **3.** 46 **5.** 11 **7.** 75 **9.** 30

11. 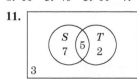 **13.** 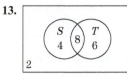 **15.** 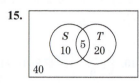 **17.**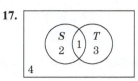

19. 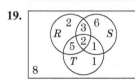 **21.** 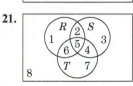 **23.** 25 **25.** 4; 2; 1 **27.** 55 **29.** 5 **31. (a)**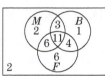

(b) 2 **(c)** 5 **33.** 28 **35.** 2 **37.** 51 **39.** 190 **41.** 180 **43.** 210 **45.** 100 **47.** 450 **49.** 3750 **51.** 30 **53.** 4 **55.** 49 **57.** 29
59. 26 **61.** 90 **63.** 6 **65.** 140 **67.** 30 **69.** 30 **71.** 3

Exercises 5.4, page 206

1. $4 \cdot 2 = 8$ **3.** $3 \cdot 2 = 6$ **5.** $44 \cdot 43 \cdot 42 = 79,464$ **7.** $20 \cdot 19 \cdot 18 = 6840$ **9.** 30, since $30 \cdot 29 = 870$
11. (a) $8 \cdot 7 \cdot 6 \cdot 5 \cdot 4 \cdot 3 \cdot 2 \cdot 1 = 40,320$ **(b)** $5 \cdot 4 \cdot 3 \cdot 2 \cdot 1 \cdot 3 \cdot 2 \cdot 1 = 720$ **13.** $4 \cdot 3 \cdot 2 \cdot 1 = 24$ **15.** $2 \cdot 3 = 6$
17. $3 \cdot 12 \cdot 10 \cdot 10 \cdot 10 \cdot 10 = 360,000$ **19.** $10^9 - 1 = 999,999,999$ **21.** $8 \cdot 2 \cdot 10 = 160$ **23.** $9 \cdot 10 \cdot 10 \cdot 1 \cdot 1 = 900$ **25.** $26 \cdot 26 \cdot 1 \cdot 1 = 676$
27. $15 \cdot 15 = 225$ **29.** $3200 \cdot 2 \cdot 24 \cdot 52 = 7,987,200$ **31.** Since $26 \cdot 26 \cdot 26 = 17,576 < 20,000$, two students must have the same initials.
33. $7 \cdot 5 = 35$ **35.** $5 \cdot 4 = 20$ **37.** $2^6 = 64$ **39.** $2^5 = 32$ **41.** $4^{10} = 1,048,576$ **43.** $10^5 = 100,000$
45. $8 \cdot 7 \cdot 6 \cdot 5 \cdot 4 \cdot 3 \cdot 2 \cdot 1 = 40,320$; one week **47.** $6 \cdot 7 \cdot 4 = 168$ days or 24 weeks **49.** $5 \cdot 11 \cdot (7 \cdot 2 + 1) \cdot 10 = 8250$ **51.** $2^4 = 16$
53. $2 \cdot 38 \cdot 38 = 2888$ **55. (a)** $9 \cdot 8 \cdot 7 \cdot 6 \cdot 5 \cdot 4 \cdot 3 \cdot 2 \cdot 1 = 362,880$ **(b)** $8 \cdot 7 \cdot 6 \cdot 5 \cdot 4 \cdot 3 \cdot 2 \cdot 1 \cdot 1 = 40,320$
(c) $1 \cdot 6 \cdot 5 \cdot 4 \cdot 3 \cdot 2 \cdot 1 \cdot 1 \cdot 1 = 720$ **57.** $\frac{10 \cdot 9}{2} + 10 \cdot 10 = 145$ **59.** $4 \cdot 3 \cdot 3 \cdot 3 \cdot 3 \cdot 3 = 972$ **61.** $7 \cdot 4 \cdot 2^6 = 1792$;
$8 \cdot 5 \cdot 3^6 = 29,160$ **63.** $2^4 = 16$

Exercises 5.5, page 213

1. 12 **3.** 120 **5.** 120 **7.** 5 **9.** 7 **11.** n **13.** 1 **15.** $\frac{n(n-1)}{2}$ **17.** 720 **19.** 72 **21.** Permutation **23.** Combination
25. Neither **27.** $4! = 24$ **29.** $C(9, 7) = 36$ **31.** $C(8, 4) = 70$ **33.** $P(65, 5) = 991,186,560$ **35.** $C(10, 5) = 252$
37. $C(100, 3) = 161,700$; $C(7, 3) = 35$ **39.** $P(150, 3) = 3,307,800$ **41.** $C(52, 5) = 2,598,960$ **43.** $C(13, 5) = 1287$ **45.** $5! = 120$
47. (a) $C(10, 4) = 210$ **(b)** $C(10, 6) = 210$ **(c)** Selecting four sweaters to take is equivalent to selecting six sweaters to leave.
49. $C(8, 2) = 28$ **51.** $C(69, 5) \cdot 26 = 292,201,338$ **53. (b)** $\dfrac{C(59, 6)}{C(49, 6)} \approx 3.22$ **55.** Yes; Moe: $C(9, 2) = 36$; Joe: $C(7, 4) = 35$
57. $4! \cdot P(4, 3) \cdot P(5, 3) \cdot P(6, 3) \cdot P(7, 3) = 870,912,000$ **59.** $3! \cdot 3!^3 = 1296$ **61.** 10 **63.** $C(15, 3) + 15 \cdot 14 + 15 = 680$
65. $720 - 3! - 5! = 594$ **67. (a)** $C(45, 5) = 1,221,759$ **(b)** $C(100, 4) = 3,921,225$ **(c)** Lottery (a)
69. Yes; the number of ways to shuffle a deck of 52 cards is $52! \approx 8 \times 10^{67}$.

Exercises 5.6, page 219

1. (a) $2^8 = 256$ **(b)** $C(8, 4) = 70$ **3. (a)** $C(7, 5) + C(7, 6) + C(7, 7) = 29$ **(b)** $2^7 - 29 = 99$ **5.** $2 \cdot C(6, 3) = 40$; $2 \cdot C(5, 3) = 20$
7. $C(11, 5) \cdot C(6, 5) \cdot 1 = 2772$ **9.** $C(8, 5) = 56$ **11.** $C(7, 2) = 21$ **13.** $C(5, 2) \cdot C(4, 2) = 60$ **15.** $C(6, 2) = 15$
17. (d) $56 + 70 = 126$ **19.** $C(8, 3) = 56$ **21.** $C(10, 6) - C(7, 4) = 175$ **23. (a)** $C(12, 5) = 792$ **(b)** $C(7, 5) = 21$
(c) $C(7, 2) \cdot C(5, 3) = 210$ **(d)** $C(7, 4) \cdot 5 + 21 = 196$ **25. (a)** $C(10, 3) = 120$ **(b)** $C(8, 3) = 56$ **(c)** $120 - 56 = 64$
27. $C(4, 2) \cdot C(6, 2) = 90$ **29.** $C(4, 3) \cdot C(4, 2) = 24$ **31.** $13 \cdot C(4, 3) \cdot 12 \cdot C(4, 2) = 3744$ **33.** $C(7, 5) \cdot 5! \cdot 21 \cdot 20 = 1,058,400$
35. $C(10, 5) \cdot P(21, 5) = 615,353,760$ **37.** $6! \cdot 7 \cdot 3! = 30,240$ **39.** $C(9, 4) = 126$ **41.** $C(26, 22) \cdot C(10, 7) = 1,794,000$
43. $C(12, 6) = 924$ **45.** $C(100, 50)/2^{100} \approx 7.96\%$ **47.** $\dfrac{C(50, 10) \cdot C(50, 10)}{C(100, 20)} \approx 19.7\%$

Exercises 5.7, page 225

1. 153 **3.** 15 **5.** 8 **7.** 1 **9.** 1 **11.** n **13.** 1 **15.** $n!$ **17.** 64 **19.** 20 **21.** $x^{10} + 10x^9y + 45x^8y^2$ **23.** $105x^2y^{13} + 15xy^{14} + y^{15}$
25. $184,756x^{10}y^{10}$ **27.** 6 **29.** 330 **31.** $x^9 + 18x^8y + 144x^7y^2$ **33.** $673,596x^6y^6$ **35.** $-945x^4y^3$ **37.** $2^8 = 256$ **39.** $2^4 = 16$
41. $2^5 = 32$ **43.** $2^8 - 1 = 255$ **45.** $2 \cdot 3 \cdot 2^{13} = 49,152$ **47.** $2^7 - C(7, 6) - C(7, 7) = 120$ **49.** $2^8 - C(8, 1) - C(8, 0) = 247$
51. $[2^9 - C(9, 1) - C(9, 0)] \cdot [2^{10} - C(10, 1) - C(10, 0)] = 508,526$ **53.** No **57.** $2^{10} - 2 = 1022$

Exercises 5.8, page 230

1. 20 **3.** 180 **5.** 210 **7.** 34,650 **9.** 166,320 **11.** 1,401,400 **13.** 2,858,856 **15.** $\binom{20}{7, 5, 8} = 99,768,240$ **17.** $\binom{8}{2, 1, 4, 1} = 840$
19. $\binom{9}{3, 2, 4} = 1260$ **21.** $\dfrac{1}{5!} \cdot \dfrac{20!}{(4!)^5} = 2,546,168,625$ **23.** $\binom{30}{10, 2, 18} = 5,708,552,850$ **25.** $\binom{4}{1, 1, 2} = 12$ **27.** $\dfrac{1}{7!} \cdot \dfrac{14!}{(2!)^7} = 135,135$
29. $\dfrac{1}{2!} \cdot \dfrac{10!}{(5!)^2} = 126$ **31.** $n!$ **33.** $\binom{38}{10, 12, 10, 6} = 115,166,175,166,136,334,240$ **35.** Greater than; $\binom{52}{13, 13, 13, 13} \approx 5.4$ octillion

Chapter 5: Review Exercises, page 234

1. $\varnothing, \{a\}, \{b\}, \{a, b\}$ **2.**   **3.** $C(16, 2) = 120$ **4.** $2 \cdot 5! = 240$ **5.**   **6.** $x^{12} - 24x^{11}y + 264x^{10}y^2$

7. $C(8, 3) \cdot C(6, 2) = 840$ **8.** 15   **9.** $7 \cdot 5 = 35$ **10.** $\binom{12}{2, 4, 6} = 13{,}860$ **11.** 0   **12.** 136   **13.** 6   **14.** 49   **15.** 47   **16.** 11

17. 46   **18.** 58   **19.** 22   **20.** 23   **21.** $C(9, 4) = 126$ **22.** $2^{20} = 1{,}048{,}576$ **23.** 550   **24.** 480   **25.** $\binom{5}{1, 3, 1} = 20$

26. $9 \cdot 10^4 \cdot 1 \cdot 5 = 450{,}000$ **27.** $9 \cdot 9 \cdot 10^8 = 8{,}100{,}000{,}000$ **28.** $P(7, 3) = 210$ **29.** $5^8 = 390{,}625; 5^8 - 4^8 = 325{,}089$
30. $C(12, 5) = 792$ **31.** $C(30, 14) = 145{,}422{,}675$ **32.** 27,500   **33.** $3^{10} = 59{,}049$ **34.** $C(60, 10) = 75{,}394{,}027{,}566$

35. $5^{10} = 9{,}765{,}625$ **36.** $21^6 = 85{,}766{,}121$ **37.** $C(10, 4) = 210$ **38.** $\frac{1}{3!} \cdot \frac{21!}{(7!)^3} = 66{,}512{,}160$ **39.** $14! = 87{,}178{,}291{,}200$

40. $\frac{1}{4!} \cdot \frac{20!}{(5!)^4} = 488{,}864{,}376$ **41.** $3 \cdot 5 \cdot 4 = 60$ **42.** $3 \cdot C(10, 2) = 135$ **43.** $C(n, 2) - n$ **44.** $8 \cdot 6 = 48$ **45.** $5! \cdot 4! \cdot 3! \cdot 2! \cdot 1! = 34{,}560$

46. $P(12, 5) = 95{,}040$ **47.** $4 \cdot C(13, 5) = 5148$ **48.** $C(4, 3) \cdot C(48, 2) = 4512$ **49.** $9 \cdot 10 \cdot 10 - 9 \cdot 9 \cdot 8 - 9 = 243$
50. $9 \cdot 9 \cdot 8 = 648$ **51.** $3 \cdot 3 = 9$ **52.** $2 \cdot 3! \cdot 3! = 72$ **53.** $24 \cdot 23 + 24 \cdot 23 \cdot 22 = 12{,}696$ **54.** $3 \cdot 4! \cdot 2 = 144$ **55.** $C(10, 2) = 45$

56. Second teacher: $\frac{1}{4!} \cdot \frac{24!}{(6!)^4} = 96{,}197{,}645{,}544$ vs $\frac{1}{6!} \cdot \frac{24!}{(4!)^6} = 4{,}509{,}264{,}634{,}875$ **57.** $\binom{10}{3, 4, 3} = 4200$

58. 5, since $5! = 120$ **59.** $7 \cdot 6 \cdot 5 \cdot 1 \cdot 4 \cdot 3 \cdot 2 \cdot 1 \cdot 1 = 5040$ **60. (a)** $C(12, 2) \cdot C(12, 3) = 14{,}520$ **(b)** $C(12, 5) \cdot 2^5 = 25{,}344$
61. $C(7, 2) + C(7, 1) + C(7, 0) = 29$ **62.** $2^6 = 64$ **63.** $2 \cdot 26 \cdot 26 + 2 \cdot 26 \cdot 26 \cdot 26 = 36{,}504$ **64.** 12

65. $\binom{25}{10, 9, 6} = 16{,}360{,}143{,}800$ **66.** $C(26, 3) \cdot C(10, 3) \cdot 6! = 224{,}640{,}000$

CHAPTER 6

Exercises 6.1, page 244
1. (a) {RS, RT, RU, RV, ST, SU, SV, TU, TV, UV}   **(b)** {RS, RT, RU, RV}   **(c)** {TU, TV, UV}   **3. (a)** {HH, HT, TH, TT}
(b) {HH, HT}   **5. (a)** {(I, red), (I, white), (II, red), (II, white)}   **(b)** {(I, red), (I, white)}   **7. (a)** $S = \{$All positive numbers of minutes$\}$
(b) "More than 5 minutes but less than 8 minutes"; $\varnothing$; "5 minutes or less"; "8 minutes or more"; "5 minutes or less"; "less than 4 minutes"; S
9. (a) {(Fr, Lib), (Fr, Con), (So, Lib), (So, Con), (Jr, Lib), (Jr, Con), (Sr, Lib), (Sr, Con)}   **(b)** {(Fr, Con), (So, Con), (Jr, Con), (Sr, Con)}
(c) {(Jr, Lib)}   **(d)** {(So, Lib), (Jr, Lib), (Sr, Lib)}   **11. (a)** No   **(b)** Yes   **13.** $\varnothing, \{a\}, \{b\}, \{c\}, \{a, b\}, \{a, c\}, \{b, c\}, S$ **15.** Yes
17. (a) {0, 1, 2, 3, 4, 5, 6, 7, 8, 9, 10}   **(b)** {6, 7, 8, 9, 10}   **19. (a)** No   **(b)** Yes   **(c)** Yes   **21.** {0, 1, 2, 3, 4, 5, 6, 7, 8}   **23.** (7, 4); 81
25. {2, 6, 9, 10}; 25%   **27.** {Colonel Mustard, Miss Scarlet, Professor Plum, Mrs. White, Mr. Green, Mrs. Peacock}   **(a)** 324
(b) "The murder occurred in the library with a gun."   **(c)** "Either the murder occurred in the library or it was done with a gun."

Exercises 6.2, page 254
1. Judgmental   **3.** Logical

5.

Number of Heads	Probability
0	$\frac{1}{4}$
1	$\frac{1}{2}$
2	$\frac{1}{4}$

7. $\frac{1}{19}$ **9. (a)** $\frac{191}{4487} \approx .04257$ **(b)** $\frac{272}{4487} \approx .06062$ **(c)** $\frac{4215}{4487} \approx .9394$

11. (a) $\frac{3}{13} \approx .2308$ **(b)** $\frac{5}{26} \approx .1923$ **(c)** $\frac{9}{26} \approx .3462$

13. (a) $\frac{1}{9} \approx .1111$ **(b)** $\frac{1}{6} \approx .1667$

15.

Kind of High School	Probability
Public	.820
Private	.172
Home School	.008

17. .64   **19. (a)** .6; .7   **(b)** .4   **(c)** .5   **(d)** .8   **21. (a)**

Number of Colleges Applied To	Probability
1	.10
2	.07
3	.10
4	.13
≥ 5	.60

(b) .83

23. None. For (a), the probabilities do not add to 1. For (b), $\text{Pr}(s_3) < 0$. For (c), the probabilities do not add to 1.   **25.** $\frac{1}{12}$ **27.** .24
29. 1   **31.** .9   **33.** .45   **35.** .25   **37. (a)** .7   **(b)** .2   **39.** .6   **41.** $\frac{10}{11}$ **43.** 1 to 4   **45.** 5 to 11   **47.** $\frac{2}{11}$ **49. (a)** $\frac{3}{8}, \frac{1}{4}, \frac{2}{5}, \frac{1}{5}$
(b) $\frac{49}{40}$, which is greater than 1   **(c)** Bookies have to make a living. The payoffs are a little lower than they should be, thus allowing the
bookie to make a profit.   **51.** 1

Exercises 6.3, page 261

1. (a) $\frac{9}{17} \approx .5294$ **(b)** $\frac{8}{17} \approx .4706$ **(c)** $\frac{5}{17} \approx .2941$ **(d)** $\frac{11}{17} \approx .6471$ **3. (a)** $\frac{C(5,2)}{C(11,2)} = \frac{2}{11} \approx .1818$ **(b)** $1 - \frac{C(5,2)}{C(11,2)} = \frac{9}{11} \approx .8182$

5. (a) $\frac{C(6,4) + C(7,4)}{C(13,4)} = \frac{10}{143} \approx .0699$ **(b)** $\frac{C(6,4) + C(6,3)\cdot C(7,1)}{C(13,4)} = \frac{31}{143} \approx .2168$ **7.** $1 - \frac{C(5,3)}{C(7,3)} = \frac{5}{7} \approx .7143$

9. $1 - \frac{C(9,3)}{C(13,3)} = \frac{101}{143} \approx .7063$ **11.** $1 - \frac{C(4,3)}{C(10,3)} = \frac{29}{30} \approx .9667$ **13.** $\frac{C(10,7)}{C(22,7)} = \frac{5}{7106} \approx .0007$ **15.** $1 - \frac{C(12,7) + C(12,6)\cdot C(10,1)}{C(22,7)} = \frac{16}{17} \approx .9412$

17. $1 - \frac{7\cdot6\cdot5}{7^3} = \frac{19}{49} \approx .3878$ **19.** $1 - \frac{30\cdot29\cdot28\cdot27}{30^4} = \frac{47}{250} \approx .188$ **21.** $1 - \frac{P(20,8)}{20^8} \approx .8016$ **23.** $1 - \left(\frac{364}{365}\right)^{25} \approx .06629$

25. $\frac{6\cdot5}{6^2} = \frac{5}{6} \approx .8333$ **27.** $\frac{3^4}{6^4} = \frac{1}{16} = .0625$ **29.** $\frac{C(10,4)}{2^{10}} = \frac{105}{512} \approx .2051$ **31.** $1 - \frac{7\cdot6\cdot5\cdot4}{7^4} = \frac{223}{343} \approx .6501$

33. (a) $\frac{C(3,1)\cdot C(5,2)}{C(8,3)} = \frac{15}{28} \approx .5357$ **(b)** $\frac{C(5,1)\cdot C(3,2)}{C(8,3)} = \frac{15}{56} \approx .2679$ **(c)** $\frac{C(3,1)\cdot1\cdot C(3,2)}{C(8,3)} = \frac{9}{56} \approx .1607$ **(d)** $\frac{15}{28} + \frac{15}{56} - \frac{9}{56} = \frac{9}{14} \approx .6429$

35. $1 - \frac{4^3}{5^3} = \frac{61}{125} = .488$ **37.** Increase **39.** $\frac{5\cdot4\cdot3}{5^3} = \frac{12}{25} = .48$ **41.** $\frac{3!\cdot4\cdot2!}{5!} = \frac{2}{5} = .4$ **43.** $\frac{13\cdot C(4,3)\cdot12\cdot C(4,2)}{C(52,5)} = \frac{6}{4165} \approx .0014$

45. $\frac{C(13,2)\cdot C(4,2)\cdot C(4,2)\cdot44}{C(52,5)} = \frac{198}{4165} \approx .0475$ **47. (a)** $\frac{4\cdot C(13,4)\cdot C(13,3)\cdot C(13,3)\cdot C(13,3)}{C(52,13)} \approx .1054$

(b) $\frac{C(4,2)\cdot C(13,4)\cdot C(13,4)\cdot2\cdot C(13,3)\cdot1\cdot C(13,2)}{C(52,13)} \approx .2155$ **49.** $\frac{2}{C(40,6)} = \frac{1}{1,919,190} \approx .0000005211$ **51.** $\frac{C(5,3)\cdot C(34,2)}{C(39,5)} \approx .0097$

53. $\frac{25-20}{80} = \frac{1}{16} = 6.25\%$; $\frac{25-20}{25} = \frac{1}{5} = 20\%$; 16 **55.** $\frac{20-15}{100} = \frac{1}{20} = 5\%$; $\frac{20-15}{20} = \frac{1}{4} = 25\%$; 20 **57.** 16 **59.** $1 - \frac{P(100,15)}{100^{15}} \approx .6687$

61. $1 - \frac{P(52,5)}{52^5} \approx .1797$; 9 **63.** 253; $1 - \left(\frac{364}{365}\right)^{253} \approx .5005$ **65.** 48

Exercises 6.4, page 271

1. (a) .5 **(b)** .6 **(c)** $\frac{.2}{.6} \approx .3333$ **(d)** $\frac{.2}{.5} = .4$ **3. (a)** $\frac{.1}{.4} = \frac{1}{4}$ **(b)** $\frac{.1}{.5} = \frac{1}{5}$ **(c)** $\frac{.4}{.6} = \frac{2}{3}$ **(d)** $\frac{.2}{.6} = \frac{1}{3}$ **5. (a)** $\frac{1}{12}$ **(b)** $\frac{\frac{1}{12}}{\frac{5}{12}} = \frac{1}{5}$ **(c)** $\frac{\frac{1}{12}}{\frac{1}{3}} = \frac{1}{4}$

7. (a) .1 **(b)** .6 **(c)** $\frac{.1}{.3} = \frac{1}{3}$ **(d)** .2 **9.** $\frac{5}{36}/(1 - \frac{1}{6}) = \frac{1}{6}$ **11.** 0 **13.** $\frac{C(7,4)}{C(12,4) - C(5,4)} = \frac{1}{14} \approx .0714$ **15.** $\frac{1}{2}$ **17.** $\frac{.10}{.25} = .4$

19. (a) $\frac{851}{2898} \approx .2937$ **(b)** $\frac{1201}{2898} \approx .4144$ **(c)** $\frac{522}{851} \approx .6134$ **(d)** $\frac{93}{1697} \approx .0548$ **21. (a)** $\frac{228.6}{1291.8} \approx .1770$ **(b)** $\frac{183.2}{1291.8} \approx .1418$

(c) $\frac{20.7}{1291.8} \approx .0160$ **(d)** $\frac{20.7}{183.2} \approx .1130$ **(e)** $\frac{20.7}{228.6} \approx .0906$ **23.** $\frac{\frac{1}{3}}{\frac{1}{2}} = \frac{2}{3}$ **25.** $\frac{1}{221} \approx .004525$ **27.** $\frac{1}{2}$ **29.** $.48\cdot.09 = .0432$

31. Three-point shot; .24 vs .29 **33.** Yes **35.** .8 **37.** .6 **39.** .25 **41.** .992 **43.** Not independent **45.** Independent
47. Not Independent **49.** No **51. (a)** $.80\cdot.75\cdot.60 = .36$ **(b)** .81 **53.** $.995^5\cdot.98^5\cdot.975^3 \approx .7967$ **55.** $.3^4 = .0081$
57. (a) $1 - .7^4 = .7599$ **(b)** $.7599^{10} \approx .06420$ **(c)** $1 - .9358^{20} = .7347$ **59.** 0 points; $1 - .6 = .4$; $.6\cdot.4 = .24$; $.6\cdot.6 = .36$
63. $.6\cdot.4 = .24$; $.4\cdot.6 = .24$ **69.** 26; $1 - \left(\frac{37}{38}\right)^{26} \approx .5001$

Exercises 6.5, page 279

1. 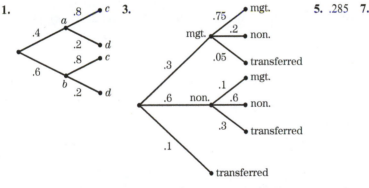 **3.** **5.** .285 **7.** .295 **9.** $\frac{7}{12}$ **11.** $\frac{1201}{5525} \approx .22$

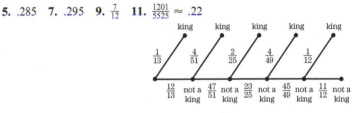

13. .14 **15.** $\frac{16}{17} \approx .9412$ **17.** .8 **19.** $\frac{4}{7}$ **21. (a)** .60 **(b)** .75 **23.** .00029997 **25.** $\frac{11}{16}$ **27. (a)** $\frac{1}{4}$; $\frac{3}{4}$ **(b)** .7 **29.** $\frac{3}{8}$ **31.** $\frac{25}{26}$
33. $\frac{3}{8}$ **35. (a)** .99 **(b)** $(.99)^{200} \approx .1340$ **37. (a)** $\frac{7}{12}$ **(b)** $\frac{7}{12}$ **(c)** $\frac{25}{36}$ **(d)** Since the red die beats the blue die more than half the time and the blue die beats the green die more than half the time, the red die appears to be the strongest of the three dice and the green die appears to be the weakest. However, paradoxically, the green die beats the red die more than half the time. **39.** True **41.** True
43. True **45.** .00473 or .473% **47.** $\frac{9}{19} \approx .474$ **49.** .84 or 84%

Exercises 6.6, page 286

1. $\frac{8}{53}$ **3.** $\frac{3}{7}$ **5.** .075 **7.** $\frac{8}{9}$ **9. (a)** .1325 **(b)** $\frac{12}{53} \approx .23$ **11.** $\frac{5}{103} \approx .049$ **13. (a)** .01 **(b)** $\frac{33}{34} \approx .971$ **15.** $\frac{31}{37} \approx .838$ **17.** .3805
19. (a) $\frac{1}{4}$ **(b)** $\frac{13}{17} \approx .765$ **(c)** .130 **21. (a)** $\frac{5}{9}$ **(b)** 11% **23.** $\frac{3}{7}$ **25.** $\frac{8}{9}$ **27.** $\frac{31}{37} \approx .838$ **29.** $\frac{5}{9}$

Chapter 6: Review Exercises, page 296

1. (a) $S = \{PN, PD, PQ, PH, ND, NQ, NH, DQ, DH, QH\}$ **(b)** $E = \{PN, PQ, NQ, DH\}$ **2. (a)** A male junior is elected.
(b) A female junior is not elected. **(c)** A male or a junior is elected. **3.** .2 **4.** .8 **5.** $\frac{1}{8}$ **6.** .5 **7.** $\frac{13}{25}$; $\frac{12}{25}$ **8.** $\frac{1}{3709}$
9. 13 to 37; 37 to 13 **10.** 1 to 3 **11.** $\frac{1}{12}$ **12.** $\frac{1}{3}$ **13.** $\approx .01163$ **14.** $\approx .3571$ **15. (a)** $\frac{2}{15}$ **(b)** $\frac{1}{3}$ **16. (a)** $\frac{2}{9}$; $\frac{1}{9}$ **(b)** $\frac{5}{36}$; $\frac{1}{6}$ **17.** $\frac{31}{32}$

18. $\frac{5}{16}$ **19.** $\frac{1}{21}$ **20.** 83,520 to 1 **21. (a)** $\left(\frac{1}{36}\right)^3$ **(b)** $\left(\frac{1}{10}\right)^4$ **(c)** $\frac{2197}{93312}$ **22. (a)** $\frac{1}{2197}$ **(b)** $\frac{469}{2197}$ **23.** $\frac{6!}{6^6} \approx .0154$ **24.** 5 to 13

25. $\approx .0271$ **26.** $\frac{223}{343} \approx .6501$ **27.** $\frac{2}{3}$ **28.** $\frac{7}{10}$ **29.** $\frac{6}{7}$ **30.** $\frac{1}{3}$ **31. (a)** $\approx .9527$ **(b)** $\approx .5215$ **(c)** $\approx .4794$ **(d)** $\approx .9545$

32. (a) $\frac{3}{10}$ **(b)** $\frac{2}{5}$ **(c)** $\frac{1}{3}$ **(d)** $\frac{2}{3}$ **33.** .04 **34.** $\frac{95}{1827} \approx .0520$ **35.** No **36.** No **37.** Yes **38.** Yes **39. (a)** $\frac{1}{12}$ **(b)** $\frac{1}{2}$ **40. (a)** .3
(b) .85 **41.** $\frac{1}{6}$ **42.** .46 **43.** $\frac{4}{9}$ **44.** $\frac{2}{3}$ **45.** Switch **46.** 56% **47.** $\frac{4}{25}$ **48.** $\frac{2}{11}$ **49.** $\frac{1}{3}$ **50.** $\frac{1}{6}$ **52.** $\Pr(E \cup F)$ **55.** True **56.** No

CHAPTER 7

Exercises 7.1, page 305

1. 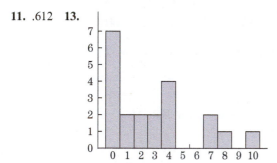 **3.** **5.** **7.** $36°$ **9.**
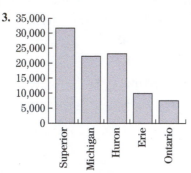
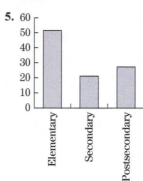
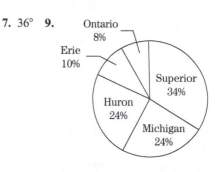

11. .612 **13.**

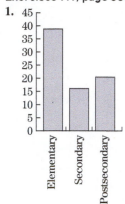

Number of Tie-Breaking Votes

15. 2.7 **17.**
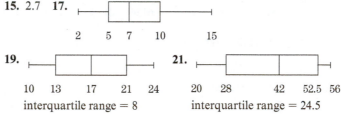

19.

21.

interquartile range = 8   interquartile range = 24.5

23. (A) c; (B) d; (C) a; (D) b **25. (a)** min = 200, $Q_1 = 400$, $Q_2 = 600$, $Q_3 = 700$, max = 800 **(b)** 25% **(c)** 25% **(d)** 50%
(e) 75% **27.**

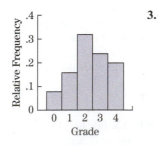

.045 .105   .200   .301 .313
Blue Jays

.239 .265 .293 .333   .429
Red Sox
The Red Sox seem more likely to win.

Exercises 7.2, page 315

1.

Grade	Relative Frequency
0	.08
1	.16
2	.32
3	.24
4	.20

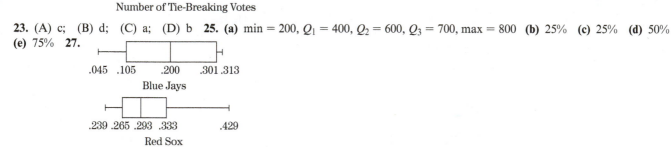

3.

Number of Visits During Minute	Relative Frequency
20	.05
21	.05
22	0
23	.10
24	.30
25	.20
26	0
27	.15
28	.10
29	.05

5.

Number of Heads	Probability
0	$\frac{1}{8}$
1	$\frac{3}{8}$
2	$\frac{3}{8}$
3	$\frac{1}{8}$

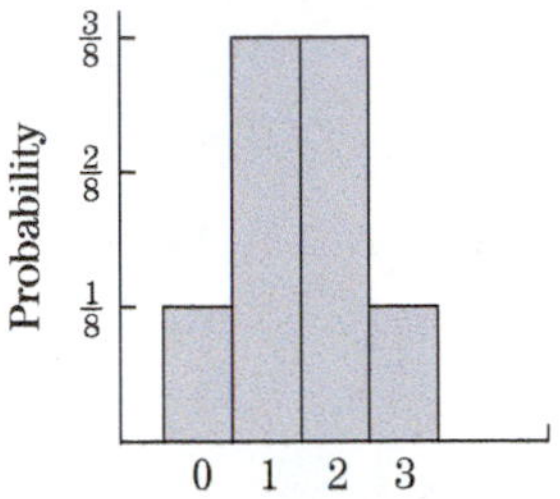

7.

Number of Red Balls	Probability
0	$\frac{4}{35}$
1	$\frac{18}{35}$
2	$\frac{12}{35}$
3	$\frac{1}{35}$

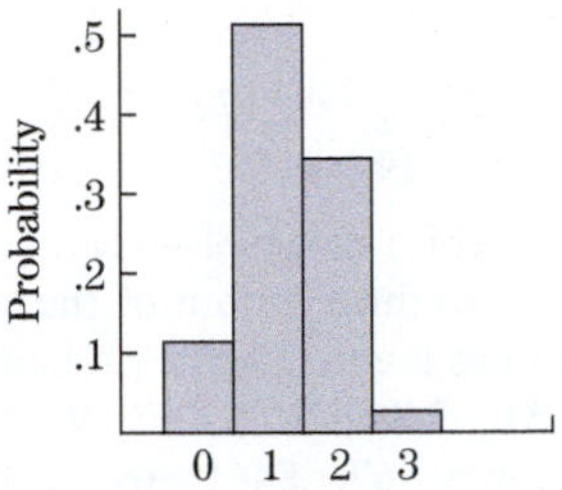

9.

Number of Red Balls	Player's Earnings	Probability
2	\$5	$\frac{1}{15}$
1	\$1	$\frac{8}{15}$
0	−\$1	$\frac{6}{15}$

11. .6

13.

k	$\Pr(X^2 = k)$
0	.1
1	.2
4	.4
9	.1
16	.2

15.

k	$\Pr(X - 1 = k)$
−1	.1
0	.2
1	.4
2	.1
3	.2

17.

k	$\Pr(\frac{1}{5}Y = k)$
1	.3
2	.4
3	.1
4	.1
5	.1

19.

k	$\Pr[(X + 1)^2 = k]$
1	.1
4	.2
9	.4
16	.1
25	.2

21.

	Relative Frequency	
Grade	9 A.M. Class	10 A.M. Class
F	.17	.17
D	.25	.20
C	.33	.17
B	.17	.20
A	.08	.26

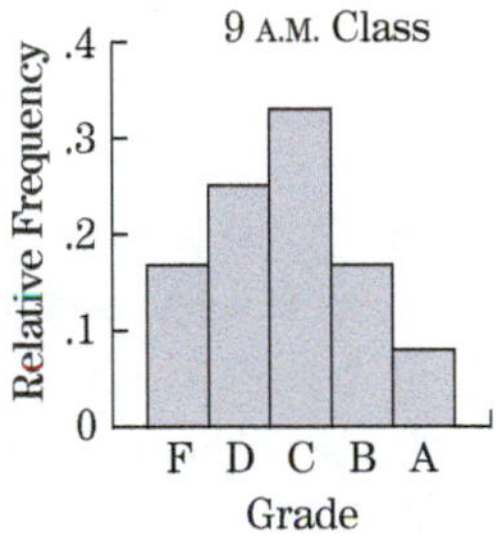

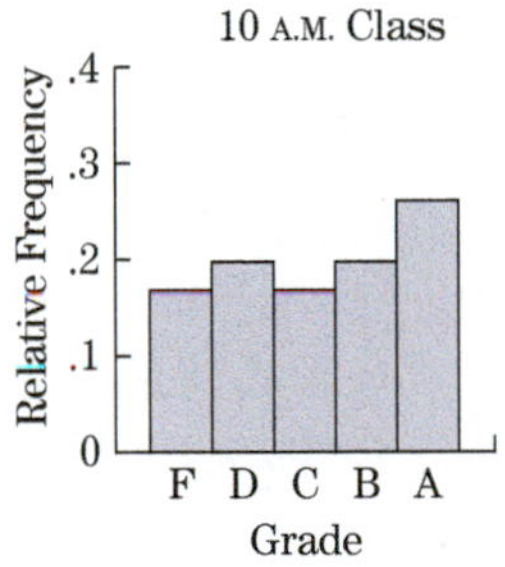

The 9 A.M. class has the distribution centered on the C grade with relatively few As. The 10 A.M. class has a large percentage of As and Bs, with fewer Cs.

23. 76% **25.** **(a)** 25% **(b)** 60% **(c)**

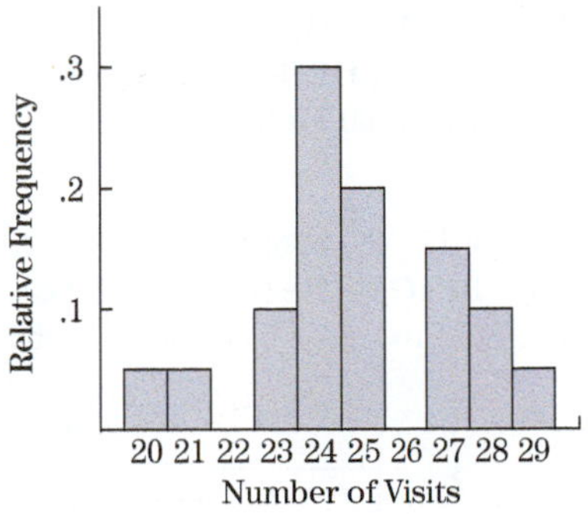

(d) approximately 25

27. **(a)** $\Pr(U = 4) = \frac{2}{15}$ **(b)** $\frac{3}{5}$ **(c)** $\frac{13}{15}$ **(d)** $\frac{2}{5}$ **(e)**

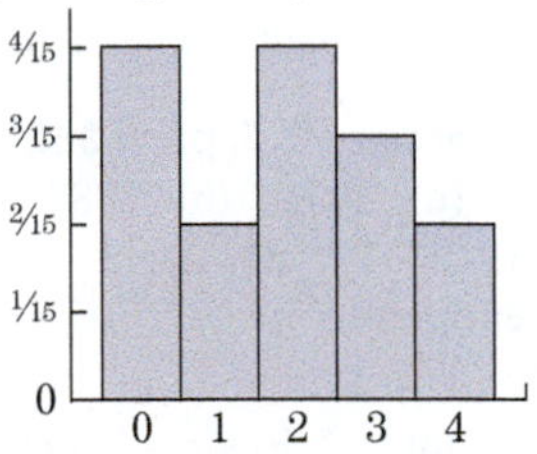

29. **(a)**

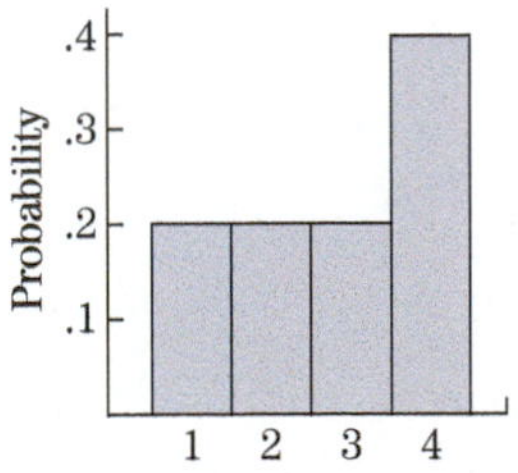

(b) .40 **(c)** .60 **(d)** 1 **(e)**

k	$\Pr[(Y + 2)^2 = k]$
9	.20
16	.20
25	.20
36	.40

Exercises 7.3, page 322

1. .1323 **3.** $\frac{8}{81} \approx .0988$ **5.** $\frac{15}{128} \approx .1172$ **7.** $\frac{165}{1024} \approx .1611$ **9.** $\frac{1023}{1024} \approx .9990$ **11.** $\frac{1}{1296} \approx .0007716$ **13.** $\frac{85}{648} \approx .1312$
15. $\frac{425}{432} \approx .9838$ **17.** .2220 **19.** .01663 **21.** .08908 **23.** .1478 **25.** .4727 **27.** (a) .375 (b) .5 (c) .125 **29.** $\Pr(X = 5)$
31. $\binom{40}{10} (.25)^{10}(.75)^{10}$; out of a group of 40 cattle, it is more likely that exactly 10 recover than exactly 9 recover. **33.** The probability
that the salesman sells cars to three or four of the customers **35.** (a) .4602 (b) .5398 **37.** Since $p = .5$, the probability of k
successes and $10 - k$ failures is equal to the probability of $10 - k$ successes and k failures. **39.** 1 **41.** .9821 **43.** .1018 **45.** .3174
47. .1198 **49.** .5781 **51.** .2401, .0756 **53.** 9 **55.** (b) $\binom{3}{0}p^0(1 - p)^3$; $\binom{3}{1}p(1 - p)^3$ **57.** 17 **59.** 114 **61.** .9648 **63.** .5433
65. (a) .2898 (b) 41 games **67.** .8441 **69.** .8301

Exercises 7.4, page 333

1. Population mean **3.** Expected value **5.** 1.95
7. (a) 2.9 (b)

Grade	Relative Frequency
4	.4
3	.2
2	.3
1	.1

(c) 2.9 **9.** $\overline{x}_A = 1.5, \overline{x}_B = 1.6$; group A had fewer cavities **11.** 2.3 **13.** 8.5

15.

Earnings	Probability
−$1	$\frac{37}{38}$
$35	$\frac{1}{38}$

$E(X) \approx -5.26¢$

17.

Earnings	Probability
−50¢	$\frac{1}{3}$
0¢	$\frac{4}{15}$
50¢	$\frac{1}{5}$
$1	$\frac{2}{15}$
$1.50	$\frac{1}{15}$

$E(X) \approx 16.67¢$

19. $1000 **21.** .322 **23.** 4.47 **25.** 0 (fair game)
27. 3.075 **29.** 10 **31.** Three free throws
33. $\frac{5}{3} \approx 1.667$; $\frac{4}{3} \approx 1.333$ **35.** $\frac{12}{7} \approx 1.714$; $\frac{12}{7} \approx 1.714$
37. 20% **39.** b **41.** 58.5° **43.** 7 **45.** 7 days
47. 3000

Exercises 7.5, page 343

1. 1.4 **3.** B **5.** (a) $\mu_A = 17, \sigma_A^2 = 186, \mu_B = 13, \sigma_B^2 = 141$ (b) A (c) B **7.** (a) $\mu_A = 103, \sigma_A^2 = 4.6, \mu_B = 104, \sigma_B^2 = 3.4$
(b) B (c) B **9.** 5, ≈ 1.581 **11.** 3, ≈ 1.719 **13.** (a) $\geq .75$ (b) $\geq .89$ (c) $\geq .31$ **15.** ≥ 4688 **17.** 8 **19.** (a) $\mu = 7, \sigma^2 = \frac{35}{6}$
(b) $\frac{5}{6}$ (c) $\geq \frac{19}{54}$ **21.** 2 **25.** $\mu = 53{,}227, \sigma \approx 4453.54$; Florida International, Texas A&M, University of Minnesota, University of
Texas, Michigan State, University of Florida **27.** (a) $\mu_A = 4.72, \sigma_A \approx 1.40$; $\mu_B = 4.8, \sigma_B \approx 1.65$ (b) B (c) A

Exercises 7.6, page 357

1. .8944 **3.** .4013 **5.** .2417 **7.** .6170 **9.** 1.75 **11.** .75 **13.** $\approx .84$ **15.** $\mu = 6, \sigma = 2$ **17.** $\mu = 9, \sigma = 1$ **19.** $-\frac{16}{3}$ **21.** $\frac{31}{2}$
23. .9772 **25.** .6247 **27.** .9544 **29.** .5 **31.** .0002 **33.** .7888 **35.** .2358 **37.** .0122 **39.** (a) 616 (b) Between 396 and 644
(c) 674 **41.** 19,750 miles **43.** True **45.** normalcdf(1,5,3,2); NORM.DIST(5,3,2, TRUE)-NORM.DIST(1,3,2,TRUE); $P(1 < x < 5)$
for x normal(3,2) **47.** normalcdf(0,1000,1200,100); NORM.DIST(1000,1200,100,TRUE); $P(x < 1000)$ for x normal(1200,100)

Exercises 7.7, page 362

1. (a) .1974 (b) .7888 (c) .9878 **3.** .0062 **5.** .0062 **7.** .0013 **9.** .6368 **11.** .1056 **13.** 2; .1469 **15.** .7498 **17.** .61
19. .2358 **21.** .0813 **23.** .0410; 1-binomcdf(300,.02,10); $1 -$ BINOM.DIST(10,300,.02,1); binomial probabilities n=300,p=.02,
endpoint=10 **25.** .0796; binompdf(100,.5,50); BINOM.DIST(50,100,.5,0); binomial probabilities n=100,p=.5,endpoint=50

Chapter 7: Review Exercises, page 366

1.

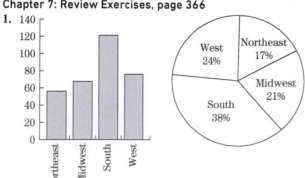

2.

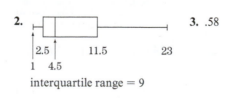

interquartile range = 9

3. .58

4. (a)

Number of Heads, k	$\Pr(X = k)$
0	.25
1	.50
2	.25

(b)

k	$\Pr(2X + 5 = k)$
5	.25
7	.50
9	.25

5. (a)

k	$\Pr(X = k)$
0	$\frac{8}{27}$
1	$\frac{12}{27}$
2	$\frac{6}{27}$
3	$\frac{1}{27}$

(b) $\mu = 1, \sigma^2 = \frac{2}{3}$

6. .2646 **7.** 52; Choose *true* for all of the questions. **8. (a)** .2961 **(b)** .6187 **(c)** 2 **9.** $\mu = 4.8, \sigma^2 = 19.76$

10.

k	$\Pr(X = k)$
0	$\frac{1}{70}$
1	$\frac{16}{70}$
2	$\frac{36}{70}$
3	$\frac{16}{70}$
4	$\frac{1}{70}$

$\mu = 2, \sigma^2 = \frac{4}{7}$

11. mean $= .4$, var $= 3.24$, std dev $= 1.8$ **12.** Lucy, about 11 cents per roll **13.** $\geq \frac{8}{9}$ **14.** at least $\frac{5}{9}$
15. .2857 **16.** .2266 **17.** 10.56% **18.** $-.75$ **19.** 22.5 **20. (a)** 1.39% **(b)** 124.75 **21.** .0122
22. .84 **27.** Yes **28.** Yes

CHAPTER 8

Exercises 8.1, page 378

1. Yes **3.** No **5.** Yes **7.** $\begin{array}{c} \\ A \\ B \end{array}\begin{array}{cc} A & B \\ \left[\begin{array}{cc} .3 & .5 \\ .7 & .5 \end{array}\right]\end{array}$ **9.** $\begin{array}{c} \\ A \\ B \\ C \end{array}\begin{array}{ccc} A & B & C \\ \left[\begin{array}{ccc} \frac{1}{3} & \frac{2}{9} & \frac{1}{3} \\ \frac{1}{3} & \frac{4}{9} & \frac{1}{6} \\ \frac{1}{3} & \frac{1}{3} & \frac{1}{2} \end{array}\right]\end{array}$ **11.** $\begin{array}{c} \\ A \\ B \\ C \end{array}\begin{array}{ccc} A & B & C \\ \left[\begin{array}{ccc} .4 & .2 & 0 \\ .5 & 0 & 0 \\ .1 & .8 & 1 \end{array}\right]\end{array}$ **13.**

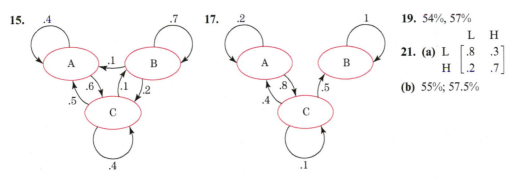

15.   **17.**   **19.** 54%, 57%

21. (a) $\begin{array}{c} \\ L \\ H \end{array}\begin{array}{cc} L & H \\ \left[\begin{array}{cc} .8 & .3 \\ .2 & .7 \end{array}\right]\end{array}$

(b) 55%; 57.5%

23. (a) $\begin{array}{c} \\ S \\ O \end{array}\begin{array}{cc} S & O \\ \left[\begin{array}{cc} .992 & .007 \\ .008 & .993 \end{array}\right]\end{array}$ **(b)** 12.52%, 13.03% **25. (a)** $\begin{array}{c} \\ L \\ R \end{array}\begin{array}{cc} L & R \\ \left[\begin{array}{cc} .9 & .7 \\ .1 & .3 \end{array}\right]\end{array}$ **(b)** $\left[\begin{array}{cc} .88 & .84 \\ .12 & .16 \end{array}\right]$ **(c)** $\left[\begin{array}{c} .8 \\ .2 \end{array}\right]_1, \left[\begin{array}{c} .86 \\ .14 \end{array}\right]_2$ **(d)** 87.5%

27. 40% will be in zone I, 40% will be in zone II, and 20% will be in zone III.
29. (a) 4% **(b)** 96% of the freshmen who held middle-of-the-road political views continued to hold those views as sophomores.

(c)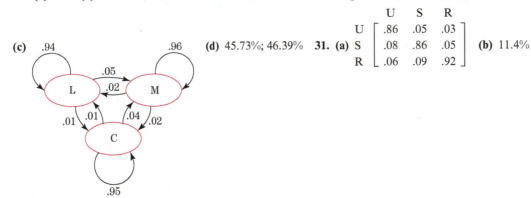

(d) 45.73%; 46.39% **31. (a)** $\begin{array}{c} \\ U \\ S \\ R \end{array}\begin{array}{ccc} U & S & R \\ \left[\begin{array}{ccc} .86 & .05 & .03 \\ .08 & .86 & .05 \\ .06 & .09 & .92 \end{array}\right]\end{array}$ **(b)** 11.4%

33. (a) $\begin{bmatrix} .34 \\ .40 \\ .26 \end{bmatrix}$ **(b)** .322; .388 **35. (a)**

$\begin{array}{c} \\ \\ 0 \\ 1 \\ 2 \\ 3 \\ 4 \end{array}\begin{array}{c}\begin{array}{ccccc} 0 & 1 & 2 & 3 & 4 \end{array} \\ \begin{bmatrix} 0 & \frac{1}{4} & 0 & 0 & 0 \\ 1 & 0 & \frac{1}{2} & 0 & 0 \\ 0 & \frac{3}{4} & 0 & \frac{3}{4} & 0 \\ 0 & 0 & \frac{1}{2} & 0 & 1 \\ 0 & 0 & 0 & \frac{1}{4} & 0 \end{bmatrix}\end{array}$

(b) .625 **37.** $\begin{bmatrix} .44 \\ .56 \end{bmatrix}_3, \begin{bmatrix} .44 \\ .56 \end{bmatrix}_4$ **39.** All powers are $\begin{bmatrix} \frac{1}{3} & \frac{1}{3} \\ \frac{2}{3} & \frac{2}{3} \end{bmatrix}$.

41. $\begin{bmatrix} .1 & .3 \\ .9 & .7 \end{bmatrix}; \begin{bmatrix} .28 & .24 \\ .72 & .76 \end{bmatrix}; \begin{bmatrix} .24 & .25 \\ .76 & .75 \end{bmatrix}; \begin{bmatrix} .25 & .25 \\ .75 & .75 \end{bmatrix}; \begin{bmatrix} .25 & .25 \\ .75 & .75 \end{bmatrix}$ **43.** All powers are the same as the original matrix. **45.** No

47. (a) $\begin{bmatrix} .35 \\ .65 \end{bmatrix}; \begin{bmatrix} .425 \\ .575 \end{bmatrix}; \begin{bmatrix} .3875 \\ .6125 \end{bmatrix}; \begin{bmatrix} .40625 \\ .59375 \end{bmatrix}$ **49.** $\begin{bmatrix} .4 \\ .6 \end{bmatrix}$ **51.** $\begin{bmatrix} .4 & .4 \\ .6 & .6 \end{bmatrix}$; Each column of this matrix is the same as the 2 × 1 matrix found in Exercise 49.

Exercises 8.2, page 387

1. Yes **3.** Yes **5.** No **7.** $\begin{bmatrix} \frac{1}{6} \\ \frac{5}{6} \end{bmatrix}$ **9.** $\begin{bmatrix} .6 \\ .4 \end{bmatrix}$ **11.** $\begin{bmatrix} \frac{5}{14} \\ \frac{3}{7} \\ \frac{3}{14} \end{bmatrix}$ **13.** 40% **15.** 87.5% **17.** 20% **19.** 40% **21. (a)**

$\begin{array}{c} \\ A \\ B \\ C \end{array}\begin{array}{c}\begin{array}{ccc} A & B & C \end{array} \\ \begin{bmatrix} .7 & .1 & .1 \\ .2 & .8 & .3 \\ .1 & .1 & .6 \end{bmatrix}\end{array}$

(b) 34%; 30.4% **(c)** 1/4 at location A, 11/20 at location B, 1/5 at location C **23.** 25% **25.** Matrix is not regular. **27.** 5/16; 3/8

29. $\begin{bmatrix} .7 & .7 \\ .3 & .3 \end{bmatrix}; \begin{bmatrix} .7 \\ .3 \end{bmatrix}$ **31.** $\begin{bmatrix} \frac{8}{35} & \frac{8}{35} & \frac{8}{35} \\ \frac{3}{7} & \frac{3}{7} & \frac{3}{7} \\ \frac{12}{35} & \frac{12}{35} & \frac{12}{35} \end{bmatrix}; \begin{bmatrix} \frac{8}{35} \\ \frac{3}{7} \\ \frac{12}{35} \end{bmatrix}$

Exercises 8.3, page 396

1. Yes **3.** No **5.** No **7.** Yes

9. $\begin{array}{c} \\ B \\ A \\ C \end{array}\begin{array}{c}\begin{array}{ccc} B & A & C \end{array} \\ \begin{bmatrix} 1 & .3 & .4 \\ 0 & .2 & .5 \\ 0 & .5 & .1 \end{bmatrix}\end{array}$ **11.** $\begin{array}{c} \\ D \\ A \\ B \\ C \end{array}\begin{array}{c}\begin{array}{cccc} D & A & B & C \end{array} \\ \begin{bmatrix} 1 & .4 & 0 & .1 \\ 0 & .1 & 1 & .6 \\ 0 & .2 & 0 & .1 \\ 0 & .3 & 0 & .2 \end{bmatrix}\end{array}$ **13.** $R = [.5]; S = \begin{bmatrix} .3 \\ .2 \end{bmatrix}; F = [2];$ $\left[\begin{array}{cc|c} 1 & 0 & .6 \\ 0 & 1 & .4 \\ \hline 0 & 0 & 0 \end{array}\right]$ **15.** $R = \begin{bmatrix} \frac{1}{4} & \frac{1}{2} \\ \frac{1}{3} & \frac{1}{3} \end{bmatrix}; S = \begin{bmatrix} \frac{1}{4} & \frac{1}{6} \\ \frac{1}{6} & 0 \end{bmatrix};$

$F = \begin{bmatrix} 2 & \frac{3}{2} \\ 1 & \frac{9}{4} \end{bmatrix}; \left[\begin{array}{cc|cc} 1 & 0 & \frac{2}{3} & \frac{3}{4} \\ 0 & 1 & \frac{1}{3} & \frac{1}{4} \\ \hline 0 & 0 & 0 & 0 \\ 0 & 0 & 0 & 0 \end{array}\right]$ **17.** $R = \begin{bmatrix} .5 & 0 \\ .1 & .6 \end{bmatrix}; S = \begin{bmatrix} .1 & .2 \\ .3 & 0 \\ 0 & .2 \end{bmatrix}; F = \begin{bmatrix} 2 & 0 \\ .5 & 2.5 \end{bmatrix}; \left[\begin{array}{ccc|cc} 1 & 0 & 0 & .3 & .5 \\ 0 & 1 & 0 & .6 & 0 \\ 0 & 0 & 1 & .1 & .5 \\ \hline 0 & 0 & 0 & 0 & 0 \\ 0 & 0 & 0 & 0 & 0 \end{array}\right]$

19. If the gambler begins with $2, he should have $1 for an expected number of .79 plays.

21. (a) 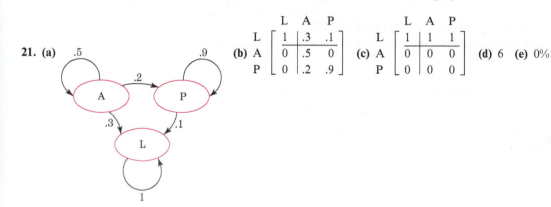 **(b)** $\begin{array}{c} \\ L \\ A \\ P \end{array}\begin{array}{c}\begin{array}{ccc} L & A & P \end{array} \\ \left[\begin{array}{c|cc} 1 & .3 & .1 \\ 0 & .5 & 0 \\ 0 & .2 & .9 \end{array}\right]\end{array}$ **(c)** $\begin{array}{c} \\ L \\ A \\ P \end{array}\begin{array}{c}\begin{array}{ccc} L & A & P \end{array} \\ \left[\begin{array}{c|cc} 1 & 1 & 1 \\ 0 & 0 & 0 \\ 0 & 0 & 0 \end{array}\right]\end{array}$ **(d)** 6 **(e)** 0%

23. (a)

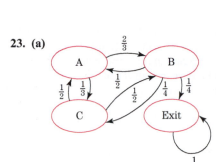

(b)

$$\begin{array}{c} \\ E \\ A \\ B \\ C \end{array} \begin{array}{cccc} \;\;E & A & B & C \\ \left[\begin{array}{c|ccc} 1 & 0 & \frac{1}{4} & 0 \\ 0 & 0 & \frac{1}{2} & \frac{1}{2} \\ 0 & \frac{2}{3} & 0 & \frac{1}{2} \\ 0 & \frac{1}{3} & \frac{1}{4} & 0 \end{array}\right] \end{array}$$

(c)

$$\begin{array}{c} \\ E \\ A \\ B \\ C \end{array} \begin{array}{cccc} \;\;E & A & B & C \\ \left[\begin{array}{c|ccc} 1 & 1 & 1 & 1 \\ 0 & 0 & 0 & 0 \\ 0 & 0 & 0 & 0 \\ 0 & 0 & 0 & 0 \end{array}\right] \end{array}$$

(d) 10.6 minutes

25. (a)

$$\begin{array}{c} \\ D \\ G \\ F \\ S \end{array} \begin{array}{cccc} \;\;D & G & F & S \\ \left[\begin{array}{cc|cc} 1 & 0 & .2 & .1 \\ 0 & 1 & 0 & .9 \\ 0 & 0 & 0 & 0 \\ 0 & 0 & .8 & 0 \end{array}\right] \end{array}$$

(b)

$$\begin{array}{c} \\ D \\ G \\ F \\ S \end{array} \begin{array}{cccc} \;\;D & G & F & S \\ \left[\begin{array}{cc|cc} 1 & 0 & .28 & .1 \\ 0 & 1 & .72 & .9 \\ 0 & 0 & 0 & 0 \\ 0 & 0 & 0 & 0 \end{array}\right] \end{array}$$

(c) .72 **(d)** 1.8 years

27. (a) $\frac{13}{14}; \frac{11}{14}$ **(b)** $\frac{20}{7}$ months **(c)** \$5786; \$1214 **29. (a)** $\frac{3}{4}; \frac{1}{2}; \frac{1}{4}$ **(b)** 4 **31.** $\frac{50}{81}; 5\frac{1}{2}$

33.

$$\begin{bmatrix} 1 & 0 & 0 & \frac{9}{37} & \frac{24}{37} \\ 0 & 1 & 0 & \frac{53}{74} & \frac{9}{37} \\ 0 & 0 & 1 & \frac{3}{74} & \frac{4}{37} \\ 0 & 0 & 0 & 0 & 0 \\ 0 & 0 & 0 & 0 & 0 \end{bmatrix}$$

Chapter 8: Review Exercises, page 400

1. Stochastic, neither **2.** Stochastic, regular **3.** Stochastic, regular **4.** Stochastic, absorbing **5.** Not stochastic

6. Stochastic, absorbing **7.** $\begin{bmatrix} \frac{5}{9} \\ \frac{4}{9} \end{bmatrix}$

8.

$$\begin{bmatrix} 1 & 0 & 0 & \frac{1}{2} & \frac{1}{2} \\ 0 & 1 & 0 & \frac{1}{4} & \frac{1}{8} \\ 0 & 0 & 1 & \frac{1}{4} & \frac{3}{8} \\ 0 & 0 & 0 & 0 & 0 \\ 0 & 0 & 0 & 0 & 0 \end{bmatrix}$$

9. (a)

$$\begin{array}{c} \\ H \\ M \\ L \end{array} \begin{array}{ccc} \;H & M & L \\ \left[\begin{array}{ccc} .5 & .4 & .3 \\ .4 & .3 & .5 \\ .1 & .3 & .2 \end{array}\right] \end{array}$$

; **(b)** 38% **(c)** $\frac{19}{97}$

10. (a)

$$\begin{array}{c} \\ P \\ N \end{array} \begin{array}{cc} \;P & N \\ \left[\begin{array}{cc} .8 & .3 \\ .2 & .7 \end{array}\right] \end{array}$$

(b) 30% **(c)** 60%

11.

$$\begin{bmatrix} 1 & 0 & \frac{11}{12} & 1 & \frac{1}{2} \\ 0 & 1 & \frac{1}{12} & 0 & \frac{1}{2} \\ 0 & 0 & 0 & 0 & 0 \\ 0 & 0 & 0 & 0 & 0 \\ 0 & 0 & 0 & 0 & 0 \end{bmatrix}$$

12. (a)

$$\begin{array}{c} \\ I \\ II \\ III \\ IV \end{array} \begin{array}{cccc} \;\;I & II & III & IV \\ \left[\begin{array}{cc|cc} 1 & 0 & 0 & \frac{1}{4} \\ 0 & 1 & \frac{1}{3} & \frac{1}{4} \\ 0 & 0 & 0 & \frac{1}{2} \\ 0 & 0 & \frac{2}{3} & 0 \end{array}\right] \end{array}$$

(b) $\frac{5}{12}$ **(c)** $\frac{5}{8}$ **(d)** $2\frac{1}{2}$ minutes **13.** c **14.** 58.5% **15. (a)** If the traffic is Moderate on a particular day, then for the next day the probability of Light is .2, the probability of Moderate is .75, and the probability of Heavy is .05. **(b)** About 37.0%, Light; 47.8%, Moderate; 15.2%, Heavy **(c)** About 3 days **16.** State G: $\approx$ 1.48 months; state S: $\approx$ 1.83 months.

17. (a)

$$\begin{bmatrix} \frac{122}{1683} \\ \frac{23}{99} \\ \frac{4}{9} \\ \frac{422}{1683} \end{bmatrix}$$

; In the long run, the probability of having 1, 2, 3, or 4 units of water in the reservoir at any given time will be $\frac{122}{1683}$, $\frac{23}{99}$, $\frac{4}{9}$, or $\frac{422}{1683}$, respectively. **(b)** About \$6881 **18.** The entries in each column are the probabilities of the transitions to the various states from the state associated with the column. These must add up to 1, just as the probabilities of all branches emanating from any node in a tree diagram must add up to 1. **19.** The probability of going from state A to state B after four periods. **20.** The stochastic matrix A represents a Markov process; A^2 represents this Markov process after two time periods, and so it is also a stochastic matrix.

21. True **22.** False

CHAPTER 9

Exercises 9.1, page 409

1. R: row 1, C: column 2; strictly determined; value $= 4$ **3.** R: row 2, C: column 2; not strictly determined   **5.** R: row 3, C: column 2; strictly determined; value $= 1$ **7.** R: row 3, C: column 2; strictly determined; value $= 0$ **9.** R: row 2, C: column 1; not strictly determined   **11.** R: row 4, C: column 2; strictly determined; value $= 3$ **13.** True   **15.** Rosa's Pizzeria should lower their price and Carlo's Pizzeria should improve their décor; strictly determined   **17.** Rigelbucks should move to mall B and Canisbucks should move to mall C. The game is strictly determined and has value 300 in Rigelbucks' favor.

19. $\begin{array}{c} \\ H \\ T \end{array}\begin{array}{cc} H & T \\ \end{array}\begin{bmatrix} 2 & -1 \\ -1 & -4 \end{bmatrix}$; strictly determined; R shows heads, C shows tails

21. $\begin{array}{c} \\ F \\ A \\ N \end{array}\begin{array}{ccc} F & A & N \\ \end{array}\begin{bmatrix} 8000 & -1000 & 1000 \\ -7000 & 4000 & -2000 \\ 3000 & 3000 & 2000 \end{bmatrix}$; strictly determined; both should be neutral   **23.** $\begin{array}{c} 5 \\ 10 \end{array}\begin{array}{ccc} 6 & 7 & 8 \\ \end{array}\begin{bmatrix} 1 & -5 & -5 \\ -6 & 3 & 2 \end{bmatrix}$; not strictly determined

Exercises 9.2, page 415

1. (a) 0   (b) 1   (c) -1.12 (d) .5; b is most advantageous to R **3.** \$29,200   **5.** 0   **7.** (a) No; no saddle point   (b) Yes; $\begin{bmatrix} \frac{11}{16} \\ \frac{5}{16} \end{bmatrix}$

9. (a) No; no saddle point   (b) Yes; $\begin{bmatrix} \frac{1}{30} \\ \frac{29}{30} \end{bmatrix}$ **11.** (a) Yes; Renée row 1, Carlos column 1   (b) Value $= 1$; no

13. (a) $\begin{array}{c} \\ \\ \text{Offense} \end{array}\begin{array}{c} \\ \\ \text{Run} \\ \text{Pass} \end{array}\begin{array}{c} \underline{Defense} \\ \begin{array}{cc} \text{Run} & \text{Pass} \end{array} \\ \begin{bmatrix} 1 & 3 \\ 7 & -4 \end{bmatrix} \end{array}$ (b) 2.44 yards   **15.** (a) $\begin{array}{c} \\ 1 \\ 2 \end{array}\begin{array}{cc} 1 & 2 \\ \end{array}\begin{bmatrix} 2 & -3 \\ -3 & 4 \end{bmatrix}$ (b) $v = .25$

Exercises 9.3, page 424

1. Minimize $y_1 + y_2$ subject to $\begin{cases} y_1 + 4y_2 \geq 1 \\ 6y_1 + 3y_2 \geq 1 \\ y_1 \geq 0, y_2 \geq 0 \end{cases}$ **3.** Maximize $z_1 + z_2$ subject to $\begin{cases} z_1 + 6z_2 \leq 1 \\ 4z_1 + 3z_2 \leq 1 \\ z_1 \geq 0, z_2 \geq 0 \end{cases}$ **5.** $v = \frac{7}{2}; \begin{bmatrix} \frac{1}{2} & \frac{1}{2} \end{bmatrix}; \begin{bmatrix} \frac{1}{4} \\ \frac{3}{4} \end{bmatrix}$

7. $v = -1; \begin{bmatrix} \frac{1}{2} & \frac{1}{2} \end{bmatrix}; \begin{bmatrix} \frac{5}{9} \\ \frac{4}{9} \end{bmatrix}$ **9.** $v = \frac{14}{5}; \begin{bmatrix} \frac{2}{5} & \frac{3}{5} \end{bmatrix}; \begin{bmatrix} \frac{3}{5} \\ \frac{2}{5} \end{bmatrix}$ **11.** $v = -\frac{1}{3}; \begin{bmatrix} \frac{1}{3} & \frac{2}{3} \end{bmatrix}; \begin{bmatrix} \frac{1}{3} \\ \frac{2}{3} \end{bmatrix}$ **13.** $v = \frac{29}{13}; \begin{bmatrix} \frac{7}{13} & \frac{6}{13} \end{bmatrix}$ **15.** $v = \frac{1}{5}; \begin{bmatrix} \frac{3}{5} & \frac{2}{5} \end{bmatrix}$

17. $v = \frac{1}{5}; \begin{bmatrix} \frac{4}{5} \\ \frac{1}{5} \end{bmatrix}$ **19.** $v = \frac{4}{7}; \begin{bmatrix} \frac{4}{7} \\ \frac{3}{7} \end{bmatrix}$ **21.** (a) $\begin{array}{c} \\ I \\ II \end{array}\begin{array}{cc} I & II \\ \end{array}\begin{bmatrix} -2 & 7 \\ 7 & -1 \end{bmatrix}$ (b) $\begin{bmatrix} \frac{8}{17} & \frac{9}{17} \end{bmatrix}$ (c) $\begin{bmatrix} \frac{8}{17} \\ \frac{9}{17} \end{bmatrix}$ (d) about \$2765

23. $v = \frac{25}{13} \approx 1.92$; offense: $\begin{bmatrix} \frac{11}{13} & \frac{2}{13} \end{bmatrix}$; defense: $\begin{bmatrix} \frac{7}{13} \\ \frac{6}{13} \end{bmatrix}$ **25.** (a) $\begin{array}{c} \\ 1 \\ 2 \end{array}\begin{array}{cc} 1 & 2 \\ \end{array}\begin{bmatrix} 2 & -3 \\ -3 & 4 \end{bmatrix}$ (b) $v = -\frac{1}{12}$; Reven: $\begin{bmatrix} \frac{7}{12} & \frac{5}{12} \end{bmatrix}$; Coddy: $\begin{bmatrix} \frac{7}{12} \\ \frac{5}{12} \end{bmatrix}$

27. $\begin{bmatrix} \frac{5}{8} & \frac{3}{8} & 0 \end{bmatrix}; \begin{bmatrix} \frac{1}{2} \\ \frac{1}{2} \end{bmatrix}$ **29.** $v = -.083331$; Reven: $[.583333 \quad .416667]$; Coddy: $\begin{bmatrix} .583333 \\ .416667 \end{bmatrix}$

Chapter 9: Review Exercises, page 427

1. Strictly determined; R 3, C 3; 2   **2.** Not strictly determined   **3.** Not strictly determined   **4.** Strictly determined; R 1, C 2; 1

5. 7   **6.** 2   **7.** 1.4   **8.** -1.3 **9.** $\begin{bmatrix} \frac{4}{11} & \frac{7}{11} \end{bmatrix}; \begin{bmatrix} \frac{6}{11} \\ \frac{5}{11} \end{bmatrix}$ **10.** $\begin{bmatrix} \frac{8}{17} & \frac{9}{17} \end{bmatrix}; \begin{bmatrix} \frac{10}{17} \\ \frac{7}{17} \end{bmatrix}$ **11.** $[0 \quad 1]$ **12.** $\begin{bmatrix} \frac{1}{4} \\ \frac{3}{4} \end{bmatrix}$ **13.** (a) Carol should play the

two $\frac{9}{14}$ of the time and the six $\frac{5}{14}$ of the time. Ruth should play the two $\frac{5}{14}$ of the time and the six $\frac{9}{14}$ of the time.   (b) Ruth

14. (a) $\begin{array}{c} \\ A \\ B \\ C \end{array}\begin{array}{ccc} S & A & W \\ \end{array}\begin{bmatrix} 3000 & 2000 & 1000 \\ 6000 & 2000 & -3000 \\ 15{,}000 & 1000 & -10{,}000 \end{bmatrix}$ (b) Buy stock A.   **15.** Consider the optimal strategies using the maxima/minima technique.

16. $h < 4$ **17.** (a) Since both lines are straight lines, the intersection of the two lines would be the only solution to the system of equations. Moving from that intersection point would increase the value of one line while decreasing the value of the other.
(b) You would use the equations $y = a_{11}r + a_{12}(1 - r)$ and $y = a_{21}r + a_{22}(1 - r)$ and find the point of intersection as in part a.
18. $r = \frac{a_{22} - a_{21}}{a_{11} - a_{21} - a_{12} + a_{22}}$

CHAPTER 10

Exercises 10.1, page 437

1. $i = .0025, n = 24$ **3.** $i = .011, n = 40$ **5.** $i = .00375, n = 42$ **7.** $i = .028, n = 4, P = \$500, F = \$558.40$
9. $i = .012, n = 19, P = \$7174.85, F = \$9000$ **11.** $i = .005, n = 360, P = \$3000, F = \$18,067.73$ **13.** \$1042.86
15. \$6505.63; \$505.63 **17.** \$12,697.35 **19.** \$54,914.06 **21.** \$9034.19 **23.** \$1400 now **25.** Better **27.** \$4; \$4.02; \$4.03
29. \$123.20; \$64.41 **31.** \$29.12 **33.** \$10,000 **35.** $\approx 4.92\%$ **37. (a)** 2.4% compounded monthly **(b)** \$20.08; \$10,060.12
(c) \$20.94; \$10,491.20 **39. (a)** $r = .015, n = \frac{1}{2}, P = \$500, F = \$503.75$ **(b)** $r = .025, n = 2, P = \$500, F = \$525$ **41.** \$1036
43. \$2500 **45.** 4.08% **47.** 40 **49.** 50 years **51.** $P = \frac{F}{1 + nr}$ **53.** \$104.06; 4.06% **55.** 4.04% **57.** 4.49% **59.** \$8874.49 **61.** a
63. 18 years **65.** d **67.** $\approx 4.6\%$ **69.** N=24, I%=2.1, PV=−1000, PMT=0, P/Y=12, Find FV; =FV(.175%,24,0,−1000);
compound interest PV=1000, i=.175%, n=24, annual **71.** N=300, I%=2.4, PMT=0, FV=−100000, P/Y=12, Find
PV; =PV(.2%,300,0,−100000); compound interest FV=100000, i=.2%, n=300, annual **73.** N=7, PMT=0, PV=1500, FV=−2100,
P/Y=1, Find I%; =RATE (7,0,1500,−2100); compound interest PV=1500, FV=2100, n=7, annual **75.** 25; 35 **77.** 37 **79.** 20

Exercises 10.2, page 446

1. $i = .00175, n = 120, R = \$100, F = \$13,340.09$ **3.** $i = .0085, n = 20, R = \$2000, P = \$36,642.08$ **5.** \$15,908.62 **7.** \$139
9. \$70,887.09 **11.** \$4048 **13. (a)** \$25,465.60 **(b)** \$24,000 **(c)** \$1465.60

(d)

Month	Interest	Balance
1		\$500.00
2	\$1.25	\$1001.25
3	\$2.50	\$1503.75

15.

Quarter	Balance at Beginning of Quarter	Interest for Quarter
1	\$10,000.00	\$55.00
2	\$9055.00	\$49.80
3	\$8104.80	\$44.58

17. \$1340.09 **19.** \$3357.92 **21.** Lump sum **23. (a)** \$2436.63 **(b)** an extra \$200 each month **25.** \$1786.57 **27.** \$9340.32
29. \$14,405.06 **31.** \$3580 **33.** a **35.** \$5600.40 **37. (a)** \$1500 **(b)** \$4151.67 **39.** \$24,649.92 **41.** \$7,886,136,075
43. \$68,404.06 **45.** \$240,000 **47. (a)** \$24,597.48 **(b)** \$6617.39 **49.** \$2,299,250.20
51. N=10, I%=2.6, PV=0, PMT=−1500, P/Y=2, Find FV; =FV(1.3%,10,−1500,0); annuity FV, n=10, i=1.3%, pmt=\$1500
53. N=12, I%=1.8, PV=0, FV=−1681.83, P/Y=12, Find PMT; =PMT(0.15%,12,0,−1681.83); annuity pmt,
n=12, i=.15%, FV=\$1681.83 **55.** N=24, I%=1.5, PMT=−3000, FV=0, P/Y=12, Find PV; =PV(0.125%,24,−3000,0);
annuity PV, i=.125%, n=24, pmt=\$3000 **57.** N=12, I%=1.6, PV=−47336.25, FV=0, P/Y=4,
Find PMT; =PMT(0.4%,12,−47336.25,0); annuity pmt, n=12, i=.4%, PV=\$47336.25 **59.** 19 years; 26 years **61.** After 33
weeks **63.** \$4734.59

Exercises 10.3, page 456

1. \$588.22 **3.** \$728.63 **5.** \$2270 **7.** \$608.13 **9.** \$143,000 **11.** \$9015.65 **13.** 7.8%

15.

Payment Number	Amount	Interest	Applied to Principal	Unpaid Balance
1	\$210.10	\$4.15	\$205.95	\$624.05
2	210.10	3.12	206.98	417.07
3	210.10	2.09	208.01	209.06
4	210.10	1.05	209.05	0.00

17. (a) \$818.80 **(b)** \$255.19 **(c)** \$204,444.81 **19. (a)** \$288.40
(b) \$10,382.40 **(c)** \$902.40 **(d)** \$6507.13 **(e)** \$3350.90 **(f)** \$304.57
(g)

Payment Number	Amount	Interest	Applied to Principal	Unpaid Balance
1	\$288.40	\$47.40	\$241.00	\$9239.00
2	288.40	46.20	242.21	8996.80
3	288.40	44.98	243.42	8753.38
4	288.40	43.77	244.63	8508.75

21. (a) \$306.91 **(b)** \$338.32 **(c)** Option (a) **23.** Option II **25.** \$14,220.61 **27.** \$1196.68 **29.** 26.56% **31.** \$111,824.09
33. \$346.81 **35.** a **37. (d)** \$101, \$102.01 **39.** N=20, I%=6.4, PV=−10000, FV=0, P/Y=4, Find PMT; =PMT(1.6%,20,
−10000,0); annuity pmt, PV=\$10000, i=1.6%, n=20 **41.** N=104, I%=7.8, PMT=−23.59, FV=0, P/Y=52, Find PV;
=PV(.15%,104,−23.59,0); annuity PV, pmt=\$23.59, i=.15%, n=104 **43.** N=300, I%=5.4, PV=−100000, FV=0, P/Y=12, Find PMT;
=PMT(0.45%,300,−100000,0); annuity pmt, PV=\$100000, i=.45%, n=300 **45.** N=360, I%=4.5, PMT=−724.56, FV=0, P/Y=12,
Find PV; =PV(.375%,360,−724.56,0); annuity PV, pmt=\$724.56, i=.375%, n=360 **47.** 46 months **49.** 25; 46; 65 **51.** 261

Exercises 10.4, page 471

1. deferred **3.** \$165,000 **5.** \$1,641,407.11 **7.** \$1,313,125.69 **9. (a)** \$412,330.99 **(b)** \$357,362.60 **(c)** Larry **(d)** Earl
11. \$366.67 **13.** \$105.83 **15.** 2.73% **17.** 3.17% **19. (a)** \$41.67 **(b)** \$41.07; The monthly payment is less. **21.** False **23.** False
25. False **27.** d **29.** a **31.** a **33.** c **35.** 12*RATE(240, −573.14, 77600, 0); 12*RATE(120, −573.14, 77600, −51624.70)
37. $(((1 + i)^{180} − 1)/(i(1 + i)^{180})) \cdot 1217.12 = 118800; 1217.12 − i \cdot 96081.51 = (1217.12 − i \cdot 118800)(1 + i)^{60}$ **39.** The salesman
should compare the present value of the loan payments with the \$1000. **41.** \$1150; \$2311.87 **43. (a)** \$1610.75 **(b)** \$224,829.73
(c) \$1729.63 **45. (a)** \$219,418.04 **(b)** \$2181.80 **(c)** \$1850.70 **(d)** \$1865.05 **(e)** \$219,432.39 **47.** 11.08% **49.** 6.83%
51. 7.08% **53.** 87.72 months **55.** 128.63 months

Exercises 10.5, page 482

1. $y_n = 1.01y_{n-1}, y_0 = 1000$ **3.** $y_n = y_{n-1} + 62.50, y_0 = 2500$ **5.** $y_n = 1.004y_{n-1} − 1073.99, y_0 = 204,700$
7. $y_n = 1.00075y_{n-1} − 100, y_0 = 25,000$ **9.** $y_n = 1.014y_{n-1} + 1000, y_0 = 2500$ **11. (a)** initial deposit = \$37,780,

annual rate of interest $= 3.2\%$, \$1196 deposited quarterly **(b)** 9 years **13. (a)** 1, 6, 11, 16, 21 **(b)** $y_n = 1 + 5n$ **15. (a)** 7, 5.8, 5.32, 5.128, 5.0512 **(b)** $y_n = 5 + 2 \cdot (.4)^n$ **17. (a)** 2, -10, 50, -250, 1250 **(b)** $y_n = 2 \cdot (-5)^n$ **19.** \$1042.86 **21.** \$35,166.61 **23.** \$1135 **25.** \$1600 **27.** \$38,566.51 **29.** \$39,184.04 **31.** \$818.20 **33.** \$35,312 **35.** \$1771.77 **37.** \$244.60 **41. (a)** $y_n = (1.01)y_{n-1} - 600$, $y_0 = 50{,}000$ **(b)** $y_n = 60{,}000 - 10{,}000 \cdot (1.01)^n$ **(c)** 48,954 people **43. (a)** $y_n = .92y_{n-1} + 8$, $y_0 = 0$ **(b)** $y_n = 100(1 - (.92)^n)$ **(c)** 60 doctors **45. (a)** $y_n = .7y_{n-1} + 3.6$, $y_0 = 0$ **(b)** $y_n = 12(1 - (.7)^n)$ **(c)** 10.6 units **47. (a)** $y_n = .8y_{n-1} + 14$, $y_0 = 40$ **(b)** $y_n = 70 - 30 \cdot (.8)^n$ **(c)** 60.2°F **49. (a)** $y_n = (1 - 2p)y_{n-1} + p$, $y_0 = 1$ **(b)** $y_n = \frac{1}{2} + \frac{1}{2}(1 - 2p)^n$ **51.** $y_n = 1.18y_{n-1}$, $y_0 = 1000$; 7 years **53. (a)** $y_n = .7y_{n-1} + 22.5$, $y_0 = 600$ **(b)** 332.25°F **(c)** after 14 minutes **55. (a)** 32.65% **(b)** 15 months **(c)** 32 months **57. (a)** $y_n = .87y_{n-1} + 130$, $y_0 = 130$ **(b)** 969.24 mg

Chapter 10: Review Exercises, page 487

1. $100(1.03)^{10} = \$134.39$ **2.** \$1744.37 **3.** \$167,069.80 **4.** \$51.11 **5.** 3% compound annually **6.** \$12,929.34 **7. (a)** \$1529.99 **(b)** \$147,627.70 **8.** \$63,678.89 **9.** \$37,054.78 **10.** \$14,086.29 **11.** \$211.37 **12.** \$12,050.34 **13.** \$556.59 **14.** \$4940.82 **15.** \$100,451.50 **16.** \$192,978.65 **17.** Investment A **18.** Yes, it is a bargain since the present value is \$1053.25. **19.** 2.2121% **20.** 2.7337% **21.** \$84,753.66

22.

Payment Number	Amount	Interest	Applied to Principal	Unpaid Balance
1	\$304.22	\$50.00	\$254.22	\$9745.78
2	304.22	48.73	255.49	9490.29
3	304.22	47.45	256.77	9233.52
4	304.22	46.17	258.05	8975.47
5	304.22	44.88	259.34	8716.13
6	304.22	43.58	260.64	8455.49

23. \$31,451.59 **24.** \$5232.12 **25.** \$787.00

26. (a) \$21,000 **(b)** \$26,095.40 **27. (a)** \$30,000 **(b)** \$40,146.77 **28.** 4.82% **29.** Consumer loan A; The monthly payment for loan B is \$265.00. **30.** $P = 88{,}200$, $n = 180$, $R = 759.47$ **31.** $P = 97{,}000$, $m = 72$, $R = 716.43$, $B = 81{,}298.32$ **32.** $i = .035/12$, $R = 63.15$, $P = 2000$ **35.** \$2185; \$3394.34 **36. (a)** \$1458.08 **(b)** \$198,690.34 **(c)** \$1582.46 **37. (a)** 5; -7; 29 **(b)** $y_n = 2 - (-3)^n$ **(c)** -79 **38. (a)** $\frac{17}{2}$; 7; $\frac{11}{2}$ **(b)** $y_n = 10 - \frac{3}{2}n$ **(c)** 1 **39.** \$2000.00 **40.** \$1109.54 **41.** \$24.79 **42.** \$237.14 **43.** $y_n = .92y_{n-1}$, $y_0 = 100$ **44.** No **45.** Higher **46.** Decrease; increase **47.** No

CHAPTER 11

Exercises 11.1, page 495

1. Statement **3.** Statement **5.** Not a statement—not a declarative sentence. **7.** Not a statement—not a declarative sentence. **9.** Statement **11.** Not a statement—x is not specified. **13.** Not a statement—not a declarative sentence. **15.** Statement **17.** $2 + 2 = 5$, $4 < 7$

19. p: The Phelps Library is in New York.
q: The Phelps Library is in Dallas.
Then we have $p \vee q$.

21. p: The Smithsonian Museum of Natural History has displays of rocks.
q: The Smithsonian Museum of Natural History has displays of bugs.
Then we have $p \wedge q$.

23. p: Amtrak trains go to Chicago.
q: Amtrak trains go to Cincinnati.
Then we have $\sim p \wedge \sim q$ or $\sim (p \vee q)$.

25. (a) Ozone is opaque to ultraviolet light, and life on Earth requires ozone.
(b) Ozone is not opaque to ultraviolet light, or life on Earth requires ozone.
(c) Ozone is not opaque to ultraviolet light, or life on Earth does not require ozone.
(d) If life on Earth does not require ozone, then ozone is opaque to ultraviolet light.

27. (a) $a \vee f$ **(b)** $a \wedge \sim f$ **(c)** $f \wedge \sim a$ **(d)** $\sim a \wedge \sim f$

Exercises 11.2, page 502

1. Since r is a statement form, so is $\sim r$. Then $p \wedge \sim r$ is a statement form, and so is $\sim (p \wedge \sim r)$. Since q is a statement form, $\sim (p \wedge \sim r) \vee q$ is a statement form. **3.** Since p is a statement form, so is $\sim p$. Since q and r are statement forms, so are $\sim p \vee r$ and $q \wedge r$, and hence also $(\sim p \vee r) \rightarrow (q \wedge r)$.

5.

p	q	p	$\wedge$	$\sim q$
T	T	T	F	F
T	F	T	T	T
F	T	F	F	F
F	F	F	F	T
(1)	(2)		(4)	(3)

7.

p	q	$(p$	$\vee$	$\sim q)$	$\wedge$	q
T	T	T	T	F	T	T
T	F	T	T	T	F	F
F	T	F	F	F	F	T
F	F	T	T	T	F	F
(1)	(2)		(5)	(3)	(6)	(4)

9.

p	q	$\sim$	$[(p$	$\vee$	$q)$	$\wedge$	$(p$	$\wedge$	$q)]$
T	T	F	T		T	T		T	T
T	F	T	T		T	F		F	F
F	T	T	T		T	F		F	F
F	F	T	F		F	F		F	F
(1)	(2)	(6)		(3)		(5)		(4)	

11.

p	q	p	⊕	(~p	∨	q)
T	T	T	**F**	F	T	T
T	F	T	**T**	F	F	F
F	T	F	**T**	T	T	T
F	F	F	**T**	T	T	F
		(1)	(2)	(5)	(3)	(4)

13.

p	q	r	(p	∧	~r)	⊕	q
T	T	T	T	F	F	**T**	T
T	T	F	T	T	T	**F**	T
T	F	T	T	F	F	**F**	F
T	F	F	T	T	T	**T**	F
F	T	T	F	F	F	**T**	T
F	T	F	F	F	T	**F**	T
F	F	T	F	F	F	**F**	F
F	F	F	F	F	T	**F**	F
			(1)	(2)	(3)	(5)	(4) (6)

15.

p	q	r	~	[(p	∧	r)	∨	q]
T	T	T	**F**	T	T	T	T	T
T	T	F	**F**	T	F	F	T	T
T	F	T	**F**	T	T	T	T	F
T	F	F	**T**	T	F	F	F	F
F	T	T	**F**	F	F	T	T	T
F	T	F	**F**	F	F	F	T	T
F	F	T	**T**	F	F	T	F	F
F	F	F	**T**	F	F	F	F	F
			(1)	(2)	(3)	(6)	(4)	(5)

17.

p	p	∨	~p
T	T	**T**	F
F	F	**T**	T
	(1)	(3)	(2)

Always true

19.

p	q	r	p	⊕	(q	⊕	r)
T	T	T	T	**T**	T	F	T
T	T	F	T	**F**	T	T	F
T	F	T	T	**F**	F	T	T
T	F	F	T	**T**	F	F	F
F	T	T	F	**F**	T	F	T
F	T	F	F	**T**	T	T	F
F	F	T	F	**T**	F	T	T
F	F	F	F	**F**	F	F	F
			(1)	(2)	(3)	(5)	(4)

21.

p	q	r	(p	∨	q)	∧	(p	∨	r)
T	T	T	T	T	T	**T**	T	T	T
T	T	F	T	T	T	**T**	T	T	F
T	F	T	T	T	F	**T**	T	T	T
T	F	F	T	T	F	**T**	T	T	F
F	T	T	F	T	T	**T**	F	T	T
F	T	F	F	T	T	**F**	F	F	F
F	F	T	F	F	F	**F**	F	T	T
F	F	F	F	F	F	**F**	F	F	F
			(1)	(2)	(3)	(4)	(6)	(5)	

23.

p	q	(p	∨	q)	∧	~	(p	∨	q)
T	T	T	T	T	**F**	F	T	T	T
T	F	T	T	F	**F**	F	T	T	F
F	T	F	T	T	**F**	F	F	T	T
F	F	F	F	F	**F**	T	F	F	F
		(1)	(2)	(3)	(6)	(5)	(4)		

25.

p	q	r	~	(p	∨	q)	∧	r
T	T	T	F	T	T	T	**F**	T
T	T	F	F	T	T	T	**F**	F
T	F	T	F	T	T	F	**F**	T
T	F	F	F	T	T	F	**F**	F
F	T	T	F	F	T	T	**F**	T
F	T	F	F	F	T	T	**F**	F
F	F	T	T	F	F	F	**T**	T
F	F	F	T	F	F	F	**F**	F
			(1)	(2)	(3)	(5)	(4)	(6)

27. Not logically equivalent
29. Not logically equivalent **31.** 16

33. (a)

p	p			p
T	T	**F**	T
F	F	**T**	F
	(1)		(2)

(b)

p	q	(p			p)			(q			q)
T	T	T	F	T	T	T	F	T
T	F	T	F	T	T	F	T	F
F	T	F	T	F	T	T	F	T
F	F	F	T	F	F	F	T	F
		(1)	(2)	(3)	(5)	(4)		

(c)

p	q	(p			q)			(p			q)
T	T	T	F	T	T	T	F	T
T	F	T	T	F	F	T	T	F
F	T	F	T	T	F	F	T	T
F	F	F	T	F	F	F	T	F
		(1)	(2)	(3)	(5)	(4)		

(d)

p	q	p			[(p | | | q)			q]
T	T	T	F	T	F	T	T	T
T	F	T	F	T	T	F	T	F
F	T	F	T	F	T	T	F	T
F	F	F	T	F	T	F	T	F
		(1)	(2)	(5)	(3)	(4)		

35. (a) T (b) F (c) T (d) F (e) F (f) F **37.** T **39.** F **41.** 1 **43.** (a) F (b) F **45.** (a) F (b) F

47.

p	q	p ∧ ~q
T	T	F
T	F	T
F	T	F
F	F	F

49.

p	q	r	p ∧ (~q ∨ r)
T	T	T	T
T	T	F	F
T	F	T	T
T	F	F	T
F	T	T	F
F	T	F	F
F	F	T	F
F	F	F	F

51. (Not p) And q, or ~p ∧ q

Exercises 11.3, page 508

1.

p	q	~p	→	~q
T	T	F	**T**	F
T	F	F	**T**	T
F	T	T	**F**	F
F	F	T	**T**	T
(1)	(2)	(3)	(5)	(4)

3.

p	q	(p	⊕	q)	→	q
T	T	T	F	T	**T**	T
T	F	T	T	F	**F**	F
F	T	F	T	T	**T**	T
F	F	F	F	F	**T**	F
(1)	(2)		(3)		(4)	

5.

p	q	r	(~	p	∧	q)	→	r
T	T	T	F	T	F	T	**T**	T
T	T	F	F	T	F	T	**T**	F
T	F	T	F	T	F	F	**T**	T
T	F	F	F	T	F	F	**T**	F
F	T	T	T	F	T	T	**T**	T
F	T	F	T	F	T	T	**F**	F
F	F	T	T	F	F	F	**T**	T
F	F	F	T	F	F	F	**T**	F
(1)	(2)	(3)	(4)		(5)		(6)	

7.

p	q	(p	→	q)	↔	(~	p	∨	q)
T	T	T	T	T	**T**	F	T	T	T
T	F	T	F	F	**T**	F	T	F	F
F	T	F	T	T	**T**	T	F	T	T
F	F	F	T	F	**T**	T	F	T	F
(1)	(2)		(4)		(6)	(3)		(5)	

9.

p	q	r	(p	→	q)	→	r
T	T	T	T	T	T	**T**	T
T	T	F	T	T	T	**F**	F
T	F	T	T	F	F	**T**	T
T	F	F	T	F	F	**T**	F
F	T	T	F	T	T	**T**	T
F	T	F	F	T	T	**F**	F
F	F	T	F	T	F	**T**	T
F	F	F	F	T	F	**F**	F
(1)	(2)	(3)		(4)		(5)	

11.

p	q	(p	∨	q)	↔	(p	∧	q)
T	T	T	T	T	**T**	T	T	T
T	F	T	T	F	**F**	T	F	F
F	T	F	T	T	**F**	F	F	T
F	F	F	F	F	**T**	F	F	F
(1)	(2)		(3)		(5)		(4)	

13. $[(\sim p) \wedge (\sim q)] \to [(\sim p) \wedge q]$ **15.** $([(\sim p) \wedge (\sim q)] \vee r) \to [(\sim q) \wedge r]$ **17.** T **19.** T **21.** T **23.** F **25.** T **27.** $p \leftrightarrow q$
29. $q \to p$; hypothesis: q; conclusion: p **31.** $q \to p$; hypothesis: q; conclusion: p **33.** $\sim p \to \sim q$; hypothesis: $\sim p$; conclusion: $\sim q$
35. $\sim p \to \sim q$; hypothesis: $\sim p$; conclusion: $\sim q$; TRUE **37.** $\sim q \to \sim p$; hypothesis: $\sim q$; conclusion: $\sim p$; TRUE **39. (a)** hyp: I will run a marathon.; con: Amy watches my dogs. **(b)** hyp: A student earns a B.; con: A student passes the course. **(c)** hyp: The football team is my favorite.; con: The football team wears green uniforms. **(d)** hyp: Isaac has wrinkled fingers.; con: Isaac was in the bathtub.
41. (a) If City Sanitation collects the garbage, then the mayor calls. **(b)** The price of beans goes down if there is no drought.
(c) Goldfish swim in Lake Erie if Lake Erie is fresh water. **(d)** Tap water is not salted if it boils slowly. **43. (a)** 4 **(b)** 4 **(c)** 4
(d) 4 **(e)** 6 **(f)** 6 **45. (a)** 7 **(b)** 7 **(c)** 0 **(d)** 7 **(e)** 0 **(f)** 0 **47. (a)** 6 **(b)** −12 **(c)** 0 **(d)** 0 **(e)** 28 **(f)** 0

Exercises 11.4, page 516

1. $[(p \to q) \wedge q] \to p$

F	T	T	T	T	F	F

When p is false and q is true, the statement is FALSE.

3. Show that the corresponding biconditional is a tautology.

p	q	(p	→	q)	↔	[~	(p	∧	~	q)]
T	T	T	T	T	**T**	T	T	F	F	T
T	F	T	F	F	**T**	F	T	T	T	F
F	T	F	T	T	**T**	T	F	F	F	T
F	F	F	T	F	**T**	T	F	F	T	F
(1)	(2)		(3)		(7)	(6)		(5)	(4)	

5.

p	q	c	(p	→	q)	↔	[(p	∧	~	q)	→	c)]
T	T	F	T	T	T	**T**	T	F	F	T	T	F
T	F	F	T	F	F	**T**	T	T	T	F	F	F
F	T	F	F	T	T	**T**	F	F	F	T	T	F
F	F	F	F	T	F	**T**	F	F	T	F	T	F
(1)	(2)	(3)		(6)		(8)		(5)	(4)		(7)	

7. False **9.** $\sim(\sim p \to q)$ **11.** $\sim[\sim(p \vee q) \vee \sim(\sim p \vee \sim q)]$
13. $\sim(p \vee q) \vee (q \wedge \sim r)$ **15.** $(p \wedge \sim q) \vee (p \vee \sim r)$
17. (a) Arizona does not border California, or Arizona does not border Nevada. **(b)** There are no tickets available, and the agency cannot get tickets. **(c)** The killer's hat was neither white nor gray.

19. $\sim p \wedge q \wedge \sim r$ **21. (a)** Jeremy does not take 12 credits and Jeremy does not take 15 credits.   **(b)** Sandra did not receive a gift from Sally, or Sandra did not receive a gift from Sacha.   **23. (a)** I have a ticket to the theater, and I did not spend a lot of money. **(b)** Basketball is played on an indoor court, and the players do not wear sneakers.   **(c)** The stock market is going up, and interest rates are not going down.   **(d)** Humans have enough water, and humans are not staying healthy.   **25. (a)** If x is an even number, then the sum $2 + x$ is even.; TRUE   **(b)** If "S" is next to "K," then the computer keyboard does not have the QWERTY layout; TRUE **27. (a)** Contrapositive: If a dog is not a Chihuahua, then it is not small; FALSE Inverse: If a dog is not small, then it is not a Chihuahua; TRUE Converse: If a dog is a Chihuahua, then it is small; TRUE   **(b)** Contrapositive: If you stop, then the traffic light is not green; TRUE Inverse: If the traffic light is not green, then you stop; FALSE Converse: If you do not stop, then the traffic light is green; FALSE   **(c)** Contrapositive: If we are not French citizens, then we do not live in France; FALSE Inverse: If we do not live in France, then we are not French citizens; FALSE Converse: If we are French citizens, then we live in France; FALSE   **29. (a)** $(q \vee r) \rightarrow \sim p$ **(b)** $(q \vee r) \wedge p$; We wait on some other person or some other time, and change will come.   **31. (a)** Inverse: If you wish to claim the premium tax credit for 2015, then you need the information in Part II. Converse: If you do not need the information in Part II, then you do not wish to claim the premium tax credit for 2015. Contrapositive: If you need the information in Part II, then you wish to claim the premium tax credit for 2015. Negation: You do not wish to claim the premium tax credit for 2015 and you do not need the information in Part II.   **(b)** Inverse: If your spouse does not itemize deductions, then you can take the standard deduction. Converse: If you can't take the standard deduction, then your spouse itemizes deductions. Contrapositive: If you can take the standard deduction, then your spouse does not itemize deductions. Negation: Your spouse itemizes deductions and you can take the standard deduction.   **(c)** Inverse: If you do not check a box, your tax will change and your refund will change. Converse: If your tax or refund don't change, then you check a box. Contrapositive: If your tax or refund change, then you don't check a box. Negation: You check a box and your tax will change and your refund will change.

33. We must show that $[p \wedge (p \rightarrow q)] \rightarrow q$ is a tautology.

p	q	$[p$	$\wedge$	$(p$	$\rightarrow$	$q)]$	$\rightarrow$	q
T	T		T		T		**T**	
T	F		F		F		**T**	
F	T		F		T		**T**	
F	F		F		T		**T**	
(1)	(2)		(4)		(3)		(5)	

Exercises 11.5, page 523

1. e: "The auto manufacturer follows environmental regulations."
l: "The auto manufacturer faces legal action."

1.	$e \vee l$	hyp.
2.	$\sim e$	hyp.
3.	l	disj. syll. (1, 2)

3. a: "My allowance comes this week."
p: "I pay the rent."
b: "My bank account will be in the black."
e: "I will be evicted."

1.	$(a \wedge p) \rightarrow b$	hyp.
2.	$\sim p \rightarrow e$	hyp.
3.	$\sim e \wedge a$	hyp.
4.	$\sim e$	subtr. (3)
5.	p	mod. tollens (2, 4)
6.	a	subtr. (3)
7.	b	mod. ponens (5, 6, 1)

5. p "The price of oil increases."
a "The OPEC countries are in agreement."
d "There is a U.N. debate."

1.	$p \rightarrow a$	hyp.
2.	$\sim d \rightarrow p$	hyp.
3.	$\sim a$	hyp.
4.	$\sim p$	mod. tollens (1, 3)
5.	d	mod. tollens (2, 4)

7. g: "The germ is present."
r: "The rash is present."
f: "The fever is present."

1.	$g \rightarrow (r \wedge f)$	hyp.
2.	f	hyp.
3.	$\sim r$	hyp.
4.	$\sim r \vee \sim f$	addition (3)
5.	$\sim (r \wedge f)$	De Morgan (4)
6.	$\sim g$	mod. tollens (1, 5)

9. c: "The material is cotton."
r: "The material is rayon."
d: "The material can be made into a dress."

1.	$(c \vee r) \rightarrow d$	hyp.
2.	$\sim d$	hyp.
3.	$\sim (c \vee r)$	mod. tollens (1, 2)
4.	$\sim c \wedge \sim r$	De Morgan (3)
5.	$\sim r$	subtraction (4)

11. Invalid   **13.** Valid   **15.** Valid   **17.** Invalid   **19.** Invalid

21. s: "Sam goes to the store."   1. s $\sim C$
  m: "Sam needs milk."   2. $s \to m$ H_1
  H_1: $s \to m$ 3. m $\sim H_2$; mod. ponens (1, 2)
  H_2: $\sim m$
  C: $\sim s$

23. n: "The newspaper reports the crime."   1. $\sim s$ $\sim C$
  t: "Television reports the crime."   2. $(n \wedge t) \to s$ H_1
  s: "The crime is serious."   3. $\sim(n \wedge t)$ mod. tollens (1, 2)
  k: "A person is killed."   4. $\sim n \vee \sim t$ De Morgan (3)
  H_1: $(n \wedge t) \to s$ 5. t H_4
  H_2: $k \to n$ 6. $\sim n$ disj. syllogism (4, 5)
  H_3: k 7. $k \to n$ H_2
  H_4: t 8. $\sim k$ $\sim H_3$; mod. tollens (6, 7)
  C: s

25. j: "Jimmy finds his keys."
  h: "He does his homework."

Direct proof   Indirect proof
1. $\sim j \to h$ H_1 1. $\sim j$ $\sim C$
2. $\sim h$ H_2 2. $\sim j \to h$ H_1
3. j mod. tollens (1, 2)   3. h mod. ponens (1, 2) $\Big\}$ contradiction
                4. $\sim h$ H_2

27. m: "Marissa goes to the movies."   1. $\sim m \to \sim i$ H_1
  k: "Marissa is in a knitting class."   2. $\sim k \to i$ H_2
  i: "Marissa is idle."   3. $\sim k$ H_3
                4. i mod. ponens (2, 3)
                5. m mod. tollens (1, 4)

Exercises 11.6, page 532

1. (a) F **(b)** T **(c)** T **(d)** T **(e)** F **3.** $\forall x \sim p(x)$, or $\sim[\exists x \, p(x)]$. This is FALSE. Abby meant to say, "Not all men cheat on their wives." **5. (a)** $\forall x \, p(x)$ **(b)** $\exists x \sim p(x)$ **(c)** $\exists x \, p(x)$ **(d)** $\sim[\forall x \, p(x)]$ **(e)** $\forall x \sim p(x)$ **(f)** $\sim[\exists x \, p(x)]$ **(g)** (b) and (d); (e) and (f) **7. (a)** T **(b)** F **9. (a)** T **(b)** F **(c)** T **(d)** F **(e)** F **(f)** T **(g)** F **(h)** T **11. (a)** Not every dog has his day. **(b)** No men fight wars. **(c)** Some women are unmarried. **(d)** There exists a pot without a cover. **(e)** All children have pets. **(f)** Every month has 30 days. **13. (a)** "The sum of any two nonnegative integers is greater than 12." FALSE: consider $x = 1$, $y = 2$. "There exist two nonnegative integers whose sum is not greater than 12." **(b)** "For any nonnegative integer, there is a nonnegative integer that, added to the first, makes a sum greater than 12." TRUE **(c)** "There is a nonnegative integer that, added to any other nonnegative integer, makes a sum greater than 12." TRUE (Try $x = 13$.) **(d)** "There are two nonnegative integers, the sum of which is greater than 12." TRUE (Try $x = 6$, $y = 7$.) **15. (a)** FALSE: let $x = 2$, $y = 3$. **(b)** TRUE: For any x, let $y = x$. **(c)** TRUE: Let $x = 1$. **(d)** FALSE: No y is divisible by every x. **(e)** TRUE: for any y, let $x = y$. **(f)** TRUE: any x divides itself. **17. (a)** $S \subseteq T$ translates as $\forall x \, [x \geq 8 \to x \leq 10]$. **(b)** No; consider $x = 11$. **19.** $S = \{2, 4, 6, 8\}$, $T = \{1, 2, 3, 4, 6, 8\}$. So $\forall x \, [x \in S \to x \in T]$.

Exercises 11.7, page 536

1. $(p \vee q) \wedge (p \vee \sim q)$ **3.** $[(p \wedge q) \wedge \sim r] \wedge (\sim q \vee r)$

9. p **11.** $r \wedge [(p \wedge q) \vee \sim(p \vee q)]$

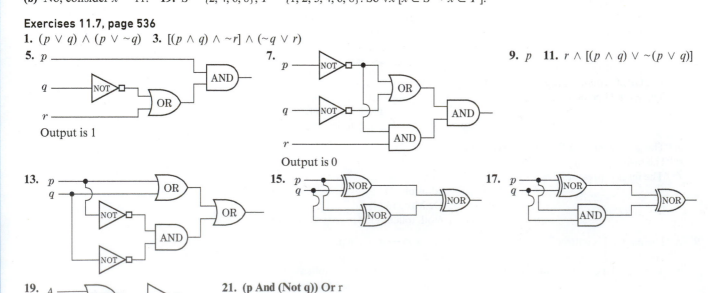

5. Output is 1

7. Output is 0

21. (p And (Not q)) Or r

Chapter 11: Review Exercises, page 540

1. (a) Statement **(b)** Not a statement—not a declarative sentence **(c)** Statement **(d)** Not a statement—"he" is not specified
(e) Statement **2. (a)** If two lines are perpendicular, then their slopes are negative reciprocals of each other. **(b)** If goldfish can live in a fishbowl, then the water is aerated. **(c)** If it rains, then Jane uses her umbrella. **(d)** If Sally gives Morris a treat, then he ate all of his food. **3. (a)** Contrapositive: If the Yankees are not playing in Yankee Stadium, then they are not in New York City.; Converse: If the Yankees are playing in Yankee Stadium, then they are in New York City. **(b)** Contrapositive: If the quake is not considered major, then the Richter scale does not indicate the earthquake is a 7.; Converse: If the quake is considered major, then the Richter scale indicates the earthquake is a 7. **(c)** Contrapositive: If a coat is not warm, then it is not made of fur.; Converse: If a coat is warm, then it is made of fur. **(d)** Contrapositive: If Jane is not in Moscow, then she is not in Russia.; Converse: If Jane is in Moscow, then she is in Russia.

4. (a) p: "Two triangles are similar."
   q: "Their sides are equal."
   $p \rightarrow q$ negated becomes $\sim(p \rightarrow q)$ or $p \wedge \sim q$, or "Two triangles are similar but their sides are unequal."
(b) $U = \{$real numbers$\}$ and $p(x) = (x^2 = 5)$.
   $\exists x\, p(x)$ negated becomes $\forall x \sim p(x)$ or "For every real number x, $x^2 \neq 5$."
(c) $U = \{$positive integers$\}$,
   $p(n) = (n$ is even$)$, and $q(n) = (n^2$ is even$)$.
   $\forall n\, [p(n) \rightarrow q(n)]$ negated becomes $\exists n \sim[p(n) \rightarrow q(n)]$ or $\exists n\, [p(n) \wedge \sim q(n)]$, or "There exists a positive integer n such that n is even and n^2 is not even."
(d) $U = \{$real numbers$\}$ and $p(x) = (x^2 + 4 = 0)$.
   $\exists x\, p(x)$ negated becomes $\forall x \sim p(x)$ or "For every real number x, $x^2 + 4 \neq 0$."

5. (a) Tautology **(b)** Tautology **(c)** Not a tautology **(d)** Not a tautology

6. (a)

p	q	r	p	$\rightarrow$	($\sim$	q	$\vee$	r)
T	T	T	T	**T**	F	T	T	T
T	T	F	T	**F**	F	T	F	F
T	F	T	T	**T**	T	F	T	T
T	F	F	T	**T**	T	F	T	F
F	T	T	F	**T**	F	T	T	T
F	T	F	F	**T**	F	T	F	F
F	F	T	F	**T**	T	F	T	T
F	F	F	F	**T**	T	F	T	F
(1)	(2)	(3)	(6)	(4)		(5)		

(b)

p	q	r	p	$\wedge$	[q	$\leftrightarrow$	(r	$\wedge$	p)]
T	T	T	T	**T**	T	T	T	T	T
T	T	F	T	**F**	T	F	F	F	T
T	F	T	T	**F**	F	F	T	T	T
T	F	F	T	**T**	F	T	F	F	T
F	T	T	F	**F**	T	F	T	F	F
F	T	F	F	**F**	T	F	F	F	F
F	F	T	F	**F**	F	T	T	F	F
F	F	F	F	**F**	F	T	F	F	F
(1)	(2)	(3)	(6)		(5)		(4)		

7. (a) TRUE **(b)** FALSE **8. (a)** TRUE **(b)** FALSE **9. (a)** FALSE **(b)** TRUE **10. (a)** 17 **(b)** 100 **(c)** 100 **(d)** 100
11. (a) 50 **(b)** -25 **(c)** -15 **(d)** 10 **12. (a)** Cannot be determined **(b)** TRUE **(c)** TRUE **(d)** Cannot be determined
(e) Cannot be determined **13. (a)** Cannot be determined **(b)** Cannot be determined **(c)** TRUE **14. (a)** TRUE **(b)** FALSE
(c) TRUE **15. (a)** Cannot be determined **(b)** Cannot be determined **(c)** TRUE

16. t: "Taxes go up."

s: "I sell the house."	1. $t \rightarrow (s \wedge m)$	hyp.
m: "I move to India."	2. $\sim m$	hyp.
	3. $\sim s \vee \sim m$	addition (2)
	4. $\sim(s \wedge m)$	De Morgan (3)
	5. $\sim t$	mod. tollens (1, 4)

17. m: "I study mathematics."

b: "I study business."	1. $m \wedge b$	hyp.
p: "I can write poetry."	2. $b \rightarrow (\sim p \vee \sim m)$	hyp.
	3. b	subtraction (1)
	4. $\sim p \vee \sim m$	mod. ponens (2, 3)
	5. m	subtraction (1)
	6. $\sim p$	disj. syllogism (4, 5)

18. d: "I shop for a dress."

h: "I wear high heels."	1. $d \rightarrow h$	hyp.
s: "I have a sore foot."	2. $s \rightarrow \sim h$	hyp.
	3. d	hyp.
	4. h	mod. ponens (1, 3)
	5. $\sim s$	mod. tollens (2, 4)

19. a: "Asters grow in the garden."

d: "Dahlias grow in the garden."	1. $a \vee d$	hyp.
s: "It is spring."	2. $s \rightarrow \sim a$	hyp.
	3. s	hyp.
	4. $\sim a$	mod. ponens (2, 3)
	5. d	disj. syllogism (1, 4)

20.
t: "The professor gives a test."
h: "Nancy studies hard."
d: "Nancy has a date."
s: "Nancy takes a shower."
H_1: $t \rightarrow h$
H_2: $d \rightarrow s$
H_3: $\sim t \rightarrow \sim s$
H_4: d
C: h

1.	$\sim h$	$\sim C$
2.	$t \rightarrow h$	H_1
3.	$\sim t$	mod. tollens (1, 2)
4.	$\sim t \rightarrow \sim s$	H_3
5.	$\sim s$	mod. ponens (3, 4)
6.	$d \rightarrow s$	H_2
7.	$\sim d$	$\sim H_4$; mod. tollens (5, 6)

21.

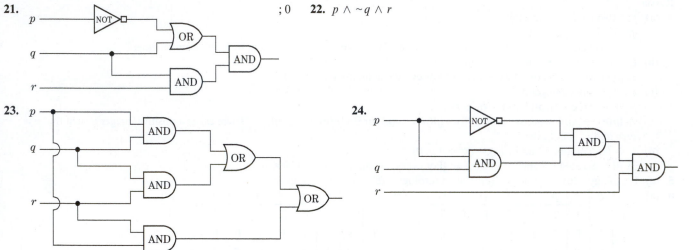

; 0   **22.** $p \wedge \sim q \wedge r$

23.

24.

25. (p NOR q) NOR (p AND q)

Index of Applications

Index

See Online Chapter 12 for references to pages beginning with D.

KEY DEFINITIONS, CONCEPTS, AND FORMULAS

1.1   Slope–Intercept Form   The equation of the line with slope m and y-intercept $(0, b)$ is $y = mx + b$.

1.4   Slope of a Line $m = \dfrac{y_2 - y_1}{x_2 - x_1}$, where (x_1, y_1) and (x_2, y_2) are points on the line with $x_1 \neq x_2$.

1.5   Perpendicular and Parallel Lines   Lines of slopes m_1 and m_2 are perpendicular if $m_1 \cdot m_2 = -1$ and are parallel if the lines are different and $m_1 = m_2$.

2.1   Elementary Row Operations for a Matrix
(a) Interchange any two rows.
(b) Multiply a row by a nonzero number.
(c) Change a row by adding to it a multiple of another row.

2.3   Product of Two Matrices   The product of an $m \times n$ matrix and an $n \times r$ matrix is the $m \times r$ matrix whose ij^{th} element is obtained by multiplying the ith row of the first matrix by the jth column of the second matrix. (The product of each row and column is calculated as the sum of the products of corresponding entries.)

2.5   Finding the Inverse of the $n \times n$ Matrix A
(a) Append I_n to the right of the original matrix to form $[A \,|\, I_n]$.
(b) Perform pivots on $[A \,|\, I_n]$ to obtain a matrix of the form $[I_n \,|\, B]$. The matrix B will then be A^{-1}.

3.2   Solving a Linear Programming Problem Graphically
(a) Assign variables to the unknown quantities.
(b) Translate the restrictions into a system of linear inequalities.
(c) Form an objective function for the quantity to be optimized.
(d) Graph the feasible set.
(E) Determine the vertices of the feasible set.
(F) Evaluate the objective function at each vertex and identify the vertex that gives the optimal value.

5.2   Inclusion–Exclusion Principle $n(S \cup T) = n(S) + n(T) - n(S \cap T)$

5.2   De Morgan's Laws $(S \cup T)' = S' \cap T' \quad \text{and} \quad (S \cap T)' = S' \cup T'$

5.5   Factorial $n! = n(n - 1)(n - 2) \cdots 2 \cdot 1$

5.5   Permutations and Combinations $P(n, r) = \dfrac{n!}{(n - r)!} \quad \text{and} \quad C(n, r) = \dfrac{n!}{r!(n - r)!}$

5.7   Binomial Theorem $(x + y)^n = \binom{n}{0}x^n + \binom{n}{1}x^{n-1}y + \binom{n}{2}x^{n-2}y^2 + \cdots + \binom{n}{n-1}xy^{n-1} + \binom{n}{n}y^n$

6.2   Inclusion–Exclusion Principle $\Pr(A \cup B) = \Pr(A) + \Pr(B) - \Pr(A \cap B)$

6.3   Probability Principle $\Pr(E) = \dfrac{n(E)}{n(S)}$, where S is a sample space with equally likely outcomes and E is an event in S.

6.3   Complement Rule $\Pr(E) = 1 - \Pr(E')$

6.4   Conditional Probability $\Pr(E \,|\, F) = \dfrac{\Pr(E \cap F)}{\Pr(F)}, \Pr(F) \neq 0.$

6.4   Product Rule $\Pr(E \cap F) = \Pr(F) \cdot \Pr(E \,|\, F), \Pr(F) \neq 0.$

6.4   Independent Events $\Pr(E \cap F) = \Pr(E) \cdot \Pr(F)$ or $\Pr(E \,|\, F) = \Pr(E)$ or $\Pr(F \,|\, E) = \Pr(F)$

6.7	**Bayes' Theorem**	$\Pr(B_i \mid A) = \dfrac{\Pr(B_i) \cdot \Pr(A \mid B_i)}{\Pr(B_1) \cdot \Pr(A \mid B_1) + \cdots + \Pr(B_n) \cdot \Pr(A \mid B_n)}$, where $B_1 \cdots, B_n$ are mutually exclusive events whose union is the entire sample space and A is an event.

6.7 **Bayes' Theorem**

$$\Pr(B_i \mid A) = \frac{\Pr(B_i) \cdot \Pr(A \mid B_i)}{\Pr(B_1) \cdot \Pr(A \mid B_1) + \cdots + \Pr(B_n) \cdot \Pr(A \mid B_n)},$$ where $B_1 \cdots, B_n$ are mutually exclusive events whose union is the entire sample space and A is an event.

7.3 **Binomial Probability**

$\Pr(k \text{ successes in } n \text{ trials}) = \binom{n}{k} p^k (1 - p)^{n-k}$, where p is the probability of success in each trial.

7.4 **Expected Value**

$E(X) = x_1 p_1 + x_2 p_2 + x_3 p_3 + \cdots + x_n p_n$, where $p_i = \Pr(x_i)$.

7.5 **Variance**

$\text{Var}(X) = (x_1 - \mu)^2 p_1 + (x_2 - \mu)^2 p_2 + \cdots + (x_N - \mu)^2 p_N$, where $\mu = E(X)$ and $p_i = \Pr(x_i)$. Alternatively, $\text{Var}(X) = E(X^2) - [E(X)]^2$.

7.4, 7.5 **Binomial Random Variable**

Expected value $= np$, standard deviation $= \sqrt{np(1 - p)}$, where p is the probability of success in each of n trials.

7.5 **Sample Variance and Standard Deviation**

Sample variance $= s^2 = \dfrac{(x_1 - \overline{x})^2 + (x_2 - \overline{x})^2 + \cdots + (x_n - \overline{x})^2}{n - 1}$, where $\overline{x}$ is the mean of the n numbers $x_1, x_2, \cdots, x_n$. Sample standard deviation $= s$.

8.2 **Regular Stochastic Matrix**

$A^n \to X$, where A is a regular stochastic matrix and $AX = X$.

10.1 **Future Value of an Investment**

$F = P(1 + i)^n$, where i is the compound interest rate per period, n is the number of interest periods, and P is the principal invested.

10.1 **Present Value of an Amount**

$P = \dfrac{F}{(1 + i)^n}$, where i is the compound interest rate per period, n is the number of interest periods, and F is the future value.

10.1 **Simple Interest**

$F = (1 + nr)P$, where n is the number of years and r is the annual rate of interest.

10.1 **Effective Rate of Interest**

$r_{\text{eff}} = (1 + i)^m - 1$, where i is the compound interest rate per period and interest is compounded m times per year.

10.2 **Future Value of Increasing Annuity**

$F = \dfrac{(1 + i)^n - 1}{i} \cdot R$, where i is the interest rate per period, n is the number of interest periods, and R is the periodic payment.

10.2 **Present Value of Decreasing Annuity**

$P = \dfrac{1 - (1 + i)^{-n}}{i} \cdot R$, where i is the interest rate per period, n is the number of interest periods, and R is the periodic payment.

10.5 **A Unifying Equation**

$y_n = a y_{n-1} + b$, y_0 given, $a \neq 1$, has solution $y_n = \dfrac{b}{1 - a} + \left(y_0 - \dfrac{b}{1 - a}\right) a^n$.

$y_n = y_{n-1} + b$, y_0 given, has solution $y_n = y_0 + bn$.

11.1–11.3 **Logical Operators**

p	q	$p \wedge q$	$p \vee q$	$\sim p$	$p \oplus q$	$p \to q$	$p \leftrightarrow q$
T	T	T	T	F	F	T	T
T	F	F	T	F	T	F	F
F	T	F	T	T	T	T	F
F	F	F	F	T	F	T	T